T0319350

Robust Optimization

Princeton Series in Applied Mathematics

Series Editors: Ingrid Daubechies (Princeton University); Weinan E (Princeton University); Jan Karel Lenstra (Eindhoven University); Endre Süli (University of Oxford)

The Princeton Series in Applied Mathematics publishes high quality advanced texts and monographs in all areas of applied mathematics. Books include those of a theoretical and general nature as well as those dealing with the mathematics of specific applications areas and real-world situations.

Chaotic Transitions in Deterministic and Stochastic Dynamical Systems: Applications of Melnikov Processes in Engineering, Physics, and Neuroscience, Emil Simiu

Self-Regularity: A New Paradigm for Primal-Dual Interior Point Algorithms, Jiming Peng, Cornelis Roos, and Tamas Terlaky

Selfsimilar Processes, Paul Embrechts and Makoto Maejima

Analytic Theory of Global Bifurcation: An Introduction, Boris Buffoni and John Toland

Entropy, Andreas Greven, Gerhard Keller, and Gerald Warnecke, editors

Auxiliary Signal Design for Failure Detection, Stephen L. Campbell and Ramine Nikoukhah

Max Plus at Work Modeling and Analysis of Synchronized Systems: A Course on Max-Plus Algebra and Its Applications, Bernd Heidergott, Geert Jan Olsder, and Jacob van der Woude

Optimization: Insights and Applications, Jan Brinkhuis and Vladimir Tikhomirov

Thermodynamics: A Dynamical Systems Approach, Wassim M. Haddad, VijaySekhar Chellaboina, and Sergey G. Nersesov

Impulsive and Hybrid Dynamical Systems Stability, Dissipativity, and Control, Wassim M. Haddad, VijaySekhar Chellaboina, and Sergey G. Nersesov

Distributed Control of Robotic Networks: A Mathematical Approach to Motion Coordination Algorithms, Francesco Bullo, Jorge Cortés, and Sonia Martínez

Genomic Signal Processing, Ilya Shmulevich and Edward Dougherty

Positive Definite Matrices, Rajendra Bhatia

The Traveling Salesman Problem: A Computational Study, David L. Applegate, Robert E. Bixby, Vasek Chvatal, and William J. Cook

Wave Scattering by Time-Dependent Perturbations: An Introduction, G. F. Roach

Algebraic Curves over a Finite Field, J.W.P. Hirschfeld, G. Korchmáros, and F. Torres

Distributed Control of Robotic Networks: A Mathematical Approach to Motion Coordination Algorithms, Francesco Bullo, Jorge Cortés, and Sonia Martínez

Robust Optimization, Aharon Ben-Tal, Laurent El Ghaoui, and Arkadi Nemirovski

Robust Optimization

Aharon Ben-Tal

Laurent El Ghaoui

Arkadi Nemirovski

Princeton University Press

Princeton and Oxford

Published by Princeton University Press, 41 William Street, Princeton, New Jersey

08540

In the United Kingdom: Princeton University Press, 6 Oxford Street, Woodstock,

Oxfordshire OX20 1TW

Library of Congress Cataloging-in-Publication Data

Ben-Tal, A..
  Robust optimization / Aharon Ben-Tal, Laurent El Ghaoui, Arkadi Nemirovski.
    p. cm.
  ISBN 978-0-691-14368-2 (hardcover : alk. paper) 1. Robust optimization. 2. Linear
programming. I. El Ghaoui, Laurent. II. Nemirovskii, Arkadii Semenovich. III. Title.
  QA402.5.B445 2009
  519.6--dc22
                              2009013229

The publishers would like to acknowledge the authors of this volume for providing the

electronic files from which this book was printed

Printed on acid-free paper. ∞

press.princeton.edu

Printed in the United States of America

10 9 8 7 6 5 4 3 2 1

Contents

Preface

This book is devoted to *Robust Optimization* — a specific and relatively novel methodology for handling optimization problems with *uncertain data*. The primary goal of this Preface is to provide the reader with a first impression of what the story is about:

• what is the phenomenon of data uncertainty and why it deserves a dedicated treatment,

• how this phenomenon is treated in Robust Optimization, and how this treatment compares to those offered by more traditional methodologies for handling data uncertainty.

The secondary, quite standard, goal is to outline the main topics of the book and describe its contents.

A. Data Uncertainty in Optimization

The very first question we intend to address here is whether the underlying phenomenon — data uncertainty — is worthy of special treatment. To answer this question, consider a simple example — problem PILOT4 from the well-known NETLIB library. This is a Linear Programming problem with 1,000 variables and 410 constraints; one of the constraints (# 372) is:

$$\begin{aligned}
a^T x \equiv &-15.79081 x_{826} - 8.598819 x_{827} - 1.88789 x_{828} - 1.362417 x_{829} - 1.526049 x_{830} \\
&-0.031883 x_{849} - 28.725555 x_{850} - 10.792065 x_{851} - 0.19004 x_{852} - 2.757176 x_{853} \\
&-12.290832 x_{854} + 717.562256 x_{855} - 0.057865 x_{856} - 3.785417 x_{857} - 78.30661 x_{858} \\
&-122.163055 x_{859} - 6.46609 x_{860} - 0.48371 x_{861} - 0.615264 x_{862} - 1.353783 x_{863} \\
&-84.644257 x_{864} - 122.459045 x_{865} - 43.15593 x_{866} - 1.712592 x_{870} - 0.401597 x_{871} \\
&+x_{880} - 0.946049 x_{898} - 0.946049 x_{916} \geq b \equiv 23.387405.
\end{aligned} \tag{C}$$

The related *nonzero* coordinates of the optimal solution x^* of the problem, as reported by CPLEX, are as follows:

$$\begin{aligned}
&x_{826}^* = 255.6112787181108 \quad &&x_{827}^* = 6240.488912232100 \quad &&x_{828}^* = 3624.613324098961 \\
&x_{829}^* = 18.20205065283259 \quad &&x_{849}^* = 174397.0389573037 \quad &&x_{870}^* = 14250.00176680900 \\
&x_{871}^* = 25910.00731692178 \quad &&x_{880}^* = 104958.3199274139.
\end{aligned}$$

Note that within machine precision x^* makes (C) an equality.

Observe that most of the coefficients in (C) are "ugly reals" like -15.79081 or -84.644257. Coefficients of this type usually (and PILOT4 is not an exception) characterize certain technological devices/processes, forecasts for future demands, etc., and as such *they could hardly be known to high accuracy*. It is quite natural to assume that the "ugly coefficients" are in fact *uncertain* — they coincide with the "true" values of the corresponding data within accuracy of 3 to 4 digits, not more. The only exception is the coefficient 1 of x_{880}; it perhaps reflects the structure of the problem and is therefore exact, that is certain.

Assuming that the uncertain entries of a are, say, 0.1%-accurate approximations of unknown entries of the "true" vector of coefficients $\tilde{a}$, let us look what would be the effect of this uncertainty on the validity of the "true" constraint $\tilde{a}^T x \geq b$ at x^*. What happens is as follows:

• Over all vectors of coefficients $\tilde{a}$ compatible with our 0.1%-uncertainty hypothesis, the minimum value of $\tilde{a}^T x^* - b$, is < -104.9; in other words, the violation of the constraint can be as large as 450% of the right hand side!

• Treating the above worst-case violation as "too pessimistic" (why should the true values of all uncertain coefficients differ from the values indicated in (C) in the "most dangerous" way?), consider a less extreme measure of violation. Specifically, assume that the true values of the uncertain coefficients in (C) are obtained from the "nominal values" (those shown in (C)) by random perturbations $a_j \mapsto \tilde{a}_j = (1 + \xi_j)a_j$ with independent and, say, uniformly distributed on $[-0.001, 0.001]$ "relative perturbations" ξ_j. What will be a "typical" relative violation,

$$V = \max \left[\frac{b - \tilde{a}^T x^*}{b}, 0 \right] \times 100\%,$$

of the "true" (now random) constraint $\tilde{a}^T x \geq b$ at x^*? The answer is nearly as bad as for the worst scenario:

Prob$\{V > 0\}$	Prob$\{V > 150\%\}$	Mean(V)
0.50	0.18	125%

Table 1. Relative violation of constraint 372 in PILOT4
(1,000-element sample of 0.1% perturbations of the uncertain data)

We see that *quite small (just 0.1%) perturbations of "obviously uncertain" data coefficients can make the "nominal" optimal solution x^* heavily infeasible and thus practically meaningless.*

A "case study" reported in [7] shows that the phenomenon we have just described is not an exception – in 13 of 90 *NETLIB* Linear Programming problems considered in this study, already 0.01%-perturbations of "ugly" coefficients result in violations of some constraints, as evaluated at the nominal optimal solutions by more than 50%. In 6 of these 13 problems the magnitude of constraint violations was over 100%, and in PILOT4 — "the champion" — it was as large as 210,000%, that is, 7 orders of magnitude larger than the relative perturbations in the data.

The techniques presented in this book as applied to the NETLIB problems allow one to eliminate the outlined phenomenon by passing out of the nominal optimal to *robust optimal* solutions. At the 0.1%-uncertainty level, the price of this "immunization against uncertainty" (the increase in the value of the objective when passing from the nominal to the robust solution), *for every one of the* NETLIB *problems*, is less than 1% (see [7] for details).

The outlined case study and many other examples lead to several observations:

A. *The data of real-world optimization problems more often than not are uncertain — not known exactly at the time the problem is being solved.* The reasons for data uncertainty include, among others:

measurement/estimation errors coming from the impossibility to measure/estimate exactly the data entries representing characteristics of physical systems/technological processes/environmental conditions, etc.

implementation errors coming from the impossibility to implement a solution exactly as it is computed. For example, whatever the entries "in reality" in the above nominal solution x^* to PILOT4 — control inputs to physical systems, resources allocated for various purposes, etc. — they clearly cannot be implemented with the same high precision to which they are computed. The effect of the implementation errors, like $x_j^* \mapsto (1 + \epsilon_j)x_j^*$, is as if there were no implementation errors, but the coefficients a_{ij} in the constraints of PILOT4 were subject to perturbations $a_{ij} \mapsto (1 + \epsilon_j)a_{ij}$.

B. *In real-world applications of Optimization one cannot ignore the possibility that even a small uncertainty in the data can make the nominal optimal solution to the problem completely meaningless from a practical viewpoint.*

C. *Consequently, in Optimization, there exists a real need of a methodology capable of detecting cases when data uncertainty can heavily affect the quality of the nominal solution, and in these cases to generate a robust solution, one that is immunized against the effect of data uncertainty.*

A methodology addressing the latter need is offered by *Robust Optimization*, which is the subject of this book.

B. Robust Optimization — The Paradigm

To explain the paradigm of Robust Optimization, we start by addressing the particular case of Linear Programming — the generic optimization problem that is perhaps the best known and the most frequently used in applications. Aside from its importance, this generic problem is especially well-suited for our current purposes, since the structure and the data of a Linear Programming program $\min_{x}\{c^T x : Ax \leq b\}$ are clear. Given the form in which we wrote the program down, the structure is the sizes of the constraint matrix A, while the data is comprised of the numerical values of the entries in (c, A, b). In Robust Optimization, an *uncertain* LP problem is defined as a collection $\left\{\min_x\{c^T x : Ax \leq b\} : (c, A, B) \in \mathcal{U}\right\}$

of LP programs of a common structure with the data (c, A, b) varying in a given *uncertainty set* $\mathcal{U}$. The latter summarizes all information on the "true" data that is available to us when solving the problem. Conceptually, the most important question is what does it mean to solve an uncertain LP problem. The answer to this question, as offered by Robust Optimization in its most basic form, rests on three implicit assumptions on the underlying "decision-making environment":

A.1. All entries in the decision vector x represent "here and now" decisions: they should get specific numerical values as a result of solving the problem *before* the actual data "reveals itself."

A.2. The decision maker is fully responsible for consequences of the decisions to be made when, and only when, the actual data is within the prespecified uncertainty set $\mathcal{U}$.

A.3. The constraints of the uncertain LP in question are "hard" — the decision maker cannot tolerate violations of constraints when the data is in $\mathcal{U}$.

These assumptions straightforwardly lead to the definition of an "immunized against uncertainty" solution to an uncertain problem. Indeed, by A.1, such a solution should be a fixed vector that, by A.2 – A.3, should remain feasible for the constraints, whatever the realization of the data within $\mathcal{U}$; let us call such a solution *robust feasible*. Thus, in our decision-making environment, meaningful solutions to an uncertain problem are exactly its robust feasible solutions. It remains to decide how to interpret the value of the objective, (which can also be uncertain), at such a solution. As applied to the objective, our "worst-case-oriented" philosophy makes it natural to quantify the quality of a robust feasible solution x by the *guaranteed* value of the original objective, that is, by its largest value $\sup \left\{ c^T x : (c, A, b) \in \mathcal{U} \right\}$. Thus, the best possible robust feasible solution is the one that solves the optimization problem

$$\min_{x} \left\{ \sup_{(c,A,b) \in \mathcal{U}} c^T x : Ax \leq b \; \forall (c, A, b) \in \mathcal{U} \right\},$$

or, which is the same, the optimization problem

$$\min_{x,t} \left\{ t : c^T x \leq t, \; Ax \leq b \; \forall (c, A, b) \in \mathcal{U} \right\}. \tag{RC}$$

The latter problem is called the *Robust Counterpart* (RC) of the original uncertain problem. The feasible/optimal solutions to the RC are called *robust feasible/robust optimal* solutions to the uncertain problem. The Robust Optimization methodology, in its simplest version, proposes to associate with an uncertain problem its Robust Counterpart and to use, as our "real life" decisions, the associated robust optimal solutions.

At this point, it is instructive to compare the RO paradigm with more traditional approaches to treating data uncertainty in Optimization, specifically, with *Stochastic Optimization* and *Sensitivity Analysis*.

Robust vs. Stochastic Optimization. In Stochastic Optimization (SO), the uncertain numerical data are assumed to be *random*. In the simplest case, these random data obey a known in advance probability distribution, while in more advanced settings, this distribution is only partially known. Here again an uncertain LP problem is associated with a deterministic counterpart, most notably with the *chance constrained* problem[1]

$$\min_{x,t} \left\{ t : \text{Prob}_{(c,A,b)\sim P} \left\{ c^T x \le t \ \& \ Ax \le b \right\} \ge 1 - \epsilon \right\}, \qquad \text{(ChC)}$$

where $\epsilon \ll 1$ is a given tolerance and P is the distribution of the data (c, A, b). When this distribution is only partially known — all we know is that P belongs to a given family $\mathcal{P}$ of probability distributions on the space of the data — the above setting is replaced with the *ambiguous chance constrained* setting,

$$\min_{x,t} \left\{ t : \text{Prob}_{(c,A,b)\sim P} \left\{ c^T x \le t \ \& \ Ax \le b \right\} \ge 1 - \epsilon \ \forall P \in \mathcal{P} \right\}. \qquad \text{(Amb)}$$

The SO approach seems to be less conservative than the worst-case-oriented RO approach. However, this is so *if* indeed the uncertain data are of a stochastic nature, *if* we are smart enough to point out the associated probability distribution (or at least a "narrow" family of distributions to which the true one belongs), and *if* indeed we are ready to accept probabilistic guarantees as given by chance constraints. The three if's above are indeed satisfied in some applications, such as Signal Processing, or analysis and synthesis of service systems[2]. At the same time, in numerous applications the three aforementioned if's are too restrictive. Think, e.g., of measurement/estimation errors for *individual* problems, like PILOT4. Even assuming that preparation of data entries for PILOT4 indeed involved something random, we perhaps could think about the distribution of the nominal data given the true ones, but not about what we actually need — the distribution of the true data given the nominal ones. The latter most probably just does not make sense — PILOT4 represents a particular decision-making problem with particular deterministic (albeit not known to us exactly) data, and all we can say about this true data given the nominal ones, is that the former data lies in given confidence intervals around the nominal data (and even this can be said under the assumption that when

---

[1] The concept of chance constraints goes back to A. Charnes, W.W. Copper, and G.H. Symonds [40], 1958. An alternative to chance constrained setting is where we want to optimize the expected value of the objective (the latter can incorporate penalty terms for violation of uncertain constraints) under the certain part of the original constraints. This approach, however, is aimed at "soft" constraints, while we are primarily interested in the case there the constraints are hard.

[2] Indeed, in these subject areas the random factors (like observation noises in Signal Processing, or interarrival/service times in service systems) are of random nature with more or less easy-to-identify distributions, especially when we have reasons to believe that different components of random data (like different entries in the observation noises, or individual inter-arrival and service times) are independent of each other. In such situations identifying the distribution of the data reduces to identifying a bunch of low-dimensional distributions, which is relatively easy. Furthermore, the systems in question are aimed at servicing many customers over long periods of time, so that here the probabilistic guarantees do make sense. For example, day by day many hundreds/thousands of users are sending/receiving e-mails or contacting a calling center, and a probabilistic description of the service level (the probability for an e-mail to be lost, or for the time to get an operator response to become unacceptably long) makes good sense — it merely says that in the long run, a certain fraction of users/custmers will be dissatisfied.

measuring the true data to get the nominal, no "rare event" took place). Further, even when the true data indeed are of a stochastic nature, it is usually difficult to properly identify the underlying distributions. Unless there are good reasons to a priori specify these distributions up to a small number of parameters that further can be estimated sufficiently well from observations[3], accurate identification of a "general type" multi-dimensional probability distribution usually requires an astronomical, completely unrealistic number of observations. As a result, Stochastic Optimization more often than not is forced to operate with oversimplified *guesses* for the actual distributions (like the log-normal factor model for stock returns), and usually it is very difficult to evaluate the influence of this new uncertainty — in the probability distribution — on the quality of the SO-based decisions.

The third of the above if's, our willingness to accept probabilistic guarantees, also can be controversial. Imagine, for the sake of argument, that we have at our disposal a perfect stochastic model of the stock market — as solid as the transparent model of a lottery played every week in many countries. Does the relevance of the stochastic model of the stock market make the associated probabilistic guarantees of the performance of a pension fund really meaningful for an individual customer, as meaningful as a similar guarantee in the lottery case? We believe that many customers will answer this question negatively, and rightfully so. People playing a lottery on a regular basis during their life span, participate in several hundreds of lotteries, and thus can refer to the Law of Large Numbers as a kind of indication that probabilistic guarantees indeed are meaningful for them. In contrast to this, every individual plays the "pension fund lottery" just once, which makes the interpretation of probabilistic guarantees much more problematic. Of course, the three if's above become less restrictive when passing from the chance constrained problem (ChC), where the distribution of the uncertain data is known exactly, to the ambiguously chance constrained problem (Amb), and become the less restrictive the wider families of distributions $\mathcal{P}$ we are ready to consider. Note, however, that passing from (ChC) to (Amb) is, conceptually, a step towards the Robust Counterpart — the latter is nothing but the ambiguously chance constrained problem associated with the family $\mathcal{P}$ of *all* distributions supported on a given set $\mathcal{U}$.

In fact the above three if's should be augmented by a fourth, even more restrictive "if" — chance constrained settings (ChC) and (Amb) can be treated as actual sources of "immunized against uncertainty" decisions only *if* these problems are computationally tractable; when that is not the case, these settings become more wishful thinking than actual decision-making tools. As a matter of fact, the computational tractability of chance constrained problems is a pretty rare commodity — aside of a number of very particular cases, it is difficult to verify (especially when ϵ is really small) whether a given candidate solution is feasible for a chance constrained problem. In addition, chance constraints more often than not result in *nonconvex* feasible sets, which make the optimization required in (ChC) and (Amb)

---

[3]For example, one can refer to the Central Limit Theorem in order to justify the standard — the Gaussian — model of noise in communications.

highly problematic. In sharp contrast to this, the Robust Counterparts of uncertain *Linear Programming* problems are *computationally tractable*, provided the underlying uncertainty sets $\mathcal{U}$ satisfy mild convexity and computability assumptions (e.g., are given by explicit systems of efficiently computable convex inequalities).

It should be added that the "conservatism" of RO as compared to SO is in certain respects an advantage rather than a disadvantage. When designing a construction, like a railroad bridge, by applying quantitative techniques, engineers usually increase the safety-related design parameters, like thicknesses of bars, by a reasonable margin, such as 30 to 50%, in order to account for modeling inaccuracies, rare but consequential environmental conditions, etc. With the Robust Optimization approach, this desire "to stay on the safe side" can be easily achieved by enlarging the uncertainty set. This is not the case in a chance constrained problem (ChC), where the total "budget of uncertainty" is fixed — the total probability mass of all realizations of the uncertain data must be one, so that when increasing the probabilities of some "scenarios" to make them more "visible," one is forced to reduce probabilities of other scenarios, and there are situations where this phenomenon is difficult to handle. Here again, in order to stay "on the safe side" one needs to pass from chance constrained problems to their ambiguously chance constrained modifications, that is, to move towards Robust Counterparts.

In our opinion, Stochastic and Robust Optimization are *complementary* approaches for handling data uncertainty in Optimization, each having its own advantages and drawbacks. For example, information on the stochastic nature of data uncertainty, if any, can be utilized in the RO framework, as a kind of a guideline for building uncertainty sets $\mathcal{U}$. It turns out that the latter can be built in such a way that by immunizing a candidate solution against *all* realizations of the data from $\mathcal{U}$, we automatically immunize it against *nearly all* (namely, up to realizations of total probability mass $\leq \epsilon$) random perturbations, thus making the solution feasible for the chance constrained problem. A naive way to achieve this goal would be to choose $\mathcal{U}$ as a computationally tractable convex set that "$(1 - \epsilon)$-supports" all distributions from $\mathcal{P}$ (that is, $P(\mathcal{U}) \geq 1 - \epsilon$ for all $P \in \mathcal{P}$). In this book, however, we show that under mild assumptions there exist less evident and *incomparably less conservative* ways to come up with uncertainty sets achieving the above goal.

Robust Optimization and Sensitivity Analysis. Along with Stochastic Optimization, another traditional body of knowledge dealing, in a sense, with data uncertainty in optimization is *Sensitivity Analysis*. Here the issues of primary importance are the continuity properties of the usual (the nominal) optimal solution as a function of the underlying nominal data. It is immediately seen that both Robust and Stochastic Optimization are aimed at answering the same question (albeit in different settings), the question of building an uncertainty-immunized solution to an optimization problem with uncertain data; Sensitivity Analysis is aimed at a completely different question.

Robust Optimization History. Robust optimization has many roots and precursors in the applied sciences. Some of these connections are explicit, while others are a way, looking backwards in time, to interpret an approach that was developed under different ideas. We mention three areas where robustness has played, and continues to play, an important role.

Robust Control. The field of *Robust Control* has evolved, mainly during the 90s, in the interest of control systems designers for some level of guarantee in terms of stability of the controlled system. The quest for robustness can be historically traced back to the concept of a stability margin developed in the early 30s by Bode and others, in the context of feedback amplifiers. Questions such as the "stability margin," which is the amount of feedback gain required to de-stabilize a controlled system, led naturally to a "worst-case" point of view, in which "bad" parameter values are too dangerous to be allowed, even with low probability. In the late 80s, the then-classical approach to control of large-scale feedback systems, which was based on stochastic optimization ideas, came under criticism as it could not be guaranteed to offer any kind of stability margin. The approach of $\mathbf{H}_\infty$ control was then developed as a multivariate generalization of the stability margin in the early 90s. Later, the approach was extended under the name μ-control, to handle more general, parametric perturbations (the $\mathbf{H}_\infty$ norm measures robustness with respect to a very special kind of perturbation). The corresponding robust control design problem turns out to be difficult, but relaxations based on conic (precisely, semidefinite) optimization were introduced under the name of Linear Matrix Inequalities.

Robust Statistics. In Statistics, robustness usually refers to insensitivity to outliers. Huber (see, e.g., [65]) has proposed a way to handle outliers by a modification of loss functions. The precise connection with Robust Optimization is yet to be made.

Machine Learning. More recently, the field of Machine Learning has witnessed great interest in Support Vector Machines, which are classification algorithms that can be interpreted as maximizing robustness to a special kind of uncertainty. We return to this topic in chapter 12.

Robust Linear and Convex Optimization. Aside of the outlined precursors, the paradigm of Robust Optimization per se, in the form considered here, goes back to A.L. Soyster [109], who was the first to consider, as early as in 1973, what now is called Robust Linear Programming. To the best of our knowledge, in two subsequent decades there were only two publications on the subject [52, 106]. The activity in the area was revived circa 1997, independently and essentially simultaneously, in the frameworks of both Integer Programming (Kouvelis and Yu [70]) and

Convex Programming (Ben-Tal and Nemirovski [3, 4], El Ghaoui et al. [49, 50]). Since 2000, the RO area is witnessing a burst of research activity in both theory and applications, with numerous researchers involved worldwide. The magnitude and diversity of the related contributions make it beyond our abilities to discuss them here. The reader can get some impression of this activity from [9, 16, 110, 89] and references therein.

C. The Scope of Robust Optimization and Our Focus in this Book

By itself, the RO methodology can be applied to every generic optimization problem where one can separate numerical data (that can be partly uncertain and are only known to belong to a given uncertainty set) from problem's structure (that is known in advance and is common for all instances of the uncertain problem). In particular, the methodology is fully applicable to

- *conic problems* — convex problems of the form

$$\min_x \left\{ c^T x : b - Ax \in \mathbf{K} \right\}, \qquad (C)$$

  where $\mathbf{K}$ is a given "well-structured" convex cone, representing, along with the sizes of A, a problem's structure, while the numerical entries (c, A, b) form problems's data. Conic problems look very similar to LP programs that are recovered when $\mathbf{K}$ is specified as the nonnegative orthant $\mathbb{R}^m_+$. Two other common choices of the cone $\mathbf{K}$ are:

  — a direct product of *Lorentz cones* of different dimensions. The k-dimensional Lorentz cone (also called the Second Order, or the Ice-Cream cone) is defined as

$$\mathbf{L}^k = \{ x \in \mathbb{R}^k : x_k \geq \sqrt{x_1^2 + \dots + x_{k-1}^2}.$$

  Problems (C) with direct products of Lorentz cones in the role of $\mathbf{K}$ are called *Conic Quadratic*, or *Second Order Conic* Programming problems;

  — a direct product of *semidefinite cones* of different sizes. The semidefinite cone of size k, denoted by $\mathbf{S}^k_+$, is the set of all symmetric positive semidefinite $k \times k$ matrices; it "lives" in the linear space $\mathbf{S}^k$ of all symmetric $k \times k$ matrices. Problems (C) with direct products of semidefinite cones in the role of $\mathbf{K}$ are called *Semidefinite programs*.

  Conic Quadratic and especially Semidefinite Programming problems possess extremely rich "expressive abilities"; in fact, Semidefinite Programming "captures" nearly all convex problems arising in applications, see, e.g., [8, 32, 33].

- *Integer and Mixed Integer Linear Programming* – Linear Programming problems where all or part of the variables are further restricted to be integers.

The broad spectrum of research questions related to Robust Optimization can be split into three main categories.

i) Extensions of the RO paradigm. It turns out that the implicit assumptions A.1, A.2, A.3 that led us to the central notion of Robust Counterpart of an uncertain optimization problem, while meaningful in numerous applications, in some other applications do not fully reflect the possibilities to handle the uncertain data. At present, two extensions addressing this added flexibility exist:

• *Globalized Robust Counterpart.* This extension of the notion of RC corresponds to the case when we revise Assumption A.2. Specifically, now we require a candidate solution x to satisfy the constraints for all instances of the uncertain data in $\mathcal{U}$ and, in addition, seek for *controlled* deterioration of the constraints evaluated at x when the uncertain data runs out of $\mathcal{U}$. The corresponding analogy to the Robust Counterpart of an uncertain (conic) problem, called *Globalized Robust Counterpart* (GRC), is the optimization program

$$\min_{x,t}\left\{t: \begin{array}{l} c^T x - t \leq \alpha_{\text{obj}}\text{dist}((c,A,b),\mathcal{U}) \\ \text{dist}(b - Ax, \mathbf{K}) \leq \alpha_{\text{cons}}\text{dist}((c,A,b),\mathcal{U}) \end{array} \right\} \forall (c,A,b)\right\}, \quad \text{(GRC)}$$

where the distances come from given norms on the corresponding spaces and α_{obj}, α_{cons} are given nonnegative *global sensitivities*.

• *Adjustable Robust Counterpart.* This extension of the notion of RC corresponds to the case when some of the decision variables x_j represent "wait and see" decisions to be made when the true data partly reveals itself, or are *analysis variables* that do not represent decisions (e.g., slack variables introduced to convert the original problem into a prescribed form, say, an LP one). It is natural to allow these *adjustable* variables to adjust themselves to the true data. Specifically, we can assume that every decision variable x_j is allowed to depend on a given "portion" $P_j(c,A,b)$ of the true data (c,A,B) of a (conic) problem:

$$x_j = X_j(P_j(c,A,b)),$$

where $X_j(\cdot)$ can be arbitrary functions. We then require from the resulting *decision rules* to satisfy the constraints of the uncertain conic problem for all realizations of the data from $\mathcal{U}$. The corresponding *Adjustable Robust Counterpart* (ARC) of an uncertain conic problem is the optimization program

$$\min_{X_1(\cdot),\ldots,X_n(\cdot),t}\left\{t: b - A\left[\begin{array}{c} X_1(P_1(c,A,b)) \\ \vdots \\ X_n(P_n(c,A,b)) \end{array}\right] \in \mathbf{K} \; \forall (c,A,b) \in \mathcal{U}\right\}. \quad \text{(ARC)}$$

It should be stressed that the optimization in (ARC) is carried out not over finite-dimensional vectors, as is the case in RC and GRC, but over infinite-dimensional *decision rules* — arbitrary functions $X_j(\cdot)$ on the corresponding finite-dimensional vector spaces. In order to cope, to some extent, with a severe computational intractability of ARCs, one can restrict the structure of decision rules, most notably, to make them affine in their arguments:

$$X_j(p_j) = q_j + r_j^T p_j.$$

When restricted to affine decision rules, the ARC becomes an optimization problem in finitely many real variables q_j, r_j, $1 \leq j \leq n$. This problem is called the *Affinely Adjustable Robust Counterpart* (AARC) of the original uncertain conic problem corresponding to the *information base* $P_1(\cdot), ..., P_n(\cdot)$.

ii) Investigating tractability issues of Robust Counterparts. Already the plain Robust Counterpart,

$$\min_{x,t} \left\{ t : c^T x \leq t, b - Ax \in \mathbf{K} \ \forall (c, A, b) \in \mathcal{U} \right\}, \qquad \text{(RC)}$$

of an uncertain conic problem,

$$\left\{ \min_x \left\{ c^T x : b - Ax \in \mathbf{K} \right\} : (c, A, b) \in \mathcal{U} \right\},$$

has a more complicated structure than an instance of the uncertain problem itself: (RC) is what is called a *semi-infinite* conic problem, one with *infinitely many* conic constraints $\begin{bmatrix} t - c^T x \\ b - Ax \end{bmatrix} \in \mathbf{K}_+ = \mathbb{R}_+ \times \mathbf{K}$ parameterized by the uncertain data (c, A, b) running through the uncertainty set. While (RC) is still convex, the semi-infinite nature makes it more difficult computationally than the instances of the associated uncertain problem. It may well happen that (RC) is computationally intractable, even when the uncertainty set $\mathcal{U}$ is a nice convex set (say, a ball, or a polytope) and the cone $\mathbf{K}$ is as simple as in the case of Conic Quadratic and Semidefinite programs. At the same time, in order for RO to be a working tool rather than wishful thinking, we need the RC to be computationally tractable; after all, what is the point in reducing something to an optimization problem that we do not know how to process computationally? This motivates what is in our opinion *the* the main theoretical challenge in Robust Optimization: *identifying the cases where the RC (GRC, AARC, ARC) of an uncertain conic problem admits a computationally tractable equivalent reformulation, or at least a computationally tractable safe approximation.* Here safety means that every feasible solution to the approximation is feasible for the "true" Robust Counterpart.

At the present level of our knowledge, the "big picture" here is as follows.
• When the cone $\mathbf{K}$ is "as simple as possible," i.e., is a nonnegative orthant (the case of uncertain Linear Programming), the Robust Counterpart (and under mild additional structural conditions, the GRC and the AARC as well) is computationally tractable, provided that the underlying (convex) uncertainty set $\mathcal{U}$ is so. The latter means that $\mathcal{U}$ is a convex set given by an explicit system of efficiently computable convex constraints (say, a polytope given by an explicit list of linear inequalities).
• When the (convex) uncertainty set $\mathcal{U}$ is "as simple as possible," i.e., a polytope given as a convex hull of a finite set of reasonable cardinality (scenario uncertainty), the RC is computationally tractable whenever $\mathbf{K}$ is a computationally tractable convex cone, as is the case in Linear, Conic Quadratic, and Semidefinite Programming.

• In between the above two extremes, for example, in the case of uncertain Conic Quadratic and Semidefinite problems with polytopes in the role of uncertainty sets, the RCs are, in general, computationally intractable. There are however particular cases, important for applications, where the RC is tractable, and even more cases where it admits safe tractable approximations that are *tight*, in a certain precise sense.

iii) *Applications.* This avenue of RO research is aimed at building and processing Robust Counterparts of specific optimization problems arising in various applications.

The position of our book with respect to these three major research areas in Robust Optimization is as follows:

> *Our primary emphasis is on presenting in full detail the Robust Optimization paradigm* (including its recent extensions mentioned in item 1, as well as links with Chance Constrained Stochastic Optimization) *and tractability issues, primarily for Uncertain Linear, Conic Quadratic, and Semidefinite Programming.*

D. Prerequisites and Contents

Prerequisites for reading this book are quite modest — essentially, all that is expected of a reader is knowledge of basic Analysis, Linear Algebra, and Probability, plus general mathematical culture. Preliminary "subject-specific" knowledge, (which in our case means knowledge of the Convex Optimization basics, primarily of Conic Programming and Conic Duality, on one hand, and of tractability issues in Convex Programming, on the other), while being highly welcomed, is not absolutely necessary. All required basics can be found in the Appendix augmenting the main body of the book.

The contents. The main body of the book is split into four parts:

• Part I is the basic theory of the "here and now" (i.e., the non-adjustable) Robust Linear Programming, which starts with detailed discussion of the concepts of an uncertain Linear Programming problem and its Robust/Generalized Robust Counterparts. Along with other results, we demonstrate that the RC/GRC of an uncertain LP problem is computationally tractable, provided that the uncertainty set is so. As it was already mentioned, such a general tractability result is a specific feature of uncertain LP. Another major theme of Part I is that of computationally tractable safe approximations of chance constrained uncertain LP problems with randomly perturbed data.

Part I, perhaps with chapter 4 skipped, can be used as a stand-alone graduate-level textbook on Robust Linear Programming, or as a base of a semester-long graduate course on Robust Optimization.

• Part II can be treated as a "conic version" of Part I, where the main concepts of non-adjustable Robust Optimization are extended to uncertain Convex Programming problems in the conic form, with emphasis on uncertain Conic Quadratic and Semidefinite Programming problems. As it was already mentioned, aside of the (in fact, trivial) case of scenario uncertainty, Robust/Generalized Robust Counterparts of uncertain CQP/SDP problems are, in general, computationally intractable. This is why the emphasis is on identifying, and illustrating the importance of generic situations where the RCs/GRCs of uncertain Conic Quadratic/Semidefinite problems admit tractable reformulation, or a tight safe tractable approximation. Another theme considered in Part II is that of safe tractable approximation of chance constrained uncertain Conic Quadratic and Semidefinite problems with randomly perturbed data. As compared to its "LP predecessor" from Part I, this theme now has an unexpected twist: it turns out that safe tractable approximations of the chance constrained Conic Quadratic/Semidefinite inequalities are easier to build and to process than the tight safe tractable approximations to the RCs of these conic inequalities. This is completely opposite of what happens in the case of uncertain LP problems, where it is easy to process *exactly* the RCs, but not the chance constrained versions of uncertain linear inequality constraints.

We conclude Part II investigating Robust Counterparts of specific "well structured" uncertain convex constraints arising in Machine Learning and Linear Regression models. Since the most interesting uncertain constraints arising in this context are neither Conic Quadratic nor Semidefinite, the tractability-related questions associated with the RCs of these constraints need a dedicated treatment, and this is our major goal in the corresponding chapter.

• Part III is devoted to *Robust Multi-Stage Decision Making*, specifically, to Robust Dynamic Programming, and to Adjustable (with emphasis on *Affinely* Adjustable) Robust Counterparts of uncertain conic problems, primarily uncertain multi-stage LPs. As always, our emphasis is on the tractability issues. We demonstrate, in particular, that the AARC methodology allows for efficient handling of the finite-horizon synthesis of linear controllers for uncertainty-affected Linear Dynamical systems with certain (and known in advance) dynamics. The design specifications in this synthesis can be given by general-type systems of linear constraints on states and controls, to be satisfied in a robust with respect to the initial state and the external inputs fashion.

• A short, single-chapter Part IV presents three realistic examples, worked out in full detail, of application of the RO methodology. While not pretending to give an impression of a wide and diverse range of existing applications of RO, these examples, we believe, add a "bit of reality" to our primarily theoretical treatment of the subject.

Reading modes. We believe that acquaintance with Part I is a natural prerequisite for reading Parts II and III; however, the latter two parts can be read independently of each other. In addition, those not interested in the theme of chance constraints, may skip the related chapters 2, 4 and 10; those interested in this theme may in the first reading skip chapter 4.

Acknowledgments. In our decade-long research on Robust Optimization, summarized in this book, we collaborated with many excellent colleagues and excellent students. This collaboration contributed significantly to the contents of this book. We are greatly indebted to those with whom we had the honor and privilege to collaborate in various ways in our RO-related research: A. Beck, D. Bertsimas, S. Boyd, O. Boni, D. Brown, G. Calafiore, M. Campi, F. Dan Barb, Y. Eldar, B. Golany, A. Goryashko, E. Guslitzer, R. Hildenbrand, G. Iyengar, A. Juditsky, M. Kočvara, H. Lebret, T. Margalit, Yu. Nesterov, A. Nilim, F. Oustry, C. Roos, A. Shapiro, S. Shtern, M. Sim, T. Terlaky, U. Topcu, L. Tuncel, J.-Ph. Vial, J. Zowe. Special gratitude goes to D. den Hertog and E. Stinstra for permitting us to use the results of their research on robust models in TV tube manufacturing (Section 15.1).

We gratefully acknowledge the relevant financial support from several agencies: Israeli Ministry of Science (grant 0200–1–98), Israeli Science Foundation (grants 306/94–3, 683/99–10.0), Germany-Israel Foundation for R&D (grant I0455–214.06/95), Minerva Foundation, Germany, US National Science Foundation (grants DMS-0510324, DMI-0619977, MSPA-MCS-0625371, SES-0835531), and US-Israel Binational Science Foundation (grant 2002038).

Aharon Ben-Tal
Laurent El Ghaoui
Arkadi Nemirovski
November 2008

Part I

Robust Linear Optimization

Chapter One

Uncertain Linear Optimization Problems and their Robust Counterparts

In this chapter, we introduce the concept of the uncertain Linear Optimization problem and its Robust Counterpart, and study the computational issues associated with the emerging optimization problems.

1.1 DATA UNCERTAINTY IN LINEAR OPTIMIZATION

Recall that the Linear Optimization (LO) *problem* is of the form

$$\min_x \left\{ c^T x + d : Ax \leq b \right\}, \tag{1.1.1}$$

where $x \in \mathbb{R}^n$ is the vector of *decision variables*, $c \in \mathbb{R}^n$ and $d \in \mathbb{R}$ form the *objective*, A is an $m \times n$ *constraint matrix*, and $b \in \mathbb{R}^m$ is the *right hand side vector*.

> Clearly, the constant term d in the objective, while affecting the optimal value, does not affect the optimal solution, this is why it is traditionally skipped. As we shall see, when treating the LO problems with *uncertain data* there are good reasons not to neglect this constant term.

The *structure* of problem (1.1.1) is given by the number m of constraints and the number n of variables, while the *data* of the problem are the collection (c, d, A, b), which we will arrange into an $(m+1) \times (n+1)$ *data matrix*

$$D = \left[\begin{array}{c|c} c^T & d \\ \hline A & b \end{array} \right].$$

> Usually not all constraints of an LO program, as it arises in applications, are of the form $a^T x \leq \text{const}$; there can be linear "$\geq$" inequalities and linear equalities as well. Clearly, the constraints of the latter two types can be represented equivalently by linear "$\leq$" inequalities, and we will assume henceforth that these are the only constraints in the problem.

Typically, the data of real world LOs (Linear Optimization problems) is not known exactly. The most common reasons for data uncertainty are as follows:

- Some of data entries (future demands, returns, etc.) do not exist when the problem is solved and hence are replaced with their forecasts. These data entries are thus subject to *prediction errors*;

- Some of the data (parameters of technological devices/processes, contents associated with raw materials, etc.) cannot be measured exactly – in reality their values drift around the measured "nominal" values; these data are subject to *measurement errors*;

- Some of the decision variables (intensities with which we intend to use various technological processes, parameters of physical devices we are designing, etc.) cannot be implemented exactly as computed. The resulting *implementation errors* are equivalent to appropriate artificial data uncertainties.

  Indeed, the contribution of a particular decision variable x_j to the left hand side of constraint i is the product $a_{ij}x_j$. Hence the consequences of an additive implementation error $x_j \mapsto x_j + \epsilon$ are as if there were no implementation error at all, but the left hand side of the constraint got an extra additive term $a_{ij}\epsilon$, which, in turn, is equivalent to the perturbation $b_i \mapsto b_i - a_{ij}\epsilon$ in the right hand side of the constraint. The consequences of a more typical *multiplicative* implementation error $x_j \mapsto (1 + \epsilon)x_j$ are as if there were no implementation error, but each of the data coefficients a_{ij} was subject to perturbation $a_{ij} \mapsto (1 + \epsilon)a_{ij}$. Similarly, the influence of additive and multiplicative implementation error in x_j on the value of the objective can be mimicked by appropriate perturbations in d or c_j.

In the traditional LO methodology, a small data uncertainty (say, 1% or less) is just ignored; the problem is solved *as if* the given ("nominal") data were exact, and the resulting *nominal* optimal solution is what is recommended for use, in hope that small data uncertainties will not affect significantly the feasibility and optimality properties of this solution, or that small adjustments of the nominal solution will be sufficient to make it feasible. We are about to demonstrate that these hopes are not necessarily justified, and sometimes even small data uncertainty deserves significant attention.

1.1.1 Introductory Example

Consider the following very simple linear optimization problem:

Example 1.1.1. A company produces two kinds of drugs, DrugI and DrugII, containing a specific active agent A, which is extracted from raw materials purchased on the market. There are two kinds of raw materials, RawI and RawII, which can be used as sources of the active agent. The related production, cost, and resource data are given in table 1.1. The goal is to find the production plan that maximizes the profit of the company.

Parameter	DrugI	DrugII
Selling price, $ per 1000 packs	6,200	6,900
Content of agent A, g per 1000 packs	0.500	0.600
Manpower required, hours per 1000 packs	90.0	100.0
Equipment required, hours per 1000 packs	40.0	50.0
Operational costs, $ per 1000 packs	700	800

(a) Drug production data

Raw material	Purchasing price, $ per kg	Content of agent A, g per kg
RawI	100.00	0.01
RawII	199.90	0.02

(b) Contents of raw materials

Budget, $	Manpower, hours	Equipment, hours	Capacity of raw materials storage, kg
100,000	2,000	800	1,000

(c) Resources

Table 1.1   Data for Example 1.1.1.

The problem can be immediately posed as the following linear programming program:

(Drug):

$$\text{Opt} = \min \left\{ \overbrace{[100 \cdot RawI + 199.90 \cdot RawII + 700 \cdot DrugI + 800 \cdot DrugII]}^{\text{purchasing and operational costs}} \\ \underbrace{- [6200 \cdot DrugI + 6900 \cdot DrugII]}_{\text{income from selling the drugs}} \right\} \quad \text{[minus total profit]}$$

subject to

$$
\begin{array}{ll}
0.01 \cdot RawI + 0.02 \cdot RawII - 0.500 \cdot DrugI - 0.600 \cdot DrugII \geq 0 & \text{[balance of active agent]} \\
RawI + RawII \leq 1000 & \text{[storage constraint]} \\
90.0 \cdot DrugI + 100.0 \cdot DrugII \leq 2000 & \text{[manpower constraint]} \\
40.0 \cdot DrugI + 50.0 \cdot DrugII \leq 800 & \text{[equipment constraint]} \\
100.0 \cdot RawI + 199.90 \cdot RawII + 700 \cdot DrugI + 800 \cdot DrugII \leq 100000 & \text{[budget constraint]} \\
RawI, RawII, DrugI, DrugII \geq 0 &
\end{array}
$$

The problem has four variables — the amounts $RawI$, $RawII$ (in kg) of raw materials to be purchased and the amounts $DrugI$, $DrugII$ (in 1000 of packs) of drugs to be produced.

The optimal solution of our LO problem is

$$\text{Opt} = -8819.658; RawI = 0, RawII = 438.789, DrugI = 17.552, DrugII = 0.$$

Note that both the budget and the balance constraints are active (that is, the production process utilizes the entire 100,000 budget and the full amount of active agent contained in the raw materials). The solution promises the company a modest, but quite respectable profit of 8.8%.

1.1.2 Data Uncertainty and its Consequences

Clearly, even in our simple problem some of the data cannot be "absolutely reliable"; e.g., one can hardly believe that the contents of the active agent in the raw materials are exactly 0.01 g/kg for RawI and 0.02 g/kg for RawII. In reality, these contents vary around the indicated values. A natural assumption here is that the actual contents of active agent aI in RawI and aII in RawII are realizations of random variables somehow distributed around the "nominal contents" $anI = 0.01$ and $anII = 0.02$. To be more specific, assume that aI drifts in a 0.5% margin of anI, thus taking values in the segment $[0.00995, 0.01005]$. Similarly, assume that aII drifts in a 2% margin of $anII$, thus taking values in the segment $[0.0196, 0.0204]$. Moreover, assume that aI, aII take the two extreme values in the respective segments with probabilities 0.5 each. How do these perturbations of the contents of the active agent affect the production process? The optimal solution prescribes to purchase 438.8 kg of RawII and to produce 17.552K packs of DrugI (K stands for "thousand"). With the above random fluctuations in the content of the active agent in RawII, this production plan will be infeasible with probability 0.5, i.e., the actual content of the active agent in raw materials will be less than the one required to produce the planned amount of DrugI. This difficulty can be resolved in the simplest way: when the actual content of the active agent in raw materials is insufficient, the output of the drug is reduced accordingly. With this policy, the actual production of DrugI becomes a random variable that takes with equal probabilities the nominal value of 17.552K packs and the (2% less) value of 17.201K packs. These 2% fluctuations in the production affect the profit as well; it becomes a random variable taking, with probabilities 0.5, the nominal value 8,820 and the 21% (!) less value 6,929. The expected profit is 7,843, which is 11% less than the nominal profit 8,820 promised by the optimal solution of the nominal problem.

We see that in our simple example a pretty small (and unavoidable in reality) perturbation of the data may make the nominal optimal solution infeasible. Moreover, a straightforward adjustment of the nominally optimal solution to the actual data may heavily affect the quality of the solution.

Similar phenomenon can be met in many practical linear programs where at least part of the data are not known exactly and can vary around their nominal values. The consequences of data uncertainty can be much more severe than in our toy example. The analysis of linear optimization problems from the NETLIB collection[1] reported in [7] reveals that for 13 of 94 NETLIB problems, random 0.01% perturbations of the uncertain data can make the nominal optimal solution severely infeasible: with a non-negligible probability, it violates some of the constraints by

---

[1]A collection of LP programs, including those of real world origin, used as a standard benchmark for testing LP solvers.

50% and more. It should be added that in the general case (in contrast to our toy example) there is no evident way to adjust the optimal solution to the actual values of the data by a small modification, and there are cases when such an adjustment is in fact impossible; in order to become feasible for the perturbed data, the nominal optimal solution should be "completely reshaped."

The conclusion is as follows:

> *In applications of LO, there exists a real need of a technique capable of detecting cases when data uncertainty can heavily affect the quality of the nominal solution, and in these cases to generate a "reliable" solution, one that is immunized against uncertainty.*

We are about to introduce the *Robust Counterpart* approach to uncertain LO problems aimed at coping with data uncertainty.

1.2 UNCERTAIN LINEAR PROBLEMS AND THEIR ROBUST COUNTERPARTS

Definition 1.2.1. An uncertain Linear Optimization problem is a collection

$$\left\{\min_x \left\{c^T x + d : Ax \le b\right\}\right\}_{(c,d,A,b) \in \mathcal{U}} \tag{LO$_\mathcal{U}$}$$

of LO problems (instances) $\min_x \left\{c^T x + d : Ax \le b\right\}$ of common structure (i.e., with common numbers m of constraints and n of variables) with the data varying in a given <u>uncertainty set</u> $\mathcal{U} \subset \mathbb{R}^{(m+1)\times(n+1)}$.

We always assume that the uncertainty set is parameterized, in an affine fashion, by *perturbation vector* ζ varying in a given *perturbation set* $\mathcal{Z}$:

$$\mathcal{U} = \left\{ \left[\begin{array}{c|c} c^T & d \\ \hline A & b \end{array}\right] = \underbrace{\left[\begin{array}{c|c} c_0^T & d_0 \\ \hline A_0 & b_0 \end{array}\right]}_{\substack{\text{nominal} \\ \text{data } D_0}} + \sum_{\ell=1}^{L} \zeta_\ell \underbrace{\left[\begin{array}{c|c} c_\ell^T & d_\ell \\ \hline A_\ell & b_\ell \end{array}\right]}_{\substack{\text{basic} \\ \text{shifts } D_\ell}} : \zeta \in \mathcal{Z} \subset \mathbb{R}^L \right\}. \tag{1.2.1}$$

For example, the story told in section 1.1.2 makes (Drug) an uncertain LO problem as follows:

- *Decision vector:* $x = [RawI; RawII; DrugI; DrugII]$;

- *Nominal data:* $D_0 = \left[\begin{array}{cccc|c} 100 & 199.9 & -5500 & -6100 & 0 \\ \hline -0.01 & -0.02 & 0.500 & 0.600 & 0 \\ 1 & 1 & 0 & 0 & 1000 \\ 0 & 0 & 90.0 & 100.0 & 2000 \\ 0 & 0 & 40.0 & 50.0 & 800 \\ 100.0 & 199.9 & 700 & 800 & 100000 \\ -1 & 0 & 0 & 0 & 0 \\ 0 & -1 & 0 & 0 & 0 \\ 0 & 0 & -1 & 0 & 0 \\ 0 & 0 & 0 & -1 & 0 \end{array}\right]$

- *Two basic shifts:*

$$D_1 = 5.0\text{e-}5 \cdot \begin{bmatrix} 0 & 0 & 0 & 0 & | & 0 \\ \hline 1 & 0 & 0 & 0 & | & 0 \\ 0 & 0 & 0 & 0 & | & 0 \\ 0 & 0 & 0 & 0 & | & 0 \\ 0 & 0 & 0 & 0 & | & 0 \\ 0 & 0 & 0 & 0 & | & 0 \\ 0 & 0 & 0 & 0 & | & 0 \\ 0 & 0 & 0 & 0 & | & 0 \\ 0 & 0 & 0 & 0 & | & 0 \end{bmatrix}, \ D_2 = 4.0\text{e-}4 \cdot \begin{bmatrix} 0 & 0 & 0 & 0 & | & 0 \\ \hline 0 & 1 & 0 & 0 & | & 0 \\ 0 & 0 & 0 & 0 & | & 0 \\ 0 & 0 & 0 & 0 & | & 0 \\ 0 & 0 & 0 & 0 & | & 0 \\ 0 & 0 & 0 & 0 & | & 0 \\ 0 & 0 & 0 & 0 & | & 0 \\ 0 & 0 & 0 & 0 & | & 0 \\ 0 & 0 & 0 & 0 & | & 0 \end{bmatrix}$$

- *Perturbation set:*

$$\mathcal{Z} = \left\{ \zeta \in \mathbb{R}^2 : -1 \leq \zeta_1, \zeta_2 \leq 1 \right\}.$$

This description says, in particular, that the only uncertain data in (Drug) are the coefficients anI, $anII$ of the variables $RawI$, $RawII$ in the balance inequality, (which is the first constraint in (Drug)), and that these coefficients vary in the respective segments $[0.01 \cdot (1 - 0.005), 0.01 \cdot (1 + 0.005)]$, $[0.02 \cdot (1 - 0.02), 0.02 \cdot (1 + 0.02)]$ around the nominal values 0.01, 0.02 of the coefficients, which is exactly what was stated in section 1.1.2.

Remark 1.2.2. If the perturbation set $\mathcal{Z}$ in (1.2.1) itself is represented as the image of another set $\widehat{\mathcal{Z}}$ under affine mapping $\xi \mapsto \zeta = p + P\xi$, then we can pass from perturbations ζ to perturbations ξ:

$$\begin{aligned}
\mathcal{U} &= \left\{ \begin{bmatrix} c^T & d \\ \hline A & b \end{bmatrix} = D_0 + \sum_{\ell=1}^{L} \zeta_\ell D_\ell : \zeta \in \mathcal{Z} \right\} \\
&= \left\{ \begin{bmatrix} c^T & d \\ \hline A & b \end{bmatrix} = D_0 + \sum_{\ell=1}^{L} [p_\ell + \sum_{k=1}^{K} P_{\ell k}\xi_k] D_\ell : \xi \in \widehat{\mathcal{Z}} \right\} \\
&= \left\{ \begin{bmatrix} c^T & d \\ \hline A & b \end{bmatrix} = \underbrace{\left[D_0 + \sum_{\ell=1}^{L} p_\ell D_\ell \right]}_{\widehat{D}_0} + \sum_{k=1}^{K} \xi_k \underbrace{\left[\sum_{\ell=1}^{L} P_{\ell k} D_\ell \right]}_{\widehat{D}_k} : \xi \in \widehat{\mathcal{Z}} \right\}.
\end{aligned}$$

It follows that when speaking about perturbation sets with simple geometry (parallelotopes, ellipsoids, etc.), we can normalize these sets to be "standard." For example, a parallelotope is by definition an affine image of a unit box $\{\xi \in \mathbb{R}^k : -1 \leq \xi_j \leq 1, j = 1, ..., k\}$, which gives us the possibility to work with the unit box instead of a general parallelotope. Similarly, an ellipsoid is by definition the image of a unit Euclidean ball $\{\xi \in \mathbb{R}^k : \|x\|_2^2 \equiv x^T x \leq 1\}$ under affine mapping, so that we can work with the standard ball instead of the ellipsoid, etc. We will use this normalization whenever possible.

Note that a *family* of optimization problems like $(\text{LO}_{\mathcal{U}})$, in contrast to a single optimization problem, is not associated by itself with the concepts of feasible/optimal solution and optimal value. How to define these concepts depends of course on the underlying "decision environment." Here we focus on an environment with the following characteristics:

A.1. All decision variables in (LO$_\mathcal{U}$) represent "here and now" decisions; they should be assigned specific numerical values as a result of solving the problem *before* the actual data "reveals itself."

A.2. The decision maker is fully responsible for consequences of the decisions to be made when, and only when, the actual data is within the prespecified uncertainty set $\mathcal{U}$ given by (1.2.1).

A.3. The constraints in (LO$_\mathcal{U}$) are "hard" — we cannot tolerate violations of constraints, even small ones, when the data is in $\mathcal{U}$.

The above assumptions determine, in a more or less unique fashion, what are the meaningful feasible solutions to the uncertain problem (LO$_\mathcal{U}$). By A.1, these should be fixed vectors; by A.2 and A.3, they should be *robust feasible*, that is, they should satisfy all the constraints, whatever the realization of the data from the uncertainty set. We have arrived at the following definition.

Definition 1.2.3. A vector $x \in \mathbb{R}^n$ is a <u>robust feasible</u> solution to (LO$_\mathcal{U}$), if it satisfies all realizations of the constraints from the uncertainty set, that is,

$$Ax \leq b \quad \forall (c, d, A, b) \in \mathcal{U}. \tag{1.2.2}$$

As for the objective value to be associated with a meaningful (i.e., robust feasible) solution, assumptions A.1 — A.3 do not prescribe it in a unique fashion. However, "the spirit" of these worst-case-oriented assumptions leads naturally to the following definition:

Definition 1.2.4. Given a candidate solution x, the <u>robust</u> value $\widehat{c}(x)$ of the objective in (LO$_\mathcal{U}$) at x is the largest value of the "true" objective $c^T x + d$ over all realizations of the data from the uncertainty set:

$$\widehat{c}(x) = \sup_{(c,d,A,b) \in \mathcal{U}} [c^T x + d]. \tag{1.2.3}$$

After we agree what are meaningful candidate solutions to the uncertain problem (LO$_\mathcal{U}$) and how to quantify their quality, we can seek the best robust value of the objective among all robust feasible solutions to the problem. This brings us to the central concept of this book, *Robust Counterpart* of an uncertain optimization problem, which is defined as follows:

Definition 1.2.5. The Robust Counterpart of the uncertain LO problem (LO$_\mathcal{U}$) is the optimization problem

$$\min_x \left\{ \widehat{c}(x) = \sup_{(c,d,A,b) \in \mathcal{U}} [c^T x + d] : Ax \leq b \; \forall (c, d, A, b) \in \mathcal{U} \right\} \tag{1.2.4}$$

of minimizing the robust value of the objective over all robust feasible solutions to the uncertain problem.

An optimal solution to the Robust Counterpart is called a robust optimal solution to (LO$_\mathcal{U}$), and the optimal value of the Robust Counterpart is called the robust optimal value of (LO$_\mathcal{U}$).

In a nutshell, the robust optimal solution is simply "the best uncertainty-immunized" solution we can associate with our uncertain problem.

Example 1.1.1 continued. Let us find the robust optimal solution to the uncertain problem (Drug). There is exactly one uncertainty-affected "block" in the data, namely, the coefficients of *RawI*, *RawII* in the balance constraint. A candidate solution is thus robust feasible if and only if it satisfies all constraints of (Drug), except for the balance constraint, *and* it satisfies the "worst" realization of the balance constraint. Since *RawI*, *RawII* are nonnegative, the worst realization of the balance constraint is the one where the uncertain coefficients *anI*, *anII* are set to their minimal values in the uncertainty set (these values are 0.00995 and 0.0196, respectively). Since the objective is not affected by the uncertainty, the robust objective values are the same as the original ones. Thus, the RC (Robust Counterpart) of our uncertain problem is the LO problem

RC(Drug):

RobOpt = min $\{-100 \cdot RawI - 199.9 \cdot RawII + 5500 \cdot DrugI + 6100 \cdot DrugII\}$
subject to

$$
\begin{array}{rcl}
0.00995 \cdot RawI + 0.0196 \cdot RawII - 0.500 \cdot DrugI - 0.600 \cdot DrugII & \geq & 0 \\
RawI + RawII & \leq & 1000 \\
90.0 \cdot DrugI + 100.0 \cdot DrugII & \leq & 2000 \\
40.0 \cdot DrugI + 50.0 \cdot DrugII & \leq & 800 \\
100.0 \cdot RawI + 199.90 \cdot RawII + 700 \cdot DrugI + 800 \cdot DrugII & \leq & 100000 \\
RawI, RawII, DrugI, DrugII & \geq & 0
\end{array}
$$

Solving this problem, we get

RobOpt $= -8294.567$; $RawI = 877.732, RawII = 0, DrugI = 17.467, DrugII = 0$.

The "price" of robustness is the reduction in the promised profit from its nominal optimal value 8819.658 to its robust optimal value 8294.567, that is, by 5.954%. This is much less than the 21% reduction of the actual profit to 6,929 which we may suffer when sticking to the nominal optimal solution when the "true" data are "against" it. Note also that the structure of the robust optimal solution is quite different from the one of the nominal optimal solution: with the robust solution, we shall buy only raw materials RawI, while with the nominal one, only raw materials RawII. The explanation is clear: with the nominal data, RawII as compared to RawI results in a bit smaller per unit price of the active agent (9,995 \$/g vs. 10,000 \$/g). This is why it does not make sense to use RawI with the nominal data. The robust optimal solution takes into account that the uncertainty in *anI* (i.e., the variability of contents of active agent in RawI) is 4 times smaller than that of *anII* (0.5% vs. 2%), which ultimately makes it better to use RawI.

1.2.1 More on Robust Counterparts

We start with several useful observations.

A. The Robust Counterpart (1.2.4) of LO$_{\mathcal{U}}$ can be rewritten equivalently as the problem

$$
\min_{x,t} \left\{ t : \begin{array}{rcl} c^T x - t & \leq & -d \\ Ax & \leq & b \end{array} \right\} \ \forall (c, d, A, b) \in \mathcal{U} \right\}. \tag{1.2.5}
$$

Note that we can arrive at this problem in another fashion: we first introduce the extra variable t and rewrite instances of our uncertain problem $(\mathrm{LO}_\mathcal{U})$ equivalently as

$$\min_{x,t}\left\{t: \begin{array}{rcl} c^T x - t & \leq & -d \\ Ax & \leq & b \end{array}\right\},$$

thus arriving at an equivalent to $(\mathrm{LO}_\mathcal{U})$ uncertain problem in variables x, t with the objective t that is not affected by uncertainty at all. The RC of the reformulated problem is exactly (1.2.5). We see that

> An uncertain LO problem can always be reformulated as an uncertain LO problem with certain objective. The Robust Counterpart of the reformulated problem has the same objective as this problem and is equivalent to the RC of the original uncertain problem.

As a consequence, *we lose nothing when restricting ourselves with uncertain LO programs with certain objectives* and we shall frequently use this option in the future.

> We see now why the constant term d in the objective of (1.1.1) should not be neglected, or, more exactly, should not be neglected if it is uncertain. When d is certain, we can account for it by the shift $t \mapsto t - d$ in the slack variable t which affects only the optimal value, but not the optimal solution to the Robust Counterpart (1.2.5). When d is uncertain, there is no "universal" way to eliminate d without affecting the optimal solution to the Robust Counterpart (where d plays the same role as the right hand sides of the original constraints).

B. Assuming that $(\mathrm{LO}_\mathcal{U})$ is with certain objective, the Robust Counterpart of the problem is

$$\min_x\left\{c^T x + d : Ax \leq b, \ \forall (A, b) \in \mathcal{U}\right\} \tag{1.2.6}$$

(note that the uncertainty set is now a set in the space of the constraint data $[A, b]$). We see that

> The Robust Counterpart of an uncertain LO problem with a certain objective is a purely "constraint-wise" construction: to get RC, we act as follows:
>
> - *preserve the original certain objective as it is, and*
> - *replace every one of the original constraints*
>
> $$(Ax)_i \leq b_i \Leftrightarrow a_i^T x \leq b_i \tag{C$_i$}$$
>
> *(a_i^T is i-th row in A) with its Robust Counterpart*
>
> $$a_i^T x \leq b_i \ \forall [a_i; b_i] \in \mathcal{U}_i, \tag{RC(C$_i$)}$$
>
> *where $\mathcal{U}_i$ is the projection of $\mathcal{U}$ on the space of data of i-th constraint:*
>
> $$\mathcal{U}_i = \{[a_i; b_i] : [A, b] \in \mathcal{U}\}.$$

In particular,

> The RC of an uncertain LO problem with a certain objective remains intact when the original uncertainty set $\mathcal{U}$ is extended to the direct product
> $$\widehat{\mathcal{U}} = \mathcal{U}_1 \times \ldots \times \mathcal{U}_m$$
> of its projections onto the spaces of data of respective constraints.

Example 1.2.6. The RC of the system of uncertain constraints

$$\{x_1 \geq \zeta_1,\ x_2 \geq \zeta_2\} \tag{1.2.7}$$

with $\zeta \in \mathcal{U} := \{\zeta_1 + \zeta_2 \leq 1, \zeta_1, \zeta_2 \geq 0\}$ is the infinite system of constraints

$$x_1 \geq \zeta_1,\ x_1 \geq \zeta_2 \ \forall \zeta \in \mathcal{U};$$

on variables x_1, x_2. The latter system is clearly equivalent to the pair of constraints

$$x_1 \geq \max_{\zeta \in \mathcal{U}} \zeta_1 = 1,\ x_2 \geq \max_{\zeta \in \mathcal{U}} \zeta_2 = 1. \tag{1.2.8}$$

The projections of $\mathcal{U}$ to the spaces of data of the two uncertain constraints (1.2.7) are the segments $\mathcal{U}_1 = \{\zeta_1 : 0 \leq \zeta_1 \leq 1\}$, $\mathcal{U}_2 = \{\zeta_2 : 0 \leq \zeta_2 \leq 1\}$, and the RC of (1.2.7) w.r.t.[2] the uncertainty set $\widehat{\mathcal{U}} = \mathcal{U}_1 \times \mathcal{U}_2 = \{\zeta \in \mathbb{R}^2 : 0 \leq \zeta_1, \zeta_2 \leq 1\}$ clearly is (1.2.8).

> The conclusion we have arrived at seems to be counter-intuitive: it says that it is immaterial whether the perturbations of data in different constraints are or are not linked to each other, while intuition says that such a link should be important. We shall see later (chapter 14) that this intuition is valid when a more advanced concept of *Adjustable* Robust Counterpart is considered.

C. If x is a robust feasible solution of (C_i), then x remains robust feasible when we extend the uncertainty set $\mathcal{U}_i$ to its convex hull $\mathrm{Conv}(\mathcal{U}_i)$. Indeed, if $[\bar{a}_i; \bar{b}_i] \in \mathrm{Conv}(\mathcal{U}_i)$, then

$$[\bar{a}_i; \bar{b}_i] = \sum_{j=1}^{J} \lambda_j [a_i^j; b_i^j],$$

with appropriately chosen $[a_i^j; b_i^j] \in \mathcal{U}_i$, $\lambda_j \geq 0$ such that $\sum_j \lambda_j = 1$. We now have

$$\bar{a}_i^T x = \sum_{j=1}^{J} \lambda_j [a_i^j]^T x \leq \sum_j \lambda_j b_i^j = \bar{b}_i,$$

where the inequality is given by the fact that x is feasible for $\mathrm{RC}(C_i)$ and $[a_i^j; b_i^j] \in \mathcal{U}_i$. We see that $\bar{a}_i^T x \leq \bar{b}_i$ for all $[\bar{a}_i; \bar{b}_i] \in \mathrm{Conv}(\mathcal{U}_i)$, QED.

By similar reasons, the set of robust feasible solutions to (C_i) remains intact when we extend $\mathcal{U}_i$ to the closure of this set. Combining these observations with **B.**, we arrive at the following conclusion:

---

[2]abbr. for "with respect to"

> *The Robust Counterpart of an uncertain LO problem with a certain objective remains intact when we extend the sets $\mathcal{U}_i$ of uncertain data of respective constraints to their closed convex hulls, and extend $\mathcal{U}$ to the direct product of the resulting sets.*
>
> *In other words, we lose nothing when assuming from the very beginning that the sets $\mathcal{U}_i$ of uncertain data of the constraints are closed and convex, and $\mathcal{U}$ is the direct product of these sets.*

In terms of the parameterization (1.2.1) of the uncertainty sets, the latter conclusion means that

> *When speaking about the Robust Counterpart of an uncertain LO problem with a certain objective, we lose nothing when assuming that the set $\mathcal{U}_i$ of uncertain data of i-th constraint is given as*
>
> $$\mathcal{U}_i = \left\{ [a_i; b_i] = [a_i^0; b_i^0] + \sum_{\ell=1}^{L_i} \zeta_\ell [a_i^\ell; b_i^\ell] : \zeta \in \mathcal{Z}_i \right\}, \qquad (1.2.9)$$
>
> *with a closed and convex perturbation set $\mathcal{Z}_i$.*

D. An important modeling issue. In the usual — with certain data — Linear Optimization, constraints can be modeled in various equivalent forms. For example, we can write:

$$
\begin{array}{ll}
(a) & a_1 x_1 + a_2 x_2 \le a_3 \\
(b) & a_4 x_1 + a_5 x_2 = a_6 \\
(c) & x_1 \ge 0, x_2 \ge 0
\end{array}
\qquad (1.2.10)
$$

or, equivalently,

$$
\begin{array}{ll}
(a) & a_1 x_1 + a_2 x_2 \le a_3 \\
(b.1) & a_4 x_1 + a_5 x_2 \le a_6 \\
(b.2) & -a_5 x_1 - a_5 x_2 \le -a_6 \\
(c) & x_1 \ge 0, x_2 \ge 0.
\end{array}
\qquad (1.2.11)
$$

Or, equivalently, by adding a slack variable s,

$$
\begin{array}{ll}
(a) & a_1 x_1 + a_2 x_2 + s = a_3 \\
(b) & a_4 x_1 + a_5 x_2 = a_6 \\
(c) & x_1 \ge 0, x_2 \ge 0, s \ge 0.
\end{array}
\qquad (1.2.12)
$$

However, when (part of) the data $a_1, ..., a_6$ become *uncertain*, not all of these equivalences remain valid: the RCs of our now uncertainty-affected systems of constraints are not equivalent to each other. Indeed, denoting the uncertainty set by $\mathcal{U}$, the RCs read, respectively,

$$
\left.
\begin{array}{ll}
(a) & a_1 x_1 + a_2 x_2 \le a_3 \\
(b) & a_4 x_1 + a_5 x_2 = a_6 \\
(c) & x_1 \ge 0, x_2 \ge 0
\end{array}
\right\} \forall a = [a_1; ...; a_6] \in \mathcal{U}.
\qquad (1.2.13)
$$

$$
\left.
\begin{array}{ll}
(a) & a_1 x_1 + a_2 x_2 \leq a_3 \\
(b.1) & a_4 x_1 + a_5 x_2 \leq a_6 \\
(b.2) & -a_5 x_1 - a_5 x_2 \leq -a_6 \\
(c) & x_1 \geq 0, x_2 \geq 0
\end{array}
\right\} \quad \forall a = [a_1; ...; a_6] \in \mathcal{U}. \qquad (1.2.14)
$$

$$
\left.
\begin{array}{ll}
(a) & a_1 x_1 + a_2 x_2 + s = a_3 \\
(b) & a_4 x_1 + a_5 x_2 = a_6 \\
(c) & x_1 \geq 0, x_2 \geq 0, s \geq 0
\end{array}
\right\} \quad \forall a = [a_1; ...; a_6] \in \mathcal{U}. \qquad (1.2.15)
$$

It is immediately seen that while the first and the second RCs are equivalent to each other,[3] they are *not* equivalent to the third RC. The latter RC is more conservative than the first two, meaning that whenever (x_1, x_2) can be extended, by a properly chosen s, to a feasible solution of (1.2.15), (x_1, x_2) is feasible for (1.2.13)≡(1.2.14) (this is evident), but not necessarily vice versa. In fact, the gap between (1.2.15) and (1.2.13)≡(1.2.14) can be quite large. To illustrate the latter claim, consider the case where the uncertainty set is

$$
\mathcal{U} = \{a = a_\zeta := [1 + \zeta; 2 + \zeta; 4 - \zeta; 4 + \zeta; 5 - \zeta; 9] : -\rho \leq \zeta \leq \rho\},
$$

where ζ is the data perturbation. In this situation, $x_1 = 1$, $x_2 = 1$ is a feasible solution to (1.2.13)≡(1.2.14), provided that the uncertainty level ρ is $\leq 1/3$:

$$
(1 + \zeta) \cdot 1 + (2 + \zeta) \cdot 1 \leq 4 - \zeta \, \forall (\zeta : |\zeta| \leq \rho \leq 1/3) \ \& (4 + \zeta) \cdot 1 + (5 - \zeta) \cdot 1 = 9 \, \forall \zeta.
$$

At the same time, when $\rho > 0$, our solution $(x_1 = 1, x_2 = 1)$ cannot be extended to a feasible solution of (1.2.15), since the latter system of constraints is infeasible and remains so even after eliminating the equality (1.2.15.b).

Indeed, in order for x_1, x_2, s to satisfy (1.2.15.a) for all $a \in \mathcal{U}$, we should have

$$
x_1 + 2x_2 + s + \zeta[x_1 + x_2] = 4 - \zeta \ \forall (\zeta : |\zeta| \leq \rho);
$$

when $\rho > 0$, we therefore should have $x_1 + x_2 = -1$, which contradicts (1.2.15.c)

The origin of the outlined phenomenon is clear. Evidently the inequality $a_1 x_1 + a_2 x_2 \leq a_3$, where all a_i and x_i are fixed reals, holds true if and only if we can "certify" the inequality by pointing out a real $s \geq 0$ such that $a_1 x_1 + a_2 x_2 + s = a_3$. When the data a_1, a_2, a_3 become uncertain, the restriction on (x_1, x_2) to be robust feasible for the uncertain inequality $a_1 x_1 + a_2 x_2 \leq a_3$ for all $a \in \mathcal{U}$ reads, "in terms of certificate," as

$$
\forall a \in \mathcal{U} \, \exists s \geq 0 : a_1 x_1 + a_2 x_2 + s = a_3,
$$

that is, the certificate s should be allowed to depend on the true data. In contrast to this, in (1.2.15) we require from both the decision variables x *and the slack variable ("the certificate")* s to be independent of the true data, which is by far too conservative.

What can be learned from the above examples is that when modeling an uncertain LO problem one should avoid whenever possible converting inequality

---

[3] Clearly, this always is the case when an equality constraint, certain or uncertain alike, is replaced with a pair of opposite inequalities.

constraints into equality ones, unless all the data in the constraints in question are certain. Aside from avoiding slack variables,[4] this means that restrictions like "total expenditure cannot exceed the budget," or "supply should be at least the demand," which in LO problems with certain data can harmlessly be modeled by equalities, in the case of uncertain data should be modeled by inequalities. This is in full accordance with common sense saying, e.g., that when the demand is uncertain and its satisfaction is a must, it would be unwise to forbid surplus in supply. Sometimes a good for the RO methodology modeling requires eliminating "state variables" — those which are readily given by variables representing actual decisions — via the corresponding "state equations." For example, time dynamics of an inventory is given in the simplest case by the state equations

$$x_0 = c$$
$$x_{t+1} = x_t + q_t - d_t, \, t = 0, 1, ..., T,$$

where x_t is the inventory level at time t, d_t is the (uncertain) demand in period $[t, t+1)$, and variables q_t represent actual decisions – replenishment orders at instants $t = 0, 1, ..., T$. A wise approach to the RO processing of such an inventory problem would be to eliminate the state variables x_t by setting

$$x_t = c + \sum_{\tau=1}^{t-1} q_\tau, \, t = 0, 1, 2, ..., T + 1,$$

and to get rid of the state equations. As a result, typical restrictions on state variables (like "x_t should stay within given bounds" or "total holding cost should not exceed a given bound") will become uncertainty-affected inequality constraints on the actual decisions q_t, and we can process the resulting inequality-constrained uncertain LO problem via its RC.[5]

1.2.2 What is Ahead

After introducing the concept of the Robust Counterpart of an uncertain LO problem, we confront two major questions:

i) What is the "computational status" of the RC? When is it possible to process the RC efficiently?

ii) How to come-up with meaningful uncertainty sets?

The first of these questions, to be addressed in depth in section 1.3, is a "structural" one: what should be the structure of the uncertainty set in order to make the RC computationally tractable? Note that the RC as given by (1.2.5) or (1.2.6) is a *semi-infinite* LO program, that is, an optimization program with simple linear

---

[4]Note that slack variables do not represent actual decisions; thus, their presence in an LO model contradicts assumption A.1, and thus can lead to too conservative, or even infeasible, RCs.
[5]For more advanced robust modeling of uncertainty-affected multi-stage inventory, see chapter 14.

objective and *infinitely many* linear constraints. In principle, such a problem can be "computationally intractable" — NP-hard.

Example 1.2.7. Consider an uncertain "essentially linear" constraint

$$\{\|Px - p\|_1 \leq 1\}_{[P;p]\in\mathcal{U}}, \tag{1.2.16}$$

where $\|z\|_1 = \sum_j |z_j|$, and assume that the matrix P is certain, while the vector p is uncertain and is parameterized by perturbations from the unit box:

$$p \in \{p = B\zeta : \|\zeta\|_\infty \leq 1\},$$

where $\|\zeta\|_\infty = \max_\ell |\zeta_\ell|$ and B is a given positive semidefinite matrix. To check whether $x = 0$ is robust feasible is exactly the same as to verify whether $\|B\zeta\|_1 \leq 1$ whenever $\|\zeta\|_\infty \leq 1$; or, due to the evident relation $\|u\|_1 = \max_{\|\eta\|_\infty \leq 1} \eta^T u$, the same as to check whether $\max_{\eta,\zeta} \{\eta^T B\zeta : \|\eta\|_\infty \leq 1, \|\zeta\|_\infty \leq 1\} \leq 1$. The maximum of the bilinear form $\eta^T B\zeta$ with positive semidefinite B over η, ζ varying in a convex symmetric neighborhood of the origin is always achieved when $\eta = \zeta$ (you may check this by using the polarization identity $\eta^T B\zeta = \frac{1}{4}(\eta + \zeta)^T B(\eta + \zeta) - \frac{1}{4}(\eta - \zeta)^T B(\eta - \zeta)$). Thus, to check whether $x = 0$ is robust feasible for (1.2.16) is the same as to check whether the maximum of a given nonnegative quadratic form $\zeta^T B\zeta$ over the unit box is ≤ 1. The latter problem is known to be NP-hard,[6] and therefore so is the problem of checking robust feasibility for (1.2.16).

The second of the above is a modeling question, and as such, goes beyond the scope of purely theoretical considerations. However, theory, as we shall see in section 2.1, contributes significantly to this modeling issue.

1.3 TRACTABILITY OF ROBUST COUNTERPARTS

In this section, we investigate the "computational status" of the RC of uncertain LO problem. The situation here turns out to be as good as it could be: we shall see, essentially, that *the RC of the uncertain LO problem with uncertainty set $\mathcal{U}$ is computationally tractable whenever the convex uncertainty set $\mathcal{U}$ itself is computationally tractable.* The latter means that we know in advance the affine hull of $\mathcal{U}$, a point from the relative interior of $\mathcal{U}$, and we have access to an efficient *membership oracle* that, given on input a point u, reports whether $u \in \mathcal{U}$. This can be reformulated as a precise mathematical statement; however, we will prove a slightly restricted version of this statement that does not require long excursions into complexity theory.

1.3.1 The Strategy

Our strategy will be as follows. First, we restrict ourselves to uncertain LO problems with a certain objective — we remember from item **A** in Section 1.2.1 that we lose

---

[6]In fact, it is NP-hard to compute the maximum of a nonnegative quadratic form over the unit box with inaccuracy less than 4% [61].

nothing by this restriction. Second, all we need is a "computationally tractable" representation of the RC of a *single* uncertain linear constraint, that is, an equivalent representation of the RC by an explicit (and "short") system of efficiently verifiable convex inequalities. Given such representations for the RCs of every one of the constraints of our uncertain problem and putting them together (cf. item **B** in Section 1.2.1), we reformulate the RC of the problem as the problem of minimizing the original linear objective under a finite (and short) system of explicit convex constraints, and thus — as a computationally tractable problem.

To proceed, we should explain first what does it mean to represent a constraint by a system of convex inequalities. Everyone understands that the system of 4 constraints on 2 variables,

$$x_1 + x_2 \leq 1, x_1 - x_2 \leq 1, -x_1 + x_2 \leq 1, -x_1 - x_2 \leq 1, \qquad (1.3.1)$$

represents the nonlinear inequality

$$|x_1| + |x_2| \leq 1 \qquad (1.3.2)$$

in the sense that both (1.3.2) and (1.3.1) define the same feasible set. Well, what about the claim that the system of 5 linear inequalities

$$-u_1 \leq x_1 \leq u_1, -u_2 \leq x_2 \leq u_2, u_1 + u_2 \leq 1 \qquad (1.3.3)$$

represents the same set as (1.3.2)? Here again everyone will agree with the claim, although we cannot justify the claim in the former fashion, since the feasible sets of (1.3.2) and (1.3.3) live in different spaces and therefore cannot be equal to each other!

What actually is meant when speaking about "equivalent representations of problems/constraints" in Optimization can be formalized as follows:

Definition 1.3.1. A set $X^+ \subset \mathbb{R}^n_x \times \mathbb{R}^k_u$ is said to represent a set $X \subset \mathbb{R}^n_x$, if the projection of X^+ onto the space of x-variables is exactly X, i.e., $x \in X$ if and only if there exists $u \in \mathbb{R}^k_u$ such that $(x, u) \in X^+$:

$$X = \{x : \exists u : (x, u) \in X^+\}.$$

A system of constraints $\mathcal{S}^+$ in variables $x \in \mathbb{R}^n_x$, $u \in \mathbb{R}^k_u$ is said to represent a system of constraints $\mathcal{S}$ in variables $x \in \mathbb{R}^n_x$, if the feasible set of the former system represents the feasible set of the latter one.

With this definition, it is clear that the system (1.3.3) indeed represents the constraint (1.3.2), and, more generally, that the system of $2n + 1$ linear inequalities

$$-u_j \leq x_j \leq u_j, j = 1, ..., n, \sum_j u_j \leq 1$$

in variables x, u represents the constraint

$$\sum_j |x_j| \leq 1.$$

To understand how powerful this representation is, note that to represent the same constraint in the style of (1.3.1), that is, without extra variables, it would take as much as 2^n linear inequalities.

Coming back to the general case, assume that we are given an optimization problem

$$\min_x \left\{ f(x) \text{ s.t. } x \text{ satisfies } \mathcal{S}_i, \; i = 1, ..., m \right\}, \tag{P}$$

where $\mathcal{S}_i$ are systems of constraints in variables x, and that we have in our disposal systems $\mathcal{S}_i^+$ of constraints in variables x, v^i which represent the systems $\mathcal{S}_i$. Clearly, the problem

$$\min_{x, v^1, ..., v^m} \left\{ f(x) \text{ s.t. } (x, v^i) \text{ satisfies } \mathcal{S}_i^+, \; i = 1, ..., m \right\} \tag{P$^+$}$$

is equivalent to (P): the x component of every feasible solution to (P$^+$) is feasible for (P) with the same value of the objective, and the optimal values in the problems are equal to each other, so that the x component of an ϵ-optimal (in terms of the objective) feasible solution to (P$^+$) is an ϵ-optimal feasible solution to (P). We shall say that (P$^+$) represents equivalently the original problem (P). What is important here, is that a representation can possess desired properties that are absent in the original problem. For example, an appropriate representation can convert the problem of the form $\min_x \{\|Px - p\|_1 : Ax \leq b\}$ with n variables, m linear constraints, and k-dimensional vector p, into an LO problem with $n + k$ variables and $m + 2k + 1$ linear inequality constraints, etc. Our goal now is to build a representation capable of expressing equivalently a semi-infinite linear constraint (specifically, the robust counterpart of an uncertain linear inequality) as a finite system of explicit convex constraints, with the ultimate goal to use these representations in order to convert the RC of an uncertain LO problem into an explicit (and as such, computationally tractable) convex program.

The outlined strategy allows us to focus on a *single* uncertainty-affected linear inequality — a family

$$\left\{ a^T x \leq b \right\}_{[a;b] \in \mathcal{U}}, \tag{1.3.4}$$

of linear inequalities with the data varying in the uncertainty set

$$\mathcal{U} = \left\{ [a; b] = [a^0; b^0] + \sum_{\ell=1}^{L} \zeta_\ell [a^\ell; b^\ell] : \zeta \in \mathcal{Z} \right\} \tag{1.3.5}$$

— and on "tractable representation" of the RC

$$a^T x \leq b \quad \forall \left([a; b] = [a^0; b^0] + \sum_{\ell=1}^{L} \zeta_\ell [a^\ell; b^\ell] : \zeta \in \mathcal{Z} \right) \tag{1.3.6}$$

of this uncertain inequality.

By reasons indicated in item **C** of Section 1.2.1, we assume from now on that the associated perturbation set $\mathcal{Z}$ is convex.

1.3.2 Tractable Representation of (1.3.6): Simple Cases

We start with the cases where the desired representation can be found by "bare hands," specifically, the cases of *interval* and *simple ellipsoidal* uncertainty.

Example 1.3.2. Consider the case of *interval uncertainty*, where $\mathcal{Z}$ in (1.3.6) is a box. W.l.o.g.[7] we can normalize the situation by assuming that

$$\mathcal{Z} = \mathrm{Box}_1 \equiv \{\zeta \in \mathbb{R}^L : \|\zeta\|_\infty \leq 1\}.$$

In this case, (1.3.6) reads

$$[a^0]^T x + \sum_{\ell=1}^L \zeta_\ell [a^\ell]^T x \leq b^0 + \sum_{\ell=1}^L \zeta_\ell b^\ell \qquad \forall(\zeta : \|\zeta\|_\infty \leq 1)$$

$$\Leftrightarrow \quad \sum_{\ell=1}^L \zeta_\ell [[a^\ell]^T x - b^\ell] \leq b^0 - [a^0]^T x \qquad \forall(\zeta : |\zeta_\ell| \leq 1, \ell = 1, ..., L)$$

$$\Leftrightarrow \quad \max_{-1 \leq \zeta_\ell \leq 1} \left[\sum_{\ell=1}^L \zeta_\ell [[a^\ell]^T x - b^\ell] \right] \leq b^0 - [a^0]^T x$$

The concluding maximum in the chain is clearly $\sum_{\ell=1}^L |[a^\ell]^T x - b^\ell|$, and we arrive at the representation of (1.3.6) by the explicit convex constraint

$$[a^0]^T x + \sum_{\ell=1}^L |[a^\ell]^T x - b^\ell| \leq b^0, \tag{1.3.7}$$

which in turn admits a representation by a system of linear inequalities:

$$\begin{cases} -u_\ell \leq [a^\ell]^T x - b^\ell \leq u_\ell, \ \ell = 1, ..., L, \\ [a^0]^T x + \sum_{\ell=1}^L u_\ell \leq b^0. \end{cases} \tag{1.3.8}$$

Example 1.3.3. Consider the case of *ellipsoidal uncertainty* where $\mathcal{Z}$ in (1.3.6) is an ellipsoid. W.l.o.g. we can normalize the situation by assuming that $\mathcal{Z}$ is merely the ball of radius Ω centered at the origin:

$$\mathcal{Z} = \mathrm{Ball}_\Omega = \{\zeta \in \mathbb{R}^L : \|\zeta\|_2 \leq \Omega\}.$$

In this case, (1.3.6) reads

$$[a^0]^T x + \sum_{\ell=1}^L \zeta_\ell [a^\ell]^T x \leq b^0 + \sum_{\ell=1}^L \zeta_\ell b^\ell \qquad \forall(\zeta : \|\zeta\|_2 \leq \Omega)$$

$$\Leftrightarrow \quad \max_{\|\zeta\|_2 \leq \Omega} \left[\underline{\sum_{\ell=1}^L \zeta_\ell [[a^\ell]^T x - b^\ell]} \right] \leq b^0 - [a^0]^T x$$

$$\Leftrightarrow \quad \Omega \sqrt{\sum_{\ell=1}^L ([a^\ell]^T x - b^\ell)^2} \leq b^0 - [a^0]^T x,$$

and we arrive at the representation of (1.3.6) by the explicit convex constraint ("conic quadratic inequality")

$$[a^0]^T x + \Omega \sqrt{\sum_{\ell=1}^L ([a^\ell]^T x - b^\ell)^2} \leq b^0. \tag{1.3.9}$$

---

[7]abbr. for "without loss of generality."

1.3.3 Tractable Representation of (1.3.6): General Case

Now consider a rather general case when the perturbation set $\mathcal{Z}$ in (1.3.6) is given by a *conic representation* (cf. section A.2.4 in Appendix):

$$\mathcal{Z} = \left\{ \zeta \in \mathbb{R}^L : \exists u \in \mathbb{R}^K : P\zeta + Qu + p \in \mathbf{K} \right\}, \tag{1.3.10}$$

where $\mathbf{K}$ is a closed convex pointed cone in $\mathbb{R}^N$ with a nonempty interior, P, Q are given matrices and p is a given vector. In the case when $\mathbf{K}$ is *not* a polyhedral cone, assume that this representation is strictly feasible:

$$\exists(\bar{\zeta}, \bar{u}) : P\bar{\zeta} + Q\bar{u} + p \in \operatorname{int} K. \tag{1.3.11}$$

Theorem 1.3.4. Let the perturbation set $\mathcal{Z}$ be given by (1.3.10), and in the case of non-polyhedral $\mathbf{K}$, let also (1.3.11) take place. Then the semi-infinite constraint (1.3.6) can be represented by the following system of conic inequalities in variables $x \in \mathbb{R}^n, y \in \mathbb{R}^N$:

$$\begin{aligned} & p^T y + [a^0]^T x \leq b^0, \\ & Q^T y = 0, \\ & (P^T y)_\ell + [a^\ell]^T x = b^\ell, \ell = 1, ..., L, \\ & y \in \mathbf{K}_*, \end{aligned} \tag{1.3.12}$$

where $\mathbf{K}_* = \{y : y^T z \geq 0 \,\forall z \in \mathbf{K}\}$ is the cone dual to $\mathbf{K}$.

Proof. We have

$$x \text{ is feasible for } (1.3.6)$$
$$\Leftrightarrow \quad \sup_{\zeta \in \mathcal{Z}} \Big\{ \underbrace{[a^0]^T x - b^0}_{d[x]} + \sum_{\ell=1}^L \zeta_\ell \underbrace{\left[[a^\ell]^T x - b^\ell \right]}_{c_\ell[x]} \Big\} \leq 0$$
$$\Leftrightarrow \quad \sup_{\zeta \in \mathcal{Z}} \left\{ c^T[x]\zeta + d[x] \right\} \leq 0$$
$$\Leftrightarrow \quad \sup_{\zeta \in \mathcal{Z}} c^T[x]\zeta \leq -d[x]$$
$$\Leftrightarrow \quad \max_{\zeta, v} \left\{ c^T[x]\zeta : P\zeta + Qv + p \in \mathbf{K} \right\} \leq -d[x].$$

The concluding relation says that x is feasible for (1.3.6) if and only if the optimal value in the conic program

$$\max_{\zeta, v} \left\{ c^T[x]\zeta : P\zeta + Qv + p \in \mathbf{K} \right\} \tag{CP}$$

is $\leq -d[x]$. Assume, first, that (1.3.11) takes place. Then (CP) is strictly feasible, and therefore, applying the Conic Duality Theorem (Theorem A.2.1), the optimal value in (CP) is $\leq -d[x]$ if and only if the optimal value in the conic dual to the (CP) problem

$$\min_{y} \left\{ p^T y : Q^T y = 0, P^T y = -c[x], y \in \mathbf{K}_* \right\}, \tag{CD}$$

is attained and is $\leq -d[x]$. Now assume that $\mathbf{K}$ is a polyhedral cone. In this case the usual LO Duality Theorem, (which does not require the validity of (1.3.11)), yields exactly the same conclusion: the optimal value in (CP) is $\leq -d[x]$ if and only if the optimal value in (CD) is achieved and is $\leq -d[x]$. In other words, under the premise of the Theorem, x is feasible for (1.3.6) if and only if (CD) has a feasible solution y with $p^T y \leq -d[x]$. □

Observing that nonnegative orthants, Lorentz and Semidefinite cones are self-dual, we derive from Theorem 1.3.4 the following corollary:

Corollary 1.3.5. Let the nonempty perturbation set in (1.3.6) be:

(i) polyhedral, i.e., given by (1.3.10) with a nonnegative orthant $\mathbb{R}_+^N$ in the role of $\mathbf{K}$, or

(ii) conic quadratic representable, i.e., given by (1.3.10) with a direct product $\mathbf{L}^{k_1} \times \ldots \times \mathbf{L}^{k_m}$ of Lorentz cones $\mathbf{L}^k = \{x \in \mathbb{R}^k : x_k \geq \sqrt{x_1^2 + \ldots + x_{k-1}^2}\}$ in the role of $\mathbf{K}$, or

(iii) semidefinite representable, i.e., given by (1.3.10) with the positive semidefinite cone $\mathbf{S}_+^k$ in the role of $\mathbf{K}$.

In the cases of (ii), (iii) assume in addition that (1.3.11) holds true. Then the Robust Counterpart (1.3.6) of the uncertain linear inequality (1.3.4) — (1.3.5) with the perturbation set $\mathcal{Z}$ admits equivalent reformulation as an explicit system of

— linear inequalities, in the case of (i),

— conic quadratic inequalities, in the case of (ii),

— linear matrix inequalities, in the case of (iii).

In all cases, the size of the reformulation is polynomial in the number of variables in (1.3.6) and the size of the conic description of $\mathcal{Z}$, while the data of the reformulation is readily given by the data describing, via (1.3.10), the perturbation set $\mathcal{Z}$.

Remark 1.3.6. A. Usually, the cone $\mathbf{K}$ participating in (1.3.10) is the direct product of simpler cones $\mathbf{K}^1, \ldots, \mathbf{K}^S$, so that representation (1.3.10) takes the form

$$\mathcal{Z} = \{\zeta : \exists u^1, \ldots, u^S : P_s\zeta + Q_s u^s + p_s \in \mathbf{K}^s, s = 1, \ldots, S\}. \tag{1.3.13}$$

In this case, (1.3.12) becomes the system of conic constraints in variables $x, y^1, \ldots, y^S$ as follows:

$$\begin{aligned}
&\sum_{s=1}^S p_s^T y^s + [a^0]^T x \leq b^0, \\
&Q_s^T y^s = 0, \; s = 1, \ldots, S, \\
&\sum_{s=1}^S (P_s^T y^s)_\ell + [a^\ell]^T x = b^\ell, \ell = 1, \ldots, L, \\
&y^s \in \mathbf{K}_*^s, s = 1, \ldots, S,
\end{aligned} \tag{1.3.14}$$

where K_*^s is the cone dual to K^s.

B. Uncertainty sets given by LMIs seem "exotic"; however, they can arise under quite realistic circumstances, see section 1.4.

1.3.3.1 Examples

We are about to apply Theorem 1.3.4 to build tractable reformulations of the semi-infinite inequality (1.3.6) in two particular cases. While at a first glance no natural "uncertainty models" lead to the "strange" perturbation sets we are about to consider, it will become clear later that these sets are of significant importance — they allow one to model *random* uncertainty.

Example 1.3.7. $\mathcal{Z}$ *is the intersection of concentric co-axial box and ellipsoid, specifically,*

$$\mathcal{Z} = \{\zeta \in \mathbb{R}^L : -1 \leq \zeta_\ell \leq 1, \ell \leq L, \sqrt{\sum_{\ell=1}^{L} \zeta_\ell^2/\sigma_\ell^2} \leq \Omega\}, \tag{1.3.15}$$

where $\sigma_\ell > 0$ and $\Omega > 0$ are given parameters.

Here representation (1.3.13) becomes

$$\mathcal{Z} = \{\zeta \in \mathbb{R}^L : P_1\zeta + p_1 \in \mathbf{K}^1, P_2\zeta + p_2 \in \mathbf{K}^2\},$$

where

• $P_1\zeta \equiv [\zeta; 0]$, $p_1 = [0_{L \times 1}; 1]$ and $\mathbf{K}^1 = \{(z, t) \in \mathbb{R}^L \times \mathbb{R} : t \geq \|z\|_\infty\}$, whence $\mathbf{K}_*^1 = \{(z, t) \in \mathbb{R}^L \times \mathbb{R} : t \geq \|z\|_1\}$;

• $P_2\zeta = [\Sigma^{-1}\zeta; 0]$ with $\Sigma = \text{Diag}\{\sigma_1, ..., \sigma_L\}$, $p_2 = [0_{L \times 1}; \Omega]$ and $\mathbf{K}^2$ is the Lorentz cone of the dimension $L + 1$ (whence $\mathbf{K}_*^2 = \mathbf{K}^2$)

Setting $y^1 = [\eta_1; \tau_1]$, $y^2 = [\eta_2; \tau_2]$ with one-dimensional τ_1, τ_2 and L-dimensional η_1, η_2, (1.3.14) becomes the following system of constraints in variables τ, η, x:

$$
\begin{array}{rll}
(a) & \tau_1 + \Omega\tau_2 + [a^0]^T x & \leq \quad b^0, \\
(b) & (\eta_1 + \Sigma^{-1}\eta_2)_\ell & = \quad b^\ell - [a^\ell]^T x, \ell = 1, ..., L, \\
(c) & \|\eta_1\|_1 & \leq \quad \tau_1 \quad [\Leftrightarrow [\eta_1; \tau_1] \in \mathbf{K}_*^1], \\
(d) & \|\eta_2\|_2 & \leq \quad \tau_2 \quad [\Leftrightarrow [\eta_2; \tau_2] \in \mathbf{K}_*^2].
\end{array}
$$

We can eliminate from this system the variables τ_1, τ_2 — for every feasible solution to the system, we have $\tau_1 \geq \bar{\tau}_1 \equiv \|\eta_1\|_1$, $\tau_2 \geq \bar{\tau}_2 \equiv \|\eta_2\|_2$, and the solution obtained when replacing τ_1, τ_2 with $\bar{\tau}_1$, $\bar{\tau}_2$ still is feasible. The reduced system in variables x, $z = \eta_1$, $w = \Sigma^{-1}\eta_2$ reads

$$
\begin{array}{rll}
\sum_{\ell=1}^{L} |z_\ell| + \Omega\sqrt{\sum_\ell \sigma_\ell^2 w_\ell^2} + [a^0]^T x & \leq & b^0, \\
z_\ell + w_\ell & = & b^\ell - [a^\ell]^T x, \ell = 1, ..., L,
\end{array} \tag{1.3.16}
$$

which is also a representation of (1.3.6), (1.3.15).

Example 1.3.8. [*"budgeted uncertainty"*] Consider the case where $\mathcal{Z}$ is the intersection of $\|\cdot\|_\infty$- and $\|\cdot\|_1$-balls, specifically,

$$\mathcal{Z} = \{\zeta \in \mathbb{R}^L : \|\zeta\|_\infty \leq 1, \|\zeta\|_1 \leq \gamma\}, \tag{1.3.17}$$

where γ, $1 \leq \gamma \leq L$, is a given "uncertainty budget."

Here representation (1.3.13) becomes

$$\mathcal{Z} = \{\zeta \in \mathbb{R}^L : P_1\zeta + p_1 \in \mathbf{K}^1, P_2\zeta + p_2 \in \mathbf{K}^2\},$$

where

- $P_1\zeta \equiv [\zeta; 0]$, $p_1 = [0_{L \times 1}; 1]$ and $\mathbf{K}^1 = \{[z; t] \in \mathbb{R}^L \times \mathbb{R} : t \geq \|z\|_\infty\}$, whence $\mathbf{K}_*^1 = \{[z; t] \in \mathbb{R}^L \times \mathbb{R} : t \geq \|z\|_1\}$;

- $P_2\zeta = [\zeta; 0]$, $p_2 = [0_{L \times 1}; \gamma]$ and $\mathbf{K}^2 = \mathbf{K}_*^1 = \{[z; t] \in \mathbb{R}^L \times \mathbb{R} : t \geq \|z\|_1\}$, whence $\mathbf{K}_*^2 = \mathbf{K}^1$.

Setting $y^1 = [z; \tau_1]$, $y^2 = [w; \tau_2]$ with one-dimensional τ and L-dimensional z, w, system (1.3.14) becomes the following system of constraints in variables τ_1, τ_2, z, w, x:

$$
\begin{array}{rlrl}
(a) & \tau_1 + \gamma\tau_2 + [a^0]^T x & \leq & b^0, \\
(b) & (z + w)_\ell & = & b^\ell - [a^\ell]^T x, \ \ell = 1, ..., L, \\
(c) & \|z\|_1 & \leq & \tau_1 \quad [\Leftrightarrow [\eta_1; \tau_1] \in \mathbf{K}_*^1], \\
(d) & \|w\|_\infty & \leq & \tau_2 \quad [\Leftrightarrow [\eta_2; \tau_2] \in \mathbf{K}_*^2].
\end{array}
$$

Same as in Example 1.3.7, we can eliminate the τ-variables, arriving at a representation of (1.3.6), (1.3.17) by the following system of constraints in variables x, z, w:

$$
\begin{array}{rcl}
\sum_{\ell=1}^{L} |z_\ell| + \gamma \max_\ell |w_\ell| + [a^0]^T x & \leq & b^0, \\
z_\ell + w_\ell & = & b^\ell - [a^\ell]^T x, \ \ell = 1, ..., L,
\end{array}
\tag{1.3.18}
$$

which can be further converted into the system of linear inequalities in z, w and additional variables.

1.4 NON-AFFINE PERTURBATIONS

In the first reading this section can be skipped.

So far we have assumed that the uncertain data of an uncertain LO problem are *affinely* parameterized by a perturbation vector ζ varying in a closed convex set $\mathcal{Z}$. We have seen that this assumption, combined with the assumption that $\mathcal{Z}$ is computationally tractable, implies tractability of the RC. What happens when the perturbations enter the uncertain data in a nonlinear fashion? Assume w.l.o.g. that every entry a in the uncertain data is of the form

$$
a = \sum_{k=1}^{K} c_k^a f_k(\zeta),
$$

where c_k^a are given coefficients (depending on the data entry in question) and $f_1(\zeta), ..., f_K(\zeta)$ are certain basic functions, perhaps non-affine, defined on the perturbation set $\mathcal{Z}$. Assuming w.l.o.g. that the objective is certain, we still can define the RC of our uncertain problem as the problem of minimizing the original objective over the set of robust feasible solutions, those which remain feasible for all values of the data coming from $\zeta \in \mathcal{Z}$, but what about the tractability of this RC? An immediate observation is that the case of nonlinearly perturbed data can be immediately reduced to the one where the data are affinely perturbed. To this end, it suffices to pass from the original perturbation vector ζ to the new vector

$$
\widehat{\zeta}[\zeta] = [\zeta_1; ...; \zeta_L; f_1(\zeta); ...; f_K(\zeta)].
$$

As a result, the uncertain data become *affine* functions of the new perturbation vector $\widehat{\zeta}$ which now runs through the image $\widetilde{\mathcal{Z}} = \widehat{\zeta}[\mathcal{Z}]$ of the original uncertainty set $\mathcal{Z}$ under the mapping $\zeta \mapsto \widehat{\zeta}[\zeta]$. As we know, in the case of affine data perturbations the RC remains intact when replacing a given perturbation set with its closed convex hull. Thus, we can think about our uncertain LO problem as an affinely perturbed problem where the perturbation vector is $\widehat{\zeta}$, and this vector runs through the closed convex set $\widehat{\mathcal{Z}} = \mathrm{cl}\,\mathrm{Conv}(\widehat{\zeta}[\mathcal{Z}])$. We see that formally speaking, the case of general-type perturbations can be reduced to the one of affine perturbations. This, unfortunately, does not mean that non-affine perturbations do not cause difficulties. Indeed, in order to end up with a computationally tractable RC, we need more than affinity of perturbations and convexity of the perturbation set — we need this set to be computationally tractable. And the set $\widehat{\mathcal{Z}} = \mathrm{cl}\,\mathrm{Conv}(\widehat{\zeta}[\mathcal{Z}])$ may fail to satisfy this requirement even when both $\mathcal{Z}$ and the *nonlinear* mapping $\zeta \mapsto \widehat{\zeta}[\zeta]$ are simple, e.g., when $\mathcal{Z}$ is a box and $\widehat{\zeta} = [\zeta; \{\zeta_\ell \zeta_r\}_{\ell,r=1}^L]$, (i.e., when the uncertain data are quadratically perturbed by the original perturbations ζ).

We are about to present two generic cases where the difficulty just outlined does not occur (for justification and more examples, see section 14.3.2).

Ellipsoidal perturbation set $\mathcal{Z}$, quadratic perturbations. Here $\mathcal{Z}$ is an ellipsoid, and the basic functions f_k are the constant, the coordinates of ζ and the pairwise products of these coordinates. This means that the uncertain data entries are quadratic functions of the perturbations. W.l.o.g. we can assume that the ellipsoid $\mathcal{Z}$ is centered at the origin: $\mathcal{Z} = \{\zeta : \|Q\zeta\|_2 \leq 1\}$, where $\mathrm{Ker}Q = \{0\}$. In this case, representing $\widehat{\zeta}[\zeta]$ as the matrix $\left[\begin{array}{c|c} & \zeta^T \\ \hline \zeta & \zeta\zeta^T \end{array}\right]$, we have the following semidefinite representation of $\widehat{\mathcal{Z}} = \mathrm{cl}\,\mathrm{Conv}(\widehat{\zeta}[\mathcal{Z}])$:

$$\widehat{\mathcal{Z}} = \left\{ \left[\begin{array}{c|c} & w^T \\ \hline w & W \end{array}\right] : \left[\begin{array}{c|c} 1 & w^T \\ \hline w & W \end{array}\right] \succeq 0, \mathrm{Tr}(QWQ^T) \leq 1 \right\}$$

(for proof, see Lemma 14.3.7).

Separable polynomial perturbations. Here the structure of perturbations is as follows: ζ runs through the box $\mathcal{Z} = \{\zeta \in \mathbb{R}^L : \|\zeta\|_\infty \leq 1\}$, and the uncertain data entries are of the form

$$a = p_1^a(\zeta_1) + \ldots + p_L^a(\zeta_L),$$

where $p_\ell^a(s)$ are given algebraic polynomials of degrees not exceeding d; in other words, the basic functions can be split into L groups, the functions of ℓ-th group being $1 = \zeta_\ell^0, \zeta_\ell, \zeta_\ell^2, \ldots, \zeta_\ell^d$. Consequently, the function $\widehat{\zeta}[\zeta]$ is given by

$$\widehat{\zeta}[\zeta] = [[1; \zeta_1; \zeta_1^2; \ldots; \zeta_1^d]; \ldots; [1; \zeta_L; \zeta_L^2; \ldots; \zeta_L^d]].$$

Setting $P = \{\widehat{s} = [1; s; s^2; \ldots; s^d] : -1 \leq s \leq 1\}$, we conclude that $\widetilde{\mathcal{Z}} = \widehat{\zeta}[\mathcal{Z}]$ can be identified with the set $\underbrace{P^L = P \times \ldots \times P}_{L}$, so that $\widehat{\mathcal{Z}}$ is nothing but the set

$\underbrace{\mathcal{P} \times \ldots \times \mathcal{P}}_{L}$, where $\mathcal{P} = \text{Conv}(P)$. It remains to note that the set $\mathcal{P}$ admits an explicit semidefinite representation, see Lemma 14.3.4.

1.5 EXERCISES

Exercise 1.1. Consider an uncertain LO problem with instances

$$\min_x \left\{ c^T x : Ax \leq b \right\} \qquad [A : m \times n]$$

and with simple interval uncertainty:

$$\mathcal{U} = \{(c, A, b) : |c_j - c_j^{\mathrm{n}}| \leq \sigma_j, |A_{ij} - A_{ij}^{\mathrm{n}}| \leq \alpha_{ij}, |b_i - b_i^{\mathrm{n}}| \leq \beta_i \forall i, j\}$$

(^{n} marks the nominal data). Reduce the RC of the problem to an LO problem with m constraints (not counting the sign constraints on the variables) and $2n$ nonnegative variables.

Exercise 1.2. Represent the RCs of every one of the uncertain linear constraints given below:

$$a^T x \leq b, [a; b] \in \mathcal{U} = \{[a; b] = [a^{\mathrm{n}}; b^{\mathrm{n}}] + P\zeta : \|\zeta\|_p \leq \rho\}$$
$$[p \in [1, \infty]] \quad (a)$$
$$a^T x \leq b, [a; b] \in \mathcal{U} = \{[a; b] = [a^{\mathrm{n}}; b^{\mathrm{n}}] + P\zeta : \|\zeta\|_p \leq \rho, \zeta \geq 0\}$$
$$[p \in [1, \infty]] \quad (b)$$
$$a^T x \leq b, [a; b] \in \mathcal{U} = \{[a; b] = [a^{\mathrm{n}}; b^{\mathrm{n}}] + P\zeta : \|\zeta\|_p \leq \rho\}$$
$$[p \in (0, 1)] \quad (c)$$

as explicit convex constraints.

Exercise 1.3. Represent in tractable form the RC of uncertain linear constraint

$$a^T x \leq b$$

with $\cap$-ellipsoidal uncertainty set

$$\mathcal{U} = \{[a, b] = [a^{\mathrm{n}}; b^{\mathrm{n}}] + P\zeta : \zeta^T Q_j \zeta \leq \rho^2, 1 \leq j \leq J\},$$

where $Q_j \succeq 0$ and $\sum_j Q_j \succ 0$.

1.6 NOTES AND REMARKS

NR 1.1. The paradigm of Robust Linear Optimization in the form considered here goes back to A.L. Soyster [109], 1973. To the best of our knowledge, in two subsequent decades there were only two publications on the subject [52, 106]. The activity in the area was revived circa 1997, independently and essentially simultaneously, in the frameworks of both Integer Programming (Kouvelis and Yu [70]) and Convex Programming (Ben-Tal and Nemirovski [3, 4], El Ghaoui et al. [49, 50]). Since 2000, the RO area is witnessing a burst of research activity in both theory and applications, with numerous researchers involved worldwide. The magnitude and diversity of the related contributions make it beyond our abilities to discuss them here. The reader can get some impression of this activity from [9, 16, 110, 89] and references therein.

NR 1.2. By itself, the RO methodology can be applied to every optimization problem where one can separate numerical data (that can be partly uncertain) from a problem's structure (that is known in advance and common for all instances of the uncertain problem). In particular, the methodology is fully applicable to uncertain *mixed integer* LO problems, where part of the decision variables are restricted to be integer. Note, however, that tractability issues, (which are our main focus in this book), in Uncertain LO with real variables and Uncertain Mixed-Integer LO need quite different treatment. While Theorem 1.3.4 is fully applicable to the mixed integer case and implies, in particular, that the RC of an uncertain mixed-integer LO problem $\mathcal{P}$ with a polyhedral uncertainty set is an explicit mixed-integer LO program with exactly the same integer variables as those of the instances of $\mathcal{P}$, the "tractability consequences" of this fact are completely different from those we made in the main body of this chapter. With no integer variables, the fact that the RC is an LO program straightforwardly implies tractability of the RC, while in the presence of integer variables no such conclusion can be made. Indeed, in the mixed integer case already the instances of the uncertain problem $\mathcal{P}$ typically are intractable, which, of course, implies intractability of the RC. In the case when the instances of $\mathcal{P}$ are tractable, the "fine structure" of the instances responsible for this rare phenomenon usually is destroyed when passing to the mixed-integer reformulation of the RC. There are some remarkable exceptions to this rule (see, e.g., [25]); however, in general the Uncertain Mixed-Integer LO is incomparably more complex computationally than the Uncertain LO with real variables. As it was already stated, our book is primarily focused on tractability issues of RO, and in order to get positive results in this direction, we restrict ourselves to uncertain problems with well-structured convex (and thus tractable) instances.

NR 1.3. Tractability of the RC of an uncertain LO problem with a tractable uncertainty set was established in the very first papers on convex RO. Theorem 1.3.4 and Corollary 1.3.5 are taken from [5].

Chapter Two
Robust Counterpart Approximations of Scalar Chance Constraints

2.1 HOW TO SPECIFY AN UNCERTAINTY SET

The question posed in the title of this section goes beyond general-type theoretical considerations — this is mainly a modeling issue that should be resolved on the basis of application-driven considerations. There is however a special case where this question makes sense and can, to some extent, be answered — this is the case where our goal is not to build an uncertainty model "from scratch," but rather to *translate* an already existing uncertainty model, namely, a stochastic one, to the language of "uncertain-but-bounded" perturbation sets and the associated robust counterparts. By exactly the same reasons as in the previous section, we can restrict our considerations to the case of a *single* uncertainty-affected linear inequality (1.3.4), (1.3.5).

Probabilistic vs. "uncertain-but-bounded" perturbations. When building the RC (1.3.6) of uncertain linear inequality (1.3.4), we worked with the so called "uncertain-but-bounded" data model (1.3.5) — one where all we know about the possible values of the data $[a; b]$ is their domain $\mathcal{U}$ defined in terms of a given affine parameterization of the data by perturbation vector ζ varying in a given perturbation set $\mathcal{Z}$. It should be stressed that we did not assume that the perturbations are of a stochastic nature and therefore used the only approach meaningful under the circumstances, namely, we looked for solutions that remain feasible whatever the data perturbation from $\mathcal{Z}$. This approach has its advantages:

i) More often than not there are no reasons to assign the perturbations a stochastic nature.

Indeed, stochasticity makes sense only when one repeats a certain action many times, or executes many similar actions in parallel; here it might be reasonable to think of frequencies of successes, etc. Probabilistic considerations become, methodologically, much more problematic when applied to a unique action, with no second attempt possible.

ii) Even when the unknown data can be thought of as stochastic, it might be difficult, especially in the large-scale case, to specify reliably data distribution. Indeed, the mere fact that the data are stochastic does not help unless we possess at least a partial knowledge of the underlying distribution.

Of course, the uncertain-but-bounded models of uncertainty also require a priori knowledge, namely, to know what is the uncertainty set (a probabilistically
oriented person could think about this set as the *support* of data distribution,
that is, the smallest closed set in the space of the data such that the probability for the data to take a value outside of this set is zero). Note, however,
that it is much easier to point out the support of the relevant distribution
than the distribution itself.

With the uncertain-but-bounded model of uncertainty, we can make clear predic-
tions like "with such and such behavior, we definitely will survive, provided that
the unknown parameters will differ from their nominal values by no more than
15%, although we may die when the variation will be as large as 15.1%." In case
we do believe that 15.1% variations are also worthy to worry about, we have an
option to increase the perturbation set to take care of 30% perturbations in the
data. With luck, we will be able to find a robust feasible solution for the increased
perturbation set. This is a typical engineering approach — after the required thick-
ness of a bar supporting certain load is found, a civil engineer will increase it by
factor like 1.2 or 1.5 "to be on the safe side" — to account for model inaccuracies, material imperfections, etc. With a stochastic uncertainty model, this "being
on the safe side" is impossible — increasing the probability of certain events, one
must decrease simultaneously the probability of certain other events, since the
"total probability budget" is once and for ever fixed. While all these arguments
demonstrate that there are situations in reality when the uncertain-but-bounded
model of data perturbations possesses significant methodological advantages over
the stochastic models of uncertainty, there are, of course, applications (like communications, weather forecasts, mass production, and, to some extent, finance) where
one can rely on probabilistic models of uncertainty. Whenever this is the case,
the much less informative uncertain-but-bounded model and associated worst-case-
oriented decisions can be too conservative and thus impractical. The bottom line is
that *while the stochastic models of data uncertainty are by far not the only meaningful ones, they definitely deserve attention.* Our goal in this chapter is to *develop
techniques that are capable to utilize, to some extent, knowledge of the stochastic
nature of data perturbations when building uncertainty-immunized solutions.* This
goal will be achieved via a specific "translation" of stochastic models of uncertain
data to the language of uncertain-but-bounded perturbations and the associated
robust counterparts. Before developing the approach in full detail, we will explain
why we choose such an implicit way to treat stochastic uncertainty models instead
of treating them directly.

2.2 CHANCE CONSTRAINTS AND THEIR SAFE TRACTABLE APPROXIMATIONS

The most direct way to treat stochastic data uncertainty in the context of uncertain
Linear Optimization is offered by an old concept (going back to 50s [40]) of *chance*

constraints. Consider an uncertain linear inequality

$$a^T x \le b, \ [a; b] = [a^0; b^0] + \sum_{\ell=1}^{L} \zeta_\ell [a^\ell; b^\ell] \tag{2.2.1}$$

(cf. (1.3.4), (1.3.5)) and assume that the perturbation vector ζ is random with, say, completely known probability distribution P. Ideally, we would like to work with candidate solutions x that make the constraint valid with probability 1. This "ideal goal," however, means coming back to the uncertain-but-bounded model of perturbations; indeed, it is easily seen that a given x satisfies (2.2.1) for almost all realizations of ζ if and only if x is robust feasible w.r.t. the perturbation set that is the closed convex hull of the support of P. The only meaningful way to utilize the stochasticity of perturbations is to require a candidate solution x to satisfy the constraint for "nearly all" realizations of ζ, specifically, to satisfy the constraint with probability at least $1 - \epsilon$, where $\epsilon \in (0, 1)$ is a prespecified small tolerance. This approach associates with the randomly perturbed constraint (2.2.1) the *chance constraint*

$$p(x) \equiv \mathrm{Prob}_{\zeta \sim P} \left\{ \zeta : [a^0]^T x + \sum_{\ell=1}^{L} \zeta_\ell [a^\ell]^T x > b^0 + \sum_{\ell=1}^{L} \zeta_\ell b^\ell \right\} \le \epsilon, \tag{2.2.2}$$

where $\mathrm{Prob}_{\zeta \sim P}$ is the probability associated with the distribution P. Note that (2.2.2) is a usual certain constraint. Replacing all uncertainty-affected constraints in an uncertain LO problem with their chance constrained versions and minimizing the objective function, (which we, w.l.o.g., may assume to be certain) under these constraints, we end up with the *chance constrained* version of (LO$_\mathcal{U}$), which is a deterministic optimization problem.

While the outlined approach seems to be quite natural, it suffers from a severe drawback — *typically, it results in a severely computationally intractable problem.* The reason is twofold:

i) Usually, it is difficult to evaluate with high accuracy the probability in the left hand side of (2.2.2), even in the case when P is simple.

For example, it is known [68] that computing the left hand side in (2.2.2) is NP-hard already when ζ_ℓ are independent and uniformly distributed in $[-1, 1]$. This means that unless P=NP, there is no algorithm that, given on input a rational x, rational data $\{[a^\ell; b^\ell]\}_{\ell=0}^{L}$ and rational $\delta \in (0, 1)$, allows to evaluate $p(x)$ within accuracy δ in time polynomial in the bit size of the input. Unless ζ takes values in a finite set of moderate cardinality, the only known general method to evaluate $p(x)$ is based on Monte-Carlo simulations; this method, however, requires samples with cardinality of order of $1/\delta$, where δ is the required accuracy of evaluation. Since the meaningful values of this accuracy are $\le \epsilon$, we conclude that in reality the Monte-Carlo approach can hardly be used when ϵ is like 0.0001 or less.

ii) More often than not the feasible set of (2.2.2) is non-convex, which makes optimization under chance constraints a highly problematic task.

Note that while the first difficulty becomes an actual obstacle only when ϵ is small enough, the second difficulty makes chance constrained optimization highly problematic for "large" ϵ as well.

Essentially, the only known case when none of the outlined difficulties occur is the case where ζ is a Gaussian random vector and $\epsilon < 1/2$.

Due to the severe computational difficulties associated with chance constraints, a natural course of action is to replace a chance constraint with its *computationally tractable safe approximation*. The latter notion is defined as follows:

Definition 2.2.1. Let $\{[a^\ell; b^\ell]\}_{\ell=0}^{L}$, P, ϵ be the data of chance constraint (2.2.2), and let $\mathcal{S}$ be a system of convex constraints on x and additional variables v. We say that $\mathcal{S}$ is a safe convex approximation of chance constraint (2.2.2), if the x component of every feasible solution (x, v) of $\mathcal{S}$ is feasible for the chance constraint.

A safe convex approximation $\mathcal{S}$ of (2.2.2) is called computationally tractable, if the convex constraints forming $\mathcal{S}$ are efficiently computable.

It is clear that by replacing the chance constraints in a given chance constrained optimization problem with their safe convex approximations, we end up with a convex optimization problem in x and additional variables that is a "safe approximation" of the chance constrained problem: the x component of every feasible solution to the approximation is feasible for the chance constrained problem. If the safe convex approximation in question is tractable, then the above approximating program is a convex program with efficiently computable constraints and as such it can be processed efficiently.

In the sequel, when speaking about safe convex approximations, we omit for the sake of brevity the adjective "convex," which should always be added "by default."

2.2.1 Ambiguous Chance Constraints

Chance constraint (2.2.2) is associated with randomly perturbed constraint (2.2.1) *and a given distribution P of random perturbations*, and it is reasonable to use this constraint when we do know this distribution. In reality we usually have only *partial* information on P, that is, we know only that P belongs to a given family $\mathcal{P}$ of distributions. When this is the case, it makes sense to pass from (2.2.2) to the *ambiguous* chance constraint

$$\forall (P \in \mathcal{P}) : \text{Prob}_{\zeta \sim P} \left\{ \zeta : [a^0]^T x + \sum_{\ell=1}^{L} \zeta_\ell [a^\ell]^T x > b^0 + \sum_{\ell=1}^{L} \zeta_\ell b^\ell \right\} \leq \epsilon. \quad (2.2.3)$$

Of course, the definition of a safe tractable approximation of chance constraint extends straightforwardly to the case of ambiguous chance constraint. In the sequel, we usually skip the adjective "ambiguous"; what exactly is meant depends on whether we are speaking about a partially or a fully known distribution P.

Next we present a simple scheme for the safe approximation of chance constraints.

2.3 SAFE TRACTABLE APPROXIMATIONS OF SCALAR CHANCE CONSTRAINTS: BASIC EXAMPLES

Consider the case of chance constraint (2.2.3) where all we know about the random variables ζ_ℓ is that

$$\mathbf{E}\{\zeta_\ell\} = 0 \ \& \ |\zeta_\ell| \leq 1, \ \ell = 1, ..., L \ \& \ \{\zeta_\ell\}_{\ell=1}^L \text{ are independent} \qquad (2.3.1)$$

(that is, $\mathcal{P}$ is comprised of all distributions satisfying (2.3.1)). Note that a more general case of independent random variables ζ_ℓ taking values in given finite segments centered at the expectations of ζ_ℓ by "scalings" $\zeta_\ell \mapsto \xi_\ell = \alpha_\ell \zeta_\ell + \beta_\ell$ with deterministic α_ℓ, β_ℓ can be reduced to (2.3.1) (cf. Remark 1.2.2).

Observe that the body of chance constraint (2.2.2) can be rewritten as

$$\eta \equiv \sum_{\ell=1}^L [[a^\ell]^T x - b^\ell] \zeta_\ell \leq b^0 - [a^0]^T x. \qquad (2.3.2)$$

In the case of (2.3.1), for x fixed, η is a random variable with zero mean and standard deviation

$$\mathrm{StD}[\eta] = \sqrt{\sum_{\ell=1}^L ([a^\ell]^T x - b^\ell)^2 \mathbf{E}\{\zeta_\ell^2\}} \leq \sqrt{\sum_{\ell=1}^L ([a^\ell]^T x - b^\ell)^2}.$$

The chance constraint requires for (2.3.2) to be satisfied with probability $\geq 1 - \epsilon$. An engineer would respond to this requirement arguing that a random variable is "never" greater than its mean plus 3 times the standard deviation, so that η is "never" greater than the quantity $3\sqrt{\sum_{\ell=1}^L ([a^\ell]^T x - b^\ell)^2}$. We need not be as specific as an engineer and say that η is "nearly never" greater than the quantity $\Omega\sqrt{\sum_{\ell=1}^L ([a^\ell]^T x - b^\ell)^2}$, where Ω is a "safety parameter" of order of 1; the larger Ω, the less the chances for η to be larger than the outlined quantity. We thus arrive at a parametric "safe" version

$$\Omega\sqrt{\sum_{\ell=1}^L ([a^\ell]^T x - b^\ell)^2} \leq b^0 - [a^0]^T x \qquad (2.3.3)$$

of the randomly perturbed constraint (2.3.2). It seems that with properly defined Ω, every feasible solution to this constraint satisfies, with probability at least $1 - \epsilon$, the inequality in (2.3.2). This indeed is the case; a simple analysis, which we will carry later on, demonstrates that our "engineering reasoning" can be justified.

Specifically, the following is true (for proof see Remark 2.4.10 and Proposition 2.4.2):

Proposition 2.3.1. Let z_ℓ, $\ell = 1, ..., L$, be deterministic coefficients and ζ_ℓ, $\ell = 1, ..., L$, be independent random variables with zero mean taking values in $[-1, 1]$. Then for every $\Omega \geq 0$ it holds that

$$\text{Prob}\left\{\zeta : \sum_{\ell=1}^{L} z_\ell \zeta_\ell > \Omega \sqrt{\sum_{\ell-1}^{L} z_\ell^2}\right\} \leq \exp\{-\Omega^2/2\}. \tag{2.3.4}$$

As an immediate conclusion, we get

$$(2.3.1) \Rightarrow \text{Prob}\left\{\eta > \Omega \sqrt{\sum_{\ell=1}^{L} ([a^\ell]^T x - b^\ell)^2}\right\} \leq \exp\{-\Omega^2/2\} \; \forall \Omega \geq 0, \tag{2.3.5}$$

and we have arrived at the result as follows.

Corollary 2.3.2. In the case of (2.3.1), the conic quadratic constraint (2.3.3) is a computationally tractable safe approximation of the chance constraint

$$\text{Prob}\left\{[a^0]^T x + \sum_{\ell=1}^{L} \zeta_\ell [a^\ell]^T x > b^0 + \sum_{\ell=1}^{L} \zeta_\ell b^\ell\right\} \leq \exp\{-\Omega^2/2\}. \tag{2.3.6}$$

In particular, with $\Omega \geq \sqrt{2\ln(1/\epsilon)}$, the constraint (2.3.3) is a tractable safe approximation of the chance constraint (2.2.2).

Now let us make the following important observation:

In view of Example 1.3.3, *inequality* (2.3.3) *is nothing but the RC of the uncertain linear inequality* (2.2.1), (1.3.5), *with the perturbation set* $\mathcal{Z}$ *in* (1.3.5) *specified as the ball*

$$\text{Ball}_\Omega = \{\zeta : \|\zeta\|_2 \leq \Omega\}. \tag{2.3.7}$$

This observation is worthy of in-depth discussion.

A. By itself, the assumption that ζ_ℓ vary in $[-1, 1]$, (which is a part of the assumptions in (2.3.1)), suggests to consider, as the perturbation set $\mathcal{Z}$ in (1.3.5), the box

$$\text{Box}_1 = \{\zeta : -1 \leq \zeta_\ell \leq 1, \; \ell = 1, ..., L\}.$$

For this $\mathcal{Z}$, the associated RC of the uncertain linear inequality (2.2.1), (1.3.5) is

$$\sum_{\ell=1}^{L} |[a^\ell]^T x - b^\ell| \leq b^0 - [a^0]^T x \tag{2.3.8}$$

(see Example 1.3.2). In the case of (2.3.1), this "box RC" guarantees "100% immunization against perturbations," meaning that every feasible solution to the box RC is feasible for the randomly perturbed inequality in question with probability 1. With the same stochastic model of uncertainty (2.3.1), the "ball RC," that is, the conic constraint (2.3.3), guarantees less, namely, "$(1 - \exp\{-\Omega^2/2\}) \cdot 100\%$-immunization." Note that with quite a moderate Ω, the "unreliability" $\exp\{-\Omega^2/2\}$ is negligible: it is less than 10^{-6} for $\Omega = 5.26$ and less than 10^{-12} for $\Omega = 7.44$. For all practical purposes, probability like 10^{-12} is the same as zero probability, so that there are all reasons to claim that the ball RC with $\Omega = 7.44$ is as "practically reliable" as the box RC.[1] Given that the "immunization power" of both RCs is essentially the same, it is very instructive to compare the "sizes" of the underlying perturbation sets Box_1 and Ball_Ω. This comparison leads to a great surprise: *when the dimension L of the perturbation set is not too small, the ball Ball_Ω with Ω "of order of one," say, $\Omega = 7.44$, is incomparably smaller that the unit box Box_1 with respect to all natural size measures such as diameter, volume, etc.* For example,

- the Euclidean diameters of Ball_Ω and Box_1 are respectively, 2Ω and $2\sqrt{L}$; with $\Omega = 7.44$, the second diameter is larger than the first starting with $L = 56$, and the ratio of the second diameter to the first one blows up to ∞ as L grows;

- the ratio of volumes of the ball and the box is

$$\frac{\text{Vol}(\text{Ball}_\Omega)}{\text{Vol}(\text{Box}_1)} = \frac{(\Omega\sqrt{\pi})^L}{2^L \Gamma(L/2 + 1)} \leq \left(\frac{\Omega\sqrt{e\pi/2}}{\sqrt{L}}\right)^L,$$

Γ being the Euler Gamma function. For $\Omega = 7.44$, this ratio is < 1 starting with $L = 237$ and goes to 0 super-exponentially fast at $L \to \infty$.

B. As a counter-argument to what was said in **A**, one can argue that for small L the uncertainty set $\text{Ball}_{7.44}$ is essentially larger than the uncertainty set Box_1. Well, here is a "rectification," interesting by its own right, of the ball RC which nullifies this counter-argument. Consider the case when the perturbation set $\mathcal{Z}$ is the intersection of the unit box and the ball of radius Ω centered at the origin:

$$\mathcal{Z} = \{\zeta \in \mathbb{R}^L : \|\zeta\|_\infty \leq 1, \|\zeta\|_2 \leq \Omega\} = \text{Box}_1 \cap \text{Ball}_\Omega. \tag{2.3.9}$$

Proposition 2.3.3. *The RC of the uncertain linear constraint (2.2.1) with the uncertainty set (2.3.9) is equivalent to the system of conic quadratic constraints*

$$\begin{aligned} (a) \quad & z_\ell + w_\ell = b^\ell - [a^\ell]^T x, \; \ell = 1, ..., L; \\ (b) \quad & \sum_\ell |z_\ell| + \Omega\sqrt{\sum_\ell w_\ell^2} \leq b^0 - [a^0]^T x. \end{aligned} \tag{2.3.10}$$

In the case of (2.3.1), the x component of every feasible solution to this system satisfies the randomly perturbed inequality (2.2.1) with probability at least $1 - \exp\{-\Omega^2/2\}$.

Proof. The fact that (2.3.10) represents the RC of (2.2.1), the perturbation set being (2.3.9), is readily given by Example 1.3.7 where one should set $\sigma_\ell \equiv 1$. Now let us prove that if (2.3.1) takes place and x, z, w is feasible for (2.3.10), then x is feasible for (2.2.1) with probability at least $1 - \exp\{-\Omega^2/2\}$. Indeed, when

---

[1]This conclusion tacitly assumes that the underlying stochastic uncertainty model is accurate enough to be trusted even when speaking about probabilities as small as 1.e-12; concerns of this type seem to be the inevitable price for using stochastic models of uncertainty.

$\|\zeta\|_\infty \leq 1$, we have

$$\sum_{\ell=1}^{L} [[a^\ell]^T x - b^\ell]\zeta_\ell > b^0 - [a^0]^T x$$

$$\Rightarrow \quad -\sum_{\ell=1}^{L} z_\ell \zeta_\ell - \sum_{\ell=1}^{L} w_\ell \zeta_\ell > b^0 - [a^0]^T x \quad [\text{by } (2.3.10.a)]$$

$$\Rightarrow \quad \sum_{\ell=1}^{L} |z_\ell| - \sum_{\ell=1}^{L} w_\ell \zeta_\ell > b^0 - [a^0]^T x \quad [\text{since } \|\zeta\|_\infty \leq 1]$$

$$\Rightarrow \quad -\sum_{\ell=1}^{L} w_\ell \zeta_\ell > \Omega \sqrt{\sum_{\ell=1}^{L} w_\ell^2} \quad [\text{by } (2.3.10.b)]$$

Therefore for every distribution P compatible with (2.3.1) we have

$$\text{Prob}_{\zeta \sim P} \{x \text{ is ineasible for } (2.2.1)\} \leq \text{Prob}_{\zeta \sim P} \left\{ -\sum_{\ell=1}^{L} w_\ell \zeta_\ell > \Omega \sqrt{\sum_{\ell=1}^{L} w_\ell^2} \right\}$$

$$\leq \exp\{-\Omega^2/2\},$$

where the last inequality is due to Proposition 2.3.1. $\qquad\square$

Note that perturbation set (2.3.9) is *never* greater than the perturbation set Box_1 and, as was explained in **A**, for every fixed Ω is incomparably smaller than the latter set when the dimension L of the perturbation vector ζ is large. Nevertheless, Proposition 2.3.3 says that when the perturbation vector is random and obeys (2.3.1), the "immunization power" of the RC associated with the small perturbation set (2.3.9), where $\Omega = 7.44$, is essentially as strong as that of the 100% reliable box RC (2.3.8). This phenomenon becomes even more striking when we consider the following special case of (2.2.1): ζ_ℓ are independent and each of them takes values ± 1 with probabilities $1/2$. In this case, when $L > \Omega^2$, the perturbation set (2.3.9) does not contain even a *single* realization of the random perturbation vector! Thus, the "immunization power" of the RC (2.3.10) cannot be explained by the fact that the underlying perturbation set contains "nearly all" realizations of the random perturbation vector.

C. Our considerations justify the use of "strange" perturbation sets like ellipsoids and intersections of ellipsoids and parallelotopes: while it may seem difficult to imagine a natural perturbation mechanism that produces perturbations from such sets, our analysis demonstrates that these sets do emerge naturally when "immunizing" solutions against random perturbations of the type described in (2.3.1). The same is true for the "budgeted" perturbation set considered in Example 1.3.8:

Proposition 2.3.4. Consider the RC of uncertain linear constraint (2.2.1) in the case of budgeted uncertainty:

$$\mathcal{Z} = \{\zeta \in \mathbb{R}^L : -1 \leq \zeta_\ell \leq 1, \ell = 1, ..., L, \sum_{\ell=1}^{L} |\zeta_\ell| \leq \gamma\}. \qquad (2.3.11)$$

This RC, according to Example 1.3.8, can be represented by the system of constraints

$$(a) \quad \sum_{\ell=1}^{L} |z_\ell| + \gamma \max_\ell |w_\ell| + [a^0]^T x \leq b^0,$$
$$(b) \quad z_\ell + w_\ell = b^\ell - [a^\ell]^T x, \ \ell = 1, ..., L \tag{2.3.12}$$

in variables x, z, w. In the case of (2.3.1), the x component of every feasible solution to this system satisfies the randomly perturbed inequality (2.2.1) with probability at least $1 - \exp\{-\frac{\gamma^2}{2L}\}$.

Thus, the quantity $\frac{\gamma}{\sqrt{L}}$ in our present situation plays the same role as the quantity Ω plays in the situation of Proposition 2.3.3.

Proof. Let (x, z, w) be feasible for (2.3.12). We have

$$\|w\|_2^2 = \sum_{\ell=1}^{L} w_\ell^2 \leq \sum_{\ell=1}^{L} |w_\ell| \|w\|_\infty \leq \|w\|_\infty \sum_{\ell=1}^{L} |w_\ell| \leq \|w\|_\infty \sqrt{L} \|w\|_2,$$

where the last $\leq$ is by the Cauchy inequality. Thus, $\|w\|_2 \leq \sqrt{L} \|w\|_\infty$; since x, z, w satisfy (2.3.12), we have $\sum_{\ell=1}^{L} |z_\ell| + \frac{\gamma}{\sqrt{L}} \|w\|_2 \leq b^0 - [a^0]^T x$, which combines with (2.3.12) to imply that x, z, w satisfy (2.3.10) with $\Omega = \frac{\gamma}{\sqrt{L}}$. Now we can apply Proposition 2.3.3 to conclude that in the case of (2.3.1) x satisfies (2.2.1) with probability $\geq 1 - \exp\{-\frac{\gamma^2}{2L}\}$. $\quad\square$

Remark 2.3.5. The proof of Proposition 2.3.4 shows that the "budgeted" RC (2.3.11) is more conservative (that is, associated with a larger perturbation set) than the ball RC (2.3.3), provided that the uncertainty budget γ in the budgeted RC is linked to the safety parameter Ω in the ball RC according to $\Omega = \frac{\gamma}{\sqrt{L}}$. The question arises: Why should we be interested in the budgeted RC at all, given that the only "good news" about this RC, expressed in Proposition 2.3.4, holds true for the less conservative ball RC? The answer is, that the budgeted RC can be represented by a system of *linear* constraints, that is, it is of the same "level of complexity" as the instances of the underlying uncertain constraint (2.2.1). As a result, when using budgeted uncertainty models for every one of the uncertain constraints in an uncertain LO problem, the RC of the problem is itself an LO problem and as such can be processed by well-developed commercial LO solvers. In contrast to this, the ball RC (2.3.3) leads to a conic quadratic problem, which is more computationally demanding (although still efficiently tractable).

2.3.1 Illustration: A Single-Period Portfolio Selection

Example 2.3.6. Let us apply the outlined techniques to the following single-period portfolio selection problem:

There are 200 assets. Asset # 200 ("money in the bank") has yearly return $r_{200} = 1.05$ and zero variability. The yearly returns r_ℓ, $\ell = 1, ..., 199$ of the remaining assets are independent random variables taking values in the segments $[\mu_\ell - \sigma_\ell, \mu_\ell + \sigma_\ell]$ with expected values μ_ℓ; here

$$\mu_\ell = 1.05 + 0.3\frac{(200 - \ell)}{199}, \ \sigma_\ell = 0.05 + 0.6\frac{(200 - \ell)}{199}, \ \ell = 1, ..., 199.$$

The goal is to distribute \$1 between the assets in order to maximize the value-at-risk of the resulting portfolio, the required risk level being $\epsilon = 0.5\%$.

We want to solve the uncertain LO problem

$$\max_{y,t}\left\{t : \sum_{\ell=1}^{199} r_\ell y_\ell + r_{200}y_{200} - t \geq 0, \ \sum_{\ell=0}^{200} y_\ell = 1, y_\ell \geq 0 \, \forall \ell\right\},$$

where y_ℓ is the capital to be invested in asset # ℓ. The uncertain data are the returns r_ℓ, $\ell = 1, ..., 199$; their natural parameterization is

$$r_\ell = \mu_\ell + \sigma_\ell \zeta_\ell,$$

where ζ_ℓ, $\ell = 1, ..., 199$, are independent random perturbations with zero mean varying in the segments $[-1, 1]$. Setting $x = [y; -t] \in \mathbb{R}^{201}$, the problem becomes

$$
\begin{array}{ll}
\text{minimize} & x_{201} \\
\text{subject to} &
\end{array}
$$

$$(a) \qquad [a^0 + \sum_{\ell=1}^{199} \zeta_\ell a^\ell]^T x - [b^0 + \sum_{\ell=1}^{199} \zeta_\ell b^\ell] \leq 0$$

$$(b) \qquad \sum_{j=1}^{200} x_\ell = 1 \tag{2.3.13}$$

$$(c) \qquad x_\ell \geq 0, \ \ell = 1, ..., 200$$

where

$$
\begin{aligned}
a^0 &= [-\mu_1; -\mu_2; ...; -\mu_{199}; -r_{200}; -1]; \\
a^\ell &= \sigma_\ell \cdot [0_{\ell-1,1}; 1; 0_{201-\ell,1}], \ 1 \leq \ell \leq 199; \\
b^\ell &= 0, \ 0 \leq \ell \leq 199.
\end{aligned}
\tag{2.3.14}
$$

The only uncertain constraint in the problem is the inequality $(2.3.13.a)$. We consider 3 perturbation sets along with the associated robust counterparts of $(2.3.13)$:

i) *Box RC* that ignores the information on the stochastic nature of the perturbations affecting the uncertain inequality and uses the only fact that these perturbations vary in $[-1, 1]$. The underlying perturbation set $\mathcal{Z}$ for $(2.3.13.a)$ is

$$\{\zeta : \|\zeta\|_\infty \leq 1\};$$

ii) *Ball-box RC* given by Proposition 2.3.3, with the safety parameter

$$\Omega = \sqrt{2\ln(1/\epsilon)} = 3.255,$$

which ensures that the robust optimal solution satisfies the uncertainty-affected constraint with probability at least $1 - \epsilon = 0.995$. The underlying perturbation set $\mathcal{Z}$ for $(2.3.13.a)$ is

$$\{\zeta : \|\zeta\|_\infty \leq 1, \|\zeta\|_2 \leq 3.255\};$$

iii) Budgeted RC given by Proposition 2.3.4, with the uncertainty budget

$$\gamma = \sqrt{2\ln(1/\epsilon)}\sqrt{199} = 45.921,$$

which results in the same probabilistic guarantees as for the ball-box RC. The underlying perturbation set $\mathcal{Z}$ for $(2.3.13.a)$ is

$$\{\zeta : \|\zeta\|_\infty \leq 1, \|\zeta\|_1 \leq 45.921\}.$$

Box RC. The associated RC for the uncertain inequality is given by (2.3.8); after straightforward computations, the resulting RC of (2.3.13) becomes the LO problem

$$\max_{y,t}\left\{t : \begin{array}{l} \sum_{\ell=1}^{199}(\mu_\ell - \sigma_\ell)y_\ell + 1.05y_{200} \geq t \\ \sum_{\ell=1}^{200} y_\ell = 1,\ y \geq 0 \end{array}\right\}; \qquad (2.3.15)$$

as it should be expected, this is nothing but the instance of our uncertain problem corresponding to the worst possible values $r_\ell = \mu_\ell - \sigma_\ell$, $\ell = 1,...,199$, of the uncertain returns. Since these values are less than the guaranteed return for money, the robust optimal solution prescribes to keep our initial capital in the bank, with a guaranteed yearly return of 1.05, that is, a guaranteed profit of 5%.

Ball-box RC. The associated RC for the uncertain inequality is given by Proposition 2.3.3. The resulting RC of (2.3.13) is the conic quadratic problem

$$\max_{y,z,w,t}\left\{t : \begin{array}{l} \sum_{\ell=1}^{199}\mu_\ell y_\ell + 1.05y_{200} - \sum_{\ell=1}^{199}|z_\ell| - 3.255\sqrt{\sum_{\ell=1}^{199}w_\ell^2} \geq t \\ z_\ell + w_\ell = \sigma_\ell y_\ell,\ \ell = 1,...,199,\ \sum_{\ell=1}^{200} y_\ell = 1,\ y \geq 0 \end{array}\right\}. \qquad (2.3.16)$$

The robust optimal value is 1.1200, meaning 12.0% profit with risk as low as $\epsilon = 0.5\%$. The distribution of capital between assets is depicted in figure 2.1.

Budgeted RC. The associated RC for the uncertain inequality is given by Proposition 2.3.4. The resulting RC of (2.3.13) is the LO problem

$$\max_{y,z,w,t}\left\{t : \begin{array}{l} \sum_{\ell=1}^{199}\mu_\ell y_\ell + 1.05y_{200} - \sum_{\ell=1}^{199}|z_\ell| - 45.921\max_{1\leq\ell\leq199}|w_\ell| \geq t \\ z_\ell + w_\ell = \sigma_\ell y_\ell,\ \ell = 1,...,199,\ \sum_{\ell=1}^{200} y_\ell = 1,\ y \geq 0 \end{array}\right\}. \qquad (2.3.17)$$

The robust optimal value is 1.1014, meaning 10.1% profit with risk as low as $\epsilon = 0.5\%$. The distribution of capital between assets is depicted in figure 2.1.

Discussion. First, we see how useful stochastic information might be — with risk as low as 0.5%, the value-at-risk of the portfolio profits yielded by the ball-box RC (12%) and the Budgeted RC (10%) are twice as large as the profit guaranteed by the box RC (5%). Note also that both the ball-box and the Budgeted RCs suggest "active" investment decisions, while the box RC suggests keeping the initial capital in bank. Second, the Budgeted RC, as it should be, is more conservative than the ball-box one. Finally, we should remember that the actual risk associated with

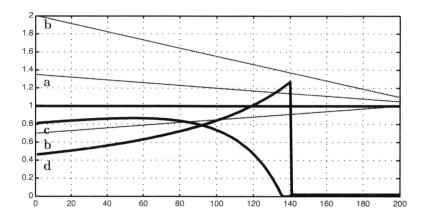

Figure 2.1 Robust solutions to portfolio selection problem from Example 2.3.6. Along the x-axis: indices 1,2,...,200 of the assets. **a**: expected returns, **b**: upper and lower endpoints of the return ranges, **c**: invested capital for ball-box RC, %, **d**: invested capital for Budgeted RC, %.

the portfolio designs offered by the ball-box and the Budgeted RCs (that is, the probability for the actual total yearly return to be less than the corresponding robust optimal value) is *at most* the required 0.5%, and is likely to be less than this amount; indeed, both RCs in question utilize conservative approximations of the chance constraint

$$\text{Prob}\{\sum_{\ell=1}^{199} r_\ell y_\ell + r_{200} y_{200} < t\} \leq \epsilon.$$

It is interesting to find out how small the actual risk is. The answer, of course, depends on the actual probability distributions of uncertain returns (recall that in our model, we postulated only partial knowledge of these distributions, specifically, knowledge of their supports and expectations). Assuming that "in reality" ζ_ℓ, $\ell = 1, ..., 199$, take only their extreme values ± 1, with probability $1/2$ each, and carrying out a Monte-Carlo simulation with a sample of 1,000,000 realizations, we found that the actual risk for the "ball-box" portfolio is less than the required risk 0.5% by factor 10, and for the "Budgeted" portfolio, by factor 50. Based on this observation, it seems plausible that we can reduce our conservatism by "tuning," that is, by replacing the required risk in the RCs with a larger quantity, in hope that the resulting actual risk, (which can be evaluated via simulation), will still be below the required level. With this tuning, reducing the safety parameter $\Omega = 3.255$ in (2.3.16) to $\Omega = 2.589$, one ends up with the robust optimal value 1.1470 (that is, with a profit of 14.7% instead of the initial 12.0%), while keeping the empirical risk (as evaluated over 500,000 realization sample) still as low as 0.47%. Similarly, reducing the uncertainty budget $\gamma = 45.921$ in (2.3.17) to $\gamma = 30.349$, we increase the robust optimal value from 1.1012 to 1.1395 (i.e., increase profit from 10.12% to 13.95%), with the empirical risk as low as 0.42%.

2.3.2 Illustration: Cellular Communication

Example 2.3.7. Consider the following problem:

Signal Recovery: *Given indirect observations*

$$u = As + \rho\xi \qquad (2.3.18)$$

of a signal $s \in \mathcal{S} = \{s \in \mathbb{R}^n : s_i = \pm 1, i = 1, ...n\}$ *(A is a given $m \times n$ matrix, $\xi \sim \mathcal{N}(0, I_m)$ is observation noise, $\rho \geq 0$ is a deterministic noise level), find an estimate $\hat{s}$ of the signal of the form*

$$\hat{s} = \text{sign}[Gu] \qquad (2.3.19)$$

such that the requirement

$$\forall(s \in \mathcal{S}, i \leq n) : \text{Prob}\{\hat{s}_i \neq s_i\} \equiv \text{Prob}\{(\text{sign}[GAs + \rho G\xi])_i \neq s_i\} \leq \epsilon \qquad (2.3.20)$$

is satisfied. Here $\epsilon \ll 1$ is a given tolerance, and $\text{sign}[v]$ acts in the coordinate-wise fashion: $\text{sign}[[v_1; ...; v_n]] = [\text{sign}(v_1); ...; \text{sign}(v_n)]$.

The situation when a signal $s \in \mathcal{S}$ is observed according to (2.3.18) can be regarded as a meaningful (although somehow simplified) model of cellular communication. Note also that estimates of the form (2.3.19) are practical — while not the best from the viewpoint of their sensitivity to noise, they are frequently used in reality due to their computational simplicity. Finally, let us explain what is the rationale behind (2.3.19). Assuming s to be random and Gaussian with zero mean (and independent of ξ), the best recovery, in the sense of the mean square error, is indeed a linear one: $\hat{s} = Gu$ with a properly defined matrix G (the so called *Wiener filter*). Engineers often use optimal solutions to simple problems as "practical solutions" to more complicated problems, the Wiener filter not being an exception. A linear estimator Gu of a signal observed according to (2.3.18) is frequently used in situations when s is not necessarily Gaussian with zero mean. Now, when we know in advance that s is a ± 1 vector, we can try to improve the purely linear estimator Gu as follows: assuming that Gu is not too far from s (namely, the typical Euclidean distance from Gu to s is < 1), the vector $\text{sign}[Gu]$ "equally typically" will be *exactly* s. All this being said, let us focus on the Signal Recovery problem as a mathematical beast. Our first observation is immediate:

A necessary and sufficient condition for (2.3.20) is

$$\forall i \leq n : \sum_{j \neq i} |(GA)_{ij}| - (GA)_{ii} + \rho\|g_i\|_2 \text{ErfInv}(\epsilon) \leq 0 \ \& \ g_i \neq 0, \ (2.3.21)$$

where g_i^T is i-th row of G and ErfInv is the inverse error function defined by the relation

$$0 < \delta < 1 \Rightarrow \text{Erf}(\text{ErfInv}(\delta)) = \delta,$$
$$\left[\text{Erf}(s) = \int\limits_s^\infty \frac{1}{\sqrt{2\pi}} \exp\{-r^2/2\}dr \text{ is the error function.} \right] \qquad (2.3.22)$$

Indeed, assume that G satisfies (2.3.20). Then for every $i \leq n$ we have

$$\forall(s \in \mathcal{S}, s_i = -1) : \mathrm{Prob}\{\sum_{j \neq i}(GA)_{ij}s_j - (GA)_{ii} + \rho(g_i^T \xi) \geq 0\} \leq \epsilon.$$

The latter is possible only when $g_i \neq 0$ (since otherwise the probability in the left hand side is 1) and is equivalent to

$$\forall(s \in \mathcal{S}, s_i = -1) : \mathrm{Prob}\{-\rho(g_i^T \xi) \leq \sum_{j \neq i}(GA)_{ij}s_j - (GA)_{ii}\} \leq \epsilon,$$

which, due to $\rho(g_i^T \xi) \sim \mathcal{N}(0, \rho^2 \|g_i\|_2^2)$, is equivalent to

$$\forall(s \in \mathcal{S}, s_i = -1) : \sum_{j \neq i}(GA)_{ij}s_j - (GA)_{ii} + \rho\mathrm{ErfInv}(\epsilon)\|g_i\|_2 \leq 0$$
$$\Updownarrow$$
$$\sum_{j \neq i}|(GA)_{ij}| - (GA)_{ii} + \rho\|g_i\|_2\mathrm{ErfInv}(\epsilon)$$
$$\equiv \max_{s \in \mathcal{S}, s_i = -1}\left[\sum_{j \neq i}(GA)_{ij}s_j - (GA)_{ii}\right] + \rho\mathrm{ErfInv}(\epsilon)\|g_i\|_2 \leq 0,$$

and we see that (2.3.21) takes place. Vice versa, if the latter relation takes place, then, inverting the above reasoning, we see that

$$\forall(i \leq n, s \in \mathcal{S} : s_i = -1) : \mathrm{Prob}\left\{(\mathrm{sign}[GAs + \rho G\xi])_i \neq s_i\right\} \leq \epsilon.$$

Since ξ is symmetrically distributed, the latter relation is equivalent to (2.3.20). Observe that when $\rho > 0$, (2.3.21) clearly implies $(GA)_{ii} > 0$. Multiplying the rows in G by appropriate positive constants, we can normalize G to have $(GA)_{ii} = 1$ for all i, and this normalization clearly does not affect the validity of (2.3.21). It follows that the problem of interest is equivalent to the optimization problem

$$\max_{\rho, G}\left\{\rho : \begin{array}{l} \sum_j |(GA - I)_{ij}| + \rho\sqrt{\sum_j G_{ij}^2}\mathrm{ErfInv}(\epsilon) \leq 1 = (GA)_{ii}, \\ 1 \leq i \leq n \end{array}\right\}. \qquad (2.3.23)$$

Note that this problem, while not being exactly convex, is nevertheless computationally tractable: for every positive ρ, the system of constraints in the right hand side is a system of efficiently computable convex constraints in G, and we can check efficiently whether it is feasible. If it is feasible for a given ρ, it is feasible for all smaller ρ as well, so that the largest ρ for which the system is feasible (and that is exactly the ρ we want to find) can be easily approximated to a high accuracy by bisection. We are about to show that in fact no bisection is necessary — our problem admits a closed form solution. Specifically, the following is true:

Proposition 2.3.8. Problem (2.3.23) has a feasible solution with $\rho > 0$ if and only if the rank of the matrix A is equal to n (the dimension of the signal s), and in this case an optimal solution to (2.3.23) is as follows:
• G is the pseudo-inverse of A, that is, the $n \times m$ matrix with the transposed rows belonging to the image space of A and such that $GA = I$ (these conditions uniquely define G);
• $\rho = (\mathrm{ErfInv}(\epsilon) \max_i \sqrt{\sum_j G_{ij}^2})^{-1}$.

Proof. Observe, first, that if $\mathrm{Rank}(A) < n$, the problem (2.3.23) has no feasible solutions with $\rho > 0$. Indeed, assume that $\mathrm{Rank}(A) < n$ and that $(\rho > 0, G)$ is a feasible solution to the problem. Then the image $L \subset \mathbb{R}^n$ of the matrix GA is a proper linear subspace of $\mathbb{R}^n$ and as such it does not intersect the interior of at least one of the 2^n orthants $\mathbb{R}_\kappa = \{s : \kappa_i s_i \geq 0, 1 \leq i \leq n\}$, $\kappa_i = \pm 1$. Indeed, there exists a nonzero vector e which is orthogonal to L; setting $\kappa_i = \text{sign}(e_i)$ when $e_i \neq 0$ and choosing, say, $\kappa_i = 1$ when $e_i = 0$, we ensure that $e^T f > 0$ for all $f \in \text{int}\mathbb{R}_\kappa$, so that L cannot intersect $\text{int}\mathbb{R}_\kappa$. Thus, there exists $\kappa \in \mathcal{S}$ such that L does not intersect $\text{int}\mathbb{R}_\kappa$, meaning that $(\kappa - GA\kappa)_i \geq 1$ for at least one i. On the other hand, (G, ρ) is feasible for (2.3.23), meaning that $(GA)_i = 1$ and $\sum_{j \neq i} |GA_{ij}| < 1$; these relations clearly imply that $(\kappa - GA\kappa)_i = -\sum_{j \neq i}(GA)_{ij}\kappa_j < 1$, the desired contradiction.

Now assume that $\text{Rank}(A) = n$, so that there exists the pseudo-inverse of A, let it be denoted $G^\dagger$. Setting $\rho_* = (\text{ErfInv}(\epsilon) \max_i \sqrt{\sum_j (G_{ij}^\dagger)^2})^{-1}$, we get a feasible solution $(\rho_*, G^\dagger)$ to (2.3.23). Let us prove that this solution is optimal. To this end, assume that there exists a feasible solution $(\rho, \widehat{G})$ to (2.3.23) with $\rho > \rho_*$, and let us lead this assumption to a contradiction. Let a_i, $i = 1, ..., n$, be the columns of A, and g_i^T, $i = 1, ..., n$, be the rows of $G^\dagger$, so that

$$g_i^T a_j = \delta_{ij} \equiv \begin{cases} 1, & i = j \\ 0, & i \neq j \end{cases}$$

due to $G^\dagger A = I$. Let, further, $\widehat{g}_i^T$ be the rows of $\widehat{G}$. Observe that we can, w.l.o.g., assume that $\widehat{g}_i$ belong to the image space of A. Indeed, replacing the (transposed of the) rows in $\widehat{G}$ by their orthogonal projections on the image space of A, we do not change $\widehat{G}A$ and do not increase the Euclidean norms of the rows in $\widehat{G}$ and thus preserve the feasibility of $(\rho, \widehat{G})$ for (2.3.23). Further, from $(\widehat{G}A)_{ii} = 1$, $i = 1, ..., n$, it follows that $\widehat{g}_i^T a_i = 1$, so that $\widehat{g}_i \neq 0$ for all i. Now comes the final step: we have $\underbrace{\rho_* \text{ErfInv}(\epsilon)}_{\nu_*} \|g_i\|_2 \leq 1$ for every i, with the inequality being equality for some i; w.l.o.g. we may assume that

$$\nu_* \|g_i\|_2 \leq \nu_* \|g_1\|_2 = 1, \, i = 1, ..., n.$$

Now let

$$f(g) = \sum_{j \neq 1} |g^T a_j| + \nu_* \|g\|_2.$$

Then $f(g_1) = 1$. We claim that $f(\widehat{g}_1) < 1$. Indeed, setting $\nu = \rho \text{ErfInv}(\epsilon)$, we have

$$f(\widehat{g}_1) = \sum_{j \neq 1} |(\widehat{G}A)_{1j}| + \nu_* \|\widehat{g}_1\|_2 < \sum_{j \neq 1} |(\widehat{G}A)_{1j}| + \nu \|\widehat{g}_1\|_2 \leq 1,$$

where the strict inequality is due to the fact that $\widehat{g}_1 \neq 0$ and $\nu > \nu_*$, and the concluding inequality follows from the fact that $(\nu, \widehat{G})$ is feasible for (2.3.23). Now, the vector $\widehat{g}_1$ belongs to the image space of A, and this image is spanned by the vectors $g_1, ..., g_n$. Indeed, by the definition of pseudo-inverse the vectors g_i belong to this space; if their linear span were less than the image of A, there would exist vector As, $s \neq 0$, orthogonal to $g_1, ..., g_n$, whence $G^\dagger As = 0$ instead of $G^\dagger As = s \neq 0$. It follows that

$$\widehat{g}_1 = g_1 + \sum_k r_k g_k$$

cond(A)	$\rho(10^{-3})$	$\rho(10^{-6})$
3.7e2	0.0122	0.009219
1.46e4	5.5e-4	4.16e-4

Table 2.1 Critical noise levels for two instances of the Signal Recovery problem.

for some r_k. Since $(\widehat{G}A)_{11} = \widehat{g}_1^T a_1 = 1$ and $g_k^T a_1 = \delta_{k1}$, we have $r_1 = 0$, that is, $\widehat{g}_1 = g_1 + \sum_{k \neq 1} r_k g_k$. We now have

$$
\begin{aligned}
1 > f(\widehat{g}_1) = f(g_1 + \sum_{k \neq 1} r_k g_k) &= \sum_{j \neq 1} |(g_1 + \sum_{k \neq 1} r_k g_k)^T a_j| \\
+\nu_* \|g_1 + \sum_{k \neq 1} r_k g_k\|_2 &= \sum_{j \neq 1} |\sum_{k \neq 1} r_k \underbrace{g_k^T a_j}_{\delta_{kj}}| + \nu_* \|g_1 + \sum_{k \neq 1} r_k g_k\|_2 \\
&= \sum_{k \neq 1} |r_k| + \nu_* \|g_1 + \sum_{k \neq 1} r_k g_k\|_2 \geq \sum_{k \neq 1} |r_k| + \underbrace{\nu_* \|g_1\|_2}_{1} - \sum_{k \neq 1} \underbrace{\nu_* \|g_k\|_2}_{\leq 1} |r_k| \geq 1,
\end{aligned}
$$

which gives the desired contradiction.   □

Proposition 2.3.8 brings us good and bad news. The good news is that the solution to our problem is simple and natural; the bad news is that when A is ill-conditioned, the optimal recovery suggested by the proposition will be optimal, but nevertheless of a very poor performance, since the "straightforward recovery" $u \mapsto G^\dagger u$ will amplify the noise. In table 2.1, we present numerical results for two 32×32 randomly generated matrices A: in the first, the entries are sampled from $\mathcal{N}(0,1)$ distribution, in the second we generate the matrix in the same fashion, but then multiply one of its columns by a small number to make the matrix ill-conditioned. This experiment demonstrates, as a byproduct, how tight our approximation is: for the first matrix and $\rho = 0.0122$, (which provably guarantees precise recovery with probability at least 0.999), in a sample of 100,000 experiments there were 3 recovery errors, meaning that the true error probability at this level of noise is hardly much less than 10^{-5}; at the same time, at a level of noise $0.75\rho(10^{-3})$, the error probability is provably less than 10^{-6}.

Can we "beat" the straightforward recovery given by the pseudo-inverse of A? The answer is yes, provided that we slightly restrict the set of tentative signals S. Assume, e.g., that our signals s are vectors with coordinates ± 1 *and such that among the entries of s there are at least k ones and k minus ones*; here $k \leq n/2$ is a given integer. Let S_k be the set of all signals of this type (in this notation, what used to be S is nothing but S_0). The derivation that led us to (2.3.23) can be carried out with S_k in the role of S, and the resulting equivalent reformulation of the problem is

$$
\max_{\rho, G} \left\{ \max_{\substack{s \in S_k: \\ s_i = -1}} \sum_{j \neq i} (GA)_{ij} s_j + \rho \mathrm{ErfInv}(\epsilon) \sqrt{\sum_j G_{ij}^2} \leq 1 = (GA)_{ii}, 1 \leq i \leq n \right\}.
$$

$$(2.3.24)$$

The only difference with the case of $k = 0$ is in computing

$$\Phi_i(G) = \max_{s \in \mathcal{S}_k : s_i = -1} \sum_{j \neq i} (GA)_{ij} s_j,$$

the quantity that in the case of $k = 0$ was $\sum_{j \neq i} |(GA)_{ij}|$. Now this quantity is of the form

$$F(z_1, ..., z_{n-1}) = \max_{s \in \mathcal{S}_k^*} \sum_{j=1}^{n-1} z_j s_j,$$

where $\mathcal{S}_k^*$ is the set of all vectors in $\mathbb{R}^{n-1}$ with coordinates ± 1 that have at least $k - 1$ coordinates equal to -1 and at least k coordinates equal to 1; indeed, we have

$$\Phi_i(G) = F((GA)_{i1}, ..., (GA)_{i,i-1}, (GA)_{i,i+1}, ..., (GA)_{in}).$$

In order to compute F, note that $\mathcal{S}_k^*$ is exactly the set of extreme points of the polytope $\{s \in \mathbb{R}^{n-1} : -1 \leq s_j \leq 1 \, \forall j, \sum_j s_j \leq n + 1 - 2k, \sum_j s_j \geq -n + 1 + 2k\}$, so that

$$F(z) = \max_{s \in \mathbb{R}^{n-1}} \left\{ \sum_{j=1}^{n-1} z_j s_j : \begin{array}{c} -1 \leq s_j \leq 1, \sum_j s_j \leq n + 1 - 2k, \\ \sum_j s_j \geq -n + 1 + 2k \end{array} \right\}$$

$$= \min_{\{\mu_j \geq 0, \nu_j \geq 0\}_{j=0}^{n-1}} \left\{ \sum_{j=1}^{k} [\mu_j + \nu_j] + (n + 1 - 2k)\mu_0 + (n - 1 - 2k)\nu_0 : \right.$$

$$\left. \mu_j - \nu_j + \mu_0 - \nu_0 = z_j, 1 \leq j \leq n - 1 \right\},$$

where the concluding equality is given by the Linear Programming Duality Theorem. It follows that (2.3.24) is equivalent to the explicit computationally tractable optimization program

$$\max_{\rho, G} \left\{ \rho : \begin{array}{l} \sum_{\substack{1 \leq j \leq n \\ j \neq i}} [\mu_{ij} + \nu_{ij}] + (n + 1 - 2k)\mu_{i0} + (n - 1 - 2k)\nu_{i0} \\ \quad + \rho \text{ErfInv}(\epsilon) \sqrt{\sum_j G_{ij}^2} \leq 1 = (GA)_{ii}, \, i = 1, ..., n \\ \mu_{ij} - \nu_{ij} + \mu_{i0} - \nu_{i0} = (GA)_{ij}, 1 \leq i \leq n, 1 \leq j \leq n, j \neq i \\ \mu_{ij}, \nu_{ij} \geq 0, 1 \leq i \leq n, 0 \leq j \leq n, i \neq j \end{array} \right\}. \quad (2.3.25)$$

For this problem, a closed form solution is seemingly out of the question; the problem, however, can be solved efficiently by the bisection-based strategy presented before Proposition 2.3.8. Whether the recovering routine yielded by the optimal solution to this problem "beats" the straightforward recovery based on the pseudo-inverse of A depends on the "geometry" of A. Experiments demonstrate that for randomly generated matrices A, both procedures are essentially of the same power. At the same time, for special matrices A the recovering routine based upon the optimal solution to (2.3.25) can beat significantly the "straightforward" one. As an example, consider the case when A is close to the orthonormal projector P onto the hyperplane $\sum_i s_i = 0$, specifically,

$$A = A_\gamma = P + \gamma \frac{\mathbf{1} \cdot \mathbf{1}^T}{n},$$

where $\mathbf{1}$ is the vector of ones. The closeness of this matrix to P is controlled by γ — the closer this parameter is to 0, the closer A is to P. In table 2.2, we present results of numerical experiments with $n = 32$, $k = 1$, and $A = A_\gamma$ for $\gamma = 0.005$

γ	$\rho(10^{-4})$, G is given by (2.3.25)	$\rho(10^{-4})$, $G = G^{\dagger}$
0.005	0.0146	0.00606
0.001	0.0135	0.00121

Table 2.2 Experiments with A being near-projector onto the hyperplane $\{s \in \mathbb{R}^{32} : \sum_i s_i = 0\}$ and $k = 1$.

and $\gamma = 0.001$. We see that in order to ensure 0.9999-reliable recovery, the routine based on $G^{\dagger} = A^{-1}$ requires an essentially smaller level of noise than the one based on the optimal solution to (2.3.25). For example, with $\gamma = 0.001$ and the noise level 0.0135, the probability of wrong recovery by the procedure based on $G^{\dagger}$, evaluated on a 10,000-element sample is as large as 0.68 (and remains 0.42 even when the level of noise is reduced by a factor of 2). In contrast to this, the procedure based on (2.3.25) at the noise level 0.0135 provably guarantees 0.9999-reliable recovery. Note that this significant advantage in performance "is bought" by forbidding just two of the $2^{32} = 429,4967,296$ signals with coordinates ± 1, namely, the signal with all coordinates 1 and the one with all coordinates -1.

It should be added that by itself, the problem of highly reliable recovery of a signal $s \in \mathcal{S}_0$ from "deficient" observations (2.3.18), (e.g., those with $\mathrm{Rank}(A) < n$) is not necessarily ill-posed. Consider, e.g., the case when $m = \mathrm{Rank}(A) = n - 1$. Then the observations are obtained from s by the following sequence of transformations: (a) projecting on a hyperplane (the orthogonal complement to the null space of A), (b) applying an invertible linear transformation to the projection, and (c) adding noise to the result. The possibility to recover $s \in \mathcal{S}_0$ from the resulting observation depends on whether the projection in (a) *when restricted onto the set of vertices of the unit cube* is a one to one mapping. Whenever this is the case (as it indeed happens when A is "in general position"), we can recover the signal from the observations *exactly*, provided that there is no noise, and, consequently, can recover it with arbitrarily high reliability, provided that the noise level is small enough. What does become impossible — independently of the level of noise! — is an errorless recovery *of the form* (2.3.19).

2.4 EXTENSIONS

In the preceding section, we were focusing on building a safe approximation of the chance version (2.2.3) of a randomly perturbed linear constraint (2.2.1). Under specific assumptions, expressed by (2.3.1), we have built such an approximation in the form of the RC of (2.2.1) with the properly chosen perturbation set $\mathcal{Z}$. We are about to extend this construction to wider families of random perturbations than those captured by (2.3.1). Specifically, let us assume that the random perturbation ζ affecting (2.2.1) possesses the following properties:

P.1. ζ_ℓ, $\ell = 1, ..., L$, are independent random variables;

P.2. The distributions P_ℓ of the components ζ_ℓ are such that

$$\int \exp\{ts\} dP_\ell(s) \le \exp\{\max[\mu_\ell^+ t, \mu_\ell^- t] + \frac{1}{2}\sigma_\ell^2 t^2\} \ \forall t \in \mathbb{R} \qquad (2.4.1)$$

with known constants $\mu_\ell^- \le \mu_\ell^+$ and $\sigma_\ell \ge 0$.

Property P.2 can be validated in several interesting cases to be considered later. Right now, let us derive some consequences of P.1–P.2.

Given P.1–2, consider the problem of bounding from above the probability $p(z)$ of the event $z_0 + \sum_{\ell=1}^{L} z_\ell \zeta_\ell > 0$, where $z = [z_0; z_1; ...; z_L]$ is a given deterministic vector. Let us set

$$\Phi(w_1, ..., w_L) = \sum_{\ell=1}^{L} \left[\max[\mu_\ell^+ w_\ell, \mu_\ell^- w_\ell] + \frac{1}{2}\sigma_\ell^2 w_\ell^2 \right], \qquad (2.4.2)$$

so that by P.1–2 for all deterministic reals $w_1, ..., w_L$ one has

$$\mathbf{E}\left\{\exp\{\sum_{\ell=1}^{L} \zeta_\ell w_\ell\}\right\} \le \exp\{\Phi(w_1, ..., w_L)\}. \qquad (2.4.3)$$

We have

$$z_0 + \sum_{\ell=1}^{L} z_\ell \zeta_\ell > 0$$

$$\Leftrightarrow \quad \exp\{\alpha[z_0 + \sum_{\ell=1}^{L} z_\ell \zeta_\ell]\} > 1 \ \forall \alpha > 0$$

$$\Rightarrow \quad \forall \alpha > 0 : \mathbf{E}\left\{\exp\{\alpha[z_0 + \sum_{\ell=1}^{L} z_\ell \zeta_\ell]\}\right\} \ge p(z)$$

$$\Rightarrow \quad \forall \alpha > 0 : \exp\{\alpha z_0 + \Phi(\alpha z_1, \alpha z_2, ..., \alpha z_L)\} \ge p(z)$$

$$\Leftrightarrow \quad \forall \alpha > 0 : \alpha z_0 + \Phi(\alpha[z_1; ...; z_L]) \ge \ln p(z).$$

We have arrived at the inequality

$$\forall \alpha > 0 : \ln p(z) \le \alpha z_0 + \Phi(\alpha[z_1; ...; z_L]).$$

If, for certain $\alpha > 0$, the right hand side of this inequality is $\le \ln \epsilon$, then the inequality implies that $p(z) \le \epsilon$. We have arrived at the following conclusion:

(∗) *Whenever, for a given $\epsilon \in (0, 1)$ and given z, there exists $\alpha > 0$ such that*

$$\alpha z_0 + \Phi(\alpha[z_1; ...; z_L]) \le \ln(\epsilon), \qquad (2.4.4)$$

one has

$$\mathrm{Prob}\left\{\zeta : z_0 + \sum_{\ell=1}^{L} z_\ell \zeta_\ell > 0\right\} \le \epsilon. \qquad (2.4.5)$$

In other words, the set

$$Z_\epsilon^o = \{z = [z_0; ...; z_L] : \exists \alpha > 0 : \alpha z_0 + \Phi(\alpha[z_1; ...; z_L]) \le \ln(\epsilon)\}$$

is contained in the feasible set of the chance constraint (2.4.5).

Now, the feasible set of (2.4.5) is clearly closed; since it contains the set Z_ϵ^o, it contains the set

$$Z_\epsilon = \operatorname{cl} Z_\epsilon^o. \tag{2.4.6}$$

Let us find an explicit description of the set Z_ϵ. We should understand first when a given point z belongs to Z_ϵ^o. By definition of the latter set, this is the case if and only if

$$\begin{aligned} \exists \alpha > 0 : \ln(\epsilon) \;\geq\;& f_z(\alpha) \equiv \alpha z_0 + \Phi(\alpha[z_1;...;z_L]) \\ =\;& \underbrace{\alpha \left(z_0 + \sum_{\ell=1}^L \max[\mu_\ell^- z_\ell, \mu_\ell^+ z_\ell] \right) + \tfrac{\alpha^2}{2} \sum_{\ell=1}^L \sigma_\ell^2 z_\ell^2.}_{a(z)} \end{aligned} \tag{2.4.7}$$

Assuming $b(z) \equiv \sum_\ell \sigma_\ell^2 z_\ell^2 > 0$, the function $f_z(\alpha)$ attains its minimum on $[0,\infty)$, and this minimum is either $f_z(0) = 0$ (this is the case when $a(z) \geq 0$), or $-\tfrac{1}{2}\tfrac{a^2(z)}{b(z)}$ (this is the case when $a(z) < 0$). Since $\ln(\epsilon) < 0$, we conclude that in the case $b(z) > 0$, relation (2.4.7) holds if and only if $a(z) + \sqrt{2\ln(1/\epsilon)} b(z) \leq 0$, i.e., if and only if

$$z_0 + \sum_{\ell=1}^L \max[\mu_\ell^+ z_\ell, \mu_\ell^- z_\ell] + \sqrt{2\ln(1/\epsilon)} \sqrt{\sum_{\ell=1}^L \sigma_\ell^2 z_\ell^2} \leq 0. \tag{2.4.8}$$

In the case $b(z) = 0$, relation (2.4.7) takes place if and only if $a(z) < 0$. To summarize: $z \in Z_\epsilon^o$ if and only if z satisfies (2.4.8) *and* $a(z) < 0$. The closure Z_ϵ of this set is exactly the set of solutions to (2.4.8). We have proved the following:

Proposition 2.4.1. Under assumptions P.1–2, relation (2.4.8) is a sufficient condition for the validity of (2.4.5). In other words, the explicit convex constraint (2.4.8) in variables z is a safe approximation of the chance constraint (2.4.5).

As an immediate corollary, we get the following useful statement:

Proposition 2.4.2. Let ζ_ℓ, $\ell = 1,...,L$, be independent random variables with distributions satisfying P.2. Then, for every deterministic vector $[z_1;...;z_L]$ and constant $\Omega \geq 0$ one has

$$\operatorname{Prob}\left\{ \sum_{\ell=1}^L z_\ell \zeta_\ell > \sum_{\ell=1}^L \max[\mu_\ell^- z_\ell, \mu_\ell^+ z_\ell] + \Omega \sqrt{\sum_{\ell=1}^L \sigma_\ell^2 z_\ell^2} \right\} \leq \exp\{-\Omega^2/2\}. \tag{2.4.9}$$

Proof. Setting

$$z_0 = -\max[\mu_\ell^- z_\ell, \mu_\ell^+ z_\ell] - \Omega \sqrt{\sum_{\ell=1}^L \sigma_\ell^2 z_\ell^2}, \;\; \epsilon = \exp\{-\Omega^2/2\},$$

we ensure the validity of (2.4.8); by Proposition 2.4.1, we have therefore

$$\operatorname{Prob}\{z_0 + \sum_{\ell=1}^L z_\ell \zeta_\ell > 0\} \leq \epsilon = \exp\{-\Omega^2/2\},$$

which, in view of the origin of z_0, is nothing but (2.4.9). $\qquad \square$

Now let us make the following observation:

(∗∗) *Consider a perturbation set $\mathcal{Z}$ given by the following conic quadratic representation:*

$$\mathcal{Z} = \left\{ \eta \in \mathbb{R}^L : \exists u \in \mathbb{R}^L : \begin{array}{c} \mu_\ell^- \leq \eta_\ell - u_\ell \leq \mu_\ell^+, \ell = 1, ..., L \\ \sqrt{\sum_{\ell=1}^{L} u_\ell^2/\sigma_\ell^2} \leq \sqrt{2\ln(1/\epsilon)} \end{array} \right\},$$

(2.4.10)

where, by definition, $a^2/0^2$ is 0 or $+\infty$ depending on whether $a = 0$ or $a \neq 0$. Then for every vector $y \in \mathbb{R}^L$ one has

$$\sum_{\ell=1}^{L} \max[\mu_\ell^+ y_\ell, \mu_\ell^- y_\ell] + \sqrt{2\ln(1/\epsilon)} \sqrt{\sum_{\ell=1}^{L} \sigma_\ell^2 y_\ell^2} = \max_{\eta \in \mathcal{Z}} \eta^T y.$$

Indeed, we have $\mathcal{Z} = \{\eta = u + v : \mu_\ell^- \leq v_\ell \leq \mu_\ell^+, \sqrt{\sum_{\ell=1}^{L} u_\ell^2/\sigma_\ell^2} \leq \sqrt{2\ln(1/\epsilon)}\}$, whence

$$\begin{aligned}
&\max_{\eta \in \mathcal{Z}} \eta^T y \\
=\ &\max_{u,v} \left\{ (u+v)^T y : \mu_\ell^- \leq v_\ell \leq \mu_\ell^+ \,\forall \ell, \sqrt{\sum_{\ell=1}^{L} u_\ell^2/\sigma_\ell^2} \leq \sqrt{2\ln(1/\epsilon)} \right\} \\
=\ &\max_{v} \left\{ v^T y : \mu_\ell^- \leq v_\ell \leq \mu_\ell^+ \,\forall \ell \right\} \\
&+ \max_{u} \left\{ u^T y : \sqrt{\sum_{\ell=1}^{L} u_\ell^2/\sigma_\ell^2} \leq \sqrt{2\ln(1/\epsilon)} \right\} \\
=\ &\sum_{\ell=1}^{L} \max[\mu_\ell^- y_\ell, \mu_\ell^+ y_\ell] + \sqrt{2\ln(1/\epsilon)} \sqrt{\sum_{\ell=1}^{L} \sigma_\ell^2 y_\ell^2},
\end{aligned}$$

□

We can summarize our findings in the following

Theorem 2.4.3. Let the random perturbations affecting (2.2.1) obey P.1–2, and consider the RC of (2.2.1) corresponding to the perturbation set (2.4.10). This RC can be equivalently represented by the explicit convex inequality

$$\begin{aligned}
&[[a^0]^T x - b^0] + \sum_{\ell=1}^{L} \max[\mu_\ell^- ([a^\ell]^T x - b^\ell), \mu_\ell^+ ([a^\ell]^T x - b^\ell)] \\
&+ \sqrt{2\ln(1/\epsilon)} \sqrt{\sum_{\ell=1}^{L} \sigma_\ell^2 ([a^\ell]^T x - b^\ell)^2} \leq 0,
\end{aligned}$$

(2.4.11)

and every feasible solution to this inequality is feasible for the chance constraint (2.2.3).

Proof. By definition, x is feasible for the RC in question if and only if

$$\underbrace{[a^0]^T x - b^0}_{z_0} + \sum_{\ell=1}^{L} \eta_\ell \underbrace{([a^\ell]^T x - b^\ell)}_{z_\ell} \leq 0 \,\forall \eta \in \mathcal{Z},$$

or, which is the same,

$$z_0 + \sup_{\eta \in \mathcal{Z}} \eta^T[z_1; ...; z_L] \leq 0.$$

By (∗∗), the latter inequality is nothing but (2.4.11), and by Proposition 2.4.1, for every solution to this inequality we have

$$\text{Prob}\{z_0 + \sum_{\ell=1}^{L} \zeta_\ell z_\ell > 0\} \leq \epsilon,$$

when the random vector ζ obeys P.1–2.   □

2.4.1 Refinements in the Case of Bounded Perturbations

In addition to assumptions P.1–2, assume that ζ_ℓ have bounded ranges:

$$\text{Prob}\{a_\ell^- \leq \zeta_\ell \leq a_\ell^+\} = 1, \ \ell = 1, ..., L, \tag{2.4.12}$$

where $-\infty < a_\ell^- \leq a_\ell^+ < \infty$ are deterministic. In this case, Theorem 2.4.3 admits the following refinement (cf. Proposition 2.3.3):

Theorem 2.4.4. Assume that random perturbations affecting (2.2.1) obey P.1–2 and (2.4.12) with $a_\ell^- \leq \mu_\ell^- \leq \mu_\ell^+ \leq a_\ell^+$ for all ℓ, and consider the RC of (2.2.1) corresponding to the perturbation set (cf. (2.4.10))

$$\mathcal{Z} = \left\{ \eta \in \mathbb{R}^L : \exists u \in \mathbb{R}^L : \begin{array}{c} \mu_\ell^- \leq \eta_\ell - u_\ell \leq \mu_\ell^+, 1 \leq \ell \leq L \\ \sqrt{\sum_{\ell=1}^{L} u_\ell^2/\sigma_\ell^2} \leq \sqrt{2\ln(1/\epsilon)} \\ a_\ell^- \leq \eta_\ell \leq a_\ell^+, \ 1 \leq \ell \leq L \end{array} \right\}, \tag{2.4.13}$$

where, by definition, $a^2/0^2$ is 0 or $+\infty$ depending on whether $a = 0$ or $a \neq 0$. This RC can be equivalently represented by the explicit system of convex inequalities

$$\begin{array}{ll} (a) & [a^\ell]^T x - b^\ell = u_\ell + v_\ell, \ \ell = 0, 1, ..., L \\ (b) & u_0 + \sum_{\ell=1}^{L} \max[a_\ell^- u_\ell, a_\ell^+ u_\ell] \leq 0 \\ (c) & v_0 + \sum_{\ell=1}^{L} \max[\mu_\ell^- v_\ell, \mu_\ell^+ v_\ell] + \sqrt{2\ln(1/\epsilon)}\sqrt{\sum_{\ell=1}^{L} \sigma_\ell^2 v_\ell^2} \leq 0, \end{array} \tag{2.4.14}$$

in variables x, u, v. Moreover, every x that can be extended to a feasible solution (x, u, v) to the latter system is feasible for the chance constraint (2.2.2).

Proof. 1^0. Let us prove that with $\Omega = 2\sqrt{\ln(1/\epsilon)}$, for every vector $z = [z_0; z_1; ...; z_L]$ the equivalence

$$(a) \quad \max_{\eta \in \mathcal{Z}}[z_0 + \sum_{\ell=1}^{L} \eta_\ell z_\ell \leq 0]$$

$$\Updownarrow$$

$$(b) \quad \exists u, v : \left\{ \begin{array}{ll} u + v = z & (b.1) \\ u_0 + \sum_{\ell=1}^{L} \max[a_\ell^- u_\ell, a_\ell^+ u_\ell] \leq 0 & (b.2) \\ v_0 + \sum_{\ell=1}^{L} \max[\mu_\ell^- v_\ell, \mu_\ell^+ v_\ell] + \Omega\sqrt{\sum_{\ell=1}^{L} \sigma_\ell^2 v_\ell^2} \leq 0 & (b.3) \end{array} \right.$$

$$\tag{2.4.15}$$

holds true. It is immediately seen that the validity of this equivalence remains intact under "shifts" $(a_\ell^\pm, \mu_\ell^\pm) \mapsto (a_\ell^\pm + c_\ell, \mu_\ell^\pm + c_\ell)$ of the coefficients, so that we can assume, w.l.o.g., that

$$\forall \ell : a_\ell^- \leq \mu_\ell^- \leq 0 \leq \mu_\ell^+ \leq a_\ell^+. \tag{2.4.16}$$

We first prove that $(2.4.15.b)$ implies $(2.4.15.a)$; so, let u, v satisfy the relations in $(2.4.15.b)$, and let $\eta \in \mathcal{Z}$. Then $\eta_\ell \in [a_\ell^-, a_\ell^+]$ for every ℓ, which combines with $(2.4.15.b.2)$ to imply that

$$u_0 + \sum_{\ell=1}^{L} \eta_\ell u_\ell \leq 0. \tag{2.4.17}$$

Besides this, due to the constraints in the definition of $\mathcal{Z}$, we have $\eta = \eta^0 + \eta^1$ with $\eta_\ell^0 \in [\mu_\ell^-, \mu_\ell^+]$ and $\sum_{\ell=1}^{L}(\eta_\ell^1)^2/\sigma_\ell^2 \leq \Omega^2$. It follows that

$$\begin{aligned}
v_0 + \sum_{\ell=1}^{L} \eta_\ell v_\ell &= v_0 + \sum_{\ell=1}^{L} \eta_\ell^0 v_\ell + \sum_{\ell=1}^{L} \eta_\ell^1 v_\ell \\
&\leq v_0 + \sum_{\ell=1}^{L} \max[\mu_\ell^- v_\ell, \mu_\ell^+ v_\ell] + \sum_{\ell=1}^{L} [\sigma_\ell v_\ell][\eta_\ell^1/\sigma_\ell] \\
&\leq v_0 + \sum_{\ell=1}^{L} \max[\mu_\ell^- v_\ell, \mu_\ell^+ v_\ell] + \Omega\sqrt{\sum_{\ell=1}^{L} \sigma_\ell^2 v_\ell^2} \leq 0
\end{aligned} \tag{2.4.18}$$

where the concluding inequality is given by $(2.4.15.b.3)$. Combining $(2.4.17)$, $(2.4.18)$, and $(2.4.15.b.1)$, we arrive at $(2.4.15.a)$.

Next we prove that $(2.4.15.a)$ implies $(2.4.15.b)$. Let

$$\begin{aligned}
\mathcal{P} &= \{\eta : a_\ell^- \leq \eta_\ell \leq a_\ell^+, 1 \leq \ell \leq L\}, \\
\mathcal{Q} &= \{\eta : \exists v : \mu_\ell^- \leq \eta_\ell - v_\ell \leq \mu_\ell^+, 1 \leq \ell \leq L, \sqrt{\sum_{\ell=1}^{L} v_\ell^2/\sigma_\ell^2} \leq \Omega\},
\end{aligned}$$

so that $\mathcal{P}$, $\mathcal{Q}$ are convex compact sets and $\mathcal{Z} = \mathcal{P} \cap \mathcal{Q}$; besides this, we clearly have $\mathcal{Z} \supset \{\eta : \mu_\ell^- \leq \eta_\ell \leq \mu_\ell^+, 1 \leq \ell \leq L\}$. Assume, first, $\mu_\ell^- < \mu_\ell^+$ for all ℓ, so that $\mathrm{int}\mathcal{P} \cap \mathrm{int}\mathcal{Q} \neq \emptyset$. In this case, by well-known results of Convex Analysis, a vector $z = [z_0; ...; z_L]$ satisfies the relation

$$z_0 + \max_{\eta \in \mathcal{P} \cap \mathcal{Q}} \eta^T [z_1; ...; z_L] \leq 0$$

if and only if there exists a decomposition $z = u + v$ with

$$u_0 + \max_{\eta \in \mathcal{P}} \eta^T[u_1; ...; u_L] \leq 0, \quad v_0 + \max_{\eta \in \mathcal{Q}} \eta^T[v_1; ...; v_L] \leq 0.$$

The first inequality clearly says that u satisfies $(2.4.15.b.2)$, while the second inequality, by $(**)$, says that v satisfies $(2.4.15.b.3)$. Thus, z satisfies $(2.4.15.b)$, as claimed.

We have proved that if z satisfies $(2.4.15.a)$ and all the inequalities $\mu_\ell^- \leq \mu_\ell^+$ are strict, then z satisfies $(2.4.15.b)$; all we need to complete the proof of $(2.4.15)$ is to show that the latter conclusion remains valid when some of the inequalities $\mu_\ell^- \leq \mu_\ell^+$ are equalities. To this end, assume that z satisfies $(2.4.15.a)$, let t be a positive integer, and let $\mu_{t,\ell}^- = \mu_\ell^- - 1/t$, $\mu_{t,\ell}^+ = \mu_\ell^+ + 1/t$, and similarly for $a_{t,\ell}^\pm$. Let

$$
\mathcal{Z}^t = \left\{ \eta \in \mathbb{R}^L : \exists u \in \mathbb{R}^L : \begin{array}{c} \mu_{t,\ell}^- \le \eta_\ell - u_\ell \le \mu_{t,\ell}^+, 1 \le \ell \le L \\ \sqrt{\sum_{\ell=1}^L u_\ell^2/\sigma_\ell^2} \le \sqrt{2\ln(1/\epsilon)} \\ a_{t,\ell}^- \le \eta_\ell \le a_{t,\ell}^+, 1 \le \ell \le L \end{array} \right\}.
$$

From the fact that z satisfies $(2.4.15.a)$, by standard compactness arguments, it follows that

$$
\delta_t := \max_{\eta \in \mathcal{Z}^t}[z_0 + \eta^T[z_1; ...; z_L]] \to 0, \ t \to \infty.
$$

Let $z_0^t = z_0 - \delta_t$, $z_\ell^t = z_\ell$, $1 \le \ell \le L$, so that

$$
\max_{\eta \in \mathcal{Z}^t}[z_0^t + \eta^T[z_1^t; ...; z_L^t]] \le 0, \ t = 1, 2, ...
$$

According to the proved version of the implication $(2.4.15.a) \Rightarrow (2.4.15.b)$, the system $\mathcal{S}^t$ of constraints on variables u, v obtained from inequalities in $(2.4.15.b)$ when replacing the data $\mu_\ell^\pm$, $a_\ell^\pm$ with $\mu_{t,\ell}^\pm$, $a_{t,\ell}^\pm$ and z with z^t, admits a solution u^t, v^t. From $(2.4.16)$ it follows that if u, v solve $\mathcal{S}^t$ and u', v' are such that $u_0' = u_0$, $v_0' = v_0$, $u' + v' = u + v$ and the entries u_ℓ', v_ℓ', $1 \le \ell \le L$ are of the same signs and of smaller magnitudes then the corresponding entries in u, v, then u', v' solve $\mathcal{S}^t$ as well. It follows that the above u^t, v^t can be chosen uniformly bounded.

> Indeed, if for certain $\ell \ge 1$ we have $u_\ell^t > |z_\ell^t|$, then, replacing ℓ-th coordinates in u^t, v^t with $|z_\ell^t|$ and $z_\ell^t - |z_\ell^t|$, respectively, and keeping the remaining entries in u^t, v^t intact, we get a new feasible solution of $\mathcal{S}^t$ with ℓ-th entries in u, v bounded in magnitude by $2|z_\ell^t|$. Similar correction is possible when $u_\ell^t \le -|z_\ell^t|$; applying these corrections, we can ensure that $|u_\ell^t|, |v_\ell^t|$ do not exceed $2|z_\ell^t|$ for every $\ell \ge 1$. And of course under this normalization, a solution to $\mathcal{S}^t$ does not need entries u_0^t, v_0^t with too large magnitudes.

With u^t, v^t uniformly bounded, passing to a subsequence, we can assume that $u^t \to u$ and $v^t \to v$ as $t \to \infty$; since $z^t = u^t + v^t \to z$ as $t \to \infty$ due to $d_t \to 0$, $t \to \infty$, and the "perturbed" data $a_{t,\ell}^\pm$, $\mu_{t,\ell}^\pm$ converge to the "true" data $a_\ell^\pm$, $\mu_\ell^\pm$, u, v certify that z satisfies $(2.4.15.b)$.

2^0. By $(2.4.15)$, the system of constraints in $(2.4.14)$ equivalently represents the fact that x is robust feasible for the uncertain inequality $(2.2.1)$, the perturbation set being $(2.4.13)$. All we need to complete the proof is to demonstrate that whenever x can be extended to a feasible solution (x, u, v) of $(2.4.14)$, x is feasible for the chance constraint $(2.2.2)$. This is immediate: setting $z_\ell = [a^\ell]^T x - b^\ell$, $\ell = 0, 1, ..., L$, and invoking $(2.4.14.a)$, we get

$$
z_0 + \sum_{\ell=1}^L \zeta_\ell z_\ell = \underbrace{u_0 + \sum_{\ell=1}^L \zeta_\ell u_\ell}_{A} + \underbrace{v_0 + \sum_{\ell=1}^L \zeta_\ell v_\ell}_{B}.
$$

Since ζ takes values from the box $\{a_\ell^- \le \zeta_\ell \le a_\ell^+, 1 \le \ell \le L\}$ with probability 1, we have $A \le 0$ with probability 1 by $(2.4.14.b)$. Applying Proposition 2.4.1 to v

in the role of z and invoking (2.4.14.c), we conclude that $\text{Prob}\{B > 0\} \leq \epsilon$. Thus, the quantity $\text{Prob}\{A + B > 0\}$, which is exactly the probability for x to violate the chance constraint (2.2.2), is $\leq \epsilon$. $\square$

2.4.2 Examples

In order to make the constructions presented in Theorems 2.4.3, 2.4.4 useful, we should understand how to "translate" a priori partial knowledge on the distributions of ζ_ℓ's in the perturbation vector ζ into concrete values of the parameters $\mu_\ell^\pm, \sigma_\ell$ in P.2. We are about to present several instructive examples of such a translation (most of them originating from [83]).

2.4.2.1 Note on normalization

To avoid messy formulas, we subject the components of ζ_ℓ to suitable normalization. Observe that what we are interested in, is a randomly perturbed inequality

$$z_0 + \sum_{\ell=1}^{L} z_\ell \zeta_\ell \leq 0 \tag{2.4.19}$$

with random $\zeta = [\zeta_1; ...; \zeta_L]$ satisfying P.1–2, along with specific bounds, given by Proposition 2.4.1, for this inequality to be violated. Now assume that we subject every component ζ_ℓ to a deterministic affine transformation, setting

$$\zeta_\ell = \alpha_\ell + \beta_\ell \widehat{\zeta}_\ell \tag{2.4.20}$$

with deterministic $\beta_\ell > 0, \alpha_\ell$. With this substitution, the left hand side in (2.4.19) becomes

$$\widehat{z}_0 + \sum_{\ell=1}^{L} \widehat{z}_\ell \widehat{\zeta}_\ell, \ \widehat{z}_0 = z_0 + \sum_{\ell=1}^{L} \alpha_\ell z_\ell, \ \widehat{z}_\ell = \beta_\ell z_\ell, \ 1 \leq \ell \leq L, \tag{2.4.21}$$

and, of course,

$$\text{Prob}\left\{\zeta : z_0 + \sum_{\ell=1}^{L} z_\ell \zeta_\ell > 0\right\} = \text{Prob}\left\{\widehat{\zeta} : \widehat{z}_0 + \sum_{\ell=1}^{L} \widehat{z}_\ell \widehat{\zeta}_\ell > 0\right\}. \tag{2.4.22}$$

Now, if ζ satisfies P.1–2 with certain parameters $\{\mu_\ell^\pm, \sigma_\ell\}$, then $\widehat{\zeta}$ satisfies the same assumptions with the parameters $\{\widehat{\mu}_\ell^\pm, \widehat{\sigma}_\ell\}$ given by

$$\mu_\ell^\pm = \alpha_\ell + \beta_\ell \widehat{\mu}_\ell^\pm, \ \sigma_\ell = \beta_\ell \widehat{\sigma}_\ell. \tag{2.4.23}$$

It follows that *the machinery referred to by Propositions 2.4.1, 2.4.2 and Theorems 2.4.3, 2.4.4 "respects" substitution (2.4.20): the conclusions about probability (2.4.22) that we can make with this machinery when working with ̂ quantities are identical to those we can make when working with the original quantities.* For example, the key condition (2.4.8) in the original quantities remains exactly the same condition in $\widehat{}$ quantities, since with the correspondences (2.4.21), (2.4.23) we have

$$z_0 + \sum_{\ell=1}^{L} \max[\mu_\ell^+ z_\ell, \mu_\ell^- z_\ell] \equiv \widehat{z}_0 + \sum_{\ell=1}^{L} \max[\widehat{\mu}_\ell^+ \widehat{z}_\ell, \widehat{\mu}_\ell^- \widehat{z}_\ell],$$

$$\sum_{\ell=1}^{L} \sigma_\ell^2 z_\ell^2 \equiv \sum_{\ell=1}^{L} \widehat{\sigma}_\ell^2 \widehat{z}_\ell^2.$$

The bottom line is as follows: *we lose nothing when passing from the original random variables ζ_ℓ to their scaled versions $\widehat{\zeta}_\ell$.* Below, we mainly work with variables ζ_ℓ varying in given finite ranges $[a_\ell, b_\ell]$, $a_\ell < b_\ell$. It is convenient to scale ζ_ℓ in such a way that the induced ranges for $\widehat{\zeta}_\ell$ variables are $[-1, 1]$. We always assume that this scaling is carried out in advance, so that the ranges of the variables ζ_ℓ themselves, when finite, are the segment $[-1, 1]$.

2.4.2.2 Gaussian perturbations

Example 2.4.5. Assume that $\zeta_1, ..., \zeta_L$ are independent Gaussian random variables with partially known expectations μ_ℓ and variances s_ℓ^2; specifically, all we know is that $\mu_\ell \in [\mu_\ell^-, \mu_\ell^+]$ and $s_\ell^2 \leq \sigma_\ell^2$, with known $\mu_\ell^\pm$ and σ_ℓ, $1 \leq \ell \leq L$. For $\mu_- \leq \mu \leq \mu_+$ and $\xi \sim \mathcal{N}(\mu, \sigma^2)$ we have

$$\begin{aligned}
\mathbf{E}\{\exp\{t\xi\}\} &= \tfrac{1}{\sqrt{2\pi}\sigma} \int \exp\{ts\} \exp\{-(s-\mu)^2/(2\sigma^2)\} ds \\
&= \exp\{\mu t\} \tfrac{1}{\sqrt{2\pi}\sigma} \int \exp\{tr\} \exp\{-r^2/(2\sigma^2)\} dr && [r = s - \mu] \\
&= \exp\{\mu t\} \tfrac{1}{\sqrt{2\pi}\sigma} \int \exp\{t^2\sigma^2/2\} \exp\{-(r - t\sigma^2)^2/(2\sigma^2)\} dr \\
&= \exp\{t\mu + t^2\sigma^2/2\} \leq \exp\{\max[\mu^- t, \mu^+ t] + t^2\sigma^2/2\}.
\end{aligned}$$

We see that ζ satisfies P.1–2 with the parameters $\mu_\ell^\pm$, σ_ℓ, $\ell = 1, ..., L$. The safe tractable approximation of (2.2.3) as given by Theorem 2.4.3 is

$$\begin{aligned}
&[[a^0]^T x - b^0] + \sum_{\ell=1}^{L} \max[\mu_\ell^-([a^\ell]^T x - b^\ell), \mu_\ell^+([a^\ell]^T x - b^\ell)] \\
&+ \sqrt{2\ln(1/\epsilon)} \sqrt{\sum_{\ell=1}^{L} \sigma_\ell^2([a^\ell]^T x - b^\ell)^2} \leq 0.
\end{aligned} \tag{2.4.24}$$

When $\mu_\ell^\pm = 0$, $\sigma_\ell = 1$, $\ell = 1, ..., L$, this is nothing but the ball RC (2.3.3) of (2.2.1) with $\Omega = \sqrt{2\ln(1/\epsilon)}$.

Note that in the simple case in question the ambiguous chance constraint (2.2.3) needs no approximation: when $\epsilon \leq 1/2$, it is *exactly equivalent* to the convex constraint

$$\begin{aligned}
&[[a^0]^T x - b^0] + \sum_{\ell=1}^{L} \max[\mu_\ell^-([a^\ell]^T x - b^\ell), \mu_\ell^+([a^\ell]^T x - b^\ell)] \\
&+ \mathrm{ErfInv}(\epsilon) \sqrt{\sum_{\ell=1}^{L} \sigma_\ell^2([a^\ell]^T x - b^\ell)^2} \leq 0,
\end{aligned} \tag{2.4.25}$$

where ErfInv is the inverse error function (2.3.22). The same remains true when we assume that ζ is Gaussian, and all we know about the expectation μ and the covariance matrix Σ of ζ is that $\mu^- \leq \mu \leq \mu^+$ and $\Sigma \preceq \mathrm{Diag}\{\sigma_1^2, ..., \sigma_L^2\}$. Note that (2.4.25) is of exactly the same structure as (2.4.24), with the only difference in the factor at the $\sqrt{\cdot}$. In (2.4.24) this factor is $\Omega(\epsilon) = \sqrt{2\ln(1/\epsilon)}$, while in (2.4.25) this factor is $\mathrm{ErfInv}(\epsilon) < \Omega(\epsilon)$. The difference, however, is not that huge, as can be seen from the following comparison:

ϵ	10^{-1}	10^{-2}	10^{-3}	10^{-4}	10^{-5}	10^{-6}	$\to +0$
$\dfrac{\mathrm{ErfInv}(\epsilon)}{\Omega(\epsilon)}$	0.597	0.767	0.831	0.867	0.889	0.904	$\to 1$

2.4.2.3 Bounded perturbations

Example 2.4.6. Assume that all we know about probability distribution P is that it is supported on $[-1, 1]$. Then

$$\int \exp\{st\}dP(s) \leq \int \exp\{|t|\}dP(s) = \exp\{|t|\},$$

so that P satisfies P.2 with $\mu^- = -1$, $\mu^+ = 1$, $\sigma = 0$.

In particular, if all we know about random perturbation ζ affecting (2.2.1) is that ζ_ℓ are independent and vary in $[-1, 1]$, that is, we take $\mu_\ell^- = -1$, $\mu_\ell^+ = 1$, $\sigma_\ell = 0$ for all ℓ, then the RC (2.4.11) becomes

$$[a^0]^T x - b^0 + \sum_{\ell=1}^{L} |[a^\ell]^T x - b^\ell| \leq 0;$$

this is nothing but the box RC (2.3.15) which gives 100% immunization against uncertainty. Note that since with our a priori information, ζ can be an arbitrary deterministic perturbation from the unit box, this RC is the best we can build under the circumstances.

2.4.2.4 Bounded unimodal perturbations

Example 2.4.7. Assume that all we know about probability distribution P is that it is supported on $[-1, 1]$ and is *unimodal w.r.t. 0*, that is, possesses a density $p(s)$ that is unimodal w.r.t. 0 (i.e., is nondecreasing when $s < 0$ and is nonincreasing when $s > 0$). In this case, assuming $t \geq 0$, we have

$$\int \exp\{ts\}dP(s) = \int_{-1}^{1} \exp\{ts\}p(s)ds.$$

It is easily seen that the latter functional of $p(\cdot)$, restricted to unimodal w.r.t. 0 densities, attains its maximum when $p(s)$ vanishes on $[-1, 0]$ and is $\equiv 1$ on $[0, 1]$, so that

$$t \geq 0 \Rightarrow \int \exp\{ts\}dP(s) \leq f(t) = \int_{0}^{1} \exp\{ts\}ds = \frac{\exp\{t\} - 1}{t}. \tag{2.4.26}$$

Indeed, assume first that P has a smooth density $p(s)$ that vanishes outside of $[-1, 1]$, is nondecreasing when $s < 0$ and is nonincreasing when $s > 0$, and let $F(s) = \int_0^s \exp\{tr\}dr$. We have

$$\int \exp\{ts\}dP(s) = \int \exp\{ts\}p(s)ds = \int F'(s)p(s)ds = \int F(s)(-p'(s))ds$$
$$= \int (F(s)/s)(-sp'(s))ds,$$

where $F(0)/0 := \lim\limits_{s \to +0} F(s)/s = 1$. Since p is unimodal w.r.t. 0 and vanishes outside $[-1, 1]$, the function $q(s) = -sp'(s)$ is nonnegative and also vanishes outside of $[-1, 1]$; besides this, $\int q(s)ds = \int(-sp'(s))ds = \int p(s)ds = 1$, that is, $q(\cdot)$ is the density of a probability distribution supported on $[-1, 1]$. In addition, the function $F(s)/s$ is clearly nondecreasing in s (recall that $t \geq 0$). Therefore

$$\int(F(s)/s)(-sp'(s))ds = \int_{-1}^{1}(F(s)/s)q(s)ds \leq F(1) = \int_{0}^{1}\exp\{tr\}dr = \frac{\exp\{t\} - 1}{t},$$

as required in (2.4.26).

We have proved that (2.4.26) holds true when the density of P is a smooth unimodal w.r.t. 0 function vanishing outside of $[-1, 1]$. Now, every probability density $p(s)$ that is unimodal w.r.t. 0 and supported on $[-1, 1]$ can be approximated by a sequence of smooth unimodal w.r.t. 0 and vanishing outside of $[-1, 1]$ densities $p_i(\cdot)$ in the sense that $\int \phi(s)p_i(s)ds \to \int \phi(s)p(s)ds$, $i \to \infty$, for every continuous function $\phi(\cdot)$. Specifying $\phi(s) = \exp\{ts\}$) and noting that, as we have already seen, $\int \exp\{ts\}p_i(s) \leq t^{-1}(\exp\{t\} - 1)$, we conclude that (2.4.26) is valid for every unimodal w.r.t. 0 distribution P supported on $[-1, 1]$.

By symmetry, we derive from (2.4.26) that

$$\int \exp\{ts\}dP(s) \leq f(t) \equiv \frac{\exp\{|t|\} - 1}{|t|} \ \forall t.$$

It follows that

$$\ln\left(\int \exp\{ts\}dP(s)\right) \leq h(|t|), \ h(t) = \ln f(t).$$

Now, direct computation shows that $h(0) = 0$, $h'(0) = \frac{1}{2}$ and $h''(0) = \frac{1}{12}$. A natural guess is that

$$h(t) \leq h(0) + h'(0)t + \frac{1}{2}h''(0)t^2 \equiv \frac{1}{2}t + \frac{1}{24}t^2$$

for all $t \geq 0$. The guess is indeed true:

$$h(t) \ ? \leq? \ \tfrac{1}{2}t + \tfrac{1}{24}t^2 \ \forall t \geq 0$$

$$\Leftrightarrow \ \int_{0}^{1}\exp\{ts\}ds \ ? \leq? \ \exp\{\tfrac{1}{2}t + \tfrac{1}{24}t^2\} \ \forall t > 0$$

$$\Leftrightarrow \ \int_{-1/2}^{1/2}\exp\{tr\}dr \ ? \leq? \ \exp\{\tfrac{1}{24}t^2\}$$

$$\Leftrightarrow \ \sum_{k=0}^{\infty}\frac{t^{2k}}{(2k+1)!2^{2k}} \ ? \leq? \ \sum_{k=0}^{\infty}\frac{t^{2k}}{24^k k!} \qquad (*)$$

where the concluding reformulation is given by expanding $\exp\{ts\}$ and $\exp\{\frac{1}{24}t^2\}$ into a Taylor series. It is immediately seen that the right hand side series in $(*)$ dominates term-wise the left hand side series, so that the last " $? \leq?$ " in the chain is indeed "$\leq$."

We conclude that P satisfies P.2 with $\mu^- = -\frac{1}{2}$, $\mu^+ = \frac{1}{2}$ and $\sigma^2 = \frac{1}{12}$.

2.4.2.5 Bounded symmetric unimodal perturbations

Example 2.4.8. Assume that all we know about a probability distribution P is that it is supported on $[-1, 1]$, symmetric w.r.t. 0 and unimodal w.r.t. 0. In this case, we have

$$\int \exp\{ts\}dP(s) = \int_{0}^{1} 2\cosh(ts)p(s)ds;$$

here again it is easy to see that the latter functional attains its maximum on the set of unimodal symmetric probability densities on $[-1,1]$ when $p(s) \equiv \frac{1}{2}$, $-1 \le s \le 1$, so that

$$\int \exp\{ts\}dP(s) \le \int_0^1 \cosh(ts)ds = f(t) := \frac{\sinh(t)}{t}. \tag{2.4.27}$$

Indeed, as in the proof of (2.4.26), it suffices to prove (2.4.27) for the case when $p(s)$ is a smooth even density, nonincreasing when $s > 0$ and vanishing when $s \ge 1$. Setting $F(s) = \int_0^s \cosh(ts)ds$ and $q(s) = -2sp'(s)$, we have, as in the proof of (2.4.26), that $\int_0^1 2\cosh(ts)p(s)ds = \int_0^1 (F(s)/s)q(s)ds$, the function $F(s)/s$ is nondecreasing and $q(s)$ is a probability density on $[0,1]$, whence $\int_0^1 (F(s)/s)q(s)ds \le F(1) = f(t)$.

Direct computation shows that the function $h(t) = \ln f(t)$ satisfies $h(0) = 0$, $h'(0) = 0$, $h''(t) = \frac{1}{3}$, and a natural guess again is that

$$h(t) \le h(0) + h'(0)t + \frac{1}{2}h''(0)t^2 \equiv \frac{1}{6}t^2$$

for all t, which again is true:

$$h(t) \ ? \le? \ \tfrac{1}{6}t^2 \ \forall t$$
$$\Leftrightarrow \quad \tfrac{1}{2}\int_{-1}^1 \exp\{ts\}ds \ ? \le? \ \exp\{\tfrac{1}{6}t^2\} \ \forall t$$
$$\Leftrightarrow \quad \sum_{k=0}^\infty \frac{t^{2k}}{(2k+1)!} \ ? \le? \ \sum_{k=0}^\infty \frac{t^{2k}}{6^k k!} \quad (*)$$

and $(*)$ is indeed true by the same argument as in Example 2.4.7. We conclude that P satisfies P.2 with $\mu^- = \mu^+ = 0, \sigma^2 = \frac{1}{3}$.

2.4.2.6 Range and expectation information

Example 2.4.9. Assume that all we know about a probability distribution P is that it is supported on $[-1,1]$ and that the expectation of the associated random variable belongs to a given segment $[\mu^-, \mu^+]$; we may of course assume that $-1 \le \mu^- \le \mu^+ \le 1$. Let μ be the true mean of P. Given t, consider the function

$$\phi(s) = \exp\{ts\} - \sinh(t)s, \ -1 \le s \le 1.$$

This function is convex on $[-1,1]$ and therefore attains it maximum over this segment at an endpoint of it. Since $\phi(1) = \phi(-1) = \cosh(t)$, we have

$$\int \exp\{ts\}dP(s) = \int \phi(s)dP(s) + \mu\sinh(t) \le \max_{-1 \le s \le 1} \phi(s) + \mu\sinh(t)$$
$$= \cosh(t) + \mu\sinh(t).$$

Thus,

$$\int \exp\{ts\}dP(s) \le f_\mu(t) := \cosh(t) + \mu\sinh(t). \tag{2.4.28}$$

Note that the bound (2.4.28) is the best possible under the circumstances — the inequality (2.4.28) becomes an equality when P is a 2-point distribution assigning mass $(1 + \mu)/2$

to the point $s = 1$ and mass $(1 - \mu)/2$ to the point $s = -1$; this distribution indeed is supported on $[-1, 1]$ and has expectation μ.

Setting $h_\mu(t) = \ln f_\mu(t)$, we have $h_\mu(0) = 0$, $h'_\mu(t) = \frac{\sinh(t) + \mu \cosh(t)}{\cosh(t) + \mu \sinh(t)}$, whence $h'_\mu(0) = \mu$, and $h''_\mu(t) = 1 - \left(\frac{\sinh(t) + \mu \cosh(t)}{\cosh(t) + \mu \sinh(t)} \right)^2$, whence $h''_\mu(t) \leq 1$ for all t. We conclude that

(!) When $\mu^- \leq \mu \leq \mu^+$, we have $h_\mu(t) \leq \max[\mu^- t, \mu^+ t] + \frac{1}{2} t^2 \ \forall t$.

Now let us set

$$\Sigma_{(1)}(\mu^-, \mu^+) = \min \left\{ c \geq 0 : h_\mu(t) \leq \max[\mu^- t, \mu^+ t] + \frac{c^2}{2} t^2 \ \forall (\mu \in [\mu^-, \mu^+], t) \right\}. \quad (2.4.29)$$

The graph of $\Sigma_{(1)}(\mu^-, \mu^+)$, $-1 \leq \mu^- \leq \mu^+ \leq 1$, is plotted on figure 2.2.(a). By (!), $\Sigma_{(1)}(\mu-, \mu^+)$ is well defined and ≤ 1, and recalling (2.4.28),

$$\ln \left(\int \exp\{ts\} dP(s) \right) \leq \max[\mu^- t, \mu^+ t] + \frac{\Sigma_{(1)}^2(\mu^-, \mu^+)}{2} t^2 \ \forall t,$$

so that P satisfies P.2 with the parameters $\mu^\pm, \sigma = \Sigma_{(1)}(\mu^-, \mu^+) \leq 1$.

Remark 2.4.10. We have proved Proposition 2.3.1. Indeed, under the premise of this Proposition, Example 2.4.9 (where we should set $\mu^\pm = 0$) says that the random variables ζ_ℓ satisfy P.1–2 with $\mu_\ell^\pm = 0$, $\sigma_\ell = 1$, $\ell = 1, ..., L$, which makes Proposition 2.3.1 a particular case of Proposition 2.4.2.

2.4.3 More Examples

Example 2.4.9 is highly instructive, and we can proceed in the direction outlined in this example, utilizing more and more specific information on the distributions of ζ_ℓ.

Before passing to further examples of this type, let us clarify the main ingredient of the reasoning used in Example 2.4.9, that is, the fashion in which we have established the key inequality (2.4.28). Similar reasoning can be used in all examples to follow. The essence of the matter is: given a function $w_t(s)$ on the axis (which in Example 2.4.9 is $\exp\{ts\}$) and "moment-type" information

$$\int g_j(s) dP(s) \begin{cases} = \mu_j, & j \in J_= \\ \leq \mu_j, & j \in J_\leq \end{cases},$$

on a probability distribution P supported on a segment Δ of the axis (in Example 2.4.9, $J_= = \{1\}$, $J_\leq = \emptyset$, $\mu_1 = \mu$, $g_1(s) \equiv s$ and $\Delta = [-1, 1]$), we want to bound from above the quantity $\int w_t(s) dP(s)$. The scheme we have used is a kind of Lagrange relaxation: we observe that the information on P implies that whenever $\lambda_j, j \in J_= \cup J_\leq$ are such that $\lambda_j \geq 0$ for $j \in J_\leq$, we have

$$\begin{aligned} \int w_t(s) dP(s) &= \int [w_t(s) - \sum_j \lambda_j g_j(s)] dP(s) + \sum_j \lambda_j \int g_j(s) dP(s) \\ &\leq \max_{s \in \Delta} [w_t(s) - \sum_j \lambda_j g_j(s)] + \sum_j \lambda_j \mu_j, \end{aligned} \quad (2.4.30)$$

where the concluding inequality is given by the fact that P is a probability distribution supported on Δ and consistent with our a priori moment information.

When proving inequalities like (2.4.28) in the examples to follow, we use the outlined bounding scheme with properly chosen λ_j (in the case of (2.4.28), $\lambda_1 = \sinh(t)$). In fact, the λ's to be used are given by minimization of the resulting bound in λ, but we do not bother to justify this fact directly; we simply show that the resulting bound is unimprovable, since it becomes equality on certain distribution P consistent with our a priori information.

2.4.3.1 Information on range, mean, and variance

Example 2.4.11. Assume that all we know about a probability distribution P is that it is supported on $[-1, 1]$ with $\mathrm{Mean}[P] \in [\mu^-, \mu^+]$ and $\mathrm{Var}[P] \leq \nu^2$, where ν and $\mu^\pm$ are known in advance. W.l.o.g., we may focus on the case $|\mu^\pm| \leq \nu \leq 1$. We claim that

(i) With $\mu = \mathrm{Mean}[P]$ one has

$$\int \exp\{ts\}dP(s) \leq f_{\mu,\nu}(t) \equiv \begin{cases} \dfrac{(1-\mu)^2 \exp\{t\frac{\mu-\nu^2}{1-\mu}\}+(\nu^2-\mu^2)\exp\{t\}}{1-2\mu+\nu^2}, & t \geq 0 \\[2ex] \dfrac{(1+\mu)^2 \exp\{t\frac{\mu+\nu^2}{1+\mu}\}+(\nu^2-\mu^2)\exp\{-t\}}{1+2\mu+\nu^2}, & t \leq 0 \end{cases} \qquad (2.4.31)$$

and the bound (2.4.31) is the best possible under the circumstances: when $t > 0$, it is achieved at the 2-point distribution assigning to the points $\bar{s} = \frac{\mu-\nu^2}{1-\mu}$ and 1 the masses $(1-\mu)^2/(1-2\mu+\nu^2)$, $(\nu^2-\mu^2)/(1-2\mu+\nu^2)$, respectively. This distribution P^μ_+ is compatible with our a priori information: $\mathrm{Mean}[P^\mu_+] = \mu$, $\mathrm{Var}[P^\mu_+] = \nu^2$. When $t < 0$, the bound (2.4.31) is achieved when P is the "reflection" P^μ_- of $P^{(-\mu)}_+$ w.r.t. 0.

(ii) The function $h_{\mu,\nu}(t) = \ln f_{\mu,\nu}(t)$ satisfies $h_{\mu,\nu}(t) \leq \mu t + \frac{1}{2}t^2$ for all t. As a result, The function

$$\Sigma_{(2)}(\mu^-, \mu^+, \nu) = \min\left\{ c \geq 0: \begin{array}{c} h_{\mu,\nu}(t) \leq \max[\mu^- t, \mu^+ t] + \frac{c^2}{2}t^2 \\ \forall(\mu \in [\mu^-, \mu^+], t) \end{array} \right\} \quad (2.4.32)$$

is well defined and is ≤ 1, and P satisfies P.2 with the parameters $\mu^\pm, \sigma = \Sigma_{(2)}(\mu^-, \mu^+, \nu)$.

Note that the function $\Sigma_{(1)}(\mu^-, \mu^+)$ is nothing but $\Sigma_{(2)}(\mu^-, \mu^+, 1)$. The graph of $\Sigma_{(2)}(\mu, \mu, \nu)$ is plotted on figure 2.2.(b).

Indeed, to prove (i), it suffices, by continuity, to prove (2.4.31) in the case when $|\mu| < \nu \leq 1$; by symmetry, we may assume that $t > 0$. Let us set $\bar{s} = \frac{\mu-\nu^2}{1-\mu}$, so that $-1 < \bar{s} < 1$ due to $|\mu| < \nu < 1$. Consider the function

$$\phi(s) = \exp\{ts\} - \lambda_1 s - \lambda_2 s^2$$

where λ_1 and λ_2 are chosen in such a way that

$$\phi(\bar{s}) = \phi(1), \ \phi'(\bar{s}) = 0,$$

that is,

$$\lambda_1 = t\exp\{t\bar{s}\} - 2\lambda_2\bar{s}, \ \lambda_2 = \frac{\exp\{t\bar{s}\}[\exp\{t(1-\bar{s})\}-1-t(1-\bar{s})]}{(1-\bar{s})^2}$$

Observe that by construction $\lambda_2 \geq 0$. We claim that $\phi(s) \leq \phi(1)$ for $-1 \leq s \leq 1$, so that

$$\int \exp\{ts\}dP(s) = \int \phi(s)dP(s) + \int [\lambda_1 s + \lambda_2 s^2]dP(s) \leq \phi(1) + \lambda_1\mu + \lambda_2\nu^2$$

(cf. (2.4.30)), and the resulting bound, after substituting values of $\bar{s}$, λ_1, λ_2, becomes (2.4.31).

It remains to justify the claim that $\phi(s) \leq \phi(1)$ on $[-1,1]$. It is immediately seen that $\phi''(\bar{s}) < 0$. Thus, when s increases from $\bar{s}$ to 1, the function $\phi(s)$ first decreases, starting with the value $\phi(\bar{s}) = \phi(1)$; what happens next, we do not know exactly, except for the fact that when s reaches the value 1, ϕ recovers its initial value $\phi(\bar{s}) = \phi(1)$. It follows that $\phi'(s)$ has a zero in the open interval $(\bar{s}, 1)$. Further, assuming $\max_{\bar{s} \leq s \leq 1} \phi(s) > \phi(\bar{s})$, the function $\phi'(s)$ possesses at least 2 distinct zeros on $(\bar{s}, 1)$; besides this, $\phi'(\bar{s}) = 0$ by construction, which gives us at least 3 distinct zeros for $\phi'(s)$. But $\phi'(s)$ is a convex function of s; possessing at least 3 distinct zeros, it must be identically equal to 0 on a nontrivial segment, which definitely is not the case. Thus, $\phi(s) \leq \phi(\bar{s}) = \phi(1)$ when $\bar{s} \leq s \leq 1$. To prove that the same inequality holds true when $-1 \leq s \leq \bar{s}$, observe that as s decreases from $\bar{s}$ to -1, $\phi(s)$ first decreases (due to $\phi'(\bar{s}) = 0$, $\phi''(\bar{s}) < 0$). What happens next, we do not know exactly, but from what we have just said it follows that if $\phi(s) > \phi(\bar{s})$ somewhere on $[-1, \bar{s}]$, then $\phi'(s)$ has a zero on $(-1, \bar{s})$; with zeros at $\bar{s}$ (by construction) and somewhere in $(\bar{s}, 1)$ (we have seen that this is the case), this gives at least 3 distinct zeros of ϕ', which, as we just have explained, is impossible. Thus, $\phi(s) \leq \phi(\bar{s}) = \phi(1)$ on the entire segment $[-1, 1]$, as claimed.

To verify (ii), note that, as we have already seen, $f_{\mu,\nu}(t)$ is the maximum of the quantities $\int \exp\{ts\}dP(s)$ taken over all probability distributions P supported on $[-1, 1]$ and such that $\text{Mean}[P] = \mu$, $\text{Var}[P] \leq \nu^2$; but the latter maximum clearly is nondecreasing in ν. As about $f_{\mu,1}(t)$, this, of course, is nothing but the function $f_\mu(t)$ from Example 2.4.9; as we have seen in this Example, $\ln f_\mu(t) \leq \mu t + \frac{1}{2}t^2$, so that the same upper bound is valid for $h_{\mu,\nu}(t)$. $\square$

2.4.3.2 Range, symmetry, and variance

Example 2.4.12. Assume that all we know about a probability distribution P is that it is supported on $[-1, 1]$, is symmetric w.r.t. 0 and is such that $\text{Var}[P] \leq \nu^2$, $0 \leq \nu \leq 1$, with known ν. We claim that

(i) One has

$$\int \exp\{ts\}dP(s) \leq f(t) \equiv \nu^2 \cosh(t) + 1 - \nu^2, \tag{2.4.33}$$

and the bound is the best possible under the circumstances: it becomes precise when P is the 3-point distribution assigning masses $\nu^2/2$ to the points ± 1 and the mass $1 - \nu^2$ to the point 0.

(ii) The function $h(t) = \ln f(t)$ is convex, even and twice continuously differentiable, with the second derivative bounded by 1 on the entire real axis, and $h(0) = 0$, $h'(0) = 0$. As a result,

The function

$$\Sigma_{(3)}(\nu) \equiv \min_{c} \left\{ c \geq 0 : \frac{c^2}{2}t^2 \geq h(t) := \ln\left(\nu^2\cosh(t) + 1 - \nu^2\right) \ \forall t \right\} \leq 1$$

(2.4.34)

is well defined on $0 \leq \nu \leq 1$ and P satisfies P.2 with $\mu^\pm = 0$, $\sigma = \Sigma_{(3)}(\nu)$.

The graph of $\Sigma_{(3)}(\cdot)$ is plotted on figure 2.2.(c). The proof of our claim is left as Exercise 2.1.

2.4.3.3 Range, symmetry, unimodality, and variance

Example 2.4.13. Assume that all we know about a probability distribution P is that P is supported on $[-1, 1]$, is symmetric and unimodal w.r.t. 0, and, in addition, $\mathrm{Var}[P] \leq \nu^2 \leq 1/3$. (The upper bound on ν is natural — it can be seen that is implied by the other assumptions on P.) We claim that

(i) One has

$$\int \exp\{ts\}dP(s) \leq 1 - 3\nu^2 + 3\nu^2 \frac{\sinh(t)}{t},$$

(2.4.35)

and this bound is the best possible under the circumstances. (To see that the bound cannot be improved, look what happens when the density of P is equal to $3\nu^2/2$ everywhere on $[-1, 1]$ except for the small neighborhood $[-\epsilon, \epsilon]$ of the origin, where the density is equal to $3\nu^2/2 + (1 - 3\nu^2)/(2\epsilon)$.)

(ii) The function $h(t) = \ln\left(1 - 3\nu^2 + 3\nu^2 \frac{\sinh(t)}{t}\right)$ is even and smooth with the second derivative bounded on the entire real axis by 1, so that the function

$$\Sigma_{(4)}(\nu) = \min\left\{c \geq 0 : \ln\left(1 - 3\nu^2 + 3\nu^2 \frac{\sinh(t)}{t}\right) \leq \frac{c^2}{2}t^2 \ \forall t\right\}, \ 0 \leq \nu \leq \sqrt{1/3}, \ \ (2.4.36)$$

is well defined and is ≤ 1. As a result, P satisfies P.2 with the parameters $\mu^\pm = 0, \sigma = \Sigma_{(4)}(\nu)$.

To prove (i), it suffices to verify (2.4.35) in the case when the density $p(s)$ of P is smooth, even, unimodal w.r.t. 0 and vanishes outside $[-1, 1]$ (cf. the proof of (2.4.26)). Besides this, by continuity we may assume that $t \neq 0$. We have

$$\int \exp\{ts\}dP(s) = \int_{-1}^{1} \exp\{ts\}p(s)ds = \int_{0}^{1} \cosh(ts)(2p(s))ds.$$

(2.4.37)

Now let λ be such that the function

$$\phi(s) = \frac{\sinh(ts)}{ts} - \lambda s^2$$

satisfies $\phi(0) = \phi(1)$, that is,

$$\lambda = \frac{\sinh(t) - t}{t}.$$

Setting $F(s) = \int_{0}^{s} \cosh(tr)dr = \sinh(ts)/t$ and $q(s) = -2sp'(s)$, and following Example 2.4.8, we observe that $q(s)$, $0 \leq s \leq 1$, is a probability density and

$$\int_{0}^{1} \cosh(ts)(2p(s))ds = \int_{0}^{1} (F(s)/s)(-2sp'(s))ds = \int_{0}^{1} (F(s)/s)q(s)ds.$$

Besides this, we have

$$\int_0^1 s^2 q(s)ds = \int_0^1 (-2s^3)p'(s)ds = \int_0^1 (6s^2)p(s)ds = 3\int_{-1}^1 s^2 p(s)ds \leq 3\nu^2.$$

Observe that since q is a probability density on $[0,1]$, the equalities in the latter chain imply that $3\int_{-1}^1 s^2 p(s)ds = \int_0^1 s^2 q(s) \leq 1$, which justifies the upper bound $1/3$ on $\mathrm{Var}[P]$ and thus, the bound $\nu^2 \leq 1/3$.

We now can proceed as follows: as we have seen,

$$\int \exp\{ts\}dP(s) = \int_0^1 (\sinh(ts)/(ts))q(s)ds,$$

whence

$$\begin{aligned} \int \exp\{ts\}dP(s) &= \int_0^1 \left[\frac{\sinh ts}{ts} - \lambda s^2\right]q(s)ds + \int_0^1 \lambda s^2 q(s)ds \\ &\leq \max_{0 \leq s \leq 1} \phi(s) + 3\lambda\nu^2 \end{aligned} \tag{2.4.38}$$

(take into account that $\lambda > 0$). We now claim that $\max_{0 \leq s \leq 1} \phi(s) = \phi(0) \equiv \phi(1)$, which would combine with (2.4.38) to imply that

$$\int \exp\{ts\}dP(s) \leq \phi(0) + 3\lambda\nu^2 = 1 + 3\lambda\nu^2;$$

the latter bound, in view of the expression for λ, is exactly (2.4.35).

It remains to justify our claim that $\phi(s) \leq \phi(0) = \phi(1) = 1$ when $0 \leq s \leq 1$. It is immediately seen that $\phi'(0) = 0$, $\phi''(0) = \frac{1}{3}t^2 - 2\frac{\sinh(t)-t}{t} = -2\frac{\sinh(t)-t-t^3/6}{t} < 0$. It follows that when s grows from 0 to 1, the function $\phi(s)$ first decreases, starting with the value $\phi(0) = 1$. What happens next, we do not know exactly, except for the fact that when s reaches the value 1, $\phi(s)$ recovers its initial value $\phi(1) = \phi(0) = 1$. It follows that if $\max_{0 \leq s \leq 1} \phi(s) > 1$, then $\phi'(s)$ has at least 2 distinct zeros on $(0,1)$, which along with $\phi'(0) = 0$, gives us at least 3 distinct zeros of ϕ'. But $\phi'(s)$ is a convex function, and in order to possess 3 distinct zeros, it should vanish on a nontrivial segment, which definitely is not the case. $\square$

2.4.4 Summary

A summary of the examples we have considered is presented in table 2.3 and in figure 2.2.

2.5 EXERCISES

Exercise 2.1. Prove the claim in Example 2.4.12.

Exercise 2.2. Consider a toy chance constrained LO problem:

$$\min_{x,t} \left\{ t : \mathrm{Prob}\{\underbrace{\sum_{j=1}^n \zeta_j x_j}_{\xi^n[x]} \leq t\} \geq 1 - \epsilon, 0 \leq x_i \leq 1, \sum_j x_j = n \right\} \tag{2.5.1}$$

where $\zeta_1, ..., \zeta_n$ are independent random variables uniformly distributed in $[-1, 1]$.

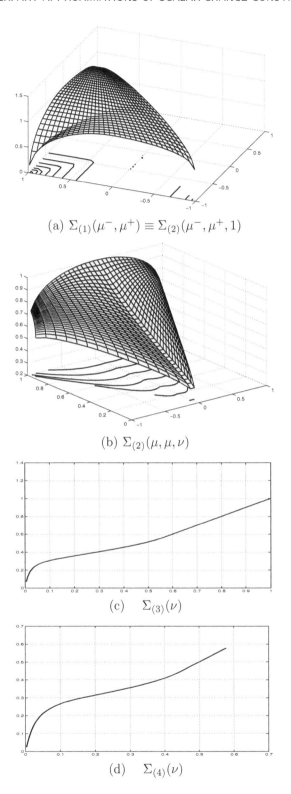

(a) $\Sigma_{(1)}(\mu^-, \mu^+) \equiv \Sigma_{(2)}(\mu^-, \mu^+, 1)$

(b) $\Sigma_{(2)}(\mu, \mu, \nu)$

(c) $\Sigma_{(3)}(\nu)$

(d) $\Sigma_{(4)}(\nu)$

Figure 2.2 Graphs of $\Sigma_{(\kappa)}$, $\kappa = 1, 2, 3, 4$.

A priori information	P satisfies P.2 with parameters			
on P	μ^-	μ^+	σ	Remark
$\operatorname{supp}(P) \subset [-1,1]$	-1	1	0	
$\operatorname{supp}(P) \subset [-1,1]$ P is unimodal w.r.t. 0	$-\frac{1}{2}$	$\frac{1}{2}$	$\sqrt{\frac{1}{12}}$	
$\operatorname{supp}(P) \subset [-1,1]$ P is unimodal w.r.t. 0 P is symmetric w.r.t. 0	0	0	$\sqrt{\frac{1}{3}}$	
$\operatorname{supp}(P) \subset [-1,1]$ $[-1 <] \ \mu^- \leq \operatorname{Mean}[P] \leq \mu^+ \ [< 1]$	μ^-	μ^+	$\Sigma_{(1)}(\mu^-, \mu^+)$	(2.4.29)
$\operatorname{supp}(P) \subset [-1,1]$ $[-\nu \leq] \ \mu^- \leq \operatorname{Mean}[P] \leq \mu^+ \ [\leq \nu]$ $\operatorname{Var}[P] \leq \nu^2 \leq 1$	μ^-	μ^+	$\Sigma_{(2)}(\mu^-, \mu^+, \nu)$	(2.4.32)
$\operatorname{supp}(P) \subset [-1,1]$ P is symmetric w.r.t. 0 $\operatorname{Var}[P] \leq \nu^2 \leq 1$	0	0	$\Sigma_{(3)}(\nu)$	(2.4.34)
$\operatorname{supp}(P) \subset [-1,1]$ P is symmetric w.r.t. 0 P is unimodal w.r.t. 0 $\operatorname{Var}[P] \leq \nu^2 \leq 1/3$	0	0	$\Sigma_{(4)}(\nu)$	(2.4.36)

Table 2.3 Summary of Examples 2.4.6 – 2.4.9 and 2.4.11 – 2.4.13. In the table for a probability distribution P on the axis, we denote by $\operatorname{Mean}[P] = \int s \, dP(s)$ and $\operatorname{Var}[P] = \int s^2 dP(s)$ the mean and the variance of the distribution.

i) Find a way to solve the problem exactly, and find the true optimal value t_{tru} of the problem for $n = 16, 256$ and $\epsilon = 0.05, 0.0005, 0.000005$.

Hint: The deterministic constraints say that $x_1 = \ldots = x_n = 1$. All we need is an efficient way to compute the probability distribution $\operatorname{Prob}\{\xi^n < t\}$ of the sum ξ^n of n independent random variables uniformly distributed on $[-1, 1]$. The density of ξ^n clearly is supported on $[-n, n]$ and is a polynomial of degree $n - 1$ in every one of the segments $[-n + 2i, -n + 2i + 2]$, $0 \leq i < n$. The coefficients of these polynomials can be computed via a simple recursion in n.

ii) For the same pairs (n, ϵ) as in i), compute the optimal values of the tractable approximations of the problem as follows:

(a) t_{Nrm} — the optimal value of the problem obtained from (2.5.1) when replacing the "true" random variable $\xi^n[x]$ with its "normal approximation" — a Gaussian random variable with the same mean and standard deviation as those of $\xi^n[x]$;

(b) t_{Bll} — the optimal value of the safe tractable approximation of (2.5.1) given by Proposition 2.3.1;

(c) t_{BllBx} — the optimal value of the safe tractable approximation of (2.5.1) given by Proposition 2.3.3;

(d) t_{Bdg} — the optimal value of the safe tractable approximation of (2.5.1) given by Proposition 2.3.4;

(e) $t_{\text{E.2.4.11}}$ — the optimal value of the safe tractable approximation of (2.5.1) suggested by Example 2.4.11, where you set $\mu^{\pm} = 0$ and $\nu = 1/\sqrt{3}$;

(f) $t_{\text{E.2.4.12}}$ — the optimal value of the safe tractable approximation of (2.5.1) suggested by Example 2.4.12, where you set $\nu = 1/\sqrt{3}$;

(g) $t_{\text{E.2.4.13}}$ — the optimal value of the safe tractable approximation of (2.5.1) suggested by Example 2.4.13, where you set $\nu = 1/\sqrt{3}$;

(h) t_{Unim} — the optimal value of the safe tractable approximation of (2.5.1) suggested by Example 2.4.7.

Think of the results as compared to each other and to those of i).

Exercise 2.3. Consider the chance constrained LO problem (2.5.1) with independent $\zeta_1, ..., \zeta_n$ taking values ± 1 with probability 0.5.

i) Find a way to solve the problem exactly, and find the true optimal value t_{tru} of the problem for $n = 16, 256$ and $\epsilon = 0.05, 0.0005, 0.000005$.

ii) For the same pairs (n, ϵ) as in i), compute the optimal values of the tractable approximations of the problem as follows:

(a) t_{Nrm} — the optimal value of the problem obtained from (2.5.1) when replacing the "true" random variable ξ^n with its "normal approximation" — a Gaussian random variable with the same mean and standard deviation as those of ξ^n;

(b) t_{Bll} — the optimal value of the safe tractable approximation of (2.5.1) given by Proposition 2.3.1;

(c) t_{BllBx} — the optimal value of the safe tractable approximation of (2.5.1) given by Proposition 2.3.3;

(d) t_{Bdg} — the optimal value of the safe tractable approximation of (2.5.1) given by Proposition 2.3.4;

(e) $t_{\text{E.2.4.11}}$ — the optimal value of the safe tractable approximation of (2.5.1) suggested by Example 2.4.11, where you set $\mu^{\pm} = 0$ and $\nu = 1$;

(f) $t_{\text{E.2.4.12}}$ — the optimal value of the safe tractable approximation of (2.5.1) suggested by Example 2.4.12, where you set $\nu = 1$.

Think of the results as compared to each other and to those of i).

Exercise 2.4. A) Verify that whenever $n = 2^k$ is an integral power of 2, one can build an $n \times n$ matrix B_n with all entries ± 1, all entries in the first column equal to 1, and with rows that are orthogonal to each other.

Hint: Use recursion $B_{2^0} = [1]$; $B_{2^{k+1}} = \begin{bmatrix} B_{2^k} & B_{2^k} \\ B_{2^k} & -B_{2^k} \end{bmatrix}$.

B) Let $n = 2^k$ and $\widehat{\zeta} \in \mathbb{R}^n$ be the random vector as follows. We fix a matrix B_n from A). To get a realization of ζ, we generate random variable $\eta \sim \mathcal{N}(0, 1)$

and pick at random (according to uniform distribution on $\{1, ..., n\}$) a column in the matrix ηB_n; the resulting vector is a realization of $\widehat{\zeta}$ that we are generating.

B.1) Prove that the marginal distributions of ζ_j and the covariance matrix of $\widehat{\zeta}$ are exactly the same as for the random vector $\widetilde{\zeta} \sim \mathcal{N}(0, I_n)$. It follows that most primitive statistical tests cannot distinguish between the distributions of $\widehat{\zeta}$ and $\widehat{\zeta}$.

B.2) Consider problem (2.5.1) with $\epsilon < 1/(2n)$ and compute the optimal values in the cases when (a) ζ is $\widetilde{\zeta}$, and (b) ζ is $\widehat{\zeta}$. Compare the results for $n = 10, \epsilon = 0.01$; $n = 100, \epsilon = 0.001$; $n = 1000, \epsilon = 0.0001$.

2.6 NOTES AND REMARKS

NR 2.1. The concept of chance constraints goes back to Charnes, Cooper, and Symonds [40], Miller and Wagner[79], and Prékopa [96]. For important convexity-related results on these constraints, see [97, 71]. For special cases where a scalar chance constraint can be processed efficiently, see [98, 45]. To the best of our knowledge, aside of these special cases, there exist only two computationally efficient techniques for handling chance constraints: the *scenario approximation* and the *safe tractable approximations* as defined in the main body of this chapter.

The scenario approximation of a chance constrained problem

$$\min_x \left\{ f_0(x) : \text{Prob}_{\zeta \sim P}\{f_i(x, \zeta) \le 0, \, i = 1, ..., m\} \ge 1 - \epsilon \right\} \qquad (*)$$

is, conceptually, very simple: one generates a sample $\zeta^1, ..., \zeta^N$ of N independent realizations of ζ and uses, as an approximation, the problem

$$\min_x \left\{ f_0(x) : f_i(x, \zeta^t) \le 0, \, i = 1, ..., m, t = 1, ..., N \right\} \qquad (!)$$

When f_i, $i = 0, 1, ..., m$, are efficiently computable convex functions of x and N is moderate, the approximation is computationally tractable; however, it is not necessarily safe — the feasible set of (!), (which by itself is random), is not necessarily contained in the feasible set of ($*$). Even a weaker property (in fact, sufficient for all our purposes) that the *optimal solution* to (!), (which again is random), is feasible for ($*$) cannot be guaranteed. However, when N is large enough, one can hope that this solution is feasible for ($*$) with probability close to 1. Deep results of Calafiore and Campi [37, 38] demonstrate that this indeed is true in the convex case; specifically, if $f_0, f_1, ..., f_m$ are convex in x, then for every $\epsilon, \delta \in (0, 1)$, the sample size

$$N \ge N^* := \text{Ceil} \left(2n\epsilon^{-1} \log\left(12/\epsilon\right) + 2\epsilon^{-1} \log\left(2/\delta\right) + 2n\right), \, n = \dim x$$

guarantees that the optimal solution $\widetilde{x}$ to (!) is feasible for ($*$) with probability $\ge 1 - \delta$. Other interesting and important results on scenario approximations of chance constrained problems can be found in [44, 66] (the latter paper addresses the case of an ambiguous chance constraint). The most attractive feature of the scenario approximation is its generality. It requires no structural assumptions on $f_i(x, \zeta)$ aside of convexity in x, and no assumptions on P (the latter by itself is not even necessary — all we need is the possibility to sample from P). On the negative side, in order for the scenario approximation to be safe (i.e., with $\tilde{x}$ feasible for $(*)$) with probability close to 1, the sample size N should be large enough, specifically, of order of $\frac{1}{\epsilon}$.[2] In reality, this means that the scenario approximation becomes impractical when ϵ is small, something like 1.e-4 or less[3].

In this book, we take another approach: *simulation-free* analytical safe tractable approximations of chance constraints. While these approximations require severe assumptions on the structure of the chance constraints in question (we hardly ever go beyond the case of bi-affine $f_i(x, \zeta)$ with independent "light tail" components of ζ), their advantage is that the complexity of the approximation is independent of the value of ϵ, (which thus can be arbitrary small).

NR 2.2. The theoretical results in section 2.3 originate from [5, 7] (Propositions 2.3.1, 2.3.3) and from [24] (Proposition 2.3.4). Nearly all results presented in section 2.4 are based on the *Bernstein approximation scheme* as presented in [83]; this scheme is investigated in more detail in chapter 4.

---

[2]It is easily seen that this requirement reflects the essence of the matter rather than being a result of "bad bounding."

[3]This shortcoming can be overcome when f_i are bi-affine in x and in ζ and the entries in ζ are "light tail" independent random variables, see [84].

Chapter Three
Globalized Robust Counterparts of Uncertain LO Problems

In this chapter we extend the concept of Robust Counterpart in order to gain certain control on what happens when the actual data perturbations run out of the postulated perturbation set.

3.1 GLOBALIZED ROBUST COUNTERPART — MOTIVATION AND DEFINITION

Let us come back to Assumptions A.1 – A.3 underlying the concept of Robust Counterpart and concentrate on **A.3**. This assumption is not a "universal truth" — in reality, there are indeed constraints that cannot be violated (e.g., you cannot order a negative supply), but also constraints whose violations, while undesirable, can be tolerated to some degree, (e.g., sometimes you can tolerate a shortage of a certain resource by implementing an "emergency measure" like purchasing it on the market, employing sub-contractors, taking out loans, etc.). Immunizing such "soft" constraints against data uncertainty should perhaps be done in a more flexible fashion than in the usual Robust Counterpart. In the latter, we ensure a constraint's validity for all realizations of the data from a given uncertainty set and do not care what happens when the data are outside of this set. For a soft constraint, we can take care of what happens in this latter case as well, namely, by ensuring *controlled deterioration* of the constraint when the data runs away from the uncertainty set. A mathematically convenient model capturing the above requirements is as follows.

Consider an uncertain linear constraint in variable x

$$\left[a^0 + \sum_{\ell=1}^{L} \zeta_\ell a^\ell \right]^T x \leq \left[b^0 + \sum_{\ell=1}^{L} \zeta_\ell b^\ell \right] \qquad (3.1.1)$$

where ζ is the perturbation vector (cf. (1.3.4), (1.3.5)). Let $\mathcal{Z}_+$ be the set of all "physically possible" perturbations, and $\mathcal{Z} \subset \mathcal{Z}_+$ be the "normal range" of the perturbations — the one for which we insist on the constraint to be satisfied. With the usual RC approach, we treat $\mathcal{Z}$ as the only set of perturbations and require a candidate solution x to satisfy the constraint for all $\zeta \in \mathcal{Z}$. With our new approach, we add the requirement that *the violation of constraint in the case when $\zeta \in \mathcal{Z}_+ \backslash \mathcal{Z}$* (that is a "physically possible" perturbation that is outside of the normal range) *should be bounded by a constant times the distance from ζ to $\mathcal{Z}$*. Both requirements — the validity of the constraint for $\zeta \in \mathcal{Z}$ and the bound on the constraint's violation when $\zeta \in \mathcal{Z}_+ \backslash \mathcal{Z}$ can be expressed by a single requirement

$$\left[a^0 + \sum_{\ell=1}^L \zeta_\ell a^\ell\right]^T x - \left[b^0 + \sum_{\ell=1}^L \zeta_\ell b^\ell\right] \leq \alpha \mathrm{dist}(\zeta, \mathcal{Z}) \quad \forall \zeta \in \mathcal{Z}_+,$$

where $\alpha \geq 0$ is a given "global sensitivity."

In order to make the latter requirement tractable, we add some structure to our setup. Specifically, let us assume that:

(G.a) The normal range $\mathcal{Z}$ of the perturbation vector ζ is a nonempty closed convex set;

(G.b) The set $\mathcal{Z}_+$ of all "physically possible" perturbations is the sum of $\mathcal{Z}$ and a closed convex cone $\mathcal{L}$:

$$\mathcal{Z}_+ = \mathcal{Z} + \mathcal{L} = \{\zeta = \zeta' + \zeta'' : \zeta' \in \mathcal{Z}, \zeta'' \in \mathcal{L}\}; \tag{3.1.2}$$

(G.c) We measure the distance from a point $\zeta \in \mathcal{Z}_+$ to the normal range $\mathcal{Z}$ of the perturbations in a way that is consistent with the structure (3.1.2) of $\mathcal{Z}_+$, specifically, by

$$\mathrm{dist}(\zeta, \mathcal{Z} \mid \mathcal{L}) = \inf_{\zeta'} \{\|\zeta - \zeta'\| : \zeta' \in \mathcal{Z}, \zeta - \zeta' \in \mathcal{L}\}, \tag{3.1.3}$$

where $\|\cdot\|$ is a fixed norm on $\mathbb{R}^L$.

In what follows, we refer to a triple $(\mathcal{Z}, \mathcal{L}, \|\cdot\|)$ arising in (G.a–c) as a *perturbation structure* for the uncertain constraint (3.1.1).

Definition 3.1.1. Given $\alpha \geq 0$ and a perturbation structure $(\mathcal{Z}, \mathcal{L}, \|\cdot\|)$, we say that a vector x is a globally robust feasible solution to uncertain linear constraint (3.1.1) with global sensitivity α, if x satisfies the semi-infinite constraint

$$\left[a^0 + \sum_{\ell=1}^L \zeta_\ell a^\ell\right]^T x \leq \left[b^0 + \sum_{\ell=1}^L \zeta_\ell b^\ell\right] + \alpha \, \mathrm{dist}(\zeta, \mathcal{Z}|\mathcal{L}) \; \forall \zeta \in \mathcal{Z}_+ = \mathcal{Z} + \mathcal{L}. \tag{3.1.4}$$

We refer to the semi-infinite constraint (3.1.4) as the Globalized Robust Counterpart (GRC) of the uncertain constraint (3.1.1).

Note that global sensitivity $\alpha = 0$ corresponds to the most conservative attitude where the constraint must be satisfied for all physically possible perturbations; with $\alpha = 0$, the GRC becomes the usual RC of the uncertain constraint with $\mathcal{Z}_+$ in the role of the perturbation set. The larger α, the less conservative the GRC.

Now, given an uncertain Linear Optimization program with affinely perturbed data

$$\left\{\min_x \left\{c^T x : Ax \leq b\right\} : [A, b] = [A^0, b^0] + \sum_{\ell=1}^L \zeta_\ell [A^\ell, b^\ell]\right\} \tag{3.1.5}$$

(w.l.o.g., we assume that the objective is certain) and a perturbation structure $(\mathcal{Z}, \mathcal{L}, \|\cdot\|)$, we can replace every one of the constraints with its Globalized Robust Counterpart, thus ending up with the GRC of (3.1.5). In this construction, we can associate different sensitivity parameters α to different constraints. Moreover, we can treat these sensitivities as design variables rather than fixed parameters, add linear constraints on these variables, and optimize both in x *and* α an objective function that is a mixture of the original objective and a weighted sum of the sensitivities.

3.2 COMPUTATIONAL TRACTABILITY OF GRC

As in the case of the usual Robust Counterpart, the central question of computational tractability of the Globalized RC of an uncertain LO reduces to a similar question for the GRC (3.1.4) of a single uncertain linear constraint (3.1.1). The latter question is resolved to a large extent by the following observation:

Proposition 3.2.1. A vector x satisfies the semi-infinite constraint (3.1.4) if and only if x satisfies the following pair of semi-infinite constraints:

$$
(a) \quad \left[a^0 + \sum_{\ell=1}^{L} \zeta_\ell a^\ell\right]^T x \leq \left[b^0 + \sum_{\ell=1}^{L} \zeta_\ell b^\ell\right] \; \forall \zeta \in \mathcal{Z}
$$

$$
(b) \quad \left[\sum_{\ell=1}^{L} \Delta_\ell a^\ell\right]^T x \leq \left[\sum_{\ell=1}^{L} \Delta_\ell b^\ell\right] + \alpha \; \forall \Delta \in \widetilde{\mathcal{Z}} \equiv \{\Delta \in \mathcal{L} : \|\Delta\| \leq 1\}.
$$

$$(3.2.1)$$

Remark 3.2.2. Proposition 3.2.1 implies that the GRC of an uncertain linear inequality is *equivalent* to a pair of semi-infinite linear inequalities of the type arising in the usual RC. Consequently, we can invoke the representation results of section 1.3 to show that *under mild assumptions on the perturbation structure, the GRC (3.1.4) can be represented by a "short" system of explicit convex constraints.*

Proof of Proposition 3.2.1. Let x satisfy (3.1.4). Then x satisfies (3.2.1.a) due to $\text{dist}(\zeta, \mathcal{Z}|\mathcal{L}) = 0$ for $\zeta \in \mathcal{Z}$. In order to demonstrate that x satisfies (3.2.1.b) as well, let $\bar{\zeta} \in \mathcal{Z}$ and $\Delta \in \mathcal{L}$ with $\|\Delta\| \leq 1$. By (3.1.4) and since $\mathcal{L}$ is a cone, for every $t > 0$ we have $\zeta_t := \bar{\zeta} + t\Delta \in \mathcal{Z} + \mathcal{L}$ and $\text{dist}(\zeta_t, \mathcal{Z}|\mathcal{L}) \leq \|t\Delta\| \leq t$; applying (3.1.4) to $\zeta = \zeta_t$, we therefore get

$$
\left[a^0 + \sum_{\ell=1}^{L} \bar{\zeta}_\ell a^\ell\right]^T x + t\left[\sum_{\ell=1}^{L} \Delta_\ell a^\ell\right]^T x \leq \left[b^0 + \sum_{\ell=1}^{L} \bar{\zeta}_\ell b^\ell\right] + t\left[\sum_{\ell=1}^{L} \Delta_\ell b^\ell\right] + \alpha t.
$$

Dividing both sides in this inequality by t and passing to limit as $t \to \infty$, we see that the inequality in (3.2.1.b) is valid at our Δ. Since $\Delta \in \widetilde{\mathcal{Z}}$ is arbitrary, x satisfies (3.2.1.b), as claimed.

It remains to prove that if x satisfies (3.2.1), then x satisfies (3.1.4). Indeed, let x satisfy (3.2.1). Given $\zeta \in \mathcal{Z} + \mathcal{L}$ and taking into account that $\mathcal{Z}$ and $\mathcal{L}$ are closed, we can find $\bar{\zeta} \in \mathcal{Z}$ and $\Delta \in \mathcal{L}$ such that $\bar{\zeta} + \Delta = \zeta$ and $t := \text{dist}(\zeta, \mathcal{Z}|\mathcal{L}) = \|\Delta\|$.

Representing $\Delta = te$ with $e \in \mathcal{L}$, $\|e\| \le 1$, we have

$$\left[a^0 + \sum_{\ell=1}^{L} \zeta_\ell a^\ell\right]^T x - \left[b^0 + \sum_{\ell=1}^{L} \zeta_\ell b^\ell\right]$$

$$= \underbrace{\left[a^0 + \sum_{\ell=1}^{L} \bar{\zeta}_\ell a^\ell\right]^T x - \left[b^0 + \sum_{\ell=1}^{L} \bar{\zeta}_\ell b^\ell\right]}_{\le 0 \text{ by } (3.2.1.a)} + \underbrace{\left[\sum_{\ell=1}^{L} \Delta_\ell a^\ell\right]^T x - \left[\sum_{\ell=1}^{L} \Delta_\ell b^\ell\right]}_{\substack{= t\left[\left[\sum\limits_{\ell=1}^{L} e_\ell a^\ell\right]^T x - \left[\sum\limits_{\ell=1}^{L} e_\ell b^\ell\right]\right] \\ \le t\alpha \text{ by } (3.2.1.b)}}$$

$$\le t\alpha = \alpha \operatorname{dist}(\zeta, \mathcal{Z}|\mathcal{L}).$$

Since $\zeta \in \mathcal{Z} + \mathcal{L}$ is arbitrary, x satisfies (3.1.4). □

Example 3.2.3. Consider the following 3 perturbation structures $(\mathcal{Z}, \mathcal{L}, \|\cdot\|)$:

(a) $\mathcal{Z}$ is a box $\{\zeta : |\zeta_\ell| \le \sigma_\ell, 1 \le \ell \le L\}$, $\mathcal{L} = \mathbb{R}^L$ and $\|\cdot\| = \|\cdot\|_1$;

(b) $\mathcal{Z}$ is an ellipsoid $\{\zeta : \sum_{\ell=1}^{L} \zeta_\ell^2 / \sigma_\ell^2 \le \Omega^2\}$, $\mathcal{L} = \mathbb{R}_+^L$ and $\|\cdot\| = \|\cdot\|_2$;

(c) $\mathcal{Z}$ is the intersection of a box and an ellipsoid: $\mathcal{Z} = \{\zeta : |\zeta_\ell| \le \sigma_\ell, 1 \le \ell \le L, \sum_{\ell=1}^{L} \zeta_\ell^2 / \sigma_\ell^2 \le \Omega^2\}$, $\mathcal{L} = \mathbb{R}^L$, $\|\cdot\| = \|\cdot\|_\infty$.

In these cases the GRC of (3.1.1) is equivalent to the finite systems of explicit convex inequalities as follows:

Case (a):

$$(a) \quad [a^0]^T x + \sum_{\ell=1}^{L} \sigma_\ell |[a^\ell]^T x - b^\ell| \le b^0$$
$$(b) \quad |[a^\ell]^T x - b^\ell| \le \alpha, \ \ell = 1, ..., L$$

Here (a) represents the constraint (3.2.1.a) (cf. Example 1.3.2), and (b) represents the constraint (3.2.1.b) (why?)

Case (b):

$$(a) \quad [a^0]^T x + \Omega \left(\sum_{\ell=1}^{L} \sigma_\ell^2 ([a^\ell]^T x - b^\ell)^2\right)^{1/2} \le b^0$$
$$(b) \quad \left(\sum_{\ell=1}^{L} \max^2[[a^\ell]^T x - b^\ell, 0]\right)^{1/2} \le \alpha.$$

Here (a) represents the constraint (3.2.1.a) (cf. Example 1.3.3), and (b) represents the constraint (3.2.1.b).

Case (c):

$$(a.1) \quad [a^0]^T x + \sum_{\ell=1}^{L} \sigma_\ell |z_\ell| + \Omega \left(\sum_{\ell=1}^{L} \sigma_\ell^2 w_\ell^2\right)^{1/2} \le b^0$$
$$(a.2) \quad z_\ell + w_\ell = [a^\ell]^T x - b^\ell, \ \ell = 1, ..., L$$
$$(b) \quad \sum_{\ell=1}^{L} |[a^\ell]^T x - b^\ell| \le \alpha.$$

Here $(a.1–2)$ represent the constraint (3.2.1.a) (cf. Example 1.3.7), and (b) represents the constraint (3.2.1.b).

3.3 EXAMPLE: SYNTHESIS OF ANTENNA ARRAYS

To illustrate the notion of Globalized Robust Counterpart, consider an example related to the problem of the synthesis of antenna arrays.

3.3.1 Building the Model

Antenna arrays. The most basic element of a transmitting antenna is an isotropic harmonic oscillator emitting spherical monochromatic electromagnetic waves of a certain wavelength λ and frequency ω. When invoked, the oscillator generates an electromagnetic field whose electric component at a point P is given by

$$d^{-1}\Re\{z\exp\{i(\omega t - 2\pi d/\lambda)\}\},$$

where the *weight* z is a complex number responsible for how the oscillator is invoked, t is time, d is the distance between P and the point where the oscillator is placed and i is the imaginary unit. The electrical component of the electromagnetic field created at P by an *array* of n coherent (that is, with the same frequency) isotropic oscillators placed at points $P_1, ..., P_n$ is

$$
\begin{aligned}
E &= \sum_{k=1}^{n} \Re\{\|P - P_k\|^{-1} z_k \exp\{i(\omega t - 2\pi\|P - P_k\|/\lambda)\}\} \\
&= \Re\{\exp\{i\omega t\} \sum_{k=1}^{n} \|P - P_k\|^{-1} z_k \exp\{-2\pi i\|P - P_k\|/\lambda)\},
\end{aligned}
\tag{3.3.1}
$$

where $z_k \in \mathbb{C}$ is the weight of k-th oscillator. When P is at a large distance r from the origin:

$$P = re, \ \|e\|_2 = 1,$$

the expression for E, neglecting terms of order of r^{-2}, becomes

$$\Re\{r^{-1}\exp\{i(\omega t - 2\pi r/\lambda)\}\underbrace{\sum_{k=1}^{n} z_k \exp\{2\pi i d_k \cos(\phi_k(e))/\lambda\}}_{D(e)}\},$$

where $d_k = \|P_k\|_2$ and $\phi_k(e)$ is the angle between the directions e_k from the origin to P_k and e from the origin to P. The complex-valued function $D(e)$ of a unit 3-D direction e is called the *diagram* of the antenna array. It turns out that squared modulus $|D(e)|^2$ of the diagram is proportional to the directional density of electromagnetic energy sent by the antenna.[1]

In a typical antenna design problem, one is given the number and the locations of isotropic monochromatic oscillators and is interested to assign them with complex weights z_k in such a way that the resulting diagram (or its modulus) is as close as possible to a desired "target." In simple cases, such a problem can be modeled as a linear optimization program.

Example 3.3.1. Consider a n-element equidistant grid of oscillators placed along the x-axis: $P_k = k\mathbf{i}$, where $\mathbf{i}$ is the basic orth of the x-axis. The diagram of such an antenna depends solely on the angle ϕ, $0 \leq \phi \leq \pi$, between a 3-D direction e and the

---

[1] For a receiving antenna, the squared modulus $|D(e)|^2$ of its diagram (mathematically, completely similar to the diagram of a transmitting antenna) is proportional to the sensitivity of the antenna to a flat wave of frequency ω incoming along a direction e.

direction $\mathbf{i}$ of the x-axis and is given by

$$D(\phi) = \sum_{k=1}^{n} z_k \exp\{2\pi \imath d_k \cos(\phi)/\lambda\}, \ d_k = k, \qquad (3.3.2)$$

(from now on, we write $D(\phi)$ instead of $D(e)$). Now consider the design problem where, given an "angle of interest" $\Delta \in (0, \pi)$, one should choose the weights z_k so that most of the energy emanating from the antenna is sent along the cone K_Δ comprised of all 3-D directions with $0 \le \phi \le \Delta$, (i.e., along the usual 3-D round cone with the nonnegative ray of the x-axis in the role of the central ray and the "angular width" 2Δ). There are many ways to model our design specification; we choose a simple way as follows. First note that when multiplying all weights by a common nonzero complex number, we do not vary the directional distribution of energy; therefore we lose nothing by normalizing the weights by the requirement that the real part of $D(0)$ is ≥ 1. We now can quantify the amount of energy sent in the *sidelobe angle* (the complement of K_Δ) by the quantity

$$\max_{\Delta < \phi \le \pi} |D(\phi)|$$

("sidelobe level"), and pose our problem as the semi-infinite optimization program

$$\min_{z_1, \ldots, z_n \in \mathbb{C}, \tau \in \mathbb{R}} \tau : \left\{ \begin{array}{c} \tau \ge |D(\phi)| \equiv |\sum_{k=1}^{n} z_k \exp\{2\pi \imath d_k \cos(\phi)/\lambda\}| \\ \forall \phi \in [\Delta, \pi] \\ \Re D(0) \equiv \Re\left\{\sum_{k=1}^{n} z_k \exp\{2\pi \imath d_k/\lambda\}\right\} \ge 1 \end{array} \right\}, \ d_k = k.$$

This (admittedly simplified) model of the Antenna Design problem has an important advantage: it is a *convex* problem, although a semi-infinite one. We can get rid of semi-infiniteness by replacing the segment $[\Delta, \pi]$ with a fine finite grid Φ of points in this segment, thus arriving at the convex program

$$\min_{z_1, \ldots, z_n \in \mathbb{C}, \tau \in \mathbb{R}} \tau : \left\{ \begin{array}{c} \tau \ge |D(\phi)| \equiv |\sum_{k=1}^{n} z_k \exp\{2\pi \imath d_k \cos(\phi)/\lambda\}| \\ \forall \phi \in \Phi \\ \Re D(0) \equiv \Re\left\{\sum_{k=1}^{n} z_k \exp\{2\pi \imath d_k/\lambda\}\right\} \ge 1 \end{array} \right\}, \ d_k = k.$$

This is not exactly a LO program, since the absolute values in question are moduli of complex numbers (that is, Euclidean norms of 2-D vectors). In order to overcome this difficulty, let us approximate the modulus $|z| = \sqrt{\Re^2(z) + \Im^2(z)}$ of a complex number z by a "polyhedral norm," specifically, by the norm

$$p_d(z) = \max_{1 \le \ell \le L} \Re\{z\mu_\ell\}, \ \mu_\ell = \exp\{\imath\ell/L\}.$$

Geometrically, we approximate the unit 2-D disc by a circumscribed L-side perfect polygon. We clearly have

$$p_L(z) \le |z| \le p_L(z)/\cos(\pi/L).$$

For example, $p_{12}(z)$ approximates $|z|$ within relative accuracy 3.5%, which is sufficiently accurate in our context. Replacing $|\cdot|$ with $p_{12}(\cdot)$, we arrive at the following LO program:

Antenna Design problem: *Given the number of oscillators n, wavelength λ, angle of interest Δ and a finite grid Φ on the segment $[\Delta, \pi]$, solve the LO* program

$$\min_{z_1,\ldots,z_n \in \mathbb{C}, \tau} \left\{ \tau : \begin{array}{c} \Re\left\{ \sum_{k=1}^{n} \mu_\ell \exp\{2\pi i d_k \cos(\phi)/\lambda\} z_k \right\} \leq \tau \\ \forall (\phi \in \Phi, \ell \leq 12) \\ \Re\left\{ -\sum_{k=1}^{n} \exp\{2\pi i d_k/\lambda\} z_k \right\} \leq -1 \end{array} \right\}. \qquad (3.3.3)$$

Sources of uncertainty. In the Antenna Design problem, there are at least two sources of data uncertainty to be accounted for.

i) *Positioning* errors. When manufacturing the antenna, the oscillators cannot be placed along an ideal equidistant grid. Moreover, their positioning varies slightly over time as a result of deformations of the antenna due to changes in temperature, wind, etc. To simplify matters, assume that the positioning errors affect only the distances from the origin to the oscillators, but not the directions from the origin to oscillators, so that the latter belong to the x-axis. We can model the positioning perturbations by collections $\{\delta d_k \in \mathbb{R}\}_{k=1}^n$, where δd_k is the deviation of the actual distance d_k of k-th oscillator to the origin from the nominal value $d_k^{\mathrm{n}} = k$ of this distance, and we assume that the positioning perturbations run through the box $\Delta_{\mathrm{p}} = \{\{\delta d_k\}_{k=1}^n : |\delta d_k| \leq \epsilon, 1 \leq k \leq n\}$.

ii) *Actuation* errors. The weights z_k are in reality characteristics of certain physical devices and as such cannot be implemented exactly as they are computed. It is natural to model these unavoidable actuation errors as multiplicative perturbations $z_k \mapsto (1+\xi_k)z_k$, where $\xi_k \in \mathbb{C}$; we refer to the quantity $\rho = \max_k |\xi_k|$ as to the *level* of actuation errors.

Clearly, both sources of uncertainty can be thought of as those of the data. Indeed, all constraints in (3.3.3) are of the form

$$\Re\{\sum_{k=1}^{n} \zeta_k z_k\} \leq p\tau + q \qquad (3.3.4)$$

with certain p, q. The consequences of a perturbation $d_k \mapsto d_k + \delta d_k$, $z_k \mapsto (1+\xi_k)z_k$ are exactly the same as if there were no positioning perturbations and actuation errors, but the coefficients $\zeta_k = \zeta_k(d_k)$ were subject to perturbations

$$\zeta_k(d_k^{\mathrm{n}}) \mapsto \zeta_k(d_k^{\mathrm{n}} + \delta d_k)(1 + \xi_k). \qquad (3.3.5)$$

3.3.2 Nominal Solution: Dream and Reality

Before thinking of how to immunize solutions against data uncertainty, it makes sense to check whether we should bother about this uncertainty in the first place. After all, both the positioning perturbations and the actuation errors are expected to be rather small (say, something like 1% of the corresponding nominal values).

Furthermore, the constraints of the problem are clearly soft — whatever the perturbations of the data, the nominal solution still makes physical sense (the designed
antenna will not explode, so to speak). The only bad thing that may happen is
that the actual value of the sidelobe level will be worse than the nominal one. If
deterioration of this level were "of the same order of magnitude" as the data perturbations (say, a 3–5% increase in the sidelobe level caused by a 1% perturbation
in the positioning of oscillators and in the weights), we could still be content with
the nominal solution as it is. Unfortunately, this is not the case — *in the Antenna
Design problem, even small (0.01%) data perturbations can result in huge (hundreds of a percent) variations of the design criteria.* Whether this is indeed the
case, depends on the nominal data. Here is an example of a setup that is really
bad:

$$n = 16, \ \lambda = 8, \ \Delta = \pi/6. \tag{3.3.6}$$

With this setup, the distances between consecutive oscillators equal to 1/8 of the
wavelength, and the spatial angle of interest — the one where we would like to send
as much energy as possible — is comprised of directions whose angular distances
from the axial direction $\mathbf{i}$ are at most 30°. Note that the relative spherical measure
of the set of these directions[2] is $(1 - \cos(\pi/6)/2 \approx 0.067$. When solving (3.3.3) with
nominal data and an equidistant 90-point grid Φ on $[0, \pi]$, we end up with a nice
nominal solution depicted on figure 3.1. For this solution *and no data perturbations*,
the sidelobe level is as low as 0.0025, and the *energy concentration* (the fraction of
total transmitted energy that goes along the desired angle) is as high as 99.99%.
Unfortunately, these nice results are nothing but a dream — with very small ($\epsilon =
0.0001$) random perturbations in the positioning, or with random actuation errors of
level as low as $\rho = 0.0001$, our design becomes a complete disaster. Let us look, e.g.,
at the sidelobe levels and energy concentrations produced by the nominal design
with randomly perturbed positions of oscillators and weights. First note that with
random perturbations of the data, the antenna diagram becomes random, and
therefore does not necessarily satisfy the normalization requirement $\Re\{D(0)\} \geq 1$.
To account for this phenomenon, we scale the sample diagrams to make $|D(0)|$
equal to 1, and look at the sidelobe levels and energy concentrations of the resulting
diagrams. Data in table 3.1 demonstrate that with $\epsilon = 0, \rho = 0.0001$, the sidelobe
level for the nominal design jumps, on the average, from its nominal value 0.0025
to 2.08, while the energy concentration drops from its nominal value of 99.99% to
just 8% — almost as if the energy were sent uniformly in all directions! Additional
evidence of how bad the nominal design is in the presence of our — unrealistically
low! — data uncertainty is given by sample diagrams depicted in figure 3.1 and
especially by "energy density" plots in the same figure.

> For a given diagram, its energy density is defined as follows. We compute the
> energy transmitted along all directions from a spatial angle K_s of directions
> forming angle at most s with $\mathbf{i}$, and treat this energy as a function of the

---

[2]Recall that directions are unit 3-D vectors, that is, points on the unit sphere in $\mathbb{R}^3$. The
relative spherical measure of a set A of directions (that is, of a subset of the unit sphere) is, by
definition, the area of the set divided by the area 4π of the entire unit sphere.

Design	ρ	ϵ	Sidelobe level	Energy concentration
Nominal Opt = 0.0025 $\alpha = 3.0e6$	0	0	$\underline{0.0025}(0.00)$	$\underline{0.9999}(0.00)$
	0	1.e-4	$\underline{1.38}(0.77)$	$\underline{0.18}(0.10)$
	1.e-4	0	$\underline{2.08}(1.54)$	$\underline{0.08}(0.07)$
	1.e-4	1.e-4	$\underline{2.18}(2.02)$	$\underline{0.09}(0.07)$
RC Opt = 0.106 $\alpha = 9.4$	0	0	$\underline{0.095}(0.00)$	$\underline{0.844}(0.00)$
	0	1.e-2	$\underline{0.099}(0.004)$	$\underline{0.837}(0.01)$
	1.e-2	0	$\underline{0.108}(0.02)$	$\underline{0.75}(0.04)$
	1.e-2	1.e-2	$\underline{0.150}(0.02)$	$\underline{0.75}(0.04)$
	3.e-2	1.e-2	$\underline{0.280}(0.07)$	$\underline{0.46}(0.14)$
GRC Opt = 0.147 $\alpha = 3.00$	0	0	$\underline{0.148}(0.00)$	$\underline{0.70}(0.00)$
	0	1.e-2	$\underline{0.149}(0.00)$	$\underline{0.70}(0.00)$
	1.e-2	0	$\underline{0.159}(0.00)$	$\underline{0.70}(0.01)$
	1.e-2	1.e-2	$\underline{0.159}(0.00)$	$\underline{0.70}(0.01)$
	3.e-2	1.e-2	$\underline{0.195}(0.02)$	$\underline{0.66}(0.04)$

Table 3.1 Performance of various designs. In the table: Opt is the optimal value in the nominal problem and its RC/GRC respectively, α is the global sensitivity $\sum_{k} |z_k|$ of the resulting solution w.r.t. actuation errors. The underlined numbers in the columns "Sidelobe level" and "Energy concentration" are averages over 100 random realizations of positioning and/or actuation errors, the numbers in parentheses are the associated standard deviations.

spherical measure of K_s, thus getting a function on the segment $[0, 4\pi]$. The energy density $p(s)$ is nothing but the derivative of the resulting function.

The plots in the second row on figure 3.1 are built as follows: we generate a sample of 100 data perturbations of the magnitude indicated under the plot and draw on a single figure the 100 resulting energy densities. The density pictures in figure 3.1 show that on the average (over 0.01% data perturbations), the energy density is nearly symmetric, meaning that under data perturbations, the nominal design does not distinguish between directions of interest and opposite directions. The bottom line is that *in the presence of even a small data uncertainty the nominal design becomes completely senseless.*

3.3.3 Immunizing Against Uncertainty

The strategy. For illustrative purposes, we intend to treat the two sources of uncertainty differently. Specifically, the perturbations in the data coefficients coming from positioning errors will be treated as the normal range of uncertainty, while the influence of the actuation errors will be controlled via the corresponding global sensitivity.

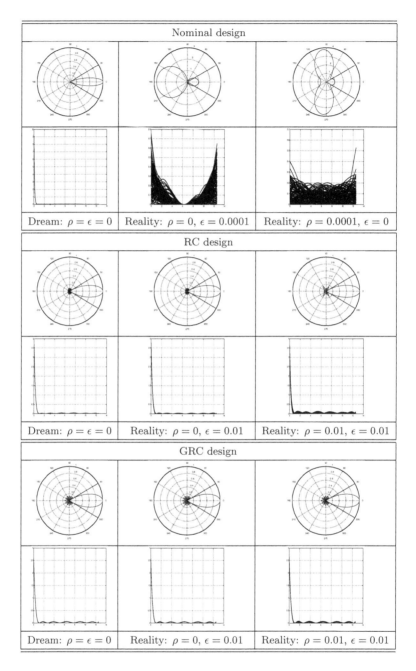

Figure 3.1 Nominal, RC and GRC antenna designs. First rows: sample plots of $|D(\phi)|$ in polar coordinates; the diagrams are normalized to have $D(0) = 1$. Second rows: bunches of 100 simulated energy densities.

Specifying perturbation structure. To implement our strategy, consider a single constraint (3.3.4) coming from (3.3.3). The actual values of the coefficients ζ_k are

$$\zeta_k = \mu \exp\{2\pi\imath \cos(\phi)(k + \delta d_k)/\lambda)\}(1 + \xi_k),$$

where μ, $|\mu| = 1$, and $\phi \in [0, \pi]$ are fixed, δd_k are the positioning errors, and ξ_k are the actuation errors. We would like to ensure that

$$\Re\left\{\sum_{k=1}^n \zeta_k z_k\right\} \le p\tau + q + \alpha\rho \ \forall(\rho \ge 0, \delta d : |\delta d_k| \le \epsilon \forall k, \xi : |\xi_k| \le \rho \forall k),$$

where p, q are given reals and $\alpha \ge 0$ is a given global sensitivity w.r.t. actuation errors. This is the same as to ensure that

$$\Re\Big\{\sum_{k=1}^n \overbrace{\mu \exp\{2\pi\imath \cos(\phi)(k + \delta d_k)/\lambda)\}}^{\zeta_k(\delta d_k)} z_k\Big\} + \Re\Big\{\sum_{k=1}^n \overbrace{\zeta_k(\delta d_k)\xi_k}^{\delta\zeta_k} z_k\Big\}$$
$$\le p\tau + q + \alpha\rho \ \ \forall(\rho \ge 0, \delta d : |\delta d_k| \le \epsilon \forall k, \xi : |\xi_k| \le \rho \forall k).$$

Taking into account that $|\zeta_k(\delta d_k)| = 1$, the latter relation clearly reads

$$\Re\Big\{\sum_{k=1}^n \zeta_k(\delta d_k)z_k\Big\} + \max_{\delta\zeta : |\delta\zeta_k| \le \rho \forall k} \Re\Big\{\sum_{k=1}^n \delta\zeta_k z_k\Big\}$$
$$\le p\tau + q + \alpha\rho \ \ \forall(\rho \ge 0, \delta d : |\delta d_k| \le \epsilon \forall k),$$

which is the same as the pair of requirements

$$\begin{aligned} (a) \quad & \Re\Big\{\sum_{k=1}^n \zeta_k(\delta d_k)z_k\Big\} \le p\tau + q \ \forall(\delta d : |\delta d_k| \le \epsilon \forall k) \\ (b) \quad & \sum_{k=1}^n |z_k| \le \alpha \end{aligned} \qquad (3.3.7)$$

on variables z_k, τ.

Note that (3.3.7.a) says that

$$\Re\Big\{\sum_{k=1}^n \zeta_k z_k\Big\} \le p\tau + q \ \forall(\zeta : \zeta_k \in \Gamma_k \forall k),$$

where Γ_k is the convex hull of the arc $\gamma_k = \{\mu \exp\{2\pi\imath \cos(\phi)(k + s)/\lambda)\} : |s| \le \epsilon\}$ of the unit circle in $\mathbb{C} = \mathbb{R}^2$. This convex hull Γ_k is a conic quadratic representable set, so that (3.3.7.a) is equivalent to an explicit finite system of conic quadratic inequalities (Theorem 1.3.4). We prefer to simplify the structure of the GRC by replacing Γ_k with the triangle Δ_k circumscribed around the arc γ_k (see figure 3.2). While increasing the conservatism slightly, this approximation allows one to rewrite (3.3.7.a) as the explicit convex constraint

$$\sum_{k=1}^n \max_{\ell=1,2,3} \Re\{v_{k\ell} z_k\} \le p\tau + q,$$

Figure 3.2 "Circle hat" Γ_k (ABC) and triangle Δ_k (ABD).

where v_{k1}, v_{k2}, v_{k3} are the vertices of the triangle Δ_k. With this approach, the GRC of (3.3.4) is given by the pair of convex constraints

$$(a) \quad \sum_{k=1}^{n} \max_{\ell=1,2,3} \Re\{v_{k\ell} z_k\} \le p\tau + q,$$
$$(b) \quad \sum_{k=1}^{n} |z_k| \le \alpha; \tag{3.3.8}$$

this is nothing but the GRC of (3.3.4) corresponding to the perturbation structure where $\mathcal{Z} = \Delta_1 \times ... \times \Delta_n$, $\mathcal{L} = \mathbb{C}^n = \mathbb{R}^{2n}$ and $\|\delta\zeta\| = \max_{1\le k\le n} |\delta\zeta_k|$.

The GRC of (3.3.3) is obtained by replacing every one of the constraints in (3.3.3) with the corresponding constraint (3.3.8.a), (which in fact gives a system of linear constraints on the design variables), and adding the constraint (3.3.7.b) "responsible" for the global sensitivity w.r.t. the actuation errors (this constraint is common for all pairs (3.3.8) coming from different constraints of (3.3.3)). The resulting GRC of (3.3.3) has a linear objective, a bunch of linear constraints and a single conic quadratic constraint (3.3.7.b).[3]

Robust design. For illustrative purposes, we built two robust designs: the first (referred to as "RC design") is obtained when immunizing against 1% positioning errors while taking no care of actuation errors, (which is equivalent to setting $\epsilon = 0.01$ and $\alpha = \infty$ in the GRC); the actual global sensitivity of the resulting design to the actuation errors is $\alpha = 9.404$. The second ("GRC") design is obtained when keeping $\epsilon = 0.01$ and setting the global sensitivity α to the value 3. The performance of all three designs (the nominal, the RC and the GRC) is presented in table 3.1, and illustrated in figure 3.1. We see that "immunizing" the design against positioning errors of magnitude 0.01 is rather costly in terms of the objective: the RC sidelobe level is 0.106 vs. the nominal sidelobe level 0.0025, and the energy concentration for the RC design, evaluated at the nominal data, is 84%, which is by 16% worse than the 99.99% concentration for the nominal design. Good news, however, is that while the nice performance of the nominal design is a matter of pure

---

[3]We could further approximate $|z_k|$ in (3.3.7.b) by $p_{12}(z_k)$, thus ending up with a GRC that is equivalent to an LO program.

imagination — it is completely destroyed by positioning/actuation errors as low as 0.01%, the performance of the RC design is fully immunized against positioning errors with $\epsilon = 0.01$ and is incomparably less sensitive to the actuation errors than the performance of the nominal design (the corresponding global sensitivities are 9.4 and 3.0e6, respectively). However, the performance of the RC design is still too sensitive to actuation errors: when the level ρ of these errors grows from 0 to 0.03, the sidelobe level jumps, on the average, from 0.11 to 0.28, and the energy concentration drops from 0.84 to 0.46. The GRC design is aimed to moderate this phenomenon; here we require the sensitivity to actuation errors to be at most 3 (vs. 9.4 yielded by the RC design). As a result, we again lose in optimality: for the GRC design, without data perturbations the sidelobe level is 0.15, and the energy concentration is 70% (vs. 0.11 and 84% for the RC design). As a compensation, the GRC design, in contrast to the RC one, is well immunized against the actuation errors: with a level of these errors 3%, (i.e., $\rho = 0.03$), the performance of the GRC design is essentially the same as when there are no actuation errors at all, (i.e., $\rho = 0$).

3.4 EXERCISES

Exercise 3.1. Consider a situation as follows. A factory consumes n types of raw materials, coming from n different suppliers, to be decomposed into m pure components. The per unit content of component i in raw material j is $p_{ij} \geq 0$, and the necessary per month amount of component i is a given quantity $b_i \geq 0$. You need to make a long-term arrangement on the amounts of raw materials x_j coming every month from each of the suppliers, and these amounts should satisfy the system of linear constraints

$$Px \geq b, \ P = [p_{ij}].$$

The current per unit price of product j is c_j; this price, however, can vary in time, and from the history you know the volatilities $v_j \geq 0$ of the prices. How to choose x_j's in order to minimize the total cost of supply at the current prices, given an upper bound α on the sensitivity of the cost to possible future drifts in prices?

Test your model on the following data:

$$n = 32, m = 8, p_{ij} \equiv 1/m, b_i \equiv 1.e3,$$
$$c_j = 0.8 + 0.2\sqrt{((j-1)/(n-1))}, v_j = 0.1(1.2 - c_j),$$

and build the tradeoff curve "supply cost with current prices vs. sensitivity."

3.5 NOTES AND REMARKS

NR 3.1. The theoretical results of this chapter originate from [15, 16]. The nominal Antenna Design model goes back to [72].

Chapter Four

More on Safe Tractable Approximations of Scalar Chance Constraints

This chapter can be treated as an "advanced extension" of chapter 2. The entity of our major interest is the chance constrained version

$$p(z) \equiv \mathrm{Prob}\left\{ z_0 + \sum_{\ell=1}^{L} z_\ell \zeta_\ell > 0 \right\} \leq \epsilon \tag{4.0.1}$$

of a randomly perturbed linear inequality

$$z_0 + \sum_{\ell=1}^{L} z_\ell \zeta_\ell \leq 0, \tag{4.0.2}$$

where ζ_ℓ are random perturbations and z_ℓ are deterministic parameters (in applications in Uncertain Linear Optimization, these parameters will be specified as affine functions of the decision variables).

4.1 ROBUST COUNTERPART REPRESENTATION OF A SAFE CONVEX APPROXIMATION TO A SCALAR CHANCE CONSTRAINT

Recall that a *safe approximation* of the chance constraint (4.0.1) is a system $\mathcal{S}$ of convex constraints in variables z and perhaps additional variables u such that the projection $Z[\mathcal{S}]$ of the feasible set of the system onto the space of z variables is contained in the feasible set of the chance constraint (cf. Definition 2.2.1). In chapter 2, we dealt with a particular approximation scheme (to be revisited in section 4.2 below) that resulted in a *Robust Counterpart* type approximation:

$$Z[\mathcal{S}] = \{ z = [z_0; ...; z_L] : z_0 + \sum_{\ell=1}^{L} \zeta_\ell z_\ell \leq 0 \quad \forall (\zeta \in \mathcal{Z}_\epsilon) \},$$

where $\mathcal{Z}_\epsilon$ is an explicitly given convex compact set. In other words, *the approximating constraint requires from z to be robust feasible for the uncertain constraint (4.0.2) equipped with a properly defined "artificial" uncertainty set $\mathcal{Z}_\epsilon$.*

We are about to demonstrate that the "Robust Counterpart representability" of a safe convex approximation to (4.0.1) is a common property of a wide spectrum of approximations of the chance constraint, rather than being a specific property of the approximations considered in chapter 2.

We start with observing that the true feasible set Z_* of the chance constraint (4.0.1) possesses the following properties:

i) Z_* is a *conic* set, meaning that $0 \in Z_*$ and $\lambda z \in Z_*$ whenever $z \in Z_*$ and $\lambda \geq 0$;

ii) Z_* is closed;

iii) the set $Z[z_0] = \{z = [z_0; z_1; ...; z_L] : \|[z_1; ...; z_L]\|_2 \leq 1\}$ with $z_0 < 0$ and large enough $|z_0|$ is contained in Z_*;

iv) the set $Z[z_0]$ with large enough $z_0 > 0$ does not intersect Z_*.

For a safe convex approximation $\mathcal{S}$ of (4.0.1), the set $Z[\mathcal{S}]$ always inherits property *iv* of Z_*. We introduce the following

Definition 4.1.1. A safe convex approximation $\mathcal{S}$ of (4.0.1) is called normal, if $Z[\mathcal{S}]$ inherits properties *i-iii* (and thus all four properties) of Z_*.

Remark 4.1.2. For a safe convex approximation $\mathcal{S}$ of (4.0.1), the set $Z = Z[\mathcal{S}]$ is convex. From this observation it immediately follows that the normality of a safe approximation $\mathcal{S}$ is equivalent to the fact that $Z = Z[\mathcal{S}] \subset Z_*$ is a closed convex cone with $e \equiv [-1; 0; ...; 0] \in \mathrm{int} Z$ and $-e \notin Z$.

Our interest in normal safe approximations of chance constraints stems from the following simple observation:

Proposition 4.1.3. Let $\mathcal{S}$ be a normal safe convex approximation of the chance constraint (4.0.1). Then the approximation is Robust Counterpart representable: there exists a convex compact uncertainty set $\mathcal{Z}$ such that

$$Z[\mathcal{S}] = \{z : z_0 + \sum_{\ell=1}^{L} \zeta_\ell z_\ell \leq 0 \,\forall \zeta \in \mathcal{Z}\}. \tag{4.1.1}$$

Proof. Let $\mathcal{S}$ be normal. As we have seen, the set $Z = Z[\mathcal{S}]$ is a closed convex cone with $e \in \mathrm{int} Z$ and $-e \notin Z$. As every closed convex cone, Z is the anti-dual of its anti-dual cone Z_-:

$$Z = \{z \in \mathbb{R}^{L+1} : z^T \zeta \leq 0 \,\forall \zeta \in Z_-\}, \quad Z_- = \{\zeta \in \mathbb{R}^{L+1} : \zeta^T z \leq 0 \,\forall z \in Z\}.$$

Since Z has a nonempty interior, Z_- is a closed pointed convex cone, and since $e \in \mathrm{int} Z$, the set $\widehat{\mathcal{Z}} = \{[y_0; ...; y_L] \in Z_- : e^T y = -1\} = \{y = [1; y_1; ...; y_L] \in Z_-\}$ is a convex compact set that intersects every nontrivial ray from Z_-. It follows that the set $\mathcal{Z} = \{\zeta \in \mathbb{R}^L : [1; \zeta] \in Z_-\}$ is a convex compact set and that

$$Z = \{z : z^T y \leq 0 \,\forall y \in Z_-\} = \{z : z^T [1; \zeta] \leq 0 \,\forall \zeta \in \widehat{\mathcal{Z}}\}$$
$$= \{z = [z_0; z_1; ...; z_L] : z_0 + \sum_{\ell=1}^{L} \zeta_\ell z_\ell \leq 0 \,\forall \zeta \in \mathcal{Z}\}. \qquad \square$$

Proposition 4.1.3 shows that a natural safe approximation of a chance constrained uncertain linear inequality is nothing but the RC of this uncertain inequality associated with the appropriate convex compact uncertainty set. When the approximation in question is tractable, so is this uncertainty set (modulo mild regularity assumptions). What is on our agenda now, is to introduce a number of specific safe approximation schemes.

4.2 BERNSTEIN APPROXIMATION OF A CHANCE CONSTRAINT

This approximation scheme is closely related to the one we used in section 2.4, and it is instructive to start our acquaintance with the Bernstein approximation by reviewing this scheme.

Our goal is to build a safe approximation of the chance constraint (4.0.1). To achieve this goal:

1) We assume that ζ_ℓ, $\ell = 1, ..., L$, are independent with distributions P_ℓ such that

$$\int \exp\{ts\}dP_\ell(s) \leq \exp\{\max[\mu_\ell^+ t, \mu_\ell^- t] + \frac{1}{2}\sigma_\ell^2 t^2\} \ \forall t \in \mathbb{R},$$

whence

$$\ln\left(\mathbf{E}\left\{\exp\{w^T\zeta\}\right\}\right) \leq \Phi(w) = \sum_{\ell=1}^{L}\left[\max[\mu_\ell^- w_\ell, \mu_\ell^+ w_\ell] + \frac{1}{2}\sum_{\ell}\sigma_\ell^2 w_\ell^2\right];$$
$$(4.2.1)$$

2) We have inferred from (4.2.1) that the validity of the relation

$$z \in Z_\epsilon = \text{cl}\left\{z : \exists \alpha > 0 : \alpha z_0 + \Phi(\alpha[z_1; ...; z_L]) \leq \ln(\epsilon)\right\} \qquad (4.2.2)$$

is a sufficient condition for the validity of (4.0.1) (statement $(*)$ in section 2.4);

3) Finally, we derived from 2) that the sufficient condition in question is equivalent to the robust feasibility of $[z_0; ...; z_L]$ for the uncertain linear inequality

$$z_0 + \sum_{\ell=1}^{L}\zeta_\ell z_\ell \leq 0$$

with an appropriately chosen perturbation set $\mathcal{Z}$ (statement $(**)$ in section 2.4).

We are about to demonstrate that the outlined approximation scheme is a particular case of a more general one, a *Bernstein approximation*.

4.2.1 Bernstein Approximation: Basic Observation

Assume that

Q.1. The distribution P of the random perturbation $\zeta = [\zeta_1; ...; \zeta_L]$ in (4.0.1) is such that

$$\ln\left(\mathbf{E}\left\{\exp\{w^T\zeta\}\right\}\right) \leq \Phi(w) \qquad (4.2.3)$$

for some known convex function Φ that is finite everywhere on $\mathbb{R}^L$ and satisfies $\Phi(0) = 0$.

Example 4.2.1. It is immediately seen that under assumptions P.1–2 from section 2.4 (or, which is the same, under the assumptions from item 1 above), relation (4.2.3) is satisfied with

$$\Phi(w) = \sum_{\ell=1}^{L} \left[\max[\mu_\ell^+ w_\ell, \mu_\ell^- w_\ell] + \frac{1}{2}\sigma_\ell^2 w_\ell^2 \right]. \tag{4.2.4}$$

Given $\epsilon \in (0, 1)$, let us set

$$\begin{aligned} Z_\epsilon^o &= \{z = [z_0; w] \in \mathbb{R}^{L+1} : \exists \alpha > 0 : \alpha z_0 + \Phi(\alpha w) \leq \ln(\epsilon)\}, \\ Z_\epsilon &= \operatorname{cl} Z_\epsilon^o. \end{aligned} \tag{4.2.5}$$

The *Bernstein approximation* scheme is given by the following statement:

Proposition 4.2.2. Under assumption (4.2.3), Z_ϵ is exactly the solution set of the convex inequality

$$\inf_{\beta > 0} \left[z_0 + \beta \Phi(\beta^{-1} w) + \beta \ln(1/\epsilon) \right] \leq 0, \tag{4.2.6}$$

and this convex inequality is a normal safe approximation of the chance constraint (4.0.1).

For the proof, see section B.1.1.

Our current goal is to develop a scheme that, under favorable circumstances, allows us to describe efficiently the set $\mathcal{Z}_\epsilon$.

4.2.2 Bernstein Approximation: Dualization

In addition to Q.1, let us assume that,

Q.2. We can represent the convex function Φ participating in (4.2.3) in the following form:

$$\Phi(w) = \sup_u \left\{ w^T (Au + a) - \phi(u) \right\}, \tag{4.2.7}$$

where

- $\phi(u)$ is a convex and lower semicontinuous function on $\mathbb{R}^M$ taking real values and the value $+\infty$,

- $u \mapsto Au + a$ is an affine mapping from $\mathbb{R}^M$ into $\mathbb{R}^L$,

- every level set $U_c = \{u : \phi(u) \leq c\}$, $c \in \mathbb{R}$, of ϕ is bounded.

Note that *every* convex and finite everywhere on $\mathbb{R}^L$ function $\Phi(\cdot)$ admits a required representation, e.g., the representation with

$$\begin{aligned} Au + a &\equiv u, \\ \phi(u) &= \sup_w \left\{ u^T w - \Phi(w) \right\}. \end{aligned} \tag{4.2.8}$$

Indeed, it is known that the function ϕ given by the latter relation — the *Legendre transformation* (or the *Fenchel dual*, or the *conjugate*) of Φ — is a convex and lower semicontinuous function such that

$$\Phi(w) = \sup_u \left\{ w^T u - \phi(u) \right\}. \tag{4.2.9}$$

As for the requirement on ϕ to have bounded level sets, this requirement, in the case of (4.2.8) is readily given by the fact that Φ is everywhere finite. This is implied by the following well-known fact:

Proposition 4.2.3. Let $\Phi(\cdot)$, $\phi(\cdot)$ be linked by (4.2.7), Φ be everywhere defined, and ϕ be lower semicontinuous. Assume also that A has trivial kernel. Then the level sets of ϕ are bounded.

For the proof, see section B.1.2.

Example 4.2.4. [Example 4.2.1 continued] The function (4.2.4) admits an explicit representation of the form (4.2.7), specifically, as follows:

$$\begin{aligned}
\Phi(w) &\equiv \sum_{\ell=1}^{L} \left[\max[\mu_\ell^- w_\ell, \mu_\ell^+ w_\ell] + \tfrac{1}{2}\sigma_\ell^2 w_\ell^2 \right] \\
&= \sup_{u=\{u^\ell, u_\ell\}_{\ell=1}^L} \left\{ \sum_\ell \left[w_\ell(u^\ell + u_\ell) - \tfrac{u_\ell^2}{2\sigma_\ell^2} \right] : \begin{array}{l} \mu_\ell^- \leq u^\ell \leq \mu_\ell^+, \\ 1 \leq \ell \leq L \end{array} \right\},
\end{aligned}$$

that is,

$$\begin{aligned}
(Au + a)_\ell &= u_\ell + u^\ell, \\
\phi(u) &= \sum_{\ell=1}^{L} \phi_\ell(u^\ell, u_\ell), \\
\phi_\ell(u^\ell, u_\ell) &= \begin{cases} \tfrac{1}{2\sigma_\ell^2} u_\ell^2, & \mu_\ell^- \leq u^\ell \leq \mu_\ell^+ \\ +\infty, & \text{otherwise}. \end{cases}
\end{aligned} \tag{4.2.10}$$

4.2.3 Bernstein Approximation: Main Result

The outlined assumptions give rise to the following result:

Theorem 4.2.5. Consider the chance constrained inequality (4.0.1) and assume that the distribution P of the random vector ζ satisfies Assumptions Q.1–2. Then the set

$$\mathcal{U}_\epsilon = \{u : \phi(u) \leq \ln(1/\epsilon)\}$$

is a nonempty convex compact set, and for the set Z_ϵ given by (4.2.5) one has

$$z \equiv [z_0; w] \in Z_\epsilon \Leftrightarrow z_0 + \zeta^T w \leq 0 \; \forall \zeta \in \mathcal{Z}_\epsilon \equiv A\mathcal{U}_\epsilon + a. \tag{4.2.11}$$

In particular, the condition

$$z \equiv [z_0; w] \in Z_\epsilon \Leftrightarrow \inf_{\beta > 0} \left[z_0 + \beta \Phi(\beta^{-1} w) + \beta \ln(1/\epsilon) \right] \leq 0$$

which, by Proposition 4.2.2, is a sufficient condition for z to satisfy the chance constraint (4.0.1), is nothing but the condition that z is a robust feasible solution to the uncertain linear constraint

$$z_0 + \sum_{\ell=1}^{L} \zeta_\ell w_\ell \leq 0,$$

in variables z_0, w, the perturbation set being the set $\mathcal{Z}_\epsilon$ given by (4.2.11).

For the proof, see section B.1.3.

The uncertainty set $\mathcal{Z}_\epsilon$ in (4.2.11) is defined in terms of two entities: the function $\Phi(\cdot)$ and its representation (4.2.7), and the data of this representation is not uniquely defined by Φ. Nevertheless, the set $\mathcal{Z}_\epsilon$ depends solely on $\Phi(\cdot)$. Indeed, the set $\mathcal{Z}_\epsilon$ is defined solely in terms of Φ, and therefore the support function $\Theta(w) = \max_{\zeta \in \mathcal{Z}_\epsilon} \zeta^T w$ of the compact convex set $\mathcal{Z}_\epsilon$ can be expressed solely in terms of Φ, since

$$\Theta(w) = -\sup\{z_0 : [z_0; w] \in Z_\epsilon\}$$

by (4.2.11). It remains to recall that the support function of a closed nonempty convex set completely determines this set [100, Section 13].

The bottom line is as follows:

Every convex upper bound $\Phi(w) : \mathbb{R}^L \to \mathbb{R}$, $\Phi(0) = 0$, *on the logarithmic moment-generating function*

$$\ln\left(\mathbf{E}\left\{\exp\{w^T \zeta\}\right\}\right)$$

of the distribution of random perturbation vector ζ *implies a safe normal approximation of the chance constrained inequality* (4.0.1). *Under Assumption Q.2, this approximation has the form* $z_0 + \max_{\zeta \in \mathcal{Z}_\epsilon} \sum_{\ell=1}^{L} \zeta_\ell z_\ell \leq 0$ *where* $\mathcal{Z}_\epsilon$ *is a properly defined nonempty convex compact set. This approximation is the RC of the uncertain inequality* $z_0 + \sum_{\ell=1}^{L} \zeta_\ell z_\ell \leq 0$ *with* $\mathcal{Z}_\epsilon$ *in the role of the perturbation set.*

The latter result extends those in section 2.4, where we restricted Φ to be of the specific form (4.2.1).

Example 4.2.6. [Example 4.2.4 continued] Observe that under assumptions P.1–2 Theorem 4.2.5 recovers Theorem 2.4.3.

4.2.4 Bernstein Approximation: Examples

Example 4.2.7. [Gaussian case] Assume that all we know about the random vector ζ is that its entries ζ_ℓ are independent Gaussian random variables with means μ_ℓ belonging to given segments $[\mu_\ell^-, \mu_\ell^+]$ and variances belonging to given segments $[(\sigma_\ell^-)^2, (\sigma_\ell^+)^2]$, $\ell = 1, ..., L$. In this case it is immediately seen that:

(i) The best (that is, the smallest) function $\Psi(w)$ satisfying Q.1 w.r.t. all distributions of ζ compatible with the above information is

$$\Phi(w) = \sum_{\ell=1}^{L} \left[\max[\mu_\ell^- w_\ell, \mu_\ell^+ w_\ell] + \frac{(\sigma_\ell^+)^2}{2} w_\ell^2 \right] ;$$

(ii) Setting

$$
\begin{aligned}
\phi(u) &= \sum_{\ell=1}^{L} \phi_\ell(u^\ell, u_\ell), \\
\phi_\ell(u^\ell, u_\ell) &= \begin{cases} \frac{1}{2(\sigma_\ell^+)^2} u_\ell^2, & \mu_\ell^- \leq u^\ell \leq \mu_\ell^+ \\ +\infty, & \text{otherwise} \end{cases}, \\
(Au + a)_\ell &= u^\ell + u_\ell
\end{aligned}
$$

we ensure the validity of Q.2 for the function Φ given in (i);

(iii) The Bernstein approximation of the chance constraint (4.0.1) associated with the outlined data is exactly the approximation given by Theorem 2.4.3, cf. Example 2.4.5.

Example 4.2.8. Assume that all we know about random perturbations ζ_ℓ is that they are independent, take values in $[-1, 1]$ and their means belong to given sub-segments $[\mu_\ell^-, \mu_\ell^+]$ of $[-1, 1]$ (cf. Example 2.4.9). As stated in Example 2.4.9, the best (the smallest) function $\Phi(\cdot)$ satisfying Q.1 w.r.t. all distributions of ζ compatible with the above information is

$$\Phi(w) = \sum_{\ell=1}^{L} \underbrace{\ln \left(\max_{\mu \in [\mu_\ell^-, \mu_\ell^+]} [\cosh(w_\ell) + \mu \sinh(w_\ell)] \right)}_{\Phi_\ell(w_\ell)}. \tag{4.2.12}$$

It is easily seen that

$$\Phi_\ell(w_\ell) = \max_{\mu_\ell^- \leq \mu \leq \mu_\ell^+} \ln \left(\exp \left\{ w_\ell + \ln \left(\frac{1+\mu}{2} \right) \right\} + \exp \left\{ -w_\ell + \ln \left(\frac{1-\mu}{2} \right) \right\} \right).$$

This relation, due to an immediate equality

$$\ln \left(\exp\{x_1\} + \ldots + \exp\{x_n\} \right) = \max_y \left\{ x^T y - \sum_{i=1}^{n} y_i \ln y_i : y \geq 0, \sum_i y_i = 1 \right\}, \tag{4.2.13}$$

implies that

$$\Phi_\ell(w_\ell) = \max_{-1 \leq u_\ell \leq 1} \left\{ w_\ell u_\ell - \phi_\ell(u_\ell) \right\}$$

with

$$
\phi_\ell(u_\ell) = \begin{cases} \frac{1}{2} \left[(1 + u_\ell) \ln \left(\frac{1+u_\ell}{1+\mu_\ell^-} \right) + (1 - u_\ell) \ln \left(\frac{1-u_\ell}{1-\mu_\ell^-} \right) \right] & , -1 \leq u_\ell \leq \mu_\ell^- \\ 0 & , \mu_\ell^- \leq u_\ell \leq \mu_\ell^+ \\ \frac{1}{2} \left[(1 + u_\ell) \ln \left(\frac{1+u_\ell}{1+\mu_\ell^+} \right) + (1 - u_\ell) \ln \left(\frac{1-u_\ell}{1-\mu_\ell^+} \right) \right] & , \mu_\ell^+ \leq u_\ell \leq 1 \end{cases}. \tag{4.2.14}
$$

It follows that setting

$$\phi(u) = \sum_\ell \phi_\ell(u_\ell), \ Au + a \equiv u,$$

with $\phi_\ell(\cdot)$ given by (4.2.14), we arrive at a representation, required in Q.2, of the function $\Phi(w)$ given by (4.2.12). The perturbation set $\mathcal{Z}_\epsilon$ given by the Bernstein approximation associated with the outlined data is

$$\mathcal{Z}_\epsilon = \left\{ \zeta : \sum_{\ell=1}^{L} \phi_\ell(\zeta_\ell) \leq \ln(1/\epsilon) \right\}.$$

Example 4.2.9. Consider the situation described in Example 4.2.8 and assume that $\mu_\ell^\pm = 0$ for all ℓ, (i.e., ζ_ℓ are independent random variables with zero means taking values in $[-1, 1]$). Let us compare the safe approximation of the chance constraint (4.0.1) given in Example 2.4.9 and the Bernstein approximation of this constraint by the following experiment:

i) We set $z_1 = ... = z_L = 1$, assume that the "true" (unknown when the approximation is built) distribution of ζ is the uniform distribution on the vertices of the unit box, and consider the chance constrained optimization problem

$$\mathrm{Opt}^+(\epsilon) = \max_{z_0} \left\{ z_0 : \mathrm{Prob}\{ z_0 + \sum_{\ell=1}^{L} \zeta_\ell z_\ell > 0 \} \leq \epsilon \right\}; \qquad (P)$$

ii) Replacing the chance constraint in (P) with its safe approximation, we replace (P) with a tractable approximating problem with easily computable optimal value, that is a lower bound on $\mathrm{Opt}(\epsilon)$. Let us plot and compare to each other and to the true optimal value of (P), (which is easily computable when $z_1 = ... = z_L$), the dependencies of the resulting lower bounds on $\mathrm{Opt}(\epsilon)$ as functions of ϵ for the following approximations:

- the one given by Example 2.4.9 and Theorem 2.4.3 (Approximation I);
- the refinement of the latter approximation given by Theorem 2.4.4 instead of Theorem 2.4.3 (Approximation II);
- the Bernstein approximation as given by Example 4.2.8 (Approximation III).

Note that Approximation I is nothing but the Ball RC approximation of our chance constraint, while Approximation II is simultaneously its Ball-Box and its Budgeted RC approximation, the budget, as defined in section 2.3, being $\sqrt{2L\ln(1/\epsilon)}$.

In the experiments, we use $L = 16$ and $L = 64$ and scan the range $10^{-12} \leq \epsilon \leq 10^{-1}$. The results of the experiments are as follows:

A. The optimal values of the approximations are given by

$$
\begin{aligned}
\mathrm{Opt}_I(\epsilon) &= -\sqrt{2L\ln(1/\epsilon)}; \\
\mathrm{Opt}_{II}(\epsilon) &= -\min_w \left\{ \sum_{\ell=1}^{L} |1 - w_\ell| + \sqrt{2\ln(1/\epsilon)} \|w\|_2 \right\} \\
&= -\min_{0 \leq s \leq 1} \left\{ L(1 - s) + \sqrt{2L\ln(1/\epsilon)}s \right\} \\
&= \max\left[-L, -\sqrt{2L\ln(1/\epsilon)} \right]; \\
\mathrm{Opt}_{III}(\epsilon) &= -\max_{u:\|u\|_\infty \leq 1} \left\{ \sum_{\ell=1}^{L} u_\ell : \right. \\
&\qquad \left. \sum_{\ell=1}^{L} [(1 + u_\ell)\ln(1 + u_\ell) + (1 - u_\ell)\ln(1 - u_\ell)] \leq 2\ln(1/\epsilon) \right\} \\
&= -\max_{0 \leq s \leq 1} \left\{ Ls : L[(1 + s)\ln(1 + s) + (1 - s)\ln(1 - s)] \right. \\
&\qquad\qquad \left. \leq 2\ln(1/\epsilon) \right\}.
\end{aligned}
$$

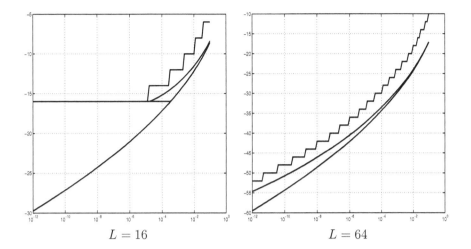

$$L = 16 \qquad\qquad L = 64$$

Figure 4.1   Plots of $\mathrm{Opt}_I(\epsilon) - \mathrm{Opt}_{III}(\epsilon)$ and $\mathrm{Opt}^+(\epsilon)$. Top to bottom: $\mathrm{Opt}^+(\epsilon)$, $\mathrm{Opt}_{III}(\epsilon)$, $\mathrm{Opt}_{II}(\epsilon)$, $\mathrm{Opt}_I(\epsilon)$ (the latter two curves are undistinguishable when $L = 64$).

B. The plots of $\mathrm{Opt}_I(\cdot) - \mathrm{Opt}_{III}(\cdot)$ and $\mathrm{Opt}^+(\cdot)$ are presented in figure 4.1. It is seen that in this example Approximation III is the best.

Example 4.2.10. Consider the case where all we know about ζ_ℓ is that these random variables are independent with zero mean, take values in $[-1, 1]$ and the variance of ζ_ℓ does not exceed ν_ℓ^2, $0 < \nu_\ell \leq 1$. Invoking Example 2.4.11 (where one should set $\mu = 0$), the best (the smallest) function Φ satisfying Q.1 is

$$\Phi(w) = \sum_{\ell=1}^{L} \Phi_\ell(w_\ell),$$

where

$$\exp\{\Phi_\ell(w_\ell)\} = \begin{cases} \frac{\exp\{-w_\ell \nu_\ell^2\} + \nu_\ell^2 \exp\{w_\ell\}}{1 + \nu_\ell^2}, & w_\ell \geq 0 \\ \frac{\exp\{w_\ell \nu_\ell^2\} + \nu_\ell^2 \exp\{-w_\ell\}}{1 + \nu_\ell^2}, & w_\ell \leq 0 \end{cases}$$

$$= \max\left[\frac{\exp\{-w_\ell \nu_\ell^2\} + \nu_\ell^2 \exp\{w_\ell\}}{1 + \nu_\ell^2}, \frac{\exp\{w_\ell \nu_\ell^2\} + \nu_\ell^2 \exp\{-w_\ell\}}{1 + \nu_\ell^2}\right]$$

$$= \max_{0 \leq \lambda \leq 1} \frac{\lambda \exp\{-w_\ell \nu_\ell^2\} + (1-\lambda) \exp\{w_\ell \nu_\ell^2\} + \nu_\ell^2 \lambda \exp\{w_\ell\} + \nu_\ell^2 (1-\lambda) \exp\{-w_\ell\}}{1 + \nu_\ell^2}$$

Let us show that

$$\Phi_\ell(w_\ell) = \sup_{u^\ell} \left\{ w_\ell \alpha_\ell^T u^\ell - \phi_\ell(u^\ell) \right\},$$

where

$$u^\ell = [u_1^\ell; ...; u_4^\ell] \in \mathbb{R}^4,$$

$$\alpha_\ell^T u^\ell = \nu_\ell^2 (u_3^\ell - u_1^\ell) + (u_2^\ell - u_4^\ell)$$

$$\phi_\ell(u^\ell) = \begin{cases} u_1^\ell \ln u_1^\ell + u_2^\ell \ln u_2^\ell + u_3^\ell \ln u_3^\ell + u_4^\ell \ln u_4^\ell \\ \quad -(u_1^\ell + u_2^\ell) \ln(u_1^\ell + u_2^\ell) - (u_3^\ell + u_4^\ell) \ln(u_3^\ell + u_4^\ell) \\ \quad - \ln(\nu_\ell^2)(u_2^\ell + u_4^\ell) + \ln(1 + \nu_\ell^2) & , u^\ell \in \Delta, \\ +\infty & , \text{otherwise} \end{cases}$$

with $\Delta = \{u \in \mathbb{R}^4 : u \geq 0, \sum_{i=1}^{4} u_i = 1\}$.

Indeed, w.l.o.g. we may assume that $L = 1$, which allows to skip index ℓ. We have

$$\exp\{\Phi(w)\} = \max_{0 \leq \lambda \leq 1} \frac{\lambda \exp\{-w\nu^2\} + (1-\lambda)\exp\{w\nu^2\} + \nu^2 \lambda \exp\{w\} + \nu^2(1-\lambda)\exp\{-w\}}{1+\nu^2}.$$

Invoking (4.2.13), we proceed as follows:

$$\begin{aligned}
\Rightarrow \Phi(w) &= \ln \max_{0 \leq \lambda \leq 1} \Big\{ \exp\{-w\nu^2 + \ln(\lambda) - \ln(1+\nu^2)\} \\
&\quad + \exp\{w + \ln(\lambda) + \ln(\nu^2) - \ln(1+\nu^2)\} \\
&\quad + \exp\{w\nu^2 + \ln(1-\lambda) - \ln(1+\nu^2)\} \\
&\quad + \exp\{-w + \ln(1-\lambda) + \ln(\nu^2) - \ln(1+\nu^2)\} \Big\}
\end{aligned}$$

$$\begin{aligned}
\Rightarrow \Phi(w) &= \sup_{0 \leq \lambda \leq 1} \sup_{u \in \Delta} \Big\{ [-w\nu^2 + \ln(\lambda) - \ln(1+\nu^2)]u_1 \\
&\quad + [w + \ln(\lambda) + \ln(\nu^2) - \ln(1+\nu^2)]u_2 \\
&\quad + [w\nu^2 + \ln(1-\lambda) - \ln(1+\nu^2)]u_3 \\
&\quad + [-w + \ln(1-\lambda) + \ln(\nu^2) - \ln(1+\nu^2)]u_4 \\
&\quad - \sum_{i=1}^{4} u_i \ln u_i \Big\} \\
&= \sup_{u \in \Delta} \sup_{0 \leq \lambda \leq 1} \Big\{ w[-\nu^2 u_1 + \nu^2 u_3 + u_2 - u_4] \\
&\quad + [u_1 + u_2]\ln(\lambda) + [u_3 + u_4]\ln(1-\lambda) \\
&\quad + \ln(\nu^2)(u_2 + u_4) - \sum_{i=1}^{4} u_i \ln u_i - \ln(1+\nu^2) \Big\} \\
&= \sup_{u \in \Delta} \Big\{ w[-\nu^2 u_1 + \nu^2 u_3 + u_2 - u_4] \\
&\quad + (u_1 + u_2)\ln(u_1 + u_2) + (u_3 + u_4)\ln(u_3 + u_4) \\
&\quad + \ln(\nu^2)(u_2 + u_4) - \sum_{i=1}^{4} u_i \ln u_i - \ln(1+\nu^2) \Big\} \\
&\quad [\text{the optimal } \lambda \text{ is } u_1 + u_2 = 1 - (u_3 + u_4)]
\end{aligned}$$

QED.

Thus, in Example 4.2.10 the uncertainty set $\mathcal{Z}_\epsilon$ associated with the Bernstein approximation is

$$\mathcal{Z}_\epsilon = \Big\{ [\nu_1^2(u_3^1 - u_1^1) + (u_2^1 - u_4^1); ...; \nu_L^2(u_3^L - u_1^L) + (u_2^L - u_4^L)] :$$
$$\sum_{\ell=1}^{L} \phi_\ell(u^\ell) \leq \ln(1/\epsilon) \Big\}.$$

4.3 FROM BERNSTEIN APPROXIMATION TO CONDITIONAL VALUE AT RISK AND BACK

The Bernstein approximation is a particular case of a conceptually simple approach to bounding the probability of a random variable to be positive. The approach is as follows.

4.3.1 Generating Function Based Approximation Scheme

Consider the random variable

$$\xi^z = z_0 + \sum_{\ell=1}^{L} \zeta_\ell z_\ell$$

(z is a deterministic vector of parameters, ζ_ℓ are random perturbations with well-defined expectations). To bound from above the quantity

$$p(z) = \mathrm{Prob}\{\xi^z > 0\},$$

we fix a *generating function* $\gamma(s)$ on the axis such that

$$\gamma(\cdot) \text{ is convex and nondecreasing, } \gamma(\cdot) \geq 0, \ \gamma(0) \geq 1, \gamma(s) \to 0, s \to -\infty. \quad (4.3.1)$$

Since γ is non-decreasing and is ≥ 1 at the origin, we have $\gamma(s) \geq 1$ for all $s \geq 0$; since, in addition, γ is nonnegative, the quantity $\Psi_*(z) \equiv \mathbf{E}\{\gamma(\xi^z)\}$ is an upper bound on $p(z)$:

$$p(z) \leq \Psi_*(z).$$

Note that $\Psi_*(z)$ is a convex function of z; from now on we assume that it is finite everywhere. Under this assumption, we have

$$\Psi_*(z + t \underbrace{[-1; 0; ...; 0]}_{e}) \to 0, t \to \infty.$$

Suppose we can find a convex function $\Psi(z)$ taking values in $\mathbb{R}$ such that

$$\forall z : \Psi_*(z) \equiv \mathbf{E}\{\gamma(z_0 + \sum_{\ell=1}^{L} \zeta_\ell z_\ell)\} \leq \Psi(z) \ \& \ \Psi(z + te) \to 0, \ t \to \infty, \quad (4.3.2)$$

so that $p(z) \leq \Psi(z)$ for all z. Since $p(z) = p(\alpha z)$ for all $\alpha > 0$, we have

$$\forall z : p(z) \leq \inf_{\alpha > 0} \Psi(\alpha z). \quad (4.3.3)$$

We have, essentially, established the following simple statement:

Proposition 4.3.1. Given $\epsilon \in (0, 1)$ and a generating function $\gamma(\cdot)$ satisfying (4.3.1), let $\Psi(z)$ be a finite convex function satisfying (4.3.2). Let us set

$$\Gamma_\epsilon^o = \{z : \exists \alpha > 0 : \Psi(\alpha z) \leq \epsilon\}, \ \Gamma_\epsilon = \mathrm{cl}\,\Gamma_\epsilon^o.$$

Then Γ_ϵ is the solution set of the convex inequality (cf. (4.2.6))

$$\inf_{\beta > 0} \left[\beta \Psi(\beta^{-1} z) - \beta \epsilon \right] \leq 0, \quad (4.3.4)$$

and this inequality is a safe normal approximation of the chance constraint (4.0.1).

For the proof, see section B.1.4.

Note that the Bernstein approximation is, essentially, a particular case of the approximation scheme associated with Proposition 4.3.1 corresponding to the choice $\gamma(s) = \exp\{s\}$. With this choice of the generating function, $\ln(\Psi_*(z))$ is a convex function, which allowed us to operate "in the logarithmic scale," that is, to start with a convex majorant $\Phi(z)$ on $\ln \Psi_*(z)$ and to work with the equivalent version

$$\forall (\alpha > 0, z) : \ln p(z) \leq \Phi(\alpha z)$$

of the bound (4.3.3).

4.3.2 Robust Counterpart Representation of Γ_ϵ

In addition to the premise of Proposition 4.3.1, assume that

$$\Psi(z) = \sup_u \left\{ z^T (Bu + b) - \psi(u) \right\}$$

with an appropriately chosen lower semicontinuous convex function $\psi(\cdot)$ possessing bounded level sets. Applying Theorem B.1.2, we conclude that for every $\epsilon \in (0, 1)$, the set

$$\mathcal{U}_\epsilon = \{ u : \psi(u) \leq -\epsilon \}$$

is a nonempty convex compact set, and

$$z \in \Gamma_\epsilon \Leftrightarrow z^T (Bu + b) \leq 0 \ \forall u \in \mathcal{U}_\epsilon.$$

In other words, Γ_ϵ is nothing but the robust feasible set of the uncertain linear inequality

$$\sum_{\ell=0}^{L} \zeta_\ell z_\ell \leq 0 \tag{4.3.5}$$

associated with the convex compact uncertainty set $\widehat{\mathcal{Z}}_\epsilon = \{ \zeta = Bu + b : u \in \mathcal{U}_\epsilon \}$.

   From an aesthetical viewpoint, a disadvantage of the Robust Counterpart representation of Γ_ϵ is that in (4.3.5) the coefficient of z_0 becomes uncertain rather than being equal to 1, as in all our previous results. This can be easily cured. Indeed, let

$$\widetilde{\mathcal{Z}}_\epsilon = \mathrm{cl} \left\{ \zeta = [1; z_1; ...; z_L] \in \mathbb{R}^{L+1} : \zeta = \alpha \zeta' \text{ with } \alpha > 0, \zeta' \in \widehat{\mathcal{Z}}_\epsilon \right\}.$$

We claim that $\widetilde{\mathcal{Z}}_\epsilon = \{ \zeta = [1; w] : w \in \mathcal{Z}_\epsilon \}$ with a compact convex set $\mathcal{Z}_\epsilon$ and that

$$\Gamma_\epsilon = \{ z : z_0 + \sum_{\ell=1}^{L} \zeta_\ell z_\ell \leq 0 \ \forall \zeta \in \mathcal{Z}_\epsilon \}, \tag{4.3.6}$$

that is, Γ_ϵ is the robust feasible set of the uncertain linear inequality

$$z_0 + \sum_{\ell=1}^{L} \zeta_\ell z_\ell \leq 0,$$

the uncertainty set being $\mathcal{Z}_\epsilon$. To justify our claim, observe, first, that the second relation in (4.3.2) combines with convexity of Ψ to imply that for every bounded set $U \subset \mathbb{R}^L$, all vectors $[-t; z] \in \mathbb{R}^{L+1}$ with $z \in U$ are contained in Γ_ϵ, provided that t is large enough. Since Γ_ϵ is a cone, it follows that $e = [-1; 0; ...; 0] \in \mathrm{int}\Gamma_\epsilon$. It follows that the only vector $\zeta \in \widehat{\mathcal{Z}}_\epsilon$ with $\zeta_0 \leq 0$, if such a vector exists, is the vector $\zeta = 0$. Indeed, we know that

$$\Gamma_\epsilon = \{ z : z^T \zeta \leq 0 \ \forall \zeta \in \widehat{\mathcal{Z}}_\epsilon \}. \tag{4.3.7}$$

Assuming that $\zeta_0 \leq 0$ for certain $0 \neq \zeta \in \widehat{\mathcal{Z}}_\epsilon$, we would conclude from (4.3.7) that $e = [-1; 0; ...; 0] \notin \mathrm{int}\Gamma_\epsilon$, which is not the case. Observe also that $\widehat{\mathcal{Z}}_\epsilon \neq \{0\}$; indeed, otherwise (4.3.7) would say that $\Gamma_\epsilon = \mathbb{R}^{L+1}$, which is not the case as we know that $-e \notin \Gamma_\epsilon$. The bottom line is that $\widehat{\mathcal{Z}}_\epsilon$ is a closed convex set contained in the half-space $\zeta_0 \geq 0$, intersecting with the boundary hyperplane $\zeta_0 = 0$ of this half-space, if at all, at the only point $\zeta = 0$ and not reducible to this point; as a result, the vectors $\zeta \in \widehat{\mathcal{Z}}_\epsilon$ with $\zeta_0 > 0$ are dense in $\widehat{\mathcal{Z}}_\epsilon$, which combines with (4.3.7) and the definition of $\widetilde{\mathcal{Z}}_\epsilon$ to imply that

$$\Gamma_\epsilon = \{z : z^T \zeta \geq 0 \; \forall \zeta \in \widetilde{\mathcal{Z}}_\epsilon\}. \tag{4.3.8}$$

Further, the set $\widetilde{\mathcal{Z}}_\epsilon$, by its construction, is convex (since $\widehat{\mathcal{Z}}_\epsilon$ is so), nonempty and closed. The only part of the claim that still is not justified is that $\widetilde{\mathcal{Z}}_\epsilon$ is bounded; but this is an immediate consequence of (4.3.8) and the fact that $e \in \text{int}\Gamma_\epsilon$. $\square$

4.3.3 Optimal Choice of the Generating Function and Conditional Value at Risk

A natural question is, how to choose the function $\gamma(\cdot)$. If the only criterion was the quality of the bound (4.3.3), the answer would be

$$\gamma(s) = \gamma_\sharp(s) \equiv \max[1 + s, 0] \tag{4.3.9}$$

or, which in our context is the same, $\gamma(s) = \max[1 + \alpha s, 0]$ with $\alpha > 0$ (note that the generating functions $\gamma(s)$ and $\gamma_\alpha(s) = \gamma(\alpha s)$, $\alpha > 0$, produce the same approximation Γ_ϵ). Indeed, let $\gamma(\cdot)$ be a generating function (that is, a function satisfying (4.3.1)), $\Psi(z)$ be a convex function such that

$$\forall z : \mathbf{E}\{\gamma(z_0 + \sum_{\ell=1}^{L} \zeta_\ell z_\ell)\} \leq \Psi(z),$$

and let

$$\Psi_\sharp(z) = \mathbf{E}\{\gamma_\sharp(z_0 + \sum_{\ell=1}^{L} \zeta_\ell z_\ell)\}.$$

We claim that $\Psi_\sharp(z)$ is a finite convex function on $\mathbb{R}^L$ and that

$$(a) \quad \inf_{\alpha>0} \Psi_\sharp(\alpha z) \leq \inf_{\alpha>0} \Psi(\alpha z), \quad (b) \quad \Gamma_\epsilon \subset \Gamma_\epsilon^\sharp,$$
$$\left[\begin{array}{l} \Gamma_\epsilon^o = \{z : \exists \alpha > 0 : \Psi(\alpha z) \leq \epsilon\}, \; \Gamma_\epsilon = \text{cl}\,\Gamma_\epsilon^o, \\ \Gamma_\epsilon^{o,\sharp} = \{z : \exists \alpha > 0 : \Psi_\sharp(\alpha z) \leq \epsilon\}, \; \Gamma_\epsilon^\sharp = \text{cl}\,\Gamma_\epsilon^{o,\sharp}, \end{array} \right] \tag{4.3.10}$$

so that the bound (4.3.3) associated with $\Psi_\sharp$ is at least as good as the bound associated with Ψ, and consequently the safe approximation $\Gamma_\epsilon^\sharp$ of the feasible set of (4.0.1) is no more conservative as the approximation Γ_ϵ.

> Indeed, ζ_ℓ have well defined expectations, so that $\Psi_\sharp$ is well defined. Since γ satisfies (4.3.1), we clearly have $\gamma'(+0) > 0$. Replacing $\gamma(s)$ with $\gamma(\beta s)$, $\beta > 0$ (and thus replacing $\Psi(z)$ with $\Psi(\beta z)$), we do not vary the right-hand side in (4.3.3); "scaling" γ in this fashion, we can enforce $\gamma'(+0) = 1$. In the latter case, we have $\gamma(s) \geq \gamma(0) + \gamma'(+0)s \geq 1 + s$ for all s (recall that γ is convex and $\gamma(0) \geq 1$). Since, in addition, $\gamma(\cdot) \geq 0$, we conclude that $\gamma(s) \geq \gamma_\sharp(s)$ for all s, whence also $\Psi_\sharp(z) \leq \Psi(z)$ for all z, which implies all the facts stated in (4.3.10).

For a given distribution P of the perturbation vector ζ, the "optimal" choice

$$\Psi_\sharp(z) = \mathbf{E}_{\zeta \sim P}\{\gamma_\sharp(z_0 + \sum_{\ell=1}^{L} \zeta_\ell z_\ell)\} \tag{4.3.11}$$

of Ψ is closely related to the *Conditional Value at Risk* $\mathrm{CVaR}_\epsilon(\xi^z)$ of the associated parametric random variable $\xi^z = z_0 + \sum_{\ell=1}^{L} \zeta_\ell z_\ell$. For a random variable ξ with well defined expectation and $\epsilon \in (0,1)$, the associated Conditional Value at Risk is defined as

$$\mathrm{CVaR}_\epsilon(\xi) = \inf_{a \in \mathbb{R}} \left[a + \frac{1}{\epsilon} \mathbf{E}\{\max[\xi - a, 0]\} \right]; \tag{4.3.12}$$

it is well-known that the inf in the right hand side of this relation is attained, and that $\mathrm{Prob}\{\xi > \mathrm{CVaR}_\epsilon(\xi)\} \leq \epsilon$. Besides this, if ξ is of the parametric form $\xi = \xi^z \equiv z_0 + \sum_{\ell=1}^{L} \zeta_\ell z_\ell$ and all ζ_ℓ have well defined expectations, then $\mathrm{CVaR}_\epsilon(\xi^z)$ is convex in z, so that the relation

$$\mathrm{CVaR}_\epsilon(\xi^z) \leq 0 \tag{4.3.13}$$

is a convex inequality in z, and its validity is a sufficient condition for $\mathrm{Prob}\{\xi^z > 0\} \leq \epsilon$. The link between this condition and our constructions is explained in the following observation:

Proposition 4.3.2. Let ζ be a random perturbation with distribution P possessing expectation, let $\epsilon \in (0,1)$, and let $\Gamma_\epsilon^\sharp$ be the associated set (4.3.10). Then

$$\Gamma_\epsilon^\sharp = \{z : \mathrm{CVaR}_\epsilon(\xi^z) \leq 0\}.$$

Proof. We have

$$
\begin{aligned}
\Gamma_\epsilon^{o,\sharp} &= \{z : \exists \alpha > 0 : \mathbf{E}\{\max[1 + \xi^{\alpha z}, 0]\} \leq \epsilon\} \\
&= \{z : \exists \alpha > 0 : \mathbf{E}\{\max[1 + \alpha \xi^z, 0]\} \leq \epsilon\} \\
&= \{z : \exists \alpha > 0 : \mathbf{E}\{\max[1 + \alpha^{-1} \xi^z, 0]\} \leq \epsilon\} \\
&= \{z : \exists \alpha > 0 : \tfrac{1}{\epsilon} \mathbf{E}\{\max[\alpha + \xi^z, 0]\} \leq \alpha\} \\
&= \{z : \exists a = -\alpha < 0 : a + \tfrac{1}{\epsilon} \mathbf{E}\{\max[\xi^z - a, 0]\} \leq 0\}.
\end{aligned}
\tag{4.3.14}
$$

From the latter relation it immediately follows that $\Gamma_\epsilon^{o,\sharp} \subset \mathcal{C} = \{z : \mathrm{CVaR}_\epsilon(\xi^z) \leq 0\}$. As we have already mentioned, $\mathrm{CVaR}_\epsilon(\xi^z)$ is a finite convex function of z, so that this function is continuous, and therefore $\mathcal{C}$ is a closed set. Thus, the above inclusion implies that $\Gamma_\epsilon^\sharp \subset \mathcal{C}$. To prove the inverse inclusion, let us fix $z \in \mathcal{C}$ and prove that $z \in \Gamma_\epsilon^\sharp$. To this end, observe that the function

$$f(a) = a + \frac{1}{\epsilon} \mathbf{E}\{\max[\xi^z - a, 0]\}$$

clearly is a convex finite function that tends to $+\infty$ as $|a| \to \infty$, so that this function attains its minimum at certain $a = a_*$. We have

$$a_* + \frac{1}{\epsilon} \mathbf{E}\{\max[\xi^z - a_*, 0]\} \leq 0 \tag{4.3.15}$$

due to $z \in \mathcal{C}$. From the latter inequality, $a_* \leq 0$. In the case of $a_* < 0$, relations (4.3.14) say that $z \in \Gamma_\epsilon^{o,\sharp}$. If $a_* = 0$, then (4.3.15) implies that $\xi^z \leq 0$ with probability 1, whence, setting $z' = [z_0 - \delta; z_1; ...; z_L]$, $\delta > 0$, we get $\xi^{z'} < -\delta$ with probability 1. In the latter case we clearly have $\frac{1}{\epsilon}\mathbf{E}\{\max[\alpha + \xi^{z'}, 0]\} = 0 \leq \alpha$ for all small positive α, which, in view of relations (4.3.14), implies that $z' \in \Gamma_\epsilon^{o,\sharp}$. Since $z' \to z$ as $\delta \to +0$, we conclude that $z \in \Gamma_\epsilon^\sharp$. $\square$

4.3.4 Tractability Issues

We have seen that the "CVaR approximation"

$$\mathrm{CVaR}_\epsilon(\xi^z) \leq 0,$$

of chance constraint (4.0.1) is the best — the least conservative — among the approximations yielded by our generating-function-based approximation scheme. Given this fact, why could we be interested in other, more conservative, approximations, e.g., the Bernstein approximation?

The answer is, that the level of conservatism is not the only consideration: we are interested in *computationally tractable* approximations, and to this end the underlying function Ψ should be efficiently computable. For the Bernstein approximation, this indeed is the case, provided that the random perturbations ζ_ℓ are independent with distributions belonging to not too complicated families (see examples in section 4.2). In contrast, the function $\Psi_\sharp$ underlying the CVaR approximation is *not* efficiently computable even in the case when ζ_ℓ are independent and possess simple distributions, (e.g., are uniformly distributed in $[-1, 1]$). Seemingly the only generic case where we have no difficulty in computing $\Psi_\sharp$ is when ζ is supported on a finite set of moderate cardinality, in that case the CVaR approximation is given by the following.

Proposition 4.3.3. Let $\zeta \in \mathbb{R}^L$ be a discrete random vector taking values $\zeta^1, ..., \zeta^N$ with probabilities $\pi_1, ..., \pi_N$. Then

$$\Psi_\sharp(z) = \sup_u \left\{ z^T(Bu + b) - \psi(u) \right\},$$

$$Bu + b = \left[\sum_i u_i; \sum_i u_i \zeta^i\right], \quad \psi(u) = \begin{cases} -\sum_i u_i, & \begin{matrix} 0 \leq u_i \leq \pi_i, \\ 1 \leq i \leq N \end{matrix} \\ +\infty, & \text{otherwise} \end{cases}$$

and the Robust Counterpart representation of the CVaR approximation is

$$\begin{cases} \Gamma_\epsilon^\sharp & = \left\{ z : z^T[1; \epsilon^{-1}\sum_i u_i \zeta^i] \leq 0 \, \forall u \in \overline{\mathcal{U}}_\epsilon \right\}, \\ \overline{\mathcal{U}}_\epsilon & \equiv \left\{ u : 0 \leq u_i \leq \pi_i \, \forall i, \sum_i u_i = \epsilon \right\} \end{cases}$$

$$\Updownarrow$$

$$\begin{cases} \Gamma_\epsilon^\sharp & = \left\{ z : z_0 + \sum_\ell \zeta_\ell z_\ell \leq 0, \, \forall \zeta \in \mathcal{Z}_\epsilon \right\}, \\ \mathcal{Z}_\epsilon & = \left\{ \zeta = \sum_i u_i \zeta^i : 0 \leq u_i \leq \pi_i/\epsilon, \sum_i u_i = 1 \right\} \end{cases}.$$

Proof. We have

$$\Psi_\sharp(z)$$

$$= \mathbf{E}\{\max[1 + z_0 + \sum_{\ell=1}^{L} \zeta_\ell z_\ell, 0]\} = \sum_{i=1}^{N} \pi_i \max[1 + z_0 + [z_1; ...; z_L]^T \zeta^i, 0]$$

$$= \sum_{i=1}^{N} \max_{0 \leq u_i \leq \pi_i} u_i[1 + z_0 + [z_1; ...; z_L]^T \zeta^i]$$

$$= \max_{u:0 \leq u_i \leq \pi_i} \left\{[z_0; ...; z_L]^T \underbrace{[\sum_i u_i; \sum_i u_i \zeta^i]}_{Bu+b} - (-\sum_i u_i)\right\}.$$

Consequently, by the results from section 4.3.2,

$$\Gamma_\epsilon^\sharp = \{z : z^T[\sum_i u_i; \sum_i u_i \zeta^i] \leq 0 \ \forall u \in \mathcal{U}_\epsilon \equiv \left\{ u : \begin{array}{l} 0 \leq u_i \leq \pi_i \ \forall i; \\ \sum_i u_i \geq \epsilon \end{array} \right\},$$

which is equivalent to what is stated in Proposition. □

4.3.5 Extensions to Vector Inequalities

The outlined approach can be applied to the chance constrained version

$$\text{Prob}\{z^0 + \sum_{\ell=1}^{L} \zeta_\ell z^\ell \notin -\mathbf{K}\} \leq \epsilon \tag{4.3.16}$$

of randomly perturbed *vector* inequality

$$\xi^z \equiv z^0 + \sum_{\ell=1}^{L} \zeta_\ell z^\ell \in -\mathbf{K}, \tag{4.3.17}$$

where $z^0, z^1, ..., z^L \in \mathbb{R}^d$ are deterministic parameters, ζ_ℓ are random perturbations, and $\mathbf{K}$ is a given closed convex cone with a nonempty interior in $\mathbb{R}^d$. To this end, it suffices to choose a convex function $\gamma : \mathbb{R}^d \to \mathbb{R}$ that is $\mathbf{K}$-monotone,

$$\gamma(y + h) \geq \gamma(y) \ \forall(h \in \mathbf{K}, y \in \mathbb{R}^d)$$

and satisfies the relations

$$\gamma(y) \geq 0 \ \forall y, \ \gamma(y) \geq 1 \ \forall(y \notin -\mathbf{K})$$

and

$$\exists e : \forall y : \gamma(y + te) \to 0, t \to \infty.$$

For example, we can choose a norm $\|\cdot\|$ on $\mathbb{R}^d$, set

$$\gamma(y) = 1 + \text{dist}(y, -\mathbf{K}), \ \text{dist}(y, -\mathbf{K}) = \min_{v \in -\mathbf{K}} \|y - v\|$$

and use in the role of e a direction from $-\text{int}\mathbf{K}$.

Given a $\gamma(\cdot)$ as above, assume that we have at our disposal an everywhere finite function $\Psi(z)$ that is convex in $z = [z^0; z^1; ...; z^L]$ and is an upper bound on $\mathbf{E}\{\gamma(\xi^z)\}$, along with a convex and lower semicontinuous function $\psi(u)$ with bounded level sets such that

$$\Psi(z) = \sup_u \left\{ z^T(Bu + b) - \psi(u) \right\}.$$

It can be easily proved that in the outlined situation:

(i) One has

$$\forall z : \text{Prob}\{\xi^z \notin -\mathbf{K}\} \leq \inf_{\alpha > 0} \Psi(\alpha z)$$

(ii) The set $Z_\epsilon = \text{cl}\left\{ z : \exists \alpha > 0 : \Psi(\alpha z) \leq \epsilon \right\}$ is such that

$$z \in Z_\epsilon \Rightarrow \text{Prob}\{\xi^z \notin -\mathbf{K}\} \leq \epsilon;$$

(iii) The set Z_ϵ is nothing but the robust feasible set of the uncertain linear constraint

$$z^T(Bu + b) \leq 0 \; \forall u \in \mathcal{U}_\epsilon = \{u : \psi(u) \leq -\epsilon\}$$

associated with the nonempty convex compact perturbation set $\mathcal{U}_\epsilon$.

4.3.6 Bridging the Gap between the Bernstein and the CVaR Approximations

We have seen that the Bernstein approximation of a chance constraint (4.0.1) is a particular case of the general generating-function-based scheme for building a safe convex approximation of the constraint, and that this particular approximation is not the best in terms of conservatism. What makes it attractive, is that under certain structural assumptions (namely, those of independence of $\zeta_1, ..., \zeta_L$ plus availability of efficiently computable convex upper bounds on the functions $\ln(\mathbf{E}\{\exp\{s\zeta_\ell\}\}))$ this approximation is computationally tractable. The question we now address is how to reduce, to some extent, the conservatism of the Bernstein approximation without sacrificing computational tractability. The idea is as follows. Assume that

A. *The random perturbations $\zeta_1, ..., \zeta_L$ are independent, and we can compute efficiently the associated moment-generating functions*

$$\Psi_\ell(s) = \mathbf{E}\left\{\exp\{s\zeta_\ell\}\right\} : \mathbb{C} \to \mathbb{C}.$$

Under this assumption, whenever $\gamma(s) = \sum\limits_{\nu=0}^{d} c_\nu \exp\{\omega_\nu s\}$ is an exponential polynomial, we can efficiently compute the function

$$\Psi(z) = \mathbf{E}\left\{\gamma(z_0 + \sum_{\ell=1}^{L} \zeta_\ell z_\ell)\right\} = \sum_{\nu=0}^{d} c_\nu \exp\{\omega_\nu z_0\} \prod_{\ell=1}^{L} \Psi_\ell(\omega_\nu z_\ell).$$

In other words,

(!) *Whenever a generating function $\gamma(\cdot) : \mathbb{R} \to \mathbb{R}$ satisfying (4.3.1) is an exponential polynomial, the associated upper bound*

$$\inf_{\alpha > 0} \Psi(\alpha z), \ \ \Psi(z) = \mathbf{E}\left\{ \gamma\left(z_0 + \sum_{\ell=1}^{L} \zeta_\ell z_\ell\right) \right\}$$

on the quantity $p(z) = \mathrm{Prob}\{z_0 + \sum_{\ell=1}^{L} \zeta_\ell z_\ell > 0\}$ is efficiently computable.

We now can utilize (!) in the following construction:

Given design parameters $T > 0$ ("window width") and d ("degree of approximation"), we build the trigonometric polynomial

$$\chi_{c_*}(s) \equiv \sum_{\nu=0}^{d} \left[c_{*\nu} \exp\{\imath \pi \nu s/T\} + \overline{c_{*\nu}} \exp\{-\imath \pi \nu s/T\} \right]$$

by solving the following problem of the best uniform approximation:

$$c_* \in \underset{c \in \mathbb{C}^{d+1}}{\mathrm{Argmin}} \left\{ \max_{-T \leq s \leq T} |\exp\{s\}\chi_c(s) - \max[1+s, 0]| : \right.$$
$$0 \leq \chi_c(s) \leq \chi_c(0) = 1 \ \forall s \in \mathbb{R}, \ \exp\{s\}\chi_c(s) \ \text{is convex}$$
$$\left. \text{and nondecreasing on } [-T, T] \right\}$$

and use in (!) the exponential polynomial

$$\gamma_{d,T}(s) = \exp\{s\}\chi_{c_*}(s). \tag{4.3.18}$$

It can be immediately verified that

(i) The outlined construction is well defined and results in generating function $\gamma_{d,T}(s)$ that is an exponential polynomial satisfying the requirements (4.3.1) and thus inducing an efficiently computable convex upper bound on $p(z)$.

(ii) The resulting upper bound on $p(z)$ is $\leq$ the Bernstein upper bound associated, according to (!), with $\gamma(s) = \exp\{s\}$.

The generating function $\gamma_{11,8}(\cdot)$ is depicted in figure 4.2.

The case of ambiguous chance constraint. A disadvantage of the improved Bernstein approximation as compared to the plain one is that the improved approximation requires *precise* knowledge of the moment-generating functions $\mathbf{E}\{\exp\{s\zeta_\ell\}\}$, $s \in \mathbb{C}$, of the independent random variables ζ_ℓ, while the original approximation requires knowledge of *upper bounds* on these functions and thus is applicable in the case of ambiguous chance constraints, those with only partially known distributions of ζ_ℓ. Such partial information is equivalent to the fact that the distribution P of ζ belongs to a given family $\mathcal{P}$ in the space of product probability distributions on $\mathbb{R}^L$. All we need in this situation is a possibility to compute efficiently the convex function

$$\Psi_{\mathcal{P}}(z) = \sup_{P \in \mathcal{P}} \mathbf{E}_{\zeta \sim P}\left\{ \gamma\left(z_0 + \sum_{\ell=1}^{L} \zeta_\ell z_\ell\right) \right\}$$

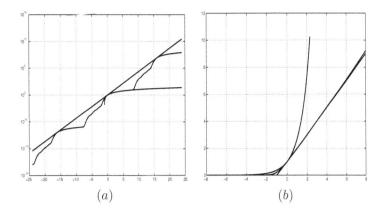

(a) (b)

Figure 4.2 Generating function $\gamma_{11,8}(s)$ (middle curve) vs. $\exp\{s\}$ (top curve) and $\max[1+ s,0]$ (bottom curve). (a): $-24 \leq s \leq 24$, logarithmic scale along the y-axis; (b): $-8 \leq s \leq 8$, natural scale along the y-axis.

associated with $\mathcal{P}$ and with a given generating function $\gamma(\cdot)$ satisfying (4.3.1). When $\Psi_{\mathcal{P}}(\cdot)$ is available, a computationally tractable safe approximation of the ambiguous chance constraint

$$\forall (P \in \mathcal{P}) : \mathrm{Prob}_{\zeta \sim P}\{z_0 + \sum_{\ell=1}^{L} \zeta_\ell z_\ell > 0\} \leq \epsilon \tag{4.3.19}$$

is

$$\mathrm{cl}\left\{z : \exists \alpha > 0 : \Psi_{\mathcal{P}}(\alpha z) \leq \epsilon\right\}.$$

Now, in all applications of the "plain" Bernstein approximation we have considered so far the family $\mathcal{P}$ was comprised of all product distributions $P = P_1 \times ... \times P_L$ with P_ℓ running through given families $\mathcal{P}_\ell$ of probability distributions on the axis, and these families $\mathcal{P}_\ell$ were "simple," specifically, allowing us to compute explicitly the functions

$$\Psi_\ell(s) = \sup_{P_\ell \in \mathcal{P}_\ell} \int \exp\{s\zeta_\ell\}dP_\ell(\zeta_\ell).$$

With these functions at our disposal and with $\gamma(s) = \exp\{s\}$, the function

$$\Psi_{\mathcal{P}}(z) = \sup_{P \in \mathcal{P}} \mathbf{E}\left\{\exp\{z_0 + \sum_{\ell=1}^{L} \zeta_\ell z_\ell\}\right\}$$

is readily available — it is merely $\exp\{z_0\} \prod_{\ell=1}^{L} \Psi_\ell(z_\ell)$. Note, however, that when $\gamma(\cdot)$ is an exponential polynomial rather than the exponent, the associated function $\Psi_{\mathcal{P}}(z)$ does *not* admit a simple representation via the functions $\Psi_\ell(\cdot)$. Thus, it is indeed unclear how to implement the improved Bernstein approximation in the case of an ambiguous chance constraint.

Our current goal is to implement the improved Bernstein approximation in the case of a particular ambiguous chance constraint (4.3.19), namely, in the case when $\mathcal{P}$ is comprised of all product distributions $P = P_1 \times ... \times P_L$ with the marginal distributions P_ℓ satisfying the restrictions

$$\operatorname{supp} P_\ell \subset [-1,1] \ \& \ \mu_\ell^- \leq \mathbf{E}_{\zeta_\ell \sim P_\ell}\{\zeta_\ell\} \leq \mu_\ell^+ \qquad (4.3.20)$$

with known $\mu_\ell^\pm \in [-1,1]$ (cf. Example 4.2.8).

The result is as follows:

Proposition 4.3.4. For the just defined family $\mathcal{P}$ and with every $\gamma(\cdot)$ satisfying (4.3.1), one has

$$\Psi_\mathcal{P}(z) = \mathbf{E}_{\zeta \sim P^z}\left\{\gamma\left(z_0 + \sum_{\ell=1}^L \zeta_\ell z_\ell\right)\right\},$$

where $P^z = P_1^{z_1} \times \dots \times P_L^{z_L}$ and P_ℓ^s is the distribution supported at the endpoints of $[-1,1]$ given by

$$P_\ell^s\{1\} = 1 - P_\ell^s\{-1\} = \begin{cases} \frac{1+\mu_\ell^+}{2}, & s \geq 0 \\ \frac{1+\mu_\ell^-}{2}, & s < 0 \end{cases} ;$$

In particular, when $\gamma(\cdot) \equiv \gamma_{d,T}(\cdot)$, the function $\Psi_\mathcal{P}(z)$ is efficiently computable.

Proof. It suffices to prove the following

<u>Claim:</u> If $P = P_1 \times \dots \times P_L$ with P_ℓ satisfying (4.3.20) and $\ell_* \in \{1, \dots, L\}$, then, passing from the distribution P to the distribution $P' = P_1 \times \dots \times P_{\ell_*-1} \times P_{\ell_*}^{z_{\ell_*}} \times P_{\ell_*+1} \times \dots \times P_L$ (which clearly belongs to $\mathcal{P}$ as well), we do not decrease the associated quantity $\mathbf{E}\left\{\gamma\left(z_0 + \sum_{\ell=1}^L \zeta_\ell z_\ell\right)\right\}$.

When proving the claim, we can assume w.l.o.g. that $\ell_* = 1$.

Let us set

$$\widehat{\gamma}(t) = \mathbf{E}_{[\zeta_2;\dots;\zeta_L] \sim P_2 \times \dots \times P_L}\left\{\gamma\left(z_0 + z_1 t + \zeta_2 z_2 + \dots + \zeta_L z_L\right)\right\}.$$

Since $\gamma(\cdot)$ satisfies (4.3.1), the function $\widehat{\gamma}(\cdot)$ is a finite convex function that is non-decreasing when $z_1 \geq 0$ and is nonincreasing when $z_1 < 0$. In terms of $\widehat{\gamma}(\cdot)$, our claim reads:

$$\int_{-1}^1 \widehat{\gamma}(\zeta_1)dP_1(\zeta_1) \leq \int_{-1}^1 \widehat{\gamma}(\zeta_1)dP_1^{z_1}(\zeta_1) = P_1^{z_1}\{1\}\widehat{\gamma}(1)+(1-P_1^{z_1}\{1\})\widehat{\gamma}(-1). \quad (4.3.21)$$

The proof of the latter relation is immediate. Let $\mu_1 = \int \zeta_1 dP_1(\zeta_1)$. Since $\widehat{\gamma}$ is convex, we have

$$\begin{aligned} \int_{-1}^1 \widehat{\gamma}(\zeta_1)dP_1(\zeta_1) &\leq \int_{-1}^1 \left[\tfrac{1+\zeta_1}{2}\widehat{\gamma}(1) + \tfrac{1-\zeta_1}{2}\widehat{\gamma}(-1)\right]dP_1(\zeta_1) \\ &= \phi(\mu_1) \equiv \tfrac{1+\mu_1}{2}\widehat{\gamma}(1) + \tfrac{1-\mu_1}{2}\widehat{\gamma}(-1). \end{aligned} \qquad (4.3.22)$$

Since $\widehat{\gamma}(\cdot)$ is nondecreasing when $z_1 \geq 0$ and is nonincreasing when $z_1 < 0$, the function $\phi(r)$ is nondecreasing on the segment $[\mu_1^-, \mu_1^+] \subset [-1,1]$ when $z_1 \geq 0$ and is nonincreasing on the same segment when $z_1 < 0$. Since μ_1 belongs to this segment by (4.3.20), we have $\phi(\mu_1) \leq \phi(\mu_1^+)$ when $z_1 \geq 0$ and $\phi(\mu_1) \leq \phi(\mu_1^-)$ when $z_1 < 0$, meaning that in both cases $\phi(\mu_1) \leq \int_{-1}^1 \widehat{\gamma}(\zeta_1)dP_1^{z_1}(\zeta_1)$ (see the concluding equality in (4.3.21)), so that (4.3.22) implies the inequality in (4.3.21). $\square$

4.3.6.1 Illustration I

To illustrate our findings, assume that all our a priori information on the random perturbations ζ_ℓ in (4.0.1) is that they are independent, supported on $[-1, 1]$ and with zero means. Let us overview the safe approximations to the corresponding ambiguous chance constraint

$$\forall((P_1, ..., P_L) \in \mathcal{P}) : \text{Prob}_{\zeta \sim P_1 \times ... \times P_L} \left\{ z_0 + \sum_{\ell=1}^{L} \zeta_\ell z_\ell > 0 \right\} \leq \epsilon, \qquad (4.3.23)$$

where $\mathcal{P}$ is the family of all collections of L probability distributions with zero mean supported on $[-1, 1]$. Note that on a closest inspection, the results yielded by all approximation schemes to be listed below remain intact when instead of the ambiguous chance constraint we were speaking about the usual one, with ζ distributed uniformly on the vertices of the unit box $\{\zeta : \|\zeta\|_\infty \leq 1\}$.

We are about to outline the approximations, ascending in their conservatism and descending in their complexity. When possible, we present approximations in both the "inequality form" (via an explicit system of convex constraints) and in the "Robust Counterpart form"

$$\{z : z_0 + \zeta^T[z_1; ...; z_L] \leq 0 \ \forall \zeta \in \mathcal{Z}\}.$$

- CVaR *approximation* [Proposition 4.3.2]

$$\inf_{\beta > 0} \left[z_0 + \max_{(P_1, ..., P_L) \in \mathcal{P}} \int \max[\beta + z_0 + \sum_{\ell=1}^{L} \zeta_\ell z_\ell, 0] dP_1(\zeta_1) ... dP_L(\zeta_L) - \beta\epsilon \right] \leq 0 \qquad (4.3.24)$$

While being the least conservative among all generation-function-based approximations, the CVaR approximation is in general intractable. It remains intractable already when passing from the ambiguous chance constraint case to the case where ζ_ℓ are, say, uniformly distributed on $[-1, 1]$ (which corresponds to replacing $\max_{(P_1, ..., P_N) \in \mathcal{P}} \int ... dP_1(\zeta_1) ... dP_L(\zeta_\ell)$ in (4.3.24) with $\int_{\|\zeta\|_\infty \leq 1} ... d\zeta$).

We have "presented" the inequality form of the CVaR approximation. By Propositions 4.1.3 and 4.3.1, this approximation admits a Robust Counterpart form; the latter "exists in the nature," but is computationally intractable, and thus of not much use.

- *Bridged Bernstein-*CVaR *approximation* [p. 98 and Proposition 4.3.4]

$$\inf_{\beta > 0} \left[\beta \Psi_{d,T}(\beta^{-1} z) - \beta\epsilon \right] \leq 0,$$
$$\Psi_{d,T}(\zeta) = \sum_{\epsilon_\ell = \pm 1, 1 \leq \ell \leq L} 2^{-L} \gamma_{d,T} \left(z_0 + \sum_{\ell=1}^{L} \epsilon_\ell z_\ell \right), \qquad (4.3.25)$$

where d, T are parameters of the construction and $\gamma_{d,T}$ is the exponential polynomial (4.3.18). Note that we used Proposition 4.3.4 to cope with the ambiguity of the chance constraint of interest.

In spite of the disastrous complexity of the representation (4.3.25), the function $\Psi_{d,T}$ is efficiently computable (via the recipe from Proposition 4.3.4, and *not* via the formula in (4.3.25)). Thus, our approximation is computationally tractable. Recall that this tractable safe approximation is less conservative than the plain Bernstein one.

Due to Propositions 4.1.3 and 4.3.1, approximation (4.3.24) admits a Robust Counterpart representation that now involves a computationally tractable uncertainty set $\mathcal{Z}^{\mathrm{BCV}}$; this set, however, seems to have no explicit representation.

- *Bernstein approximation* [Example 4.2.8]

$$
\begin{aligned}
&\inf_{\beta > 0} \left[z_0 + \sum_{\ell=1}^{L} \beta \ln \left(\cosh(\beta^{-1} z_\ell) \right) + \beta \ln(1/\epsilon) \right] \leq 0 \\
&\Leftrightarrow z_0 + \sum_{\ell=1}^{L} \zeta_\ell z_\ell \leq 0 \ \forall \zeta \in \mathcal{Z}_\epsilon^{\mathrm{Brn}} = \{ \zeta : \sum_{\ell=1}^{L} \phi(\zeta_\ell) \leq \ln(1/\epsilon) \} \\
&\left[\phi(u) = \tfrac{1}{2} \left[(1+u) \ln(1+u) + (1-u) \ln(1-u) \right], \ \mathrm{Dom} \ \phi = [-1, 1] \right].
\end{aligned}
\tag{4.3.26}
$$

-    *Robust Counterpart approximation with Ball-Box uncertainty* [Proposition 2.3.3, or, equivalently, Example 2.4.9 and Theorem 2.4.4]

$$
\begin{aligned}
&\exists u, v : z = u + v, \ v_0 + \sum_{\ell=1}^{L} |v_\ell| \leq 0, \ u_0 + \sqrt{2 \ln(1/\epsilon)} \sqrt{\sum_{\ell=1}^{L} u_\ell^2} \leq 0 \\
&\Leftrightarrow z_0 + \sum_{\ell=1}^{L} \zeta_\ell z_\ell \leq 0 \ \forall \zeta \in \mathcal{Z}^{\mathrm{BlBx}} := \left\{ \zeta \in \mathbb{R}^L : \begin{array}{l} |\zeta_\ell| \leq 1, \ \ell = 1, ..., L, \\ \sqrt{\sum_{\ell=1}^{L} \zeta_\ell^2} \leq \sqrt{2 \ln(1/\epsilon)} \end{array} \right\}.
\end{aligned}
\tag{4.3.27}
$$

It is immediately seen that (4.3.27) is a simplified conservative version of (4.3.26) that, in hindsight, can be obtained from the quadratic lower bound on the entropy $\phi(u)$:

$$
\phi(u) \geq \frac{1}{2} u^2;
$$

(to get this bound, note that $\phi(0) = \phi'(0) = 0$ and $\phi''(u) = \frac{1}{1-u^2} \geq 1$ when $|u| < 1$), whence

$$
\mathcal{Z}^{\mathrm{Brn}} = \{ \zeta : \sum_{\ell=1}^{L} \phi(\zeta_\ell) \leq \ln(1/\epsilon) \} \subset \{ \zeta : \|\zeta\|_\infty \leq 1, \sum_{\ell=1}^{L} \frac{\zeta_\ell^2}{2} \leq \ln(1/\epsilon) \} = \mathcal{Z}^{\mathrm{BlBx}}.
$$

-    *Robust Counterpart approximation with Budgeted uncertainty* [Proposition 2.3.4]

$$
\begin{aligned}
&\exists u, v : z = u + v, \ v_0 + \sum_{\ell=1}^{L} |v_\ell| \leq 0, \ u_0 + \sqrt{2L \ln(1/\epsilon)} \max_\ell |u_\ell| \leq 0 \\
&\Leftrightarrow z_0 + \sum_{\ell=1}^{L} \zeta_\ell z_\ell \leq 0 \ \forall \zeta \in \mathcal{Z}^{\mathrm{Bdg}} := \left\{ \zeta \in \mathbb{R}^L : \begin{array}{l} |\zeta_\ell| \leq 1, \ \ell = 1, ..., L, \\ \sum_{\ell=1}^{L} |\zeta_\ell| \leq \sqrt{2L \ln(1/\epsilon)} \end{array} \right\}.
\end{aligned}
\tag{4.3.28}
$$

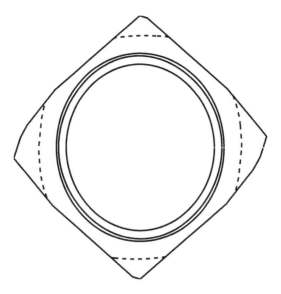

Figure 4.3 Intersection of uncertainty sets, underlying various approximation schemes in
    Illustration I, with a random 2-D plane. From inside to outside:
    – Bridged Bernstein-CVaR approximation, $d = 11$, $T = 8$;
    – Bernstein approximation;
    – Ball-Box approximation;
    – Budgeted approximation;
    – "worst-case" approximation with the support $\{\|\zeta\|_\infty \leq 1\}$ of ζ in the role of
    the uncertainty set.

Note that (4.3.28) is a simplified conservative version of (4.3.27) given by the evident
inequality

$$\sum_{\ell=1}^{L} |u_\ell| \leq \sqrt{L}\sqrt{\sum_{\ell=1}^{L} u_\ell^2},$$

which implies that $\mathcal{Z}^{\mathrm{BlBx}} \subset \mathcal{Z}^{\mathrm{Bdg}}$.

The computationally tractable uncertainty sets we have listed form a chain:

$$\mathcal{Z}^{\mathrm{BCV}} \subset \mathcal{Z}^{\mathrm{Brn}} \subset \mathcal{Z}^{\mathrm{BlBx}} \subset \mathcal{Z}^{\mathrm{Bdg}}.$$

Figure 4.3, where we plot a random 2-D cross-section of our nested uncertainty
sets, gives an impression of the "gaps" in this chain.

4.3.6.2 Illustration II

In this illustration, which is a continuation of Example 4.2.9, we use the above
approximation schemes to build safe approximations of the ambiguously chance

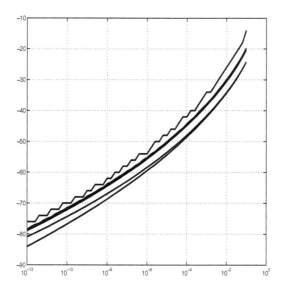

Figure 4.4 Optimal values of various approximations of (4.3.29) with $L = 128$ vs. ϵ.
From bottom to top:
– Budgeted and Ball-Box approximations
– Bernstein approximation
– Bridged Bernstein-CVaR approximation, $d = 11$, $T = 8$
– CVaR-approximation
– $\mathrm{Opt}^+(\epsilon)$

constrained problem

$$\mathrm{Opt}(\epsilon) = \max\left\{ z_0 : \max_{(P_1,\dots,P_L)\in\mathcal{P}} \mathrm{Prob}\{z_0 + \sum_{\ell=1}^{L} \zeta_\ell z_\ell > 0\} \le \epsilon, z_1 = \dots = z_L = 1 \right\}$$
$$(4.3.29)$$

where, as before, $\mathcal{P}$ is the set of L-element tuples of probability distributions supported on $[-1,1]$ and possessing zero means. Due to the simplicity of our chance constraint, here we can build efficiently the CVaR-approximation of the problem. Moreover, we can solve *exactly* the chance constrained problem

$$\mathrm{Opt}^+(\epsilon) = \max\left\{ z_0 : \max_{\zeta\sim U} \mathrm{Prob}\{z_0 + \sum_{\ell=1}^{L} \zeta_\ell z_\ell > 0\} \le \epsilon, z_1 = \dots = z_L = 1 \right\}$$

where U is the uniform distribution on the vertices of the unit box $\{\zeta : \|\zeta\|_\infty \le 1\}$; this is in fact problem (P) from Example 4.2.9. Clearly, $\mathrm{Opt}^+(\epsilon)$ is an upper bound on the true optimal value $\mathrm{Opt}(\epsilon)$ of the ambiguously chance constrained problem (4.3.29), while the optimal values of our approximations are lower bounds on $\mathrm{Opt}(\epsilon)$. In our experiment, we used $L = 128$. The results are depicted in figure 4.4 and are displayed in table 4.1.

ϵ	$\mathrm{Opt}^+(\epsilon)$	$\mathrm{Opt}_V(\epsilon)$	$\mathrm{Opt}_{IV}(\epsilon)$	$\mathrm{Opt}_{III}(\epsilon)$	$\mathrm{Opt}_{II}(\epsilon)$	$\mathrm{Opt}_I(\epsilon)$
10^{-12}	-76.00	$-78.52\ (-3.3\%)$	$-78.88\ (-0.5\%)$	$-80.92\ (-3.1\%)$	$-84.10\ (-7.1\%)$	$-84.10\ (-7.1\%)$
10^{-11}	-74.00	$-75.03\ (-1.4\%)$	$-75.60\ (-0.8\%)$	$-77.74\ (-3.6\%)$	$-80.52\ (-7.3\%)$	$-80.52\ (-7.3\%)$
10^{-10}	-70.00	$-71.50\ (-2.1\%)$	$-72.13\ (-0.9\%)$	$-74.37\ (-4.0\%)$	$-76.78\ (-7.4\%)$	$-76.78\ (-7.4\%)$
10^{-9}	-66.00	$-67.82\ (-2.8\%)$	$-68.45\ (-0.9\%)$	$-70.80\ (-4.4\%)$	$-72.84\ (-7.4\%)$	$-72.84\ (-7.4\%)$
10^{-8}	-62.00	$-63.88\ (-3.0\%)$	$-64.49\ (-1.0\%)$	$-66.97\ (-4.8\%)$	$-68.67\ (-7.5\%)$	$-68.67\ (-7.5\%)$
10^{-7}	-58.00	$-59.66\ (-2.9\%)$	$-60.23\ (-1.0\%)$	$-62.85\ (-5.4\%)$	$-64.24\ (-7.7\%)$	$-64.24\ (-7.7\%)$
10^{-6}	-54.00	$-55.25\ (-2.3\%)$	$-55.60\ (-0.6\%)$	$-58.37\ (-5.7\%)$	$-59.47\ (-7.6\%)$	$-59.47\ (-7.6\%)$
10^{-5}	-48.00	$-49.98\ (-4.1\%)$	$-50.52\ (-1.1\%)$	$-53.46\ (-7.0\%)$	$-54.29\ (-8.6\%)$	$-54.29\ (-8.6\%)$
10^{-4}	-42.00	$-44.31\ (-5.5\%)$	$-44.85\ (-1.2\%)$	$-47.97\ (-8.3\%)$	$-48.56\ (-9.6\%)$	$-48.56\ (-9.6\%)$
10^{-3}	-34.00	$-37.86(-11.4\%)$	$-38.34\ (-1.2\%)$	$-41.67(-10.1\%)$	$-42.05(-11.1\%)$	$-42.05(-11.1\%)$
10^{-2}	-26.00	$-29.99(-15.4\%)$	$-30.55\ (-1.9\%)$	$-34.13(-13.8\%)$	$-34.34(-14.5\%)$	$-34.34(-14.5\%)$
10^{-1}	-14.00	$-19.81(-41.5\%)$	$-20.43\ (-3.1\%)$	$-24.21(-22.2\%)$	$-24.28(-22.5\%)$	$-24.28(-22.5\%)$

Table 4.1   Comparing various safe approximations of the ambiguously chance constrained problem (4.3.29). $\mathrm{Opt}_I(\epsilon)$ through $\mathrm{Opt}_V(\epsilon)$ are optimal values of the Ball, Ball-Box (or, which in the case of (4.3.29) is the same, the Budgeted), Bernstein, Bridged Bernstein-CVaR and the CVaR approximations, respectively. Numbers in parentheses in column "$\mathrm{Opt}_V(\epsilon)$" refer to the conservativeness of the CVaR-approximation as compared to $\mathrm{Opt}^+(\cdot)$, and in remaining columns to the conservativeness of the corresponding approximation as compared to the CVaR approximation.

4.4 MAJORIZATION

One way to bound from above the probability

$$\mathrm{Prob}\left\{z_0 + \sum_{\ell=1}^{L} z_\ell \zeta_\ell > 0\right\}$$

for independent random variables ζ_ℓ is to replace ζ_ℓ with "more diffused" random variables ξ_ℓ (meaning that the probability in question increases when we replace ζ_ℓ with ξ_ℓ) such that the quantity $\mathrm{Prob}\left\{z_0 + \sum_{\ell=1}^{L} z_\ell \xi_\ell > 0\right\}$, (which now is an upper bound on probability in question), is easy to handle. Our goal here is to investigate the outlined approach in the case of random variables with symmetric and unimodal w.r.t. 0 probability distributions.

In contrast to chapter 2, where an unimodal w.r.t. 0 random variable was defined as a variable with probability density that is nondecreasing on $\mathbb{R}_-$ and nonincreasing on $\mathbb{R}_+$, now it is convenient to allow the variable to take the value 0 with a positive probability. Thus, in what follows a symmetric and unimodal w.r.t. 0 probability distribution P on the axis is a distribution given by $P(A) = \int_A p(s)ds + \delta(A)$, where A is a measurable subset of $\mathbb{R}$, $p(\cdot) \geq 0$ is an even and nonincreasing on $\mathbb{R}_+$ function such that $\int p(s)ds \leq 1$, and $\delta(A)$ is either $1 - \int p(s)ds$ or 0 depending on whether or not $0 \in A$. We call $p(\cdot)$ the density of P, and say that P is regular if $p(\cdot)$ is a usual probability density, that is, $\int p(s)ds = 1$ (or, equivalently, there is no nontrivial probability mass at 0).

In what follows, we denote the family of densities of all symmetric and uni-modal w.r.t. 0 random variables by $\mathcal{P}$, and the family of these random variables themselves by Π.

If we want the outlined scheme to work, the notion of a "more diffused" random variable should imply the following: If $p, q \in \mathcal{P}$ and q is "more diffused" than p, then, for every $a \geq 0$, we should have $\int_a^\infty p(s)ds \leq \int_a^\infty q(s)ds$. We make this requirement the definition of "more diffused":

Definition 4.4.1. Let $p, q \in \mathcal{P}$. We say that q is more diffused than p (notation: $q \succeq_m p$, or $p \preceq_m q$) if

$$\forall a \geq 0 : P(a) := \int_a^\infty p(s)ds \leq Q(a) := \int_a^\infty q(s)ds.$$

When $\xi, \eta \in \Pi$, we say that η is more diffuse than ξ (notation: $\eta \succeq_m \xi$), if the corresponding densities are in the same relation.

It is immediately seen that the relation $\succeq_m$ is a partial order on $\mathcal{P}$; this order is called "monotone dominance." It is well known that an equivalent description of this order is given by the following

Proposition 4.4.2. Let $\pi, \theta \in \Pi$, let ν, q be the probability distribution of θ and the density of θ, and let μ, p be the probability distribution and the density of π. Finally, let $\mathcal{M}_b$ be the family of all continuously differentiable even and bounded functions on the axis that are nondecreasing on $\mathbb{R}_+$. Then $\theta \succeq_m \pi$ if and only if

$$\int f(s)d\nu(s) \geq \int f(s)d\mu(s)ds \ \forall f \in \mathcal{M}_b, \tag{4.4.1}$$

same as if and only if

$$\int f(s)q(s)ds \geq \int f(s)p(s)ds \ \forall f \in \mathcal{M}_b. \tag{4.4.2}$$

Moreover, when (4.4.1) takes place, the inequalities in (4.4.1), (4.4.2) hold true for every even function on the axis that is nondecreasing on $\mathbb{R}_+$.

For the proof, see section B.1.5.

Example 4.4.3. Let $\xi \in \Pi$ be a random variable that is supported on $[-1, 1]$, ζ be uniformly distributed on $[-1, 1]$ and $\eta \sim \mathcal{N}(0, 2/\pi)$. We claim that $\xi \preceq_m \zeta \preceq_m \eta$.

Indeed, let $p(\cdot)$, $q(\cdot)$ be the densities of random variables $\pi, \theta \in \Pi$. Then the functions $P(t) = \int_t^\infty p(s)ds$, $Q(t) = \int_t^\infty q(s)ds$ of $t \geq 0$ are convex, and $\pi \preceq_m \theta$ iff $P(t) \leq Q(t)$ for all $t \geq 0$. Now let $\pi \in \Pi$ be supported on $[-1, 1]$ and θ be uniform on $[-1, 1]$. Then $P(t)$ is convex on $[0, \infty)$ with $P(0) \leq 1/2$ and $P(t) \equiv P(1) = 0$ for $t \geq 1$, while $Q(t) = \frac{1}{2}\max[1-t, 0]$ when $t \geq 0$. Since $Q(0) \geq P(0)$, $Q(1) = P(1)$, P is convex, and Q is linear on $[0, 1]$, we have $P(t) \leq Q(t)$ for all $t \in [0, 1]$, whence $P(t) \leq Q(t)$ for all $t \geq 0$, and thus $\pi \preceq_m \theta$. Now let π be uniform on $[-1, 1]$, so that $P(t) = \frac{1}{2}\max[1-t, 0]$, and θ be $\mathcal{N}(0, 2/\pi)$, so that $Q(t)$ is a convex function and therefore $Q(t) \geq Q(0) + Q'(0)t = (1-t)/2$

for all $t \geq 0$. This combines with $Q(t) \geq 0$, $t \geq 0$, to imply that $P(t) \leq Q(t)$ for all $t \geq 0$ and thus $\pi \preceq_m \theta$.

We start with the following observation:

Proposition 4.4.4. One has

(i) If $\xi, \eta \in \Pi$, λ is a deterministic real and $\eta \succeq_m \xi$, then $\lambda\eta \succeq_m \lambda\xi$.

(ii) If $\xi, \bar{\xi}, \eta, \bar{\eta} \in \Pi$ are independent random variables such that $\eta \succeq_m \xi$, $\bar{\eta} \succeq_m \bar{\xi}$, then $\xi + \bar{\xi} \in \Pi$, $\eta + \bar{\eta} \in \Pi$ and $\eta + \bar{\eta} \succeq_m \xi + \bar{\xi}$.

For the proof, see section B.1.5.
As a corollary of Proposition 4.4.4, we get our first majorization result:

Proposition 4.4.5. Let $z_0 \leq 0$, $z_1, ..., z_L$ be deterministic reals, $\{\zeta_\ell\}_{\ell=1}^{L}$ be independent random variables with unimodal and symmetric w.r.t. 0 distributions, and $\{\eta_\ell\}_{\ell=1}^{L}$ be a similar collection of independent random variables such that $\eta_\ell \succeq_m \zeta_\ell$ for every ℓ. Then

$$\text{Prob}\{z_0 + \sum_{\ell=1}^{L} z_\ell\zeta_\ell > 0\} \leq \text{Prob}\{z_0 + \sum_{\ell=1}^{L} z_\ell\eta_\ell > 0\}. \tag{4.4.3}$$

If, in addition, $\eta_\ell \sim \mathcal{N}(0, \sigma_\ell^2)$, $\ell = 1, ..., L$, then, for every $\epsilon \in (0, 1/2]$, one has

$$z_0 + \text{ErfInv}(\epsilon)\sqrt{\sum_{\ell=1}^{L} \sigma_\ell^2 z_\ell^2} \leq 0 \Rightarrow \text{Prob}\{z_0 + \sum_{\ell=1}^{L} \zeta_\ell z_\ell > 0\} \leq \epsilon, \tag{4.4.4}$$

where $\text{ErfInv}(\cdot)$ is the inverse error function (2.3.22).

Proof. By Proposition 4.4.4.(i) the random variables $\widehat{\zeta}_\ell = z_\ell\zeta_\ell$ and $\widehat{\eta}_\ell = z_\ell\eta_\ell$ are linked by $\widehat{\eta}_\ell \succeq_m \widehat{\zeta}_\ell$. This, by Proposition 4.4.4.(ii), implies that

$$\widehat{\eta} := \sum_{\ell=1}^{L} \widehat{\eta}_\ell \succeq_m \widehat{\zeta} := \sum_{\ell=1}^{L} \widehat{\zeta}_\ell.$$

The latter, by definition of $\succeq_m$, implies that

$$\text{Prob}\{z_0 + \sum_{\ell=1}^{L} z_\ell\zeta_\ell > 0\} = \text{Prob}\{\widehat{\zeta} > |z_0|\} \leq \text{Prob}\{\widehat{\eta} > |z_0|\}$$
$$= \text{Prob}\{z_0 + \sum_{\ell=1}^{L} z_\ell\eta_\ell > 0\}.$$

The concluding claim in Proposition 4.4.5 is readily given by the fact that under the premise of the claim we have $\widehat{\eta} \sim \mathcal{N}(0, \sum_{\ell=1}^{L} \sigma_\ell^2 z_\ell^2)$. $\qquad \square$

Relation (4.4.4) seems to be the major "yield" we can extract from Proposition 4.4.4, since the case of independent $\mathcal{N}(0, \sigma_\ell^2)$ random variables η_ℓ is, essentially, the only interesting case for which we can easily compute $\text{Prob}\{z_0 + \sum_{\ell=1}^{L} z_\ell\eta_\ell > 0\}$ and the chance constraint $\text{Prob}\{z_0 + \sum_{\ell=1}^{L} z_\ell\eta_\ell > 0\} \leq \epsilon$ for $\epsilon \leq 1/2$ is equivalent to an explicit convex constraint, specifically,

$$z_0 + \mathrm{ErfInv}(\epsilon)\sqrt{\sum_{\ell=1}^{L} \sigma_\ell^2 z_\ell^2} \leq 0. \tag{4.4.5}$$

Comparison with Proposition 2.4.1. Assume that independent random variables $\zeta_\ell \in \Pi$, $\ell = 1, ..., L$, admit "Gaussian upper bounds" $\eta_\ell \succeq_m \zeta_\ell$ with $\eta_\ell \sim \mathcal{N}(0, \sigma_\ell^2)$. Then ζ satisfies assumptions P.1–2 from section 2.4 with the parameters $\mu_\ell^\pm = 0, \sigma_\ell$.

Indeed, all we should prove is that if $\eta \succeq_m \zeta$ and $\eta \sim \mathcal{N}(0, \sigma^2)$, then

$$\int \exp\{ts\}d\mu(s) \leq \exp\{\sigma^2 t^2/2\} \ \forall t,$$

where μ is the distribution of ζ. Using the symmetry of μ w.r.t. 0, we have

$$\int \exp\{ts\}d\mu(s) = \int \cosh(ts)d\mu(s) \leq \int \cosh(ts)\frac{1}{\sqrt{2\pi}\sigma}\exp\{-s^2/(2\sigma^2)\}ds$$
$$= \exp\{\sigma^2 t^2/2\},$$

where "$\leq$" is given by Proposition 4.4.2 due to the fact that $\zeta \preceq_m \eta \sim \mathcal{N}(0, \sigma^2)$.

Since ζ satisfies P.1–2 with the parameters $\mu_\ell^\pm = 0, \sigma_\ell$, our previous results, (i.e., Proposition 2.4.1) state that

$$\left(z_0 + \sqrt{2\log(1/\epsilon)}\sqrt{\sum_{\ell=1}^{L}\sigma_\ell^2 z_\ell^2} \leq 0\right) \Rightarrow \left(\mathrm{Prob}\left\{z_0 + \sum_{\ell=1}^{L} z_\ell \zeta_\ell > 0\right\} \leq \epsilon\right). \tag{4.4.6}$$

The only disadvantage of this result as compared to (4.4.4) is in the fact that $\mathrm{ErfInv}(\epsilon) < \sqrt{2\ln(1/\epsilon)}$.

4.4.1 Majorization Theorem

Proposition 4.4.5 can be rephrased as follows:

> Let $\{\zeta_\ell\}_{\ell=1}^{L}$ be independent random variables with unimodal and symmetric w.r.t. 0 distributions, and $\{\eta_\ell\}_{\ell=1}^{L}$ be a similar collection of independent random variables such that $\eta_\ell \succeq_m \zeta_\ell$ for every ℓ. Given a deterministic vector $z \in \mathbb{R}^L$ and $z_0 \leq 0$, consider the "strip"
>
> $$S = \{x \in \mathbb{R}^L : |z^T x| \leq -z_0\}.$$
>
> Then
> $$\mathrm{Prob}\{[\zeta_1; ...; \zeta_L] \in S\} \geq \mathrm{Prob}\{[\eta_1; ...; \eta_L] \in S\}.$$

It turns out that the resulting inequality holds true for every closed convex set S that is symmetric w.r.t. the origin.

Theorem 4.4.6. [Majorization Theorem] Let $\{\zeta_\ell\}_{\ell=1}^{L}$ be independent random variables with unimodal and symmetric w.r.t. 0 distributions, and $\{\eta_\ell\}_{\ell=1}^{L}$ be a similar collection of independent random variables such that $\eta_\ell \succeq_{\mathrm{m}} \zeta_\ell$ for every ℓ. Then for every closed convex set $S \subset \mathbb{R}^L$ that is symmetric w.r.t. the origin one has

$$\mathrm{Prob}\{[\zeta_1; ...; \zeta_L] \in S\} \geq \mathrm{Prob}\{[\eta_1; ...; \eta_L] \in S\}. \tag{4.4.7}$$

For the proof, see section B.1.6.

Example 4.4.7. Let $\xi \sim \mathcal{N}(0, \Sigma)$ and $\eta \sim \mathcal{N}(0, \Theta)$ be two Gaussian random vectors taking values in $\mathbb{R}^n$ and let $\Sigma \preceq \Theta$. We claim that for every closed convex set $S \subset \mathbb{R}^n$ symmetric w.r.t. 0 one has

$$\mathrm{Prob}\{\xi \in S\} \geq \mathrm{Prob}\{\eta \in S\}.$$

Indeed, by continuity reasons, it suffices to consider the case when Θ is nondegenerate. Passing from random vectors ξ, η to random vectors $A\xi, A\eta$ with properly defined nonsingular A, we can reduce the situation to the one where $\Theta = I$ and Σ is diagonal, meaning that the densities $p(\cdot)$ of ξ and q of η are of the forms

$$p(x) = p_1(x_1)...p_n(x_n), \ q(x) = q_1(x_1)...q_n(x_n),$$

with $p_i(s)$ being the $\mathcal{N}(0, \Sigma_{ii})$ densities, and $q_i(s)$ being the $\mathcal{N}(0, 1)$ densities. Since $\Sigma \preceq \Theta = I$, we have $\Sigma_{ii} \leq 1$, meaning that $p_i \preceq_{\mathrm{m}} q_i$ for all i. It remains to apply the Majorization Theorem.

4.5 BEYOND THE CASE OF INDEPENDENT LINEAR PERTURBATIONS

So far, we dealt with a linearly perturbed chance constraint

$$p(z) \equiv \mathrm{Prob}\left\{z_0 + \sum_{\ell=1}^{L} z_\ell \zeta_\ell > 0\right\} \leq \epsilon \tag{4.0.1}$$

in the case when the random perturbations $\zeta_1, ..., \zeta_L$ are independent. In this section, we consider the case when the perturbations are dependent and/or enter nonlinearly the body of the chance constraint.

4.5.1 Dependent Linear Perturbations

Here we remove the assumption that the perturbations in (4.0.1) are independent. Instead, we make the following

> **Assumption S:** *All our a priori information on the distribution of ζ reduces to knowledge of some sets $\mathcal{P}_\ell$, $\ell = 1, ..., L$, in the space of probability distributions with well defined first moment on the axis such that the distributions P_ℓ of ζ_ℓ belong to the respective sets. In particular, we know nothing about the structure of dependence between the perturbations $\zeta_1, ..., \zeta_L$.*

We want to build a safe convex approximation of the ambiguous version of (4.0.1) associated with Assumption S, which is the constraint

Whenever the marginal distributions P_ℓ of $\zeta = [\zeta_1; ...; \zeta_L]$ belong to $\mathcal{P}_\ell$, $\ell = 1, ..., L$, one has

$$\text{Prob}\left\{z_0 + \sum_{\ell=1}^{L} z_\ell \zeta_\ell > 0\right\} \leq \epsilon.$$

(4.5.1)

To this end, we can use an approach similar to the one in section 4.2, specifically, as follows. Assume that we can point out functions $\gamma_\ell(\cdot)$ on the axis such that

$$
\begin{aligned}
(a) \quad & \gamma(u) \equiv \sum_{\ell=1}^{L} \gamma_\ell(u_\ell) \geq 0 \,\forall u \in \mathbb{R}^L \\
(b) \quad & z_0 + \sum_{\ell=1}^{L} u_\ell z_\ell > 0 \Rightarrow \gamma(u) \geq 1.
\end{aligned}
$$

(4.5.2)

Then clearly $p(z) \leq \mathbf{E}\{\gamma(\zeta)\}$, so that the condition

$$\mathbf{E}\{\gamma(\zeta)\} \leq \epsilon$$

is a sufficient condition for the validity of (4.0.1). Now, an evident *necessary and sufficient* condition for the validity of (4.5.2.a) is

$$\sum_{\ell=1}^{L} \inf_{u_\ell \in \mathbb{R}} \gamma_\ell(u_\ell) \geq 0,$$

(4.5.3)

while an evident *sufficient* condition for the validity of (4.5.2.b) is

$$\exists \lambda > 0: \inf_{u \in \mathbb{R}^L} \left\{\lambda(\gamma(u) - 1) - z_0 - \sum_{\ell=1}^{L} z_\ell u_\ell\right\} \geq 0.$$

(4.5.4)

These observations pave the road to the following.

Theorem 4.5.1. If z can be extended to a feasible solution of the system

$$
\begin{aligned}
\left[\begin{array}{c|c} \lambda & 1 \\ \hline 1 & \tau \end{array}\right] \succeq 0, \qquad & \sum_{\ell=1}^{L} \beta_\ell \geq \lambda + z_0, \\
\sum_{\ell=1}^{L} \sup_{P_\ell \in \mathcal{P}_\ell} \int \max[0, z_\ell s + \beta_\ell] dP_\ell(s) \leq \lambda \epsilon &
\end{aligned}
$$

(4.5.5)

of convex constraints in variables $z, \lambda, \beta_\ell, \tau$, then z is feasible for the ambiguous chance constraint (4.5.1) associated with Assumption S. Thus, (4.5.5) is a safe convex approximation of (4.5.1); this approximation is tractable, provided that given z_ℓ, β_ℓ, the quantities

$$\sup_{P_\ell \in \mathcal{P}_\ell} \int \max[0, z_\ell s + \beta_\ell] dP_\ell(s), \ \ell = 1, ..., L,$$

are efficiently computable.

Proof. 1^0. Observe, first, that the condition

$$
\exists \left(\begin{array}{c} \alpha \in \mathbb{R}^L, \beta \in \mathbb{R}^L, \lambda \\ \{\gamma_\ell(\cdot)\}_{\ell=1}^L \end{array} \right) : \left. \begin{array}{ll} \lambda > 0 & (a) \\ \sum_\ell \alpha_\ell \geq 0 & (b.1) \\ \gamma_\ell(s) \geq \alpha_\ell \,\forall s \in \mathbb{R} & (b.2) \\ \sum_\ell \beta_\ell \geq \lambda + z_0 & (c.1) \\ \lambda \gamma_\ell(s) \geq z_\ell s + \beta_\ell \,\forall s \in \mathbb{R} & (c.2) \\ \sum_{\ell=1}^L \mathbf{E}\{\gamma_\ell(\zeta_\ell)\} \leq \epsilon & (d) \end{array} \right\}
\qquad (4.5.6)
$$

is a sufficient condition for the validity of (4.0.1).

Indeed, let $(\{\alpha_\ell, \beta_\ell\}_{\ell=1}^L, \lambda, \{\gamma_\ell(\cdot)\}_{\ell=1}^L)$ satisfy (4.5.6.a-d), and let

$$
\gamma(u) = \sum_{\ell=1}^L \gamma_\ell(u_\ell).
$$

By (4.5.6.b) we have $\gamma(\cdot) \geq 0$, and by (4.5.6.a,c), we have

$$
\lambda \left[\gamma(u) - 1\right] - z_0 - \sum_{\ell=1}^L z_\ell u_\ell \geq \sum_{\ell=1}^L \beta_\ell - \lambda - z_0 \geq 0
$$

for all u, whence $z_0 + \sum_{\ell=1}^L u_\ell z_\ell > 0 \Rightarrow \gamma(u) \geq 1$. According to the reasoning preceding Theorem 4.5.1, the latter relation along with $\gamma(\cdot) \geq 0$ implies that $p(z) \leq \mathbf{E}\{\gamma(\zeta)\}$, which combines with (4.5.6.d) to imply that $p(z) \leq \epsilon$, as claimed.

2^0. Now let us prove that the condition (4.5.6) is equivalent to the condition

$$
\exists \left(\begin{array}{c} \theta \in \mathbb{R}^L, \beta \in \mathbb{R}^L, \lambda, \\ \{\delta_\ell(\cdot)\}_{\ell=1}^L \end{array} \right) : \left. \begin{array}{ll} \lambda > 0 & (a) \\ \sum_\ell \theta_\ell \geq 0 & (b.1) \\ \delta_\ell(s) \geq \theta_\ell \,\forall s \in \mathbb{R} & (b.2) \\ \sum_\ell \beta_\ell \geq \lambda + z_0 & (c.1) \\ \delta_\ell(s) \geq z_\ell s + \beta_\ell \,\forall s \in \mathbb{R} & (c.2) \\ \sum_\ell \mathbf{E}\{\delta_\ell(\zeta_\ell)\} \leq \lambda \epsilon & (d) \end{array} \right\},
\qquad (4.5.7)
$$

which in turn is equivalent to the condition

$$
\exists \left(\theta \in \mathbb{R}^L, \beta \in \mathbb{R}^L, \lambda \right) : \left. \begin{array}{ll} \lambda > 0 & (a) \\ \sum_\ell \theta_\ell \geq 0 & (b) \\ \sum_\ell \beta_\ell \geq \lambda + z_0 & (c) \\ \sum_\ell \mathbf{E}\{\max[\theta_\ell, z_\ell \zeta_\ell + \beta_\ell]\} \leq \lambda \epsilon & (d) \end{array} \right\}.
\qquad (4.5.8)
$$

Indeed, passing in the condition (4.5.6) from the variables $\alpha_\ell, \beta_\ell, \lambda, \gamma_\ell(\cdot)$ to the variables $\theta_\ell = \alpha_\ell \lambda, \beta_\ell, \lambda, \delta_\ell(\cdot) = \lambda \gamma_\ell(\cdot)$, the condition becomes exactly (4.5.7), so that (4.5.6) and (4.5.7) are equivalent. Now, the conditions (4.5.7.a-d) hold true for certain $\theta_\ell, \beta_\ell, \lambda, \delta_\ell(\cdot)$ if and only if these conditions remain true when $\theta_\ell, \beta_\ell, \lambda$ are kept intact and $\delta_\ell(\cdot)$ are replaced with the functions $\max[\theta_\ell, \zeta_\ell s + \beta_\ell] \leq \delta_\ell(s)$; since this transformation can only decrease the left hand side in (4.5.7.d), the condition (4.5.7) is equivalent to (4.5.8).

3^0. Now note that the condition (4.5.8) is equivalent to the condition

$$\exists \left(\beta \in \mathbb{R}^L, \lambda\right) : \quad \begin{array}{ll} \lambda > 0 & (a) \\ \sum_\ell \beta_\ell \geq \lambda + z_0 & (c) \\ \sum_\ell \mathbf{E}\{\max[0, z_\ell \zeta_\ell + \beta_\ell]\} \leq \lambda\epsilon & (d) \end{array} \Bigg\} . \qquad (4.5.9)$$

Indeed, (4.5.9) clearly implies the condition (4.5.8) (set $\theta_\ell = 0$ for all ℓ). Conversely, assume that (4.5.8) takes place, and let us prove (4.5.9). Observe, first, that we lose nothing by assuming that θ_ℓ in (4.5.8) satisfy $\sum_\ell \theta_\ell = 0$ rather than $\sum_\ell \theta_\ell \geq 0$; indeed, reducing, say, θ_1 to make the inequality $\sum_\ell \theta_\ell \geq 0$ an equality, we clearly do not violate the validity of (4.5.8.d). Now, assuming that θ_ℓ, β_ℓ satisfy all relations in (4.5.8) and $\sum_\ell \theta_\ell = 0$ and setting $\beta'_\ell = \beta_\ell - \theta_\ell$, we get $\sum_\ell \beta'_\ell = \sum_\ell \beta_\ell \geq \lambda + z_0$ and

$$\lambda\epsilon \geq \sum_\ell \mathbf{E}\{\max[\theta_\ell, z_\ell\zeta_\ell + \beta_\ell]\} = \sum_\ell \mathbf{E}\{\theta_\ell + \max[0, z_\ell\zeta_\ell + \beta'_\ell]\}$$
$$= \sum_\ell \theta_\ell + \sum_\ell \mathbf{E}\{\max[0, z_\ell\zeta_\ell + \beta'_\ell]\} = \sum_\ell \mathbf{E}\{\max[0, z_\ell\zeta_\ell + \beta'_\ell]\},$$

so that β'_ℓ, $\ell = 1, ..., d$, satisfy (4.5.9).

4^0. As a consequence of 1^0 through 3^0, the condition (4.5.9) is sufficient for the validity of (4.0.1) for any distribution of ζ compatible with Assumption S. $\quad\square$

4.5.2 A Modification

The simple idea we have used (its scientific name is "Lagrange relaxation") can be utilized in a closely related situation, specifically, as follows. Assume that we are given a piecewise linear convex function on the axis:

$$f(s) = \max_{1 \leq j \leq J} [a_j + b_j s] \qquad (4.5.10)$$

and we wish to bound from above the expectation

$$F(z) = \mathbf{E}\left\{ f\left(z_0 + \sum_{\ell=1}^{L} z_\ell\zeta_\ell \right) \right\}, \qquad (4.5.11)$$

where z is a deterministic parameter vector, ζ is a vector of random perturbations, and all our a priori information on the distribution of ζ is as stated in Assumption S.

Our starting point is the following observation: if a separable Borel function $\gamma(u) = \sum_{\ell=1}^{L} \gamma_\ell(u_\ell) : \mathbb{R}^L \to \mathbb{R}$ is everywhere $\geq$ than the function $g(u) = f(z_0 + \sum_{\ell=1}^{L} z_\ell u_\ell)$, then the quantity $\sum_{\ell=1}^{L} \mathbf{E}\{\gamma_\ell(\zeta_\ell)\}$ is an upper bound on $F(z)$:

$$F(z) \leq \Phi[f, z] \equiv \inf_{\gamma(\cdot) \in \Gamma_z} \left\{ \sum_{\ell=1}^{L} \sup_{P_\ell \in \mathcal{P}_\ell} \int \gamma_\ell(u_\ell) dP_\ell(u_\ell) \right\}$$
$$\Gamma_z = \{\gamma(u) = \sum_{\ell=1}^{L} \gamma_\ell(u_\ell) : \gamma(u) \geq f(z_0 + \sum_{\ell=1}^{L} z_\ell u_\ell) \,\forall u\} \qquad (4.5.12)$$

This leads to the following result:

Theorem 4.5.2. Relation (4.5.12) can be rewritten equivalently as

$$
F(z) \leq \Phi[f(\cdot), z]
$$
$$
= \inf_{\{\alpha_{\ell j}\}} \left\{ \sum_{\ell=1}^{L} \sup_{P_{\ell} \in \mathcal{P}_{\ell}} \int \max_{1 \leq j \leq J} [\alpha_{\ell j} + b_j z_{\ell} u_{\ell}] \, dP_{\ell}(u_{\ell}) : \begin{array}{c} \sum_{\ell=1}^{L} \alpha_{\ell j} = a_j + b_j z_0, \\ 1 \leq j \leq J \end{array} \right\}. \quad (4.5.13)
$$

The right hand side of the latter relation is a convex function of z, and this function is efficiently computable provided that we can compute efficiently the quantities of the form

$$
\sup_{P_{\ell} \in \mathcal{P}_{\ell}} \int g_{\ell}(u_{\ell}) dP_{\ell}(u_{\ell})
$$

associated with piecewise linear convex functions $g_{\ell}(\cdot)$ with at most J explicitly given linear pieces.

Proof. 1^0. Observe, first, that

$$
\Gamma_z = \left\{ \sum_{\ell} \gamma_{\ell}(u_{\ell}) : \exists \{\alpha_{\ell j}\}_{\substack{1 \leq j \leq J \\ 1 \leq \ell \leq L}} : \begin{array}{c} \sum_{\ell=1}^{L} \alpha_{\ell j} \geq a_j + b_j z_0, \\ 1 \leq j \leq J \\ \gamma_{\ell}(u_{\ell}) \geq \alpha_{\ell j} + b_j z_{\ell} u_{\ell} \, \forall u_{\ell}, \\ 1 \leq \ell \leq L, 1 \leq j \leq J \end{array} \right\}. \quad (4.5.14)
$$

Indeed, we have

$$
\sum_{\ell=1}^{L} \gamma_{\ell}(u_{\ell}) \in \Gamma_z
$$
$$
\Leftrightarrow \quad \forall j \leq J : \sum_{\ell=1}^{L} \gamma_{\ell}(u_{\ell}) \geq a_j + b_j \left[z_0 + \sum_{\ell=1}^{L} z_{\ell} u_{\ell} \right] \, \forall u
$$
$$
\Leftrightarrow \quad \forall j \leq J : \sum_{\ell=1}^{L} [\gamma_{\ell}(u_{\ell}) - b_j z_{\ell} u_{\ell}] \geq a_j + b_j z_0 \, \forall u
$$
$$
\Leftrightarrow \quad \forall j \leq J \, \exists \{\alpha_{\ell j}\}_{\ell=1}^{L} : \left\{ \begin{array}{c} \sum_{\ell=1}^{L} \alpha_{\ell j} \geq a_j + b_j z_0 \\ \gamma_{\ell}(u_{\ell}) \geq \alpha_{\ell j} + b_j z_{\ell} u_{\ell} \, \forall u_{\ell}, \, 1 \leq \ell \leq L \end{array} \right.
$$

as required in (4.5.14).

2^0. We claim that

$$
\Phi[f(\cdot), z]
$$
$$
= \inf_{\{\alpha_{\ell j}\}} \left\{ \sum_{\ell=1}^{L} \sup_{P_{\ell} \in \mathcal{P}_{\ell}} \int \max_{1 \leq j \leq J} [\alpha_{\ell j} + b_j z_{\ell} u_{\ell}] \, dP_{\ell}(u_{\ell}) : \begin{array}{c} \sum_{\ell=1}^{L} \alpha_{\ell j} \geq a_j + b_j z_0, \\ 1 \leq j \leq J \end{array} \right\} \quad (4.5.15)
$$

Indeed, by 1^0 the inf in the right hand side of (4.5.12) remains intact when we restrict the domain of minimization to functions $\gamma(u)$ of the form

$$
\sum_{\ell=1}^{L} \max_{1 \leq j \leq J} [\alpha_{\ell j} + b_j z_{\ell} u_{\ell}]
$$

with $\alpha_{\ell j}$ satisfying the constraints in (4.5.15).

3^0. Finally, we claim that the inequality constraints in (4.5.15) can be replaced with equalities without affecting the inf in the right hand side of (4.5.15), so that (4.5.15) is equivalent to (4.5.13), which would complete the proof.

The claim is evident: given a feasible solution to the optimization problem in the right hand side of (4.5.15), we can decrease appropriately the variables $\alpha_{11},...,\alpha_{1J}$ in order to make all the constraints equalities; clearly this transformation can only decrease the value of the objective of the problem in question. □

To proceed, we need the following simple observation:

Proposition 4.5.3. The functional $\Phi[f(\cdot), z]$, where $f(\cdot)$ belongs to the family $\mathcal{CL}$ of piecewise linear convex functions on the axis, possesses the following properties:

(i) Φ is well-defined: $\Phi[f, z]$ depends on f and z, but is independent of a particular representation (4.5.10) of f as the maximum of a collection of affine functions;

(ii) [homogeneity] $\Phi[\lambda f, z] = \lambda \Phi[f, z]$ whenever $\lambda \geq 0$;

(iii) [monotonicity] $\Phi[f, z] \leq \Phi[g, z]$, provided that $f \leq g$ and $f, g \in \mathcal{CL}$;

(iv) [sub-additivity] $\Phi[f + g, z] \leq \Phi[f, z] + \Phi[g, z]$, provided that $f, g \in \mathcal{CL}$.

All these facts follow immediately from the fact that $\Phi[f, z]$ is the optimal value of the optimization problem in the right hand side of (4.5.12).

The case of $\mathcal{P}_\ell = \{P_\ell\}$. In the case when all $\mathcal{P}_\ell$ are singletons, i.e., the distributions of ζ_ℓ are known exactly, we have the following nice result inspired by a remarkable paper [46] (in fact we are offering here an alternative proof to the main result of this paper):

Proposition 4.5.4. The bound $\Phi[f, z]$ is unimprovable: for $z, P_1, ..., P_L$ fixed, one can point out a collection of random variables $\zeta_1,...,\zeta_L$ with distributions $P_1, ..., P_L$ such that

$$\mathbf{E}\{f(z_0 + \sum_{\ell=1}^{L} \zeta_\ell z_\ell)\} = \Phi[f, z] \tag{4.5.16}$$

for all convex piecewise linear (and therefore for all convex) functions $f(\cdot)$.

This statement deals with *fixed* z and $P_1, ..., P_L$; passing from the distributions P_ℓ to distributions P'_ℓ of random variables $z_\ell \zeta_\ell$ with $\zeta_\ell \sim P_\ell$ and adding distribution P_{L+1} that sits at the point z_0, the situation can be reduced to the one where $z_1 = z_2 = ... = z_L = 1$ and $z_0 = 0$. In this "normalized" situation, the collection of random variables ζ_ℓ that makes (4.5.16) valid can be defined by the following construction from [46]:

As it is well-known, for every Borel probability distribution $P(t) = \mathrm{Prob}\{\xi \leq t\}$ of a scalar random variable ξ, there exists a nondecreasing continuous from the left function $\phi_P(s)$ on $(0, 1)$ (namely, $\phi_P(s) = \inf\{t : P(t) \geq s\}$) such that the distribution of the random variable $\phi_P(\nu)$ with ν uniformly distributed on $(0, 1)$ is exactly P. Let $\phi_\ell(\cdot)$, $\ell = 1, ..., L$, be nondecreasing continuous from the left functions on $(0, 1)$ that "produce" in this fashion the distributions $P_1, ..., P_L$. The desired collection of random variables $\zeta_1, ..., \zeta_L$ is nothing but

$$\zeta_1 = \phi_1(\nu), ..., \zeta_L = \phi_L(\nu) \qquad (4.5.17)$$

with ν uniformly distributed on $(0, 1)$.

Note that our ζ_ℓ are deterministic (and monotone) transformations of a *common* random variable ν — in a sense, a situation that is completely opposite to independence.

For the proof of Proposition 4.5.4 see section B.1.7.

4.5.3 Utilizing Covariance Matrix

Now let us add to (just partial) knowledge of marginal distributions of ζ_ℓ some knowledge of the covariance matrix of ζ. Specifically, let us "upgrade" Assumption S to

> **Assumption T:** *All our a priori information on the distribution of ζ reduces to knowledge of some sets $\mathcal{P}_\ell$, $\ell = 1, ..., L$, in the space of probability distributions on the axis with finite second moment such that the distributions P_ℓ of ζ_ℓ belong to the respective sets, plus knowledge of certain set $\mathcal{V} \subset \mathbf{S}_+^L$ such that $V_\zeta := \mathbf{E}\{\zeta\zeta^T\} \in \mathcal{V}$.*

We want to build a safe convex approximation of the ambiguous version of (4.0.1) associated with Assumption T, that is, of the constraint

> *Whenever the marginal distributions P_ℓ of $\zeta = [\zeta_1; ...; \zeta_L]$ belong to $\mathcal{P}_\ell$, $\ell = 1, ..., L$, and $V_\zeta \in \mathcal{V}$, one has*
> $$p(z) := \mathrm{Prob}\left\{z_0 + \textstyle\sum_{\ell=1}^{L} z_\ell\zeta_\ell > 0\right\} \leq \epsilon. \qquad (4.5.18)$$

Now we know, to some extent, expectations of functions of ζ that are more general than in the previous case, specifically, of functions of the form

$$\zeta^T\Gamma\zeta + 2\sum_{\ell=1}^{L}\gamma_\ell(\zeta_\ell), \ \Gamma \in \mathbf{S}^L.$$

We can therefore modify the previous approach as follows: whenever the condition

$$\begin{aligned}(a) & \quad \gamma(u) \equiv u^T\Gamma u + 2\textstyle\sum_{\ell=1}^{L}\gamma_\ell(u_\ell) \geq 0 \,\forall u \in \mathbb{R}^L \\ (b) & \quad z_0 + \textstyle\sum_{\ell=1}^{L} u_\ell z_\ell > 0 \Rightarrow \gamma(u) \geq 1,\end{aligned} \qquad (4.5.19)$$

holds, we clearly have $p(z) \leq \mathbf{E}\{\gamma(\zeta)\}$, so that the condition

$$\mathbf{E}\{\gamma(\zeta)\} \leq \epsilon$$

is a sufficient condition for the validity of (4.0.1). What remains is to extract from this condition a safe convex approximation of (4.0.1). We are about to derive such an approximation under additional restriction that we work only with $\Gamma \succeq 0$ and convex $\gamma_\ell(\cdot)$, $\ell = 1, ..., L$.

Theorem 4.5.5. If z can be extended to a solution of the system

(a) $\left[\begin{array}{c|c} \lambda & 1 \\ \hline 1 & \tau \end{array}\right] \succeq 0$ (b) $\left[\begin{array}{c|c} \Delta & \theta \\ \hline \theta^T & 2\sum_\ell \beta_\ell \end{array}\right] \succeq 0$

(c) $\left[\begin{array}{c|c} \Delta & \widehat{\theta} - \frac{1}{2}[z_1;...;z_L] \\ \hline (\widehat{\theta} - \frac{1}{2}[z_1;...;z_L])^T & 2\sum_\ell \widehat{\beta}_\ell - \lambda - z_0 \end{array}\right] \succeq 0$ (4.5.20)

(d) $\displaystyle\sup_{V \in \mathcal{V}} \mathrm{Tr}(V\Delta) + 2\sum_\ell \sup_{P_\ell \in \mathcal{P}_\ell} \int \max[\beta_\ell + \theta_\ell s, \widehat{\beta}_\ell + \widehat{\theta}_\ell s] dP_\ell(s) \leq \lambda\epsilon$

of convex constraints in variables z, $\Delta \in \mathbf{S}^L$, $\theta \in \mathbb{R}^L$, $\beta \in \mathbb{R}^L$, $\widehat{\theta} \in \mathbb{R}^L$, $\widehat{\beta} \in \mathbb{R}^L$, λ, τ, then z is feasible for the ambiguous chance constraint (4.5.18) associated with Assumption T. Thus, (4.5.20) is a safe convex approximation of (4.5.18); this approximation is tractable provided that given $\Delta, \beta_\ell, \theta_\ell, \widehat{\beta}_\ell, \widehat{\theta}_\ell$, the quantities $\sup_{V \in \mathcal{V}} \mathrm{Tr}(V\Delta)$ and

$$\sup_{P_\ell \in \mathcal{P}_\ell} \int \max[\beta_\ell + \theta_\ell s, \widehat{\beta}_\ell + \widehat{\theta}_\ell s] dP_\ell(s), \ \ell = 1, ..., L,$$

are efficiently computable.

Proof. 1^0. Observe, first, that the condition

$\exists(\lambda, \{\delta_\ell(\cdot)\}_{\ell=1}^L, \Delta):$

$$\begin{aligned} \lambda &> 0 & (a) \\ u^T\Delta u + 2\sum_\ell \delta_\ell(u_\ell) &\geq 0 \ \forall u \in \mathbb{R}^L & (b) \\ u^T\Delta u + 2\sum_\ell \delta_\ell(u_\ell) - \sum_\ell u_\ell z_\ell &\geq \lambda + z_0 \ \forall u \in \mathbb{R}^L & (c) \\ \mathbf{E}\{\zeta^T\Delta\zeta + 2\sum_\ell \delta_\ell(\zeta_\ell)\} &\leq \lambda\epsilon & (d) \end{aligned}\right\}$$ (4.5.21)

is a sufficient condition for the validity of the relation $p(z) \leq \epsilon$.

Indeed, let $(\lambda, \{\delta_\ell(\cdot)\}_{\ell=1}^L, \Delta)$ satisfy (4.5.21.a–d). Setting $\Gamma = \lambda^{-1}\Delta$, $\gamma_\ell(\cdot) = \lambda^{-1}\delta_\ell(\cdot)$, $\gamma(u) = u^T\Gamma u + 2\sum_\ell \gamma_\ell(u_\ell)$, we get that $\gamma(\cdot)$ satisfies (4.5.19.a) (by (4.5.21.b)) and (4.5.19.b) (by (4.5.21.c)), while $\mathbf{E}\{\gamma(\zeta)\} \leq \epsilon$ (by (4.5.21.d)).

2^0. Now let $\Delta \succ 0$, let $\phi_\ell(\cdot)$, $\ell = 1, ..., L$, be convex real-valued functions on the axis, and let $G(u) = u^T\Delta u + 2\sum_\ell \phi_\ell(u_\ell)$. We claim that

1) If $G(u) \geq 0$ for all u, then

$$\exists\left(\theta \in \mathbb{R}^L, \beta \in \mathbb{R}^L\right): \begin{array}{l} \phi_\ell(s) \geq \beta_\ell + \theta_\ell s \ \forall(s \in \mathbb{R}, \ell \leq L) \quad (a) \\ \left[\begin{array}{c|c} \Delta & \theta \\ \hline \theta^T & 2\sum_\ell \beta_\ell \end{array}\right] \succeq 0 \hspace{2.2cm} (b) \end{array}\right\}.$$ (4.5.22)

2) If condition (4.5.22) takes place then $G(u) \geq 0$ for all u, and this conclusion is valid independently of the convexity of $\phi_\ell(\cdot)$ and the assumption $\Delta \succ 0$.

To prove 1), let $G(\cdot) \geq 0$. By evident reasons, the convex problem $\min_u G(u)$ has a solution u^*, and from optimality conditions there exist $\theta_\ell \in \partial\phi_\ell(u_\ell^*)$ such that $\Delta u^* + \theta = 0$. Setting $\beta_\ell = \phi_\ell(u_\ell^*) - \theta_\ell u_\ell^*$, consider the quadratic form

$$\widehat{G}(u) = u^T\Delta u + 2\sum_\ell [\beta_\ell + \theta_\ell u_\ell].$$

By construction, $\widehat{G}$ is convex, has zero gradient at u^* and $\widehat{G}(u^*) = G(u^*) \geq 0$, so that $\widehat{G}(\cdot) \geq 0$, and therefore (4.5.22.$b$) takes place. Due to the convexity of $\phi_\ell(\cdot)$ and the origin of θ_ℓ, β_ℓ, (4.5.22.a) holds true as well. 1) is proved.

To prove 2), let $\widehat{G}(u) = u^T \Delta u + 2 \sum_\ell [\beta_\ell + \theta_\ell u_\ell]$. We have $\widehat{G}(\cdot) \geq 0$ by (4.5.22.b) and $\widehat{G}(\cdot) \leq G(\cdot)$ by (4.5.22.a), whence $G(\cdot) \geq 0$.

3^0. We are ready to complete the proof. Assuming that (4.5.20) takes place, let us set $G(u) = u^T \Delta u + 2 \sum_\ell \max[\beta_\ell + \theta_\ell u_\ell, \widehat{\beta}_\ell + \widehat{\theta}_\ell u_\ell]$. Invoking 2^0.2), this function is ≥ 0 everywhere (by (4.5.20.b)) and satisfies the relation $G(u) - \sum_\ell u_\ell z_\ell \geq \lambda + z_0$ for all u (by (4.5.20.c)). By (4.5.20.d) we have $\mathbf{E}\{G(\zeta)\} \leq \lambda \epsilon$. Finally, (4.5.20.$a$) implies that $\lambda > 0$. These relations combine with 1^0 to imply that $p(z) \leq \epsilon$. $\square$

4.5.4 Illustration

Let us compare the performance of the various approximations suggested by our developments using the following simple chance constrained problem

$$\text{Opt}(\epsilon) = \max \left\{ t : \text{Prob} \left\{ \zeta^{10} \equiv \sum_{\ell=1}^{10} \zeta_\ell > t \right\} \geq 1 - \epsilon \right\}. \tag{4.5.23}$$

The "cover story" might be as follows:

> You have a portfolio with unit investments in every one of 10 assets. The yearly return for asset #ℓ is ζ_ℓ. You should find the Value-at-Risk ϵ of the portfolio, that is, the largest t such that the value of the portfolio in a year from now is $< t$ with probability at most ϵ.

Our setups are as follows:

- the returns ζ_ℓ are log-normal random variables of the form

$$\zeta_\ell = \exp\{\mu_\ell + \sigma_\ell e_\ell^T \eta\},$$

where μ_ℓ is the deterministic *trend* (expected log of the return ζ_ℓ), $\sigma_\ell > 0$ is the deterministic variability of the log of the return, $\eta \sim \mathcal{N}(0, I_m)$ is the vector of random factors, common for all returns, underlying the actual values of the returns and e_ℓ are m-dimensional deterministic unit vectors indicating how the factors η affect individual returns.

- we consider two sets of the data:

DataI : $\mu_\ell = \sigma_\ell = \ln(1.25)$, $\ell = 1, ..., 10$, $m = 1$, $e_\ell = 1$, $\ell = 1, ..., 10$ (that is, the returns are equal to each other).

DataII : σ_ℓ, μ_ℓ are as in DataI, and the e_ℓ^T are the rows of the following matrix:

$$\begin{bmatrix}
0.7559 & -0.1997 & 0.6235 \\
0.2861 & -0.8873 & 0.3616 \\
-0.9516 & 0.2221 & -0.2124 \\
-0.5155 & -0.8472 & -0.1286 \\
0.9354 & 0.2621 & -0.2374 \\
-0.7447 & -0.3724 & 0.5538 \\
-0.9315 & -0.2806 & 0.2316 \\
0.0721 & 0.3435 & 0.9364 \\
0.2890 & -0.8465 & 0.4472 \\
-0.9159 & 0.4003 & -0.0292
\end{bmatrix}$$

The numerical results are shown in tables 4.2, 4.3. The notation in the tables is as follows:

- Bound A is the one given by Theorem 4.5.1;

- Bound B is the one given by Theorem 4.5.5;

- Bound C is the "engineering bound"

$$\text{Opt} = \mathbf{E}\{\zeta^{10}\} - \text{ErfInv}(\epsilon)\text{StD}\{\zeta^{10}\}$$

  built *as if* the random variable ζ^{10} were Gaussian, (note that in contrast to Bounds A, B that provably underestimate $\text{Opt}(\epsilon)$, Bound C can overestimate this quantity);

- Bound D is the empirical lower ϵ quantile of the distribution of $\sum_\ell \zeta_\ell$ computed over 1,000,000 realizations of ζ^{10}.

Probabilities in the tables are empirical probabilities computed over 10^6 realizations of ζ^{10}. For the case of DataI, it is easy to compute the true value of $\text{Opt}(\epsilon)$, and we provide it in table 4.2. As it could be guessed in advance, utilizing covariances does not help in the case of "degenerate" data DataI (rows "Bound A" and "Bound B" in table 4.2); good news is that it helps significantly in the case of DataII. We see also that the "engineering" bound C *on our data* is safe and most of the time outperforms the bounds A, B; it is, however, worse than these bounds on DataI with "small" ϵ.

4.5.5 Extensions to Quadratically Perturbed Chance Constraints

Consider next a chance constrained version of the randomly perturbed quadratic inequality

$$p(W, w) \equiv \text{Prob}\left\{ \zeta^T W \zeta + 2 \sum_{\ell=1}^{L} \zeta_\ell w_\ell + w_0 > 0 \right\} \leq \epsilon \qquad (4.5.24)$$

in variables $z = (W, w) \in \mathbf{S}^L \times \mathbb{R}^{L+1}$, $\zeta \in \mathbb{R}^L$ being random perturbations.

We start with the case when all we know about ζ is that the distributions of ζ_ℓ belong to given families $\mathcal{P}_\ell$, $\ell = 1, ..., L$. Exactly the same reasoning as in the proof of Theorem 4.5.5 yields the following results.

	ϵ			
	0.100	0.050	0.010	0.005
Bound A	8.484	7.915	6.912	6.570
$\mathrm{Prob}\{\zeta^{10} < \text{Bound A}\}$	0.0413	0.0190	0.0040	0.0020
Bound B	8.484	7.915	6.912	6.570
$\mathrm{Prob}\{\zeta^{10} < \text{Bound B}\}$	0.0413	0.0190	0.0040	0.0020
Bound C	9.104	8.052	6.079	5.357
$\mathrm{Prob}\{\zeta^{10} < \text{Bound C}\}$	0.0777	0.0230	0.0006	0.0001
Bound D	9.389	8.656	7.445	7.040
$\mathrm{Opt}(\epsilon)$	9.394	8.669	7.449	7.044

Table 4.2 Results for `DataI`.

	ϵ			
	0.100	0.050	0.010	0.005
Bound A	8.484	7.915	6.912	6.570
$\mathrm{Prob}\{\zeta^{10} < \text{Bound A}\}$	0.0000	0.0000	0.0000	0.0000
Bound B	10.371	9.823	8.289	7.471
$\mathrm{Prob}\{\zeta^{10} < \text{Bound B}\}$	0.0020	0.0000	0.0000	0.0000
Bound C	11.382	10.976	10.213	9.934
$\mathrm{Prob}\{\zeta^{10} < \text{Bound C}\}$	0.0810	0.0267	0.0008	0.0001
Bound D	11.478	11.185	10.702	10.545

Table 4.3 Results for `DataII`.

Theorem 4.5.6. Let ζ satisfy Assumption S and let all distributions form the sets $\mathcal{P}_\ell$, $\ell = 1, ..., L$, possess finite second moments. Then the condition

$$
\exists \left(\begin{array}{c} \theta \in \mathbb{R}^L, \beta \in \mathbb{R}^L, \lambda, \\ \widehat{\theta} \in \mathbb{R}^L, \widehat{\beta} \in \mathbb{R}^L, \mu \in \mathbb{R}^L \end{array} \right) :
$$
$$
\left[\begin{array}{c|c} \lambda & 1 \\ \hline 1 & \tau \end{array} \right] \succeq 0, \quad \left[\begin{array}{c|c} \mathrm{Diag}\{\mu\} & \theta \\ \hline \theta^T & 2\sum_\ell \beta_\ell \end{array} \right] \succeq 0
$$
$$
\left[\begin{array}{c|c} \mathrm{Diag}\{\mu\} - W & \widehat{\theta} - [w_1; ...; w_L] \\ \hline [\widehat{\theta} - [w_1; ...; w_L]]^T & 2\sum_\ell \widehat{\beta}_\ell - \lambda - w_0 \end{array} \right] \succeq 0
$$
$$
\sum_{\ell=1}^{L} \sup_{P_\ell \in \mathcal{P}_\ell} \int [\mu_\ell s^2 + 2\max[\beta_\ell + \theta_\ell s, \widehat{\beta}_\ell + \widehat{\theta}_\ell s]] dP_\ell(s) \leq \lambda \epsilon
$$

(4.5.25)

is sufficient for the validity of the ambiguous chance constraint

"Whenever the distributions P_ℓ of ζ_ℓ belong to $\mathcal{P}_\ell$, $1 \leq \ell \leq L$, one has
$\mathrm{Prob}\left\{ \zeta^T W \zeta + 2\sum_{\ell=1}^{L} w_\ell \zeta_\ell + w_0 > 0 \right\} \leq \epsilon.$"

Theorem 4.5.7. Let ζ satisfy Assumption T. Then the condition

$$\exists \left(\begin{array}{c} \theta \in \mathbb{R}^L, \beta \in \mathbb{R}^L, \lambda, \tau, \\ \widehat{\theta} \in \mathbb{R}^L, \widehat{\beta} \in \mathbb{R}^L, \Delta \in \mathbf{S}^L \end{array} \right) :$$

$$\left[\begin{array}{c|c} \lambda & 1 \\ \hline 1 & \tau \end{array} \right] \succeq 0, \quad \left[\begin{array}{c|c} \Delta & \theta \\ \hline \theta^T & 2 \sum_\ell \beta_\ell \end{array} \right] \succeq 0$$

$$\left[\begin{array}{c|c} \Delta - W & \widehat{\theta} - [w_1; ...; w_L] \\ \hline [\widehat{\theta} - [w_1; ...; w_L]]^T & 2 \sum_\ell \widehat{\beta}_\ell - \lambda - w_0 \end{array} \right] \succeq 0$$

$$\sup_{V \in \mathcal{V}} \mathrm{Tr}(V \Delta) + 2 \sum_{\ell=1}^L \sup_{P_\ell \in \mathcal{P}_\ell} \int \max[\beta_\ell + \theta_\ell s, \widehat{\beta}_\ell + \widehat{\theta}_\ell s] dP_\ell(s) \leq \lambda \epsilon$$

(4.5.26)

is sufficient for the validity of the ambiguous chance constraint

"*Whenever the distributions P_ℓ of ζ_ℓ belong to $\mathcal{P}_\ell$, $1 \leq \ell \leq L$, and $V_\zeta := \mathbf{E}\{\zeta\zeta^T\} \in \mathcal{V}$, one has $\mathrm{Prob}\left\{\zeta^T W \zeta + 2 \sum_{\ell=1}^L w_\ell \zeta_\ell + w_0 > 0\right\} \leq \epsilon$.*"

4.5.5.1 Refinements in the case of Gaussian perturbations.

Now assume that $\zeta \sim \mathcal{N}(0, I)$. Let

$$\begin{aligned} F(W, w) &= w_0 - \tfrac{1}{2} \ln \mathrm{Det}(I - 2W) + 2b^T (I - 2W)^{-1}[w_1; ...; w_L] \\ \mathrm{Dom}\, F &= \{(W, w) \in \mathbf{S}^L \times \mathbb{R}^{L+1} : 2W \prec I\} \end{aligned}$$

(4.5.27)

Our interest in this function stems from the following immediate observation:

Lemma 4.5.8. Let $\zeta \sim \mathcal{N}(0, I)$, and let

$$\xi = \xi^{W,w} = \zeta^T W \zeta + 2[w_1; ...; w_L]^T \zeta + w_0.$$

Then $\ln \left(\mathbf{E} \left\{ \exp\{\xi^{W,w}\} \right\} \right) = F(W, w)$.

Applying the Bernstein approximation scheme (section 4.2), we arrive at the following result:

Theorem 4.5.9. Let

$$\begin{aligned} \Phi(\beta, W, w) &= \beta F(\beta^{-1}(W, w)) \\ &= \beta \left[-\tfrac{1}{2} \ln \mathrm{Det}(I - 2\beta^{-1}W) \right. \\ &\quad \left. + 2\beta^{-2}[w_1; ...; w_L]^T (I - 2\beta^{-1}W)^{-1}[w_1; ...; w_L] \right] + w_0, \\ \mathrm{Dom}\, \Phi &= \{(\beta, W, w) : \beta > 0, 2W \prec \beta I\}, \\ Z_\epsilon^o &= \left\{(W, w) : \exists \beta > 0 : \Phi(\beta, W, w) + \beta \ln(1/\epsilon) \leq 0\right\}, \\ Z_\epsilon &= \mathrm{cl}\, Z_\epsilon^o. \end{aligned}$$

(4.5.28)

Then Z_ϵ is the solution set of the convex inequality

$$H(W, w) \equiv \inf_{\beta > 0} [\Phi(\beta, W, w) + \beta \ln(1/\epsilon)] \leq 0. \tag{4.5.29}$$

If $\zeta \sim \mathcal{N}(0, I)$, then this inequality is a safe tractable approximation of the chance constraint (4.5.24).

For the proof, see section B.1.8.

Application: A useful inequality. Let W be a symmetric $L \times L$ matrix and w be an L-dimensional vector. Consider the quadratic form

$$f(s) = s^T W s + 2w^T s,$$

and let $\zeta \sim \mathcal{N}(0, I)$. We clearly have $\mathbf{E}\{f(\zeta)\} = \mathrm{Tr}(W)$. Our goal is to establish a simple bound on $\mathrm{Prob}\{f(\zeta) - \mathrm{Tr}(W) > t\}$, and here is this bound:

Proposition 4.5.10. Let λ be the vector of eigenvalues of W. Then

$$\forall \Omega > 0 : \mathrm{Prob}_{\zeta \sim \mathcal{N}(0,I)} \left\{ [\zeta^T W \zeta + 2w^T \zeta] - \mathrm{Tr}(W) > \Omega \sqrt{\lambda^T \lambda + w^T w} \right\}$$
$$\leq \exp \left\{ -\frac{\Omega^2 \sqrt{\lambda^T \lambda + w^T w}}{4 \left(2\sqrt{\lambda^T \lambda + w^T w} + \|\lambda\|_\infty \Omega \right)} \right\} \tag{4.5.30}$$

(by definition, the right hand side is 0 when $W = 0$, $w = 0$).

Proof. The claim is clearly true in the trivial case of $W = 0$, $w = 0$, thus assume that f is not identically zero. Passing to the orthonormal eigenbasis of W, we can w.l.o.g. assume that W is diagonal with diagonal entries $\lambda_1, ..., \lambda_L$. Given $\Omega > 0$, let us set $s = \Omega \sqrt{\lambda^T \lambda + w^T w}$ and let

$$\gamma = \frac{s}{2 \left(2(\lambda^T \lambda + w^T w) + \|\lambda\|_\infty s \right)},$$

so that

$$0 < \gamma \ \& \ 2\gamma W \prec I \ \& \ \frac{4\gamma(\lambda^T \lambda + w^T w)}{1 - 2\gamma \|\lambda\|_\infty} = s. \tag{4.5.31}$$

Applying Theorem 4.5.9 with $w_0 = -[\mathrm{Tr}(W) + s]$ and specifying β as $1/\gamma$, we get

$$\mathrm{Prob}\{f(\zeta) > \mathrm{Tr}(W) + s\}$$
$$\leq \exp \left\{ -\gamma s + \sum_{\ell=1}^L \left(-\frac{1}{2} \ln(1 - 2\gamma\lambda_\ell) + 2\gamma^2 \frac{w_\ell^2}{1 - 2\gamma\lambda_\ell} - \gamma\lambda_\ell \right) \right\}$$
$$\leq \exp \left\{ -\gamma s + \sum_{\ell=1}^L \left(\frac{\gamma\lambda_\ell}{1 - 2\gamma\lambda_\ell} + 2\gamma^2 \frac{w_\ell^2}{1 - 2\gamma\lambda_\ell} - \gamma\lambda_\ell \right) \right\}$$
$$[\text{since } \ln(1 - \delta) + \frac{\delta}{1 - \delta} \geq \ln(1) = 0 \text{ by the concavity of } \ln(\cdot)]$$
$$= \exp \left\{ -\gamma s + \sum_{\ell=1}^L \left(\frac{2\gamma^2(\lambda_\ell^2 + w_\ell^2)}{1 - 2\gamma\lambda_\ell} \right) \right\} \leq \exp \left\{ -\gamma s + \frac{2\gamma^2(\lambda^T \lambda + w^T w)}{1 - 2\gamma \|\lambda\|_\infty} \right\}$$
$$\leq \exp\{-\frac{\gamma s}{2}\}$$
$$[\text{by (4.5.31)}] \ .$$

Substituting the values of γ and s, we arrive at (4.5.30). $\qquad \square$

Application: Linearly perturbed Least Squares inequality. Consider a chance constrained linearly perturbed Least Squares inequality

$$\mathrm{Prob} \left\{ \|A[x]\zeta + b[x]\|_2 \leq c[x] \right\} \geq 1 - \epsilon, \tag{4.5.32}$$

where $A[x]$, $b[x]$, $c[x]$ are affine in the variables x and $\zeta \sim \mathcal{N}(0, I)$. Taking squares of both sides in the body of the constraint, this inequality is equivalent to

$$\exists U, u, u_0:$$
$$\left[\begin{array}{c|c} U & u^T \\ \hline u & u_0 \end{array}\right] \succeq \left[\begin{array}{c|c} A^T[x]A[x] & A^T[x]b[x] \\ \hline b^T[x]A[x] & b^T[x]b[x] - c^2[x] \end{array}\right],$$
$$\text{Prob}\left\{\zeta^T U \zeta + 2\sum_{\ell=1}^{L} u_\ell \zeta_\ell + u_0 > 0\right\} \leq \epsilon.$$

Assuming $c[x] > 0$, passing from U, u variables to $W = c^{-1}[x]U$, $w = c^{-1}[x]u$, and dividing both sides of the LMI by $c[x]$, this can be rewritten equivalently as

$$\exists (W, w, w_0):$$
$$\left[\begin{array}{c|c} W & w^T \\ \hline w & w_0 + c[x] \end{array}\right] \succeq c^{-1}[x][A[x], b[x]]^T [A[x], b[x]],$$
$$\text{Prob}\left\{\zeta^T W \zeta + 2\sum_{\ell=1}^{L} w_\ell \zeta_\ell + w_0 > 0\right\} \leq \epsilon.$$

The constraint linking W, w and x is, by the Schur Complement Lemma, nothing but the Linear Matrix Inequality

$$\left[\begin{array}{c|c|c} W & [w_1; ...; w_L] & A^T[x] \\ \hline [w_1, ..., w_L] & w_0 + c[x] & b^T[x] \\ \hline A[x] & b[x] & c[x]I \end{array}\right] \succeq 0. \tag{4.5.33}$$

Invoking Theorem 4.5.9, we arrive at the following

Corollary 4.5.11. The system of convex constraints (4.5.33) and (4.5.29) in variables W, w, x is a safe tractable approximation of the chance constrained Least Squares Inequality (4.5.32).

Note that while we have derived this Corollary under the assumption that $c[x] > 0$, the result is trivially true when $c[x] = 0$, since in this case (4.5.33) already implies that $A[x] = 0$, $b[x] = 0$ and thus (4.5.32) holds true.

4.5.6 Utilizing Domain and Moment Information

We proceed with considering the chance constraint (4.5.24), which we now rewrite equivalently as

$$\text{Prob}\{A(W, w; \zeta) > 0\} \leq \epsilon,$$
$$\text{where}$$
$$A(W, w; u) := [u; 1]^T Z[W, w][u; 1], \tag{4.5.34}$$
$$Z[W, w] = \left[\begin{array}{c|c} W & [w_1; ...; w_L] \\ \hline [w_1; ...; w_L]^T & w_0 \end{array}\right]$$

In contrast to what was assumed in the previous subsection, now we make the following assumptions:

R.1) We have partial information on the expectation and the covariance matrix of ζ, specifically, we are given convex compact set $\mathcal{V} \subset \mathbf{S}_+^{L+1}$ that *contains the matrix*

$$V_\zeta = \mathbf{E}\left\{\left[\begin{array}{c|c} \zeta\zeta^T & \zeta \\ \hline \zeta^T & 1 \end{array}\right]\right\}.$$

Note that the matrix in question is $\mathbf{E}\{[\zeta; 1][\zeta; 1]^T\}$ and is therefore $\succeq 0$; besides this, $(V_\zeta)_{L+1,L+1} = 1$. This is why we lose nothing by assuming that $\mathcal{V} \subset \mathbf{S}_+^{L+1}$ and that $V_{L+1,L+1} = 1$ for all $V \in \mathcal{V}$.

R.2) ζ *is supported in a known set U given by a finite system of quadratic (not necessarily convex) constraints:*

$$U = \{u \in \mathbb{R}^L : f_j(u) = [u; 1]^T A_j [u; 1] \leq 0, \, j = 1, ..., m\},$$

where $A_j \in \mathbf{S}^{L+1}$.

We are about to build a safe tractable approximation of the chance constraint (4.5.34), and our strategy will combine the approach we have used when building Bernstein and CVaR approximations with Lagrange relaxation (see p. 112) and is very close to the strategy developed in [18]. Specifically, given a quadratic form

$$h(u) = u^T P u + 2p^T u + r = [u; 1]^T \underbrace{\left[\begin{array}{c|c} P & p \\ \hline p^T & r \end{array}\right]}_{H} [u; 1]$$

on $\mathbb{R}^{L+1}$, assumption R.1 allows us to bound from above the expectation of $h(\zeta)$:

$$\mathbf{E}\{h(\zeta)\} = \mathrm{Tr}(HV_\zeta) \leq \max_{V \in \mathcal{V}} \mathrm{Tr}(HV).$$

Now assume that $h(\cdot)$ is nonnegative everywhere on U and is > 1 at every $u \in U \backslash Q[W, w]$, where

$$Q[W, w] = \{u \in \mathbb{R}^L : A(W, w; u) \leq 0\}.$$

Then $h(u)$ everywhere in U is an upper bound on the characteristic function of the set $U \backslash Q[W, w]$; since ζ is supported in U, we have essentially proved the following

Lemma 4.5.12. Let $\mathcal{H}[W, w]$ be the set of all symmetric matrices $H \in \mathbf{S}^{L+1}$ such that

$$\begin{aligned} (a) \quad & [u; 1]^T H [u; 1] \geq 0 \, \forall u \in U \\ (b) \quad & \inf_{u \in U} \left\{ -A(W, w; u) : [u; 1]^T [H - E][u; 1] \leq 0 \right\} \geq 0, \end{aligned} \tag{4.5.35}$$

where

$$E = \left[\begin{array}{c|c} & \\ \hline & 1 \end{array}\right] \in \mathbf{S}^{L+1}.$$

Then

$$\begin{aligned} p(W, w) &\equiv \mathrm{Prob}\{A(W, w; \zeta) > 0\} \leq \inf_{H \in \mathcal{H}[W, w]} \psi(H), \\ \psi(H) &= \max_{V \in \mathcal{V}} \mathrm{Tr}(HV). \end{aligned} \tag{4.5.36}$$

Proof. Let $H \in \mathcal{H}[W, w]$ and $h(u) = [u; 1]^T H [u; 1]$. By (4.5.35.a) we have $h(u) \geq 0$ for all $u \in U$. Besides this, by (4.5.35.b), if $u \in U \backslash Q[W, w]$, that is, if

$-A(W, w; u) < 0$, we have $h(u) - 1 = [u; 1]^T [H - E][u; 1] > 0$; thus, $h(u) > 1$ everywhere on $U \backslash Q[W, w]$. It follows that

$$p(W, w) \leq \mathbf{E}\{h(\zeta)\} = \mathrm{Tr}(HV_\zeta) \leq \max_{V \in \mathcal{V}} \mathrm{Tr}(HV).$$

Thus,

$$\forall H \in \mathcal{H}[W, w] : p(W, w) \leq \psi(H),$$

and (4.5.36) follows.   □

Our local goal is to extract from Lemma 4.5.12 a safe tractable approximation of (4.5.34).

Observe, first, that by evident reasons one has

$$\left(\exists \lambda \in \mathbb{R}_+^m : H + \sum_{j=1}^{m} \lambda_j A_j \succeq 0 \right) \Rightarrow H \text{ satisfies } (4.5.35.a) \qquad (4.5.37)$$

and

$$\left(\exists (\mu \in \mathbb{R}_+^m, \gamma > 0) : \gamma(H - E) - Z[W, w] + \sum_{j=1}^{m} \mu_j A_j \succeq 0 \right) \\ \Rightarrow H \text{ satisfies } (4.5.35.b). \qquad (4.5.38)$$

Essentially, we have established the following:

Proposition 4.5.13. The condition

$\exists P, \nu, \mu, \gamma, \tau :$

$(a.1) \quad P + \sum_j \nu_j A_j \succeq 0 \qquad\qquad (a.2) \quad \nu \geq 0$

$(b.1) \quad P - \gamma E - Z[W, w] + \sum_j \mu_j A_j \succeq 0 \quad (b.2) \quad \mu \geq 0 \qquad (4.5.39)$

$(c) \quad \psi(P) := \max_{V \in \mathcal{V}} \mathrm{Tr}(PV) \leq \gamma \epsilon \qquad (d) \quad \left[\begin{array}{c|c} \gamma & 1 \\ \hline 1 & \tau \end{array} \right] \succeq 0$

is sufficient for (W, w) to satisfy the chance constraint (4.5.34) and thus defines a safe convex approximation of the constraint. This approximation is computationally tractable, provided that $\psi(\cdot)$ is efficiently computable.

Proof. Consider the condition

$\exists (H \in \mathbf{S}^{L+1}, \lambda \in \mathbb{R}_+^m, \mu \in \mathbb{R}_+^m, \gamma > 0) :$

$$\begin{array}{ll} H + \sum_j \lambda_j A_j \succeq 0 & (a) \\ \gamma(H - E) - Z[W, w] + \sum_j \mu_j A_j \succeq 0 & (b) \\ \gamma \max_{V \in \mathcal{V}} \mathrm{Tr}(VH) - \gamma \epsilon \leq 0 & (c) \end{array} \qquad (4.5.40)$$

We claim that this condition is sufficient for the validity of the chance constraint (4.5.34).

Indeed, let $(H, \lambda \geq 0, \mu \geq 0, \gamma > 0)$ satisfy the relations in (4.5.40). By (4.5.40.a) and due to (4.5.37), H satisfies (4.5.35.a). By (4.5.40.b) and due to (4.5.38), H satisfies (4.5.35.b). Combining Lemma 4.5.12 and (4.5.40.c), we derive from these observations that $p(W, w) \leq \epsilon$, as claimed.

To complete the proof, it remains to show that (4.5.40) is equivalent to (4.5.39). Indeed, let $(H, \lambda \geq 0, \mu \geq 0, \gamma > 0)$ be such that the conditions (4.5.40.a-c) take place. Let us set $P = \gamma H$, $\nu = \gamma \lambda$, $\tau = 1/\gamma$. Then $(P, \nu, \mu, \tau, \gamma)$ clearly satisfies (4.5.39.a, b, d). Since ψ clearly is homogeneous of degree 1, (4.5.39.c) is satisfied as well. Vice versa, if $(P, \nu, \mu, \tau, \gamma)$ satisfies (4.5.39.a-c), then $\gamma > 0$ by (4.5.39.d); setting $H = \gamma^{-1}P$, $\lambda = \gamma^{-1}\nu$ and taking into account that ψ is homogeneous of degree 1, we have $\lambda \geq 0, \nu \geq 0, \gamma > 0$ and $(H, \lambda, \mu, \gamma)$ satisfies (4.5.40). Thus, conditions (4.5.40) and (4.5.39) are equivalent. $\qquad \square$

Approximation in Proposition 4.5.13 admits a useful modification as follows:

Proposition 4.5.14. Given $\epsilon > 0$, consider the chance constraint (4.5.34) and assume that

$$\text{Prob}\left\{\zeta : [\zeta; 1]^T A_j [\zeta; 1] \leq 0, \ j = 1, ..., m\right\} \geq 1 - \delta$$

for some known $\delta \in [0, \epsilon]$. Then the condition

$\exists P, \mu, \gamma, \tau :$

(a) $\quad P \succeq 0$

(b.1) $\quad P - \gamma E - Z[W, w] + \sum_j \mu_j A_j \succeq 0 \quad$ (b.2) $\quad \mu \geq 0 \qquad\qquad$ (4.5.41)

(c) $\quad \psi(P) := \max_{V \in \mathcal{V}} \text{Tr}(PV) \leq \gamma[\epsilon - \delta] \quad$ (d) $\quad \left[\begin{array}{c|c} \gamma & 1 \\ \hline 1 & \tau \end{array}\right] \succeq 0$

is sufficient for (W, w) to satisfy the chance constraint (4.5.34).

Proof. Let $(P, \mu, \gamma, \tau, W, w)$ be a feasible solution to (4.5.41.a–d); note that $\gamma > 0$ by (4.5.41.d). Setting $H = \gamma^{-1}P$ and taking into account that ψ is homogeneous of degree 1, we have

$$
\begin{aligned}
&h(u) \equiv [u; 1]^T H[u, 1] \geq 0 \, \forall u && \text{[by (4.5.41.a)]} \\
&u \in G \equiv \{u : [u; 1]^T A_j[u; 1] \leq 0, \ 1 \leq j \leq m\} \\
&\quad \Rightarrow [u; 1]^T Z[W, w][u; 1] \leq \gamma[h(u) - 1] && \text{[by (4.5.41.b)]} \\
&\max_{V \in \mathcal{V}} \text{Tr}(HV) \leq \epsilon - \delta && \text{[by (4.5.41.c)]}
\end{aligned}
$$

The first and the second of these relations say that the function $h(u)$ is everywhere nonnegative and is ≥ 1 whenever $u \in G$ is such that $[u; 1]^T Z[W, w][u; 1] > 0$. Denoting by $\chi(\cdot)$ the characteristic function of the set $G \backslash \{u : [u; 1]^T Z[W, w][u; 1] \leq 0\}$, we therefore have $\chi(u) \leq h(u)$ for all u. It follows that

$$\text{Prob}\{\zeta \in G \ \& \ [\zeta; 1]^T Z[W, w][\zeta; 1] > 0\} = \mathbf{E}\{\chi(\zeta)\} \leq \mathbf{E}\{h(\zeta)\} = \text{Tr}(HV_\zeta)$$
$$\leq \max_{V \in \mathcal{V}} \text{Tr}(HV) \leq \epsilon - \delta,$$

whence

$$\text{Prob}\{[\zeta; 1]^T Z[W, w][\zeta; 1] > 0\} \leq \text{Prob}\{\zeta \in G \ \& \ [\zeta; 1]^T Z[W, w][\zeta; 1] > 0\}$$
$$+\text{Prob}\{\zeta \notin G\} \leq (\epsilon - \delta) + \delta = \epsilon. \qquad\qquad \square$$

Basic properties of the approximation (4.5.39). These properties are as follows.

1. The approximation "respects invertible affine transformations." Specifically, let the random perturbations ζ and η be linked by $\zeta = R\eta + r$ with deterministic R, r and nonsingular R. Then the chance constraints

$$\begin{aligned}
(a) \quad & p(W, w) \equiv \text{Prob}\{[\zeta; 1]^T Z[W, w][\zeta; 1] > 0\} \leq \epsilon \\
(b) \quad & \widehat{p}(\widehat{W}, \widehat{w}) \equiv \text{Prob}\{[\eta; 1]^T Z[\widehat{W}, \widehat{w}][\eta; 1] > 0\} \leq \epsilon,
\end{aligned}$$

$$Z[\widehat{W}, \widehat{w}] = \underbrace{\left[\begin{array}{c|c} R & r \\ \hline & 1 \end{array}\right]^T Z[W, w] \left[\begin{array}{c|c} R & r \\ \hline & 1 \end{array}\right]}_{\mathcal{R}} \tag{4.5.42}$$

are equivalent to each other, and information on ζ given in assumptions R.1–2 induces similar information on η, namely,

$$\begin{aligned}
(a) \quad & V_\eta \equiv \mathbf{E}\left\{[\eta; 1][\eta; 1]^T\right\} \in \widehat{\mathcal{V}} = \{\mathcal{R}^{-1} V \mathcal{R}^{-T} : V \in \mathcal{V}\}, \\
(b) \quad & \text{Prob}\{[\eta; 1]^T \widehat{A}_j[\eta; 1] > 0\} = 0, j = 1, ..., m, \\
& \text{where } \widehat{A}_j = \mathcal{R}^T A_j \mathcal{R}.
\end{aligned} \tag{4.5.43}$$

It is easily seen that the approximations, given by Proposition 4.5.13, of the chance constraints (4.5.42.a,b) are also equivalent to each other: whenever the approximation of the first chance constraint says that a pair (W, w) is feasible for it, the approximation of the second chance constraint says that the pair $(\widehat{W}, \widehat{w})$ corresponding to (W, w) according to (4.5.42) is feasible for the second chance constraint, and vice versa.

2. Observe that given a system of quadratic inequalities defining U, we can always add to it linear combinations, with nonnegative coefficients, of the original quadratic inequalities and identically true ones. Such a "linear extension" of the original description of U results in a new approximation (4.5.39) of (4.5.34). It turns out that our approximation scheme is intelligent enough to recognize that such linear extension in fact adds no new information: it is easily seen that whenever (W, w) can be extended to a feasible solution of the system of constraints in (4.5.39) corresponding to the original description of U, (W, w) can be extended to a feasible solution to the similar system associated with a linear extension of this description, and vice versa.

3. The approximation is intelligent enough to recognize that probability is always ≤ 1: the condition (4.5.39) with $\epsilon > 1$ is always satisfied.

4. Consider the case when the body of (4.5.34) is *linearly* perturbed: $W = 0$, and, in addition, $\mathcal{V} = \{V_\zeta\}$ is a singleton, and ζ is centered: $V_\zeta = \left[\begin{array}{c|c} V & \\ \hline & 1 \end{array}\right]$. In this case, assuming that $w_0 < 0$, we can bound the probability

$$p(0, w) = \text{Prob}\left\{[\zeta; 1]^T Z[0, w][\zeta; 1] > 0\right\} \equiv \text{Prob}\left\{2\sum_{p=1}^{L} \zeta_i w_i + w_0 > 0\right\}$$

by Tschebyshev inequality. Specifically, setting $\bar{w} = [w_1; \dots; w_L]$, we have $2\bar{w}^T\zeta + w_0 > 0 \Rightarrow 2\bar{w}^T\zeta > |w_0| \Rightarrow 4(\bar{w}^T\zeta)^2 \geq w_0^2$, whence

$$p(0, w) \leq \min[1, 4\mathbf{E}\{(\bar{w}^T\zeta)^2\}/w_0^2] =: \bar{\epsilon}.$$

It is easily seen that in this special case the approximation (4.5.39) is intelligent enough to be at least as good as the outlined Tschebyshev bound, that is, if ϵ_* is the infimum of those ϵ for which the system of LMIs

$(a.1)$ $P + \sum_j \nu_j A_j \succeq 0$ $\qquad\qquad$ $(a.2)$ $\nu \geq 0$

$(b.1)$ $P - \gamma E - Z[0, w] + \sum_j \mu_j A_j \succeq 0$ $\qquad$ $(b.2)$ $\mu \geq 0$ $\qquad\qquad$ (4.5.44)

(c) $\mathrm{Tr}(P\mathrm{Diag}\{V, 1\}) - \gamma\epsilon \leq 0$ $\qquad$ (d) $\begin{bmatrix} \gamma & 1 \\ \hline 1 & \tau \end{bmatrix} \succeq 0$

in variables $P, \nu, \mu, \tau, \gamma$ is feasible, then $\epsilon_* \leq \bar{\epsilon}$.

Indeed, assume first that $\bar{\epsilon} < 1$, and set

$$\nu = 0, \mu = 0, \gamma = -w_0/2, \tau = 1/\gamma, P = \begin{bmatrix} \frac{1}{\gamma}\bar{w}\bar{w}^T & \\ \hline & 0 \end{bmatrix}.$$

For these values of the variables, relations $(4.5.44.a, b.2, d)$ clearly are valid. Further, since $-\gamma - w_0 = \gamma > 0$, we have

$$P - \gamma E - Z[0, w] + \sum_j \mu_j A_j = \begin{bmatrix} \frac{1}{\gamma}\bar{w}\bar{w}^T & -\bar{w} \\ \hline -\bar{w}^T & \gamma \end{bmatrix} \succeq 0,$$

so that $(4.5.44.b.1)$ takes place as well. Let us verify that $(4.5.44.c)$ is valid with $\epsilon = \bar{\epsilon}$ (this would imply that $\epsilon_* \leq \bar{\epsilon}$). Indeed,

$$\mathrm{Tr}(P\mathrm{Diag}\{V, 1\}) = \mathrm{Tr}(\gamma^{-1}\bar{w}\bar{w}^T V) = \frac{\bar{w}^T V \bar{w}}{\gamma} = \gamma\frac{4\bar{w}^T V \bar{w}}{w_0^2} = \gamma\bar{\epsilon},$$

as claimed. The case of $\bar{\epsilon} = 1$ is resolved by item 3.

5. Assume that $\mathcal{V} = \{\mathrm{Diag}\{V, 1\} : 0 \preceq V \preceq \widehat{V}\}$ (that is, we know that ζ has zero mean and the covariance matrix of ζ is $\preceq \widehat{V}$). Given an $n \times n$ matrix $Q \succeq 0$ and a positive α, we can bound the quantity $p \equiv \mathrm{Prob}\{\zeta^T Q\zeta > \alpha\}$ from above via the Tschebyshev bound:

$$p \leq \mathbf{E}\{\zeta^T Q\zeta\}/\alpha \leq \bar{\epsilon} = \min[1, \mathrm{Tr}(\widehat{V}Q)/\alpha].$$

It turns out that the condition (4.5.39) is intelligent enough to recover this bound. Indeed, let W, w be given by $Z[W, w] = \mathrm{Diag}\{Q, -\alpha\}$, and let ϵ_* be the infimum of those ϵ for which the system of LMIs

$(a.1)$ $P + \sum_j \nu_j A_j \succeq 0$ $\qquad\qquad$ $(a.2)$ $\nu \geq 0$

$(b.1)$ $P - \gamma E - Z[W, w] + \sum_j \mu_j A_j \succeq 0$ $\qquad$ $(b.2)$ $\mu \geq 0$ $\qquad\qquad$ (4.5.45)

(c) $\max_{0 \preceq V \preceq \widehat{V}} \mathrm{Tr}(P\mathrm{Diag}\{V, 1\}) - \gamma\epsilon \leq 0$ $\qquad$ (d) $\begin{bmatrix} \tau & 1 \\ \hline 1 & \gamma \end{bmatrix} \succeq 0$

in variables $P, \nu, \mu, \tau, \gamma$ has a solution; then $\epsilon_* \le \bar{\epsilon}$.

Indeed, assume first that $\bar{\epsilon} < 1$, and let us set

$$P = \mathrm{Diag}\{Q, 0\}, \nu = 0, \mu = 0, \gamma = \alpha, \tau = 1/\gamma.$$

This choice clearly ensures (4.5.45.a, b, d), and makes the left hand side in (4.5.45.c) equal to $\mathrm{Tr}(Q\widehat{V}) - \alpha\epsilon$; thus, (4.5.45) is satisfied when $\epsilon = \bar{\epsilon} \equiv \mathrm{Tr}(Q\widehat{V})/\alpha$, and therefore $\epsilon_* \le \bar{\epsilon}$. The case of $\bar{\epsilon} = 1$ is resolved by item 3.

6. Assume that $A'_j \preceq \theta_j A_j$, $\theta_j > 0$, $j = 1, ..., m$, and $(W, w), (W', w')$ are such that $\theta Z[W, w] \preceq Z[W', w']$ with $\theta > 0$, so that

$$\{u : [u; 1]^T A_j[u; 1] \le 0, 1 \le j \le m\} \subset \{u : [u; 1]^T A'_j[u; 1] \le 0, 1 \le j \le m\},$$
$$[u; 1]^T Z[W, w][u; 1] > 0 \Rightarrow [u; 1]^T Z[W', w'][u; 1] > 0.$$

In view of these relations, the validity of assumptions R.1–2 associated with some $\mathcal{V}$ and the data $\{A_j\}$ implies the validity of assumptions R.1–2 associated with the same $\mathcal{V}$ and the data $\{A'_j\}$, and for every random vector ζ one has

$$\mathrm{Prob}\{[\zeta; 1]^T Z[W', w'][\zeta; 1] > 0\} \ge \mathrm{Prob}\{[\zeta; 1]^T Z[W, w][\zeta; 1] > 0\}.$$

Thus, if for all distributions of ζ compatible with assumptions R.1–2, the data being $\{A'_j\}$, one has $\mathrm{Prob}\{[\zeta; 1]^T Z[W', w'][\zeta; 1] > 0\} \le \epsilon$, then for all distributions of ζ compatible with assumptions R.1–2, the data being $\{A_j\}$, one has $\mathrm{Prob}\{[\zeta; 1]^T Z[W, w][\zeta; 1] > 0\} \le \epsilon$.

It is easily seen that the approximation of (4.5.34) given by (4.5.39) is intelligent enough "to understand" the above conclusion. Specifically, if ϵ is such that the system of constraints

$$
\begin{array}{ll}
(a.1) \quad P' + \sum_j \nu'_j A'_j \succeq 0 & (a.2) \quad \nu' \ge 0 \\[2mm]
(b.1) \quad P' - \gamma' E - Z[W', w'] + \sum_j \mu'_j A'_j \succeq 0 & (b.2) \quad \mu' \ge 0 \\[2mm]
(c) \quad \psi(P') - \gamma'\epsilon \le 0 & (d) \quad \left[\begin{array}{c|c} \tau' & 1 \\ \hline 1 & \gamma' \end{array}\right] \succeq 0
\end{array}
$$
$$(4.5.46)$$

in variables $P', \nu', \mu', \tau', \gamma'$ is feasible, then the system of constraints

$$
\begin{array}{ll}
(a.1) \quad P + \sum_j \nu_j A_j \succeq 0 & (a.2) \quad \nu \ge 0 \\[2mm]
(b.1) \quad P - \gamma E - Z[W, w] + \sum_j \mu_j A_j \succeq 0 & (b.2) \quad \mu \ge 0 \\[2mm]
(c) \quad \psi(P) - \gamma\epsilon \le 0 & (d) \quad \left[\begin{array}{c|c} \tau & 1 \\ \hline 1 & \gamma \end{array}\right] \succeq 0
\end{array}
$$
$$(4.5.47)$$

in variables $P, \nu, \mu, \tau, \gamma$ is feasible as well.

Indeed, let $(P', \nu', \mu', \tau', \gamma')$ be a feasible solution to (4.5.46). Setting

$$\nu_j = \theta^{-1}\theta_j \nu'_j, \; \mu_j = \theta^{-1}\theta_j \mu'_j, \; P = \theta^{-1}P', \; \gamma = \theta^{-1}\gamma', \tau = \theta\tau',$$

and taking into account that $\psi(\cdot)$ is homogeneous of degree 1, it is immediately seen that $(P, \nu, \mu, \tau, \gamma)$ is a feasible solution to (4.5.47). For example, we have

$$P - \gamma E - Z[W, w] + \sum_j \mu_j A_j = \theta^{-1} \left[P' - \gamma' E - \theta Z[W, w] + \sum_j \mu'_j \theta_j A_j \right]$$
$$\underbrace{\succeq}_{(*)} \theta^{-1} \left[P' - \gamma' E - Z[W', w'] + \sum_j \mu'_j A'_j \right] \succeq 0,$$

where $(*)$ is given by $\theta Z[W, w] \preceq Z[W', w']$ and $\theta_j A_j \succeq A'_j$.

7. Assume that the body of the chance constraint (4.5.34) is linearly perturbed: $Z = Z[0, w] = \left[\begin{array}{c|c} & p \\ \hline p^T & q \end{array} \right]$ and that $q < 0$. Then we can write

$$[\zeta; 1]^T Z[0, w][\zeta; 1] > 0 \Leftrightarrow 2p^T \zeta > -q$$
$$\Rightarrow 4\zeta^T pp^T \zeta > q^2 \Leftrightarrow [\zeta; 1]^T Z[W', w'][\zeta; 1] > 0,$$
$$Z[W', w'] = \left[\begin{array}{c|c} 4pp^T & \\ \hline & -q^2 \end{array} \right]$$

whence also

$$[\zeta; 1]^T Z[0, w][\zeta; 1] > 0 \Rightarrow [\zeta; 1]^T Z[\beta W', \alpha w + \beta w'][\zeta; 1] > 0, \; 0 \neq [\alpha; \beta] \geq 0,$$

so that
$$\forall([\alpha; \beta] \geq 0, [\alpha; \beta] \neq 0) : \text{Prob}\{[\zeta; 1]^T Z[0, w][\zeta; 1] > 0\}$$
$$\leq \text{Prob}\{[\zeta; 1]^T Z[\beta W', \alpha w + \beta w'][\zeta; 1] > 0\}.$$

It follows that for every $0 \neq [\alpha; \beta] \geq 0$, the safe convex approximation, given by (4.5.39), of the chance constraint

$$\text{Prob}\{[\zeta; 1]^T Z[\beta W', \alpha w + \beta w'][\zeta; 1] > 0\} \leq \epsilon$$

is a safe convex approximation of the chance constraint of interest (4.5.34). Thus, we end up with seemingly a two-parametric *family* of safe tractable approximations of (4.5.34). Is there the best — the least conservative — member in this family? The answer is positive, and one of the best members is the original pair $(0, w)$ (corresponding to the choice $\alpha = 1, \beta = 0$).

Indeed, invoking item 6, in order to prove that the validity of (4.5.39) with $(\widetilde{W}, \widetilde{w}) = \alpha(0, w) + \beta(W', w')$ (where $0 \neq [\alpha; \beta] \geq 0$) in the role of $(0, w)$ implies the validity of (4.5.39) as it is, it suffices to prove that there exists $\theta > 0$ such that $Z[\widetilde{W}, \widetilde{w}] \succeq \theta Z[0, w]$. To this end note that $Z[W', w'] \succeq 2|q| Z[0, w]$:

$$Z[W', w'] - 2|q| Z[0, w] = \left[\begin{array}{c|c} 4pp^T & -2|q|p \\ \hline -2|q|p^T & q^2 \end{array} \right] \succeq 0.$$

Thus, $Z[\beta W', \alpha w + \beta w'] \succeq (\alpha + 2|q|\beta) Z[0, w]$. Since $0 \neq [\alpha; \beta] \geq 0$, the quantity $\alpha + 2|q|\beta$ is positive.

Strengthening approximation (4.5.39). Assume that the system of quadratic inequalities in R.2 contains a linear inequality, say,

$$f_1(u) = 2a^T u + \alpha \Leftrightarrow A_1 = \left[\begin{array}{c|c} & a \\ \hline a^T & \alpha \end{array} \right].$$

Assume, further, that we know a constant $\beta > \alpha$ such that

$$f_j(u) \leq 0, \ i = 1, ..., m \Rightarrow 2a^T u + \beta \geq 0.$$

Then we can add to the original constraints $f_j(u) \leq 0$, $j = 1, ..., m$, specifying U the redundant constraint $f_{m+1}(u) \equiv -2a^T u - \beta \leq 0$. Another option is to replace the linear constraint $f_1(u) \leq 0$ with the quadratic one

$$\widehat{f}_1(u) \equiv \frac{1}{4} f_1(u)(-f_{m+1}(u)) \equiv u^T a a^T u + \frac{\alpha + \beta}{2} a^T u + \frac{\alpha\beta}{4} \leq 0 \qquad (*)$$

and keep remaining constraints $f_j(u) \leq 0$, $j = 1, ..., m$, intact. Note that since $(*)$ clearly is valid on U, this transformation can only increase U (in fact, it keeps U intact) and therefore it preserves the validity of R.2. A natural question is, what is wiser (that is, what results in the less conservative safe approximation (4.5.39) of the chance constraint (4.5.34)):

A. To use the original description $f_j(u) \leq 0$, $j = 1, ..., m$ of U,

B. To describe U by the constraints $f_j(u) \leq 0$, $j = 1, ..., m+1$,

C. To describe U by the constraints $\widehat{f}_1(u) \leq 0$, $f_j(u) \leq 0$, $j = 2, ..., m$,

D. To describe U by the constraints $f_j(u) \leq 0$, $j = 1, ..., m+1$, $\widehat{f}_1(u) \leq 0$.

If we were adding to the system of constraints defining U its "linear consequences," see item 2 above, the options A through C would be the same. However, the constraint $f_{m+1}(u) \leq 0$, while being a consequence of the original constraints defining U, is not necessarily their *linear* consequence, so that item 2 does not apply now.

The correct answer is that in terms of conservatism, option B clearly is not worse than option A. A less trivial observation is that option C is at least as good as option B (and in fact can be much better than B), and is exactly as good as option D (that is, option C is the best — it is not more conservative than options A, B and is simpler than the equally conservative option D).

Indeed, option B means that we extend the collection of matrices A_j, $1 \leq j \leq m$, by adding the matrix $A_{m+1} = \begin{bmatrix} & -a \\ \hline -a^T & -\beta \end{bmatrix}$, option C means that we update the original collection by replacing A_1 with the matrix

$$\widehat{A}_1 = \begin{bmatrix} aa^T & pa \\ \hline pa^T & q \end{bmatrix}, \ p = \frac{\alpha + \beta}{4}, q = \frac{\alpha\beta}{4},$$

and option D means that we add to the original collection both the matrices A_{m+1} and $\widehat{A}_1$. Let us verify that $\widehat{A}_1 \succeq \theta_1 A_1$ and $\widehat{A}_1 \succeq \theta_{m+1} A_{m+1}$ with properly chosen $\theta_1, \theta_{m+1} > 0$; in view of the result of item 6 above, this would imply that option C is not worse than options B and D. Verification is immediate: we have $\beta - \alpha > 0$,

$$\widehat{A}_1 - \frac{\beta - \alpha}{4} A_1 = \begin{bmatrix} aa^T & \frac{\alpha}{2} a \\ \hline \frac{\alpha}{2} a^T & \frac{\alpha^2}{4} \end{bmatrix} \succeq 0$$

and

$$\widehat{A}_1 - \frac{\beta - \alpha}{4} A_{m+1} = \begin{bmatrix} aa^T & \frac{\beta}{2} a \\ \hline \frac{\beta}{2} a^T & \frac{\beta^2}{4} \end{bmatrix} \succeq 0.$$

Discussion. The innocently looking question of what is the best among the alternatives A through D leads to serious and challenging research questions related to the quality of Lagrange Relaxation. Indeed, the considerations that led us to the safe approximation (4.5.39) were based on the following evident observation:

A *sufficient* condition for a function $f(u)$ to be nonnegative on the domain $U = \{u : f_j(u) \leq 0, \, j = 1, ..., m\}$ is

$$\exists \lambda \geq 0 : \forall u : f(u) + \sum_i \lambda_i f_i(u) \geq 0. \qquad (*)$$

When f and all f_j are quadratic functions, (which is the case we are concerned with), this condition is tractable — it is equivalent to the existence of a feasible solution to an explicit system of LMIs. Now, aside from the cases when (a) all f, f_j are affine, and (b) all f, f_j are convex and the system of constraints $f_j(u) \leq 0$, $j = 1, ..., m$, is strictly feasible, condition $(*)$ is only sufficient, but not necessary, for the relation $\min_U f(u) \geq 0$. Clearly, the "gap" between the validity of the latter fact and the validity of the condition $(*)$ can only shrink when we add to the list of constraints specifying U their *consequences* — (quadratic) inequalities of the form $g(u) \leq 0$ that are valid on U. The question (its importance goes far beyond the topic of chance constraints) is, how to generate these consequences in such a way that the gap indeed shrinks, that is, $(*)$ is invalid before we add the consequence and becomes valid after we add it.

There are many ways to generate consequences of a system of quadratic constraints, the simplest ones being as follows:

i) "Linear aggregation" mentioned in item 2 above: we add to the list of the original constraints weighted sums, with nonnegative weights, of the original constraints and, perhaps, identically true quadratic inequalities (say, $-x^2 - y^2 + 2xy - 1 \leq 0$). This way "to add consequences" is of no interest; it cannot convert an invalid predicate $(*)$ into a valid one (cf. item 2 above).

ii) Passing from linear constraints to quadratic ones. Specifically, assume that $f_1(u)$ is linear and that we can bound from below this linear function on U, that is, we can find c such that $f_1(u) + c \geq 0$ for $u \in U$. Then the quadratic inequality $g(u) \equiv f_1(x)(f_1(x) + c) \leq 0$ is valid everywhere on U and thus it can be added to the list of constraints defining U. This modification indeed can convert invalid predicate $(*)$ into a valid one[1]. In fact there is no need to keep both the original linear constraint and the new quadratic in the list of constraints; we lose nothing by just replacing the linear inequality $f_1(u) \leq 0$ with the quadratic inequality $g(u) \leq 0$.

---

[1] Meaning, by the way, that Lagrange Relaxation "does not understand" the rule known to every kid: the product of two nonnegative reals is nonnegative.

iii) When among the original constraints $f_j(u) \leq 0$ there is a pair of linear ones, say, f_1 and f_2, we can add to the original constraints their quadratic consequence $g(u) \equiv -f_1(u)f_2(u) \leq 0$. This again can convert an invalid predicate (∗) into a valid one.

iv) When all the functions f_j are convex, we can build a consequence g as follows: take a linear form $e^T u$ and find its maximum β and its minimum α on U by solving the corresponding convex problems, so that the quadratic inequality $g(u) := (e^T x - \alpha)(e^T x - \beta) \leq 0$ is valid on U, and we can add it to the list of inequalities defining U. This again can convert an invalid predicate (∗) into a valid one.

The bottom line contains both good and bad news. The good news is that there are simple ways to shrink the gap between the target relation $\min_U f(u) \geq 0$ and the sufficient condition (∗) for this relation. The bad news is that we do not know how to use these ways in the best possible fashion. Take, e.g., the last (*iv*) "extension procedure": we can use it several times with different linear forms. How many times should we use the procedure and which linear forms to use? Note that the trivial answer "the more, the better" is of no interest, since the computational effort required to check (∗) grows with m as m^3. As for more intelligent "universal guidelines," we are not aware of their existence...

We conclude this short visit to the topic of Lagrange Relaxation by presenting three numerical illustrations. In all of them, we want to bound from above the probability of violating the constraint

$$[\zeta; 1]^T Z[W, w][\zeta; 1] \leq 0$$

given the following data:

i) We know that ζ is with zero mean and with a given covariance matrix V;

ii) The domain of ζ is known to belong to a given polytope U.

Illustration A. In this illustration, ζ is 3-dimensional and the constraint in question is linear:

$$a + b^T \zeta \leq 0 \qquad\qquad \left[Z[W, w] = \left[\begin{array}{c|c} & b \\ \hline b^T & 2a \end{array} \right] \right].$$

The covariance matrix of ζ is

$$V = \begin{bmatrix} 1 & 1/3 & -1/3 \\ 1/3 & 1 & 1/3 \\ -1/3 & 1/3 & 1 \end{bmatrix},$$

and U is the box $U_\rho = \{u \in \mathbb{R}^3 : \|u\|_\infty \leq \rho\}$. Note that when $\rho \geq 1$, the hypothesis $\text{supp}\,\zeta \subset U_\rho$ does not contradict the assumption that ζ is with zero mean and possesses the outlined covariance matrix; this hypothesis, e.g., is valid when the

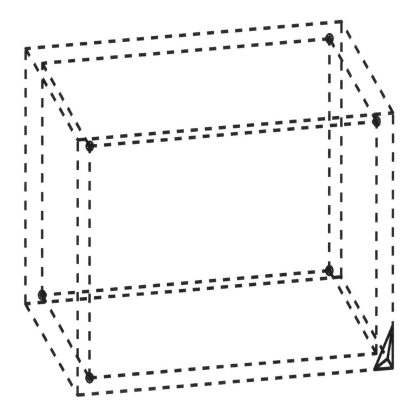

Figure 4.5 The geometry of Illustration A. Inner box: U_1; outer box: $U_{1.1}$.

distribution of ζ is the uniform distribution on the 6 vertices of U_1 marked on figure 4.5.

In our experiment, we chose a and b in such a way that the half-space $\Pi = \{u : a + b^T u > 0\}$ (that is, the half-space where the constraint is *not* satisfied) does not intersect U_1 and intersects $U_{1.1}$, cutting the small tetrahedron off the latter box as shown in figure 4.5 .

We built the following bounds on the probability for the constraint to be violated:

- The Tschebyshev bound p_{T} (see item 4).

- The bound p_{N} given by safe tractable approximation, as presented in (4.5.39), of our chance constraint, when the domain information is ignored.

- The bounds $p_{\mathrm{L}}(\rho)$ and $p_{\mathrm{Q}}(\rho)$ given by the safe approximation (4.5.39) when we do use the domain information supp $\zeta \subset U_\rho$, specifically, represent U_ρ by

  — the system of linear inequalities $-\rho \leq u_i \leq \rho$, $i = 1, 2, 3$, in the case of the bound $p_{\mathrm{L}}(\rho)$;

	$\rho = 1.000$	$\rho = 1.509$	$\rho \geq 2.018$
p_{T}	0.2771		
p_{N}	0.2170		
$p_{\mathrm{L}}(\rho)$	< 1.e-10	0.1876	0.2170
$p_{\mathrm{Q}}(\rho)$	0	0.1876	0.2170

Table 4.4 Numerical results for Illustration A.

— the system of quadratic inequalities $u_i^2 \leq \rho^2$, $i = 1, 2, 3$, in the case of the bound $p_{\mathrm{Q}}(\rho)$.

The results are presented in table 4.4. What can be concluded from this experiment is as follows:

- Even when no information on the domain is used, approximation scheme (4.5.39) produces a significantly better bound than the Tschebyshev inequality.

- Both $p_{\mathrm{L}}(\cdot)$ and $p_{\mathrm{Q}}(\cdot)$ are intelligent enough to understand that when Π does not intersect U_ρ, then the probability of the chance constraint to be violated is 0 (see what happens when $\rho = 1$).

- As ρ grows, the bounds $p_{\mathrm{L}}(\cdot)$ and $p_{\mathrm{Q}}(\cdot)$ also grow, eventually stabilizing at the level of the "no domain information" bound p_{N}.

Note that while we have reasons to expect the bound $p_{\mathrm{Q}}(\cdot)$ to be better than $p_{\mathrm{L}}(\cdot)$, we did not observe this phenomenon in our experiment.

Illustration B. Here ζ is 2-dimensional and the constraint in question is linear: $a + b^T u \leq 0$. The covariance matrix of ζ is

$$V = \left[\begin{array}{cc} 0.5 & 0 \\ 0 & 0.5 \end{array} \right]$$

and $U = U_\rho$ is the equilateral triangle with the barycenter at the origin and one of the vertices at $[\rho; 0]$. Here again in the case of $\rho \geq 1$ the assumption supp $\zeta \subset U_\rho$ is compatible with the assumption that the mean of ζ is 0 and the covariance matrix of ζ is V; indeed, V is exactly the covariance matrix of the uniform distribution on the vertices of U_1.

We chose a and b in such a way that the half-plane $\Pi = \{u : a + b^T u > 0\}$ does not intersect U_1 and intersects $U_{1.1}$, cutting off $U_{1.1}$ the small triangle shown on figure 4.6.a.

We built the same 4 bounds on the probability for our linear constraint to be violated as in Illustration A. When building $p_{\mathrm{L}}(\rho)$, we used the natural description of the triangle U_ρ by 3 linear inequalities $f_j^\rho(u) \leq 0$, $j = 1, 2, 3$.

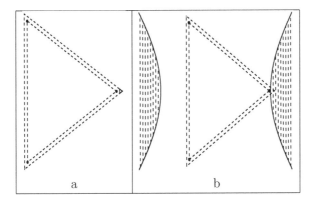

Figure 4.6 Geometries of Illustrations B (a) and C (b). Inner triangle: U_1; outer triangle: $U_{1.1}$.

	$\rho = 1.000$	$\rho \geq 1.1$
p_{T}	0.4535	
p_{N}	0.3120	
$p_{\mathrm{L}}(\rho)$	< 1.e-10	0.3120
$p_{\mathrm{Q}}(\rho)$	0	0.3120

Table 4.5 Numerical results for Illustration B.

When building $p_{\mathrm{Q}}(\rho)$, we used the representation of U_ρ by 3 quadratic inequalities $g_j^\rho(u) \equiv f_j^\rho(u)(f_j^\rho(u) + \delta_j^\rho) \leq 0$, where δ_j^ρ is given by the requirement $\min_{u \in U_\rho}(f_j^\rho(u) + \delta_j^\rho) = 0$.

The results of our experiment are presented in table 4.5. The conclusions are exactly the same as in Illustration A, with one more observation: it is easily seen that when $\rho > 1$, then our a priori information is compatible with the assumption that ζ takes just 4 values: 3 values corresponding to the vertices of U_ρ, with probability of every one of these 3 values equal to $1/(3\rho^2)$, and the value 0 taken with probability $1 - \rho^{-2}$. It follows that when $\rho \geq 1.1$, the true probability for our randomly perturbed constraint to be violated can be as large as $1/(3 \cdot 1.1^2) = 0.2755$, so that our bounds are not that bad.

Illustration C. The purpose of this experiment was to demonstrate that passing from representation of U by linear constraints to a representation of the same set by quadratic constraints can indeed be profitable — the phenomenon that we did not observe in two previous experiments. As far as the information on the mean, the covariance and the domain of ζ are concerned, our current setup is exactly the same as in Illustration B. What is different is the chance constraint; now it is the quadratically perturbed constraint

$$\mathrm{Prob}\{[\zeta;1]^T Z[W,w][\zeta;1] \equiv \zeta_1^2 - \zeta_2^2 - 1.05 > 0\} \leq \epsilon \quad [Z[W,w] = \mathrm{Diag}\{1, -1, -1.05\}]$$

	$\rho = 1.0$	$\rho = 1.1$	$\rho = 1.2$	$\rho = 1.4$	$\rho \geq 1.8$
p_{N}	0.4762				
$p_{\mathrm{L}}(\rho)$	0.4762	0.4762	0.4762	0.4762	0.4762
$p_{\mathrm{Q}}(\rho)$	0	0.3547	0.3996	0.4722	0.4762

Table 4.6 Numerical results for Illustration C.

The domain Π on the plane of ζ where the constraint is violated is the union of two domains bounded by a hyperbola (dashed areas on figure 4.6.b). As we see, this domain does not intersect the triangle U_1 and does intersect the triangle $U_{1.1}$. We built the same bounds as in Illustrations A, B, except for the Tschebyshev bound that now does not make sense. The results are presented in table 4.6. We see that the bound $p_{\mathrm{Q}}(\cdot)$ is indeed less conservative than the bound $p_{\mathrm{L}}(\cdot)$; in particular, this bound still understands that when Π does not intersect U_ρ, then the probability for our randomly perturbed constraint to be violated is 0.

4.6 EXERCISES

Exercise 4.1. Let ζ_ℓ, $1 \leq \ell \leq L$, be independent Poisson random variables with parameters λ_ℓ, (i.e., ζ_ℓ takes nonnegative integer value k with probability $\frac{\lambda_\ell^k}{k!}e^{-\lambda_\ell}$). Build Bernstein approximation of the chance constraint

$$\mathrm{Prob}\{z_0 + \sum_{\ell=1}^{L} w_\ell \zeta_\ell \leq 0\} \geq 1 - \epsilon.$$

What is the associated uncertainty set $\mathcal{Z}_\epsilon$ as given by Theorem 4.2.5?

Exercise 4.2. The stream of customers of an ATM can be split into L groups, according to the amounts of cash c_ℓ they are withdrawing. The per-day number of customers of type ℓ is a realization of Poisson random variable ζ_ℓ with parameter λ_ℓ, and these variables are independent of each other. What is the minimal amount of cash $w(\epsilon)$ to be loaded in the ATM in the morning in order to ensure service level $1 - \epsilon$, (i.e., the probability of the event that not all customers arriving during the day are served should be $\leq \epsilon$)?

Consider the case when

$$L = 7, c = [20; 40; 60; 100; 300; 500; 1000], \lambda_\ell = 1000/c_\ell$$

and compute and compare the following quantities:

i) The expected value of the per-day customer demand for cash.

ii) The true value of $w(\epsilon)$ and its CVaR-upper bound (utilize the integrality of c_ℓ to compute these quantities efficiently).

iii) The bridged Bernstein - CVaR, and the pure Bernstein upper bounds on $w(\epsilon)$.

$iv)$ The $(1 - \epsilon)$-reliable empirical upper bound on $w(\epsilon)$ built upon a 100,000-element simulation sample of the per day customer demands for cash.

The latter quantity is defined as follows. Assume that given an N-element sample $\{\eta_i\}_{i=1}^N$ of independent realizations of a random variable η, and a tolerance $\delta \in (0, 1)$, we want to infer from the sample a "$(1 - \delta)$-reliable" upper bound on the upper ϵ quantile $q_\epsilon = \min\{q : \text{Prob}\{\eta > q\} \leq \epsilon\}$ of η. It is natural to take, as this bound, the M-th order statistics S_M of the sample, (i.e., M-th element in the non-descending rearrangement of the sample), and the question is, how to choose M in order for S_M to be $\geq q_\epsilon$ with probability at least $1 - \delta$. Since $\text{Prob}\{\eta \geq q_\epsilon\} \geq \epsilon$, the relation $S_M < q_\epsilon$ for a given M implies that in our sample of N independent realizations η_i of η the relation $\{\eta_i \geq q_\epsilon\}$ took place at most $N - M$ times, and the probability of this event is at most $p_M = \sum_{k=0}^{N-M} \binom{N}{k} \epsilon^k (1 - \epsilon)^{N-k}$. It follows that if M is such that $p_M \leq \delta$, then the event in question takes place with probability at most δ, i.e., S_M is an upper bound on q_ϵ with probability at least $1 - \delta$. Thus, it is natural to choose M as the smallest integer $\leq N$ such that $p_M \leq \delta$. Note that such an integer not necessarily exists — it may happen that already $p_N > \delta$, meaning that the sample size N is insufficient to build a $(1-\delta)$-reliable upper bound on q_ϵ.

Carry out the computation for $\epsilon = 10^{-k}$, $1 \leq k \leq 6$. [2]

Exercise 4.3. Consider the same situation as in Exercise 4.2, with the only difference that now we do not assume the Poisson random variables ζ_ℓ to be independent, and make no assumptions whatsoever on how they relate to each other. Now the minimal amount of "cash input" to the ATM that guarantees service level $1 - \epsilon$ is the optimal value $\widehat{w}(\epsilon)$ of the "ambiguously chance constrained" problem

$$\min\left\{w_0 : \text{Prob}_{\zeta \sim P}\{\sum_\ell c_\ell \zeta_\ell \leq w_0\} \geq 1 - \epsilon \,\forall P \in \mathcal{P}\right\},$$

where $\mathcal{P}$ is the set of all distributions P on $\mathbb{R}^L$ with Poisson distributions with parameters $\lambda_1, ..., \lambda_L$ as their marginals.

By which margin can $\widehat{w}(\epsilon)$ be larger than $w(\epsilon)$? To check your intuition, use the same data as in Exercise 4.2 to compute
– the upper bound on $\widehat{w}(\epsilon)$ given by Theorem 4.5.1;
– the lower bound on $\widehat{w}(\epsilon)$ corresponding to the case where $\zeta_1, ..., \zeta_L$ are comonotone, (i.e., are deterministic nondecreasing functions of the same random variable η uniformly distributed on $[0, 1]$, cf. p. 114).
Carry out computations for $\epsilon = 10^{-k}$, $1 \leq k \leq 6$.

Exercise 4.4. 1) Consider the same situation as in Exercise 4.2, but assume that the nonnegative vector $\lambda = [\lambda_1; ...; \lambda_L]$ is known to belong to a given convex

---

[2]Of course, in our ATM story the values of ϵ like 0.001 and less make no sense. Well, you can think about an emergency center and requests for blood transfusions instead of an ATM and dollars.

compact set $\Lambda \subset \{\lambda \geq 0\}$. Prove that with

$$B_\Lambda(\epsilon) = \max_{\lambda \in \Lambda} \inf \left\{ w_0 : \inf_{\beta > 0} \left[-w_0 + \beta \sum_\ell \lambda_\ell (\exp\{c_\ell/\beta\} - 1) - \beta \ln(1/\epsilon) \right] \leq 0 \right\}$$

one has

$$\forall \lambda \in \Lambda : \mathrm{Prob}_{\zeta \sim P_{\lambda_1} \times \dots \times P_{\lambda_L}} \left\{ \sum_\ell c_\ell \zeta_\ell > B_\Lambda(\epsilon) \right\} \leq \epsilon,$$

where P_μ stands for the Poisson distribution with parameter μ. In other words, initial charge of $B_\Lambda(\epsilon)$ dollars is enough to ensure service level $1 - \epsilon$, whatever be the vector $\lambda \in \Lambda$ of parameters of the (independent of each other) Poisson streams of customers of different types.

2) In 1), we have considered the case when λ runs through a given "uncertainty set" Λ, and we want the service level to be at least $1 - \epsilon$, whatever be $\lambda \in \Lambda$. Now consider the case when we impose a chance constraint on the service level, specifically, assume that λ is picked at random every morning, according to a certain distribution P on the nonnegative orthant, and we want to find a once and forever fixed morning cash charge w_0 of the ATM such that the probability for a day to be "bad" (such that the service level in this day drops below the desired level $1 - \epsilon$) is at most a given $\delta \in (0, 1)$. Now consider the chance constraint

$$\mathrm{Prob}_{\lambda \sim P}\{z_0 + \sum_\ell \lambda_\ell z_\ell > 0\} \leq \delta$$

in variables $z_0, ..., z_L$, and assume that we have in our disposal a Robust Counterpart type safe convex approximation of this constraint, i.e., we know a convex compact set $\Lambda \subset \{\lambda \geq 0\}$ such that

$$\forall(z_0, ..., z_L) : z_0 + \max_{\lambda \in \Lambda} \sum_\ell \lambda_\ell z_\ell \leq 0 \Rightarrow \mathrm{Prob}_{\lambda \sim P}\{z_0 + \sum_\ell \lambda_\ell z_\ell > 0\} \leq \delta.$$

Prove that by loading the ATM with $B_\Lambda(\epsilon)$ dollars in the morning, we ensure that the probability of a day to be bad is $\leq \delta$.

4.6.1 Mixed Uncertainty Model

To motivate what follows, let us revisit the Portfolio Selection problem (Example 2.3.6). According to our analysis, the yearly return R of the portfolio given by the optimal solution to (2.3.16) is less than 1.12 with probability ≤ 0.005. Replacing the constant $3.255 = \sqrt{2\ln(1/0.005)}$ with $4.799 = \sqrt{2\ln(1/1.\mathrm{e}\text{-}5)}$, a similar conclusion is that the return of the optimized portfolio is < 1.0711 with probability $< 1.\mathrm{e}\text{-}5$. Now ask yourself whether *in real life* you would indeed bet 100,000:1 that the return of the latter portfolio will be at least 1.0711. We believe that on a close inspection, such a bet would be somehow risky, in spite of the theoretical solidity of the conclusion in question. The reason is, that *in order for our conclusion to be applicable to a real life portfolio, the stochastic model of uncertainty underlying the conclusion should describe the "real life" returns of assets fairly well – so well that we could trust it even when it predicts the "event of interest" to happen with probability as small as 1.e-5."* And common sense says that as far as a real market is concerned, such a precise model definitely is out of question. Assume, e.g., that the "true" model of returns is as follows: in the beginning of a year, "nature" flips a coin and, depending on the result, decides whether in the year the returns of our 199 market assets will be $\mu_\ell + \sigma_\ell \zeta_\ell$ with *independent* $\zeta_1, ..., \zeta_{199}$ taking values ± 1 (such a decision is made by nature with probability 0.99), or all $\zeta_1, ..., \zeta_{199}$ will be equal to each other and take values ± 1 with equal probabilities (such a decision is made by nature with probability 0.01). Experiment shows that with this new distribution of returns, the probability for the portfolio associated with the old ("the nominal") uncertainty model to have a return < 1.05, (i.e., less than the return guaranteed by the bank) is about 5.e-3, 500 times larger than the probability 1.e-5 promised by the nominal uncertainty model! And of course the existing market data do not allow one to distinguish reliably between the two models in question. As a matter of fact, the only seemingly solid conclusion from market data is that *there hardly exists a "tractable" probabilistic model of the real life market capable of reliably predicting probabilities like 0.001 and less.* A similar conclusion holds true for basically all probabilistic uncertainty models in optimization under uncertainty, although in some cases, (e.g., in Signal Processing), "the scope of predictability" can include probabilities like 1.e-5 to 1.e-6. Thus, in real-life applications of optimization under uncertainty one typically arrives at the dilemma as follows: on one hand, the uncertain-but-bounded model of data perturbations allows to reliably immunize solutions against data perturbations of a desired magnitude, but seems to be too conservative when the uncertainty is of a stochastic nature. On the other hand, stochastic uncertainty models available in real life usually are not accurate enough to ensure reliably that "bad things" happen with probabilities as low as 1.e-5 and less.[3] A possible resolution of the outlined dilemma could be in *synthesis* of the two uncertainty models in question, specifically, in utilizing the *combined uncertainty model* as follows: the actual perturbation ζ is of the form

$$\zeta = \xi + \eta,$$

where ξ is a "deterministic" perturbation known to belong to a given perturbation set $\mathcal{Z}_\xi$, and η is random perturbation with distribution known to belong to a given family $\mathcal{P}$. For example, we can assume the vector r of returns r_ℓ of market assets to be the sum of a random vector η with independent coordinates distributed in given segments $[\mu_\ell - \sigma_\ell, \mu_\ell + \sigma_\ell]$, $\mu_\ell = \mathbf{E}\{\eta_\ell\}$, and a deterministic vector ξ with $|\xi_\ell| \leq \alpha \sigma_\ell$, where $\alpha \ll 1$ (say, $\alpha = 0.1$). The first of these components represents the "internal noises" in asset returns, while the second accounts for unavoidable inaccuracies in the estimated mean returns, small inter-asset dependencies between the returns, etc. Since in our example the magnitude of entries in ξ is significantly less than those in η, this uncertainty model is essentially less conservative than the one where the only information on r is the range of returns (in our case, the induced range of r_ℓ is $[\mu_\ell - (1 + \alpha)\sigma_\ell, \mu_\ell + (1 + \alpha)\sigma_\ell]$); at the same time, it, as we have mentioned, allows to account to some extent for inaccuracies of the "purely stochastic" uncertainty model.

In order to utilize the combined model of uncertainty in LO, we should define the notion of a "safe" version of a linear constraint affected by this uncertainty, that is, the constraint

$$[a^0 + \sum_{\ell=1}^{L}[\xi_\ell + \eta_\ell]a^\ell]^T x \leq b^0 + \sum_{\ell=1}^{L}[\xi_\ell + \eta_\ell]b^\ell \atop [\xi \in \mathcal{Z}_\xi, \eta \sim P \in \mathcal{P}]. \qquad (4.6.1)$$

---

[3]Note that while really low probabilities are of no actual interest in finance, they are a must in many other applications — think of the reliability you expect from the steering mechanism of your car or from the airplane you are boarding.

The most natural answer is to associate with the uncertain constraint (4.6.1) its "chance constrained version"

$$(\forall \xi \in \mathcal{Z}_\xi, P \in \mathcal{P}):$$
$$\mathrm{Prob}_{\eta \sim P}\left\{[a^0 + \sum_{\ell=1}^{L}[\xi_\ell + \eta_\ell]a^\ell]^T x \leq b^0 + \sum_{\ell=1}^{L}[\xi_\ell + \eta_\ell]b^\ell\right\} \geq 1 - \epsilon, \qquad (4.6.2)$$

where $\epsilon < 1$ is a given tolerance. Note that when $\mathcal{Z}_\xi = \{0\}$, this definition recovers the chance constrained version of (4.6.1). The opposite extreme — $\mathcal{P}$ contains a single trivial distribution (mass 1 at the origin) and $\mathcal{Z}_\xi$ is nontrivial — recovers the RC of (4.6.1) with $\mathcal{Z}_\xi$ in the role of the perturbation set.

After the notion of a "safe version" of an uncertain constraint with combined uncertainty is defined, the question of primary importance is how to process this safe version. The next exercise demonstrates that this can be done by combining the techniques for tractable processing RCs and for building safe tractable approximations of the "plain" chance constraints.

Exercise 4.5. Prove the following fact:

Theorem 4.6.1. Consider the uncertain constraint (4.6.1) and assume that the set $\mathcal{Z}_\xi$ admits a strictly feasible conic representation (cf. Theorem 1.3.4):

$$\mathcal{Z}_\xi = \{\xi : \exists u : P\zeta + Qu + p \in \mathbf{K}\}$$

(when $\mathbf{K}$ is a polyhedral cone, strict feasibility can be reduced to feasibility). Let, further, the ambiguous chance constraint

$$\forall(P \in \mathcal{P}) : \mathrm{Prob}_{\eta \sim P}\left\{[a^0 + \sum_{\ell=1}^{L} \eta_\ell a^\ell]^T x \leq b^0 + \sum_{\ell=1}^{L} \eta_\ell b^\ell\right\} \geq 1 - \epsilon,$$

associated with $\mathcal{P}$ and (4.6.1) admit a safe tractable Robust Counterpart type approximation, i.e., we can point out a computationally tractable convex compact set $\mathcal{Z}_\eta^\epsilon$ such that with

$$f(z) = \max_{\eta \in \mathcal{Z}_\eta^\epsilon} \sum_{\ell=1}^{L} \eta_\ell z_\ell$$

the implication

$$\forall(z_0, z) : z_0 + f(z) \leq 0 \Rightarrow \forall(P \in \mathcal{P}) : \mathrm{Prob}_{\eta \sim P}\left\{z_0 + \sum_{\ell=1}^{L} \zeta_\ell z_\ell \leq 0\right\} \geq 1 - \epsilon \qquad (4.6.3)$$

holds true (cf. Proposition 4.1.3). Then the system of explicit convex constraints

$$\begin{array}{ll}
(a) & p^T y + [a^0]^T x - b_0 \leq t, \\
(b) & Q^T y = 0, \\
(c) & (P^T y)_\ell + [a^\ell]^T x = b^\ell, \ell = 1, ..., L, \\
(d) & y \in \mathbf{K}_* \equiv \{y : y^T z \geq 0 \forall z \in \mathbf{K}\}, \\
(e) & t + f([a^1]^T x - b^1, ..., [a^L]^T x - b^l) \leq 0
\end{array} \qquad (4.6.4)$$

in variables x, y, t is a safe tractable approximation of (4.6.2): whenever x can be extended to a solution to this system, x is feasible for (4.6.2).

Thus, all techniques for building safe tractable approximations of chance constraints developed in this chapter and in chapter 2 can be used to process efficiently the safe versions of linear constraints with combined uncertainty.

Mixed uncertainty model: illustration.   The next exercise is aimed at experimentation with the Mixed uncertainty model in the following (admittedly oversimplified, but still instructive) situation (cf. Example 2.3.6):

Portfolio selection revisited:
There are n assets; asset # 1 ("money in safe") has a deterministic yearly return $\zeta_1 \equiv 1$, and the remaining assets have independent Gaussian yearly returns $\zeta_\ell \sim \mathcal{N}(\mu_\ell, \sigma_\ell^2)$, $2 \leq \ell \leq n$. Setting $\mu_1 = 1$, $\sigma_1 = 0$, we can say that the returns of all n assets are independent Gaussian random variables. We assume further that the variances σ_ℓ^2 are known for all ℓ and are positive when $\ell \geq 2$; w.l.o.g. we can assume that $0 = \sigma_1 < \sigma_2 \leq ... \leq \sigma_n$. The expected returns μ_ℓ, $\ell \geq 2$, are *not* known in advance; the only related information is the "historical data" — a sample $\zeta^1, ..., \zeta^N$ of N independent realizations of the vectors of yearly returns. Given a risk level $\epsilon \in (0, 1)$, our goal is to distribute \$1 among the assets in order to maximize the value at risk of the resulting portfolio a year from now.

The precise formulation of our goal is as follows. A portfolio can be identified with vector $x = [x_1; ...; x_n]$ from the standard simplex $\Delta_n = \{x \in \mathbb{R}^n : x \geq 0, \sum_\ell x_\ell = 1\}$; x_ℓ is the capital invested in i-th asset. The value of our portfolio a year from now is the random variable

$$V^x = \sum_{\ell=1}^n \zeta_\ell x_\ell = \underbrace{\sum_{\ell=1}^n \mu_\ell x_\ell}_{\mu(x)} + \underbrace{\sqrt{\sum_{\ell=1}^n \sigma_\ell^2 x_\ell^2}}_{\sigma(x)} \xi. \qquad [\xi \sim \mathcal{N}(0, 1)]$$

The *value at risk* $\epsilon \, \mathrm{VaR}_\epsilon[\eta]$ of a random variable η is the (lower) ϵ quantile of this variable — the largest real a such that $\mathrm{Prob}\{\eta < a\} \leq \epsilon$; for the Gaussian random variable V^x, we have $\mathrm{VaR}[V^x] = \mu(x) - \mathrm{ErfInv}(\epsilon)\sigma(x)$, so that our ideal goal is to solve the optimization problem

$$\mathrm{Opt} = \max_x \Big\{ \sum_{\ell=1}^n \mu_\ell x_\ell - \mathrm{ErfInv}(\epsilon)\sigma(x) : x \in \Delta_n \Big\}. \qquad (4.6.5)$$

This goal is indeed an ideal one, since we do not know what exactly is the objective we should maximize, since the coefficients μ_ℓ are not exactly known. Moreover, aside for the trivial case where $\sigma_\ell = 0$ for all ℓ, we cannot localize $\mu = [\mu_1; ...; \mu_n]$ with 100% reliability in a whatever large, but *bounded* set. As a result, the only portfolio x for which we can guarantee a finite lower bound on $\mathrm{VaR}[V^x]$ is the trivial portfolio $x_1 = 1, x_2 = ... = x_n = 0$. To allow for selecting nontrivial portfolios, we should replace "100%-reliable" guarantees with "$(1 - \delta)$-reliable" ones. Thus, assume that we are given a tolerance δ, $0 < \delta \ll 1$, and want to find a procedure that, given on input random historical data $\widetilde\zeta = [\zeta^1; ...; \zeta^N]$, converts this data into a portfolio $x = X(\widetilde\zeta)$ and a *guessed* lower bound $\mathrm{VaR} = \mathrm{VaR}(\widetilde\zeta)$ on $\mathrm{VaR}_\epsilon[V^x]$, which should indeed be a lower bound on $\mathrm{VaR}_\epsilon[V^x]$ with probability $\geq 1 - \delta$:

$$\mathrm{Prob}\Big\{ \mathrm{VaR}(\widetilde\zeta) \leq \mu^T X(\widetilde\zeta) - \mathrm{ErfInv}(\epsilon)\sigma(X(\widetilde\zeta)) \Big\} \geq 1 - \epsilon, \qquad (4.6.6)$$

whatever be the true vector μ of expected returns with $\mu_1 = 1$. Under this restriction, we want the "typical" values of $\mathrm{VaR}(\widetilde\zeta)$ to be as large as possible. The latter informal goal can be formalized as maximization of the expected value of $\mathrm{VaR}(\widetilde\zeta)$, by imposing a chance constrained lower bound on $\mathrm{VaR}(\widetilde\zeta)$ and maximizing this bound, etc. Whatever the formalization, the resulting problem seems to be severely computationally intractable; indeed, already the constraint (4.6.6) seems to be a disaster — this is a *semi-infinite* chance constraint in *decision rules*. However, there are ways to get efficiently if not optimal in terms of a given criterion, but at least feasible solutions, and we are about to consider several of these ways. The goal of the exercises to follow is to investigate the proposed approaches numerically, and here are the three recommended data sets:

Type	n	δ	ϵ	$\mu_\ell, \ell \geq 2 \ (\mu_1 = 1)$	$\sigma_\ell, \ell \geq 2 \ (\sigma_1 = 0)$	N
(a)	500	0.01	0.01	$1 + 0.1\frac{\ell-1}{n-1}$	$0.21\frac{\ell-1}{n-1}$	40
(b)	500	0.01	0.01	$\begin{cases} 1, & \ell < n \\ 1.1, & \ell = n \end{cases}$	$0.02, \ 2 \leq \ell \leq n$	40
(c)	200	0.01	0.01	$1 + 0.1\frac{\ell-1}{n-1}$	$0.3\sqrt{\frac{\ell-1}{n-1}}$	100

$$(4.6.7)$$

These data sets include the true values of μ_ℓ; these values are not used by the portfolio selection routines we are about to consider, but can be used when evaluating these routines by simulation. Note that with our data, the expectations of all market assets are $\geq$ than the guaranteed return $\mu = 1$ of the "money in safe" policy, and that our data possess the natural property that the more promising an asset (the larger is μ_ℓ), the more risky it is (the larger is σ_ℓ).

Exercise 4.6. Find the true optimal solutions to (4.6.5) corresponding to every one of the data sets (4.6.7.*a*–*c*).

RC approximation: the strategy. With the RC approximation,

• We use the historical data $\widetilde{\zeta}$ to build a set $\mathcal{M} = \mathcal{M}(\widetilde{\zeta})$ in the μ space in such a way that whatever the true vector μ of expected returns with $\mu_1 = 1$, the probability for $\mathcal{M}$ to contain a lower bound for μ is at least $1 - \delta$:

$$\forall(\mu : \mu_1 = 1) : \mathrm{Prob}_\mu \left\{ \exists \bar{\mu} \in \mathcal{M}(\widetilde{\zeta}) : \bar{\mu} \leq \mu \right\} \geq 1 - \delta, \qquad (4.6.8)$$

where Prob_μ is the probability w.r.t. the distribution of the historical data associated with μ, (i.e., the distribution where $\zeta^i \sim \mathcal{N}(\mu, \mathrm{Diag}\{\sigma_1^2, ..., \sigma_n^2\})$ and $\zeta^1, .., \zeta^N$ are independent).

• After $\mathcal{M}$ is built, we treat (4.6.5) as an uncertain optimization problem, where the only uncertain data is μ, and the uncertainty set is $\mathcal{M}$. We solve the RC of this uncertain problem and take the robust optimal solution as the recommended portfolio $x = X(\widetilde{\zeta})$, and the robust optimal value — as the guess $\mathrm{VaR}(\widetilde{\zeta})$.

Exercise 4.7. Prove that the RC approximation is safe — it indeed ensures (4.6.6).

RC approximation: implementations. An implementation of the just outlined strategy depends on how we choose the set $\mathcal{M}$. For the sake of simplicity, assume that we choose the this set as

$$\mathcal{M}(\widetilde{\zeta}) = \widehat{\mu} + \mathcal{O}, \qquad (4.6.9)$$

where $\mathcal{O}$ is a deterministic convex body and $\widehat{\mu}$ is the empirical expected return given by the historical data:

$$\widehat{\mu} = \frac{1}{N} \sum_{t=1}^{N} \zeta^t = \mu + \underbrace{[\mathrm{Diag}\{\sigma\}N^{-1/2}]}_{\Sigma} \eta, \ \eta \sim \mathcal{N}(0, I_n).$$

Exercise 4.8. Prove that in order for the random set (4.6.9) to satisfy the requirement (4.6.8), it suffices to have

$$\mathrm{Prob}_{\eta \sim \mathcal{N}(0, I_n)} \left\{ -\Sigma \eta \in \mathcal{O} + \mathbb{R}_+^n \right\} \geq 1 - \delta. \tag{4.6.10}$$

Let us stick to the sets (4.6.9) with $\mathcal{O}$ satisfying (4.6.10). This still leaves us with "uncountably many" choices. We start by exploring the simplest of them, specifically,

Ball: $\mathcal{O} = \Sigma B_2^{\rho_2}$, where $B_2^{\rho_2} = \{u \in \mathbb{R}^n : u \leq 0, \|u\|_2 \leq \rho_2\}$ and ρ_2 is such that

$$\mathrm{Prob}_{\eta \sim \mathcal{N}(0, I_n)} \{ -\eta \in B_2^{\rho_2} + \mathbb{R}_+^n \} \geq 1 - \delta. \tag{4.6.11}$$

Note that we are interested in as small ρ_2 as possible, since the less is $\mathcal{O}$, the less conservative is the RC associated with the uncertainty set (4.6.9).

Box: $\mathcal{O} = \Sigma B_\infty^{\rho_\infty}$, where $B_\infty^{\rho_\infty} = \{u \in \mathbb{R}^n : u \leq 0, \|u\|_\infty \leq \rho_\infty\}$ and ρ_∞ is such that

$$\mathrm{Prob}_{\eta \sim \mathcal{N}(0, I_n)} \{ -\eta \in B_\infty^{\rho_\infty} + \mathbb{R}_+^n \} \geq 1 - \delta. \tag{4.6.12}$$

Exercise 4.9. 1) Prove that the RCs associated with the Ball and the Box choices of $\mathcal{O}$ are, respectively, the optimization problems

$$\max_x \left\{ \sum_{\ell=1}^{n} \widehat{\mu}_\ell x_\ell - [\rho_2 N^{-1/2} + \mathrm{ErfInv}(\epsilon)]\sigma(x) : x \in \Delta_n \right\} \tag{Bl}$$

and

$$\max_x \left\{ \sum_{\ell=1}^{n} [\widehat{\mu}_\ell - \rho_\infty N^{-1/2}\sigma_\ell]x_\ell - \mathrm{ErfInv}(\epsilon)\sigma(x) : x \in \Delta_n \right\}. \tag{Bx}$$

2) Find a way to bound from above ρ_2 and ρ_∞.
Hint: To bound ρ_2, you can use the Bernstein approximation scheme.

3) Use the bounds from 2) to implement the RC approach and test it on the data sets (4.6.7.a–c). Compare the results with each other and with the "ideal results" yielded by Exercise 4.6. Is it possible to say in advance what is better — the Ball or the Box choice of $\mathcal{O}$? Where the difficulty comes from?
Hint: Verify that the minimal $\rho_2 = \rho_2(n, \delta)$ satisfying (4.6.11) as a function of n grows with n as $O(\sqrt{n})$, while the minimal $\rho_\infty = \rho_\infty(n, \delta)$ satisfying (4.6.12) is $O(\sqrt{\ln(n/\delta)})$.

4) Combine the Ball and the Box RC approximations into a single RC approximation that is provably nearly as good as the best of the Ball and the Box approximations. Implement this approximation for the data sets (4.6.7.a–c) and compare the results with the those of the pure Ball and Box RC approximations.

Hint: Take $\mathcal{O} = \Sigma(B_2^{\rho_2} \cap B_\infty^{\rho_\infty})$, where ρ_2 and ρ_∞ satisfy the respective relations (4.6.11), (4.6.12) with δ replaced with $\delta/2$.

Exercise 4.10. Consider the following "soft" RC approximation. For a given portfolio $x \in \Delta_n$, the difference between its "true" expected return $\mu^T x$ and the estimate $\widehat{\mu}^T x$ of this expected return is a Gaussian random variable with zero mean and the variance $N^{-1}\sigma^2(x)$. Consequently, the random quantity $\widehat{\mu}^T x - \mathrm{ErfInv}(\delta)N^{-1/2}\sigma(x)$ is, with probability $\geq 1 - \delta$, a lower bound on the true expected return $\mu^T x$. Given this observation, let us approximate the problem of interest (4.6.5) with the random problem

$$\mathrm{VaR} = \max_x \left\{ \sum_{\ell=1}^n \widehat{\mu}_\ell x_\ell - \mathrm{ErfInv}(\delta)N^{-1/2}\sigma(x) - \mathrm{ErfInv}(\epsilon)\sigma(x) : x \in \Delta_n \right\} \quad (4.6.13)$$

with the objective that underestimates, with probability $1 - \delta$, the objective of the 'true' problem.

1) Is the just outlined "soft" RC approximation safe, i.e., is it true that taking the optimal value of the problem as $\mathrm{VaR} = \mathrm{VaR}(\widetilde{\zeta})$, and the optimal solution as $X(\widetilde{\zeta})$, we ensure the validity of (4.6.6)?

2) Implement the soft RC approximation and, using the data sets (4.6.7), empirically find the probability for VaR, X to violate the relation $\mathrm{VaR} \leq \mu^T X - \mathrm{ErfInv}(\epsilon)\sigma(X)$.

Correcting a soft RC approximation. Now consider a conceptual approximation as follows.

- We fix in advance a finite number M of "basic portfolios" $x^1, ..., x^M \in \Delta_n$.

- Given the empirical average return $\widehat{\mu}$, we build lower bounds L^i on the expected returns of the basic portfolios x^i, $i = 1, ..., M$, such that whatever the true vector μ of expected returns with $\mu_1 = 1$, we have

$$\mathrm{Prob} \left\{ L^i \leq \mu^T x^i, \, 1 \leq i \leq M \right\} \geq 1 - \delta. \quad (4.6.14)$$

The simplest way to ensure this relation is to set

$$L^i = \widehat{\mu}^T x^i - \mathrm{ErfInv}(\delta/M)N^{-1/2}\sigma(x^i). \quad (4.6.15)$$

- We now restrict ourselves to the portfolios that are convex combinations of the basic ones:

$$x = x(\lambda) = \sum_{i=1}^M \lambda_i x^i, \qquad\qquad [\lambda \geq 0, \sum_i \lambda_i = 1]$$

estimate the expected return of such a portfolio as

$$L(\lambda) = \sum_{i=1}^M \lambda_i L^i,$$

and then look for the portfolio with the largest possible estimated VaR_ϵ. That is, we solve the optimization problem

$$\lambda_* \in \underset{\lambda}{\text{Argmin}} \left\{ f(\lambda) = L(\lambda) - \text{ErfInv}(\epsilon)\sigma(x(\lambda)) : \lambda \geq 0, \sum_i \lambda_i = 1 \right\}, \quad (4.6.16)$$

and take $x(\lambda_*)$ as the resulting portfolio, and $f(\lambda_*)$ as the guessed lower bound VaR on $\text{VaR}_\epsilon[x(\lambda_*)]$.

Exercise 4.11. 1) Prove that the outlined approximation is safe, so that the resulting portfolio and the guessed lower bound on its value at risk do satisfy (4.6.6).

2) Verify that when $M = n$ and $x^1, ..., x^n$ are the standard basic orths in $\mathbb{R}^n$, the outlined approximation is nothing but the Box RC approximation. Would it make sense to use richer sets of basic portfolios? Does it make sense to use "very large" sets of this type?

3) Look what happens when we use $M = 2n$ basic portfolios, namely, n standard basic orths and n portfolios

$$x^{(k)} : x^{(k)}_\ell = \begin{cases} 0, & \ell < k \\ \frac{1}{n-k+1}, & k \leq \ell \leq n \end{cases},$$

$k = 1, ..., n$. Use the data sets (4.6.7) and compare the results with those yielded by other approximations we have considered.

4.7 NOTES AND REMARKS

NR 4.1. The Bernstein approximation scheme (section 4.2) goes back to J. Pinter [92], where, however, the "scale" parameter (parameter β in Proposition 4.2.2) was considered as a chosen a priori constant rather than an adjustable parameter of the bounding routine. The advanced form of the Bernstein approximation scheme as presented in section 4.2 was proposed in [83]. The latter paper underlies the results of section 4.3 (except for those related to the bridged Bernstein-CVaR approximation; these results are new).

NR 4.2. The monotone dominance considered in section 4.4 is the symmetrized version of the *first order stochastic dominance* well studied in econometrics [47, 60, 101, 102]. The main result of the section, Theorem 4.4.6, while being close to the Uniformity Principle of Barmish and Lagoa [1], seems to be new; note that the Uniformity Principle is an immediate consequence of this Theorem and the result of Example 4.4.3.

NR 4.3. The idea of Lagrange relaxation implicitly underlying the developments of section 4.5 is now quite standard and is one of the most powerful sources (if not *the* most powerful source) of efficiently computable bounds on "difficult to compute" optimization-related quantities, (e.g., optimal values of NP-hard combinatorial problems). The developments in section 4.5.5 can be traced to [26, 27] and are based upon specific implementation of Lagrange relaxation, the so called *semidefinite relaxation scheme* that goes back to Naum Shor and Laslo Lovacz.[4]

---

[4] A more detailed presentation of the semidefinite relaxation scheme can be found, among many other sources, in [8, Chapter 4] and [33].

Part II

Robust Conic Optimization

Chapter Five
Uncertain Conic Optimization: The Concepts

In this chapter, we extend the RO methodology onto *non-linear* convex optimization problems, specifically, *conic* ones.

5.1 UNCERTAIN CONIC OPTIMIZATION: PRELIMINARIES

5.1.1 Conic Programs

A *conic* optimization (CO) problem (also called *conic program*) is of the form

$$\min_x \left\{ c^T x + d : Ax - b \in \mathbf{K} \right\}, \tag{5.1.1}$$

where $x \in \mathbb{R}^n$ is the decision vector, $\mathbf{K} \subset \mathbb{R}^m$ is a closed pointed convex cone with a nonempty interior, and $x \mapsto Ax - b$ is a given affine mapping from $\mathbb{R}^n$ to $\mathbb{R}^m$. Conic formulation is one of the universal forms of a Convex Programming problem; among the many advantages of this specific form is its "unifying power." An extremely wide variety of convex programs is covered by just three types of cones:

i) Direct products of nonnegative rays, i.e., $\mathbf{K}$ is a non-negative orthant $\mathbb{R}^m_+$. These cones give rise to Linear Optimization problems

$$\min_x \left\{ c^T x : a_i^T x - b_i \geq 0, \ 1 \leq i \leq m \right\}.$$

ii) Direct products of *Lorentz* (or *Second-order*, or Ice-cream) cones $\mathbf{L}^k = \{x \in \mathbb{R}^k : x_k \geq \sqrt{\sum_{j=1}^{k-1} x_j^2}\}$. These cones give rise to Conic Quadratic Optimization (called also Second Order Conic Optimization). The Mathematical Programming form of a CO problem is

$$\min_x \left\{ c^T x : \|A_i x - b_i\|_2 \leq c_i^T x - d_i, \ 1 \leq i \leq m \right\};$$

here i-th scalar constraint (called *Conic Quadratic Inequality*) (CQI) expresses the fact that the vector $[A_i x; c_i^T x] - [b_i; d_i]$ that depends affinely on x belongs to the Lorentz cone $\mathbf{L}_i$ of appropriate dimension, and the system of all constraints says that the affine mapping

$$x \mapsto \left[[A_1 x; c_1^T x]; ...; [A_m x; c_m^T x] \right] - \left[[b_1; d_1]; ...; ; [b_m; d_m] \right]$$

maps x into the direct product of the Lorentz cones $\mathbf{L}_1 \times ... \times \mathbf{L}_m$.

iii) Direct products of *semidefinite* cones $\mathbf{S}^k_+$.

$\mathbf{S}_+^k$ is the cone of positive semidefinite $k \times k$ matrices; it "lives" in the space $\mathbf{S}^k$ of symmetric $k \times k$ matrices. We treat $\mathbf{S}^k$ as Euclidean space equipped with the *Frobenius* inner product $\langle A, B \rangle = \text{Tr}(AB) = \sum_{i,j=1}^{k} A_{ij} B_{ij}$.

The family of semidefinite cones gives rise to *Semidefinite Optimization* (SDO) — optimization programs of the form

$$\min_{x} \left\{ c^T x + d : \mathcal{A}_i x - B_i \succeq 0, \, 1 \leq i \leq m \right\},$$

where

$$x \mapsto \mathcal{A}_i x - B_i \equiv \sum_{j=1}^{n} x_j A^{ij} - B_i$$

is an affine mapping from $\mathbb{R}^n$ to $\mathbf{S}^{k_i}$ (so that A^{ij} and B_i are symmetric $k_i \times k_i$ matrices), and $A \succeq 0$ means that A is a symmetric positive semidefinite matrix. The constraint of the form "a symmetric matrix affinely depending on the decision vector should be positive semidefinite" is called an LMI — Linear Matrix Inequality. Thus, a Semidefinite Optimization problem (called also *semidefinite program*) is the problem of minimizing a linear objective under finitely many LMI constraints. One can rewrite an SDO program in the Mathematical Programming form, e.g., as

$$\min_{x} \left\{ c^T x + d : \lambda_{\min}(\mathcal{A}_i x - B_i) \geq 0, \, 1 \leq i \leq m \right\},$$

where $\lambda_{\min}(A)$ stands for the minimal eigenvalue of a symmetric matrix A, but this reformulation usually is of no use.

Keeping in mind our future needs related to Globalized Robust Counterparts, it makes sense to modify slightly the format of a conic program, specifically, to pass to programs of the form

$$\min_{x} \left\{ c^T x + d : A_i x - b_i \in \mathbf{Q}_i, \, 1 \leq i \leq m \right\}, \tag{5.1.2}$$

where $\mathbf{Q}_i \subset \mathbb{R}^{k_i}$ are nonempty closed convex sets given by finite lists of conic inclusions:

$$\mathbf{Q}_i = \{ u \in \mathbb{R}^{k_i} : Q_{i\ell} u - q_{i\ell} \in \mathbf{K}_{i\ell}, \, \ell = 1, ..., L_i \}, \tag{5.1.3}$$

with closed convex pointed cones $\mathbf{K}_{i\ell}$. We will restrict ourselves to the cases where $\mathbf{K}_{i\ell}$ are nonnegative orthants, or Lorentz, or Semidefinite cones. Clearly, a problem in the form (5.1.2) is equivalent to the conic problem

$$\min_{x} \left\{ c^T x + d : Q_{i\ell} A_i x - [Q_{i\ell} b_i + q_{i\ell}] \in \mathbf{K}_{i\ell} \, \forall (i, \ell \leq L_i) \right\}$$

We treat the collection $(c, d, \{A_i, b_i\}_{i=1}^{m})$ as *natural data* of problem (5.1.2). The collection of sets $\mathbf{Q}_i$, $i = 1, ..., m$, is interpreted as the *structure* of problem (5.1.2), and thus the quantities $Q_{i\ell}, q_{i\ell}$ specifying these sets are considered as certain data.

5.1.2 Uncertain Conic Problems and their Robust Counterparts

Uncertain conic problem (5.1.2) is a problem with fixed structure and uncertain natural data *affinely* parameterized by a *perturbation* vector $\zeta \in \mathbb{R}^L$

$$(c, d, \{A_i, b_i\}_{i=1}^m) = (c^0, d^0, \{A_i^0, b_i^0\}_{i=1}^m) + \sum_{\ell=1}^L \zeta_\ell (c^\ell, d^\ell, \{A_i^\ell, b_i^\ell\}_{i=1}^m). \qquad (5.1.4)$$

running through a given perturbation set $\mathcal{Z} \subset \mathbb{R}^L$.

5.1.2.1 Robust Counterpart of an uncertain conic problem

The notions of a robust feasible solution and the *Robust Counterpart* (RC) of uncertain problem (5.1.2) are defined exactly as in the case of an uncertain LO problem (see Definition 1.2.5):

   Definition 5.1.1. Let an uncertain problem (5.1.2), (5.1.4) be given and let $\mathcal{Z} \subset \mathbb{R}^L$ be a given perturbation set.

   (i) A candidate solution $x \in \mathbb{R}^n$ is robust feasible, if it remains feasible for all realizations of the perturbation vector from the perturbation set:

$$x \text{ is robust feasible}$$
$$\Updownarrow$$
$$[A_i^0 + \sum_{\ell=1}^L \zeta_\ell A_i^\ell] x - [b_i^0 + \sum_{\ell=1}^L \zeta_\ell b_i^\ell] \in \mathbf{Q}_i \; \forall (i, 1 \le i \le m, \zeta \in \mathcal{Z}).$$

   (ii) The Robust Counterpart of (5.1.2), (5.1.4) is the problem

$$\min_{x,t} \left\{ t : \begin{array}{l} [c^0 + \sum_{\ell=1}^L \zeta_\ell c^\ell]^T x + [d^0 + \sum_{\ell=1}^L \zeta_\ell d^\ell] - t \in \mathbf{Q}_0 \equiv \mathbb{R}_-, \\ [A_i^0 + \sum_{\ell=1}^L \zeta_\ell A_i^\ell] x - [b_i^0 + \sum_{\ell=1}^L \zeta_\ell b_i^\ell] \in \mathbf{Q}_i, \; 1 \le i \le m \end{array} \right\} \forall \zeta \in \mathcal{Z} \right\} \quad (5.1.5)$$

of minimizing the guaranteed value of the objective over the robust feasible solutions.

   As in the LO case, it is immediately seen that the RC remains intact when the perturbation set $\mathcal{Z}$ is replaced with its closed convex hull; so, from now on we assume the perturbation set to be closed and convex. Note also that the case when the entries of the uncertain data $[A; b]$ are affected by perturbations in a *non-affine* fashion in principle could be reduced to the case of affine perturbations (see section 1.4); however, we do not know meaningful cases beyond uncertain LO where such a reduction leads to a tractable RC.

5.2 ROBUST COUNTERPART OF UNCERTAIN CONIC PROBLEM: TRACTABILITY

In contrast to uncertain LO, where the RC/GRC turn out to be computationally tractable whenever the perturbation set is so, uncertain conic problems with computationally tractable RCs are a "rare commodity." The ultimate reason for this phenomenon is rather simple: the RC (5.1.5) of an uncertain conic problem (5.1.2), (5.1.4) is a convex problem with linear objective and constraints of the generic form

$$P(y, \zeta) = \pi(y) + \Phi(y)\zeta = \phi(\zeta) + \Phi(\zeta)y \in \mathbf{Q}, \qquad (5.2.1)$$

where $\pi(y), \Phi(y)$ are affine in the vector y of the decision variables, $\phi(\zeta), \Phi(\zeta)$ are affine in the perturbation vector ζ, and $\mathbf{Q}$ is a "simple" closed convex set. For such a problem, its computational tractability is, essentially, equivalent to the possibility to check efficiently whether a given candidate solution y is or is not feasible. The latter question, in turn, is whether the image of the perturbation set $\mathcal{Z}$ under an affine mapping $\zeta \mapsto \pi(y) + \Phi(y)\zeta$ is or is not contained in a given convex set $\mathbf{Q}$. This question is easy when $\mathbf{Q}$ is a polyhedral set given by an explicit list of scalar linear inequalities $a_i^T u \leq b_i$, $i = 1, ..., I$ (in particular, when $\mathbf{Q}$ is a nonpositive ray, that is what we deal with in LO), in which case the required verification consists in checking whether the maxima of I affine functions $a_i^T(\pi(y) + \Phi(y)\zeta) - b_i$ of ζ over $\zeta \in \mathcal{Z}$ are or are not nonnegative. Since the maximization of an affine (and thus concave!) function over a computationally tractable convex set $\mathcal{Z}$ is easy, so is the required verification. When $\mathbf{Q}$ is given by *nonlinear* convex inequalities $a_i(u) \leq 0$, $i = 1, ..., I$, the verification in question requires checking whether the *maxima* of *convex* functions $a_i(\pi(y) + \Phi(y)\zeta)$ over $\zeta \in \mathcal{Z}$ are or are not nonpositive. A problem of maximizing a convex function $f(\zeta)$ over a convex set $\mathcal{Z}$ can be computationally intractable already in the case of $\mathcal{Z}$ as simple as the unit box and f as simple as a convex quadratic form $\zeta^T Q\zeta$. Indeed, it is known that the problem

$$\max_{\zeta} \left\{ \zeta^T B\zeta : \|\zeta\|_\infty \leq 1 \right\}$$

with positive semidefinite matrix B is NP-hard; in fact, it is already NP-hard to approximate the optimal value in this problem within a relative accuracy of 4%, even when probabilistic algorithms are allowed [61]. This example immediately implies that the RC of a generic uncertain conic quadratic problem with a perturbation set as simple as a box is computationally intractable.

Indeed, consider a simple-looking uncertain conic quadratic inequality

$$\|0 \cdot y + Q\zeta\|_2 \leq 1$$

(Q is a given square matrix) along with its RC, the perturbation set being the unit box:

$$\|0 \cdot y + Q\zeta\|_2 \leq 1 \ \forall(\zeta : \|\zeta\|_\infty \leq 1). \qquad \text{(RC)}$$

The feasible set of the RC is either the entire space of y-variables, or is empty, which depends on whether or not one has

$$\max_{\|\zeta\|_\infty \leq 1} \zeta^T B\zeta \leq 1. \qquad [B = Q^T Q]$$

Varying Q, we can get, as B, an arbitrary positive semidefinite matrix of a given size. Now, assuming that we can process (RC) efficiently, we can check efficiently whether the feasible set of (RC) is or is not empty, that is, we can compare efficiently the maximum of a positive semidefinite quadratic form over the unit box with the value 1. If we can do it, we can compute the maximum of a general-type positive semidefinite quadratic form $\zeta^T B \zeta$ over the unit box within relative accuracy ϵ in time polynomial in the dimension of ζ and $\ln(1/\epsilon)$ (by comparing $\max_{\|\zeta\|_\infty \leq 1} \lambda \zeta^T B \zeta$ with 1 and applying bisection in $\lambda > 0$). Thus, the NP-hard problem of computing $\max_{\|\zeta\|_\infty \leq 1} \zeta^T B \zeta$, $B \succ 0$, within relative accuracy $\epsilon = 0.04$ reduces to checking feasibility of the RC of a CQI with a box perturbation set, meaning that it is NP-hard to process the RC in question.

s This unpleasant phenomenon we have just outlined leaves us with only two options:

A. To identify meaningful particular cases where the RC of an uncertain conic problem is computationally tractable; and

B. To develop *tractable approximations* of the RC in the remaining cases.

Note that the RC, same as in the LO case, is a "constraint-wise" construction, so that investigating tractability of the RC of an uncertain conic problem reduces to the same question for the RCs of the conic constraints constituting the problem. Due to this observation, from now on we focus on tractability of the RC

$$\forall(\zeta \in \mathcal{Z}) : A(\zeta)x + b(\zeta) \in \mathbf{Q}$$

of a *single* uncertain conic inequality.

5.3 SAFE TRACTABLE APPROXIMATIONS OF RCS OF UNCERTAIN CONIC INEQUALITIES

In chapters 6, 8 we will present a number of special cases where the RC of an uncertain CQI/LMI is computationally tractable; these cases have to do with rather specific perturbation sets. The question is, what to do when the RC is *not* computationally tractable. A natural course of action in this case is to look for a *safe tractable approximation* of the RC, defined as follows:

Definition 5.3.1. Consider the RC

$$\underbrace{A(\zeta)x + b(\zeta)}_{\equiv \alpha(x)\zeta + \beta(x)} \in \mathbf{Q} \; \forall \zeta \in \mathcal{Z} \tag{5.3.1}$$

of an uncertain constraint

$$A(\zeta)x + b(\zeta) \in \mathbf{Q}. \tag{5.3.2}$$

$(A(\zeta) \in \mathbb{R}^{k \times n}, b(\zeta) \in \mathbb{R}^k$ are affine in ζ, so that $\alpha(x), \beta(x)$ are affine in the decision vector x). We say that a system $\mathcal{S}$ of convex constraints in variables x and, perhaps, additional variables u is a safe approximation of the RC (5.3.1), if the projection of the feasible set of $\mathcal{S}$ on the space of x variables is contained in the feasible set of the RC:

$$\forall x : (\exists u : \; (x, u) \text{ satisfies } \mathcal{S}) \Rightarrow x \text{ satisfies } (5.3.1).$$

This approximation is called tractable, provided that $\mathcal{S}$ is so, (e.g., $\mathcal{S}$ is an explicit system of CQIs/LMIs or, more generally, the constraints in $\mathcal{S}$ are efficiently computable).

The rationale behind the definition is as follows: assume we are given an uncertain conic problem (5.1.2) with vector of design variables x and a certain objective $c^T x$ (as we remember, the latter assumption is w.l.o.g.) and we have at our disposal a safe tractable approximation $\mathcal{S}_i$ of i-th constraint of the problem, $i = 1, ..., m$. Then the problem

$$\min_{x, u^1, ..., u^m} \left\{ c^T x : (x, u^i) \text{ satisfies } \mathcal{S}_i, \, 1 \leq i \leq m \right\}$$

is a computationally tractable *safe approximation* of the RC, meaning that the x-component of every feasible solution to the approximation is feasible for the RC, and thus an optimal solution to the approximation is a *feasible* suboptimal solution to the RC.

In principle, there are many ways to build a safe tractable approximation of an uncertain conic problem. For example, assuming $\mathcal{Z}$ bounded, which usually is the case, we could find a simplex $\Delta = \text{Conv}\{\zeta^1, ..., \zeta^{L+1}\}$ in the space $\mathbb{R}^L$ of perturbation vectors that is large enough to contain the actual perturbation set $\mathcal{Z}$. The RC of our uncertain problem, the perturbation set being Δ, is computationally tractable (see section 6.1) and is a safe approximation of the RC associated with the actual perturbation set $\mathcal{Z}$ due to $\Delta \supset \mathcal{Z}$. The essence of the matter is, of course, how conservative an approximation is: how much it "adds" to the built-in conservatism of the worst-case-oriented RC. In order to answer the latter question, we should quantify the "conservatism" of an approximation. There is no evident way to do it. One possible way could be to look by how much the optimal value of the approximation is larger than the optimal value of the true RC, but here we run into a severe difficulty. It may well happen that the feasible set of an approximation is empty, while the true feasible set of the RC is not so. Whenever this is the case, the optimal value of the approximation is "infinitely worse" than the true optimal value. It follows that comparison of optimal values makes sense only when the approximation scheme in question guarantees that the approximation inherits the feasibility properties of the true problem. On a closer inspection, such a requirement is, in general, not less restrictive than the requirement for the approximation to be precise.

The way to quantify the conservatism of an approximation to be used in this book is as follows. Assume that $0 \in \mathcal{Z}$ (this assumption is in full accordance with the interpretation of vectors $\zeta \in \mathcal{Z}$ as data perturbations, in which case $\zeta = 0$ corresponds to the nominal data). With this assumption, we can embed our closed convex perturbation set $\mathcal{Z}$ into a single-parametric family of perturbation sets

$$\mathcal{Z}_\rho = \rho \mathcal{Z}, \, 0 < \rho \leq \infty, \tag{5.3.3}$$

thus giving rise to a single-parametric family

$$\underbrace{A(\zeta)x + b(\zeta)}_{\equiv \alpha(x)\zeta + \beta(x)} \in \mathbf{Q} \;\; \forall \zeta \in \mathcal{Z}_\rho \tag{RC$_\rho$}$$

of RCs of the uncertain conic constraint (5.3.2). One can think about ρ as *perturbation level*; the original perturbation set $\mathcal{Z}$ and the associated RC (5.3.1) correspond to the perturbation level 1. Observe that the feasible set X_ρ of (RC$_\rho$) shrinks as ρ grows. This allows us to quantify the conservatism of a safe approximation to (RC) by "positioning" the feasible set of $\mathcal{S}$ with respect to the scale of "true" feasible sets X_ρ, specifically, as follows:

Definition 5.3.2. Assume that we are given an approximation scheme that puts into correspondence to (5.3.3), (RC$_\rho$) a finite system $\mathcal{S}_\rho$ of efficiently computable convex constraints on variables x and, perhaps, additional variables u, depending on $\rho > 0$ as on a parameter, in such a way that for every ρ the system $\mathcal{S}_\rho$ is a safe tractable approximation of (RC$_\rho$), and let $\widehat{X}_\rho$ be the projection of the feasible set of $\mathcal{S}_\rho$ onto the space of x variables.

We say that the conservatism (or "tightness factor") of the approximation scheme in question does not exceed $\vartheta \geq 1$ if, for every $\rho > 0$, we have

$$X_{\vartheta\rho} \subset \widehat{X}_\rho \subset X_\rho.$$

Note that the fact that $\mathcal{S}_\rho$ is a safe approximation of (RC$_\rho$) tight within factor ϑ is equivalent to the following pair of statements:

i) [safety] *Whenever a vector x and $\rho > 0$ are such that x can be extended to a feasible solution of $\mathcal{S}_\rho$, x is feasible for (RC$_\rho$);*

ii) [tightness] *Whenever a vector x and $\rho > 0$ are such that x cannot be extended to a feasible solution of $\mathcal{S}_\rho$, x is not feasible for (RC$_{\vartheta\rho}$).*

Clearly, a tightness factor equal to 1 means that the approximation is precise: $\widehat{X}_\rho = X_\rho$ for all ρ. In many applications, especially in those where the level of perturbations is known only "up to an order of magnitude," a safe approximation of the RC with a moderate tightness factor is almost as useful, from a practical viewpoint, as the RC itself.

An important observation is that *with a bounded perturbation set $\mathcal{Z} = \mathcal{Z}_1 \subset \mathbb{R}^L$ that is symmetric w.r.t. the origin, we can always point out a safe computationally tractable approximation scheme for (5.3.3), (RC$_\rho$) with tightness factor $\leq L$.*

Indeed, w.l.o.g. we may assume that $\mathrm{int}\mathcal{Z} \neq \emptyset$, so that $\mathcal{Z}$ is a closed and bounded convex set symmetric w.r.t. the origin. It is known that for such a set, there always exist two similar ellipsoids, centered at the origin, with the similarity ratio at most $\sqrt{L}$, such that the smaller ellipsoid is contained in $\mathcal{Z}$, and the larger one contains $\mathcal{Z}$. In particular, one can choose, as the smaller ellipsoid, the largest volume ellipsoid contained in $\mathcal{Z}$; alternatively, one can choose, as the larger ellipsoid, the smallest volume ellipsoid containing $\mathcal{Z}$. Choosing coordinates in which the smaller ellipsoid is the unit Euclidean ball B, we conclude that $B \subset \mathcal{Z} \subset \sqrt{L}B$. Now observe that B, and therefore $\mathcal{Z}$, contains the convex hull $\underline{\mathcal{Z}} = \{\zeta \in \mathbb{R}^L : \|\zeta\|_1 \leq 1\}$ of the $2L$ vectors $\pm e_\ell$, $\ell = 1, ..., L$, where e_ℓ are the basic orths of the axes in question. Since $\underline{\mathcal{Z}}$ clearly contains $L^{-1/2}B$, the convex hull $\widehat{\mathcal{Z}}$ of the vectors $\pm Le_\ell$, $\ell = 1, ..., L$, contains $\mathcal{Z}$ and is contained in $L\mathcal{Z}$. Taking, as $\mathcal{S}_\rho$, the RC of our uncertain constraint, the perturbation set being $\rho\widehat{\mathcal{Z}}$, we clearly get an L-tight safe approximation of (5.3.3), (RC$_\rho$), and this approximation is merely the system of constraints

$$A(\rho Le_\ell)x + b(\rho Le_\ell) \in \mathbf{Q}, \; A(-\rho Le_\ell)x + b(-\rho Le_\ell) \in \mathbf{Q}, \; \ell = 1, ..., L,$$

that is, our approximation scheme is computationally tractable.

5.4 EXERCISES

Exercise 5.1. Find and try to close a logical gap in the proof of the statement

With a bounded perturbation set $\mathcal{Z} = \mathcal{Z}_1 \subset \mathbb{R}^L$ symmetric w.r.t. the origin, we always can point out a safe computationally tractable approximation scheme for (5.3.3), (RC$_\rho$) with tightness factor $\leq L$.

concluding the previous section.

Exercise 5.2. Consider a semi-infinite conic constraint

$$\forall(\zeta \in \rho\mathcal{Z}) : a_0[x] + \sum_{\ell=1}^{L} \zeta_i a_\ell[x] \in \mathbf{Q} \qquad (C_{\mathcal{Z}}[\rho])$$

Assume that for certain ϑ and some closed convex set $\mathcal{Z}_*$, $0 \in \mathcal{Z}_*$, the constraint $(C_{\mathcal{Z}_*}[\cdot])$ admits a safe tractable approximation tight within the factor ϑ. Now let $\mathcal{Z}$ be a closed convex set that can be approximated, up to a factor λ, by $\mathcal{Z}_*$, meaning that for certain $\gamma > 0$ we have

$$\gamma\mathcal{Z}_* \subset \mathcal{Z} \subset (\lambda\gamma)\mathcal{Z}_*.$$

Prove that $(C_{\mathcal{Z}}[\cdot])$ admits a safe tractable approximation, tight within the factor $\lambda\vartheta$.

Exercise 5.3. Let $\vartheta \geq 1$ be given, and consider the semi-infinite conic constraint $(C_{\mathcal{Z}}[\cdot])$ "as a function of $\mathcal{Z}$," meaning that $a_\ell[\cdot]$, $0 \leq \ell \leq L$, and $\mathbf{Q}$ are once and forever fixed. In what follows, $\mathcal{Z}$ always is a solid (convex compact set with a nonempty interior) symmetric w.r.t. 0.

Assume that whenever $\mathcal{Z}$ is an ellipsoid centered at the origin, $(C_{\mathcal{Z}}[\cdot])$ admits a safe tractable approximation tight within factor ϑ (as it is the case for $\vartheta = 1$ when $\mathbf{Q}$ is the Lorentz cone, see section 6.5).

i) Prove that when $\mathcal{Z}$ is the intersection of M centered at the origin ellipsoids:

$$\mathcal{Z} = \{\zeta : \zeta^T Q_i \zeta \leq 1, \; i = 1, ..., M\} \qquad [Q_i \succeq 0, \sum_i Q_i \succ 0]$$

$(C_{\mathcal{Z}}[\cdot])$ admits a safe tractable approximation tight within the factor $\sqrt{M}\vartheta$.

ii) Prove that if $\mathcal{Z} = \{\zeta : \|\zeta\|_\infty \leq 1\}$, then $(C_{\mathcal{Z}}[\cdot])$ admits a safe tractable approximation tight within the factor $\vartheta\sqrt{\dim \zeta}$.

iii) Assume that $\mathcal{Z}$ is the intersection of M ellipsoids not necessarily centered at the origin. Prove that then $(C_{\mathcal{Z}}[\cdot])$ admits a safe tractable approximation tight within a factor $\sqrt{2M}\vartheta$.

5.5 NOTES AND REMARKS

NR 5.1. The central role played by Uncertain CO in Robust Optimization stems from the following reasons:

- The conic form $\min_x \left\{ c^T x : Ax - b \in \mathbf{K} \right\}$ of a convex problem allows one to naturally separate the problem's structure (represented by the cone $\mathbf{K}$) from the problem's data (represented by (c, A, b)), which is vitally important for investigating issues related to data uncertainty. Technically speaking, the main advantage of this format is that the left hand side of conic inequality $Ax - b \in \mathbf{K}$ is bi-affine in the data and the decision variables; this fact, essentially, is the starting point of all our tractability-related results on the RC. This is in sharp contrast to the usual "Mathematical Programming" format of an optimization problem $\min_x \left\{ f_0(x, \zeta) : f_i(x, \zeta) \leq 0, \, i = 1, ..., m \right\}$, where ζ stands for the data; without additional structural assumptions on how the data enters the objective, this format, while allowing to define the notion of RC, is poorly suited for investigating the related tractability issues.

- The conic form of a convex program is not only "structure and data revealing"; it allows also for unified treatment of a wide variety of convex programs, since just 3 "generic" cones with well understood geometry — (direct products of) rays, Lorentz and Semidefinite cones — are responsible for "nearly all" convex problems arising in applications.

The outlined "exceptional" role of the conic format in Robust Optimization was understood already in the very first papers on Convex RO [3, 4, 49, 50, 18].

NR 5.2. The concept of *safe* tractable approximation (as opposed to "more aggressive" approximations, where a solution to the approximating problem not necessarily is feasible for the problem of interest) in the RO context is very natural — finally, the entire RO methodology is about safety. A general scheme for building safe tractable approximations of uncertain conic problems is proposed in [22]; this scheme, however, does not admit good bounds on its conservatism. The quantification of conservatism of a safe tractable approximation used in this book was introduced in [4]. Its "flavor" resembles that of *approximation algorithms* aimed at finding suboptimal feasible solutions to difficult optimization problems. An efficient, (i.e., polynomial time) algorithm for a generic optimization problem $\mathcal{P}$ is called α-approximating, if, as applied to every instance p of the problem, it produces a feasible solution with the value of the objective by a factor at most

α greater than the optimal value of the instance, where α is independent of the instance's data. (In order for this definition to make sense, the optimal values of all instances should be positive.) The concept of an approximating algorithm is "tailored" to the situations where a feasible solution to an instance can be found efficiently; what is difficult, is to find a feasible solution that is near-optimal. The concept of a tight tractable approximation is of the same spirit, but it is adjusted to the case where the difficulties primarily come from the necessity to satisfy the constraints.

Chapter Six
Uncertain Conic Quadratic Problems with Tractable RCs

In this chapter we focus on uncertain conic quadratic problems (that is, the sets $\mathbf{Q}_i$ in (5.1.2) are given by explicit lists of conic quadratic inequalities) for which the RCs are computationally tractable.

6.1 A GENERIC SOLVABLE CASE: SCENARIO UNCERTAINTY

We start with a simple case where the RC of an uncertain conic problem (not necessarily a conic quadratic one) is computationally tractable — the case of *scenario uncertainty*.

Definition 6.1.1. We say that a perturbation set $\mathcal{Z}$ is scenario generated, if $\mathcal{Z}$ is given as the convex hull of a given finite set of scenarios $\zeta^{(\nu)}$:

$$\mathcal{Z} = \text{Conv}\{\zeta^{(1)}, ..., \zeta^{(N)}\}. \tag{6.1.1}$$

Theorem 6.1.2. The RC (5.1.5) of uncertain problem (5.1.2), (5.1.4) with scenario perturbation set (6.1.1) is equivalent to the explicit conic problem

$$\min_{x,t}\left\{ t : \begin{array}{c} [c^0 + \sum_{\ell=1}^{L} \zeta_\ell^{(\nu)} c^\ell]^T x + [d^0 + \sum_{\ell=1}^{L} \zeta_\ell^{(\nu)} d^\ell] - t \leq 0 \\ [A_i^0 + \sum_{\ell=1}^{L} \zeta_\ell^{(\nu)} A_i^\ell]^T x - [b^0 + \sum_{\ell=1}^{L} \zeta_\ell^{(\nu)} b^\ell] \in \mathbf{Q}_i, \\ 1 \leq i \leq m \end{array}, 1 \leq \nu \leq N \right\} \tag{6.1.2}$$

with a structure similar to the one of the instances of the original uncertain problem.

Proof. This is evident due to the convexity of $\mathbf{Q}_i$ and the affinity of the left hand sides of the constraints in (5.1.5) in ζ. □

The situation considered in Theorem 6.1.2 is "symmetric" to the one considered in chapter 1, where we spoke about problems (5.1.2) with the simplest possible sets $\mathbf{Q}_i$ — just nonnegative rays, and the RC turns out to be computationally tractable whenever the perturbation set is so. Theorem 6.1.2 deals with another extreme case of the tradeoff between the geometry of the right hand side sets $\mathbf{Q}_i$ and that of the perturbation set. Here the latter is as simple as it could be — just the convex hull of an explicitly listed finite set, which makes the RC computationally tractable for rather general (just computationally tractable) sets $\mathbf{Q}_i$. Unfortunately, the second extreme is not too interesting: in the large scale case, a "scenario approximation" of a reasonable quality for typical perturbation sets, like boxes, requires an astronomically large number of scenarios, thus preventing listing them explicitly and making problem (6.1.2) computationally intractable. This is in sharp contrast with the first extreme, where the simple sets were $\mathbf{Q}_i$ — Linear Optimization is definitely interesting and has a lot of applications.

In what follows, we consider a number of less trivial cases where the RC of an uncertain conic quadratic problem is computationally tractable. As always with RC, which is a constraint-wise construction, we may focus on computational tractability of the RC of a *single* uncertain CQI

$$\| \underbrace{A(\zeta)y + b(\zeta)}_{\equiv \alpha(y)\zeta + \beta(y)} \|_2 \leq \underbrace{c^T(\zeta)y + d(\zeta)}_{\equiv \sigma^T(y)\zeta + \delta(y)}, \tag{6.1.3}$$

where $A(\zeta) \in \mathbb{R}^{k \times n}, b(\zeta) \in \mathbb{R}^k, c(\zeta) \in \mathbb{R}^n, d(\zeta) \in \mathbb{R}$ are affine in ζ, so that $\alpha(y), \beta(y), \sigma(y), \delta(y)$ are affine in the decision vector y.

6.2 SOLVABLE CASE I: SIMPLE INTERVAL UNCERTAINTY

Consider uncertain conic quadratic constraint (6.1.3) and assume that:

i) The uncertainty is *side-wise*: the perturbation set $\mathcal{Z} = \mathcal{Z}^{\text{left}} \times \mathcal{Z}^{\text{right}}$ is the direct product of two sets (so that the perturbation vector $\zeta \in \mathcal{Z}$ is split into blocks $\eta \in \mathcal{Z}^{\text{left}}$ and $\chi \in \mathcal{Z}^{\text{right}}$), with the left hand side data $A(\zeta), b(\zeta)$ depending solely on η and the right hand side data $c(\zeta), d(\zeta)$ depending solely on χ, so that (6.1.3) reads

$$\| \underbrace{A(\eta)y + b(\eta)}_{\equiv \alpha(y)\eta + \beta(y)} \|_2 \leq \underbrace{c^T(\chi)y + d(\chi)}_{\equiv \sigma^T(y)\chi + \delta(y)}, \tag{6.2.1}$$

and the RC of this uncertain constraint reads

$$\|A(\eta)y + b(\eta)\|_2 \leq c^T(\chi)y + d(\chi) \quad \forall(\eta \in \mathcal{Z}^{\text{left}}, \chi \in \mathcal{Z}^{\text{right}}); \tag{6.2.2}$$

ii) The right hand side perturbation set is as described in Theorem 1.3.4, that is,

$$\mathcal{Z}^{\text{right}} = \{\chi : \exists u : P\chi + Qu + p \in \mathbf{K}\},$$

where either $\mathbf{K}$ is a closed convex pointed cone, and the representation is strictly feasible, or $\mathbf{K}$ is a polyhedral cone given by an explicit finite list of linear inequalities;

iii) The left hand side uncertainty is a simple interval one:

$$\begin{aligned} \mathcal{Z}^{\text{left}} &= \{\eta = [\delta A, \delta b] : |(\delta A)_{ij}| \leq \delta_{ij}, 1 \leq i \leq k, 1 \leq j \leq n, \\ &\qquad\qquad |(\delta b)_i| \leq \delta_i, \ 1 \leq i \leq k\}, \\ [A(\zeta), b(\zeta)] &= [A^n, b^n] + [\delta A, \delta b]. \end{aligned}$$

In other words, every entry in the left hand side data $[A, b]$ of (6.1.3), independently of all other entries, runs through a given segment centered at the nominal value of the entry.

Proposition 6.2.1. Under assumptions $1-3$ on the perturbation set $\mathcal{Z}$, the RC of the uncertain CQI (6.1.3) is equivalent to the following explicit system of conic quadratic and linear constraints in variables y, z, τ, v:

$$
\begin{aligned}
(a) \quad & \tau + p^T v \le \delta(y), \ P^T v = \sigma(y), \\
& Q^T v = 0, \ v \in \mathbf{K}_*
\end{aligned}
$$

$$
\begin{aligned}
(b) \quad & z_i \ge |(A^{\mathrm{n}} y + b^{\mathrm{n}})_i| + \delta_i + \sum_{j=1}^n |\delta_{ij} y_j|, \ i = 1, ..., k \\
& \|z\|_2 \le \tau
\end{aligned}
\qquad (6.2.3)
$$

where $\mathbf{K}_*$ is the cone dual to $\mathbf{K}$.

Proof. Due to the side-wise structure of the uncertainty, a given y is robust feasible if and only if there exists τ such that

$$
\begin{aligned}
(a) \quad \tau & \le \min_{\chi \in \mathcal{Z}^{\mathrm{right}}} \left\{ \sigma^T(y)\chi + \delta(y) \right\} \\
& = \min_{\chi, u} \left\{ \sigma^T(y)\chi : P\chi + Qu + p \in \mathbf{K} \right\} + \delta(y), \\
(b) \quad \tau & \ge \max_{\eta \in \mathcal{Z}^{\mathrm{left}}} \|A(\eta)y + b(\eta)\|_2 \\
& = \max_{\delta A, \delta b} \left\{ \|[A^{\mathrm{n}} y + b^{\mathrm{n}}] + [\delta A y + \delta b]\|_2 : |\delta A|_{ij} \le \delta_{ij}, |\delta b_i| \le \delta_i \right\}.
\end{aligned}
$$

By Conic Duality, a given τ satisfies (a) if and only if τ can be extended, by properly chosen v, to a solution of $(6.2.3.a)$; by evident reasons, τ satisfies (b) if and only if there exists z satisfying $(6.2.3.b)$. $\qquad \square$

6.3 SOLVABLE CASE II: UNSTRUCTURED NORM-BOUNDED UNCERTAINTY

Consider the case where the uncertainty in (6.1.3) is still side-wise ($\mathcal{Z} = \mathcal{Z}^{\mathrm{left}} \times \mathcal{Z}^{\mathrm{right}}$) with the right hand side uncertainty set $\mathcal{Z}^{\mathrm{right}}$ as in section 6.2, while the left hand side uncertainty is *unstructured norm-bounded*, meaning that

$$
\mathcal{Z}^{\mathrm{left}} = \left\{ \eta \in \mathbb{R}^{p \times q} : \|\eta\|_{2,2} \le 1 \right\} \qquad (6.3.1)
$$

and either

$$
A(\eta)y + b(\eta) = A^{\mathrm{n}} y + b^{\mathrm{n}} + L^T(y)\eta R \qquad (6.3.2)
$$

with $L(y)$ affine in y and $R \ne 0$, or

$$
A(\eta)y + b(\eta) = A^{\mathrm{n}} y + b^{\mathrm{n}} + L^T \eta R(y) \qquad (6.3.3)
$$

with $R(y)$ affine in y and $L \ne 0$. Here

$$
\|\eta\|_{2,2} = \max_u \left\{ \|\eta u\|_2 : u \in \mathbb{R}^q, \|u\|_2 \le 1 \right\}
$$

is the usual matrix norm of a $p \times q$ matrix η (the maximal singular value),

Example 6.3.1.

(i) Imagine that some $p \times q$ submatrix P of the left hand side data $[A, b]$ of (6.2.1) is uncertain and differs from its nominal value P^{n} by an additive perturbation $\Delta P = M^T \Delta N$ with Δ having matrix norm at most 1, and all entries in $[A, b]$ outside of P are certain. Denoting by I the set of indices of the rows in P and by J the set of indices of the columns in P, let U be the natural projector of $\mathbb{R}^{n+1}$ on the coordinate subspace in $\mathbb{R}^{n+1}$ given by J, and V be the natural projector of $\mathbb{R}^k$ on the subspace of $\mathbb{R}^k$ given by I (e.g., with $I = \{1, 2\}$ and $J = \{1, 5\}$, $Uu = [u_1; u_5] \in \mathbb{R}^2$ and $Vu = [u_1; u_2] \in \mathbb{R}^2$). Then the outlined perturbations of $[A, b]$ can be represented as

$$[A(\eta), b(\eta)] = [A^{\mathrm{n}}, b^{\mathrm{n}}] + \underbrace{V^T M^T}_{L^T} \eta \underbrace{(NU)}_{R}, \ \|\eta\|_{2,2} \le 1,$$

whence, setting $Y(y) = [y; 1]$,

$$A(\eta)y + b(\eta) = [A^{\mathrm{n}}y + b^{\mathrm{n}}] + L^T \eta \underbrace{[RY(y)]}_{R(y)},$$

and we are in the situation (6.3.1), (6.3.3).

(ii) [Simple ellipsoidal uncertainty] Assume that the left hand side perturbation set $\mathcal{Z}^{\text{left}}$ is a p-dimensional ellipsoid; w.l.o.g. we may assume that this ellipsoid is just the unit Euclidean ball $B = \{\eta \in \mathbb{R}^p : \|\eta\|_2 \le 1\}$. Note that for vectors $\eta \in \mathbb{R}^p = \mathbb{R}^{p \times 1}$ their usual Euclidean norm $\|\eta\|_2$ and their matrix norm $\|\eta\|_{2,2}$ are the same. We now have

$$A(\eta)y + b(\eta) = [A^0 y + b^0] + \sum_{\ell=1}^{p} \eta_\ell [A^\ell y + b^\ell] = [A^{\mathrm{n}}y + b^{\mathrm{n}}] + L^T(y)\eta R,$$

where $A^{\mathrm{n}} = A^0$, $b^{\mathrm{n}} = b^0$, $R = 1$ and $L(y)$ is the matrix with the rows $[A^\ell y + b^\ell]^T$, $\ell = 1, ..., p$. Thus, we are in the situation (6.3.1), (6.3.2).

Theorem 6.3.2. The RC of the uncertain CQI (6.2.1) with unstructured norm-bounded uncertainty is equivalent to the following explicit system of LMIs in variables y, τ, u, λ:

(i) In the case of left hand side perturbations (6.3.1), (6.3.2):

$$(a) \quad \tau + p^T v \le \delta(y), \ P^T v = \sigma(y), \ Q^T v = 0, \ v \in \mathbf{K}_*$$

$$(b) \quad \left[\begin{array}{c|c|c} \tau I_k & L^T(y) & A^{\mathrm{n}}y + b^{\mathrm{n}} \\ \hline L(y) & \lambda I_p & \\ \hline [A^{\mathrm{n}}y + b^{\mathrm{n}}]^T & & \tau - \lambda R^T R \end{array} \right] \succeq 0. \qquad (6.3.4)$$

(ii) In the case of left hand side perturbations (6.3.1), (6.3.3):

$$(a) \quad \tau + p^T v \le \delta(y), \ P^T v = \sigma(y), \ Q^T v = 0, \ v \in \mathbf{K}_*$$

$$(b) \quad \left[\begin{array}{c|c|c} \tau I_k - \lambda L^T L & & A^{\mathrm{n}}y + b^{\mathrm{n}} \\ \hline & \lambda I_q & R(y) \\ \hline [A^{\mathrm{n}}y + b^{\mathrm{n}}]^T & R^T(y) & \tau \end{array} \right] \succeq 0. \qquad (6.3.5)$$

Here $\mathbf{K}_*$ is the cone dual to $\mathbf{K}$.

Proof. Same as in the proof of Proposition 6.2.1, y is robust feasible for (6.2.1) if and only if there exists τ such that

$$
\begin{aligned}
(a) \quad \tau &\leq \min_{\chi \in \mathcal{Z}^{\text{right}}} \left\{ \sigma^T(y)\chi + \delta(y) \right\} \\
&= \min_{\chi, u} \left\{ \sigma^T(y)\chi : P\chi + Qu + p \in \mathbf{K} \right\}, \qquad (6.3.6)
\end{aligned}
$$

$$
(b) \quad \tau \geq \max_{\eta \in \mathcal{Z}^{\text{left}}} \| A(\eta)y + b(\eta) \|_2,
$$

and a given τ satisfies (a) if and only if it can be extended, by a properly chosen v, to a solution of $(6.3.4.a) \Leftrightarrow (6.3.5.a)$. It remains to understand when τ satisfies (b). This requires two basic facts.

Lemma 6.3.3. [Semidefinite representation of the Lorentz cone] A vector $[y; t] \in \mathbb{R}^k \times \mathbb{R}$ belongs to the Lorentz cone $\mathbf{L}^{k+1} = \{ [y; t] \in \mathbb{R}^{k+1} : t \geq \|y\|_2 \}$ if and only if the "arrow matrix"

$$
\text{Arrow}(y, t) = \left[\begin{array}{c|c} t & y^T \\ \hline y & tI_k \end{array} \right]
$$

is positive semidefinite.

<u>Proof of Lemma 6.3.3:</u> We use the following fundamental fact:

Lemma 6.3.4. [**Schur Complement Lemma**] A symmetric block matrix

$$
A = \left[\begin{array}{c|c} P & Q^T \\ \hline Q & R \end{array} \right]
$$

with $R \succ 0$ is positive (semi)definite if and only if the matrix

$$
P - Q^T R^{-1} Q
$$

is positive (semi)definite.

<u>Schur Complement Lemma $\Rightarrow$ Lemma 6.3.3:</u> When $t = 0$, we have $[y; t] \in \mathbf{L}^{k+1}$ iff $y = 0$, and $\text{Arrow}(y, t) \succeq 0$ iff $y = 0$, as claimed in Lemma 6.3.3. Now let $t > 0$. Then the matrix tI_k is positive definite, so that by the Schur Complement Lemma we have $\text{Arrow}(y, t) \succeq 0$ if and only if $t \geq t^{-1}y^T y$, or, which is the same, iff $[y; t] \in \mathbf{L}^{k+1}$. When $t < 0$, we have $[y; t] \notin \mathbf{L}^{k+1}$ and $\text{Arrow}(y, t) \not\succeq 0$. $\square$

<u>Proof of the Schur Complement Lemma:</u> Matrix $A = A^T$ is $\succeq 0$ iff $u^T P u + 2u^T Q^T v + v^T R v \geq 0$ for all u, v, or, which is the same, iff

$$
\forall u : 0 \leq \min_v \left\{ u^T P u + 2u^T Q^T v + v^T R v \right\} = u^T P u - u^T Q^T R^{-1} Q u
$$

(indeed, since $R \succ 0$, the minimum in v in the last expression is achieved when $v = R^{-1}Qu$). The concluding relation $\forall u : u^T [P - Q^T R^{-1} Q] u \geq 0$ is valid iff $P - Q^T R^{-1} Q \succeq 0$. Thus, $A \succeq 0$ iff $P - Q^T R^{-1} Q \succeq 0$. The same reasoning implies that $A \succ 0$ iff $P - Q^T R^{-1} Q \succ 0$. $\square$

We further need the following fundamental result:

Lemma 6.3.5. [S-Lemma]

(i) [homogeneous version] Let A, B be symmetric matrices of the same size such that $\bar{x}^T A \bar{x} > 0$ for some $\bar{x}$. Then the implication

$$x^T A x \geq 0 \Rightarrow x^T B x \geq 0$$

holds true if and only if

$$\exists \lambda \geq 0 : B \succeq \lambda A.$$

(ii) [inhomogeneous version] Let A, B be symmetric matrices of the same size, and let the quadratic form $x^T A x + 2a^T x + \alpha$ be strictly positive at some point. Then the implication

$$x^T A x + 2a^T x + \alpha \geq 0 \Rightarrow x^T B x + 2b^T x + \beta \geq 0$$

holds true if and only if

$$\exists \lambda \geq 0 : \left[\begin{array}{c|c} B - \lambda A & b^T - \lambda a^T \\ \hline b - \lambda a & \beta - \lambda \alpha \end{array} \right] \succeq 0.$$

For proof of this fundamental Lemma, see Appendix B.2.

Coming back to the proof of Theorem 6.3.2, we can now understand when a given pair τ, y satisfies (6.3.6.b). Let us start with the case (6.3.2). We have

(y, τ) satisfies (6.3.6.b)

$$\Leftrightarrow \quad [\overbrace{[A^{\mathbf{n}}y + b^{\mathbf{n}}]}^{\widehat{y}} + L^T(y)\eta R; \tau] \in \mathbf{L}^{k+1} \quad \forall(\eta : \|\eta\|_{2,2} \leq 1) \qquad \text{[by (6.3.2)]}$$

$$\Leftrightarrow \quad \left[\begin{array}{c|c} \tau & \widehat{y}^T + R^T \eta^T L(y) \\ \hline \widehat{y} + L^T(y)\eta R & \tau I_k \end{array} \right] \succeq 0 \quad \forall(\eta : \|\eta\|_{2,2} \leq 1) \qquad \text{[by Lemma 6.3.3]}$$

$$\Leftrightarrow \quad \tau s^2 + 2sr^T[\widehat{y} + L^T(y)\eta R] + \tau r^T r \geq 0 \quad \forall[s; r] \ \forall(\eta : \|\eta\|_{2,2} \leq 1)$$

$$\Leftrightarrow \quad \tau s^2 + 2s\widehat{y}^T r + 2 \min_{\eta : \|\eta\|_{2,2} \leq 1} \left[s(\eta^T L(y)r)^T R \right] + \tau r^T r \geq 0 \quad \forall[s; r]$$

$$\Leftrightarrow \quad \tau s^2 + 2s\widehat{y}^T r - 2\|L(y)r\|_2 \|sR\|_2 + \tau r^T r \geq 0 \quad \forall[s; r]$$

$$\Leftrightarrow \quad \tau r^T r + 2(L(y)r)^T \xi + 2sr^T \widehat{y} + \tau s^2 \geq 0 \quad \forall(s, r, \xi : \xi^T \xi \leq s^2 R^T R)$$

$$\Leftrightarrow \quad \exists \lambda \geq 0 : \left[\begin{array}{c|c|c} \tau I_k & L^T(y) & \widehat{y} \\ \hline L(y) & \lambda I_p & \\ \hline \widehat{y}^T & & \tau - \lambda R^T R \end{array} \right] \succeq 0$$

[by the homogeneous S-Lemma; note that $R \neq 0$].

The requirement $\lambda \geq 0$ in the latter relation is implied by the LMI in the relation and is therefore redundant. Thus, in the case of (6.3.2) relation (6.3.6.b) is equivalent to the possibility to extend (y, τ) to a solution of (6.3.4.b).

Now let (6.3.3) be the case. We have

(y, τ) satisfies (6.3.6.b)

$$\Leftrightarrow \quad [\overbrace{[A^{\mathrm{n}}y + b^{\mathrm{n}}]}^{\widehat{y}} + L^T \eta R(y); \tau] \in \mathbf{L}^{k+1} \quad \forall (\eta : \|\eta\|_{2,2} \le 1) \; [\text{by } (6.3.3)]$$

$$\Leftrightarrow \quad \left[\begin{array}{c|c} \tau & \widehat{y}^T + R^T(y)\eta^T L \\ \hline \widehat{y} + L^T \eta R(y) & \tau I_k \end{array} \right] \succeq 0 \quad \forall (\eta : \|\eta\|_{2,2} \le 1)$$
$$[\text{by Lemma } 6.3.3]$$

$$\Leftrightarrow \quad \tau s^2 + 2sr^T[\widehat{y} + L^T \eta R(y)] + \tau r^T r \ge 0 \quad \forall [s; r] \; \forall (\eta : \|\eta\|_{2,2} \le 1)$$

$$\Leftrightarrow \quad \tau s^2 + 2s\widehat{y}^T r + 2 \min_{\eta: \|\eta\|_{2,2} \le 1} \left[s(\eta^T Lr)^T R(y) \right] + \tau r^T r \ge 0 \quad \forall [s; r]$$

$$\Leftrightarrow \quad \tau s^2 + 2s\widehat{y}^T r - 2\|Lr\|_2 \|sR(y)\|_2 + \tau r^T r \ge 0 \quad \forall [s; r]$$

$$\Leftrightarrow \quad \tau r^T r + 2sR^T(y)\xi + 2sr^T \widehat{y} + \tau s^2 \ge 0 \quad \forall (s, r, \xi : \xi^T \xi \le r^T L^T Lr)$$

$$\Leftrightarrow \quad \exists \lambda \ge 0 : \left[\begin{array}{c|c|c} \tau I_k - \lambda L^T L & & \widehat{y} \\ \hline & \lambda I_q & R(y) \\ \hline \widehat{y}^T & R^T(y) & \tau \end{array} \right] \succeq 0$$
$$[\text{by the homogeneous } \mathcal{S}\text{-Lemma; note that } L \ne 0].$$

As above, the restriction $\lambda \ge 0$ is redundant. We see that in the case of (6.3.3) relation (6.3.6.b) is equivalent to the possibility to extend (y, τ) to a solution of (6.3.5.b). □

6.4 SOLVABLE CASE III: CONVEX QUADRATIC INEQUALITY WITH UNSTRUCTURED NORM-BOUNDED UNCERTAINTY

A special case of an uncertain conic quadratic constraint (6.1.3) is a convex quadratic constraint

$$\begin{array}{ll} (a) & y^T A^T(\zeta)A(\zeta)y \le 2y^T b(\zeta) + c(\zeta) \\ & \quad\quad\quad\quad\quad\quad \Updownarrow \\ (b) & \|[2A(\zeta)y; 1 - 2y^T b(\zeta) - c(\zeta)]\|_2 \le 1 + 2y^T b(\zeta) + c(\zeta). \end{array} \quad (6.4.1)$$

Here $A(\zeta)$ is $k \times n$.

Assume that the uncertainty affecting this constraint is an unstructured norm-bounded one, meaning that

$$\begin{array}{ll} (a) & \mathcal{Z} = \{\zeta \in \mathbb{R}^{p \times q} : \|\zeta\|_{2,2} \le 1\}, \\ (b) & \left[\begin{array}{c} A(\zeta)y \\ y^T b(\zeta) \\ c(\zeta) \end{array} \right] = \left[\begin{array}{c} A^{\mathrm{n}}y \\ y^T b^{\mathrm{n}} \\ c^{\mathrm{n}} \end{array} \right] + L^T(y)\zeta R(y), \end{array} \quad (6.4.2)$$

where $L(y)$, $R(y)$ are matrices of appropriate sizes affinely depending on y and such that at least one of the matrices is constant. We are about to prove that the RC of (6.4.1), (6.4.2) is computationally tractable. Note that the just defined unstructured norm-bounded uncertainty in the data of convex quadratic constraint (6.4.1.a) implies similar uncertainty in the left hand side data of the equivalent uncertain CQI (6.4.1.a). Recall that Theorem 6.3.2 ensures that the RC of a general-type uncertain CQI with side-wise uncertainty and unstructured norm-bounded perturbations in the left hand side data is tractable. The result to follow removes the requirement of "side-wiseness" of the uncertainty at the cost of restricting the structure of the CQI in question — now it should come from an uncertain convex quadratic constraint. Note also that the case we are about to consider covers in particular the one when the data $(A(\zeta), b(\zeta), c(\zeta))$ of (6.4.1.a) are affinely parameterized by ζ varying in an ellipsoid (cf. Example 6.3.1.(ii)).

Proposition 6.4.1. Let us set $L(y) = [L_A(y), L_b(y), L_c(y)]$, where $L_b(y)$, $L_c(y)$ are the last two columns in $L(y)$, and let

$$\widehat{L}^T(y) = \left[L_b^T(y) + \tfrac{1}{2} L_c^T(y); L_A^T(y) \right], \quad \widehat{R}(y) = [R(y), 0_{q \times k}],$$

$$\mathcal{A}(y) = \left[\begin{array}{c|c} 2y^T b^n + c^n & [A^n y]^T \\ \hline A^n y & I_k \end{array} \right], \tag{6.4.3}$$

so that $\mathcal{A}(y)$, $\widehat{L}(y)$ and $\widehat{R}(y)$ are affine in y and at least one of the latter two matrices is constant.

The RC of (6.4.1), (6.4.2) is equivalent to the explicit LMI $\mathcal{S}$ in variables y, λ as follows:

(i) In the case when $\widehat{L}(y)$ is independent of y and is nonzero, $\mathcal{S}$ is

$$\left[\begin{array}{c|c} \mathcal{A}(y) - \lambda \widehat{L}^T \widehat{L} & \widehat{R}^T(y) \\ \hline \widehat{R}(y) & \lambda I_q \end{array} \right] \succeq 0; \tag{6.4.4}$$

(ii) In the case when $\widehat{R}(Y)$ is independent of y and is nonzero, $\mathcal{S}$ is

$$\left[\begin{array}{c|c} \mathcal{A}(y) - \lambda \widehat{R}^T \widehat{R} & \widehat{L}^T(y) \\ \hline \widehat{L}(y) & \lambda I_p \end{array} \right] \succeq 0; \tag{6.4.5}$$

(iii) In all remaining cases (that is, when either $\widehat{L}(y) \equiv 0$, or $\widehat{R}(y) \equiv 0$, or both), $\mathcal{S}$ is

$$\mathcal{A}(y) \succeq 0. \tag{6.4.6}$$

Proof. We have

$$y^T A^T(\zeta) A(\zeta) y \leq 2y^T b(\zeta) + c(\zeta) \quad \forall \zeta \in \mathcal{Z}$$

$$\Leftrightarrow \quad \left[\begin{array}{c|c} 2y^T b(\zeta) + c(\zeta) & [A(\zeta)y]^T \\ \hline A[\zeta]y & I_k \end{array} \right] \succeq 0 \quad \forall \zeta \in \mathcal{Z}$$

[Schur Complement Lemma]

$$\Leftrightarrow \quad \overbrace{\left[\begin{array}{c|c} 2y^T b^{\mathrm{n}} + c^{\mathrm{n}} & [A^{\mathrm{n}}y]^T \\ \hline A^{\mathrm{n}}y & I \end{array} \right]}^{\mathcal{A}(y)}$$

$$+ \overbrace{\left[\begin{array}{c|c} 2L_b^T(y)\zeta R(y) + L_c^T(y)\zeta R(y) & R^T(y)\zeta^T L_A(y) \\ \hline L_A^T(y)\zeta R(y) & \end{array} \right]}^{\mathcal{B}(y,\zeta)} \succeq 0 \quad \forall(\zeta : \|\zeta\|_{2,2} \leq 1)$$

[by (6.4.2)]

$$\Leftrightarrow \quad \mathcal{A}(y) + \widehat{L}^T(y)\zeta\widehat{R}(y) + \widehat{R}^T(y)\zeta^T\widehat{L}(y) \succeq 0 \quad \forall(\zeta : \|\zeta\|_{2,2} \leq 1) \qquad \text{[by (6.4.3)]}.$$

Now the reasoning can be completed exactly as in the proof of Theorem 6.3.2. Consider, e.g., the case of (i). We have

$$y^T A^T(\zeta) A(\zeta) y \leq 2y^T b(\zeta) + c(\zeta) \quad \forall \zeta \in \mathcal{Z}$$

$$\Leftrightarrow \quad \mathcal{A}(y) + \widehat{L}^T\zeta\widehat{R}(y) + \widehat{R}^T(y)\zeta^T\widehat{L} \succeq 0 \; \forall(\zeta : \|\zeta\|_{2,2} \leq 1) \text{ [already proved]}$$

$$\Leftrightarrow \quad \xi^T\mathcal{A}(y)\xi + 2(\widehat{L}\xi)^T\zeta\widehat{R}(y)\xi \geq 0 \quad \forall\xi \;\; \forall(\zeta : \|\zeta\|_{2,2} \leq 1)$$

$$\Leftrightarrow \quad \xi^T\mathcal{A}(y)\xi - 2\|\widehat{L}\xi\|_2\|\widehat{R}(y)\xi\|_2 \geq 0 \quad \forall\xi$$

$$\Leftrightarrow \quad \xi^T\mathcal{A}(y)\xi + 2\eta^T\widehat{R}(y)\xi \geq 0 \quad \forall(\xi,\eta : \eta^T\eta \leq \xi^T\widehat{L}^T\widehat{L}\xi)$$

$$\Leftrightarrow \quad \exists\lambda \geq 0 : \left[\begin{array}{c|c} \mathcal{A}(y) - \lambda\widehat{L}^T\widehat{L} & \widehat{R}^T(y) \\ \hline \widehat{R}(y) & \lambda I_q \end{array} \right] \succeq 0 \; [\mathcal{S}\text{-Lemma}]$$

$$\Leftrightarrow \quad \exists\lambda : \left[\begin{array}{c|c} \mathcal{A}(y) - \lambda\widehat{L}^T\widehat{L} & \widehat{R}^T(y) \\ \hline \widehat{R}(y) & \lambda I_q \end{array} \right] \succeq 0,$$

and we arrive at (6.4.4). $\qquad\square$

6.5 SOLVABLE CASE IV: CQI WITH SIMPLE ELLIPSOIDAL UNCERTAINTY

The last solvable case we intend to present is of uncertain CQI (6.1.3) with an ellipsoid as the perturbation set. Now, unlike the results of Theorem 6.3.2 and Proposition 6.4.1, we neither assume the uncertainty side-wise, nor impose specific structural restrictions on the CQI in question. However, whereas in all tractability results stated so far we ended up with a "well-structured" tractable reformulation of the RC (mainly in the form of an explicit system of LMIs), now the reformulation will be less elegant: we shall prove that the feasible set of the RC admits an efficiently computable *separation oracle* — an efficient computational routine that, given on input a candidate decision vector y, reports whether this vector is robust feasible, and if it is not the case, returns a *separator* — a linear form $e^T z$ on the space of decision vectors such that

$$e^T y > \sup_{z \in Y} e^T z,$$

where Y is the set of all robust feasible solutions. Good news is that equipped with such a routine, one can optimize efficiently a linear form over the intersection of Y with any convex compact set Z that is itself given by an efficiently computable separation oracle. On the negative side, the family of "theoretically efficient" optimization algorithms available in this situation is much more restricted than the family of algorithms available in the situations we encountered so far. Specifically, in these past situations, we could process the RC by high-performance Interior Point polynomial time methods, while in our present case we are forced to use slower black-box-oriented methods, like the Ellipsoid algorithm. As a result, the design dimensions that can be handled in a realistic time can drop considerably.

We are about to describe an efficient separation oracle for the feasible set

$$Y = \{y : \|\alpha(y)\zeta + \beta(y)\|_2 \le \sigma^T(y)\zeta + \delta(y) \;\; \forall(\zeta : \zeta^T\zeta \le 1)\} \tag{6.5.1}$$

of the uncertain CQI (6.1.3) with the unit ball in the role of the perturbation set; recall that $\alpha(y)$, $\beta(y)$, $\sigma(y)$, $\delta(y)$ are affine in y.

Observe that $y \in Y$ if and only if the following two conditions hold true:

$$
\begin{array}{|ll|c|}
\hline
 & 0 \le \sigma^T(y)\zeta + \delta(y) \;\; \forall(\zeta : \|\zeta\|_2 \le 1) & \\
\Leftrightarrow & \|\sigma(y)\|_2 \le \delta(y) & (a) \\
\hline
 & (\sigma^T(y)\zeta + \delta(y))^2 - [\alpha(y)\zeta + \beta(y)]^T[\alpha(y)\zeta + \beta(y)] \ge 0 & \\
 & \hspace{4cm} \forall(\zeta : \zeta^T\zeta \le 1) & \\
\Leftrightarrow & \exists \lambda \ge 0 : & \\
 & A_y(\lambda) \equiv \left[\begin{array}{c|c} \begin{array}{c} \lambda I_L + \sigma(y)\sigma^T(y) \\ -\alpha^T(y)\alpha(y) \end{array} & \begin{array}{c} \delta(y)\sigma^T(y) \\ -\beta^T(y)\alpha(y) \end{array} \\ \hline \begin{array}{c} \delta(y)\sigma(y) \\ -\alpha^T(y)\beta(y) \end{array} & \begin{array}{c} \delta^2(y) - \beta^T(y)\beta(y) \\ -\lambda \end{array} \end{array} \right] \succeq 0 & (b) \\
\hline
\end{array}
$$
$$\tag{6.5.2}$$

where the second $\Leftrightarrow$ is due to the inhomogeneous $\mathcal{S}$-Lemma. Observe that *given y, it is easy to verify the validity of* (6.5.2). Indeed,

i) Verification of (6.5.2.*a*) is trivial.

ii) To verify (6.5.2.*b*), we can use bisection in λ as follows.
First note that any $\lambda \ge 0$ satisfying the matrix inequality (MI) in (6.5.2.*b*) clearly should be $\le \lambda_+ \equiv \delta^2(y) - \beta^T(y)\beta(y)$. If $\lambda_+ < 0$, then (6.5.2.*b*) definitely does not take place, and we can terminate our verification. When $\lambda_+ \ge 0$, we can build a shrinking sequence of localizers $\Delta_t = [\underline{\lambda}_t, \overline{\lambda}_t]$ for the set Λ_* of solutions to our MI, namely, as follows:

- We set $\underline{\lambda}_0 = 0$, $\overline{\lambda}_0 = \lambda_+$, thus ensuring that $\Lambda_* \subset \Delta_0$.

- Assume that after $t-1$ steps we have in our disposal a segment Δ_{t-1}, $\Delta_{t-1} \subset \Delta_{t-2} \subset ... \subset \Delta_0$, such that $\Lambda_* \subset \Delta_{t-1}$. Let λ_t be the midpoint of Δ_{t-1}. At step t, we check whether the matrix $A_y(\lambda_t)$ is $\succeq 0$; to this end we can use any one from the well-known Linear Algebra routines capable to check in $O(k^3)$ operations positive semidefiniteness of a $k \times k$ matrix A, and if it is not the case, to produce a "certificate" for the fact that $A \not\succeq 0$ — a vector z such that $z^T A z < 0$. If $A_y(\lambda_t) \succeq 0$, we are done, otherwise we get a vector z_t such that the affine function $f_t(\lambda) \equiv z_t^T A_y(\lambda) z_t$ is negative when $\lambda = \lambda_t$. Setting $\Delta_t = \{\lambda \in \Delta_{t-1} : f_t(\lambda) \geq 0\}$, we clearly get a new localizer for Λ_* that is at least twice shorter than Δ_{t-1}; if this localizer is nonempty, we pass to step $t+1$, otherwise we terminate with the claim that (6.5.2.b) is not valid.

Since the sizes of subsequent localizers shrink at each step by a factor of at least 2, the outlined procedure rapidly converges: for all practical purposes[1] we may assume that the procedure terminates after a small number of steps with either a λ that makes the MI in (6.5.2) valid, or with an empty localizer, meaning that (6.5.2.b) is invalid.

So far we built an efficient procedure that checks whether or not y is robust feasible (i.e., whether or not $y \in Y$). To complete the construction of a separation oracle for Y, it remains to build a separator of y and Y when $y \notin Y$. Our "separation strategy" is as follows. Recall that $y \in Y$ if and only if all vectors $v_y(\zeta) = [\alpha(y)\zeta + \beta(y); \sigma^T(y)\zeta + \delta(y)]$ with $\|\zeta\|_2 \leq 1$ belong to the Lorentz cone $\mathbf{L}^{k+1}$, where $k = \dim \beta(y)$. Thus, $y \notin Y$ if there exists $\bar{\zeta}$ such that $\|\bar{\zeta}\|_2 \leq 1$ and $v_y(\bar{\zeta}) \notin \mathbf{L}^{k+1}$. Given such a $\bar{\zeta}$, we can immediately build a separator of y and Y as follows:

i) Since $v_y(\bar{\zeta}) \notin \mathbf{L}^{k+1}$, we can easily separate $v_y(\bar{\zeta})$ and $\mathbf{L}^{k+1}$. Specifically, setting $v_y(\bar{\zeta}) = [a; b]$, we have $b < \|a\|_2$, so that setting $e = [a/\|a\|_2; -1]$, we have $e^T v_y(\bar{\zeta}) = \|a\|_2 - b > 0$, while $e^T u \leq 0$ for all $u \in \mathbf{L}^{k+1}$.

ii) After a separator e of $v_y(\bar{\zeta})$ and $\mathbf{L}^{k+1}$ is built, we look at the function $\phi(z) = e^T v_z(\bar{\zeta})$. This is an affine function of z such that

$$\sup_{z \in Y} \phi(z) \leq \sup_{u \in \mathbf{L}^{k+1}} e^T u < e^T v_y(\bar{\zeta}) = \phi(y)$$

where the first $\leq$ is given by the fact that $v_z(\bar{\zeta}) \in \mathbf{L}^{k+1}$ when $z \in Y$. Thus, the homogeneous part of $\phi(\cdot)$, (which is a linear form readily given by e), separates y and Y.

In summary, all we need is an efficient routine that, in the case when $y \notin Y$, i.e.,

$$\widehat{\mathcal{Z}}_y \equiv \{\bar{\zeta} : \|\bar{\zeta}\|_2 \leq 1, v_y(\bar{\zeta}) \notin \mathbf{L}^{k+1}\} \neq \emptyset,$$

---

[1]We could make our reasoning precise, but it would require going into tedious technical details that we prefer to skip.

finds a point $\bar{\zeta} \in \widehat{\mathcal{Z}}_y$ ("an infeasibility certificate"). Here is such a routine. First, recall that our algorithm for verifying robust feasibility of y reports that $y \notin Y$ in two situations:

• $\|\sigma(y)\|_2 > \delta(y)$. In this case we can without any difficulty find a $\bar{\zeta}$, $\|\bar{\zeta}\|_2 \leq 1$, such that $\sigma^T(y)\bar{\zeta} + \delta(y) < 0$. In other words, the vector $v_y(\bar{\zeta})$ has a negative last coordinate and therefore it definitely does not belong to $\mathbf{L}^{k+1}$. Such a $\bar{\zeta}$ is an infeasibility certificate.

• We have discovered that (a) $\lambda_+ < 0$, or (b) got $\Delta_t = \emptyset$ at a certain step t of our bisection process. In this case building an infeasibility certificate is more tricky.

Step 1: Separating the positive semidefinite cone and the "matrix ray" $\{A_y(\lambda) : \lambda \geq 0\}$. Observe that with z_0 defined as the last basic orth in $\mathbb{R}^{L+1}$, we have $f_0(\lambda) \equiv z_0^T A_y(\lambda) z_0 < 0$ when $\lambda > \lambda_+$. Recalling what our bisection process is, we conclude that in both cases (a), (b) we have at our disposal a collection $z_0, ..., z_t$ of $(L+1)$-dimensional vectors such that with $f_s(\lambda) = z_s^T A_y(\lambda) z_s$ we have $f(\lambda) \equiv \min[f_0(\lambda), f_1(\lambda), ..., f_t(\lambda)] < 0$ for all $\lambda \geq 0$. By construction, $f(\lambda)$ is a piecewise linear concave function on the nonnegative ray; looking at what happens at the maximizer of f over $\lambda \geq 0$, we conclude that an appropriate convex combination of just two of the "linear pieces" $f_0(\lambda), ..., f_t(\lambda)$ of f is negative everywhere on the nonnegative ray. That is, with properly chosen and easy-to-find $\alpha \in [0, 1]$ and $\tau_1, \tau_2 \leq t$ we have

$$\phi(\lambda) \equiv \alpha f_{\tau_1}(\lambda) + (1 - \alpha) f_{\tau_2}(\lambda) < 0 \quad \forall \lambda \geq 0.$$

Recalling the origin of $f_\tau(\lambda)$ and setting $z^1 = \sqrt{\alpha} z_{\tau_1}, z^2 = \sqrt{1 - \alpha} z_{\tau_2}, Z = z^1[z^1]^T + z^2[z^2]^T$, we have

$$0 > \phi(\lambda) = [z^1]^T A_y(\lambda) z^1 + [z^2]^T A_y(\lambda) z^2 = \text{Tr}(A_y(\lambda) Z) \quad \forall \lambda \geq 0. \qquad (6.5.3)$$

This inequality has a simple interpretation: the function $\Phi(X) = \text{Tr}(XZ)$ is a linear form on $\mathbf{S}^{L+1}$ that is nonnegative on the positive semidefinite cone (since $Z \succeq 0$ by construction) and is negative everywhere on the "matrix ray" $\{A_y(\lambda) : \lambda \geq 0\}$, thus certifying that this ray does not intersect the positive semidefinite cone (the latter is exactly the same as the fact that (6.5.2.b) is false).

Step 2: from Z to $\bar{\zeta}$. Relation (6.5.3) says that an affine function $\phi(\lambda)$ is negative everywhere on the nonnegative ray, meaning that the slope of the function is nonpositive, and the value at the origin is negative. Taking into account (6.5.2), we get

$$Z_{L+1,L+1} \geq \sum_{i=1}^{L} Z_{ii}, \ \text{Tr}(Z \underbrace{\left[\begin{array}{c|c} \begin{array}{c} \sigma(y)\sigma^T(y) \\ -\alpha^T(y)\alpha(y) \end{array} & \begin{array}{c} \delta(y)\sigma^T(y) \\ -\beta^T(y)\alpha(y) \end{array} \\ \hline \begin{array}{c} \delta(y)\sigma(y) \\ -\alpha^T(y)\beta(y) \end{array} & \delta^2(y) - \beta^T(y)\beta(y) \end{array} \right]}_{A_y(0)}) < 0.$$

$$(6.5.4)$$

Besides this, we remember that Z is given as $z^1[z^1]^T + z^2[z^2]^T$. We claim that

> (!) *We can efficiently find a representation $Z = ee^T + ff^T$ such that $e, f \in \mathbf{L}^{L+1}$.*

Taking for the time being (!) for granted, let us build an infeasibility certificate. Indeed, from the second relation in (6.5.4) it follows that either $\text{Tr}(A_y(0)ee^T) < 0$, or $\text{Tr}(A_y(0)ff^T) < 0$, or both. Let us check which one of these inequalities indeed holds true; w.l.o.g., let it be the first one. From this inequality, in particular, $e \neq 0$, and since $e \in \mathbf{L}^{L+1}$, we have $e_{L+1} > 0$. Setting $\bar{e} = e/e_{L+1} = [\bar{\zeta}; 1]$, we have $\text{Tr}(A_y(0)\bar{e}\bar{e}^T) = \bar{e}^T A_y(0)\bar{e} < 0$, that is,

$$\delta^2(y) - \beta^T(y)\beta(y) + 2\delta(y)\sigma^T(y)\bar{\zeta} - 2\beta^T(y)\alpha(y)\bar{\zeta} + \bar{\zeta}^T\sigma(y)\sigma^T(y)\bar{\zeta}$$
$$-\bar{\zeta}^T\alpha^T(y)\alpha(y)\bar{\zeta} < 0,$$

or, which is the same,

$$(\delta(y) + \sigma^T(y)\bar{\zeta})^2 < (\alpha(y)\bar{\zeta} + \beta(y))^T(\alpha(y)\bar{\zeta} + \beta(y)).$$

We see that the vector $v_y(\bar{\zeta}) = [\alpha(y)\bar{\zeta}+\beta(y); \sigma^T(y)\bar{\zeta}+\delta(y)]$ does not belong to $\mathbf{L}^{L+1}$, while $\bar{e} = [\bar{\zeta}; 1] \in \mathbf{L}^{L+1}$, that is, $\|\bar{\zeta}\|_2 \leq 1$. We have built a required infeasibility certificate.

It remains to justify (!). Replacing, if necessary, z^1 with $-z^1$ and z^2 with $-z^2$, we can assume that $Z = z^1[z^1]^T + z^2[z^2]^T$ with $z^1 = [p; s]$, $z^2 = [q; r]$, where $s, r \geq 0$. It may happen that $z^1, z^2 \in \mathbf{L}^{L+1}$ — then we are done. Assume now that not both z^1, z^2 belong to $\mathbf{L}^{L+1}$, say, $z^1 \notin \mathbf{L}^{L+1}$, that is, $0 \leq s < \|p\|_2$. Observe that $Z_{L+1,L+1} = s^2 + r^2$ and $\sum_{i=1}^{L} Z_{ii} = p^Tp + q^Tq$; therefore the first relation in (6.5.4) implies that $s^2 + r^2 \geq p^Tp + q^Tq$. Since $0 \leq s < \|p\|_2$ and $r \geq 0$, we conclude that $r > \|q\|_2$. Thus, $s < \|p\|_2$, $r > \|q\|_2$, whence there exists (and can be easily found) $\alpha \in (0,1)$ such that for the vector $e = \sqrt{\alpha}z^1 + \sqrt{1-\alpha}z^2 = [u; t]$ we have $e_{L+1} = \sqrt{e_1^2 + ... + e_L^2}$. Setting $f = -\sqrt{1-\alpha}z^1 + \sqrt{\alpha}z^2$, we have $ee^T + ff^T = z^1[z^1]^T + z^2[z^2]^T = Z$. We now have

$$0 \leq Z_{L+1,L+1} - \sum_{i=1}^{L} Z_{ii} = e_{L+1}^2 + f_{L+1}^2 - \sum_{i=1}^{L}[e_i^2 + f_i^2] = f_{L+1}^2 - \sum_{i=1}^{L} f_i^2;$$

thus, replacing, if necessary, f with $-f$, we see that $e, f \in \mathbf{L}^{L+1}$ and $Z = ee^T + ff^T$, as required in (!).

6.5.1 Semidefinite Representation of the RC of an Uncertain CQI with Simple Ellipsoidal Uncertainty

This book was nearly finished when the topic considered in this section was significantly advanced by R. Hildebrand [62, 63] who discovered an explicit SDP representation of the cone of "Lorentz-positive" $n \times m$ matrices (real $m \times n$ matrices that map the Lorentz cone $\mathbf{L}^m$ into the Lorentz cone $\mathbf{L}^n$). Existence of such a representation was a long-standing open question. As a byproduct of answering this question, the construction of Hildebrand offers an explicit SDP reformulation of the RC of an uncertain conic quadratic inequality with ellipsoidal uncertainty.

The RC of an uncertain conic quadratic inequality with ellipsoidal uncertainty and Lorentz-positive matrices. Consider the RC of an uncertain conic quadratic inequality with simple ellipsoidal uncertainty; w.l.o.g., we assume that the uncertainty set $\mathcal{Z}$ is the unit Euclidean ball in some $\mathbb{R}^{m-1}$, so that the RC is the semi-infinite constraint of the form

$$B[x]\zeta + b[x] \in \mathbf{L}^n \ \forall(\zeta \in \mathbb{R}^{m-1} : \zeta^T\zeta \leq 1), \tag{6.5.5}$$

with $B[x]$, $b[x]$ affinely depending on x. This constraint is clearly exactly the same as the constraint

$$B[x]\xi + \tau b[x] \in \mathbf{L}^n \ \forall([\xi;\tau] \in \mathbf{L}^m).$$

We see that x is feasible for the RC in question if and only if the $n \times m$ matrix $M[x] = [B[x], b[x]]$ affinely depending on x is Lorentz-positive, that is, maps the cone $\mathbf{L}^m$ into the cone $\mathbf{L}^n$. It follows that in order to get an explicit SDP representation of the RC, is suffices to know an explicit SDP representation of the set $P_{n,m}$ of $n \times m$ matrices mapping $\mathbf{L}^m$ into $\mathbf{L}^n$.

SDP representation of $P_{n,m}$ as discovered by R. Hildebrand (who used tools going far beyond those used in this book) is as follows.

A. Given m, n, we define a linear mapping $A \mapsto \mathcal{W}(A)$ from the space $\mathbb{R}^{n \times m}$ of real $n \times m$ matrices into the space $\mathbf{S}^N$ of symmetric $N \times N$ matrices with $N = (n-1)(m-1)$, namely, as follows.

Let $W_n[u] = \begin{bmatrix} u_n + u_1 & u_2 & \cdots & u_{n-1} \\ u_2 & u_n - u_1 & & \\ \vdots & & \ddots & \\ u_{n-1} & & & u_n - u_1 \end{bmatrix}$, so that W_n is a sym-

metric $(n-1) \times (n-1)$ matrix depending on a vector u of n real variables. Now consider the *Kronecker product* $W[u,v] = W_n[u] \bigotimes W_m[v]$. [2] W is a symmetric $N \times N$ matrix with entries that are bilinear functions of u and v variables, so that an entry is of the form "weighted sum of pair products of the u and the v-variables." Now, given an $n \times m$ matrix A, let us replace pair products $u_i v_k$ in the representation of the entries in $W[u,v]$ with the entries A_{ik} of A. As a result of this formal substitution, W will become a symmetric $(n-1) \times (m-1)$ matrix $\mathcal{W}(A)$ that depends linearly on A.

B. We define a linear subspace $\mathcal{L}_{m,n}$ in the space $\mathbf{S}^N$ as the linear span of the Kronecker products $S \bigotimes T$ of all skew-symmetric real $(n-1) \times (n-1)$ matrices S and skew-symmetric real $(m-1) \times (m-1)$ matrices T. Note that the Kronecker product of two skew-symmetric matrices is a symmetric matrix, so that the definition makes sense. Of course, we can easily build a basis in $\mathcal{L}_{m,n}$ — it is comprised of pairwise Kronecker products of the basic $(n-1)$-dimensional and $(m-1)$-dimensional skew-symmetric matrices.

---

[2]Recall that the Kronecker product $A \bigotimes B$ of a $p \times q$ matrix A and an $r \times s$ matrix B is the $pr \times qs$ matrix with rows indexed by pairs (i,k), $1 \leq i \leq p$, $1 \leq k \leq r$, and columns indexed by pairs (j,ℓ), $1 \leq j \leq q$, $1 \leq \ell \leq s$, and the $((i,k),(j,\ell))$-entry equal to $A_{ij}B_{k\ell}$. Equivalently, $A \bigotimes B$ is a $p \times q$ block matrix with $r \times s$ blocks, the (i,j)-th block being $A_{ij}B$.

The Hildebrand SDP representation of $P_{n,m}$ is given by the following:

Theorem 6.5.1. [Hildebrand [63, Theorem 5.6]] Let $\min[m, n] \geq 3$. Then an $n \times m$ matrix A maps $\mathbf{L}^m$ into $\mathbf{L}^n$ if and only if A can be extended to a feasible solution to the explicit system of LMIs

$$\mathcal{W}(A) + X \succeq 0, \ X \in \mathcal{L}_{m,n}$$

in variables A, X.

As a corollary,

When $m - 1 := \dim \zeta \geq 2$ and $n := \dim b[x] \geq 3$, the explicit $(n-1)(m-1) \times (n-1)(m-1)$ LMI

$$\mathcal{W}([B[x], b[x]]) + X \succeq 0 \tag{6.5.6}$$

in variables x and $X \in \mathcal{L}_{m,n}$ is an equivalent SDP representation of the semi-infinite conic quadratic inequality (6.5.5) with ellipsoidal uncertainty set.

The lower bounds on the dimensions of ζ and $b[x]$ in the corollary do not restrict generality — we can always ensure their validity by adding zero columns to $B[x]$ and/or adding zero rows to $[B[x], b[x]]$.

6.6 ILLUSTRATION: ROBUST LINEAR ESTIMATION

Consider the situation as follows: we are given noisy observations

$$w = (I_p + \Delta)z + \xi \tag{6.6.1}$$

of a signal z that, in turn, is the result of passing an unknown input signal v through a given linear filter: $z = Av$ with known $p \times q$ matrix A. The measurements contain errors of two kinds:

- bias Δz linearly depending on z, where the only information on the bias matrix Δ is given by a bound $\|\Delta\|_{2,2} \leq \rho$ on its norm;

- random noise ξ with zero mean and known covariance matrix $\Sigma = \mathbf{E}\{\xi\xi^T\}$.

The goal is to estimate a given linear functional $f^T v$ of the input signal. We restrict ourselves with estimators that are linear in w:

$$\widehat{f} = x^T w,$$

where x is a fixed weight vector. For a linear estimator, the mean squares error is

$$\begin{aligned} \text{EstErr} &= \sqrt{\mathbf{E}\{(x^T[(I + \Delta)Av + \xi] - f^T v)^2\}} \\ &= \sqrt{([A^T(I + \Delta^T)x - f]^T v)^2 + x^T \Sigma x}. \end{aligned}$$

Now assume that our a priori knowledge of the true signal is that $v^T Q v \leq R^2$, where $Q \succ 0$ and $R > 0$. In this situation it makes sense to look for the *minimax optimal* weight vector x that minimizes the worst, over v and Δ compatible with our a priori information, mean squares estimation error. In other words, we choose

x as the optimal solution to the following optimization problem

$$\min_{x} \max_{\substack{v:v^T Qv \leq R^2 \\ \Delta:\|\Delta\|_{2,2} \leq \rho}} \left(\left([\underbrace{A^T(I+\Delta^T)}_{S} x - f]^T v\right)^2 + x^T \Sigma x\right)^{1/2}. \tag{P}$$

Now,

$$\max_{v:v^T Qv \leq R^2} [Sx - f]^T v = \max_{u:u^T u \leq 1} [Sx - f]^T (RQ^{-1/2} u)$$
$$= R\|Q^{-1/2}Sx - \underbrace{Q^{-1/2}f}_{\widehat{f}}\|_2,$$

so that (P) reduces to the problem

$$\min_{x} \sqrt{x^T \Sigma x + R^2 \max_{\|\Delta\|_{2,2} \leq \rho} \|\underbrace{Q^{-1/2}A^T(I+\Delta^T)}_{B} x - \widehat{f}\|_2^2},$$

which is exactly the RC of the uncertain conic quadratic program

$$\min_{x,t,r,s} \left\{ t : \begin{array}{l} \sqrt{r^2 + s^2} \leq t, \ \|\Sigma^{1/2}x\|_2 \leq r, \\ \|Bx - \widehat{f}\|_2 \leq R^{-1}s \end{array} \right\}, \tag{6.6.2}$$

where the only uncertain element of the data is the matrix $B = Q^{-1/2}A^T(I+\Delta^T)$ running through the uncertainty set

$$\mathcal{U} = \{B = \underbrace{Q^{-1/2}A^T}_{B_n} + \rho Q^{-1/2}A^T \zeta, \zeta \in \mathcal{Z} = \{\zeta \in \mathbb{R}^{p \times p} : \|\zeta\|_{2,2} \leq 1\}\}. \tag{6.6.3}$$

The uncertainty here is the unstructured norm-bounded one; the RC of (6.6.2), (6.6.3) is readily given by Theorem 6.3.2 and Example 6.3.1.(i). Specifically, the RC is the optimization program

$$\min_{x,t,r,s,\lambda} \left\{ t : \begin{array}{c} \sqrt{r^2 + s^2} \leq t, \ \|\Sigma^{1/2}x\|_2 \leq r, \\ \left[\begin{array}{c|c|c} R^{-1}sI_q - \lambda\rho^2 B_n B_n^T & & B_n x - \widehat{f} \\ \hline & \lambda I_p & x \\ \hline [B_n x - \widehat{f}]^T & x^T & R^{-1}s \end{array} \right] \succeq 0 \end{array} \right\}, \tag{6.6.4}$$

which can further be recast as an SDP.

Next we present a numerical illustration.

Example 6.6.1. Consider the problem as follows:

A thin homogeneous iron plate occupies the 2-D square $D = \{(x,y) : 0 \leq x, y \leq 1\}$. At time $t = 0$ it was heated to temperature $T(0,x,y)$ such that $\int_D T^2(0,x,y)dxdy \leq T_0^2$ with a given T_0, and then was left to cool; the temperature along the perimeter of the plate is kept at the level 0° all the time. At a given time 2τ we measure the temperature $T(2\tau,x,y)$ along the 2-D grid

$$\Gamma = \{(u_\mu, u_\nu) : 1 \leq \mu, \nu \leq N\}, \ u_k = \frac{k - 1/2}{N}$$

The vector w of measurements is obtained from the vector

$$z = \{T(2\tau, u_\mu, u_\nu) : 1 \leq \mu, \nu \leq N\}$$

according to (6.6.1), where $\|\Delta\|_{2,2} \leq \rho$ and $\xi_{\mu\nu}$ are independent Gaussian random variables with zero mean and standard deviation σ. Given the measurements, we need to estimate the temperature $T(\tau, 1/2, 1/2)$ at the center of the plate at time τ.

It is known from physics that the evolution in time of the temperature $T(t, x, y)$ of a homogeneous plate occupying a 2-D domain Ω, with no sources of heat in the domain and heat exchange solely via the boundary, is governed by the *heat equation*

$$\frac{\partial}{\partial t}T = \left(\frac{\partial^2}{\partial x^2} + \frac{\partial^2}{\partial y^2}\right).T$$

(In fact, in the right hand side there should be a factor γ representing material's properties, but by an appropriate choice of the time unit, this factor can be made equal to 1.) For the case of $\Omega = D$ and zero boundary conditions, the solution to this equation is as follows:

$$T(t, x, y) = \sum_{k,\ell=1}^{\infty} a_{k\ell} \exp\{-(k^2 + \ell^2)\pi^2 t\} \sin(\pi k x) \sin(\pi \ell y), \qquad (6.6.5)$$

where the coefficients $a_{k\ell}$ can be obtained by expanding the initial temperature into a series in the orthogonal basis $\phi_{k\ell}(x, y) = \sin(\pi k x) \sin(\pi \ell y)$ in $L_2(D)$:

$$a_{k\ell} = 4 \int_D T(0, x, y)\phi_{k\ell}(x, y)dxdy.$$

In other words, the Fourier coefficients of $T(t, \cdot, \cdot)$ in an appropriate orthogonal spatial basis decrease exponentially as t grows, with the "decay time" (the smallest time in which every one of the coefficients is multiplied by factor ≤ 0.1) equal to

$$\Delta = \frac{\ln(10)}{2\pi^2}.$$

Setting $v_{k\ell} = a_{k\ell} \exp\{-(k^2 + \ell^2)\pi^2\tau\}$, the problem in question becomes to estimate

$$T(\tau, 1/2, 1/2) = \sum_{k,\ell} v_{k\ell}\phi_{k\ell}(1/2, 1/2)$$

given observations

$$w = (I + \Delta)z + \xi, \ z = \{T(2\tau, u_\mu, u_\nu) : 1 \leq \mu, \nu \leq N\},$$
$$\xi = \{\xi_{\mu\nu} \sim \mathcal{N}(0, \sigma^2) : 1 \leq \mu, \nu \leq N\}$$

($\xi_{\mu\nu}$ are independent).

Finite-dimensional approximation. Observe that

$$a_{k\ell} = \exp\{\pi^2(k^2 + \ell^2)\tau\}v_{k\ell}$$

and that

$$\sum_{k,\ell} v_{k\ell}^2 \exp\{2\pi^2(k^2 + \ell^2)\tau\} = \sum_{k,\ell} a_{k\ell}^2 = 4 \int_D T^2(0, x, y)dxdy \leq 4T_0^2. \qquad (6.6.6)$$

It follows that

$$|v_{k\ell}| \leq 2T_0 \exp\{-\pi^2(k^2 + \ell^2)\tau\}.$$

Now, given a tolerance $\epsilon > 0$, we can easily find L such that

$$\sum_{k,\ell:k^2+\ell^2>L^2} \exp\{-\pi^2(k^2+\ell^2)\tau\} \leq \frac{\epsilon}{2T_0},$$

meaning that when replacing by zeros the actual (unknown!) $v_{k\ell}$ with $k^2+\ell^2 > L^2$, we change temperature at time τ (and at time 2τ as well) at every point by at most ϵ. Choosing ϵ really small (say, $\epsilon = 1.\text{e-}16$), we may assume for all practical purposes that $v_{k\ell} = 0$ when $k^2+\ell^2 > L^2$, which makes our problem a finite-dimensional one, specifically, as follows:

Given the parameters L, N, ρ, σ, T_0 and observations

$$w = (I+\Delta)z + \xi, \tag{6.6.7}$$

where $\|\Delta\|_{2,2} \leq \rho$, $\xi_{\mu\nu} \sim \mathcal{N}(0,\sigma^2)$ are independent, $z = Av$ is defined by the relations

$$z_{\mu\nu} = \sum_{k^2+\ell^2\leq L^2} \exp\{-\pi^2(k^2+\ell^2)\tau\}v_{k\ell}\phi_{k\ell}(u_\mu,u_\nu), \ 1\leq\mu,\nu\leq N,$$

and $v = \{v_{k\ell}\}_{k^2+\ell^2\leq L^2}$ is known to satisfy the inequality

$$v^T Qv \equiv \sum_{k^2+\ell^2\leq L^2} v_{k\ell}^2 \exp\{2\pi^2(k^2+\ell^2)\tau\} \leq 4T_0^2,$$

estimate the quantity

$$\sum_{k^2+\ell^2\leq L^2} v_{k\ell}\phi_{k\ell}(1/2,1/2),$$

where $\phi_{k\ell}(x,y) = \sin(\pi kx)\sin(\pi\ell y)$ and $u_\mu = \frac{\mu-1/2}{N}$.

The latter problem fits the framework of robust estimation we have built, and we can recover $T = T(\tau,1/2,1/2)$ by a linear estimator

$$\widehat{T} = \sum_{\mu,\nu} x_{\mu\nu}w_{\mu\nu}$$

with weights $x_{\mu\nu}$ given by an optimal solution to the associated problem (6.6.4).

Assume, for example, that τ is half of the decay time of our system:

$$\tau = \frac{1}{2}\frac{\ln(10)}{2\pi^2} \approx 0.0583,$$

and let

$$T_0 = 1000, N = 4.$$

With $\epsilon = 1.\text{e-}15$, we get $L = 8$ (this corresponds to just 41-dimensional space for v's). Now consider four options for ρ and σ:

(a) $\rho = 1.\text{e-}9,$ $\sigma = 1.\text{e-}9$
(b) $\rho = 0,$ $\sigma = 1.\text{e-}3$
(c) $\rho = 1.\text{e-}3,$ $\sigma = 1.\text{e-}3$
(d) $\rho = 1.\text{e-}1,$ $\sigma = 1.\text{e-}1$

In the case of (a), the optimal value in (6.6.4) is 0.0064, meaning that the expected squared error of the minimax optimal estimator never exceeds $(0.0064)^2$. The minimax optimal weights are

$$\begin{bmatrix} 6625.3 & -2823.0 & -2.8230 & 6625.3 \\ -2823.0 & 1202.9 & 1202.9 & -2823.0 \\ -2823.0 & 1202.9 & 1202.9 & -2823.0 \\ 6625.3 & -2823.0 & -2823.0 & 6625.3 \end{bmatrix} \quad (A)$$

(we represent the weights as a 2-D array, according to the natural structure of the observations).

In the case of (b), the optimal value in (6.6.4) is 0.232, and the minimax optimal weights are

$$\begin{bmatrix} -55.6430 & -55.6320 & -55.6320 & -55.6430 \\ -55.6320 & 56.5601 & 56.5601 & -55.6320 \\ -55.6320 & 56.5601 & 56.5601 & -55.6320 \\ -55.6430 & -55.6320 & -55.6320 & -55.6430 \end{bmatrix}. \quad (B)$$

In the case of (c), the optimal value in (6.6.4) is 8.92, and the minimax optimal weights are

$$\begin{bmatrix} -0.4377 & -0.2740 & -0.2740 & -0.4377 \\ -0.2740 & 1.2283 & 1.2283 & -0.2740 \\ -0.2740 & 1.2283 & 1.2283 & -0.2740 \\ -0.4377 & -0.2740 & -0.2740 & -0.4377 \end{bmatrix}. \quad (C)$$

In the case of (d), the optimal value in (6.6.4) is 63.9, and the minimax optimal weights are

$$\begin{bmatrix} 0.1157 & 0.2795 & 0.2795 & 0.1157 \\ 0.2795 & 0.6748 & 0.6748 & 0.2795 \\ 0.2795 & 0.6748 & 0.6748 & 0.2795 \\ 0.1157 & 0.2795 & 0.2795 & 0.1157 \end{bmatrix}. \quad (D)$$

Now, in reality we can hardly know exactly the bounds ρ, σ on the measurement errors. What happens when we under- or over-estimate these quantities? To get an orientation, let us use every one of the weights given by (A), (B), (C), (D) in every one of the situations (a), (b), (c), (d). This is what happens with the errors (obtained as the average of observed errors over 100 random simulations using the "nearly worst-case" signal v and "nearly worst-case" perturbation matrix Δ):

	(a)	(b)	(c)	(d)
(A)	0.001	18.0	6262.9	6.26e5
(B)	0.063	0.232	89.3	8942.7
(C)	8.85	8.85	8.85	108.8
(D)	8.94	8.94	8.94	63.3

We clearly see that, first, in our situation taking into account measurement errors, even pretty small ones, is a must (this is so in all *ill-posed* estimation problems —

those where the condition number of B_n is large). Second, we see that underestimating the magnitude of measurement errors seems to be much more dangerous than overestimating them.

6.7 EXERCISES

Exercise 6.1. Consider the situation as follows (cf. section 6.6). We are given an observation

$$y = Ax + b \in \mathbb{R}^m$$

of unknown signal $x \in \mathbb{R}^n$. The matrix $B \equiv [A; b]$ is not known exactly; all we know is that $B \in \mathcal{B} = \{B = B_n + L^T \Delta R : \Delta \in \mathbb{R}^{p \times q}, \|\Delta\|_{2,2} \leq \rho\}$. Build an estimate v of the vector Qx, where Q is a given $k \times n$ matrix, that minimizes the worst-case, over all possible true values of x, $\|\cdot\|_2$ estimation error.

6.8 NOTES AND REMARKS

NR 6.1. Tractable reformulation of an uncertain LMI with unstructured norm-bounded perturbation underlying Theorem 6.3.2 and Proposition 6.4.1 was discovered in [32]. $\mathcal{S}$-Lemma, along with the (much simpler) Schur Complement Lemma, form *the* two most powerful tools in Semidefinite Optimization and in Control Theory. The $\mathcal{S}$-Lemma was discovered by V.A. Yakubovich in 1971; for a comprehensive "optimization-oriented" survey of the related issues, see [94]. Tractability of the RC of uncertain CQI with simple ellipsoidal uncertainty (section 6.5) was established independently in [3] and [49].

Chapter Seven
Approximating RCs of Uncertain Conic Quadratic Problems

In this chapter we focus on *tight* tractable approximations of uncertain CQIs — those with tightness factor independent (or nearly so) of the "size" of the description of the perturbation set. Known approximations of this type deal with side-wise uncertainty and two types of the left hand side perturbations: the first is the case of *structured norm-bounded perturbations* to be considered in section 7.1, while the second is the case of ∩-*ellipsoidal* left hand side perturbation sets to be considered in section 7.2.

7.1 STRUCTURED NORM-BOUNDED UNCERTAINTY

Consider the case where the uncertainty in CQI (6.1.3) is side-wise with the right hand side uncertainty as in section 6.2, and with *structured norm-bounded* left hand side uncertainty, meaning that

i) The left hand side perturbation set is

$$
\mathcal{Z}_\rho^{\text{left}} = \rho \mathcal{Z}_1^{\text{left}} = \left\{ \eta = (\eta^1, ..., \eta^N) : \begin{array}{c} \eta^\nu \in \mathbb{R}^{p_\nu \times q_\nu} \, \forall \nu \leq N \\ \|\eta^\nu\|_{2,2} \leq \rho \, \forall \nu \leq N \\ \eta^\nu = \theta_\nu I_{p_\nu}, \theta_\nu \in \mathbb{R}, \, \nu \in \mathcal{I}_{\text{S}} \end{array} \right\} \tag{7.1.1}
$$

Here $\mathcal{I}_{\text{S}}$ is a given subset of the index set $\{1, ..., N\}$ such that $p_\nu = q_\nu$ for $\nu \in \mathcal{I}_{\text{S}}$.

Thus, the left hand side perturbations $\eta \in \mathcal{Z}_1^{\text{left}}$ are block-diagonal matrices with $p_\nu \times q_\nu$ diagonal blocks $\eta^\nu, \nu = 1, ..., N$. All of these blocks are of matrix norm not exceeding 1, and, in addition, prescribed blocks should be proportional to the unit matrices of appropriate sizes. The latter blocks are called *scalar*, and the remaining — *full* perturbation blocks.

ii) We have

$$
A(\eta)y + b(\eta) = A^{\text{n}}y + b^{\text{n}} + \sum_{\nu=1}^N L_\nu^T(y)\eta^\nu R_\nu(y), \tag{7.1.2}
$$

where all matrices $\mathbf{L}_\nu(y) \not\equiv 0$, $R_\nu(y) \not\equiv 0$ are affine in y and for every ν, either $L_\nu(y)$, or $R_\nu(y)$, or both are independent of y.

Remark 7.1.1. W.l.o.g., we assume from now on that all scalar perturbation blocks are of the size 1×1: $p_\nu = q_\nu = 1$ for all $\nu \in \mathcal{I}_{\text{S}}$.

To see that this assumption indeed does not restrict generality, note that if $\nu \in \mathcal{I}_s$, then in order for (7.1.2) to make sense, $R_\nu(y)$ should be a $p_\nu \times 1$ vector, and $L_\nu(y)$ should be a $p_\nu \times k$ matrix, where k is the dimension of $b(\eta)$. Setting $\bar{R}_\nu(y) \equiv 1$, $\bar{L}_\nu(y) = R_\nu^T(y)L_\nu(y)$, observe that $\bar{L}_\nu(y)$ is affine in y, and the contribution $\theta_\nu L_\nu^T(y)R_\nu(y)$ of the ν-th scalar perturbation block to $A(\eta)y + b(\eta)$ is exactly the same as if this block were of size 1×1, and the matrices $L_\nu(y)$, $R_\nu(y)$ were replaced with $\bar{L}_\nu(y)$, $\bar{R}_\nu(y)$, respectively.

Note that Remark 7.1.1 is equivalent to the assumption that *there are no scalar perturbation blocks at all* — indeed, 1×1 scalar perturbation blocks can be thought of as full ones as well. [1]

Recall that we have already considered the particular case $N = 1$ of the uncertainty structure. Indeed, with a single perturbation block, that, as we just have seen, we can treat as a full one, we find ourselves in the situation of side-wise uncertainty with unstructured norm-bounded left hand side perturbation (section 6.3). In this situation the RC of the uncertain CQI in question is computationally tractable. The latter is not necessarily the case for general $(N > 1)$ structured norm-bounded left hand side perturbations. To see that the general structured norm-bounded perturbations are difficult to handle, note that they cover, in particular, the case of *interval uncertainty*, where $\mathcal{Z}_1^{\text{left}}$ is the box $\{\eta \in \mathbb{R}^L : \|\eta\|_\infty \leq 1\}$ and $A(\eta)$, $b(\eta)$ are arbitrary affine functions of η.

Indeed, the interval uncertainty

$$
\begin{aligned}
A(\eta)y + b(\eta) &= [A^{\text{n}}y + b^{\text{n}}] + \sum_{\nu=1}^{N} \eta_\nu[A^\nu y + b^\nu] \\
&= [A^{\text{n}}y + b^{\text{n}}] + \sum_{\nu=1}^{N} \underbrace{[A^\nu y + b^\nu]}_{L_\nu^T(y)} \cdot \eta_\nu \cdot \underbrace{1}_{R_\nu(y)},
\end{aligned}
$$

is nothing but the structured norm-bounded perturbation with 1×1 perturbation blocks.

From the beginning of section 5.2 we know that the RC of uncertain CQI with side-wise uncertainty and interval uncertainty in the left hand side in general is computationally intractable, meaning that structural norm-bounded uncertainty can be indeed difficult.

7.1.1 Approximating the RC of Uncertain Least Squares Inequality

We start with deriving a safe tractable approximation of the RC of an *uncertain Least Squares constraint*

$$\|A(\eta)y + b(\eta)\|_2 \leq \tau, \tag{7.1.3}$$

---

[1] A reader could ask, why do we need the scalar perturbation blocks, given that finally we can get rid of them without loosing generality. The answer is, that we intend to use the same notion of structured norm-bounded uncertainty in the case of uncertain LMIs, where Remark 7.1.1 does not work.

with structured norm-bounded perturbation (7.1.1), (7.1.2).

Step 1: reformulating the RC of (7.1.3), (7.1.1), (7.1.2) as a semi-infinite LMI. Given a k-dimensional vector u (k is the dimension of $b(\eta)$) and a real τ, let us set

$$\text{Arrow}(u,t) = \left[\begin{array}{c|c} \tau & u^T \\ \hline u & \tau I_k \end{array}\right].$$

Recall that by Lemma 6.3.3 $\|u\|_2 \leq \tau$ if and only if $\text{Arrow}(u,\tau) \succeq 0$. It follows that the RC of (7.1.3), (7.1.1), (7.1.2), which is the semi-infinite Least Squares inequality

$$\|A(\eta)y + b(\eta)\|_2 \leq \tau \; \forall \eta \in \mathcal{Z}_\rho^{\text{left}},$$

can be rewritten as

$$\text{Arrow}(A(\eta)y + b(\eta), \tau) \succeq 0 \; \forall \eta \in \mathcal{Z}_\rho^{\text{left}}. \tag{7.1.4}$$

Introducing $k \times (k+1)$ matrix $\mathcal{L} = [0_{k \times 1}, I_k]$ and $1 \times (k+1)$ matrix $\mathcal{R} = [1, 0, ..., 0]$, we clearly have

$$\begin{aligned} \text{Arrow}(A(\eta)y + b(\eta), \tau) &= \text{Arrow}(A^n y + b^n, \tau) \\ &+ \sum_{\nu=1}^N \left[\mathcal{L}^T L_\nu^T(y) \eta^\nu R_\nu(y) \mathcal{R} + \mathcal{R}^T R_\nu^T(y)[\eta^\nu]^T L_\nu(y) \mathcal{L}\right]. \end{aligned} \tag{7.1.5}$$

Now, since for every ν, either $L_\nu(y)$, or $R_\nu(y)$, or both, are independent of y, renaming, if necessary $[\eta^\nu]^T$ as η^ν, and swapping $L_\nu(y)\mathcal{L}$ and $R_\nu(y)\mathcal{R}$, we may assume w.l.o.g. that in the relation (7.1.5) all factors $L_\nu(y)$ are independent of y, so that the relation reads

$$\begin{aligned} \text{Arrow}(A(\eta)y + b(\eta), \tau) &= \text{Arrow}(A^n y + b^n, \tau) \\ &+ \sum_{\nu=1}^N \left[\underbrace{\mathcal{L}^T L_\nu^T}_{\widehat{L}_\nu^T} \eta^\nu \underbrace{R_\nu(y)\mathcal{R}}_{\widehat{R}_\nu(y)} + \widehat{R}_\nu^T(y)[\eta^\nu]^T \widehat{L}_\nu\right] \end{aligned}$$

where $\widehat{R}_\nu(y)$ are affine in y and $\widehat{L}_\nu \neq 0$. Observe also that all the symmetric matrices

$$B_\nu(y, \eta^\nu) = \widehat{L}_\nu^T \eta^\nu \widehat{R}_\nu(y) + \widehat{R}_\nu^T(y)[\eta^\nu]^T \widehat{L}_\nu$$

are differences of two matrices of the form $\text{Arrow}(u, \tau)$ and $\text{Arrow}(u', \tau)$, so that these are matrices of rank at most 2. The intermediate summary of our observations is as follows:

(#): *The RC of (7.1.3), (7.1.1), (7.1.2) is equivalent to the semi-infinite LMI*

$$\underbrace{\text{Arrow}(A^n y + b^n, \tau)}_{B_0(y,\tau)} + \sum_{\nu=1}^N B_\nu(y, \eta^\nu) \succeq 0 \; \forall \left(\eta : \begin{array}{l} \eta^\nu \in \mathbb{R}^{p_\nu \times q_\nu}, \\ \|\eta^\nu\|_{2,2} \leq \rho \; \forall \nu \leq N \end{array}\right)$$

$$\left[\begin{array}{c} B_\nu(y, \eta^\nu) = \widehat{L}_\nu^T \eta^\nu \widehat{R}_\nu(y) + \widehat{R}_\nu^T(y)[\eta^\nu]^T \widehat{L}_\nu, \nu = 1, ..., N \\ p_\nu = q_\nu = 1 \; \forall \nu \in \mathcal{I}_S \end{array}\right] \tag{7.1.6}$$

Here $\widehat{R}(y)$ are affine in y, and for all y, all $\nu \geq 1$ and all η^ν the ranks of the matrices $B_\nu(y, \eta^\nu)$ do not exceed 2.

Step 2. Approximating (7.1.6). Observe that an evident *sufficient* condition for the validity of (7.1.6) for a given y is the existence of symmetric matrices Y_ν, $\nu = 1, ..., N$, such that

$$Y_\nu \succeq B_\nu(y, \eta^\nu) \,\forall\, (\eta^\nu \in \mathcal{Z}_\nu = \{\eta^\nu : \|\eta^\nu\|_{2,2} \leq 1; \nu \in \mathcal{I}_S \Rightarrow \eta^\nu \in \mathbb{R}I_{p_\nu}\}) \qquad (7.1.7)$$

and

$$B_0(y, \tau) - \rho \sum_{\nu=1}^{N} Y_\nu \succeq 0. \qquad (7.1.8)$$

We are about to demonstrate that the semi-infinite LMIs (7.1.7) in variables Y_ν, y, τ can be represented by explicit finite systems of LMIs, so that the system $\mathcal{S}^0$ of semi-infinite constraints (7.1.7), (7.1.8) on variables $Y_1, ..., Y_N, y, \tau$ is equivalent to an explicit finite system $\mathcal{S}$ of LMIs. Since $\mathcal{S}^0$, due to its origin, is a safe approximation of (7.1.6), so will be $\mathcal{S}$, (which, in addition, is tractable). Now let us implement our strategy.

1^0. Let us start with $\nu \in \mathcal{I}_S$. Here (7.1.7) clearly is equivalent to just two LMIs

$$Y_\nu \succeq B_\nu(y) \equiv \widehat{L}_\nu^T \widehat{R}_\nu(y) + \widehat{R}_\nu^T(y)\widehat{L}_\nu \quad \& \quad Y_\nu \succeq -B_\nu(y). \qquad (7.1.9)$$

2^0. Now consider relation (7.1.7) for the case $\nu \notin \mathcal{I}_S$. Here we have

$$(Y_\nu, y) \text{ satisfies } (7.1.7)$$
$$\Leftrightarrow \qquad u^T Y_\nu u \geq u^T B_\nu(y, \eta^\nu)u \,\forall u \forall (\eta^\nu : \|\eta^\nu\|_{2,2} \leq 1)$$
$$\Leftrightarrow \quad u^T Y_\nu u \geq u^T \widehat{L}_\nu^T \eta^\nu \widehat{R}_\nu(y)u + u^T \widehat{R}_\nu^T(y)[\eta^\nu]^T \widehat{L}_\nu u \,\forall u \forall (\eta^\nu : \|\eta^\nu\|_{2,2} \leq 1)$$
$$\Leftrightarrow \qquad u^T Y_\nu u \geq 2u^T \widehat{L}_\nu^T \eta^\nu \widehat{R}_\nu(y)u \,\forall u \forall (\eta^\nu : \|\eta^\nu\|_{2,2} \leq 1)$$
$$\Leftrightarrow \qquad u^T Y_\nu u \geq 2\|\widehat{L}_\nu u\|_2 \|\widehat{R}(y)u\|_2 \,\forall u$$
$$\Leftrightarrow \qquad u^T Y_\nu u - 2\xi^T \widehat{R}_\nu(y)u \,\forall (u, \xi : \xi^T \xi \leq u^T \widehat{L}_\nu^T \widehat{L}_\nu u)$$

Invoking the $\mathcal{S}$-Lemma, the concluding condition in the latter chain is equivalent to

$$\exists \lambda_\nu \geq 0 : \left[\begin{array}{c|c} Y_\nu - \lambda_\nu \widehat{L}_\nu^T \widehat{L}_\nu & -\widehat{R}_\nu^T(y) \\ \hline -\widehat{R}_\nu(y) & \lambda_\nu I_{k_\nu} \end{array}\right] \succeq 0, \qquad (7.1.10)$$

where k_ν is the number of rows in $\widehat{R}_\nu(y)$.

We have proved the first part of the following statement:

Theorem 7.1.2. The explicit system of LMIs

$$Y_\nu \succeq \pm(\widehat{L}_\nu^T \widehat{R}_\nu(y) + \widehat{R}_\nu^T(y)\widehat{L}_\nu), \; \nu \in \mathcal{I}_S$$

$$\left[\begin{array}{c|c} Y_\nu - \lambda_\nu \widehat{L}_\nu^T \widehat{L}_\nu & \widehat{R}_\nu^T(y) \\ \hline \widehat{R}_\nu(y) & \lambda_\nu I_{k_\nu} \end{array}\right] \succeq 0, \; \nu \notin \mathcal{I}_S \qquad (7.1.11)$$

$$\text{Arrow}(A^n y + b^n, \tau) - \rho \sum_{\nu=1}^{N} Y_\nu \succeq 0$$

(for notation, see (7.1.6)) in variables $Y_1, ..., Y_N, \lambda_\nu, y, \tau$ is a safe tractable approximation of the RC of the uncertain Least Squares inequality (7.1.3), (7.1.1), (7.1.2).

The tightness factor of this approximation never exceeds $\pi/2$, and equals to 1 when $N = 1$.

Proof. By construction, (7.1.11) indeed is a safe tractable approximation of the RC of (7.1.3), (7.1.1), (7.1.2) (note that a matrix of the form $\left[\begin{array}{c|c} A & B \\ \hline B^T & A \end{array}\right]$ is $\succeq 0$ if and only if the matrix $\left[\begin{array}{c|c} A & -B \\ \hline -B^T & A \end{array}\right]$ is so). By Remark and Theorem 6.3.2, our approximation is exact when $N = 1$. The fact that the tightness factor never exceeds $\pi/2$ is an immediate corollary of the following Theorem (to be proved in Appendix B.4)

Theorem 7.1.3. [Matrix Cube Theorem, real case.] Let $B_0, B_1, ..., B_p$ be symmetric $m \times m$ matrices, and let $L_j \in \mathbb{R}^{p_j \times m}$, $R_j \in \mathbb{R}^{q_j \times m}$, $j = 1, ..., q$. Consider the predicates

$$B_0 + \sum_{i=1}^{p} \theta_i B_i + \sum_{j=1}^{q} [L_j^T \Theta^j R_j + R_j^T [\Theta^j]^T L_j] \succeq 0 \ \forall \left(\begin{array}{c} \theta_i : |\theta_i| \leq \rho \\ \Theta^j : \|\Theta^j\|_{2,2} \leq \rho \end{array} \right) \qquad \mathcal{A}(\rho)$$

and

$$\begin{array}{c} \exists U_1, ..., U_p, V_1, ..., V_q : \ U_i \succeq \pm B_i, \ 1 \leq i \leq p, \\ V_j \succeq [L_j^T \Theta^j R_j + R_j^T [\Theta^j]^T L_j] \ \forall (\Theta^j : \|\Theta^j\|_{2,2} \leq 1), \ 1 \leq j \leq q, \\ B_0 - \rho \sum_{i=1}^{p} U_i - \rho \sum_{j=1}^{q} V_j \succeq 0. \end{array} \qquad \mathcal{B}(\rho)$$

Then

(i) $\mathcal{B}(\rho)$ is a sufficient condition for $\mathcal{A}(\rho)$: whenever $\mathcal{B}(\rho)$ is valid, so is $\mathcal{A}(\rho)$. When $p + q = 1$, $\mathcal{B}(\rho)$ is a necessary and sufficient condition for $\mathcal{A}(\rho)$;

(ii) If the ranks of the matrices $B_1, ..., B_p$ do not exceed an integer $\mu \geq 2$, then the "tightness factor" of the sufficient condition in question does not exceed $\vartheta(\mu)$, meaning that whenever $\mathcal{B}(\rho)$ is not valid, neither is $(\mathcal{A}(\vartheta(\mu)\rho))$. Here $\vartheta(\mu)$ is an universal nondecreasing function of μ such that

$$\vartheta(2) = \frac{\pi}{2}; \ \vartheta(4) = 2; \ \vartheta(\mu) \leq \pi\sqrt{\mu/2}.$$

To complete the proof of Theorem 7.1.2, observe that a given pair (y, τ) is robust feasible for (7.1.3), (7.1.1), (7.1.2) if and only if the matrices $B_0 = B_0(y, \tau)$, $B_i = B_{\nu_i}(y, 1)$, $i = 1, ..., p$, $L_j = \widehat{L}_{\mu_j}$, $R_j = \widehat{R}_{\mu_j}(y)$, $j = 1, ..., q$, satisfy $\mathcal{A}(\rho)$; here $\mathcal{I}_S = \{\nu_1 < ... < \nu_p\}$ and $\{1, ..., L\} \backslash \mathcal{I}_S = \{\mu_1 < ... < \mu_q\}$. At the same time, the validity of the corresponding predicate $\mathcal{B}(\rho)$ is equivalent to the possibility to extend y to a solution of (7.1.11) due to the origin of the latter system. Since all matrices B_i, $i = 1, ..., p$, are of rank at most 2 by (#), the Matrix Cube Theorem implies that if (y, τ) cannot be extended to a feasible solution to (7.1.11), then (y, τ) is not robust feasible for (7.1.3), (7.1.1), (7.1.2) when the uncertainty level is increased by the factor $\vartheta(2) = \frac{\pi}{2}$. $\qquad \square$

7.1.2 Least Squares Inequality with Structured Norm-Bounded Uncertainty, Complex Case

The uncertain Least Squares inequality (7.1.3) with structured norm-bounded perturbations makes sense in the case of complex left hand side data as well as in the case of real data. Surprisingly, in the complex case the RC admits a better in tightness factor safe tractable approximation than in the real case (specifically, the tightness factor $\frac{\pi}{2} = 1.57...$ stated in Theorem 7.1.2 in the complex case improves to $\frac{4}{\pi} = 1.27...$). Consider an uncertain Least Squares inequality (7.1.3) where $A(\eta) \in \mathbb{C}^{m \times n}$, $b(\eta) \in \mathbb{C}^m$ and the perturbations are structured norm-bounded and *complex*, meaning that (cf. (7.1.1), (7.1.2))

$$(a) \quad \mathcal{Z}_\rho^{\text{left}} = \rho \mathcal{Z}_1^{\text{left}} = \left\{ \eta = (\eta^1, ..., \eta^N) : \begin{array}{l} \eta^\nu \in \mathbb{C}^{p_\nu \times q_\nu}, \ \nu = 1, ..., N \\[4pt] \|\eta^\nu\|_{2,2} \le \rho, \ \nu = 1, ..., N \\[4pt] \eta^\nu = \theta_\nu I_{p_\nu}, \theta_\nu \in \mathbb{C}, \ \nu \in \mathcal{I}_{\mathrm{S}} \end{array} \right\},$$

$$(b) \quad A(\zeta)y + b(\zeta) = [A^{\mathrm{n}}y + b^{\mathrm{n}}] + \sum_{\nu=1}^N L_\nu^H(y)\eta^\nu R_\nu(y),$$

$$(7.1.12)$$

where $L_\nu(y)$, $R_\nu(y)$ are affine in $[\Re(y); \Im(y)]$ matrices with complex entries such that for every ν at least one of these matrices is independent on y and is nonzero, and B^H denotes the Hermitian conjugate of a complex-valued matrix B: $(B^H)_{ij} = \overline{B_{ji}}$, where $\bar{z}$ is the complex conjugate of a complex number z.

Observe that by exactly the same reasons as in the real case, we can assume w.l.o.g. that all scalar perturbation blocks are 1×1, or, equivalently, that there are no scalar perturbation blocks at all, so that from now on we assume that $\mathcal{I}_{\mathrm{S}} = \emptyset$.

The derivation of the approximation is similar to the one in the real case. Specifically, we start with the evident observation that for a complex k-dimensional vector u and a real t the relation

$$\|u\|_2 \le t$$

is equivalent to the fact that the Hermitian matrix

$$\text{Arrow}(u, t) = \left[\begin{array}{c|c} t & u^H \\ \hline u & tI_k \end{array} \right]$$

is $\succeq 0$; this fact is readily given by the complex version of the Schur Complement Lemma: *a Hermitian block matrix* $\left[\begin{array}{c|c} P & Q^H \\ \hline Q & R \end{array} \right]$ *with* $R \succ 0$ *is positive semidefinite if and only if the Hermitian matrix* $P - Q^H R^{-1} Q$ *is positive semidefinite* (cf. the proof of Lemma 6.3.3). It follows that (y, τ) is robust feasible for the uncertain

Least Squares inequality in question if and only if

$$\underbrace{\text{Arrow}(A^{\text{n}}y + b^{\text{n}}, \tau)}_{B_0(y,\tau)} + \sum_{\nu=1}^{N} B_\nu(y, \eta^\nu) \succeq 0 \forall (\eta : \|\eta^\nu\|_{2,2} \leq \rho \forall \nu \leq N) \tag{7.1.13}$$

$$\left[B_\nu(y, \eta^\nu) = \widehat{L}_\nu^H \eta^\nu \widehat{R}_\nu(y) + \widehat{R}_\nu^H(y)[\eta^\nu]^H \widehat{L}_\nu, \ \nu = 1, ..., N \right]$$

where $\widehat{L}_\nu$ are constant matrices, and $\widehat{R}(y)$ are affine in $[\Re(y); \Im(y)]$ matrices readily given by $L_\nu(y), R_\nu(y)$ (cf. (7.1.6) and take into account that we are in the situation $\mathcal{I}_{\text{S}} = \emptyset$). It follows that whenever, for a given (y, τ), one can find Hermitian matrices Y_ν such that

$$Y_\nu \succeq B_\nu(y, \eta^\nu) \ \forall (\eta^\nu \in \mathbb{C}^{p_\nu \times q_\nu} : \|\eta^\nu\|_{2,2} \leq 1), \ \nu = 1, ..., N, \tag{7.1.14}$$

and $B_0(y, \tau) \succeq \rho \sum_{\nu=1}^{N} Y_\nu$, the pair (y, τ) is robust feasible.

Same as in the real case, applying the $\mathcal{S}$-Lemma, (which works in the complex case as well as in the real one), a matrix Y_ν satisfies (7.1.14) if and only if

$$\exists \lambda_\nu \geq 0 : \left[\begin{array}{c|c} Y_\nu - \lambda_\nu \widehat{L}_\nu^H \widehat{L}_\nu & -\widehat{R}_\nu^H(y) \\ \hline -\widehat{R}_\nu(y) & \lambda_\nu I_{k_\nu} \end{array} \right],$$

where k_ν is the number of rows in $\widehat{R}_\nu(y)$. We have arrived at the first part of the following statement:

Theorem 7.1.4. The explicit system of LMIs

$$\left[\begin{array}{c|c} Y_\nu - \lambda_\nu \widehat{L}_\nu^H \widehat{L}_\nu & \widehat{R}_\nu^H(y) \\ \hline \widehat{R}_\nu(y) & \lambda_\nu I_{k_\nu} \end{array} \right] \succeq 0, \ \nu = 1, ..., N, \tag{7.1.15}$$

$$\text{Arrow}(A^{\text{n}}y + b^{\text{n}}, \tau) - \rho \sum_{\nu=1}^{N} Y_\nu \succeq 0$$

(for notation, see (7.1.13)) in the variables $\{Y_i = Y_i^H\}, \lambda_\nu, y, \tau$ is a safe tractable approximation of the RC of the uncertain Least Squares inequality (7.1.3), (7.1.12). The tightness factor of this approximation never exceeds $4/\pi$, and is equal to 1 when $N = 1$.

Proof is completely similar to the one of Theorem 7.1.2, modulo the following statement (to be proved in Appendix B.4) replacing the Real case Matrix Cube Theorem:

Theorem 7.1.5. [Matrix Cube Theorem, complex case with no scalar perturbations] Let B_0 be a Hermitian $m \times m$ matrix, and let $L_j \in \mathbb{C}^{p_j \times m}, R_j \in \mathbb{C}^{q_j \times m}$, $j = 1, ..., q$. Consider the predicates

$$B_0 + \sum_{j=1}^{q} [L_j^H \Theta^j R_j + R_j^H [\Theta^j]^H L_j] \succeq 0 \ \forall (\Theta^j \in \mathbb{C}^{p_j \times q_j} : \|\Theta^j\|_{2,2} \leq \rho) \qquad \mathcal{A}(\rho)$$

and

$$\exists V_1, ..., V_q : V_j \succeq \left(L_j^H \Theta^j R_j + R_j^H [\Theta^j]^H L_j\right) \ \forall (\Theta^j \in \mathbb{C}^{p_j \times q_j} : \|\Theta^j\|_{2,2} \leq 1),$$

$$1 \leq j \leq q, \text{ and } B_0 - \rho \sum_{j=1}^{q} V_j \succeq 0. \hspace{3cm} \mathcal{B}(\rho)$$

Then

    (i) $\mathcal{B}(\rho)$ is a sufficient condition for $\mathcal{A}(\rho)$: whenever $\mathcal{B}(\rho)$ is valid, so is $\mathcal{A}(\rho)$. When $q = 1$, $\mathcal{B}(\rho)$ is a necessary and sufficient condition for $\mathcal{A}(\rho)$;

    (ii) The "tightness factor" of the sufficient condition in question does not exceed $\frac{4}{\pi}$, meaning that when $\mathcal{B}(\rho)$ is not valid, neither is $\mathcal{A}(\frac{4}{\pi}\rho)$.

Illustration: Antenna Design revisited. Consider the "Least Squares" version of the Antenna Design problem from section 3.3. As in the original problem, we consider an array of n harmonic oscillators placed at the points $k\mathbf{i}$, $k = 1, ..., n$, $\mathbf{i}$ being the orth of the X-axis in $\mathbb{R}^3$, and normalize the weights $z_k \in \mathbb{C}$ of the oscillators by the requirement

$$\Re\{ \underbrace{\sum_{k=1}^{n} z_k D_k(\phi)}_{D(\phi)} \Big|_{\phi=0} \} \geq 1,$$

where $D_k(\phi) = \exp\{2\pi\imath\cos(\phi)k/\lambda\}$ is the diagram of the k-th oscillator. In section 3.3, our goal was to minimize, under this normalization, the uniform norm $\max_{\Delta \leq \phi \leq \pi} |D(\phi)|$ of the diagram $D(\cdot)$ in the sidelobe angle. Here we want to minimize, under the same normalization restriction, the weighted L_2 norm of the diagram $D(\cdot)$ in the sidelobe angle, specifically, the quantity

$$\|D(\cdot)\|_{\mathcal{SA}} = \left(\int_{z \in \mathcal{SA}} |D(\phi(z))|^2 dS(z) \right)^{1/2}$$

$$= \left(\frac{1}{1+\cos(\Delta)} \int_{\Delta}^{\pi} |D(\phi)|^2 \sin(\phi) d\phi \right)^{1/2},$$

where $\mathcal{SA}$ is the sidelobe angle treated as the part of the unit sphere $S_2 \subset \mathbb{R}^3$ comprised of all directions forming angle $\geq \Delta$ with the direction $\mathbf{i}$ of the antenna array, and $dS(z)$ is the element of area of S_2 normalized by the area of the entire $\mathcal{SA}$ (so that $\int_{\mathcal{SA}} dS(z) = 1$). The associated optimization problem is

$$\min_{z_1,...,z_n \in \mathbb{C}, \tau \in \mathbb{R}} \left\{ \tau : \begin{array}{l} \left(\frac{1}{1+\cos(\Delta)} \int_{\Delta}^{\pi} |\sum_{k=1}^{n} z_k D_k(\phi)|^2 \sin(\phi) d\phi \right)^{1/2} \leq \tau \\ \Re\{\sum_{k=1}^{n} z_k D_k(0)\} \geq 1 \end{array} \right\}. \quad (7.1.16)$$

In section 3.3 we allowed for perturbations in the positions of the oscillators and for actuation errors affecting the weights. Here, for the sake of simplicity, we assume that the positioning of oscillators is precise, and the only source of uncertainty is given by actuation errors

$$z_k \mapsto (1 + \zeta^k)z_k,$$

with perturbations $\zeta^k \in \mathbb{C}$ subject to the bounds $|\zeta^k| \leq \epsilon_k \rho$. As always, we lose nothing when assuming that there are no actuation errors, but the diagrams $D_k(\cdot)$ are subject to perturbations $D_k(\cdot) \mapsto (1 + \zeta^k)D_k(\cdot)$. Now, we can easily find an $n \times n$ complex-valued matrix A^{n} such that

$$\|A^{\mathrm{n}}z\|_2 = \left(\frac{1}{1 + \cos(\Delta)} \int_{\Delta}^{\pi} |\sum_{k=1}^{n} z_k D_k(\phi)|^2 \sin(\phi)d\phi \right)^{1/2} \quad \forall z;$$

to this end, it suffices to compute the positive semidefinite Hermitian matrix with the entries $H_{pq} = \frac{1}{1+\cos(\Delta)} \int_{\Delta}^{\pi} D_p(\phi)\overline{D_q(\phi)} \sin(\phi)d\phi$ and to set $A = H^{1/2}$. Doing so, we can reformulate the uncertain problem (7.1.16) equivalently as

$$\min_{z,\tau} \left\{ \tau : \begin{array}{ll} \|A(\eta)z\|_2 \leq \tau & (a) \\ \Re\{\sum_{k=1}^{n} (1 + \epsilon_k \eta^k)z_k D_k(0)\} \geq 1 & (b) \end{array} \right\},$$

$$\eta \in \mathcal{Z}_\rho^{\text{left}} = \{\eta \in \mathbb{C}^n : |\eta^k| \leq \rho, \ k = 1, ..., n\} \tag{7.1.17}$$

$$\left[\begin{array}{l} A(\eta)z = A^{\mathrm{n}}z + \sum_{k=1}^{n} L_k^H \eta^k R_k(z) \\ L_k^H \text{ is } k\text{-th column of } A^{\mathrm{n}}, R_k(z) = \epsilon_k z_k \in \mathbb{C}^{1 \times 1} \end{array} \right]$$

Taking into account that $|D_k(\cdot)| \equiv 1$, the RC of the uncertain constraint (7.1.17.b) is equivalent to the explicit convex constraint

$$\Re\{\sum_{k=1}^{n} z_k D_k(0)\} - \rho \sum_{k=1}^{n} \epsilon_k |z_k| \geq 1. \tag{7.1.18}$$

Constraint (7.1.17.a) is an uncertain Least Squares inequality with complex data and structured norm-bounded perturbations (n full 1×1 complex perturbation blocks). Theorem 7.1.4 provides us with $\frac{4}{\pi}$-tight safe tractable approximation of this constraint, which is the system

$$\left[\begin{array}{c|c} Y_k - \lambda_k \widehat{L}_k^H \widehat{L}_k & \widehat{R}_k^H(z) \\ \hline \widehat{R}_k(z) & \lambda_k \end{array} \right] \succeq 0, \ k = 1, ..., n$$

$$\text{Arrow}(A^{\mathrm{n}}z, \tau) \succeq \rho \sum_{k=1}^{n} Y_k \tag{7.1.19}$$

of LMIs in variables $Y_k = Y_k^H, \lambda_k \in \mathbb{R}, \tau \in \mathbb{R}, z \in \mathbb{C}^n$; here

$$\begin{array}{ll} \widehat{L}_k & = [0, \overline{(A_{1k}^{\mathrm{n}})}, \overline{(A_{2k}^{\mathrm{n}})}, ..., \overline{(A_{nk}^{\mathrm{n}})}] \in \mathbb{C}^{1 \times (n+1)}, \\ \widehat{R}_k(z) & = [\epsilon_k z_k, 0, ..., 0] \in \mathbb{C}^{1 \times (n+1)}. \end{array}$$

The explicit convex problem

$$\min_{z,\tau,\{Y_k,\lambda_k\}} \{\tau : (z, \tau, \{Y_k, \lambda_k\}) \text{ satisfies } (7.1.19), (7.1.18)\} \tag{7.1.20}$$

is a safe tractable approximation, tight within the factor $\frac{4}{\pi}$, of the RC of (7.1.17).

Design	ρ	Sidelobe level	$\|D(\cdot)\|_{\mathcal{SA}}$	Energy concentration
Nominal (Opt = 1.5e-5)	0	0.01(0.00)	1.5e-5(0.00)	0.9998(0.00)
	1.e-4	1.16(0.79)	0.728(0.48)	0.114(0.09)
	1.e-3	1.83(1.02)	1.193(0.68)	0.083(0.07)
RC (Opt = 0.053)	0	0.24(0.00)	0.040(0.00)	0.955(0.00)
	1.e-2	0.24(0.01)	0.043(0.00)	0.954(0.00)
	3.e-2	0.25(0.03)	0.063(0.01)	0.882(0.03)
	5.e-2	0.26(0.04)	0.091(0.01)	0.780(0.03)
	1.e-1	0.33(0.05)	0.170(0.04)	0.517(0.13)

Table 7.1 Performance of nominal and robust designs.
In the table: Opt is the optimal value in the nominal problem and its RC, respectively. The underlined numbers are averages over 100 random realizations of actuation errors, the numbers in parentheses are the associated standard deviations.

Example 7.1.6. Consider the same design data as in section 3.3, that is,

$$n = 16; \lambda = 8; \Delta = \pi/6; \epsilon_1 = ... = \epsilon_n = 1$$

(thus, the magnitude of the actuation errors is ρ). Solving the nominal problem and the (approximate) RC of the uncertain problem at the uncertainty level $\rho = 1.e\text{-}2$, we get, respectively, the nominal and the robust designs. The characteristics of these designs are presented in table 7.1 and are depicted in figure 7.1. The conclusions are, essentially, the same as those in section 3.3 — the nominal design is completely senseless already for 0.01% actuation errors, while the robust design seems completely meaningful even with 5% (and perhaps even with 10%) actuation errors. It is instructive to compare the GRC design obtained in section 3.3 (the one that is immunized against actuation errors) and the RC design we have built now. Under the same circumstances, namely, for 3% actuation errors, the former design exhibits sidelobe level about 0.19, which is essentially better than the sidelobe level 0.23 we now have; however, the performance characteristic of primary importance, that is, the energy concentration (fraction of total energy sent in the spatial angle of interest) for the new design is much better than for the old one (0.88 vs. 0.66). The conclusion is that at least in our example *the Least Squares setting of the Antenna Design problem is much better suited for robust concentration of energy in the angle of interest than the sidelobe level setting considered in section 3.3.* An immediate question is: if all we are interested in is the energy concentration, (which essentially, is the case in actual Antenna Design), why not optimize this quantity directly? Why control this concentration implicitly, by normalizing the diagram in the direction of interest and minimizing its norm in the sidelobe angle? The answer is, that direct minimization of energy concentration, even in the nominal setting, is a nonconvex problem, so that it is unclear how to solve it efficiently — a difficulty that does not arise in the models we have considered.

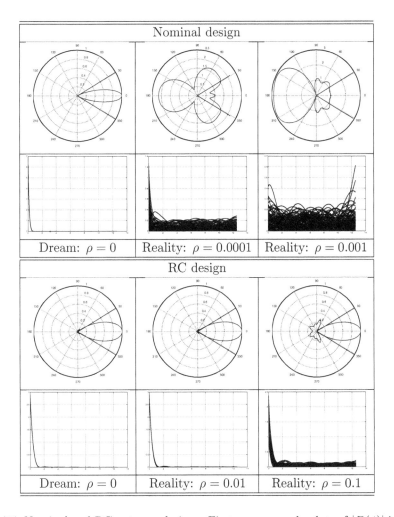

Figure 7.1 Nominal and RC antenna designs. First rows: sample plots of $|D(\phi)|$ in polar coordinates; the diagrams are normalized to have $D(0) = 1$. Second rows: bunches of 100 simulated energy densities, cf. section 3.3.

7.1.3 From Uncertain Least Squares to Uncertain CQI

Let us come back to the real case. We have already built a tight approximation for the RC of a Least Squares inequality with structured norm-bounded uncertainty in the left hand side data. Our next goal is to extend this approximation to the case of uncertain CQI with side-wise uncertainty.

Theorem 7.1.7. Consider the uncertain CQI (6.1.3) with side-wise uncertainty, where the left hand side uncertainty is the structured norm-bounded one given by (7.1.1), (7.1.2), and the right hand side perturbation set is given by a conic representation (cf. Theorem 1.3.4)

$$\mathcal{Z}_\rho^{\text{right}} = \rho \mathcal{Z}_1^{\text{right}}, \; \mathcal{Z}_1^{\text{right}} = \{\chi : \exists u : P\chi + Qu + p \in \mathbf{K}\}, \qquad (7.1.21)$$

where $0 \in \mathcal{Z}_1^{\text{right}}$, $\mathbf{K}$ is a closed convex pointed cone and the representation is strictly feasible unless $\mathbf{K}$ is a polyhedral cone given by an explicit finite list of linear inequalities, and $0 \in \mathcal{Z}_1^{\text{right}}$.

For $\rho > 0$, the explicit system of LMIs

$$
\begin{aligned}
(a) \quad & \tau + \rho p^T v \leq \delta(y), \; P^T v = \sigma(y), \; Q^T v = 0, \; v \in \mathbf{K}_* \\[4pt]
(b.1) \quad & Y_\nu \succeq \pm(\widehat{L}_\nu^T \widehat{R}_\nu(y) + \widehat{R}_\nu^T(y)\widehat{L}_\nu), \; \nu \in \mathcal{I}_{\mathrm{S}} \\[4pt]
(b.2) \quad & \left[\begin{array}{c|c} Y_\nu - \lambda_\nu \widehat{L}_\nu^T \widehat{L}_\nu & \widehat{R}_\nu^T(y) \\ \hline \widehat{R}_\nu(y) & \lambda_\nu I_{k_\nu} \end{array} \right] \succeq 0, \; \nu \notin \mathcal{I}_{\mathrm{S}} \\[4pt]
(b.3) \quad & \mathrm{Arrow}(A^{\mathrm{n}}y + b^{\mathrm{n}}, \tau) - \rho \sum_{\nu=1}^{N} Y_\nu \succeq 0
\end{aligned}
\qquad (7.1.22)
$$

(for notation, see (7.1.6)) in variables $Y_1, ..., Y_N, \lambda_\nu, y, \tau, v$ is a safe tractable approximation of the RC of (6.2.1). This approximation is exact when $N = 1$, and is tight within the factor $\frac{\pi}{2}$ otherwise.

Proof. Since the uncertainty is side-wise, y is robust feasible for (6.2.1), (7.1.1), (7.1.2), (7.1.21), the uncertainty level being $\rho > 0$, if and only if there exists τ such that

$$
\begin{aligned}
(c) \quad & \sigma^T(\chi)y + \delta(\chi) \geq \tau \; \forall \chi \in \rho \mathcal{Z}_1^{\text{right}}, \\
(d) \quad & \|A(\eta)y + b(\eta)\|_2 \leq \tau \; \forall \eta \in \rho \mathcal{Z}_1^{\text{left}}.
\end{aligned}
$$

When $\rho > 0$, we have

$$\rho \mathcal{Z}_1^{\text{right}} = \{\chi : \exists u : P(\chi/\rho) + Qu + p \in \mathbf{K}\} = \{\chi : \exists u' : P\chi + Qu' + \rho p \in \mathbf{K}\};$$

from the resulting conic representation of $\rho \mathcal{Z}_1^{\text{right}}$, same as in the proof of Theorem 1.3.4, we conclude that the relations (7.1.22.a) represent equivalently the requirement (c), that is, (y, τ) satisfies (c) if and only if (y, τ) can be extended, by properly chosen v, to a solution of (7.1.22.a). By Theorem 7.1.2, the possibility to extend (y, τ) to a feasible solution of (7.1.22.b) is a sufficient condition for the validity of (d). Thus, the (y, τ) component of a feasible solution to (7.1.22) satisfies (c), (d), meaning that y is robust feasible at the level of uncertainty ρ. Thus, (7.1.22) is a safe approximation of the RC in question.

The fact that the approximation is precise when there is only one left hand side perturbation block is readily given by Theorem 6.3.2 and Remark 7.1.1 allowing us to treat this block as full. It remains to verify that the tightness factor of the approximation is at most $\frac{\pi}{2}$, that is, to check that if a given y cannot be extended to a feasible solution of the approximation for the uncertainty level ρ, then y is not robust feasible for the uncertainty level $\frac{\pi}{2}\rho$ (see comments after Definition 5.3.2). To this end, let us set

$$\tau_y(r) = \inf_{\chi} \left\{ \sigma^T(\chi)y + \delta(\chi) : \chi \in r\mathcal{Z}_1^{\text{right}} \right\}.$$

Since $0 \in \mathcal{Z}_1^{\text{right}}$ by assumption, $\tau_y(r)$ is nonincreasing in r. Clearly, y is robust feasible at the uncertainty level r if and only if

$$\|A(\eta)y + b(\eta)\|_2 \le \tau_y(r) \; \forall \eta \in r\mathcal{Z}_1^{\text{left}}. \tag{7.1.23}$$

Now assume that a given y cannot be extended to a feasible solution of (7.1.22) for the uncertainty level ρ. Let us set $\tau = \tau_y(\rho)$; then (y, τ) can be extended, by a properly chosen v, to a feasible solution of (7.1.22.a). Indeed, the latter system expresses equivalently the fact that (y, τ) satisfies (c), which indeed is the case for our (y, τ). Now, since y cannot be extended to a feasible solution to (7.1.22) at the uncertainty level ρ, and the pair (y, τ) can be extended to a feasible solution of (7.1.22.a), we conclude that (y, τ) cannot be extended to a feasible solution of (7.1.22.b). By Theorem 7.1.2, the latter implies that y is *not* robust feasible for the semi-infinite Least Squares constraint

$$\|A(\eta)y + b(\eta)\|_2 \le \tau = \tau_y(\rho) \; \forall \eta \in \frac{\pi}{2}\rho\mathcal{Z}_1^{\text{left}}.$$

Since $\tau_y(r)$ is nonincreasing in r, we conclude that y does *not* satisfy (7.1.23) when $r = \frac{\pi}{2}\rho$, meaning that y is not robust feasible at the level of uncertainty $\frac{\pi}{2}\rho$. □

7.1.4 Convex Quadratic Constraint with Structured Norm-Bounded Uncertainty

Consider an uncertain convex quadratic constraint

$$\begin{array}{ll} (a) & y^T A^T(\zeta)A(\zeta)y \le 2y^T b(\zeta) + c(\zeta) \\ & \quad\quad\quad \Updownarrow \\ (b) & \|[2A(\zeta)y; 1 - 2y^T b(\zeta) - c(\zeta)]\|_2 \le 1 + 2y^T b(\zeta) + c(\zeta), \end{array} \tag{6.4.1}$$

where $A(\zeta)$ is $k \times n$ and the uncertainty is structured norm-bounded (cf. (6.4.2)), meaning that

$$
(a) \quad \mathcal{Z}_\rho = \rho \mathcal{Z}_1 = \left\{ \zeta = (\zeta^1, ..., \zeta^N) : \begin{array}{c} \zeta^\nu \in \mathbb{R}^{p_\nu \times q_\nu} \\ \|\zeta^\nu\|_{2,2} \le \rho, 1 \le \nu \le N \\ \zeta^\nu = \theta_\nu I_{p_\nu}, \theta_\nu \in \mathbb{R}, \nu \in \mathcal{I}_{\mathrm{s}} \end{array} \right\},
$$
$$
(b) \quad \begin{bmatrix} A(\zeta)y \\ y^T b(\zeta) \\ c(\zeta) \end{bmatrix} = \begin{bmatrix} A^{\mathrm{n}} y \\ y^T b^{\mathrm{n}} \\ c^{\mathrm{n}} \end{bmatrix} + \sum_{\nu=1}^{N} L_\nu^T(y)\zeta^\nu R_\nu(y) \tag{7.1.24}
$$

where, for every ν, $L_\nu(y)$, $R_\nu(y)$ are matrices of appropriate sizes depending affinely on y and such that at least one of the matrices is constant. Same as above, we can assume w.l.o.g. that all scalar perturbation blocks are 1×1: $p_\nu = k_\nu = 1$ for all $\nu \in \mathcal{I}_{\mathrm{s}}$.

Note that the equivalence in (6.4.1) means that we still are interested in an uncertain CQI with structured norm-bounded left hand side uncertainty. The uncertainty, however, is *not* side-wise, that is, we are in the situation we could not handle before. We can handle it now due to the fact that the uncertain CQI possesses a favorable structure inherited from the original convex quadratic form of the constraint.

We are about to derive a tight tractable approximation of the RC of (6.4.1), (7.1.24). The construction is similar to the one we used in the unstructured case $N = 1$, see section 6.4. Specifically, let us set $L_\nu(y) = [L_{\nu,A}(y), L_{\nu,b}(y), L_{\nu,c}(y)]$, where $L_{\nu,b}(y)$, $L_{\nu,c}(y)$ are the last two columns in $L_\nu(y)$, and let

$$
\tilde{L}_\nu^T(y) = \left[L_{\nu,b}^T(y) + \tfrac{1}{2}L_{\nu,c}^T(y); L_{\nu,A}^T(y) \right], \quad \tilde{R}_\nu(y) = [R_\nu(y), 0_{q_\nu \times k}],
$$
$$
\mathcal{A}(y) = \left[\begin{array}{c|c} 2y^T b^{\mathrm{n}} + c^{\mathrm{n}} & [A^{\mathrm{n}} y]^T \\ \hline A^{\mathrm{n}} y & I \end{array} \right], \tag{7.1.25}
$$

so that $\mathcal{A}(y)$, $\tilde{L}_\nu(y)$ and $\tilde{R}_\nu(y)$ are affine in y and at least one of the latter two matrices is constant.

We have

$$
y^T A^T(\zeta)A(\zeta)y \le 2y^T b(\zeta) + c(\zeta) \ \forall \zeta \in \mathcal{Z}_\rho
$$

$$
\Leftrightarrow \left[\begin{array}{c|c} 2y^T b(\zeta) + c(\zeta) & [A(\zeta)y]^T \\ \hline \underbrace{A(\zeta)y}_{\mathcal{A}(y)} & I \end{array} \right] \succeq 0 \ \forall \zeta \in \mathcal{Z}_\rho \ \text{[Schur Complement Lemma]}
$$

$$
\Leftrightarrow \overbrace{\left[\begin{array}{c|c} 2y^T b^{\mathrm{n}} + c^{\mathrm{n}} & [A^{\mathrm{n}} y]^T \\ \hline A^{\mathrm{n}} y & I \end{array} \right]}
$$
$$
+ \sum_{\nu=1}^{N} \underbrace{\left[\begin{array}{c|c} [2L_{\nu,b}(y) + L_{\nu,c}(y)]^T \zeta^\nu R_\nu(y) & [L_{\nu,A}^T(y)\zeta^\nu R_\nu(y)]^T \\ \hline L_{\nu,A}^T(y)\zeta^\nu R_\nu(y) & \end{array} \right]}_{=\tilde{L}_\nu^T(y)\zeta^\nu \tilde{R}_\nu(y) + \tilde{R}_\nu^T(y)[\zeta^\nu]^T \tilde{L}_\nu(y)} \succeq 0 \ \forall \zeta \in \mathcal{Z}_\rho
$$
$$
\text{[by (7.1.24)]}
$$

$$
\Leftrightarrow \mathcal{A}(y) + \sum_{\nu=1}^{N} \left[\tilde{L}_\nu^T(y)\zeta^\nu \tilde{R}_\nu(y) + \tilde{R}_\nu^T(y)[\zeta^\nu]^T \tilde{L}_\nu(y) \right] \succeq 0 \ \forall \zeta \in \mathcal{Z}_\rho.
$$

Taking into account that for every ν at least one of the matrices $\widetilde{L}_\nu(y)$, $\widetilde{R}_\nu(y)$ is independent of y and swapping, if necessary, ζ^ν and $[\zeta^\nu]^T$, we can rewrite the last condition in the chain as

$$\mathcal{A}(y) + \sum_{\nu=1}^{N} \left[\widehat{L}_\nu^T \zeta^\nu \widehat{R}_\nu(y) + \widehat{R}_\nu^T(y)[\zeta^\nu]^T \widehat{L}_\nu \right] \succeq 0 \; \forall (\zeta : \|\zeta^\nu\|_{2,2} \leq \rho) \tag{7.1.26}$$

where $\widehat{L}_\nu$, $\widehat{R}_\nu(y)$ are readily given matrices and $\widehat{R}_\nu(y)$ is affine in y. (Recall that we are in the situation where all scalar perturbation blocks are 1×1 ones, and we can therefore skip the explicit indication that $\zeta^\nu = \theta_\nu I_{p_\nu}$ for $\nu \in \mathcal{I}_\mathrm{S}$). Observe also that similarly to the case of a Least Squares inequality, all matrices $\left[\widehat{L}_\nu^T \zeta^\nu \widehat{R}_\nu(y) + \widehat{R}_\nu^T(y)[\zeta^\nu]^T \widehat{L}_\nu \right]$ are of rank at most 2. Finally, we lose nothing by assuming that $\widehat{L}_\nu$ are nonzero for all ν.

Proceeding exactly in the same fashion as in the case of the uncertain Least Squares inequality with structured norm-bounded perturbations, we arrive at the following result (cf. Theorem 7.1.2):

Theorem 7.1.8. The explicit system of LMIs

$$Y_\nu \succeq \pm(\widehat{L}_\nu^T \widehat{R}_\nu(y) + \widehat{R}_\nu^T(y)\widehat{L}_\nu), \; \nu \in \mathcal{I}_\mathrm{S}$$

$$\left[\begin{array}{c|c} Y_\nu - \lambda_\nu \widehat{L}_\nu^T \widehat{L}_\nu & \widehat{R}_\nu^T(y) \\ \hline \widehat{R}_\nu(y) & \lambda_\nu I_{k_\nu} \end{array} \right] \succeq 0, \; \nu \notin \mathcal{I}_\mathrm{S} \tag{7.1.27}$$

$$\mathcal{A}(y) - \rho \sum_{\nu=1}^{L} Y_\nu \succeq 0$$

(k_ν is the number of rows in $\widehat{R}_\nu$) in variables $Y_1, ..., Y_N, \lambda_\nu, y$ is a safe tractable approximation of the RC of the uncertain convex quadratic constraint (6.4.1), (7.1.24). The tightness factor of this approximation never exceeds $\pi/2$, and equals 1 when $N = 1$.

7.1.4.1 Complex case

The situation considered in section 7.1.4 admits a complex data version as well. Consider a convex quadratic constraint with complex-valued variables and a complex-valued structured norm-bounded uncertainty:

$$y^H A^H(\zeta)A(\zeta)y \leq \Re\{2y^H b(\zeta) + c(\zeta)\}$$

$$\zeta \in \mathcal{Z}_\rho = \rho \mathcal{Z}_1 = \left\{ \zeta = (\zeta^1, ..., \zeta^N) : \begin{array}{l} \zeta^\nu \in \mathbb{C}^{p_\nu \times q_\nu}, \; 1 \leq \nu \leq N \\ \|\zeta^\nu\|_{2,2} \leq \rho, \; 1 \leq \nu \leq N \\ \nu \in \mathcal{I}_\mathrm{S} \Rightarrow \zeta^\nu = \theta_\nu I_{p_\nu}, \theta_\nu \in \mathbb{C} \end{array} \right\} \tag{7.1.28}$$

$$\left[\begin{array}{c} A(\zeta)y \\ y^H b(\zeta) \\ c(\zeta) \end{array} \right] = \left[\begin{array}{c} A^\mathrm{n} y \\ y^H b^\mathrm{n} \\ c^\mathrm{n} \end{array} \right] + \sum_{\nu=1}^{N} L_\nu^H(y)\zeta^\nu R_\nu(y),$$

where $A^\mathrm{n} \in \mathbb{C}^{k \times m}$ and the matrices $L_\nu(y)$, $R_\nu(y)$ are affine in $[\Re(y); \Im(y)]$ and such that for every ν, either $L_\nu(y)$, or $R_\nu(y)$ are independent of y. Same as in the real case we have just considered, we lose nothing when assuming that all scalar perturbation blocks are 1×1, which allows us to treat these blocks as full. Thus, the general case can be reduced to the case where $\mathcal{I}_S = \emptyset$, which we assume from now on (cf. section 7.1.2).

In order to derive a safe approximation of the RC of (7.1.28), we can act exactly in the same fashion as in the real case to arrive at the equivalence

$$y^H A^H(\zeta) A(\zeta) y \le \Re\{2y^H b(\zeta) + c(\zeta)\} \ \forall \zeta \in \mathcal{Z}_\rho$$

$$\Leftrightarrow \overbrace{\left[\begin{array}{c|c} \Re\{2y^H b^n + c^n\} & [A^n y]^H \\ \hline A^n y & I \end{array} \right]}^{\mathcal{A}(y)}$$
$$+ \sum_{\nu=1}^{N} \left[\begin{array}{c|c} \Re\{2y^H L_{\nu,b}(y)\zeta^\nu R_\nu(y) + L_{\nu,c}(y)\zeta^\nu R_\nu(y)\} & R_\nu^H[\zeta^\nu]^H L_{\nu,A}(y) \\ \hline L_{\nu,A}^H(y)\zeta^\nu R_\nu(y) & \end{array} \right] \succeq 0$$
$$\forall(\zeta : \|\zeta^\nu\|_{2,2} \le \rho, 1 \le \nu \le N)$$

where $L_\nu(y) = [L_{\nu,A}(y), L_{\nu,b}(y), L_{\nu,c}(y)]$ and $L_{\nu,b}(y)$, $L_{\nu,c}(y)$ are the last two columns in $L_\nu(y)$.

Setting

$$\widetilde{L}_\nu^H(y) = \left[L_{\nu,b}^H(y) + \frac{1}{2} L_{\nu,c}^H(y); L_{\nu,A}^H(y) \right], \quad \widetilde{R}_\nu(y) = [R_\nu(y), 0_{q_\nu \times k}]$$

(cf. (7.1.25)), we conclude that the RC of (7.1.28) is equivalent to the semi-infinite LMI

$$\mathcal{A}(y) + \sum_{\nu=1}^{N} \left[\widetilde{L}_\nu^H(y)\zeta^\nu \widetilde{R}_\nu(y) + \widetilde{R}_\nu^H(y)[\zeta^\nu]^H \widetilde{L}_\nu(y) \right] \succeq 0 \tag{7.1.29}$$
$$\forall(\zeta : \|\zeta^\nu\|_{2,2} \le \rho, 1 \le \nu \le N).$$

As always, swapping, if necessary, ζ^ν and $[\zeta^\nu]^H$ we may rewrite the latter semi-infinite LMI equivalently as

$$\mathcal{A}(y) + \sum_{\nu=1}^{N} \left[\widehat{L}_\nu^H \zeta^\nu \widehat{R}_\nu(y) + \widehat{R}_\nu^H(y)[\zeta^\nu]^H \widehat{L}_\nu \right] \succeq 0$$
$$\forall(\zeta : \|\zeta^\nu\|_{2,2} \le \rho, 1 \le \nu \le N),$$

where $\widehat{R}_\nu(y)$ are affine in $[\Re(y); \Im(y)]$ and $\widehat{L}_\nu$ are nonzero. Applying the Complex case Matrix Cube Theorem (see the proof of Theorem 7.1.4), we finally arrive at the following result:

Theorem 7.1.9. The explicit system of LMIs

$$\left[\begin{array}{c|c} Y_\nu - \lambda_\nu \widehat{L}_\nu^H \widehat{L}_\nu & \widehat{R}_\nu^H(y) \\ \hline \widehat{R}_\nu(y) & \lambda_\nu I_{k_\nu} \end{array} \right] \succeq 0, \nu = 1, ..., N,$$

$$\left[\begin{array}{c|c} \Re\{2y^H b^n + c^n\} & [A^n y]^H \\ \hline A^n y & I \end{array} \right] - \rho \sum_{\nu=1}^{N} Y_\nu \succeq 0$$
$$\tag{7.1.30}$$

(k_ν is the number of rows in $\widehat{R}_\nu(y)$) in variables $Y_1 = Y_1^H, ..., Y_N = Y_N^H, \lambda_\nu \in \mathbb{R}, y \in \mathbb{C}^m$ is a safe tractable approximation of the RC of the uncertain convex quadratic inequality (7.1.28). The tightness of this approximation is $\le \frac{4}{\pi}$, and is equal to 1 when $N = 1$.

7.2 THE CASE OF ∩-ELLIPSOIDAL UNCERTAINTY

Consider the case where the uncertainty in CQI (6.1.3) is side-wise with the right hand side uncertainty exactly as in section 6.2, and with ∩-*ellipsoidal left hand side perturbation set*, that is,

$$\mathcal{Z}_\rho^{\text{left}} = \left\{ \eta : \eta^T Q_j \eta \leq \rho^2, \ j = 1, ..., J \right\}, \tag{7.2.1}$$

where $Q_j \succeq 0$ and $\sum_{j=1}^{J} Q_j \succ 0$. When $Q_j \succ 0$ for all j, $\mathcal{Z}_\rho^{\text{left}}$ is the intersection of J ellipsoids centered at the origin. When $Q_j = a_j a_j^T$ are rank 1 matrices, $\mathcal{Z}^{\text{left}}$ is a polyhedral set symmetric w.r.t. origin and given by J inequalities of the form $|a_j^T \eta| \leq \rho, \ j = 1, ..., J$. The requirement $\sum_{j=1}^{J} Q_j \succ 0$ implies that $\mathcal{Z}_\rho^{\text{left}}$ is bounded (indeed, every $\eta \in \mathcal{Z}_\rho^{\text{left}}$ belongs to the ellipsoid $\eta^T (\sum_j Q_j) \eta \leq J \rho^2$).

We have seen in section 6.3 that the case $J = 1$, (i.e., of an ellipsoid $\mathcal{Z}_\rho^{\text{left}}$ centered at the origin), is a particular case of unstructured norm-bounded perturbation, so that in this case the RC is computationally tractable. The case of general ∩-ellipsoidal uncertainty includes the situation when $\mathcal{Z}_\rho^{\text{left}}$ is a box, where the RC is computationally intractable. However, we intend to demonstrate that with ∩-ellipsoidal left hand side perturbation set, the RC of (6.2.1) admits a safe tractable approximation tight within the "nearly constant" factor $\sqrt{(O(\ln J))}$.

7.2.1 Approximating the RC of Uncertain Least Squares Inequality

Same as in section 7.1, the side-wise nature of uncertainty reduces the task of approximating the RC of uncertain CQI (6.2.1) to a similar task for the RC of the uncertain Least Squares inequality (7.1.3). Representing

$$A(\zeta)y + b(\zeta) = \underbrace{[A^{\text{n}} y + b^{\text{n}}]}_{\beta(y)} + \underbrace{\sum_{\ell=1}^{L} \eta_\ell [A^\ell y + b^\ell]}_{\alpha(y)\eta} \tag{7.2.2}$$

where $L = \dim \eta$, observe that the RC of (7.1.3), (7.2.1) is equivalent to the system of constraints

$$\tau \geq 0 \ \& \ \|\beta(y) + \alpha(y)\eta\|_2^2 \leq \tau^2 \ \forall (\eta : \eta^T Q_j \eta \leq \rho^2, \ j = 1, ..., J)$$

or, which is clearly the same, to the system

$$(a) \quad \mathcal{A}_\rho \equiv \max_{\eta, t} \left\{ \eta^T \alpha^T(y)\alpha(y)\eta + 2t\beta^T(y)\alpha(y)\eta : \eta^T Q_j \eta \leq \rho^2 \ \forall j, \ t^2 \leq 1 \right\}$$
$$\leq \tau^2 - \beta^T(y)\beta(y)$$
$$(b) \quad \tau \geq 0.$$
$$\tag{7.2.3}$$

Next we use Lagrangian relaxation to derive the following result:

(!) Assume that for certain nonnegative reals γ, γ_j, $j = 1, ..., J$, the homogeneous quadratic form in variables η, t

$$\gamma t^2 + \sum_{j=1}^{J} \gamma_j \eta^T Q_j \eta - [\eta^T \alpha^T(y)\alpha(y)\eta + 2t\beta^T(y)\alpha(y)\eta] \qquad (7.2.4)$$

is nonnegative everywhere. Then

$$\mathcal{A}_\rho \equiv \max_{\eta,t} \left\{ \eta^T \alpha^T(y)\alpha(y)\eta + 2t\beta^T(y)\alpha(y)\eta : \eta^T Q_j \eta \le \rho^2, \, t^2 \le 1 \right\}$$
$$\le \gamma + \rho^2 \sum_{j=1}^{J} \gamma_j. \qquad (7.2.5)$$

Indeed, let $F = \{(\eta, t) : \eta^T Q_j \eta \le \rho^2, j = 1, ..., J, t^2 \le 1\}$. We have

$$\mathcal{A}_\rho \;=\; \max_{(\eta,t) \in F} \left\{ \eta^T \alpha^T(y)\alpha(y)\eta + 2t\beta^T(y)\alpha(y)\eta \right\}$$
$$\le\; \max_{(\eta,t) \in F} \left\{ \gamma t^2 + \sum_{j=1}^{J} \gamma_j \eta^T Q_j \eta \right\}$$

[since the quadratic form (7.2.4) is nonnegative everywhere]

$$\le\; \gamma + \rho^2 \sum_{j=1}^{J} \gamma_j$$

[due to the origin of F and to $\gamma \ge 0$, $\gamma_j \ge 0$].

From (!) it follows that if $\gamma \ge 0$, $\gamma_j \ge 0$, $j = 1, ..., J$ are such that the quadratic form (7.2.4) is nonnegative everywhere, or, which is the same, such that

$$\left[\begin{array}{c|c} \gamma & -\beta^T(y)\alpha(y) \\ \hline -\alpha^T(y)\beta(y) & \sum_{j=1}^{J} \gamma_j Q_j - \alpha^T(y)\alpha(y) \end{array} \right] \succeq 0$$

and

$$\gamma + \rho^2 \sum_{j=1}^{J} \gamma_j \le \tau^2 - \beta^T(y)\beta(y),$$

then (y, τ) satisfies (7.2.3.a). Setting $\nu = \gamma + \beta^T(y)\beta(y)$, we can rewrite this conclusion as follows: if there exist ν and $\gamma_j \ge 0$ such that

$$\left[\begin{array}{c|c} \nu - \beta^T(y)\beta(y) & -\beta^T(y)\alpha(y) \\ \hline -\alpha^T(y)\beta(y) & \sum_{j=1}^{J} \gamma_j Q_j - \alpha^T(y)\alpha(y) \end{array} \right] \succeq 0$$

and

$$\nu + \rho^2 \sum_{j=1}^{J} \gamma_j \le \tau^2,$$

then (y, τ) satisfies (7.2.3.a).

Assume for a moment that $\tau > 0$. Setting $\lambda_j = \gamma_j/\tau$, $\mu = \nu/\tau$, the above conclusion can be rewritten as follows: if there exist μ and $\lambda_j \geq 0$ such that

$$\left[\begin{array}{c|c} \mu - \tau^{-1}\beta^T(y)\beta(y) & -\tau^{-1}\beta^T(y)\alpha(y) \\ \hline -\tau^{-1}\alpha^T(y)\beta(y) & \sum\limits_{j=1}^{J} \lambda_j Q_j - \tau^{-1}\alpha^T(y)\alpha(y) \end{array}\right] \succeq 0$$

and

$$\mu + \rho^2 \sum_{j=1}^{J} \lambda_j \leq \tau,$$

then (y, τ) satisfies $(7.2.3.a)$.

By the Schur Complement Lemma, the latter conclusion can further be reformulated as follows: if $\tau > 0$ and there exist μ, λ_j satisfying the relations

$$(a) \quad \left[\begin{array}{c|c|c} \mu & & \beta^T(y) \\ \hline & \sum\limits_{j=1}^{J} \lambda_j Q_j & \alpha^T(y) \\ \hline \beta(y) & \alpha(y) & \tau I \end{array}\right] \succeq 0 \tag{7.2.6}$$

$$(b) \quad \mu + \rho^2 \sum_{j=1}^{J} \lambda_j \leq \tau \quad (c)\ \lambda_j \geq 0,\ j = 1, ..., J$$

then (y, τ) satisfies $(7.2.3.a)$. Note that in fact our conclusion is valid for $\tau \leq 0$ as well. Indeed, assume that $\tau \leq 0$ and μ, λ_j solve $(7.2.6)$. Then clearly $\tau = 0$ and therefore $\alpha(y) = 0$, $\beta(y) = 0$, and thus $(7.2.3.a)$ is valid. We have proved the first part of the following statement:

Theorem 7.2.1. The explicit system of constraints $(7.2.6)$ in variables $y, \tau, \mu,$ $\lambda_1, ..., \lambda_J$ is a safe tractable approximation of the RC of the uncertain Least Squares constraint $(7.1.3)$ with $\cap$-ellipsoidal perturbation set $(7.2.1)$. The approximation is exact when $J = 1$, and in the case of $J > 1$ the tightness factor of this approximation does not exceed

$$\Omega(J) \leq 9.19\sqrt{\ln(J)}. \tag{7.2.7}$$

Proof. The fact that $(7.2.6)$ is a safe approximation of the RC of $(7.1.3)$, $(7.2.1)$ is readily given by the reasoning preceding Theorem 7.2.1. To prove that the approximation is tight within the announced factor, observe that the Approximate $\mathcal{S}$-Lemma (see Appendix B.3) as applied to the quadratic forms in variables $x = [\eta; t]$

$$x^T A x \equiv \left\{\eta^T \alpha^T(y)\alpha(y)\eta + 2t\beta^T(y)\alpha(y)\eta\right\}, \quad x^T B x \equiv t^2,$$

$$x^T B_j x \equiv \eta^T Q_j \eta,\ 1 \leq j \leq J,$$

states that if $J = 1$, then (y, τ) can be extended to a solution of $(7.2.6)$ if and only if (y, τ) satisfies $(7.2.3)$, that is, if and only if (y, τ) is robust feasible; thus, our approximation of the RC of $(7.1.3)$, $(7.2.1)$ is exact when $J = 1$. Now let $J > 1$, and suppose that (y, τ) cannot be extended to a feasible solution of $(7.2.6)$. Due to the origin of this system, it follows that

$$\text{SDP}(\rho) \equiv \min_{\lambda,\{\lambda_j\}} \left\{ \lambda + \rho^2 \sum_{j=1}^{J} \lambda_j : \lambda B + \sum_j \lambda_j B_j \succeq A, \lambda \geq 0, \lambda_j \geq 0 \right\}$$
$$> \tau^2 - \beta^T(y)\beta(y). \tag{7.2.8}$$

By the Approximate $\mathcal{S}$-Lemma (Appendix B.3), with appropriately chosen $\Omega(J) \leq 9.19\sqrt{\ln(J)}$ we have $\mathcal{A}_{\Omega(J)\rho} \geq \text{SDP}(\rho)$, which combines with (7.2.8) to imply that $\mathcal{A}_{\Omega(J)\rho} > \tau^2 - \beta^T(y)\beta(y)$, meaning that (y, τ) is not robust feasible at the uncertainty level $\Omega(J)\rho$ (cf. (7.2.3)). Thus, the tightness factor of our approximation does not exceed $\Omega(J)$. $\qquad\square$

7.2.2 From Uncertain Least Squares to Uncertain CQI

The next statement can obtained from Theorem 7.2.1 in the same fashion as Theorem 7.1.7 has been derived from Theorem 7.1.2.

Theorem 7.2.2. Consider uncertain CQI (6.1.3) with side-wise uncertainty, where the left hand side perturbation set is the $\cap$-ellipsoidal set (7.2.1), and the right hand side perturbation set is as in Theorem 7.1.7. For $\rho > 0$, the explicit system of LMIs

$$(a) \quad \tau + \rho p^T v \leq \delta(y), \; P^T v = \sigma(y), \; Q^T v = 0, \; v \in \mathbf{K}_*$$

$$(b.1) \quad \left[\begin{array}{c|cc} \mu & & \beta^T(y) \\ \hline & \sum_{j=1}^{J} \lambda_j Q_j & \alpha^T(y) \\ \hline \beta(y) & \alpha(y) & I \end{array} \right] \succeq 0 \tag{7.2.9}$$

$$(b.2) \quad \mu + \rho^2 \sum_{j=1}^{J} \lambda_j \leq \tau, \; \lambda_j \succeq 0 \, \forall j$$

in variables $y, v, \mu, \lambda_j, \tau$ is a safe tractable approximation of the RC of the uncertain CQI. This approximation is exact when $J = 1$ and is tight within the factor $\Omega(J) \leq 9.19\sqrt{\ln(J)}$ when $J > 1$.

7.2.3 Convex Quadratic Constraint with $\cap$-Ellipsoidal Uncertainty

Now consider approximating the RC of an uncertain convex quadratic inequality

$$y^T A^T(\zeta) A(\zeta) y \leq 2y^T b(\zeta) + c(\zeta)$$
$$\left[(A(\zeta), b(\zeta), c(\zeta)) = (A^n, b^n, c^n) + \sum_{\ell=1}^{L} \zeta_\ell (A^\ell, b^\ell, c^\ell) \right] \tag{7.2.10}$$

with $\cap$-ellipsoidal uncertainty:

$$\mathcal{Z}_\rho = \rho \mathcal{Z}_1 = \{\zeta \in \mathbb{R}^L : \zeta^T Q_j \zeta \leq \rho^2\} \quad [Q_j \succeq 0, \sum_j Q_j \succ 0] \tag{7.2.11}$$

Observe that

$$A(\zeta)y = \alpha(y)\zeta + \beta(y),$$
$$\alpha(y)\zeta = [A^1 y, ..., A^L y], \ \beta(y) = A^{\mathrm{n}} y$$

$$2y^T b(\zeta) + c(\zeta) = 2\sigma^T(y)\zeta + \delta(y),$$
$$\sigma(y) = [y^T b^1 + c^1; ...; y^T b^L + c^L], \ \delta(y) = y^T b^{\mathrm{n}} + c^{\mathrm{n}}$$
$$(7.2.12)$$

so that the RC of (7.2.10), (7.2.11) is the semi-infinite inequality

$$\zeta^T \alpha^T(y)\alpha(y)\zeta + 2\zeta^T \left[\alpha^T(y)\beta(y) - \sigma(y)\right] \leq \delta(y) - \beta^T(y)\beta(y) \ \forall \zeta \in \mathcal{Z}_\rho,$$

or, which is the same, the semi-infinite inequality

$$\mathcal{A}_\rho(y) \equiv \max_{\zeta \in \mathcal{Z}_\rho, t^2 \leq 1} \zeta^T \alpha^T(y)\alpha(y)\zeta + 2t\zeta^T \left[\alpha^T(y)\beta(y) - \sigma(y)\right]$$
$$\leq \delta(y) - \beta^T(y)\beta(y).$$
$$(7.2.13)$$

Same as in section 7.2.1, we have

$$\mathcal{A}_\rho(y) \leq \inf_{\lambda, \{\lambda_j\}} \left\{ \lambda + \rho^2 \sum_{j=1}^{J} \lambda_j : \begin{array}{l} \lambda \geq 0, \lambda_j \geq 0, j = 1, ..., J \\ \forall (t, \zeta) : \\ \lambda t^2 + \zeta^T (\sum_{j=1}^{J} \lambda_j Q_j)\zeta \geq \zeta^T \alpha^T(y)\alpha(y)\zeta \\ \qquad\qquad + 2t\zeta^T \left[\alpha^T(y)\beta(y) - \sigma(y)\right] \end{array} \right\}$$

$$= \inf_{\lambda, \{\lambda_j\}} \left\{ \lambda + \rho^2 \sum_{j=1}^{J} \lambda_j : \lambda \geq 0, \lambda_j \geq 0, j = 1, ..., J, \right.$$
$$\left. \left[\begin{array}{c|c} \lambda & -[\beta^T(y)\alpha(y) - \sigma^T(y)] \\ \hline -[\alpha^T(y)\beta(y) - \sigma(y)] & \sum_j \lambda_j Q_j - \alpha^T(y)\alpha(y) \end{array} \right] \succeq 0 \right\}.$$
$$(7.2.14)$$

We conclude that the condition

$$\exists(\lambda \geq 0, \{\lambda_j \geq 0\}) :$$
$$\left\{ \begin{array}{l} \lambda + \rho^2 \sum_{j=1}^{J} \lambda_j \leq \delta(y) - \beta^T(y)\beta(y) \\[2mm] \left[\begin{array}{c|c} \lambda & -[\beta^T(y)\alpha(y) - \sigma^T(y)] \\ \hline -[\alpha^T(y)\beta(y) - \sigma(y)] & \sum_j \lambda_j Q_j - \alpha^T(y)\alpha(y) \end{array} \right] \succeq 0 \end{array} \right.$$

is sufficient for y to be robust feasible. Setting $\mu = \lambda + \beta^T(y)\beta(y)$, this sufficient condition can be rewritten equivalently as

$$\exists(\{\lambda_j \geq 0\}, \mu) : \left\{ \begin{array}{l} \mu + \rho^2 \sum_{j=1}^{J} \lambda_j \leq \delta(y) \\[2mm] \left[\begin{array}{c|c} \mu - \beta^T(y)\beta(y) & -[\beta^T(y)\alpha(y) - \sigma^T(y)] \\ \hline -[\alpha^T(y)\beta(y) - \sigma(y)] & \sum_j \lambda_j Q_j - \alpha^T(y)\alpha(y) \end{array} \right] \succeq 0 \end{array} \right.$$
$$(7.2.15)$$

We have

$$
\begin{bmatrix}
\mu - \beta^T(y)\beta(y) & -[\beta^T(y)\alpha(y) - \sigma^T(y)] \\
-[\alpha^T(y)\beta(y) - \sigma(y)] & \sum_j \lambda_j Q_j - \alpha^T(y)\alpha(y)
\end{bmatrix}
$$
$$
=
\begin{bmatrix}
\mu & \sigma^T(y) \\
\sigma(y) & \sum_{j=1}^{J} \lambda_j Q_j
\end{bmatrix}
-
\begin{bmatrix}
\beta^T(y) \\
\alpha^T(y)
\end{bmatrix}
\begin{bmatrix}
\beta^T(y) \\
\alpha^T(y)
\end{bmatrix}^T ,
$$

so that the Schur Complement Lemma says that

$$
\begin{bmatrix}
\mu - \beta^T(y)\beta(y) & -[\beta^T(y)\alpha(y) - \sigma^T(y)] \\
-[\alpha^T(y)\beta(y) - \sigma(y)] & \sum_j \lambda_j Q_j - \alpha^T(y)\alpha(y)
\end{bmatrix} \succeq 0
$$
$$
\Leftrightarrow
\begin{bmatrix}
\mu & \sigma^T(y)] & \beta^T(y) \\
\sigma(y) & \sum_j \lambda_j Q_j & \alpha^T(y) \\
\beta(y) & \alpha(y) & I
\end{bmatrix} \succeq 0.
$$

The latter observation combines with the fact that (7.2.15) is a sufficient condition for the robust feasibility of y to yield the first part of the following statement:

Theorem 7.2.3. The explicit system of LMIs in variables y, μ, λ_j:

$$
(a) \quad
\begin{bmatrix}
\mu & \sigma^T(y)] & \beta^T(y) \\
\sigma(y) & \sum_j \lambda_j Q_j & \alpha^T(y) \\
\beta(y) & \alpha(y) & I
\end{bmatrix} \succeq 0
\tag{7.2.16}
$$
$$
(b) \quad \mu + \rho^2 \sum_{j=1}^{J} \lambda_j \leq \delta(y) \quad (c) \; \lambda_j \geq 0, \; j = 1, ..., J
$$

(for notation, see (7.2.12)) is a safe tractable approximation of the RC of (7.2.10), (7.2.11). The tightness factor of this approximation equals 1 when $J = 1$ and does not exceed $\Omega(J) \leq 9.19\sqrt{\ln(J)}$ when $J > 1$.

The proof of this theorem is completely similar to the proof of Theorem 7.2.1.

Remark 7.2.4. The tightness factor $\Omega(J) = O(\sqrt{\ln(J)})$ of the approximate RCs we have built in the case of $\cap$-ellipsoidal uncertainty (Theorems 7.2.1, 7.2.2, 7.2.3)) is not an absolute constant, as it was in the case of structured norm-bounded uncertainty, but grows, although very slowly, with the number J of ellipsoids participating in the description of the perturbation set. Of course, for all practical purposes, $\sqrt{\ln J}$ is a moderate constant, and what should be of primary importance, is the absolute constant factor hidden in the above $O(\cdot)$. As stated in the Theorems, this factor (≈ 9.2) is rather big. In fact the precise values of $\Omega(J)$ as given by the proof of Approximate $\mathcal{S}$-Lemma (Appendix B.3) are not that disastrous:

J	2	8	32	128	512	2048	8192	32678	131072
$\Omega(J)$	7.65	9.26	10.58	11.72	12.75	13.69	14.56	15.37	16.14
$\frac{\Omega(J)}{\sqrt{\ln(J)}}$	9.19	6.42	5.68	5.32	5.10	4.96	4.85	4.77	4.70

It should be added that there exists a slightly different proof of the Approximate $\mathcal{S}$-Lemma [11] that guarantees that the tightness factor does not exceed

$$\Omega = \sqrt{2\ln(6\sum_j \text{Rank}(Q_j))}.$$

Academically speaking, this bound is worse than $\Omega \leq O(\sqrt{\ln(J)})$ we have used — the total rank of the matrices Q_j can be much larger than the number J of these matrices. However, the better absolute constants in the "bad" bound imply, e.g., that the tightness of the approximation in question is at most 6, provided that the total rank of all matrices Q_j is $\leq 65,000,000$, which, for all practical purposes, is the same as to say that the tightness factor of our approximation "never" exceeds 6.

7.3 EXERCISES

Exercise 7.1. Consider an uncertain Least Squares inequality

$$\|A(\eta)x + b(\eta)\|_2 \leq \tau, \ \eta \in \rho\mathcal{Z}$$

where $\mathcal{Z}$, $0 \in \text{int}\,\mathcal{Z}$, is a symmetric w.r.t. the origin convex compact set that is the intersection of $J > 1$ ellipsoids not necessarily centered at the origin:

$$\mathcal{Z} = \{\eta : (\eta - a_j)^T Q_j(\eta - a_j) \leq 1, \ 1 \leq j \leq J\} \qquad [Q_j \succeq 0, \textstyle\sum_j Q_j \succ 0]$$

Prove that the RC of the uncertain inequality in question admits a safe tractable approximation tight within the factor $O(1)\sqrt{\ln J}$ (cf. Theorem 7.2.1).

7.4 NOTES AND REMARKS

NR 7.1. The Matrix Cube Theorem underlying Theorems 7.1.2, 7.1.4, 7.1.8 originates from [10], where only scalar perturbation blocks were considered; the more advanced version of this theorem used in the main body of the chapter is due to [12]. The Approximate $\mathcal{S}$-Lemma (Lemma B.3) underlying Theorems 7.2.1, 7.2.2, 7.2.3, in a slightly weaker form (with the total rank of the matrices Q_j instead of the number of these matrices in the bound for the tightness factor), was proved in [11]; the main ingredients of the proof go back to [81].

Chapter Eight
Uncertain Semidefinite Problems with Tractable RCs

In this chapter, we focus on uncertain Semidefinite Optimization (SDO) problems for which tractable Robust Counterparts can be derived.

8.1 UNCERTAIN SEMIDEFINITE PROBLEMS

Recall that a *semidefinite program* (SDP) is a conic optimization program

$$\min_x \left\{ c^T x + d : \mathcal{A}_i(x) \equiv \sum_{j=1}^n x_j A^{ij} - B_i \in \mathbf{S}_+^{k_i}, \ i = 1, ..., m \right\}$$

$$\Updownarrow \qquad\qquad (8.1.1)$$

$$\min_x \left\{ c^T x + d : \mathcal{A}_i(x) \equiv \sum_{j=1}^n x_j A^{ij} - B_i \succeq 0, \ i = 1, ..., m \right\}$$

where A^{ij}, B_i are symmetric matrices of sizes $k_i \times k_i$, $\mathbf{S}_+^k$ is the cone of real symmetric positive semidefinite $k \times k$ matrices, and $A \succeq B$ means that A, B are symmetric matrices of the same sizes such that the matrix $A - B$ is positive semidefinite. A constraint of the form $\mathcal{A}x - B \equiv \sum_j x_j A^j - B \succeq 0$ with symmetric A^j, B is called a *Linear Matrix Inequality* (LMI); thus, an SDP is the problem of minimizing a linear objective under finitely many LMI constraints. Another, sometimes more convenient, setting of a semidefinite program is in the form of (5.1.2), that is,

$$\min_x \left\{ c^T x + d : A_i x - b_i \in \mathbf{Q}_i, \ i = 1, ..., m \right\}, \qquad (8.1.2)$$

where nonempty sets $\mathbf{Q}_i$ are given by explicit finite lists of LMIs:

$$\mathbf{Q}_i = \{ u \in \mathbb{R}^{p_i} : \mathcal{Q}_{i\ell}(u) \equiv \sum_{s=1}^{p_i} u_s Q^{si\ell} - Q^{i\ell} \succeq 0, \ \ell = 1, ..., L_i \}.$$

Note that (8.1.1) is a particular case of (8.1.2) where $\mathbf{Q}_i = \mathbf{S}_+^{k_i}$, $i = 1, ..., m$.

The notions of the *data* of a semidefinite program, of an *uncertain* semidefinite problem and of its (exact or approximate) *Robust Counterparts* are readily given by specializing the general descriptions from sections 5.1, 5.3, to the case when the underlying cones are the cones of positive semidefinite matrices. In particular,

- The *natural data* of a semidefinite program (8.1.2) is the collection

$$(c, d, \{A_i, b_i\}_{i=1}^m),$$

while the right hand side sets $\mathbf{Q}_i$ are treated as the problem's structure;

• An *uncertain* semidefinite problem is a collection of problems (8.1.2) with common structure and natural data running through an *uncertainty set*; we always assume that the data are affinely parameterized by *perturbation vector* $\zeta \in \mathbb{R}^L$ running through a given closed and convex *perturbation set* $\mathcal{Z}$ such that $0 \in \mathcal{Z}$:

$$
\begin{aligned}
[c; d] &= [c^{\mathrm{n}}; d^{\mathrm{n}}] + \sum_{\ell=1}^{L} \zeta_\ell [c^\ell; d^\ell]; \\
[A_i, b_i] &= [A_i^{\mathrm{n}}, b_i^{\mathrm{n}}] + \sum_{\ell=1}^{L} \zeta_\ell [A_i^\ell, b_i^\ell], \ i = 1, ..., m
\end{aligned}
\tag{8.1.3}
$$

• The Robust Counterpart of uncertain SDP (8.1.2), (8.1.3) at a perturbation level $\rho > 0$ is the semi-infinite optimization program

$$
\min_{y=(x,t)} \left\{ t : \begin{array}{l} [[c^{\mathrm{n}}]^T x + d^{\mathrm{n}}] + \sum_{\ell=1}^{L} \zeta_\ell [[c^\ell]^T x + d^\ell] \leq t \\[2mm] [A_i^{\mathrm{n}} x + b_i^{\mathrm{n}}] + \sum_{\ell=1}^{L} \zeta_\ell [A_i^\ell x + b_i^\ell] \in \mathbf{Q}_i, \ i = 1, ..., m \end{array} \right\} \forall \zeta \in \rho \mathcal{Z} \right\} \tag{8.1.4}
$$

• A *safe tractable approximation* of the RC of uncertain SDP (8.1.2), (8.1.3) is a finite system $\mathcal{S}_\rho$ of explicitly computable convex constraints in variables $y = (x, t)$ (and possibly additional variables u) depending on $\rho > 0$ as a parameter, such that the projection $\widehat{Y}_\rho$ of the solution set of the system onto the space of y variables is contained in the feasible set Y_ρ of (8.1.4). Such an approximation is called *tight* within factor $\vartheta \geq 1$, if $Y_\rho \supset \widehat{Y}_\rho \supset Y_{\vartheta\rho}$. In other words, $\mathcal{S}_\rho$ is a ϑ-tight safe approximation of (8.1.4), if:

i) Whenever $\rho > 0$ and y are such that y can be extended, by a properly chosen u, to a solution of $\mathcal{S}_\rho$, y is robust feasible at the uncertainty level ρ, (i.e., y is feasible for (8.1.4)).

ii) Whenever $\rho > 0$ and y are such that y can<u>not</u> be extended to a feasible solution to $\mathcal{S}_\rho$, y is <u>not</u> robust feasible at the uncertainty level $\vartheta\rho$, (i.e., y violates some of the constraints in (8.1.4) when ρ is replaced with $\vartheta\rho$).

8.2 TRACTABILITY OF RCS OF UNCERTAIN SEMIDEFINITE PROBLEMS

Building the RC of an uncertain semidefinite problem reduces to building the RCs of the uncertain constraints constituting the problem, so that the tractability issues in Robust Semidefinite Optimization reduce to those for the Robust Counterpart

$$
\mathcal{A}_\zeta(y) \equiv \mathcal{A}^{\mathrm{n}}(y) + \sum_{\ell=1}^{L} \zeta_\ell \mathcal{A}_\ell(y) \succeq 0 \ \forall \zeta \in \rho \mathcal{Z} \tag{8.2.1}
$$

of a single uncertain LMI

$$
\mathcal{A}_\zeta(y) \equiv \mathcal{A}^{\mathrm{n}}(y) + \sum_{\ell=1}^{L} \zeta_\ell \mathcal{A}_\ell(y) \succeq 0; \tag{8.2.2}
$$

here $\mathcal{A}^n(x)$, $\mathcal{A}_\ell(x)$ are symmetric matrices affinely depending on the design vector y.

More often than not the RC of an uncertain LMI is computationally intractable. Indeed, we saw in chapter 5 that intractability is typical already for the RCs of uncertain conic quadratic inequalities, and the latter are very special cases of uncertain LMIs (due to the fact that Lorentz cones are cross-sections of semidefinite cones, see Lemma 6.3.3). In the relatively simple case of uncertain CQIs, we met just 3 generic cases where the RCs were computationally tractable, specifically, the cases of

   i) Scenario perturbation set (section 6.1);

  ii) Unstructured norm-bounded uncertainty (section 6.3);

 iii) Simple ellipsoidal uncertainty (section 6.5).

The RC associated with a scenario perturbation set is tractable for an arbitrary uncertain conic problem on a tractable cone; in particular, the RC of an uncertain LMI with scenario perturbation set is computationally tractable. Specifically, if $\mathcal{Z}$ in (8.2.1) is given as $\mathrm{Conv}\{\zeta^1,...,\zeta^N\}$, then the RC (8.2.1) is nothing but the explicit system of LMIs

$$\mathcal{A}^n(y) + \sum_{\ell=1}^{L} \zeta_\ell^i \mathcal{A}_\ell(y) \succeq 0,\ i = 1, ..., N. \tag{8.2.3}$$

The fact that the simple ellipsoidal uncertainty ($\mathcal{Z}$ is an ellipsoid) results in a tractable RC is specific for Conic Quadratic Optimization. In the LMI case, (8.2.1) can be NP-hard even with an ellipsoid in the role of $\mathcal{Z}$. In contrast to this, the case of unstructured norm-bounded perturbations remains tractable in the LMI situation. This is the only nontrivial tractable case we know. We are about to consider this case in full details.

8.2.1 Unstructured Norm-Bounded Perturbations

Definition 8.2.1. We say that uncertain LMI (8.2.2) is with unstructured norm-bounded perturbations, if

   i) The perturbation set $\mathcal{Z}$ (see (8.1.3)) is the set of all $p \times q$ matrices ζ with the usual matrix norm $\|\cdot\|_{2,2}$ not exceeding 1;

  ii) "The body" $\mathcal{A}_\zeta(y)$ of (8.2.2) can be represented as

$$\mathcal{A}_\zeta(y) \equiv \mathcal{A}^n(y) + \left[L^T(y)\zeta R(y) + R^T(y)\zeta^T L(y)\right], \tag{8.2.4}$$

where both $L(\cdot)$, $R(\cdot)$ are affine and at least one of these matrix-valued functions is in fact independent of y.

Example 8.2.2. Consider the situation where $\mathcal{Z}$ is the unit Euclidean ball in $\mathbb{R}^L$ (or, which is the same, the set of $L \times 1$ matrices of $\| \cdot \|_{2,2}$-norm not exceeding 1), and

$$\mathcal{A}_{\zeta}(y) = \left[\begin{array}{c|c} a(y) & \zeta^T B^T(y) + b^T(y) \\ \hline B(y)\zeta + b(y) & A(y) \end{array} \right], \tag{8.2.5}$$

where $a(\cdot)$ is an affine scalar function, and $b(\cdot)$, $B(\cdot)$, $A(\cdot)$ are affine vector- and matrix-valued functions with $A(\cdot) \in \mathbf{S}^M$. Setting $R(y) \equiv R = [1, 0_{1 \times M}]$, $L(y) = [0_{L \times 1}, B^T(y)]$, we have

$$\mathcal{A}_{\zeta}(y) = \underbrace{\left[\begin{array}{c|c} a(y) & b^T(y) \\ \hline b(y) & A(y) \end{array} \right]}_{\mathcal{A}^{\mathbf{n}}(y)} + L^T(y)\zeta R(y) + R^T(y)\zeta^T L(y),$$

thus, we are in the case of an unstructured norm-bounded uncertainty.

A closely related example is given by the LMI reformulation of an uncertain Least Squares inequality with unstructured norm-bounded uncertainty, see section 6.3.

Let us derive a tractable reformulation of an uncertain LMI with unstructured norm-bounded uncertainty. W.l.o.g. we may assume that $R(y) \equiv R$ is independent of y (otherwise we can swap ζ and ζ^T, swapping simultaneously L and R) and that $R \neq 0$. We have

y is robust feasible for (8.2.2), (8.2.4) at uncertainty level ρ

$$\Leftrightarrow \quad \xi^T[\mathcal{A}^{\mathbf{n}}(y) + L^T(y)\zeta R + R^T\zeta^T L(y)]\xi \geq 0 \quad \forall \xi \ \ \forall(\zeta : \|\zeta\|_{2,2} \leq \rho)$$

$$\Leftrightarrow \quad \xi^T \mathcal{A}^{\mathbf{n}}(y)\xi + 2\xi^T L^T(y)\zeta R\xi \geq 0 \quad \forall \xi \ \ \forall(\zeta : \|\zeta\|_{2,2} \leq \rho)$$

$$\Leftrightarrow \quad \xi^T \mathcal{A}^{\mathbf{n}}(y)\xi + 2\underbrace{\min_{\|\zeta\|_{2,2} \leq \rho} \xi^T L^T(y)\zeta R\xi}_{= -\rho\|L(y)\xi\|_2 \|R\xi\|_2} \geq 0 \quad \forall \xi$$

$$\Leftrightarrow \quad \xi^T \mathcal{A}^{\mathbf{n}}(y)\xi - 2\rho\|L(y)\xi\|_2 \|R\xi\|_2 \geq 0 \ \forall \xi$$

$$\Leftrightarrow \quad \xi^T \mathcal{A}^{\mathbf{n}}(y)\xi + 2\rho\eta^T L(y)\xi \geq 0 \quad \forall(\xi, \eta : \eta^T\eta \leq \xi^T R^T R\xi)$$

$$\Leftrightarrow \quad \exists \lambda \geq 0 : \left[\begin{array}{c|c} \rho L(y) \\ \hline \rho L^T(y) & \mathcal{A}^{\mathbf{n}}(y) \end{array} \right] \succeq \lambda \left[\begin{array}{c|c} -I_p \\ \hline & R^T R \end{array} \right] \quad [\mathcal{S}\text{-Lemma}]$$

$$\Leftrightarrow \quad \exists \lambda : \left[\begin{array}{c|c} \lambda I_p & \rho L(y) \\ \hline \rho L^T(y) & \mathcal{A}^{\mathbf{n}}(y) - \lambda R^T R \end{array} \right] \succeq 0.$$

We have proved the following statement:

Theorem 8.2.3. The RC

$$\mathcal{A}^{\mathbf{n}}(y) + L^T(y)\zeta R + R^T\zeta^T L(y) \succeq 0 \quad \forall(\zeta \in \mathbb{R}^{p \times q} : \|\zeta\|_{2,2} \leq \rho) \tag{8.2.6}$$

of uncertain LMI (8.2.2) with unstructured norm-bounded uncertainty (8.2.4) (where, w.l.o.g., we assume that $R \neq 0$) can be represented equivalently by the LMI

$$\left[\begin{array}{c|c} \lambda I_p & \rho L(y) \\ \hline \rho L^T(y) & \mathcal{A}^{\mathbf{n}}(y) - \lambda R^T R \end{array} \right] \succeq 0 \tag{8.2.7}$$

in variables y, λ.

8.2.2 Application: Robust Structural Design

8.2.2.1 Structural Design problem

Consider a "linearly elastic" mechanical system S that, mathematically, can be characterized by:

 i) A linear space $\mathbb{R}^M$ of *virtual displacements* of the system.

 ii) A symmetric positive semidefinite $M \times M$ matrix A, called the *stiffness matrix* of the system.

The potential energy capacitated by the system when its displacement from the equilibrium is v is
$$E = \frac{1}{2} v^T A v.$$

An external load applied to the system is given by a vector $f \in \mathbb{R}^M$. The associated *equilibrium displacement* v of the system solves the linear equation
$$A v = f.$$
If this equation has no solutions, the load destroys the system — no equilibrium exists; if the solution is not unique, so is the equilibrium displacement. Both these "bad phenomena" can occur only when A is not positive definite.

The *compliance* of the system under a load f is the potential energy capacitated by the system in the equilibrium displacement v associated with f, that is,
$$\mathrm{Compl}_f(A) = \frac{1}{2} v^T A v = \frac{1}{2} v^T f.$$
An equivalent way to define compliance is as follows. Given external load f, consider the concave quadratic form
$$f^T v - \frac{1}{2} v^T A v$$
on the space $\mathbb{R}^M$ of virtual displacements. It is easily seen that this form either is unbounded above, (which is the case when no equilibrium displacements exist), or attains its maximum. In the latter case, the compliance is nothing but the maximal value of the form:
$$\mathrm{Compl}_f(A) = \sup_{v \in \mathbb{R}^M} \left[f^T v - \frac{1}{2} v^T A v \right],$$
and the equilibrium displacements are exactly the maximizers of the form.

There are good reasons to treat the compliance as the measure of rigidity of the construction with respect to the corresponding load — the less the compliance, the higher the rigidity. A typical *Structural Design* problem is as follows:

Structural Design: *Given*

- *the space $\mathbb{R}^M$ of virtual displacements of the construction,*
- *the stiffness matrix $A = A(t)$ affinely depending on a vector t of design parameters restricted to reside in a given convex compact set $\mathcal{T} \subset \mathbb{R}^N$,*
- *a set $\mathcal{F} \subset \mathbb{R}^M$ of external loads,*

find a construction t_* that is as rigid as possible w.r.t. the "most dangerous" load from $\mathcal{F}$, that is,

$$t_* \in \operatorname*{Argmin}_{T \in \mathcal{T}} \left\{ \mathrm{Compl}_{\mathcal{F}}(t) \equiv \sup_{f \in \mathcal{F}} \mathrm{Compl}_f(A(t)) \right\}.$$

Next we present three examples of Structural Design.

Example 8.2.4. Truss Topology Design. A *truss* is a mechanical construction, like railroad bridge, electric mast, or the Eiffel Tower, comprised of thin elastic *bars* linked to each other at *nodes*. Some of the nodes are partially or completely fixed, so that their virtual displacements form proper subspaces in $\mathbb{R}^2$ (for planar constructions) or $\mathbb{R}^3$ (for spatial ones). An external load is a collection of external forces acting at the nodes. Under such a load, the nodes move slightly, thus causing elongations and compressions in the bars, until the construction achieves an equilibrium, where the tensions caused in the bars as a result of their deformations compensate the external forces. The compliance is the potential energy capacitated in the truss at the equilibrium as a result of deformations of the bars.

A mathematical model of the outlined situation is as follows.

• *Nodes and the space of virtual displacements.* Let $\mathcal{M}$ be the nodal set, that is, a finite set in $\mathbb{R}^d$ ($d = 2$ for planar and $d = 3$ for spatial trusses), and let $V_i \subset \mathbb{R}^d$ be the linear space of virtual displacements of node i. (This set is the entire $\mathbb{R}^d$ for non-supported nodes, is $\{0\}$ for fixed nodes and is something in-between these two extremes for partially fixed nodes.) The space $V = \mathbb{R}^M$ of virtual displacements of the truss is the direct product $V = V_1 \times ... \times V_m$ of the spaces of virtual displacements of the nodes, so that a virtual displacement of the truss is a collection of "physical" virtual displacements of the nodes.

Now, an external load applied to the truss can be thought of as a collection of external physical forces $f_i \in \mathbb{R}^d$ acting at nodes i from the nodal set. We lose nothing when assuming that $f_i \in V_i$ for all i, since the component of f_i orthogonal to V_i is fully compensated by the supports that make the directions from V_i the only possible displacements of node i. Thus, we can always assume that $f_i \in V_i$ for all i, which makes it possible to identify a load with a vector $f \in V$. Similarly, the collection of nodal reaction forces caused by elongations and compressions of the bars can be thought of as a vector from V.

• *Bars and the stiffness matrix.* Every bar j, $j = 1, ..., N$, in the truss links two nodes from the nodal set $\mathcal{M}$. Denoting by t_j the volume of the j-th bar, a simple analysis, (where one assumes that the nodal displacements are small and neglects all terms of order of squares of these displacements), demonstrates that the collection of the reaction forces caused by a nodal displacement $v \in V$ can be represented as $A(t)v$, where

$$A(t) = \sum_{j=1}^{N} t_j b_j b_j^T \qquad (8.2.8)$$

is the stiffness matrix of the truss. Here $b_j \in V$ is readily given by the characteristics of the material of the j-th bar and the "nominal," (i.e., in the unloaded truss), positions of the nodes linked by this bar.

In a typical Truss Topology Design (TTD) problem, one is given a *ground structure* — a set $\mathcal{M}$ of tentative nodes along with the corresponding spaces V_i of virtual displacements and the list $\mathcal{J}$ of N tentative bars, (i.e., a list of pairs of nodes that could be linked by bars), and the characteristics of the bar's material; these data determine, in particular, the vectors b_j. The design variables are the volumes t_j of the tentative bars. The design specifications always include the natural restrictions $t_j \geq 0$ and an upper bound w on $\sum_j t_j$, (which, essentially, is an upper bound on the total weight of the truss). Thus, $\mathcal{T}$ is always a subset of the standard simplex $\{t \in \mathbb{R}^N : t \geq 0, \sum_j t_j \leq w\}$. There could be other design specifications, like upper and lower bounds on the volumes of some bars. The scenario set $\mathcal{F}$ usually is either a singleton (*single-load TTD*) or a small collection of external loads (*multi-load TTD*). With this setup, one seeks for a design $t \in \mathcal{T}$, that results in the smallest possible worst case, i.e., maximal over the loads from $\mathcal{F}$ compliance.

When formulating a TTD problem, one usually starts with a dense nodal set and allows for all pair connections of the tentative nodes by bars. At an optimal solution to the associated TTD problem, usually a pretty small number of bars get positive volumes, so that the solution recovers not only the optimal bar sizing, but also the optimal topology of the construction.

Example 8.2.5. Free Material Optimization. In Free Material Optimization (FMO) one seeks to design a mechanical construction comprised of material continuously distributed over a given 2-D or 3-D domain Ω, and the mechanical properties of the material are allowed to vary from point to point. The ultimate goal of the design is to build a construction satisfying a number of constraints (most notably, an upper bound on the total weight) and most rigid w.r.t. loading scenarios from a given sample.

After finite element discretization, this (originally infinite-dimensional) optimization problem becomes a particular case of the aforementioned Structural Design problem where:

- the space $V = \mathbb{R}^M$ of virtual displacements is the space of "physical displacements" of the vertices of the finite element cells, so that a displacement $v \in V$ is a collection of displacements $v_i \in \mathbb{R}^d$ of the vertices ($d = 2$ for planar and $d = 3$ for spatial constructions). Same as in the TTD problem, displacements of some of the vertices can be restricted to reside in proper linear subspaces of $\mathbb{R}^d$;

- external loads are collections of physical forces applied at the vertices of the finite element cells; same as in the TTD case, these collections can be identified with vectors $f \in V$;

- the stiffness matrix is of the form

$$A(t) = \sum_{j=1}^{N} \sum_{s=1}^{S} b_{js} t_j b_{js}^T, \tag{8.2.9}$$

where N is the number of finite element cells and t_j is the *stiffness tensor* of the material in the j-th cell. This tensor can be identified with a $p \times p$ symmetric positive semidefinite matrix, where $p = 3$ for planar constructions and $p = 6$ for spatial ones. The number S and the $M \times p$ matrices b_{is} are readily given by the geometry of the finite element cells and the type of finite element discretization.

In a typical FMO problem, one is given the number of the finite element cells along with the matrices b_{ij} in (8.2.9), and a collection $\mathcal{F}$ of external loads of interest. The design vectors are collections $t = (t_1, ..., t_N)$ of positive semidefinite $p \times p$ matrices, and the design specifications always include the natural restrictions $t_j \succeq 0$ and an upper bound $\sum_j c_j \mathrm{Tr}(t_j) \leq w$, $c_j > 0$, on the total weighted trace of t_j; this bound reflects, essentially, an upper bound on the total weight of the construction. Along with these restrictions, the description of the feasible design set $\mathcal{T}$ can include other constraints, such as bounds on the spectra of t_j, (i.e., lower bounds on the minimal and upper bounds on the maximal eigenvalues of t_j). With this setup, one seeks for a design $t \in \mathcal{T}$ that results in the smallest worst case, (i.e., the maximal over the loads from $\mathcal{F}$) compliance.

The design yielded by FMO usually cannot be implemented "as it is" — in most cases, it would be either impossible, or too expensive to use a material with mechanical properties varying from point to point. The role of FMO is in providing an engineer with an "educated guess" of what the optimal construction could possibly be; given this guess, engineers produce something similar from composite materials, applying existing design tools that take into account finer design specifications, (which may include nonconvex ones), than those taken into consideration by the FMO design model.

Our third example, due to C. Roos, has nothing in common with mechanics — it is about design of electrical circuits. Mathematically, however, it is modeled as a Structural Design problem.

Example 8.2.6. Consider an electrical circuit comprised of resistances and sources of current. Mathematically, such a circuit can be thought of as a graph with nodes $1, ..., n$ and a set E of oriented arcs. Every arc γ is assigned with its *conductance* $\sigma_\gamma \geq 0$ (so that $1/\sigma_\gamma$ is the resistance of the arc). The nodes are equipped with external sources of current, so every node i is assigned with a real number f_i — the current supplied by the source. The steady state functioning of the circuit is characterized by currents J_γ in the arcs and potentials v_i at the nodes, (these potentials are defined up to a common additive constant). The potentials and the currents can be found from the Kirchhoff laws, specifically, as follows. Let G be the node-arc incidence matrix, so that the columns in G are indexed by the nodes, the rows are indexed by the arcs, and $G_{\gamma i}$ is 1, -1 or 0, depending on whether the arc γ starts at node i, ends at this node, or is not incident to the node, respectively. The first Kirchhoff law states that sum of all currents in the arcs leaving a given node minus the sum of all currents in the arcs entering the node is equal to the external current at the node. Mathematically, this law reads

$$G^T J = f,$$

where $f = (f_1, ..., f_n)$ and $J = \{J_\gamma\}_{\gamma \in E}$ are the vector of external currents and the vector of currents in the arcs, respectively. The second law states that the current in an arc γ is σ_γ times the arc voltage — the difference of potentials at the nodes linked by the arc. Mathematically, this law reads

$$J = \Sigma G v, \Sigma = \mathrm{Diag}\{\sigma_\gamma, \gamma \in E\}.$$

Thus, the potentials are given by the relation

$$G^T \Sigma G v = f.$$

Now, the heat H dissipated in the circuit is the sum, over the arcs, of the products of arc currents and arc voltages, that is,

$$H = \sum_\gamma \sigma_\gamma ((Gv)_\gamma)^2 = v^T G^T \Sigma G v.$$

In other words, the heat dissipated in the circuit, the external currents forming a vector f, is the maximum of the convex quadratic form

$$2v^T f - v^T G^T \Sigma G v$$

over all $v \in \mathbb{R}^n$, and the steady state potentials are exactly the maximizers of this quadratic form. In other words, the situation is as if we were speaking about a mechanical system with stiffness matrix $A(\sigma) = G^T \Sigma G$ affinely depending on the vector $\sigma \geq 0$ of arc conductances subject to external load f, with the steady-state potentials in the role of equilibrium displacements, and the dissipated heat in this state in the role of (twice) the compliance.

> It should be noted that the "stiffness matrix" in our present situation is degenerate — indeed, we clearly have $G\mathbf{1} = 0$, where $\mathbf{1}$ is the vector of ones, ("when the potentials of all nodes are equal, the currents in the arcs should be zero"), whence $A(\sigma)\mathbf{1} = 0$ as well. As a result, the necessary condition for the steady state to exist is $f^T \mathbf{1} = 0$, that is, the total sum of all external currents should be zero — a fact we could easily foresee. Whether this necessary condition is also sufficient depends on the topology of the circuit.

A straightforward "electrical" analogy of the Structural Design problem would be to build a circuit of a given topology, (i.e., to equip the arcs of a given graph with nonnegative conductances forming a design vector σ), satisfying specifications $\sigma \in \mathcal{S}$ in a way that minimizes the maximal steady-state dissipated heat, the maximum being taken over a given family $\mathcal{F}$ of vectors of external currents.

8.2.2.2 Structural Design as an uncertain Semidefinite problem

The aforementioned Structural Design problem can be easily posed as an SDP. The key element in the transformation of the problem is the following semidefinite representation of the compliance:

$$\mathrm{Compl}_f(A) \le \tau \Leftrightarrow \left[\begin{array}{c|c} 2\tau & f^T \\ \hline f & A \end{array}\right] \succeq 0. \qquad (8.2.10)$$

Indeed,

$$\mathrm{Compl}_f(A) \le \tau$$

$$\Leftrightarrow \quad f^T v - \tfrac{1}{2} v^T A v \ge \tau \quad \forall v \in \mathbb{R}^M$$

$$\Leftrightarrow \quad 2\tau s^2 - 2s f^T v + v^T A v \ge 0 \quad \forall([v,s] \in \mathbb{R}^{M+1})$$

$$\Leftrightarrow \quad \left[\begin{array}{c|c} 2\tau & -f^T \\ \hline -f & A \end{array}\right] \succeq 0$$

$$\Leftrightarrow \quad \left[\begin{array}{c|c} 2\tau & f^T \\ \hline f & A \end{array}\right] \succeq 0$$

where the last $\Leftrightarrow$ follows from the fact that

$$\left[\begin{array}{c|c} 2\tau & -f^T \\ \hline -f & A \end{array}\right] = \left[\begin{array}{c|c} 1 & \\ \hline & -I \end{array}\right]\left[\begin{array}{c|c} 2\tau & f^T \\ \hline f & A \end{array}\right]\left[\begin{array}{c|c} 1 & \\ \hline & -I \end{array}\right]^T.$$

Thus, the Structural Design problem can be posed as

$$\min_{\tau,t}\left\{\tau : \left[\begin{array}{c|c} 2\tau & f^T \\ \hline f & A \end{array}\right] \succeq 0 \quad \forall f \in \mathcal{F}, t \in \mathcal{T}\right\}. \qquad (8.2.11)$$

Assuming that the set $\mathcal{T}$ of feasible designs is LMI representable, problem (8.2.11) is nothing but the RC of the uncertain semidefinite problem

$$\min_{\tau,t}\left\{\tau : \left[\begin{array}{c|c} 2\tau & f^T \\ \hline f & A(t) \end{array}\right] \succeq 0, t \in \mathcal{T}\right\}, \qquad (8.2.12)$$

where the only uncertain data is the load f, and this data varies in a given set $\mathcal{F}$ (or, which is the same, in its closed convex hull $\mathrm{cl\,Conv}(\mathcal{F})$). Thus, in fact we are speaking about the RC of a *single-load* Structural Design problem, with the load in the role of uncertain data varying in the uncertainty set $\mathcal{U} = \mathrm{cl\,Conv}(\mathcal{F})$.

In actual design the set $\mathcal{F}$ of loads of interest is finite and usually quite small. For example, when designing a bridge for cars, an engineer is interested in a quite restricted family of scenarios, primarily in the load coming from many cars uniformly distributed along the bridge (this is, essentially, what happens in rush hours), and, perhaps, in a few other scenarios (like loads coming from a single heavy car in various positions). With finite $\mathcal{F} = \{f^1, ..., f^k\}$, we are in the situation of a scenario uncertainty, and the RC of (8.2.12) is the explicit semidefinite program

$$\min_{\tau,t}\left\{\tau : \left[\begin{array}{c|c} 2\tau & [f^i]^T \\ \hline f^i & A(t) \end{array}\right] \succeq 0, i = 1, ..., k, t \in \mathcal{T}\right\}.$$

Note, however, that in reality the would-be construction will be affected by small "occasional" loads (like side wind in the case of a bridge), and the construction should be stable with respect to these loads. It turns out, however, that the latter requirement is not necessarily satisfied by the "nominal" construction that takes into consideration only the loads of primary interest. As an instructive example, consider the design of a console.

Example 8.2.7. Figure 8.1.(c) represents optimal single-load design of a console with a 9×9 nodal grid on 2-D plane; nodes from the very left column are fixed, the remaining nodes are free, and the single scenario load is the unit force f acting down and applied at the mid-node of the very right column (see figure 8.1.(a)). We allow nearly all tentative bars (numbering 2,039), except for (clearly redundant) bars linking fixed nodes or long bars that pass through more than two nodes and thus can be split into shorter ones (figure 8.1.(b)). The set $\mathcal{T}$ of admissible designs is given solely by the weight restriction:

$$\mathcal{T} = \{t \in \mathbb{R}^{2039} : t \geq 0, \sum_{i=1}^{2039} t_i \leq 1\}$$

(compliance is homogeneous of order 1 w.r.t. t: $\mathrm{Compl}_f(\lambda t) = \lambda \mathrm{Compl}_f(t)$, $\lambda > 0$, so we can normalize the weight bound to be 1).

The compliance, in an appropriate scale, of the resulting nominally optimal truss (12 nodes, 24 bars) w.r.t. the scenario load f is 1.00. At the same time, the construction turns out to be highly unstable w.r.t. small "occasional" loads distributed along the 10 free nodes used by the nominal design. For example, the mean compliance of the nominal design w.r.t. a random load $h \sim \mathcal{N}(0, 10^{-9} I_{20})$ is 5.406 (5.4 times larger than the nominal compliance), while the "typical" norm $\|h\|_2$ of this random load is $10^{-4.5}\sqrt{20}$ — more than three orders of magnitude less than the norm $\|f\|_2 = 1$ of the scenario load. The compliance of the nominally optimal truss w.r.t. a "bad" load g that is 10^4 times smaller than f ($\|g\|_2 = 10^{-4}\|f\|_2$) is 27.6 — by factor 27 larger than the compliance w.r.t. f! Figure 8.1.(e) shows the deformation of the nominal design under the load $10^{-4}g$ (that is, the load that is 10^8 (!) times smaller than the scenario load). One can compare this deformation with the one under the load f (figure 8.1.(d)). Figure 8.1.(f) depicts shifts of the nodes under a sample of 100 random loads $h \sim \mathcal{N}(0, 10^{-16} I_{20})$ — loads of norm by 7 plus orders of magnitude less than $\|f\|_2 = 1$.

To prevent the optimal design from being crushed by a small load that is outside of the set $\mathcal{F}$ of loading scenarios, it makes sense to extend $\mathcal{F}$ to a more "massive" set, primarily by adding to $\mathcal{F}$ all loads of magnitude not exceeding a given "small" uncertainty level ρ. A challenge here is to decide where the small loads can be applied. In problems like TTD, it does not make sense to require the would-be construction to be capable of carrying small loads distributed along *all* nodes of the ground structure; indeed, not all of these nodes should be present in the final design, and of course there is no reason to bother about forces acting at non-existing nodes. The difficulty is that we do not know in advance which nodes will be present in the final design. One possibility to resolve this difficulty to some extent is to use a two-stage procedure as follows:

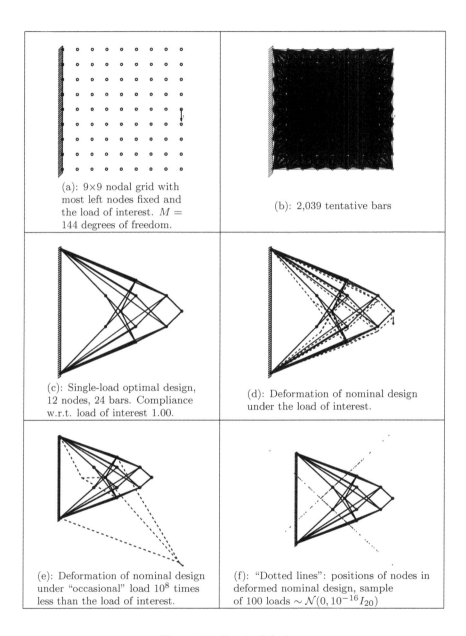

(a): 9×9 nodal grid with most left nodes fixed and the load of interest. $M = 144$ degrees of freedom.

(b): 2,039 tentative bars

(c): Single-load optimal design, 12 nodes, 24 bars. Compliance w.r.t. load of interest 1.00.

(d): Deformation of nominal design under the load of interest.

(e): Deformation of nominal design under "occasional" load 10^8 times less than the load of interest.

(f): "Dotted lines": positions of nodes in deformed nominal design, sample of 100 loads $\sim \mathcal{N}(0, 10^{-16} I_{20})$

Figure 8.1 Nominal design.

• at the first stage, we seek for the "nominal" design — the one that is optimal w.r.t. the "small" set $\mathcal{F}$ comprised of the scenario loads and, perhaps, all loads of magnitude $\leq \rho$ acting along the same nodes as the scenario loads — these nodes definitely will be present in the resulting design;

• at the second stage, we solve the problem again, with the nodes actually used by the nominal design in the role of our new nodal set $\mathcal{M}^+$, and extend $\mathcal{F}$ to the set $\mathcal{F}^+$ by taking the union of $\mathcal{F}$ and the Euclidean ball B_ρ of all loads g, $\|g\|_2 \leq \rho$, acting along $\mathcal{M}^+$.

We have arrived at the necessity to solve (8.2.11) in the situation where $\mathcal{F}$ is the union of a finite set $\{f^1, ..., f^k\}$ and a Euclidean ball. This is a particular case of the situation when $\mathcal{F}$ is the union of $S < \infty$ ellipsoids

$$E_s = \{f = f^s + B_s \zeta^s : \zeta^s \in \mathbb{R}^{k_s}, \|\zeta^s\|_2 \leq 1\}$$

or, which is the same, $\mathcal{Z}$ is the convex hull of the union of S ellipsoids $E_1, ..., E_S$. The associated "uncertainty-immunized" Structural Design problem (8.2.11) — the RC of (8.2.12) with $\mathcal{Z}$ in the role of $\mathcal{F}$ — is clearly equivalent to the problem

$$\min_{t,\tau} \left\{ \tau : \left[\begin{array}{c|c} 2\tau & f^T \\ \hline f & A(t) \end{array} \right] \succeq 0 \quad \forall f \in E_s, s = 1, ..., S; t \in \mathcal{T} \right\}. \tag{8.2.13}$$

In order to build a tractable equivalent of this semi-infinite semidefinite problem, we need to build a tractable equivalent to a semi-infinite LMI of the form

$$\left[\begin{array}{c|c} 2\tau & \zeta^T B^T + f^T \\ \hline B\zeta + f & A(t) \end{array} \right] \succeq 0 \quad \forall (\zeta \in \mathbb{R}^k : \|\zeta\|_2 \leq \rho). \tag{8.2.14}$$

But such an equivalent is readily given by Theorem 8.2.3 (cf. Example 8.2.2). Applying the recipe described in this Theorem, we end up with a representation of (8.2.14) as the following LMI in variables τ, t, λ:

$$\left[\begin{array}{c|c|c} \lambda I_k & & \rho B^T \\ \hline & 2\tau - \lambda & f^T \\ \hline \rho B & f & A(t) \end{array} \right] \succeq 0. \tag{8.2.15}$$

Observe that when $f = 0$, (8.2.15) simplifies to

$$\left[\begin{array}{c|c} 2\tau I_k & \rho B^T \\ \hline \rho B & A(t) \end{array} \right] \succeq 0. \tag{8.2.16}$$

Example 8.2.7 continued. Let us apply the outlined methodology to the Console example (Example 8.2.7). In order to immunize the design depicted on figure 8.1.(c) against small occasional loads, we start with reducing the initial 9×9 nodal set to the set of 12 nodes $\mathcal{M}^+$ (figure 8.2.(a)) used by the nominal design, and allow for $N = 54$ tentative bars on this reduced nodal set (figure 5.1.(b)) (we again allow for all pair connections of nodes, except for connections of two fixed nodes and for long bars passing through more than two nodes). According to the outlined methodology, we should then extend the original singleton $\mathcal{F} = \{f\}$ of scenario loads to the larger set $\mathcal{F}^+ = \{f\} \cup B_\rho$, where B_ρ is the Euclidean ball of radius ρ, centered at the origin in the ($M = 20$)-dimensional space of virtual displacements of the reduced planar nodal set. With this approach, an immediate question would be how to specify ρ. In order to avoid an ad hoc choice of ρ, we modify our approach as follows. Recalling that the compliance of the nominally optimal design w.r.t. the scenario load is 1.00, let us impose on our would-be "immunized" design the restriction that its worst case compliance w.r.t. the extended scenario set $\mathcal{F}_\rho = \{f\} \cup B_\rho$ should be at most $\tau_* = 1.025$, (i.e., 2.5% more than the optimal nominal compliance), and maximize under this restriction the radius ρ. In other words, we seek for a truss of the same unit weight as the nominally optimal one with "nearly optimal" rigidity w.r.t. the scenario load f and as large as possible worst-case rigidity w.r.t. occasional loads of a given magnitude. The resulting problem is the semi-infinite semidefinite program

$$\max_{t,\rho} \left\{ \rho : \begin{array}{l} \left[\begin{array}{c|c} 2\tau_* & f^T \\ \hline f & A(t) \end{array} \right] \succeq 0 \\ \left[\begin{array}{c|c} 2\tau_* & \rho h^T \\ \hline \rho h & A(t) \end{array} \right] \succeq 0 \quad \forall(h : \|h\|_2 \leq 1) \\ t \succeq 0, \sum_{i=1}^{N} t_i \leq 1 \end{array} \right\}.$$

This semi-infinite program is equivalent to the usual semidefinite program

$$\max_{t,\rho} \left\{ \rho : \begin{array}{l} \left[\begin{array}{c|c} 2\tau_* & f^T \\ \hline f & A(t) \end{array} \right] \succeq 0 \\ \left[\begin{array}{c|c} 2\tau_* I_M & \rho I_M \\ \hline \rho I_M & A(t) \end{array} \right] \succeq 0 \\ t \succeq 0, \sum_{i=1}^{N} t_i \leq 1 \end{array} \right\} \qquad (8.2.17)$$

(cf. (8.2.16)).

Computation shows that for Example 8.2.7, the optimal value in (8.2.17) is $\rho_* = 0.362$; the *robust design* yielded by the optimal solution to the problem is depicted in figure 8.2.(c). Along with the differences in sizing of bars, note the difference in the structures of the robust and the nominal design (figure 8.3). Observe that passing from the nominal to the robust design, we lose just 2.5% in the rigidity w.r.t. the scenario load and gain a dramatic improvement in the capability to carry occasional loads. Indeed, the compliance of the robust truss w.r.t. *every* load g of the magnitude $\|g\|_2 = 0.36$ (36% of the magnitude of the load of interest) is at most 1.025; the similar quantity for the nominal design is as large as 1.65×10^9 ! An additional evidence of the dramatic advantages of the robust design as compared to the nominal one can be obtained by comparing the pictures (d) through (f) in figure 8.1 with their counterparts in figure 8.2.

8.2.3 Applications in Robust Control

A major source of uncertain Semidefinite problems is Robust Control. An instructive example is given by Lyapunov Stability Analysis/Synthesis.

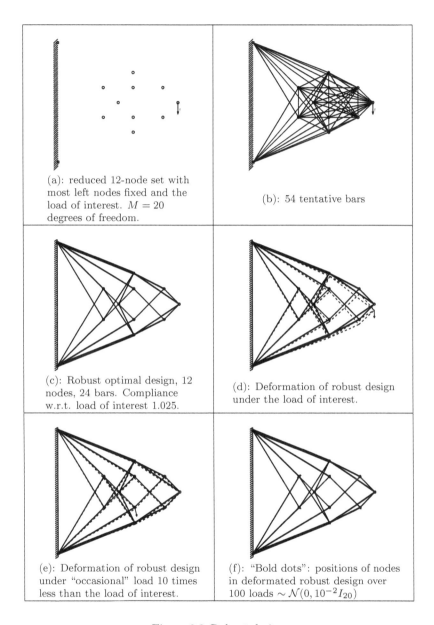

(a): reduced 12-node set with most left nodes fixed and the load of interest. $M = 20$ degrees of freedom.

(b): 54 tentative bars

(c): Robust optimal design, 12 nodes, 24 bars. Compliance w.r.t. load of interest 1.025.

(d): Deformation of robust design under the load of interest.

(e): Deformation of robust design under "occasional" load 10 times less than the load of interest.

(f): "Bold dots": positions of nodes in deformated robust design over 100 loads $\sim \mathcal{N}(0, 10^{-2}I_{20})$

Figure 8.2 Robust design.

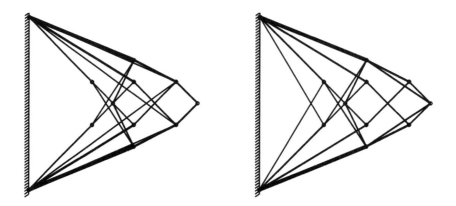

Figure 8.3 Nominal (left) and robust (right) designs.

8.2.3.1 Lyapunov Stability Analysis

Consider a time-varying linear dynamical system "closed" by a linear output-based feedback:

$$
\begin{array}{rlll}
(a) & \dot{x}(t) & = & A_t x(t) + B_t u(t) + R_t d_t \ [\text{open loop system, or } plant] \\
(b) & y(t) & = & C_t x(t) + D_t d_t \ [\text{output}] \\
(c) & u(t) & = & K_t y(t) \ [\text{output-based feedback}] \\
& & \Downarrow & \\
(d) & \dot{x}(t) & = & [A_t + B_t K_t C_t] x(t) + [R_t + B_t K_t D_t] d_t \ [\text{closed loop system}]
\end{array}
\tag{8.2.18}
$$

where $x(t) \in \mathbb{R}^n$, $u(t) \in \mathbb{R}^m$, $d_t \in \mathbb{R}^p$, $y(t) \in \mathbb{R}^q$ are respectively, the state, the control, the external disturbance, and the output at time t, A_t, B_t, R_t, C_t, D_t are matrices of appropriate sizes specifying the dynamics of the system; and K_t is the feedback matrix. We assume that the dynamical system in question is *uncertain*, meaning that we do not know the dependencies of the matrices $A_t,...,K_t$ on t; all we know is that the collection $M_t = (A_t, B_t, C_t, D_t, R_t, K_t)$ of all these matrices stays all the time within a given compact uncertainty set $\mathcal{M}$. For our further purposes, it makes sense to think that there exists an underlying time-invariant "nominal" system corresponding to known nominal values $A^n,...,K^n$ of the matrices $A_t, ..., K_t$, while the actual dynamics corresponds to the case when the matrices drift (perhaps, in a time-dependent fashion) around their nominal values.

An important desired property of a linear dynamical system is its *stability* — the fact that every state trajectory $x(t)$ of (every realization of) the closed loop system converges to 0 as $t \to \infty$, provided that the external disturbances d_t are identically zero. For a time-invariant linear system

$$
\dot{x} = Q^n x,
$$

the necessary and sufficient stability condition is that all eigenvalues of A have negative real parts or, equivalently, that there exists a *Lyapunov Stability Certificate*

(LSC) — a positive definite symmetric matrix X such that

$$[Q^{\mathbf{n}}]^T X + X Q^{\mathbf{n}} \prec 0.$$

For uncertain system (8.2.18), a *sufficient* stability condition is that all matrices

$$Q \in \mathcal{Q} = \{Q = A^M + B^M K^M C^M : M \in \mathcal{M}\}$$

have a common LSC X, that is, there exists $X \succ 0$ such that

(a) $\qquad\qquad\qquad Q^T X + X Q^T \prec 0 \quad \forall Q \in \mathcal{Q}$

$\qquad\qquad\qquad\qquad\qquad\qquad \Updownarrow$ $\qquad\qquad\qquad\qquad$ (8.2.19)

(b) $\quad [A^M + B^M K^M C^M]^T X + X[A^M + B^M K^M C^M] \prec 0 \quad \forall M \in \mathcal{M};$

here $A^M, ..., K^M$ are the components of a collection $M \in \mathcal{M}$.

The fact that the existence of a common LSC for all matrices $Q \in \mathcal{Q}$ is sufficient for the stability of the closed loop system is nearly evident. Indeed, since $\mathcal{M}$ is compact, for every feasible solution $X \succ 0$ of the semi-infinite LMI (8.2.19) one has

$$\forall M \in \mathcal{M} : [A^M + B^M K^M C^M]^T X + X[A^M + B^M K^M C^M] \prec -\alpha X \quad (*)$$

with appropriate $\alpha > 0$. Now let us look what happens with the quadratic form $x^T X x$ along a state trajectory $x(t)$ of (8.2.18). Setting $f(t) = x^T(t) X x(t)$ and invoking (8.2.18.d), we have

$$
\begin{aligned}
f'(t) &= \dot{x}^T(t) X x(t) + x(t) X \dot{x}(t) \\
&= x^T(t) \left[[A_t + B_t K_t C_t]^T X + X[A_t + B_t K_t C_t] \right] x(t) \\
&\leq -\alpha f(t),
\end{aligned}
$$

where the concluding inequality is due to $(*)$. From the resulting differential inequality

$$f'(t) \leq -\alpha f(t)$$

it follows that

$$f(t) \leq \exp\{-\alpha t\} f(0) \to 0, \ t \to \infty.$$

Recalling that $f(t) = x^T(t) X x(t)$ and X is positive definite, we conclude that $x(t) \to 0$ as $t \to \infty$.

Observe that the set $\mathcal{Q}$ is compact along with $\mathcal{M}$. It follows that X is an LSC if and only if $X \succ 0$ and

$$\exists \beta > 0 : Q^T X + X Q \preceq -\beta I \quad \forall Q \in \mathcal{Q}$$

$$\Leftrightarrow \quad \exists \beta > 0 : Q^T X + X Q \preceq -\beta I \quad \forall Q \in \mathrm{Conv}(\mathcal{Q}).$$

Multiplying such an X by an appropriate positive real, we can ensure that

$$X \succeq I \ \& \ Q^T X + X Q \preceq -I \quad \forall Q \in \mathrm{Conv}(\mathcal{Q}). \qquad (8.2.20)$$

Thus, we lose nothing when requiring from an LSC to satisfy the latter system of (semi-infinite) LMIs, and from now on LSCs in question will be exactly the solutions of this system.

Observe that (8.2.20) is nothing but the RC of the uncertain system of LMIs

$$X \succeq I \ \& \ Q^T X + X Q \preceq -I, \qquad (8.2.21)$$

the uncertain data being Q and the uncertainty set being $\text{Conv}(\mathcal{Q})$. Thus, RCs arise naturally in the context of Robust Control.

Now let us apply the results on tractability of the RCs of uncertain LMI in order to understand when the question of existence of an LSC for a given uncertain system (8.2.18) can be posed in a computationally tractable form. There are, essentially, two such cases — *polytopic* and *unstructured norm-bounded* uncertainty.

Polytopic uncertainty. By definition, polytopic uncertainty means that the set $\text{Conv}(\mathcal{Q})$ is given as a convex hull of an explicit list of "scenarios" Q^i, $i = 1, ..., N$:

$$\text{Conv}(\mathcal{Q}) = \text{Conv}\{Q^1, ..., Q^N\}.$$

In our context this situation occurs when the components A^M, B^M, C^M, K^M of $M \in \mathcal{M}$ run, independently of each other, through convex hulls of respective scenarios

$$S_A = \text{Conv}\{A^1, ..., A^{N_A}\}, S_B = \text{Conv}\{B^1, ..., B^{N_B}\},$$
$$S_C = \text{Conv}\{C^1, ..., C^{N_C}\}, S_K = \text{Conv}\{K^1, ..., K^{N_K}\};$$

in this case, the set $\text{Conv}(\mathcal{Q})$ is nothing but the convex hull of $N = N_A N_B N_C N_K$ "scenarios" $Q^{ijk\ell} = A^i + B^j K^\ell C^k$, $1 \leq i \leq N_A, ..., 1 \leq \ell \leq N_K$.

Indeed, $\mathcal{Q}$ clearly contains all matrices $Q^{ijk\ell}$ and therefore $\text{Conv}(\mathcal{Q}) \supset \text{Conv}(\{Q^{ijk\ell}\})$. On the other hand, the mapping $(A, B, C, K) \mapsto A + BKC$ is polylinear, so that the image $\mathcal{Q}$ of the set $S_A \times S_B \times S_C \times S_K$ under this mapping is contained in the convex set $\text{Conv}(\{Q^{ijk\ell}\})$, whence $\text{Conv}(\{Q^{ijk\ell}\}) \supset \text{Conv}(\mathcal{Q})$.

In the case in question we are in the situation of scenario perturbations, so that (8.2.21) is equivalent to the explicit system of LMIs

$$X \succeq I, [Q^i]^T X + X Q^i \preceq -I, i = 1, ..., N.$$

Unstructured norm-bounded uncertainty. Here

$$\text{Conv}(\mathcal{Q}) = \{Q = Q^{\mathrm{n}} + U\zeta V : \zeta \in \mathbb{R}^{p \times q}, \|\zeta\|_{2,2} \leq \rho\}.$$

In our context this situation occurs, e.g., when 3 of the 4 matrices A^M, B^M, C^M, K^M, $M \in \mathcal{M}$, are in fact certain, and the remaining matrix, say, A^M, runs through a set of the form $\{A^{\mathrm{n}} + G\zeta H : \zeta \in \mathbb{R}^{p \times q}, \|\zeta\|_{2,2} \leq \rho\}$.

In the case of unstructured norm-bounded uncertainty, the semi-infinite LMI in (8.2.21) is of the form

$$Q^T X + X Q \preceq -I \quad \forall Q \in \text{Conv}(\mathcal{Q})$$
$$\Updownarrow$$
$$\underbrace{-I - [Q^{\mathrm{n}}]^T X - X Q^{\mathrm{n}}}_{\mathcal{A}^{\mathrm{n}}(X)} + [\underbrace{-XU}_{L^T(X)} \zeta \underbrace{V}_{R} + R^T \zeta^T L(X)] \succeq 0$$
$$\forall (\zeta \in \mathbb{R}^{p \times q}, \|\zeta\|_{2,2} \leq \rho).$$

Invoking Theorem 8.2.3, (8.2.21) is equivalent to the explicit system of LMIs

$$X \succeq I, \left[\begin{array}{c|c} \lambda I_p & \rho U^T X \\ \hline \rho X U & -I - [Q^n]^T X - X Q^n - \lambda V^T V \end{array} \right] \succeq 0. \qquad (8.2.22)$$

in variables X, λ.

8.2.3.2 Lyapunov Stability Synthesis

We have considered the *Stability Analysis* problem, where one, given an uncertain closed-loop dynamical system along with the associated uncertainty set $\mathcal{M}$, seeks to verify a sufficient stability condition. A more challenging problem is *Stability Synthesis*: given an uncertain open loop system (8.2.18.*a–b*) along with the associated compact uncertainty set $\widehat{\mathcal{M}}$ in the space of collections $\widehat{M} = (A, B, C, D, R)$, find a linear output-based feedback

$$u(t) = Ky(t)$$

and an LSC for the resulting closed loop system.

The Synthesis problem has a nice solution, due to [21], in the case of *state-based* feedback (that is, $C_t \equiv I$) and under the assumption that the feedback is implemented exactly, so that the state dynamics of the closed loop system is given by

$$\dot{x}(t) = [A_t + B_t K] x(t) + [R_t + B_t K D_t] d_t. \qquad (8.2.23)$$

The pairs (K, X) of "feedback – LSC" that we are looking for are exactly the feasible solutions to the system of semi-infinite matrix inequalities in variables X, K:

$$X \succ 0 \ \& \ [A + BK]^T X + X[A + BK] \prec 0 \ \ \forall [A, B] \in \mathcal{AB}; \qquad (8.2.24)$$

here $\mathcal{AB}$ is the projection of $\widehat{M}$ on the space of $[A, B]$ data. The difficulty is that the system is *nonlinear* in the variables. As a remedy, let us carry out the nonlinear substitution of variables $X = Y^{-1}$, $K = ZY^{-1}$. With this substitution, (8.2.24) becomes a system in the new variables Y, Z:

$$Y \succ 0 \ \& \ [A + BZY^{-1}]^T Y^{-1} + Y^{-1}[A + BZY^{-1}] \prec 0 \ \ \forall [A, B] \in \mathcal{AB};$$

multiplying both sides of the second matrix inequality from the left and from the right by Y, we convert the system to the equivalent form

$$Y \succ 0, \ \& \ AY + YA^T + BZ + Z^T B^T \prec 0 \ \ \forall [A, B] \in \mathcal{AB}.$$

Since $\mathcal{AB}$ is compact along with $\widehat{\mathcal{M}}$, the solutions to the latter system are exactly the pairs (Y, Z) that can be obtained by scaling $(Y, Z) \mapsto (\lambda Y, \lambda Z)$, $\lambda > 0$, from the solutions to the system of semi-infinite LMIs

$$Y \succeq I \ \& \ AY + YA^T + BZ + Z^T B^T \preceq -I \ \ \forall [A, B] \in \mathcal{AB} \qquad (8.2.25)$$

in variables Y, Z. When the uncertainty $\mathcal{AB}$ can be represented either as a polytopic, or as unstructured norm-bounded, the system (8.2.25) of semi-infinite LMIs admits an equivalent tractable reformulation.

8.3 EXERCISES

Exercise 8.1. [Robust Linear Estimation, see [48]] Let a signal $v \in \mathbb{R}^n$ be observed according to
$$y = Av + \xi,$$
where A is an $m \times n$ matrix, known up to "unstructured norm-bounded perturbation":
$$A \in \mathcal{A} = \{A = A_\mathrm{n} + L^T \Delta R : \Delta \in \mathbb{R}^{p \times q}, \|\Delta\|_{2,2} \leq \rho\},$$
and ξ is a zero mean random noise with a known covariance matrix Σ. Our a priori information on v is that
$$v \in V = \{v : v^T Q v \leq 1\},$$
where $Q \succ 0$. We are looking for a linear estimate
$$\widehat{v} = Gy$$
with the smallest possible worst-case mean squared error
$$\mathrm{EstErr} = \sup_{v \in V, A \in \mathcal{A}} \left(\mathbf{E} \left\{ \|G[Av + \xi] - v\|_2^2 \right\} \right)^{1/2}$$
(cf. section 6.6).

1) Reformulate the problem of building the optimal estimate equivalently as the RC of uncertain semidefinite program with unstructured norm-bounded uncertainty and reduce this RC to an explicit semidefinite program.

2) Assume that $m = n$, $\Sigma = \sigma^2 I_n$, and the matrices $A_\mathrm{n}^T A_\mathrm{n}$ and Q commute, so that $A_\mathrm{n} = V \mathrm{Diag}\{a\} U^T$ and $Q = U \mathrm{Diag}\{q\} U^T$ for certain orthogonal matrices U, V and certain vectors $a \geq 0$, $q > 0$. Let, further, $\mathcal{A} = \{A_\mathrm{n} + \Delta : \|\Delta\|_{2,2} \leq \rho\}$. Prove that in the situation in question we lose nothing when looking for G in the form of
$$G = U \mathrm{Diag}\{g\} V^T,$$
and build an explicit convex optimization program with just two variables specifying the optimal choice of G.

8.4 NOTES AND REMARKS

NR 8.1. Theorem 8.2.3 was discovered in [32]. The uncertain Truss Topology Design problem (Example 8.2.4) was considered in [3]; this problem partly inspired our initial activity on Convex RO. The Free Material Optimization methodology in Structural Design was proposed by M. Bendsøe [2] and Ringertz [99]; for more detailed derivation and analysis of SDO models in Structural Design, see [6] and [8, section 4.8].

NR 8.2. The material of Section 8.2.3 is now a standard component of the LMI-based Robust Control Theory; our presentation of this material follows [32]. Along with stability analysis/synthesis for uncertain linear dynamical systems, uncertain LMIs have many other applications in Robust Control. Indeed, not only the stability, but many other "desirable properties" of certain time-invariant linear systems are "LMI-representable" — they can be certified by a solution to an appropriate system $\mathcal{S}$ of LMIs with the data readily given by the data of the dynamical system. When allowing the latter data to vary in time, staying within a given uncertainty set, that is, when passing from a certain time-invariant linear system to its uncertain time-varying counterpart, the data in $\mathcal{S}$ also become uncertain. Typically, the existence of a robust feasible solution to the resulting uncertain system of LMIs is a *sufficient* condition for the dynamical system to enjoy the desirable property in question in a robust fashion, which makes uncertain LMIs an important integral part of Robust Control. For more details on this subject, see [32].

Chapter Nine
Approximating RCs of Uncertain Semidefinite Problems

9.1 TIGHT TRACTABLE APPROXIMATIONS OF RCS OF UNCERTAIN SDPS WITH STRUCTURED NORM-BOUNDED UNCERTAINTY

We have seen that the possibility to reformulate the RC of an uncertain semidefinite program in a computationally tractable form is a "rare commodity," so that there are all reasons to be interested in the second best thing — in situations where the RC admits a tight tractable approximation. To the best of our knowledge, just one such case is known — the case of *structured norm-bounded uncertainty* we are about to consider in this chapter.

9.1.1 Uncertain LMI with Structured Norm-Bounded Perturbations

Consider an uncertain LMI

$$\mathcal{A}_\zeta(y) \succeq 0 \tag{8.2.2}$$

where the "body" $\mathcal{A}_\zeta(y)$ is bi-linear in the design vector y and the perturbation vector ζ. The definition of a structured norm-bounded perturbation follows the path we got acquainted with in chapter 5:

Definition 9.1.1. We say that the uncertain constraint (8.2.2) is affected by structured norm-bounded uncertainty with uncertainty level ρ, if

1. The perturbation set $\mathcal{Z}_\rho$ is of the form

$$\mathcal{Z}_\rho = \left\{ \zeta = (\zeta^1, ..., \zeta^L) : \begin{array}{l} \zeta^\ell \in \mathbb{R}, |\zeta^\ell| \leq \rho, \ell \in \mathcal{I}_\mathrm{s} \\ \zeta^\ell \in \mathbb{R}^{p_\ell \times q_\ell} : \|\zeta^\ell\|_{2,2} \leq \rho, \ell \notin \mathcal{I}_\mathrm{s} \end{array} \right\} \tag{9.1.1}$$

2. The body $\mathcal{A}_\zeta(y)$ of the constraint can be represented as

$$\begin{aligned}
\mathcal{A}_\zeta(y) &= \mathcal{A}^\mathrm{n}(y) + \sum_{\ell \in \mathcal{I}_\mathrm{s}} \zeta^\ell \mathcal{A}_\ell(y) \\
&+ \sum_{\ell \notin \mathcal{I}_\mathrm{s}} \left[L_\ell^T(y) \zeta^\ell R_\ell + R_\ell^T [\zeta^\ell]^T L_\ell(y) \right],
\end{aligned} \tag{9.1.2}$$

where $\mathcal{A}_\ell(y)$, $\ell \in \mathcal{I}_\mathrm{s}$, and $L_\ell(y)$, $\ell \notin \mathcal{I}_\mathrm{s}$, are affine in y, and R_ℓ, $\ell \notin \mathcal{I}_\mathrm{s}$, are nonzero.

Theorem 9.1.2. Given uncertain LMI (8.2.2) with structured norm-bounded uncertainty (9.1.1), (9.1.2), let us associate with it the following system of LMIs in variables Y_ℓ, $\ell = 1, ..., L$, λ_ℓ, $\ell \notin \mathcal{I}_s$, y:

$$(a) \quad Y_\ell \succeq \pm \mathcal{A}_\ell(y), \ \ell \in \mathcal{I}_s$$

$$(b) \quad \begin{bmatrix} \lambda_\ell I_{p_\ell} & L_\ell(y) \\ \hline L_\ell^T(y) & Y_\ell - \lambda_\ell R_\ell^T R_\ell \end{bmatrix} \succeq 0, \ \ell \notin \mathcal{I}_s \qquad (9.1.3)$$

$$(c) \quad \mathcal{A}^{\mathrm{n}}(y) - \rho \sum_{\ell=1}^{L} Y_\ell \succeq 0$$

Then system (9.1.3) is a safe tractable approximation of the RC

$$\mathcal{A}_\zeta(y) \succeq 0 \ \forall \zeta \in \mathcal{Z}_\rho \qquad (9.1.4)$$

of (8.2.2), (9.1.1), (9.1.2), and the tightness factor of this approximation does not exceed $\vartheta(\mu)$, where μ is the smallest integer ≥ 2 such that $\mu \geq \max\limits_{y} \mathrm{Rank}(\mathcal{A}_\ell(y))$ for all $\ell \in \mathcal{I}_s$, and $\vartheta(\cdot)$ is a universal function of μ such that

$$\vartheta(2) = \frac{\pi}{2}, \ \vartheta(4) = 2, \ \vartheta(\mu) \leq \pi \sqrt{\mu/2}, \ \mu > 2.$$

The approximation is exact, if either $L = 1$, or all perturbations are scalar, (i.e., $\mathcal{I}_s = \{1, ..., L\}$) and all $\mathcal{A}_\ell(y)$ are of ranks not exceeding 1.

Proof. Let us fix y and observe that a collection $y, Y_1, ..., Y_L$ can be extended to a feasible solution of (9.1.3) if and only if

$$\forall \zeta \in \mathcal{Z}_\rho : \begin{cases} -\rho Y_\ell \preceq \zeta^\ell \mathcal{A}_\ell(y), \ \ell \in \mathcal{I}_s, \\ -\rho Y_\ell \preceq L_\ell^T(y) \zeta^\ell R_\ell + R_\ell^T [\zeta^\ell]^T L_\ell(y), \ \ell \notin \mathcal{I}_s \end{cases}$$

(see Theorem 8.2.3). It follows that if, in addition, Y_ℓ satisfy (9.1.3.c), then y is feasible for (9.1.4), so that (9.1.3) is a safe tractable approximation of (9.1.4). The fact that this approximation is tight within the factor $\vartheta(\mu)$ is readily given by the Real Case Matrix Cube Theorem, see Appendix B.4.6. The fact that the approximation is exact when $L = 1$ is evident when $\mathcal{I}_s = \{1\}$ and is readily given by Theorem 8.2.3 when $\mathcal{I}_s = \emptyset$. The fact that the approximation is exact when all perturbations are scalar and all matrices $\mathcal{A}_\ell(y)$ are of ranks not exceeding 1 is evident. $\qquad \square$

9.1.2 Application: Lyapunov Stability Analysis/Synthesis Revisited

We start with the Analysis problem. Consider the uncertain time-varying dynamical system (8.2.18) and assume that the uncertainty set $\mathrm{Conv}(\mathcal{Q}) = \mathrm{Conv}(\{A^M + B^M K^M C^M\} : M \in \mathcal{M})$ in (8.2.20) is an *interval* uncertainty, meaning that

$$\mathrm{Conv}(\mathcal{Q}) = Q^{\mathrm{n}} + \rho \mathcal{Z}, \ \mathcal{Z} = \{\sum_{\ell=1}^{L} \zeta_\ell U_\ell : \|\zeta\|_\infty \leq 1\}, \qquad (9.1.5)$$
$$\mathrm{Rank}(U_\ell) \leq \mu, \ 1 \leq \ell \leq L.$$

Such a situation (with $\mu = 1$) arises, e.g., when two of the 3 matrices B_t, C_t, K_t are certain, and the remaining one of these 3 matrices, say, K_t, and the matrix

A_t are affected by entry-wise uncertainty:

$$\{(A^M, K^M) : M \in \mathcal{M}\} = \left\{(A, K) : \begin{array}{l} |A_{ij} - A_{ij}^{\mathrm{n}}| \le \rho\alpha_{ij}\forall(i,j) \\ |K_{pq} - K_{pq}^{\mathrm{n}}| \le \rho\kappa_{pq}\,\forall(p,q) \end{array} \right\},$$

In this case, denoting by B^{n}, C^{n} the (certain!) matrices B_t, C_t, we clearly have

$$\mathrm{Conv}(\mathcal{Q}) = \underbrace{A^{\mathrm{n}} + B^{\mathrm{n}}K^{\mathrm{n}}C^{\mathrm{n}}}_{Q^{\mathrm{n}}} + \rho\Bigg\{\Big[\sum_{i,j}\xi_{ij}[\alpha_{ij}e_ie_j^T]$$

$$+ \sum_{p,q}\eta_{pq}[\kappa_{pq}B^{\mathrm{n}}f_pg_q^TC^{\mathrm{n}}]\Big] : |\xi_{ij}| \le 1, |\eta_{pq}| \le 1\Bigg\},$$

where e_i, f_p, g_q are the standard basic orths in the spaces $\mathbb{R}^{\dim x}$, $\mathbb{R}^{\dim u}$ and $\mathbb{R}^{\dim y}$, respectively. Note that the matrix coefficients at the "elementary perturbations" ξ_{ij}, η_{pq} are of rank 1, and these perturbations, independently of each other, run through $[-1,1]$ — exactly as required in (9.1.5) for $\mu = 1$.

In the situation of (9.1.5), the semi-infinite Lyapunov LMI

$$Q^TX + XQ \preceq -I \;\forall Q \in \mathrm{Conv}(\mathcal{Q})$$

in (8.2.20) reads

$$\underbrace{-I - [Q^{\mathrm{n}}]^TX - XQ^{\mathrm{n}}}_{\mathcal{A}^{\mathrm{n}}(X)} + \rho\sum_{\ell=1}^{L}\zeta_\ell\underbrace{[-U_\ell^TX - XU_\ell]}_{\mathcal{A}_\ell(X)} \tag{9.1.6}$$

$$\succeq 0 \;\forall(\zeta : |\zeta_\ell| \le 1, \ell = 1, ..., L).$$

We are in the case of structured norm-bounded perturbations with $\mathcal{I}_{\mathrm{s}} = \{1, ..., L\}$. Noting that the ranks of all matrices $\mathcal{A}_\ell(X)$ never exceed 2μ (since all U_ℓ are of ranks $\le \mu$), the safe tractable approximation of (9.1.6) given by Theorem 9.1.2 is tight within the factor $\vartheta(2\mu)$. It follows, in particular, that *in the case of (9.1.5) with $\mu = 1$, we can find efficiently a lower bound, tight within the factor $\pi/2$, on the Lyapunov Stability Radius of the uncertain system (8.2.18)* (that is, on the supremum of those ρ for which the stability of our uncertain dynamical system can be certified by an LSC). The lower bound in question is the supremum of those ρ for which the approximation is feasible, and this supremum can be easily approximated to whatever accuracy by bisection.

We can process in the same fashion the Lyapunov Stability Synthesis problem in the presence of interval uncertainty. Specifically, assume that $C_t \equiv I$ and the uncertainty set $\mathcal{AB} = \{[A^M, B^M] : M \in \mathcal{M}\}$ underlying the Synthesis problem is an interval uncertainty:

$$\mathcal{AB} = [A^{\mathrm{n}}, B^{\mathrm{n}}] + \rho\{\sum_{\ell=1}^{L}\zeta_\ell U_\ell : \|\zeta\|_\infty \le 1\}, \quad \mathrm{Rank}(U_\ell) \le \mu\,\forall\ell. \tag{9.1.7}$$

We arrive at the situation of (9.1.7) with $\mu = 1$, e.g., when $\mathcal{AB}$ corresponds to entry-wise uncertainty:

$$\mathcal{AB} = [A^{\mathrm{n}}, B^{\mathrm{n}}] + \rho\{H \equiv [\delta A, \delta B] : |H_{ij}| \le h_{ij}\,\forall i, j\}.$$

In the case of (9.1.7) the semi-infinite LMI in (8.2.25) reads

$$\underbrace{-I - [A^{\mathrm{n}}, B^{\mathrm{n}}][Y; Z] - [Y; Z]^T [A^{\mathrm{n}}, B^{\mathrm{n}}]^T}_{\mathcal{A}^{\mathrm{n}}(Y, Z)}$$

$$+\rho \sum_{\ell=1}^{L} \zeta_\ell \underbrace{[-U_\ell[Y; Z] - [Y; Z]^T U_\ell^T]}_{\mathcal{A}_\ell(Y, Z)} \succeq 0 \,\forall (\zeta : |\zeta_\ell| \leq 1, \ \ell = 1, ..., L). \tag{9.1.8}$$

We again reach a situation of structured norm-bounded uncertainty with $\mathcal{I}_{\mathrm{s}} = \{1, ..., L\}$ and all matrices $\mathcal{A}_\ell(\cdot)$, $\ell = 1, ..., L$, being of ranks at most 2μ. Thus, Theorem 9.1.2 provides us with a tight, within factor $\vartheta(2\mu)$, safe tractable approximation of the Lyapunov Stability Synthesis problem.

Illustration: Controlling a multiple pendulum. Consider a multiple pendulum ("a train") depicted in figure 9.1. Denoting by m_i, $i = 1, ..., 4$, the masses of the "engine" ($i = 1$) and the "cars" ($i = 2, 3, 4$, counting from right to left), Newton's laws for the dynamical system in question read

$$\begin{array}{llll}
m_1 \frac{d^2}{dt^2} x_1(t) = & -\kappa_1 x_1(t) & +\kappa_1 x_2(t) & & +u(t) \\
m_2 \frac{d^2}{dt^2} x_2(t) = & \kappa_1 x_1(t) & -(\kappa_1 + \kappa_2) x_2(t) & +\kappa_2 x_3(t) & \\
m_3 \frac{d^2}{dt^2} x_3(t) = & & \kappa_2 x_2(t) & -(\kappa_2 + \kappa_3) x_3(t) & +\kappa_3 x_4(t) \\
m_4 \frac{d^2}{dt^2} x_4(t) = & & & \kappa_3 x_3(t) & -\kappa_3 x_4(t),
\end{array} \tag{9.1.9}$$

where $x_i(t)$ are shifts of the engine and the cars from their respective positions in the state of rest (where nothing moves and the springs are neither shrunk nor expanded), and κ_i are the elasticity constants of the springs (counted from right to left). Passing from masses m_i to their reciprocals $\mu_i = 1/m_i$ and adding to the coordinates of the cars their velocities $v_i(t) = \dot{x}_i(t)$, we can rewrite (9.1.9) as the system of 8 linear differential equations:

$$\dot{x}(t) = \underbrace{\left[\begin{array}{cccc|cccc}
 & & & & 1 & & & \\
 & & & & & 1 & & \\
 & & & & & & 1 & \\
 & & & & & & & 1 \\
\hline
-\kappa_1 \mu_1 & \kappa_1 \mu_1 & & & & & & \\
\kappa_1 \mu_2 & -[\kappa_1 + \kappa_2]\mu_2 & \kappa_2 \mu_2 & & & & & \\
 & \kappa_2 \mu_3 & -[\kappa_2 + \kappa_3]\mu_3 & \kappa_3 \mu_3 & & & & \\
 & & \kappa_3 \mu_4 & -\kappa_3 \mu_4 & & & &
\end{array}\right]}_{A_\mu} x(t)$$

$$+ \underbrace{\left[\begin{array}{c}
\\
\\
\\
\\
\hline
\mu_1 \\
\\
\\
\\
\end{array}\right]}_{B_\mu} u(t) \tag{9.1.10}$$

where $x(t) = [x_1(t); x_2(t); x_3(t); x_4(t); v_1(t); v_2(t); v_3(t); v_4(t)]$. System (9.1.10) "as

Figure 9.1 "Train": 4 masses (3 "cars" and "engine") linked by elastic springs and sliding without friction (aside of controlled force u) along "rail" AA.

it is" (i.e., with trivial control $u(\cdot) \equiv 0$) is unstable; not only it has a solution that does not converge to 0 as $t \to \infty$, it has even an unbounded solution (specifically, one where $x_i(t) = vt$, $v_i(t) \equiv v$, which corresponds to uniform motion of the cars and the engine with no tensions in the springs). Let us look for a stabilizing state-based linear feedback controller

$$u(t) = Kx(t), \qquad (9.1.11)$$

that is robust w.r.t. the masses of the cars and the engine when they vary in given segments Δ_i, $i = 1, ..., 4$. To this end we can apply the Lyapunov Stability Synthesis machinery. Observe that to say that the masses m_i run, independently of each other, through given segments is exactly the same as to say that their reciprocals μ_i run, independently of each other, through other given segments Δ'_i; thus, our goal is as follows:

Stabilization: *Given elasticity constants κ_i and segments $\Delta'_i \subset \{\mu > 0\}$, $i = 1, ..., 4$, find a linear feedback* (9.1.11) *and a Lyapunov Stability Certificate X for the corresponding closed loop system* (9.1.10), (9.1.11), *with the uncertainty set for the system being*

$$\mathcal{AB} = \{[A_\mu, B_\mu] : \mu_i \in \Delta'_i, i = 1, ..., 4\}.$$

Note that in our context the Lyapunov Stability Synthesis approach is, so to speak, "doubly conservative." First, the existence of a common LSC for all matrices Q from a given compact set $\mathcal{Q}$ is only a *sufficient* condition for the stability of the uncertain dynamical system

$$\dot{x}(t) = Q_t x(t), \quad Q_t \in \mathcal{Q} \, \forall t,$$

and as such this condition is conservative. Second, in our train example there are reasons to think of m_i as of uncertain data (in reality the loads of the cars and the mass of the engine could vary from trip to trip, and we would not like to re-adjust the controller as long as these changes are within a reasonable range), but there is absolutely no reason to think of these masses as varying in time. Indeed, we could perhaps imagine a mechanism that makes the masses m_i time-dependent, but with this mechanism our original model (9.1.9) becomes invalid — Newton's laws in the form of (9.1.9) are not applicable to systems with varying masses and at the very best they offer a reasonable approximation of the true model, provided that the changes in masses are slow. Thus, in our train example a common LSC for all matrices $Q = A + BK$, $[A, B] \in \mathcal{AB}$, would guarantee much more than required, namely, that all trajectories of the closed loop system "train plus feedback controller" converge to 0 as $t \to \infty$ even in the case when the parameters $\mu_i \in \Delta_i'$ vary in time at a high speed. This is much more than what we actually need — convergence to 0 of all trajectories in the case when $\mu_i \in \Delta_i'$ do not vary in time.

The system of semi-infinite LMIs we are about to process in the connection of the Lyapunov Stability Synthesis is

$$
\begin{array}{ll}
(a) & [A, B][Y; Z] + [Y; Z]^T [A, B]^T \preceq -\alpha Y, \ \forall [A, B] \in \mathcal{AB} \\
(b) & Y \succeq I \\
(c) & Y \leq \chi I,
\end{array}
\qquad (9.1.12)
$$

where $\alpha > 0$ and $\chi > 1$ are given. This system differs slightly from the "canonical" system (8.2.25), and the difference is twofold:

- [major] in (8.2.25), the semi-infinite Lyapunov LMI is written as

$$
[A, B][Y; Z] + [Y; Z]^T [A, B]^T \preceq -I,
$$

which is just a convenient way to express the relation

$$
[A, B][Y; Z] + [Y; Z]^T [A, B]^T \prec 0, \forall [A, B] \in \mathcal{AB}.
$$

Every feasible solution $[Y; Z]$ to this LMI with $Y \succ 0$ produces a stabilizing feedback $K = ZY^{-1}$ and the common LSC $X = Y^{-1}$ for all instances of the matrix $Q = A + BK$, $[A, B] \in \mathcal{AB}$, of the closed loop system, i.e.,

$$
[A + BK]^T X + X[A + BK] \prec 0 \ \forall [A, B] \in \mathcal{AB}.
$$

The latter condition, however, says nothing about the corresponding decay rate. In contrast, when $[Y; Z]$ is feasible for (9.1.12.a, b), the associated stabilizing feedback $K = ZY^{-1}$ and LSC $X = Y^{-1}$ satisfy the relation

$$
[A + BK]^T X + X[A + BK] \prec -\alpha X \ \forall [A, B] \in \mathcal{AB},
$$

and this relation, as we have seen when introducing the Lyapunov Stability Certificate, implies that

$$
x^T(t) X x(t) \leq \exp\{-\alpha t\} x^T(0) X x(0), \ t \geq 0,
$$

which guarantees that the decay rate in the closed loop system is at least α. In our illustration (same as in real life), we prefer to deal with this "stronger" form of the Lyapunov Stability Synthesis requirement, in order to have a control over the decay rate associated with the would-be controller.

- [minor] In (9.1.12) we impose an upper bound on the condition number (ratio of the maximal and minimal eigenvalues) of the would-be LSC; with normalization of Y given by (9.1.12.b), this bound is ensured by (9.1.12.c) and is precisely χ. The only purpose of this bound is to avoid working with extremely ill-conditioned positive definite matrices, which can cause numerical problems.

Now let us use Theorem 9.1.2 to get a tight safe tractable approximation of the semi-infinite system of LMIs (9.1.12). Denoting by $\mu_i^{\rm n}$ the midpoints of the segments Δ'_i and by δ_i the half-width of these segments, we have

$$
\begin{aligned}
\mathcal{AB} &\equiv \{[A_\mu, B_\mu] : \mu_i \in \Delta'_i, i = 1, ..., 4\} \\
&= \{[A_\mu{\rm n}, B_\mu{\rm n}] + \sum_{\ell=1}^{4} \zeta_\ell U_\ell : |\zeta_\ell| \leq 1, \ell = 1, ..., 4\}, \\
U_\ell &= \delta_\ell p_\ell q_\ell^T,
\end{aligned}
$$

where $p_\ell \in \mathbb{R}^8$ has the only nonzero entry, equal to 1, in the position $4 + \ell$, and

$$
\begin{bmatrix} q_1^T \\ q_2^T \\ q_3^T \\ q_4^T \end{bmatrix} =
\begin{bmatrix}
-\kappa_1 & \kappa_1 & & & & & & & 1 \\
\kappa_1 & -[\kappa_1 + \kappa_2] & \kappa_2 & & & & & & \\
& \kappa_2 & -[\kappa_2 + \kappa_3] & \kappa_3 & & & & & \\
& & \kappa_3 & -\kappa_3 & & & & &
\end{bmatrix}
$$

Consequently, the analogy of (9.1.12) with uncertainty level ρ ((9.1.12) itself corresponds to $\rho = 1$) is the semi-infinite system of LMIs

$$
\underbrace{-\alpha Y - [A_\mu{\rm n}, B_\mu{\rm n}][Y; Z] - [Y; Z]^T [A_\mu{\rm n}, B_\mu{\rm n}]^T}_{\mathcal{A}^{\rm n}(Y, Z)}
$$
$$
+ \rho \sum_{\ell=1}^{4} \zeta_\ell \underbrace{(-\delta_\ell [p_\ell q_\ell^T [Y; Z] + [Y; Z]^T q_\ell p_\ell^T])}_{\mathcal{A}_\ell(Y, Z)} \succeq 0 \, \forall (\zeta : |\zeta_\ell| \leq 1, \ell = 1, ..., 4) \quad (9.1.13)
$$
$$
Y \succeq I_8, \; Y \preceq \chi I_8
$$

in variables Y, Z (cf. (9.1.8)). The safe tractable approximation of this semi-infinite system of LMIs as given by Theorem 9.1.2 is the system of LMIs

$$
\begin{aligned}
Y_\ell &\succeq \pm \mathcal{A}_\ell(Y, Z), \; \ell = 1, ..., 4 \\
\mathcal{A}^{\rm n}(Y, Z) &- \rho \sum_{\ell=1}^{4} Y_\ell \succeq 0 \quad (9.1.14) \\
Y &\succeq I_8, \; Y \preceq \chi I_8
\end{aligned}
$$

in variables $Y, Z, Y_1, ..., Y_4$. Since all U_ℓ are of rank 1 and therefore all $\mathcal{A}_\ell(Y, Z)$ are of rank ≤ 2, Theorem 9.1.2 states that this safe approximation is tight within the factor $\pi/2$.

Of course, in our toy example no approximation is needed — the set $\mathcal{AB}$ is a polytopic uncertainty with just $2^4 = 16$ vertices, and we can straightforwardly convert (9.1.13) into an exactly equivalent system of 18 LMIs

$$
\begin{aligned}
\mathcal{A}^{\rm n}(Y, Z) &\succeq \rho \sum_{\ell=1}^{4} \epsilon_\ell \mathcal{A}_\ell(Y, Z), \; \epsilon_\ell = \pm 1, \ell = 1, ..., 4 \\
Y &\succeq I_8, \; Y \preceq \chi I_8
\end{aligned}
$$

in variables Y, Z. The situation would change dramatically if there were, say, 30 cars in our train rather than just 3. Indeed, in the latter case the precise "polytopic" approach would require solving a system of $2^{31} + 2 = 2,147,483,650$ LMIs of the size 62×62 in variables $Y \in \mathbf{S}^{62}, Z \in \mathbb{R}^{1 \times 63}$, which is a bit too much... In contrast, the approximation (9.1.14) is a system of just $31 + 2 = 33$ LMIs of the size 62×62 in variables $\{Y_\ell \in \mathbf{S}^{62}\}_{\ell=1}^{31}$, $Y \in \mathbf{S}^{62}$, $Z \in \mathbb{R}^{1 \times 63}$ (totally $(31 + 1)\frac{62 \cdot 63}{2} + 63 = 60606$ scalar decision variables). One can argue that the latter problem still is too large from a practical perspective. But in fact it can be shown (see Exercise 9.1) that in this problem, one can easily eliminate the matrices Y_ℓ (every one of them can be replaced with a *single* scalar decision variable), which reduces the design dimension of the approximation to $31 + \frac{62 \cdot 63}{2} + 63 = 2047$. A convex problem of this size can be solved pretty routinely.

We are about to present numerical results related to stabilization of our toy 3-car train. The setup in our computations is as follows:

$$\kappa_1 = \kappa_2 = \kappa_3 = 100.0; \alpha = 0.01; \chi = 10^8;$$
$$\Delta_1' = [0.5, 1.5], \Delta_2' = \Delta_3' = \Delta_4' = [1.5, 4.5],$$

which corresponds to the mass of the engine varying in $[2/3, 2]$ and the masses of the cars varying in $[2/9, 2/3]$.

We computed, by a kind of bisection, the largest ρ for which the approximation (9.1.14) is feasible; the optimal feedback we have found is

$$u = \ 10^7 \big[- 0.2892x_1 - 2.5115x_2 + 6.3622x_3 - 3.5621x_4$$
$$-0.0019v_1 - 0.0912v_2 - 0.0428v_3 + 0.1305v_4 \big],$$

and the (lower bound on the) Lyapunov Stability radius of the closed loop system as yielded by our approximation is $\widehat{\rho} = 1.05473$. This bound is > 1, meaning that our feedback stabilizes the train in the above ranges of the masses of the engine and the cars (and in fact, even in slightly larger ranges $0.65 \leq m_1 \leq 2.11$, $0.22 \leq m_2, m_3, m_4 \leq 0.71$). An interesting question is by how much the *lower bound* $\widehat{\rho}$ is less than the Lyapunov Stability radius ρ_* of the closed loop system. Theory guarantees that the ratio $\rho_*/\widehat{\rho}$ should be $\leq \pi/2 = 1.570....$ In our small problem we can compute ρ_* by applying the polytopic uncertainty approach, that results in $\rho_* = 1.05624$. Thus, in reality $\rho_*/\widehat{\rho} \approx 1.0014$, much better than the theoretical bound $1.570....$ In figure 9.2, we present sample trajectories of the closed loop system yielded by our design, the level of perturbations being 1.054 — pretty close to $\widehat{\rho} = 1.05473$.

9.2 EXERCISES

Exercise 9.1.

1) Let $p, q \in \mathbb{R}^n$ and $\lambda > 0$. Prove that $\lambda pp^T + \frac{1}{\lambda}qq^T \succeq \pm[pq^T + qp^T]$.

2) Let p, q be as in 1) with $p, q \neq 0$, and let $Y \in \mathbf{S}^n$ be such that $Y \succeq \pm[pq^T + qp^T]$. Prove that there exists $\lambda > 0$ such that $Y \succeq \lambda pp^T + \frac{1}{\lambda}qq^T$.

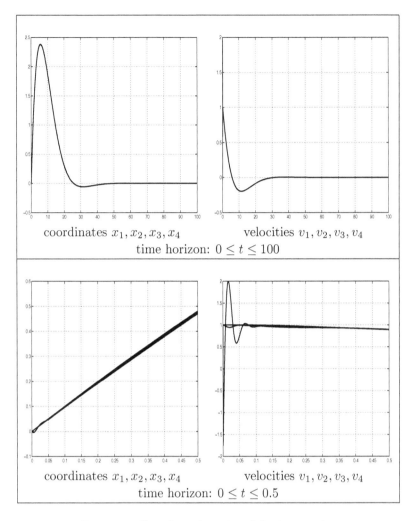

coordinates x_1, x_2, x_3, x_4 velocities v_1, v_2, v_3, v_4

time horizon: $0 \leq t \leq 100$

coordinates x_1, x_2, x_3, x_4 velocities v_1, v_2, v_3, v_4

time horizon: $0 \leq t \leq 0.5$

Figure 9.2 Sample trajectories of the 3-car train.

3) Consider the semi-infinite LMI of the following specific form:

$$\forall (\zeta \in \mathbb{R}^L : \|\zeta\|_\infty \leq 1) : \mathcal{A}_n(x) + \rho \sum_{\ell=1}^{L} \zeta_\ell \left[L_\ell^T(x) R_\ell + R_\ell^T L_\ell(x) \right] \succeq 0, \qquad (9.2.1)$$

where $L_\ell^T(x), R_\ell^T \in \mathbb{R}^n$, $R_\ell \neq 0$ and $L_\ell(x)$ are affine in x, as is the case in Lyapunov Stability Analysis/Synthesis under interval uncertainty (9.1.7) with $\mu = 1$.

Prove that the safe tractable approximation, tight within the factor $\pi/2$, of (9.2.1), that is, the system of LMIs

$$\begin{array}{l} Y_\ell \succeq \pm \left[L_\ell^T(x) R_\ell + R_\ell^T L_\ell(x) \right], \, 1 \leq \ell \leq L \\ \mathcal{A}_n(x) - \rho \sum_{\ell=1}^{L} Y_\ell \succeq 0 \end{array} \qquad (9.2.2)$$

in x and in matrix variables $Y_1, ..., Y_L$ is equivalent to the LMI

$$
\left[
\begin{array}{c|cccc}
\mathcal{A}_n(x) - \rho \sum_{\ell=1}^{L} \lambda_\ell R_\ell^T R_\ell & L_1^T(x) & L_2^T(x) & \cdots & L_L^T(x) \\
\hline
L_1(x) & \lambda_1/\rho & & & \\
L_2(x) & & \lambda_2/\rho & & \\
\vdots & & & \ddots & \\
L_L(x) & & & & \lambda_L/\rho
\end{array}
\right] \succeq 0 \qquad (9.2.3)
$$

in x and real variables $\lambda_1 ..., \lambda_L$. Here the equivalence means that x can be extended to a feasible solution of (9.2.2) if and only if it can be extended to a feasible solution of (9.2.3).

Exercise 9.2. Consider the Signal Processing problem as follows. We are given uncertainty-affected observations

$$y = Av + \xi$$

of a signal v known to belong to a set V. Uncertainty "sits" in the "measurement error" ξ, known to belong to a given set Ξ, and in A — all we know is that $A \in \mathcal{A}$. We assume that V and Ξ are intersections of ellipsoids centered at the origin:

$$V = \{v \in \mathbb{R}^n : v^T P_i v \le 1, 1 \le i \le I\}, [P_i \succeq 0, \textstyle\sum_i P_i \succ 0]$$

$$\Xi = \{\xi \in \mathbb{R}^m : \xi^T Q_j \xi \le \rho_\xi^2, 1 \le j \le J\}, [Q_j \succeq 0, \textstyle\sum_j Q_j \succeq 0]$$

and $\mathcal{A}$ is given by structured norm-bounded perturbations:

$$\mathcal{A} = \{A = A_n + \sum_{\ell=1}^{L} L_\ell^T \Delta_\ell R_\ell, \Delta_\ell \in \mathbb{R}^{p_\ell \times q_\ell}, \|\Delta_\ell\|_{2,2} \le \rho_A\}.$$

We are interested to build a linear estimate $\widehat{v} = Gy$ of v via y. The $\|\cdot\|_2$ error of such an estimate at a particular v is

$$\|Gy - v\|_2 = \|G[Av + \xi] - v\|_2 = \|(GA - I)v + G\xi\|_2,$$

and we want to build G that minimizes the worst, over all v, A, ξ compatible with our a priori information, estimation error

$$\max_{\xi \in \Xi, v \in V, A \in \mathcal{A}} \|(GA - I)v + G\xi\|_2.$$

Build a safe tractable approximation of this problem that seems reasonably tight when ρ_ξ and ρ_A are small.

9.3 NOTES AND REMARKS

NR 9.1. The model of structured norm-bounded perturbations is taken from the famous μ-theory in Robust Control [91]. On the origin of the results underlying Theorem 9.1.2, see section 7.4.

Chapter Ten

Approximating Chance Constrained CQIs and LMIs

In this chapter, we develop safe tractable approximations of *chance constrained* randomly perturbed Conic Quadratic and Linear Matrix Inequalities.

10.1 CHANCE CONSTRAINED LMIS

In previous chapters we have considered the Robust/Approximate Robust Counterparts of uncertain conic quadratic and semidefinite programs. Now we intend to consider *randomly perturbed* CQPs and SDPs and to derive safe approximations of their chance constrained versions (cf. section 2.1). From this perspective, it is convenient to treat chance constrained CQPs as particular cases of chance constrained SDPs (such an option is given by Lemma 6.3.3), so that in the sequel we focus on chance constrained SDPs. Thus, we are interested in a randomly perturbed semidefinite program

$$
\min_{y} \left\{ c^T y : \mathcal{A}^{\mathrm{n}}(y) + \rho \sum_{\ell=1}^{L} \zeta_\ell \mathcal{A}^\ell(y) \succeq 0, y \in \mathcal{Y} \right\}, \tag{10.1.1}
$$

where $\mathcal{A}^{\mathrm{n}}(y)$ and all $\mathcal{A}^\ell(y)$ are affine in y, $\rho \geq 0$ is the "perturbation level," $\zeta = [\zeta_1; ...; \zeta_L]$ is a random perturbation, and $\mathcal{Y}$ is a semidefinite representable set. We associate with this problem its *chance constrained* version

$$
\min_{y} \left\{ c^T y : \mathrm{Prob} \left\{ \mathcal{A}^{\mathrm{n}}(y) + \rho \sum_{\ell=1}^{L} \zeta_\ell \mathcal{A}^\ell(y) \succeq 0 \right\} \geq 1 - \epsilon, y \in \mathcal{Y} \right\} \tag{10.1.2}
$$

where $\epsilon \ll 1$ is a given positive tolerance. Our goal is to build a computationally tractable safe approximation of (10.1.2). We start with assumptions on the random variables ζ_ℓ, which will be in force everywhere in the following:

> *Random variables ζ_ℓ, $\ell = 1, ..., L$, are independent with zero mean satisfying either*
> **A.I** ["bounded case"] $|\zeta_\ell| \leq 1$, $\ell = 1, ..., L$,
> *or*
> **A.II** ["Gaussian case"] $\zeta_\ell \sim \mathcal{N}(0, 1)$, $\ell = 1, ..., L$.

Note that most of the results to follow can be extended to the case when ζ_ℓ are independent with zero means and "light tail" distributions. We prefer to require more in order to avoid too many technicalities.

10.1.1 Approximating Chance Constrained LMIs: Preliminaries

The problem we are facing is basically as follows:

(?) *Given symmetric matrices* $A, A_1, ..., A_L$, *find a verifiable sufficient condition for the relation*

$$\text{Prob}\{\sum_{\ell=1}^{L} \zeta_\ell A_\ell \preceq A\} \geq 1 - \epsilon. \tag{10.1.3}$$

Since ζ is with zero mean, it is natural to require $A \succeq 0$ (this condition clearly is necessary when ζ is symmetrically distributed w.r.t. 0 and $\epsilon < 0.5$). Requiring a bit more, namely, $A \succ 0$, we can reduce the situation to the case when $A = I$, due to

$$\text{Prob}\{\sum_{\ell=1}^{L} \zeta_\ell A_\ell \preceq A\} = \text{Prob}\{\sum_{\ell=1}^{L} \zeta_\ell \underbrace{A^{-1/2} A_\ell A^{-1/2}}_{B_\ell} \preceq I\}. \tag{10.1.4}$$

Now let us try to guess a verifiable sufficient condition for the relation

$$\text{Prob}\{\sum_{\ell=1}^{L} \zeta_\ell B_\ell \preceq I\} \geq 1 - \epsilon. \tag{10.1.5}$$

First of all, we do not lose much when strengthening the latter relation to

$$\text{Prob}\{\|\sum_{\ell=1}^{L} \zeta_\ell B_\ell\| \leq 1\} \geq 1 - \epsilon \tag{10.1.6}$$

(here and in what follows, $\|\cdot\|$ stands for the standard matrix norm $\|\cdot\|_{2,2}$). Indeed, the latter condition is nothing but

$$\text{Prob}\{-I \preceq \sum_{\ell=1}^{L} \zeta_\ell B_\ell \preceq I\} \geq 1 - \epsilon,$$

so that it implies (10.1.5). In the case of ζ symmetrically distributed w.r.t. the origin, we have a "nearly inverse" statement: the validity of (10.1.5) implies the validity of (10.1.6) with ϵ increased to 2ϵ.

The central observation is that whenever (10.1.6) holds true and the distribution of the random matrix

$$S = \sum_{\ell=1}^{L} \zeta_\ell B_\ell$$

is not pathological, we should have

$$\mathbf{E}\{\|S^2\|\} \leq O(1),$$

whence, by Jensen's Inequality,

$$\|\mathbf{E}\{S^2\}\| \leq O(1)$$

as well. Taking into account that $\mathbf{E}\{S^2\} = \sum_{\ell=1}^{L} \mathbf{E}\{\zeta_\ell^2\} B_\ell^2$, we conclude that when all quantities $\mathbf{E}\{\zeta_\ell^2\}$ are of order of 1, we should have $\|\sum_{\ell=1}^{L} B_\ell^2\| \leq O(1)$, or, which is the same,

$$\sum_{\ell=1}^{L} B_\ell^2 \preceq O(1)I. \tag{10.1.7}$$

By the above reasoning, (10.1.7) is a kind of a necessary condition for the validity of the chance constraint (10.1.6), at least for random variables ζ_ℓ that are symmetrically distributed w.r.t. the origin and are "of order of 1." To some extent, this condition can be treated as nearly sufficient, as is shown by the following two theorems.

Theorem 10.1.1. Let $B_1, ..., B_L \in \mathbf{S}^m$ be deterministic matrices such that

$$\sum_{\ell=1}^{L} B_\ell^2 \preceq I \tag{10.1.8}$$

and $\Upsilon > 0$ be a deterministic real. Let, further, ζ_ℓ, $\ell = 1, ..., L$, be independent random variables taking values in $[-1, 1]$ such that

$$\chi \equiv \mathrm{Prob}\left\{ \| \sum_{\ell=1}^{L} \zeta_\ell B_\ell \| \leq \Upsilon \right\} > 0. \tag{10.1.9}$$

Then

$$\forall \Omega > \Upsilon : \mathrm{Prob}\left\{ \| \sum_{\ell=1}^{L} \zeta_\ell B_\ell \| > \Omega \right\} \leq \frac{1}{\chi} \exp\{-(\Omega - \Upsilon)^2/16\}. \tag{10.1.10}$$

Proof. Let $Q = \{z \in \mathbb{R}^L : \| \sum_\ell z_\ell B_\ell \| \leq 1\}$. Observe that

$$\|[\sum_\ell z_\ell B_\ell]u\|_2 \leq \sum_\ell |z_\ell| \|B_\ell u\|_2 \leq \left(\sum_\ell z_\ell^2\right)^{1/2} \left(\sum_\ell u^T B_\ell^2 u\right)^{1/2} \leq \|z\|_2 \|u\|_2,$$

where the concluding relation is given by (10.1.8). It follows that $\| \sum_\ell z_\ell B_\ell \| \leq \|z\|_2$, whence Q contains the unit $\|\cdot\|_2$-ball B centered at the origin in $\mathbb{R}^L$. Besides this, Q is clearly closed, convex and symmetric w.r.t. the origin. Invoking the Talagrand Inequality (see the proof of Lemma B.3.3 in section B.3), we have

$$\mathbf{E}\left\{\exp\{\mathrm{dist}^2_{\|\cdot\|_2}(\zeta, \Upsilon Q)/16\}\right\} \leq (\mathrm{Prob}\{\zeta \in \Upsilon Q\})^{-1} = \frac{1}{\chi}. \tag{10.1.11}$$

Now, when ζ is such that $\| \sum_{\ell=1}^{L} \zeta_\ell B_\ell \| > \Omega$, we have $\zeta \notin \Omega Q$, whence, due to symmetry and convexity of Q, the set $(\Omega - \Upsilon)Q + \zeta$ does not intersect the set ΥQ. Since Q contains B, the set $(\Omega - \Upsilon)Q + \zeta$ contains $\|\cdot\|_2$-ball, centered at ζ, of the radius $\Omega - \Upsilon$, and therefore this ball does not intersect ΥQ either, whence $\mathrm{dist}_{\|\cdot\|_2}(\zeta, \Upsilon Q) > \Omega - \Upsilon$. The resulting relation

$$\| \sum_{\ell=1}^{L} \zeta_\ell B_\ell \| > \Omega \Leftrightarrow \zeta \notin \Omega Q \Rightarrow \mathrm{dist}_{\|\cdot\|_2}(\zeta, \Upsilon Q) > \Omega - \Upsilon$$

combines with (10.1.11) and the Tschebyshev Inequality to imply that

$$\text{Prob}\{\|\sum_{\ell=1}^{L} \zeta_\ell B_\ell\| > \Omega\} \leq \frac{1}{\chi}\exp\{-(\Omega - \Upsilon)^2/16\}. \qquad \Box$$

Theorem 10.1.2. Let $B_1, ..., B_L \in \mathbf{S}^m$ be deterministic matrices satisfying (10.1.8) and $\Upsilon > 0$ be a deterministic real. Let, further, ζ_ℓ, $\ell = 1, ..., L$, be independent $\mathcal{N}(0,1)$ random variables such that (10.1.9) holds true with $\chi > 1/2$.

Then

$$\forall \Omega \geq \Upsilon : \text{Prob}\{\|\sum_{\ell=1}^{L} \zeta_\ell B_\ell\| > \Omega\}$$
$$\leq \text{Erf}\left(\text{ErfInv}(1-\chi) + (\Omega - \Upsilon)\max[1, \Upsilon^{-1}\text{ErfInv}(1-\chi)]\right) \qquad (10.1.12)$$
$$\leq \exp\{-\frac{\Omega^2 \Upsilon^{-2}\text{ErfInv}^2(1-\chi)}{2}\},$$

where $\text{Erf}(\cdot)$ and $\text{ErfInv}(\cdot)$ are the error and the inverse error functions, see (2.3.22).

Proof. Let $Q = \{z \in \mathbb{R}^L : \|\sum_\ell z_\ell B_\ell\| \leq \Upsilon\}$. By the same argument as in the beginning of the proof of Theorem 10.1.1, Q contains the centered at the origin $\|\cdot\|_2$-ball of the radius Υ. Besides this, by definition of Q we have $\text{Prob}\{\zeta \in Q\} \geq \chi$. Invoking item (i) of Theorem B.5.1, Q contains the centered at the origin $\|\cdot\|_2$-ball of the radius $r = \max[\text{ErfInv}(1-\chi), \Upsilon]$, whence, by item (ii) of this Theorem, (10.1.12) holds true. $\qquad \Box$

The last two results are stated next in a form that is better suited for our purposes.

Corollary 10.1.3. Let $A, A_1, ..., A_L$ be deterministic matrices from $\mathbf{S}^m$ such that

$$\exists \{Y_\ell\}_{\ell=1}^{L} : \begin{cases} \left[\begin{array}{c|c} Y_\ell & A_\ell \\ \hline A_\ell & A \end{array}\right] \succeq 0, 1 \leq \ell \leq L \\ \sum_{\ell=1}^{L} Y_\ell \preceq A \end{cases}, \qquad (10.1.13)$$

let $\Upsilon > 0$, $\chi > 0$ be deterministic reals and $\zeta_1, ..., \zeta_L$ be independent random variables satisfying either **A.I**, or **A.II**, and such that

$$\text{Prob}\{-\Upsilon A \preceq \sum_{\ell=1}^{L} \zeta_\ell A_\ell \preceq \Upsilon A\} \geq \chi. \qquad (10.1.14)$$

Then

(i) When ζ_ℓ satisfy **A.I**, we have

$$\forall \Omega > \Upsilon : \text{Prob}\{-\Omega A \preceq \sum_{\ell=1}^{L} \zeta_\ell A_\ell \preceq \Omega A\} \geq 1 - \frac{1}{\chi}\exp\{-(\Omega - \Upsilon)^2/16\}; \quad (10.1.15)$$

(ii) When ζ_ℓ satisfy **A.II**, and, in addition, $\chi > 0.5$, we have

$$\forall \Omega > \Upsilon : \mathrm{Prob}\{-\Omega A \preceq \sum_{\ell=1}^{L} \zeta_\ell A_\ell \preceq \Omega A\}$$
$$\geq 1 - \mathrm{Erf}\left(\mathrm{ErfInv}(1 - \chi) + (\Omega - \Upsilon)\max\left[1, \tfrac{\mathrm{ErfInv}(1-\chi)}{\Upsilon}\right]\right), \tag{10.1.16}$$

with $\mathrm{Erf}(\cdot)$, $\mathrm{ErfInv}(\cdot)$ given by (2.3.22).

Proof. Let us prove (i). Given positive δ, let us set $A^\delta = A + \delta I$. Observe that the premise in (10.1.14) clearly implies that $A \succeq 0$, whence $A^\delta \succ 0$. Now let Y_ℓ be such that the conclusion in (10.1.13) holds true. Then $\left[\begin{array}{c|c} Y_\ell & A_\ell \\ \hline A_\ell & A^\delta \end{array}\right] \succeq 0$, whence, by the Schur Complement Lemma, $Y_\ell \succeq A_\ell[A^\delta]^{-1}A_\ell$, so that

$$\sum_\ell A_\ell[A^\delta]^{-1}A_\ell \preceq \sum_\ell Y_\ell \preceq A \preceq A^\delta.$$

We see that

$$\sum_\ell \big[\underbrace{[A^\delta]^{-1/2}A_\ell[A^\delta]^{-1/2}}_{B_\ell^\delta}\big]^2 \preceq I.$$

Further, relation (10.1.14) clearly implies that

$$\mathrm{Prob}\{-\Upsilon A^\delta \preceq \sum_\ell \zeta_\ell A_\ell \preceq \Upsilon A^\delta\} \geq \chi,$$

or, which is the same,

$$\mathrm{Prob}\{-\Upsilon I \preceq \sum_\ell \zeta_\ell B_\ell^\delta \preceq \Upsilon I\} \geq \chi.$$

Applying Theorem 10.1.1, we conclude that

$$\Omega > \Upsilon \Rightarrow \mathrm{Prob}\{-\Omega I \preceq \sum_\ell \zeta_\ell B_\ell^\delta \preceq \Omega I\} \geq 1 - \frac{1}{\chi}\exp\{-(\Omega - \Upsilon)^2/16\},$$

which in view of the structure of B_ℓ^δ is the same as

$$\Omega > \Upsilon \Rightarrow \mathrm{Prob}\{-\Omega A^\delta \preceq \sum_\ell \zeta_\ell A_\ell \preceq \Omega A^\delta\} \geq 1 - \frac{1}{\chi}\exp\{-(\Omega - \Upsilon)^2/16\}. \tag{10.1.17}$$

For every $\Omega > \Upsilon$, the sets $\{\zeta : -\Omega A^{1/t} \preceq \sum_\ell \zeta_\ell A_\ell \preceq \Omega A^{1/t}\}$, $t = 1, 2, ...$, shrink as t grows, and their intersection over $t = 1, 2, ...$ is the set $\{\zeta : -\Omega A \preceq \sum_\ell \zeta_\ell A_\ell \preceq \Omega A\}$, so that (10.1.17) implies (10.1.15), and (i) is proved. The proof of (ii) is completely similar, with Theorem 10.1.2 in the role of Theorem 10.1.1. $\square$

Comments. When $A \succ 0$, invoking the Schur Complement Lemma, the condition (10.1.13) is satisfied iff it is satisfied with $Y_\ell = A_\ell A^{-1}A_\ell$, which in turn is the case iff $\sum_\ell A_\ell A^{-1}A_\ell \preceq A$, or which is the same, iff $\sum_\ell [A^{-1/2}A_\ell A^{-1/2}]^2 \preceq I$. Thus, *condition* (10.1.4), (10.1.7) *introduced in connection with Problem* (?), *treated as a condition on the variable symmetric matrices* $A, A_1, ..., A_L$, *is LMI-representable,* (10.1.13) *being the representation.* Further, (10.1.13) can be written as the following explicit LMI on the matrices $A, A_1, ..., A_L$:

$$\mathrm{Arrow}(A, A_1, ..., A_L) \equiv \left[\begin{array}{c|ccc} A & A_1 & ... & A_L \\ \hline A_1 & A & & \\ \vdots & & \ddots & \\ A_L & & & A \end{array} \right] \succeq 0. \qquad (10.1.18)$$

Indeed, when $A \succ 0$, the Schur Complement Lemma says that the "block-arrow" matrix $\mathrm{Arrow}(A, A_1, ..., A_L)$ is $\succeq 0$ if and only if

$$\sum_\ell A_\ell A^{-1} A_\ell \preceq A,$$

and this is the case if and only if (10.1.13) holds. Thus, (10.1.13) and (10.1.18) are equivalent to each other when $A \succ 0$, which, by standard approximation argument, implies the equivalence of these two properties in the general case (that is, when $A \succeq 0$). It is worthy of noting that the set of matrices $(A, A_1, ..., A_L)$ satisfying (10.1.18) form a cone that can be considered as the matrix analogy of the Lorentz cone (look what happens when all the matrices are 1×1 ones).

10.2 THE APPROXIMATION SCHEME

To utilize the outlined observations and results in order to build a safe/"almost safe" tractable approximation of a chance constrained LMI in (10.1.2), we proceed as follows.

    1) We introduce the following:

    Conjecture 10.1. Under assumptions **A.I** or **A.II**, condition (10.1.13) implies the validity of (10.1.14) with known in advance $\chi > 1/2$ and "a moderate" (also known in advance) $\Upsilon > 0$.

    With properly chosen χ and Υ, this Conjecture indeed is true, see below. We, however, prefer not to stick to the corresponding worst-case-oriented values of χ and Υ and consider $\chi > 1/2$, $\Upsilon > 0$ as somehow chosen parameters of the construction to follow, and we proceed *as if* we know in advance that our conjecture, with the chosen Υ, χ, is true. Eventually we shall explain how to justify this tactics.

    2) Trusting in Conjecture 10.1, we have at our disposal constants $\Upsilon > 0$, $\chi \in (0.5, 1]$ such that (10.1.13) implies (10.1.14). We claim that *modulo Conjecture 10.1, the following systems of LMIs in variables $y, U_1, ..., U_L$ are safe tractable approximations of the chance constrained LMI in* (10.1.2):

In the case of **A.I**:

$$(a) \quad \left[\begin{array}{c|c} U_\ell & \mathcal{A}^\ell(y) \\ \hline \mathcal{A}^\ell(y) & \mathcal{A}^{\mathrm{n}}(y) \end{array}\right] \succeq 0, \ 1 \le \ell \le L$$

$$(b) \quad \rho^2 \sum_{\ell=1}^{L} U_\ell \preceq \Omega^{-2} \mathcal{A}^{\mathrm{n}}(y), \ \Omega = \Upsilon + 4\sqrt{\ln(\chi^{-1}\epsilon^{-1})};$$

(10.2.1)

In the case of **A.II**:

$$(a) \quad \left[\begin{array}{c|c} U_\ell & \mathcal{A}^\ell(y) \\ \hline \mathcal{A}^\ell(y) & \mathcal{A}^{\mathrm{n}}(y) \end{array}\right] \succeq 0, \ 1 \le \ell \le L$$

$$(b) \quad \rho^2 \sum_{\ell=1}^{L} U_\ell \preceq \Omega^{-2} \mathcal{A}^{\mathrm{n}}(y), \ \Omega = \Upsilon + \frac{\max[\mathrm{ErfInv}(\epsilon) - \mathrm{ErfInv}(1-\chi), 0]}{\max[1, \Upsilon^{-1}\mathrm{ErfInv}(1-\chi)]}$$

$$\le \Upsilon + \max[\mathrm{ErfInv}(\epsilon) - \mathrm{ErfInv}(1-\chi), 0].$$

(10.2.2)

Indeed, assume that y can be extended to a feasible solution $(y, U_1, ..., U_L)$ of (10.2.1). Let us set $A = \Omega^{-1}\mathcal{A}^{\mathrm{n}}(y)$, $A_\ell = \rho \mathcal{A}^\ell(y)$, $Y_\ell = \Omega \rho^2 U_\ell$. Then $\left[\begin{array}{c|c} Y_\ell & A_\ell \\ \hline A_\ell & A \end{array}\right] \succeq 0$ and $\sum_\ell Y_\ell \preceq A$ by (10.2.1). Applying Conjecture 10.1 to the matrices $A, A_1, ..., A_L$, we conclude that (10.1.14) holds true as well. Applying Corollary 10.1.3.(i), we get

$$\mathrm{Prob}\left\{\rho \sum_\ell \zeta_\ell \mathcal{A}^\ell(y) \npreceq \mathcal{A}^{\mathrm{n}}(y)\right\} = \mathrm{Prob}\left\{\sum_\ell \zeta_\ell A_\ell \npreceq \Omega A\right\}$$

$$\le \chi^{-1}\exp\{-(\Omega - \Upsilon)^2/16\} = \epsilon,$$

as claimed.

Relation (10.2.2) can be justified, modulo the validity of Conjecture 10.1, in the same fashion, with item (ii) of Corollary 10.1.3 in the role of item (i).

3) We replace the chance constrained LMI problem (10.1.2) with the outlined safe (modulo the validity of Conjecture 10.1) approximation, thus arriving at the approximating problem

$$\min_{y, \{U_\ell\}} \left\{ c^T y : \begin{array}{l} \left[\begin{array}{c|c} U_\ell & \mathcal{A}^\ell(y) \\ \hline \mathcal{A}^\ell(y) & \mathcal{A}^{\mathrm{n}}(y) \end{array}\right] \succeq 0, \ 1 \le \ell \le L \\ \\ \rho^2 \sum_\ell U_\ell \preceq \Omega^{-2} \mathcal{A}^{\mathrm{n}}(y), \ y \in \mathcal{Y} \end{array} \right\},$$

(10.2.3)

where Ω is given by the required tolerance *and our guesses for* Υ *and* χ according to (10.2.1) or (10.2.2), depending on whether we are in the case of a bounded random perturbation model (Assumption **A.I**) or a Gaussian one (Assumption **A.II**).

We solve the approximating SDO problem and obtain its optimal solution y_*. If (10.2.3) were indeed a safe approximation of (10.1.2), we would be done: y_* would be a *feasible* suboptimal solution to the chance constrained problem of interest. However, since we are not sure of the validity of Conjecture 10.1, we need an additional phase — *post-optimality analysis* — aimed at justifying the feasibility of y_* for the chance constrained problem. Note that *at this phase, we should not bother about the validity of Conjecture 10.1 in full generality — all we need is to* *justify the validity of the relation*

$$\text{Prob}\{-\Upsilon A \preceq \sum_\ell \zeta_\ell A_\ell \preceq \Upsilon A\} \geq \chi \qquad (10.2.4)$$

for specific matrices

$$A = \Omega^{-1}\mathcal{A}^{\mathrm{n}}(y_*), \ A_\ell = \rho\mathcal{A}^\ell(y_*), \ \ell = 1, ..., L, \qquad (10.2.5)$$

which we have in our disposal after y_* is found, and which indeed satisfy (10.1.13) (cf. "justification" of approximations (10.2.1), (10.2.2) in item 2)).

In principle, there are several ways to justify (10.2.4):

i) Under certain structural assumptions on the matrices A, A_ℓ and with properly chosen χ, Υ, our Conjecture 10.1 is provably true. Specifically, we shall see in section 10.4 that:

(a) when A, A_ℓ are diagonal, (which corresponds to the semidefinite reformulation of a Linear Optimization problem), Conjecture 10.1 holds true with $\chi = 0.75$ and $\Upsilon = \sqrt{3\ln(8m)}$ (recall that m is the size of the matrices $A, A_1, ..., A_L$);

(b) when A, A_ℓ are arrow matrices, (which corresponds to the semidefinite reformulation of a conic quadratic problem), Conjecture 10.1 holds true with $\chi = 0.75$ and $\Upsilon = 4\sqrt{2}$.

ii) Utilizing deep results from Functional Analysis, it can be proved (see Proposition B.5.2) that Conjecture 10.1 is true *for all matrices* $A, A_1, ..., A_L$ when $\chi = 0.75$ and $\Upsilon = 4\sqrt{\ln\max[m, 3]}$. It should be added that in order for our Conjecture 10.1 to be true for all L and all $m \times m$ matrices $A, A_1, ..., A_L$ with χ not too small, Υ should be *at least* $O(1)\sqrt{\ln m}$ with appropriate positive absolute constant $O(1)$.

In view of the above facts, we could *in principle* avoid the necessity to rely on any conjecture. However, the "theoretically valid" values of Υ, χ are *by definition* worst-case oriented and can be too conservative for the particular matrices we are interested in. The situation is even worse: these theoretically valid values reflect not the worst case "as it is," but rather our abilities to analyze this worst case and therefore are conservative estimates of the "true" (and already conservative) Υ, χ. This is why we prefer to use a technique that is based on *guessing* Υ, χ and a subsequent "verification of the guess" by a *simulation-based* justification of (10.2.4).

Comments. Note that our proposed course of action is completely similar to what we did in section 2.2. The essence of the matter there was as follows: we were interested in building a safe approximation of the chance constraint

$$\sum_{\ell=1}^{L} \zeta_\ell a_\ell \leq a \qquad (10.2.6)$$

with deterministic $a, a_1, ..., a_L \in \mathbb{R}$ and random ζ_ℓ satisfying Assumption **A.I**. To this end, we used the *provable fact* expressed by Proposition 2.3.1:

*Whenever random variables $\zeta_1, ..., \zeta_L$ satisfy **A.I** and deterministic reals $b, a_1, ..., a_L$ are such that*

$$\sqrt{\sum_{\ell=1}^{L} a_\ell^2} \leq b,$$

or, which is the same,

$$\mathrm{Arrow}(b, a_1, ..., a_L) \equiv \begin{bmatrix} b & a_1 & ... & a_L \\ \hline a_1 & b & & \\ \vdots & & \ddots & \\ a_L & & & b \end{bmatrix} \succeq 0,$$

one has

$$\forall \Omega > 0 : \mathrm{Prob}\left\{ \sum_{\ell=1}^{L} \zeta_\ell a_\ell \leq \Omega b \right\} \geq 1 - \psi(\Omega),$$
$$\psi(\Omega) = \exp\{-\Omega^2/2\}.$$

As a result, the condition

$$\mathrm{Arrow}(\Omega^{-1}a, a_1, ..., a_L) \equiv \begin{bmatrix} \Omega^{-1}a & a_1 & ... & a_L \\ \hline a_1 & \Omega^{-1}a & & \\ \vdots & & \ddots & \\ a_L & & & \Omega^{-1}a \end{bmatrix} \succeq 0$$

is sufficient for the validity of the chance constraint

$$\mathrm{Prob}\left\{ \sum_{\ell} \zeta_\ell a_\ell \leq a \right\} \geq 1 - \psi(\Omega).$$

What we are doing under Assumption **A.I** now can be sketched as follows: we are interested in building a safe approximation of the chance constraint

$$\sum_{\ell=1}^{L} \zeta_\ell A_\ell \preceq A \tag{10.2.7}$$

with deterministic $A, A_1, ..., A_L \in \mathbf{S}^m$ and random ζ_ℓ satisfying Assumption **A.I**. To this end, we use the following *provable fact* expressed by Theorem 10.1.1:

*Whenever random variables $\zeta_1, ..., \zeta_L$ satisfy **A.I** and deterministic symmetric matrices $B, A_1, ..., A_L$ are such that*

$$\mathrm{Arrow}(B, A_1, ..., A_L) \equiv \begin{bmatrix} B & A_1 & ... & A_L \\ \hline A_1 & B & & \\ \vdots & & \ddots & \\ A_L & & & B \end{bmatrix} \succeq 0, \tag{!}$$

<u>and</u>

$$\mathrm{Prob}\{-\Upsilon B \preceq \sum_{\ell} \zeta_\ell A_\ell \preceq \Upsilon B\} \geq \chi \tag{$*$}$$

with certain $\chi, \Upsilon > 0$, *one has*

$$\forall \Omega > \Upsilon : \text{Prob}\left\{\sum_{\ell=1}^{L} \zeta_\ell A_\ell \preceq \Omega B\right\} \geq 1 - \psi_{\Upsilon,\chi}(\Omega),$$

$$\psi_{\Upsilon,\chi}(\Omega) = \chi^{-1} \exp\{-(\Omega - \Upsilon)^2/16\}.$$

As a result, the condition

$$\text{Arrow}(\Omega^{-1}A, A_1, ..., A_L) \equiv \left[\begin{array}{c|ccc} \Omega^{-1}A & A_1 & \cdots & A_L \\ \hline A_1 & \Omega^{-1}A & & \\ \vdots & & \ddots & \\ A_L & & & \Omega^{-1}A \end{array}\right] \succeq 0$$

is a sufficient condition for the validity of the chance constraint

$$\text{Prob}\left\{\sum_{\ell} \zeta_\ell A_\ell \preceq A\right\} \geq 1 - \psi_{\Upsilon,\chi}(\Omega),$$

provided that $\Omega > \Upsilon$ *and* $\chi > 0$, $\Upsilon > 0$ *are such that the matrices* $B, A_1, ..., A_L$ *satisfy* (∗).

The constructions are pretty similar; the only difference is that in the matrix case we need an additional "provided that," which is absent in the scalar case. In fact, it is automatically present in the scalar case: from the Tschebyshev Inequality it follows that when $B, A_1, ..., A_L$ are scalars, condition (!) implies the validity of (∗) with, say, $\chi = 0.75$ and $\Upsilon = 2$. We now could apply the matrix-case result to recover the scalar-case, at the cost of replacing $\psi(\Omega)$ with $\psi_{2,0.75}(\Omega)$, which is not that big a loss.

Conjecture 10.1 *suggests* that in the matrix case we also should not bother much about "provided that" — it is automatically implied by (!), perhaps with a somehow worse value of Υ, but still not too large. As it was already mentioned, we can prove certain versions of the Conjecture, and we can also verify its validity, for *guessed* χ, Υ and matrices $B, A_1, ..., A_L$ that we are interested in, by simulation. The latter is the issue we consider next.

10.2.1 Simulation-Based Justification of (10.2.4)

Let us start with the following simple situation: there exists a random variable ξ taking value 1 with probability p and value 0 with probability $1 - p$; we can simulate ξ, that is, for every sample size N, observe realizations $\xi^N = (\xi_1, ..., \xi_N)$ of N independent copies of ξ. We do not know p, and our goal is to infer a reliable lower bound on this quantity from simulations. The simplest way to do this is as follows: given "reliability tolerance" $\delta \in (0, 1)$, a sample size N and an integer L, $0 \leq L \leq N$, let

$$\widehat{p}_{N,\delta}(L) = \min\left\{q \in [0, 1] : \sum_{k=L}^{N} \binom{N}{k} q^k (1 - q)^{N-k} \geq \delta\right\}.$$

The interpretation of $\widehat{p}_{N,\delta}(L)$ is as follows: imagine we are flipping a coin, and let q be the probability to get heads. We restrict q to induce chances at least δ to get L or more heads when flipping the coin N times, and $\widehat{p}_{N,\delta}(L)$ is exactly the smallest of these probabilities q. Observe that

$$(L > 0, \widehat{p} = \widehat{p}_{N,\delta}(L)) \Rightarrow \sum_{k=L}^{N} \binom{N}{k} \widehat{p}^k (1 - \widehat{p})^{N-k} = \delta \qquad (10.2.8)$$

and that $\widehat{p}_{N,\delta}(0) = 0$.

An immediate observation is as follows:

Lemma 10.2.1. For a fixed N, let $L(\xi^N)$ be the number of ones in a sample ξ^N, and let

$$\widehat{p}(\xi^N) = \widehat{p}_{N,\delta}(L(\xi^N)).$$

Then

$$\text{Prob}\{\widehat{p}(\xi^N) > p\} \leq \delta. \qquad (10.2.9)$$

Proof. Let

$$M(p) = \min \left\{ \mu \in \{0, 1, ..., N\} : \sum_{k=\mu+1}^{N} \binom{N}{k} p^k (1 - p)^{N-k} \leq \delta \right\}$$

(as always, a sum over empty set of indices is 0) and let Θ be the event $\{\xi^N : L(\xi^N) > M(p)\}$, so that by construction

$$\text{Prob}\{\Theta\} \leq \delta.$$

Now, the function

$$f(q) = \sum_{k=M(p)}^{N} \binom{N}{k} q^k (1 - q)^{N-k}$$

is a nondecreasing function of $q \in [0, 1]$, and by construction $f(p) > \delta$; it follows that if ξ^N is such that $\widehat{p} \equiv \widehat{p}(\xi^N) > p$, then $f(\widehat{p}) > \delta$ as well:

$$\sum_{k=M(p)}^{N} \binom{N}{k} \widehat{p}^k (1 - \widehat{p})^{N-k} > \delta \qquad (10.2.10)$$

and, besides this, $L(\xi^N) > 0$ (since otherwise $\widehat{p} = \widehat{p}_{N,\delta}(0) = 0 \leq p$). Since $L(\xi^N) > 0$, we conclude from (10.2.8) that

$$\sum_{k=L(\xi^N)}^{N} \binom{N}{k} \widehat{p}^k (1 - \widehat{p})^{N-k} = \delta,$$

which combines with (10.2.10) to imply that $L(\xi^N) > M(p)$, that is, ξ^N in question is such that the event Θ takes place. The bottom line is: the probability of the event $\widehat{p}(\xi^N) > p$ is at most the probability of Θ, and the latter, as we remember, is $\leq \delta$. $\square$

Lemma 10.2.1 says that the simulation-based (and thus random) quantity $\widehat{p}(\xi^N)$ is, with probability at least $1 - \delta$, a *lower bound* for unknown probability

$p \equiv \text{Prob}\{\xi = 1\}$. When p is not small, this bound is reasonably good already for moderate N, even when δ is extremely small, say, $\delta = 10^{-10}$. For example, here are simulation results for $p = 0.8$ and $\delta = 10^{-10}$:

N	10	100	1,000	10,000	100,000
$\widehat{p}$	0.06032	0.5211	0.6992	0.7814	0.7908

Coming back to our chance constrained problem (10.1.2), we can now use the outlined bounding scheme in order to carry out post-optimality analysis, namely, as follows:

Acceptance Test: *Given a reliability tolerance $\delta \in (0,1)$, guessed Υ, χ and a solution y_* to the associated problem (10.2.3), build the matrices (10.2.5). Choose an integer N, generate a sample of N independent realizations $\zeta^1, ..., \zeta^N$ of the random vector ζ, compute the quantity*

$$L = \text{Card}\{i : -\Upsilon A \preceq \sum_{\ell=1}^{L} \zeta_\ell^i A_\ell \preceq \Upsilon A\}$$

and set

$$\widehat{\chi} = \widehat{p}_{N,\delta}(L).$$

If $\widehat{\chi} \geq \chi$, accept y_*, that is, claim that y_* is a feasible solution to the chance constrained problem of interest (10.1.2).

By the above analysis, the random quantity $\widehat{\chi}$ is, with probability $\geq 1 - \delta$, a lower bound on $p \equiv \text{Prob}\{-\Upsilon A \preceq \sum_\ell \zeta_\ell A_\ell \preceq \Upsilon A\}$, so that the probability to accept y_* in the case when $p < \chi$ is at most δ. When this "rare event" does not occur, the relation (10.2.4) is satisfied, and therefore y_* is indeed feasible for the chance constrained problem. In other words, the probability to accept y_* when it is *not* a feasible solution to the problem of interest is at most δ.

The outlined scheme does not say what to do if y_* does *not* pass the Acceptance Test. A naive approach would be to check whether y_* satisfies the chance constraint by direct simulation. This approach indeed is workable when ϵ is not too small (say, $\epsilon \geq 0.001$); for small ϵ, however, it would require an unrealistically large simulation sample. A practical alternative is to resolve the approximating problem with Υ increased by a reasonable factor (say, 1.1 or 2), and to repeat this "trial and error" process until the Acceptance Test is passed.

10.2.2 A Modification

The outlined approach can be somehow streamlined when applied to a slightly modified problem (10.1.2), specifically, to the problem

$$\max_{\rho, y} \left\{ \rho : \text{Prob} \left\{ \mathcal{A}^n(y) + \rho \sum_{\ell=1}^{L} \zeta_\ell \mathcal{A}^\ell(y) \succeq 0 \right\} \geq 1 - \epsilon, c^T y \leq \tau_*, y \in \mathcal{Y} \right\} \quad (10.2.11)$$

where τ_* is a given upper bound on the original objective. Thus, now we want to maximize the level of random perturbations under the restrictions that $y \in \mathcal{Y}$ satisfies the chance constraint and is not too bad in terms of the original objective.

Approximating this problem by the method we have developed in the previous section, we end up with the problem

$$\min_{\beta,y,\{U_\ell\}} \left\{ \beta : \begin{array}{c} \left[\begin{array}{c|c} U_\ell & \mathcal{A}^\ell(y) \\ \hline \mathcal{A}^\ell(y) & \mathcal{A}^{\mathrm{n}}(y) \end{array} \right] \succeq 0, \, 1 \leq \ell \leq L \\[3mm] \sum_\ell U_\ell \preceq \beta \mathcal{A}^{\mathrm{n}}(y), \, c^T y \leq \tau_*, \, y \in \mathcal{Y} \end{array} \right\} \tag{10.2.12}$$

(cf. (10.2.3); in terms of the latter problem, $\beta = (\Omega\rho)^{-2}$, so that maximizing ρ is equivalent to minimizing β). Note that this problem remains the same whatever our guesses for Υ, χ. Further, (10.2.12) is a so called *GEVP* — Generalized Eigenvalue problem; while not being exactly a semidefinite program, it can be reduced to a "short sequence" of semidefinite programs via bisection in β and thus is efficiently solvable. Solving this problem, we arrive at a solution $\beta_*, y_*, \{U_\ell^*\}$; all we need is to understand what is the "feasibility radius" $\rho_*(y_*)$ of y_* — the largest ρ for which (y_*, ρ) satisfies the chance constraint in (10.2.11). As a matter of fact, we cannot compute this radius efficiently; what we will actually build is a reliable *lower bound* on the feasibility radius. This can be done by a suitable modification of the Acceptance Test. Let us set

$$A = \mathcal{A}^{\mathrm{n}}(y_*), A_\ell = \beta_*^{-1/2} \mathcal{A}^\ell(y_*), \ell = 1, ..., L; \tag{10.2.13}$$

note that these matrices satisfy (10.1.13). We apply to the matrices $A, A_1, ..., A_L$ the following procedure:

Randomized r-procedure:

Input: A collection of symmetric matrices $A, A_1, ..., A_L$ satisfying (10.1.13) and $\epsilon, \delta \in (0, 1)$.

Output: A random $r \geq 0$ such that with probability at least $1 - \delta$ one has

$$\mathrm{Prob}\{\zeta : -A \preceq r \sum_{\ell=1}^L \zeta_\ell A_\ell \preceq A\} \geq 1 - \epsilon. \tag{10.2.14}$$

Description:

i) We choose a K-point grid $\Gamma = \{\omega_1 < \omega_2 < ... < \omega_K\}$ with $\omega_1 \geq 1$ and a reasonably large ω_K, e.g., the grid

$$\omega_k = 1.1^k$$

and choose K large enough to ensure that Conjecture 10.1 holds true with $\Upsilon = \omega_K$ and $\chi = 0.75$; note that $K = O(1)\ln(\ln m)$ will do;

ii) We simulate N independent realizations $\zeta^1, ..., \zeta^N$ of ζ and compute the integers

$$L_k = \text{Card}\{i : -\omega_k A \preceq \sum_{\ell=1}^{L} \zeta_\ell^i A_\ell \preceq \omega_k A\}.$$

We then compute the quantities

$$\widehat{\chi}_k = \widehat{p}_{N,\delta/K}(L_k), \ k = 1, ..., K,$$

where $\delta \in (0,1)$ is the chosen in advance "reliability tolerance."
Setting

$$\chi_k = \text{Prob}\{-\omega_k A \preceq \sum_{\ell=1}^{L} \zeta_\ell A_\ell \preceq \omega_k A\},$$

we infer from Lemma 10.2.1 that

$$\widehat{\chi}_k \leq \chi_k, \ k = 1, ..., K \tag{10.2.15}$$

with probability at least $1 - \delta$.

iii) We define a function $\psi(s)$, $s \geq 0$, as follows.

*In the bounded case (Assumption **A.I**)*, we set

$$\psi_k(s) = \begin{cases} 1, \ s \leq \omega_k \\ \min\left[1, \widehat{\chi}_k^{-1} \exp\{-(s-\omega_k)^2/16\}\right], \ s > \omega_k; \end{cases}$$

*In the Gaussian case (Assumption **A.II**)*, we set

$$\psi_k(s) = \begin{cases} 1, \ \text{if} \ \widehat{\chi}_k \leq 1/2 \ \text{or} \ s \leq \omega_k, \\ \text{Erf}(\text{ErfInv}(1 - \widehat{\chi}_k) \\ \quad +(s - \omega_k)\max[1, \omega_k^{-1}\text{ErfInv}(1 - \widehat{\chi}_k)]), \\ \qquad\qquad\qquad\qquad\qquad\qquad \text{otherwise.} \end{cases}$$

In both cases, we set

$$\psi(s) = \min_{1 \leq k \leq K} \psi_k(s).$$

We claim that
(!) *When* (10.2.15) *takes place* (recall that this happens with probability at least $1 - \delta$), $\psi(s)$ *is, for all $s \geq 0$, an upper bound on*
$1 - \text{Prob}\{-sA \preceq \sum_{\ell=1}^{L} \zeta_\ell A_\ell \preceq sA\}$.
Indeed, in the case of (10.2.15), the matrices $A, A_1, ..., A_L$ (they from the very beginning are assumed to satisfy (10.1.13)) satisfy (10.1.14) with $\Upsilon = \omega_k$ and $\chi = \widehat{\chi}_k$; it remains to apply Corollary 10.1.3.

iv) We set

$$s_* = \inf\{s \geq 0 : \psi(s) \leq \epsilon\}, \quad r = \frac{1}{s_*}$$

and claim that with this r, (10.2.14) holds true.

Let us justify the outlined construction. Assume that (10.2.15) takes place. Then, by (!), we have

$$\text{Prob}\{-sA \preceq \sum_\ell \zeta_\ell \preceq sA\} \geq 1 - \psi(s).$$

Now, the function $\psi(s)$ is clearly continuous; it follows that when s_* is finite, we have $\psi(s_*) \leq \epsilon$, and therefore (10.2.14) holds true with $r = 1/s_*$. If $s_* = +\infty$, then $r = 0$, and the validity of (10.2.14) follows from $A \succeq 0$ (the latter is due to the fact that $A, A_1, ..., A_L$ satisfy (10.1.13)).

When applying the Randomized r-procedure to matrices (10.2.13), we end up with $r = r_*$ satisfying, with probability at least $1 - \delta$, the relation (10.2.14), and with our matrices $A, A_1, ..., A_L$ this relation reads

$$\text{Prob}\{-\mathcal{A}^{\mathrm{n}}(y_*) \preceq r_* \beta_*^{-1/2} \sum_{\ell=1}^L \zeta_\ell \mathcal{A}^\ell(y_*) \preceq \mathcal{A}^{\mathrm{n}}(y_*)\} \geq 1 - \epsilon.$$

Thus, setting

$$\widehat{\rho} = \frac{r_*}{\sqrt{\beta_*}},$$

we get, with probability at least $1 - \delta$, a valid lower bound on the feasibility radius $\rho_*(y_*)$ of y_*.

10.2.3 Illustration: Example 8.2.7 Revisited

Let us come back to the robust version of the Console Design problem (section 8.2.2, Example 8.2.7), where we were looking for a console capable (i) to withstand in a nearly optimal fashion a given load of interest, and (ii) to withstand equally well (that is, with the same or smaller compliance) every "occasional load" g from the Euclidean ball $B_\rho = \{g : \|g\|_2 \leq \rho\}$ of loads distributed along the 10 free nodes of the construction. Formally, our problem was

$$\max_{t,r} \left\{ r : \begin{array}{l} \left[\begin{array}{c|c} 2\tau_* & f^T \\ \hline f & A(t) \end{array}\right] \succeq 0 \\ \left[\begin{array}{c|c} 2\tau_* & rh^T \\ \hline rh & A(t) \end{array}\right] \succeq 0 \ \forall(h : \|h\|_2 \leq 1) \\ t \geq 0, \sum_{i=1}^N t_i \leq 1 \end{array} \right\}, \tag{10.2.16}$$

where $\tau_* > 0$ and the load of interest f are given and $A(t) = \sum_{i=1}^N t_i b_i b_i^T$ with $N = 54$ and known ($\mu = 20$)-dimensional vectors b_i. Note that what is now called r was called ρ in section 8.2.2.

Speaking about a console, it is reasonable to assume that in reality the "occasional load" vector is random $\sim \mathcal{N}(0, \rho^2 I_\mu)$ and to require that the construction should be capable of carrying such a load with the compliance $\leq \tau_*$ with probability at least $1 - \epsilon$, with a very small value of ϵ, say, $\epsilon = 10^{-10}$. Let us now look for a console that satisfies these requirements with the largest possible value of ρ. The corresponding chance constrained problem is

$$
\max_{t,\rho}\left\{\rho: \begin{array}{l} \left[\begin{array}{c|c} 2\tau_* & f^T \\ \hline f & A(t) \end{array}\right] \succeq 0 \\ \mathrm{Prob}_{h\sim\mathcal{N}(0,I_{20})}\left\{\left[\begin{array}{c|c} 2\tau_* & \rho h^T \\ \hline \rho h & A(t) \end{array}\right] \succeq 0\right\} \geq 1-\epsilon \\ t \geq 0, \sum_{i=1}^{N} t_i \leq 1 \end{array}\right\}, \qquad (10.2.17)
$$

and its approximation (10.2.12) is

$$
\min_{t,\beta,\{U_\ell\}_{\ell=1}^{20}}\left\{\beta: \begin{array}{l} \left[\begin{array}{c|c} 2\tau_* & f^T \\ \hline f & A(t) \end{array}\right] \succeq 0 \\ \left[\begin{array}{c|c} U_\ell & E_\ell \\ \hline E_\ell & Q(t) \end{array}\right] \succeq 0, \ 1 \leq \ell \leq \mu = 20 \\ \sum_{\ell=1}^{\mu} U_\ell \preceq \beta Q(t), \ t \geq 0, \sum_{i=1}^{N} t_i \leq 1 \end{array}\right\}, \qquad (10.2.18)
$$

where $E_\ell = e_0 e_\ell^T + e_\ell e_0^T$, $e_0, ..., e_\mu$ are the standard basic orths in $\mathbb{R}^{\mu+1} = \mathbb{R}^{21}$, and $Q(t)$ is the matrix $\mathrm{Diag}\{2\tau_*, A(t)\} \in \mathbf{S}^{\mu+1} = \mathbf{S}^{21}$.

Note that the matrices participating in this problem are simple enough to allow us to get without much difficulty a "nearly optimal" description of theoretically valid values of Υ, χ (see section 10.4). Indeed, here Conjecture 10.1 is valid with every $\chi \in (1/2, 1)$ provided that $\Upsilon \geq O(1)(1-\chi)^{-1/2}$. Thus, after the optimal solution t_{ch} to the approximating problem is found, we can avoid the simulation-based identification of a lower bound $\widehat{\rho}$ on $\rho_*(t_{\mathrm{ch}})$ (that is, on the largest ρ such that (t_{ch}, ρ) satisfies the chance constraint in (10.2.17)) and can get a 100%-reliable lower bound on this quantity, while the simulation-based technique is capable of providing no more than a $(1-\delta)$-reliable lower bound on $\rho_*(t_{\mathrm{ch}})$ with perhaps small, but positive δ. It turns out, however, that in our particular problem this 100%-reliable lower bound on $\rho_*(y_*)$ is significantly (by factor about 2) smaller than the $(1-\delta)$-reliable bound given by the outlined approach, even when δ is as small as 10^{-10}. This is why in the experiment we are about to discuss, we used the simulation-based lower bound on $\rho_*(t_{\mathrm{ch}})$.

The results of our experiment are as follows. The console given by the optimal solution to (10.2.18), let it be called the *chance constrained* design, is presented in figure 10.1 (cf. figures 8.1, 8.2 representing the nominal and the robust designs, respectively). The lower bounds on the feasibility radius for the chance constrained design associated with $\epsilon = \delta = 10^{-10}$ are presented in table 10.1; the plural ("bounds") comes from the fact that we worked with three different sample sizes N shown in table 10.1. Note that we can apply the outlined techniques to bound from below the feasibility radius of the *robust* design t_{rb} — the one given by the optimal solution to (10.2.16), see figure 8.2; the resulting bounds are presented in table 10.1.

Finally, we note that we can exploit the specific structure of the particular problem in question to get alternative lower bounds on the feasibility radii of the

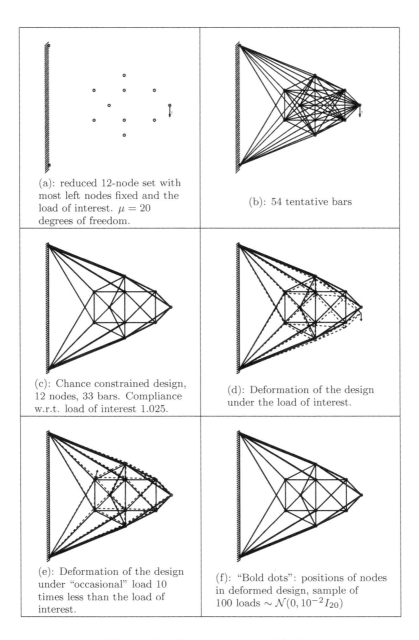

(a): reduced 12-node set with most left nodes fixed and the load of interest. $\mu = 20$ degrees of freedom.

(b): 54 tentative bars

(c): Chance constrained design, 12 nodes, 33 bars. Compliance w.r.t. load of interest 1.025.

(d): Deformation of the design under the load of interest.

(e): Deformation of the design under "occasional" load 10 times less than the load of interest.

(f): "Bold dots": positions of nodes in deformed design, sample of 100 loads $\sim \mathcal{N}(0, 10^{-2}I_{20})$

Figure 10.1 Chance constrained design.

Design	Lower bound on feasibility radius		
	$N = 10,000$	$N = 100,000$	$N = 1,000,000$
chance constrained t_{ch}	0.0354	0.0414	0.0431
robust t_{rb}	0.0343	0.0380	0.0419

Table 10.1 $(1 - 10^{-10})$-confident lower bounds on feasibility radii for the chance constrained and the robust designs.

chance constrained and the robust designs. Recall that the robust design ensures
that the compliance of the corresponding console w.r.t. *any* load g of Euclidean
norm $\leq r_*$ is at most τ_*; here $r_* \approx 0.362$ is the optimal value in (10.2.16). Now, if
ρ is such that $\mathrm{Prob}_{h \sim \mathcal{N}(0, I_{20})}\{\rho\|h\|_2 > r_*\} \leq \epsilon = 10^{-10}$, then clearly ρ is a 100%-
reliable lower bound on the feasibility radius of the robust design. We can easily
compute the largest ρ satisfying the latter condition; it turns out to be 0.0381, 9%
less than the best simulation-based lower bound. Similar reasoning can be applied
to the chance constrained design t_{ch}: we first find the largest $r = r_+$ for which
(t_{ch}, r) is feasible for (10.2.16) (it turns out that $r_+ = 0.321$), and then find the
largest ρ such that $\mathrm{Prob}_{h \sim \mathcal{N}(0, I_{20})}\{\rho\|h\|_2 > r_+\} \leq \epsilon = 10^{-10}$, ending up with the
lower bound 0.0337 on the feasibility radius of the chance constrained design (25.5%
worse than the best related bound in table 10.1).

10.3 GAUSSIAN MAJORIZATION

Under favorable circumstances, we can apply the outlined approximation scheme
to random perturbations that do not fit exactly neither Assumption **A.I**, nor Assumption **A.II**. As an instructive example, consider the case where the random
perturbations ζ_ℓ, $\ell = 1, ..., L$, in (10.1.1) are independent and symmetrically and
unimodally distributed w.r.t. 0. Assume also that we can point out scaling factors $\sigma_\ell > 0$ such that the distribution of each ζ_ℓ is less diffuse than the Gaussian
$\mathcal{N}(0, \sigma_\ell^2)$ distribution (see Definition 4.4.1). Note that in order to build a safe
tractable approximation of the chance constrained LMI

$$\mathrm{Prob}\left\{\mathcal{A}^{\mathrm{n}}(y) + \sum_{\ell=1}^{L} \zeta_\ell \mathcal{A}_\ell(y) \succeq 0\right\} \geq 1 - \epsilon, \qquad (10.1.2)$$

or, which is the same, the constraint

$$\mathrm{Prob}\left\{\mathcal{A}^{\mathrm{n}}(y) + \sum_{\ell=1}^{L} \widetilde{\zeta}_\ell \widetilde{\mathcal{A}}^\ell(y) \succeq 0\right\} \geq 1 - \epsilon \qquad \left[\begin{array}{c} \widetilde{\zeta}_\ell = \sigma_\ell^{-1}\zeta_\ell \\ \widetilde{\mathcal{A}}^\ell(y) = \sigma_\ell \mathcal{A}^\ell(y) \end{array}\right]$$

it suffices to build such an approximation for the symmetrized version

$$\mathrm{Prob}\{-\mathcal{A}^{\mathrm{n}}(y) \preceq \sum_{\ell=1}^{L} \widetilde{\zeta}_\ell \widetilde{\mathcal{A}}^\ell(y) \preceq \mathcal{A}^{\mathrm{n}}(y)\} \geq 1 - \epsilon \qquad (10.3.1)$$

of the constraint. Observe that the random variables $\widetilde{\zeta}_\ell$ are independent and possess symmetric and unimodal w.r.t. 0 distributions that are less diffuse than the $\mathcal{N}(0,1)$ distribution. Denoting by η_ℓ, $\ell = 1, ..., L$, independent $\mathcal{N}(0,1)$ random variables and invoking the Majorization Theorem (Theorem 4.4.6), we see that the validity of the chance constraint

$$\text{Prob}\{-\mathcal{A}^{\mathrm{n}}(y) \preceq \sum_{\ell=1}^{L} \eta_\ell \widetilde{\mathcal{A}}^\ell(y) \preceq \mathcal{A}^{\mathrm{n}}(y)\} \geq 1 - \epsilon$$

— and this is the constraint we do know how to handle — is a sufficient condition for the validity of (10.3.1). Thus, in the case of unimodally and symmetrically distributed ζ_ℓ admitting "Gaussian majorants," we can act, essentially, as if we were in the Gaussian case **A.II**.

It is worth noticing that we can apply the outlined "Gaussian majorization" scheme even in the case when ζ_ℓ are symmetrically and unimodally distributed in $[-1, 1]$ (a case that we know how to handle even without the unimodality assumption), and this could be profitable. Indeed, by Example 4.4.3 (section 4.4), in the case in question ζ_ℓ are less diffuse than the random variables $\eta_\ell \sim \mathcal{N}(0, 2/\pi)$, and we can again reduce the situation to Gaussian. The advantage of this approach is that the absolute constant factor $\frac{1}{16}$ in the exponent in (10.1.15) is rather small. Therefore replacing (10.1.15) with (10.1.16), even after replacing our original variables ζ_ℓ with their less concentrated "Gaussian majorants" η_ℓ, can lead to better results. To illustrate this point, here is a report on a numerical experiment:

1) We generated $L = 100$ matrices $A_\ell \in \mathbf{S}^{40}$, $\ell = 1, ..., L$, such that $\sum_\ell A_\ell^2 \preceq I$, (which clearly implies that $A = I, A_1, ..., A_L$ satisfy (10.1.13));

2) We applied the bounded case version of the Randomized r procedure to the matrices $A, A_1, ..., A_L$ and the independent random variables ζ_ℓ uniformly distributed on $[-1, 1]$, setting δ and ϵ to 10^{-10};

3) We applied the Gaussian version of the same procedure, with the same ϵ, δ, to the matrices $A, A_1, ..., A_L$ and independent $\mathcal{N}(0, 2/\pi)$ random variables η_ℓ in the role of ζ_ℓ.

In both 2) and 3), we used the same grid $\omega_k = 0.01 \cdot 10^{0.1k}$, $0 \leq k \leq 40$.

By the above arguments, both in 2) and in 3) we get, with probability at least $1 - 10^{-10}$, lower bounds on the largest ρ such that

$$\text{Prob}\{-I \preceq \rho \sum_{\ell=1}^{L} \zeta_\ell A_\ell \preceq I\} \geq 1 - 10^{-10}.$$

Here are the bounds obtained:

Bounding	Lower Bound	
scheme	$N = 1000$	$N = 10000$
2)	0.0489	0.0489
3)	0.185	0.232

We see that while we can process the case of uniformly distributed ζ_ℓ "as it is," it is better to process it via Gaussian majorization.

To conclude this section, we present another "Gaussian Majorization" result. Its advantage is that it does not require the random variables ζ_ℓ to be symmetrically or unimodally distributed; what we need, essentially, is just independence plus zero means. We start with some definitions. Let $\mathcal{R}_n$ be the space of Borel probability distributions on $\mathbb{R}^n$ with zero mean. For a random variable η taking values in $\mathbb{R}^n$, we denote by P_η the corresponding distribution, and we write $\eta \in \mathcal{R}_n$ to express that $P_\eta \in \mathcal{R}_n$. Let also $\mathcal{CF}_n$ be the set of all *convex* functions f on $\mathbb{R}^n$ with linear growth, meaning that there exists $c_f < \infty$ such that $|f(u)| \le c_f(1+\|u\|_2)$ for all u.

Definition 10.3.1. Let $\xi, \eta \in \mathcal{R}_n$. We say that η dominates ξ (notation: $\xi \preceq_{\mathrm{c}} \eta$, or $P_\xi \preceq_{\mathrm{c}} P_\eta$, or $\eta \succeq_{\mathrm{c}} \xi$, or $P_\eta \succeq_{\mathrm{c}} P_\xi$) if

$$\int f(u)dP_\xi(u) \le \int f(u)dP_\eta(u)$$

for every $f \in \mathcal{CF}_n$.

Note that in the literature the relation $\succeq_{\mathrm{c}}$ is called "convex dominance." The properties of the relation $\succeq_{\mathrm{c}}$ we need are summarized as follows:

Proposition 10.3.2.

i) $\preceq_{\mathrm{c}}$ is a partial order on $\mathcal{R}_n$.

ii) If $P_1, ..., P_k, Q_1, ..., Q_k \in \mathcal{R}_n$, and $P_i \preceq_{\mathrm{c}} Q_i$ for every i, then $\sum_i \lambda_i P_i \preceq_{\mathrm{c}} \sum_i \lambda_i Q_i$ for all nonnegative λ_i with unit sum.

iii) If $\xi \in \mathcal{R}_n$ and $t \ge 1$ is deterministic, then $t\xi \succeq_{\mathrm{c}} \xi$.

iv) Let $P_1, Q_1 \in \mathcal{R}_r$, $P_2, Q_2 \in \mathcal{R}_s$ be such that $P_i \preceq_{\mathrm{c}} Q_i$, $i = 1, 2$. Then $P_1 \times P_2 \preceq_{\mathrm{c}} Q_1 \times Q_2$. In particular, if $\xi_1, ..., \xi_n, \eta_1, ..., \eta_n \in \mathcal{R}_1$ are independent and $\xi_i \preceq_{\mathrm{c}} \eta_i$ for every i, then $[\xi_1; ...; \xi_n] \preceq_{\mathrm{c}} [\eta_1; ...; \eta_n]$.

v) If $\xi_1, ..., \xi_k, \eta_1, ..., \eta_k \in \mathcal{R}_n$ are independent random variables, $\xi_i \preceq_{\mathrm{c}} \eta_i$ for every i, and $S_i \in \mathbb{R}^{m \times n}$ are deterministic matrices, then $\sum_i S_i \xi_i \preceq_{\mathrm{c}} \sum_i S_i \eta_i$.

vi) Let $\xi \in \mathcal{R}_1$ be supported on $[-1, 1]$ and $\eta \sim \mathcal{N}(0, \pi/2)$. Then $\eta \succeq_{\mathrm{c}} \xi$.

vii) If ξ, η are symmetrically and unimodally distributed w.r.t. the origin scalar random variables with finite expectations and $\eta \succeq_{\mathrm{m}} \xi$ (see section 4.4), then $\eta \succeq_{\mathrm{c}} \xi$ as well. In particular, if ξ has unimodal w.r.t. 0 distribution and is supported on $[-1, 1]$ and $\eta \sim \mathcal{N}(0, 2/\pi)$, then $\eta \succeq_{\mathrm{c}} \xi$ (cf. Example 4.4.3).

viii) Assume that $\xi \in \mathcal{R}_n$ is supported in the unit cube $\{u : \|u\|_\infty \le 1\}$ and is "absolutely symmetrically distributed," meaning that if J is a diagonal matrix with diagonal entries ± 1, then $J\xi$ has the same distribution as ξ. Let also $\eta \sim \mathcal{N}(0, (\pi/2)I_n)$. Then $\xi \preceq_{\mathrm{c}} \eta$.

ix) Let $\xi, \eta \in \mathcal{R}_r$, $\xi \sim \mathcal{N}(0, \Sigma)$, $\eta \sim \mathcal{N}(0, \Theta)$ with $\Sigma \preceq \Theta$. Then $\xi \preceq_{\mathrm{c}} \eta$.

Our main result here is as follows.

Theorem 10.3.3. Let $\eta \sim \mathcal{N}(0, I_L)$, and let $\zeta \in \mathcal{R}_L$ be such that $\zeta \preceq_c \eta$. Let, further, $Q \subset \mathbb{R}^L$ be a closed convex set such that

$$\chi \equiv \mathrm{Prob}\{\eta \in Q\} > 1/2.$$

Then for every $\gamma > 1$, one has

$$
\begin{aligned}
\mathrm{Prob}\{\zeta \notin \gamma Q\} &\leq \inf_{1 \leq \beta < \gamma} \frac{1}{\gamma - \beta} \int_\beta^\infty \mathrm{Erf}(r\mathrm{ErfInv}(1-\chi))dr \\
&\leq \inf_{1 \leq \beta < \gamma} \frac{1}{2(\gamma - \beta)} \int_\beta^\infty \exp\{-r^2 \mathrm{ErfInv}^2(1-\chi)/2\}dr,
\end{aligned}
\tag{10.3.2}
$$

where $\mathrm{Erf}(\cdot)$, $\mathrm{ErfInv}(\cdot)$ are given by (2.3.22).

The assumption $\zeta \preceq_c \eta$ is valid, in particular, if $\zeta = [\zeta_1; ...; \zeta_L]$ with independent ζ_ℓ such that $P_{\zeta_\ell} \in \mathcal{R}_1$ and $P_{\zeta_\ell} \preceq_c \mathcal{N}(0,1)$.

The proofs are presented in section B.5.3 in the Appendix.

10.4 CHANCE CONSTRAINED LMIS: SPECIAL CASES

We intend to consider two cases where it is easy to justify Conjecture 10.1. While the structural assumptions on the matrices $A, A_1, ..., A_L$ in these two cases seem to be highly restrictive, the results are nevertheless important: they cover the situations arising in randomly perturbed Linear and Conic Quadratic Optimization. We begin with a slight relaxation of Assumptions **A.I–II**:

Assumption A.III: The random perturbations $\zeta_1, ..., \zeta_L$ are independent, zero mean and "of order of 1," meaning that

$$\mathbf{E}\{\exp\{\zeta_\ell^2\}\} \leq \exp\{1\}, \ \ell = 1, ..., L.$$

Note that Assumption **A.III** is implied by **A.I** and is "almost implied" by **A.II**; indeed, $\zeta_\ell \sim \mathcal{N}(0,1)$ implies that the random variable $\widetilde{\zeta}_\ell = \sqrt{(1-\mathrm{e}^{-2})/2}\zeta_\ell$ satisfies $\mathbf{E}\{\exp\{\widetilde{\zeta}_\ell^2\}\} \leq \exp\{1\}$.

10.4.1 The Diagonal Case: Chance Constrained Linear Optimization

Theorem 10.4.1. Let $A, A_1, ..., A_L \in \mathbf{S}^m$ be diagonal matrices satisfying (10.1.13) and let the random variables ζ_ℓ satisfy Assumption **A.III**. Then, for every $\chi \in (0,1)$, with $\Upsilon = \Upsilon(\chi) \equiv \sqrt{3 \ln\left(\frac{2m}{1-\chi}\right)}$ one has

$$\mathrm{Prob}\{-\Upsilon A \preceq \sum_{\ell=1}^L \zeta_\ell A_\ell \preceq \Upsilon A\} \geq \chi \tag{10.4.1}$$

(cf. (10.1.14)). In the case of $\zeta_\ell \sim \mathcal{N}(0,1)$, relation (10.4.1) holds true with $\Upsilon = \Upsilon(\chi) \equiv \sqrt{2 \ln\left(\frac{m}{1-\chi}\right)}$.

Proof. It is immediately seen that we lose nothing when assuming that $A \succ 0$ (cf. the proof of Corollary 10.1.3). With this assumption, passing from diagonal matrices A, A_ℓ to the diagonal matrices $B_\ell = A^{-1/2} A_\ell A^{-1/2}$, the statement to be proved reads as follows:

If $B_\ell \in \mathbf{S}^m$ are deterministic diagonal matrices such that $\sum\limits_\ell B_\ell^2 \preceq I$ and

ζ_ℓ satisfy **A.III**, *then, for every $\chi \in (0,1)$, one has*

$$\mathrm{Prob}\{\|\sum_{\ell=1}^{L} \zeta_\ell B_\ell\| \leq \underbrace{\sqrt{3\ln\left(\frac{2m}{1-\chi}\right)}}_{\Upsilon(\chi)}\} \geq \chi. \qquad (10.4.2)$$

When $\zeta_\ell \sim \mathcal{N}(0,1)$, $\ell = 1, ..., L$, the relation remains true with $\Upsilon(\chi)$ reduced to $\sqrt{2\ln(m/(1-\chi))}$.

The proof of the latter statement is based on the standard argument used in deriving results on large deviations of sums of "light-tail" independent random variables. First we need the following result.

Lemma 10.4.2. Let β_ℓ, $\ell = 1, ..., L$, $\gamma > 0$ be deterministic reals such that $\sum\limits_\ell \beta_\ell^2 \leq 1$. Then

$$\forall \Upsilon > 0 : \mathrm{Prob}\left\{|\sum_{\ell=1}^{L} \beta_\ell \zeta_\ell| > \Upsilon\right\} \leq 2\exp\{-\Upsilon^2/3\}. \qquad (10.4.3)$$

Proof of Lemma 10.4.2. Observe, first, that whenever ξ is a random variable with zero mean such that $\mathbf{E}\{\exp\{\xi^2\}\} \leq \exp\{1\}$, one has

$$\mathbf{E}\{\exp\{\gamma\xi\}\} \leq \exp\{3\gamma^2/4\}. \qquad (10.4.4)$$

Indeed, observe that by Holder Inequality the relation $\mathbf{E}\left\{\exp\{\xi^2\}\right\} \leq \exp\{1\}$ implies that $\mathbf{E}\left\{\exp\{s\xi^2\}\right\} \leq \exp\{s\}$ for all $s \in [0,1]$. It is immediately seen that $\exp\{x\} - x \leq \exp\{9x^2/16\}$ for all x. Assuming that $9\gamma^2/16 \leq 1$, we therefore have

$$\begin{aligned} \mathbf{E}\left\{\exp\{\gamma\xi\}\right\} &= \mathbf{E}\left\{\exp\{\gamma\xi\} - \gamma\xi\right\} \; [\xi \text{ is with zero mean}] \\ &\leq \mathbf{E}\left\{\exp\{9\gamma^2\xi^2/16\}\right\} \\ &\leq \exp\{9\gamma^2/16\} \; [\text{since } 9\gamma^2/16 \leq 1] \\ &\leq \exp\{3\gamma^2/4\}, \end{aligned}$$

as required in (10.4.4). Now let $9\gamma^2/16 \geq 1$. For all γ we have $\gamma\xi \leq 3\gamma^2/8 + 2\xi^2/3$, whence

$$\begin{aligned} \mathbf{E}\left\{\exp\{\gamma\xi\}\right\} &\leq \exp\{3\gamma^2/8\}\exp\{2\xi^2/3\} \leq \exp\{3\gamma^2/8 + 2/3\} \\ &\leq \exp\{3\gamma^2/4\} \; [\text{since } \gamma^2 \geq 16/9] \end{aligned}$$

We see that (10.4.4) is valid for all γ.

We now have

$$\mathbf{E}\left\{\exp\{\gamma\textstyle\sum_{\ell=1}^{L}\beta_\ell\zeta_\ell\}\right\} = \prod_{\ell=1}^{L}\mathbf{E}\left\{\exp\{\gamma\beta_\ell\zeta_\ell\}\right\} \quad [\zeta_1,...,\zeta_L \text{ are independent}]$$
$$\leq \prod_{\ell=1}^{L}\exp\{3\gamma^2\beta_\ell^2/4\} \quad [\text{by Lemma}]$$
$$\leq \exp\{3\gamma^2/4\} \quad [\text{since } \textstyle\sum_\ell \beta_\ell^2 \leq 1].$$

We now have

$$\text{Prob}\left\{\textstyle\sum_{\ell=1}^{L}\beta_\ell\zeta_\ell > \Upsilon\right\}$$
$$\leq \min_{\gamma\geq 0}\exp\{-\Upsilon\gamma\}\mathbf{E}\left\{\exp\{\gamma\textstyle\sum_\ell\beta_\ell\zeta_\ell\}\right\} \quad [\text{Tschebyshev Inequality}]$$
$$\leq \min_{\gamma\geq 0}\exp\{-\Upsilon\gamma + 3\gamma^2/4\} \quad [\text{by (10.4.4)}]$$
$$= \exp\{-\Upsilon^2/3\}.$$

Replacing ζ_ℓ with $-\zeta_\ell$, we get that $\text{Prob}\{\sum_\ell \beta_\ell\zeta_\ell < -\Upsilon\} \leq \exp\{-\Upsilon^2/3\}$ as well, and (10.4.3) follows. $\quad\square$

Proof of (10.4.1). Let s_i be the i-th diagonal entry in the random diagonal matrix $S = \sum_{\ell=1}^{L}\zeta_\ell B_\ell$. Taking into account that B_ℓ are diagonal with $\sum_\ell B_\ell^2 \preceq I$, we can apply Lemma 10.4.2 to get the bound

$$\text{Prob}\{|s_i| > \Upsilon\} \leq 2\exp\{-\Upsilon^2/3\};$$

since $\|S\| = \max_{1\leq i\leq m}|s_i|$, (10.4.2) follows.

Refinements in the case of $\zeta_\ell \sim \mathcal{N}(0,1)$ are evident: here the i-th diagonal entry s_i in the random diagonal matrix $S = \sum_\ell \zeta_\ell B_\ell$ is $\sim \mathcal{N}(0,\sigma_i^2)$ with $\sigma_i \leq 1$, whence $\text{Prob}\{|s_i| > \Upsilon\} \leq \exp\{-\Upsilon^2/2\}$ and therefore $\text{Prob}\{\|S\| > \Upsilon\} \leq m\exp\{-\Upsilon^2/2\}$, so that $\Upsilon(\chi)$ in (10.4.2) can indeed be reduced to $\sqrt{2\ln(m/(1-\chi))}$. $\quad\square$

The case of chance constrained LMI with diagonal matrices $\mathcal{A}^{\mathrm{n}}(y)$, $\mathcal{A}^\ell(y)$ has an important application — Chance Constrained Linear Optimization. Indeed, consider a randomly perturbed Linear Optimization problem

$$\min_y \left\{c^T y : A_\zeta y \geq b_\zeta\right\} \tag{10.4.5}$$

where A_ζ, b_ζ are affine in random perturbations ζ:

$$[A_\zeta, b_\zeta] = [A^{\mathrm{n}}, b^{\mathrm{n}}] + \sum_{\ell=1}^{L}\zeta_\ell[A^\ell, b^\ell];$$

as usual, we have assumed w.l.o.g. that the objective is certain. The chance constrained version of this problem is

$$\min_y \left\{c^T y : \text{Prob}\{A_\zeta y \geq b_\zeta\} \geq 1 - \epsilon\right\}. \tag{10.4.6}$$

Setting $\mathcal{A}^{\mathrm{n}}(y) = \mathrm{Diag}\{A^{\mathrm{n}}y - b^{\mathrm{n}}\}$, $\mathcal{A}^{\ell}(y) = \mathrm{Diag}\{A^{\ell}y - b^{\ell}\}$, $\ell = 1, ..., L$, we can rewrite (10.4.6) equivalently as the chance constrained semidefinite problem

$$\min_{y} \left\{ c^T y : \mathrm{Prob}\{\mathcal{A}_{\zeta}(y) \succeq 0\} \geq 1 - \epsilon \right\}, \ \mathcal{A}_{\zeta}(y) = \mathcal{A}^{\mathrm{n}}(y) + \sum_{\ell} \zeta_{\ell}\mathcal{A}^{\ell}(y), \quad (10.4.7)$$

and process this problem via the outlined approximation scheme. Note the essential difference between what we are doing now and what was done in chapter 2. There we focused on safe approximation of chance constrained *scalar* linear inequality, here we are speaking about approximating a chance constrained coordinate-wise *vector* inequality. Besides this, our approximation scheme is, in general, "semi-analytic" — it involves simulation and as a result produces a solution that is feasible for the chance constrained problem with probability close to 1, but not with probability 1.

Of course, the safe approximations of chance constraints developed in chapter 2 can be used to process coordinate-wise vector inequalities as well. The natural way to do it is to replace the chance constrained vector inequality in (10.4.6) with a bunch of chance constrained scalar inequalities

$$\mathrm{Prob}\left\{(A_{\zeta}y - b_{\zeta})_i \geq 0\right\} \geq 1 - \epsilon_i, \ i = 1, ..., m \equiv \dim b_{\zeta}, \quad (10.4.8)$$

where the tolerances $\epsilon_i \geq 0$ satisfy the relation $\sum_i \epsilon_i = \epsilon$. The validity of (10.4.8) clearly is a sufficient condition for the validity of the chance constraint in (10.4.6), so that replacing these constraints with their safe tractable approximations from chapter 2, we end up with a safe tractable approximation of the chance constrained LO problem (10.4.6). A drawback of this approach is in the necessity to "guess" the quantities ϵ_i. The ideal solution would be to treat them as additional decision variables and to optimize the safe approximation in both y and ϵ_i. Unfortunately, all approximation schemes for scalar chance constraints presented in chapter 2 result in approximations that are *not* jointly convex in $y, \{\epsilon_i\}$. As a result, joint optimization in y, ϵ_i is more wishful thinking than a computationally solid strategy. Seemingly the only simple way to resolve this difficulty is to set all ϵ_i equal to ϵ/m.

It is instructive to compare the "constraint-by-constraint" safe approximation of a chance constrained LO (10.4.6) given by the results of chapter 2 with our present approximation scheme. To this end, let us focus on the following version of the chance constrained problem:

$$\max_{\rho,y} \left\{ \rho : c^T y \leq \tau_*, \mathrm{Prob}\left\{A_{\rho\zeta}y \geq b_{\rho\zeta}\right\} \geq 1 - \epsilon \right\} \quad (10.4.9)$$

(cf. (10.2.11)). To make things as simple as possible, we assume also that $\zeta_{\ell} \sim \mathcal{N}(0,1)$, $\ell = 1, ..., L$.

The "constraint-by-constraint" safe approximation of (10.4.9) is the chance constrained problem

$$\max_{\rho,y} \left\{ \rho : c^T y \leq \tau_*, \mathrm{Prob}\left\{(A_{\rho\zeta}y - b_{\rho\zeta})_i \geq 0\right\} \geq 1 - \epsilon/m \right\},$$

where m is the number of rows in A_{ζ}. A chance constraint

$$\mathrm{Prob}\left\{(A_{\rho\zeta}y - b_{\rho\zeta})_i \geq 0\right\} \geq 1 - \epsilon/m$$

can be rewritten equivalently as

$$\text{Prob}\{[b^{\mathrm{n}} - A^{\mathrm{n}}y]_i + \rho \sum_{\ell=1}^{L} [b^\ell - A^\ell y]_i \zeta_\ell > 0\} \leq \epsilon/m.$$

Since $\zeta_\ell \sim \mathcal{N}(0,1)$ are independent, this scalar chance constraint is exactly equivalent to

$$[b^{\mathrm{n}} - A^{\mathrm{n}}y]_i + \rho \text{ErfInv}(\epsilon/m) \sqrt{\sum_\ell [b^\ell - A^\ell y]_i^2} \leq 0.$$

The associated safe tractable approximation of the problem of interest (10.4.9) is the conic quadratic program

$$\max_{\rho,y} \left\{ \rho : c^T y \leq \tau_*, \text{ErfInv}(\epsilon/m) \sqrt{\sum_\ell [b^\ell - A^\ell y]_i^2} \leq \frac{[A^{\mathrm{n}}y - b^{\mathrm{n}}]_i}{\rho}, 1 \leq i \leq m \right\}. \quad (10.4.10)$$

Now let us apply our new approximation scheme, which treats the chance constrained vector inequality in (10.4.6) "as a whole." To this end, we should solve the problem

$$\min_{\nu,y,\{U_\ell\}} \left\{ \nu : \begin{array}{c} c^T y \leq \tau_*, \left[\begin{array}{c|c} U_\ell & \text{Diag}\{A^\ell y - b^\ell\} \\ \hline \text{Diag}\{A^\ell y - b^\ell\} & \text{Diag}\{A^{\mathrm{n}}y - b^{\mathrm{n}}\} \end{array} \right] \succeq 0, \\ 1 \leq \ell \leq L \\ \sum_\ell U_\ell \preceq \nu \text{Diag}\{A^{\mathrm{n}}y - b^{\mathrm{n}}\}, c^T y \leq \tau_* \end{array} \right\}, \quad (10.4.11)$$

treat its optimal solution y_* as the y component of the optimal solution to the approximation and then bound from below the feasibility radius $\rho_*(y_*)$ of this solution, (e.g., by applying to y_* the Randomized r procedure). Observe that problem (10.4.11) is nothing but the problem

$$\min_{\nu,y} \left\{ \nu : \begin{array}{c} \sum_{\ell=1}^{L} [A^\ell y - b]_i^2 / [A^{\mathrm{n}}y - b^{\mathrm{n}}]_i \leq \nu [A^{\mathrm{n}}y - b^{\mathrm{n}}]_i, 1 \leq i \leq m, \\ A^{\mathrm{n}}y - b^{\mathrm{n}} \geq 0, c^T y \leq \tau_* \end{array} \right\},$$

where $a^2/0$ is 0 for $a = 0$ and is $+\infty$ otherwise. Comparing the latter problem with (10.4.10), we see that

Problems (10.4.11) and (10.4.10) are equivalent to each other, the optimal values being related as

$$\text{Opt}(10.4.10) = \frac{1}{\text{ErfInv}(\epsilon/m)\sqrt{\text{Opt}(10.4.11)}}.$$

Thus, the approaches we are comparing result in the same vector of decision variables y_*, the only difference being the resulting value of a lower bound on the feasibility radius of y_*. With the "constraint-by-constraint" approach originating from chapter 2, this value is the optimal value in (10.4.10), while with our new approach, which treats the vector inequality $Ax \geq b$ "as a whole," the feasibility radius is bounded from below via the provable version of Conjecture 10.1 given by Theorem 10.4.1, or by the Randomized r procedure.

A natural question is, which one of these approaches results in a less conservative lower bound on the feasibility radius of y_*. On the theoretical side of this question, it is easily seen that when the second approach utilizes Theorem 10.4.1, it results in the same (within an absolute constant factor) value of ρ as the first approach. From the practical perspective, however, it is much more interesting to consider the case where the second approach exploits the Randomized r procedure, since experiments demonstrate that this version is less conservative than the "100%-reliable" one based on Theorem 10.4.1. Thus, let us focus on comparing the "constraint-by-constraint" safe approximation of (10.4.6), let it be called Approximation I, with Approximation II based on the Randomized r procedure. Numerical experiments show that no one of these two approximations "generically dominates" the other one, so that the best thing is to choose the best — the largest — of the two respective lower bounds.

10.4.1.1 Illustration: Antenna Design revisited

Consider the Antenna Design problem (Example 3.3.1, section 3.3) in the case when there are no positioning errors, the actuation errors are Gaussian and we formulate the problem in the form of (10.4.9). Specifically, setting

$$R_{s\phi}^k = \exp\{2\pi\imath[s/12 + k\cos(\phi)/8]\},$$

the chance constrained problem of interest is

$$\max_{\rho,z}\left\{\rho : \mathrm{Prob}\left\{\begin{array}{l}\Re\left\{\sum_{k=1}^{16} R_{s\phi}^k z_k(1+\rho\eta_k)\right\} \leq \tau_*, \, \phi \in \Pi, s = 1, ..., 12 \\ \Re\left\{\sum_{k=1}^{16} R_{00}^k z_k(1+\rho\eta_k)\right\} \geq 1\end{array}\right\} \geq 1 - \epsilon\right\},$$

$$\text{(10.4.12)}$$

where Π is the equidistant grid on $[\pi/6, \pi]$ with resolution $\pi/90$. The decision variables z_k are complex numbers (so that the vector of real decision variables is $y = [\Re z; \Im z]$) and η_k are independent standard complex-valued Gaussian random variables (or, equivalently, independent $\mathcal{N}(0, I_2)$ random 2-D vectors).

Here problems (10.4.10), (10.4.11) are equivalent to

$$\min_{\mu\in\mathbb{R},z\in\mathbb{C}^{16}}\left\{\mu : \begin{array}{l}\|z\|_2 \leq \mu\left[\tau_* - \Re\left\{\sum_{k=1}^{16} R_{s\phi}^k z_k\right\}\right], \, 1 \leq s \leq 12, \phi \in \Pi \\ \|z\|_2 \leq \mu\left[\Re\left\{\sum_{k=1}^{16} R_{00}^k z_k\right\} - 1\right]\end{array}\right\}. \quad \text{(10.4.13)}$$

The only data element in this problem we did not specify yet is the quantity τ_* representing the desired upper bound on the sidelobe attenuation level. In our experiment, we set this bound to 0.15 (cf. the numbers in table 3.1). After the optimal solution (z_*, μ_*) of (10.4.12) was found, we used 3 strategies to bound from below the feasibility radius $\rho_*(z_*)$ of z_* (that is, the largest ρ for which (z_*, ρ) is feasible for the chance constrained problem of interest (10.4.12)):

i) Approximation I, which in our situation results in the lower bound

$$\rho_{\mathrm{I}} = \frac{1}{\mathrm{ErfInv}(\epsilon/m)\mu_*}.$$

ii) Approximation II, which results in the lower bound

$$\rho_{\mathrm{II}} = \frac{r}{\mu_*},$$

where r is given by the Randomized r procedure as applied to the matrices

$$A = \mathrm{Diag}\left\{ \{\tau_* - \Re\{\sum_{k=1}^{16} R_{s\phi}^k(z_*)_k\}\}_{\substack{\phi\in\Pi \\ 1\leq s\leq 12}}, \Re\{\sum_{k=1}^{16} R_{00}^k(z_*)_k\} - 1 \right\},$$

$$A_\ell = \mu_*^{-1} \cdot \begin{cases} \mathrm{Diag}\left\{ \{\Re\{R_{s\phi}^\ell(z_*)_\ell\}\}_{\substack{\phi\in\Pi \\ 1\leq s\leq 12}}, -\Re\{R_{00}^\ell(z_*)_\ell\} \right\}, \\ \hfill 1\leq \ell \leq 16 \\ \mathrm{Diag}\left\{ \{\Im\{R_{s\phi}^{\ell-16}(z_*)_{\ell-16}\}\}_{\substack{\phi\in\Pi \\ 1\leq s\leq 12}}, -\Im\{R_{00}^{\ell-16}(z_*)_{\ell-16}\} \right\}, \\ \hfill 17\leq \ell \leq 32 \end{cases}$$

iii) A version of Approximation II based on Theorem 10.4.1 rather than on simulation.

Note that Theorem 10.4.1 combines with Theorem 10.1.2 to imply the following result:

Theorem 10.4.3. Let $A, A_1, ..., A_L$ be diagonal deterministic matrices satisfying (10.1.13), and let $\zeta_1, ..., \zeta_L$ be $\sim \mathcal{N}(0,1)$ and independent. Then

$$\forall s > 0 : \mathrm{Prob}\{-sA \preceq \sum_{\ell=1}^{L} \zeta_\ell A_\ell \preceq sA\} \geq 1 - \gamma,$$

$$\gamma = \gamma(s) \equiv \inf_\theta \left\{ \mathrm{Erf}\left(\Gamma(s,\theta)\right) : 0 < \theta < 1/2, \sqrt{2\ln(m\theta^{-1})} < s \right\},$$

$$\Gamma(s,\theta) = \mathrm{ErfInv}(\theta) + (s - \sqrt{2\ln(m\theta^{-1})})\max[1, \mathrm{ErfInv}(\theta)/\sqrt{2\ln(m\theta^{-1})}]$$

As an immediate corollary, we get that

$$\rho_*(z_*) \geq \rho_{\mathrm{III}} = \frac{1}{s(\epsilon)\mu_*},$$

where $s(\epsilon)$ is the root of the equation $\gamma(s) = \epsilon$.

The results of our experiment are presented in table 10.2. We see that Approximation II is less conservative than Approximation I.

10.4.2 The Arrow Case: Chance Constrained Conic Quadratic Optimization

We are about to justify Conjecture 10.1 in the *arrow-type* case, that is, when the matrices $A_\ell \in \mathbf{S}^m$, $\ell = 1, ..., L$, are of the form

$$A_\ell = [ef_\ell^T + f_\ell e^T] + \lambda_\ell G, \tag{10.4.14}$$

Lower bound	$\epsilon = 10^{-2}$	$\epsilon = 10^{-4}$	$\epsilon = 10^{-6}$
ρ_{I}	0.00396	0.00325	0.00282
ρ_{II}	0.00504	0.00360	0.00294
ρ_{III}	0.00322	0.00288	0.00245

Table 10.2 Lower bounds on $\rho_*(z_*)$ yielded by various approximation schemes. When computing ρ_{II}, the confidence parameter δ was set to $10^{-2}\epsilon$, and the sample size N in the Randomized r-procedure was set to 100,000.

where $e, f_\ell \in \mathbb{R}^m$, $\lambda_\ell \in \mathbb{R}$ and $G \in \mathbf{S}^m$. We encounter this case in the Chance Constrained Conic Quadratic Optimization. Indeed, a Chance Constrained CQI

$$\mathrm{Prob}\{\|A(y)\zeta + b(y)\|_2 \leq c^T(y)\zeta + d(y)\} \geq 1 - \epsilon, \qquad [A(\cdot): p \times q]$$

can be reformulated equivalently as the chance constrained LMI

$$\mathrm{Prob}\left\{ \left[\begin{array}{c|c} c^T(y)\zeta + d(y) & \zeta^T A^T(y) + b^T(y) \\ \hline A(y)\zeta + b(y) & (c^T(y)\zeta + d(y))I \end{array} \right] \succeq 0 \right\} \geq 1 - \epsilon \qquad (10.4.15)$$

(see Lemma 6.3.3). In the notation of (10.1.1), for this LMI we have

$$\mathcal{A}^{\mathrm{n}}(y) = \left[\begin{array}{c|c} d(y) & b^T(y) \\ \hline b(y) & d(y)I \end{array} \right], \ \mathcal{A}^\ell(y) = \left[\begin{array}{c|c} c_\ell(y) & a_\ell^T(y) \\ \hline a_\ell(y) & c_\ell(y)I \end{array} \right],$$

where $a_\ell(y)$ in (10.4.14) is ℓ-th column of $A(y)$. We see that the matrices $\mathcal{A}^\ell(y)$ are arrow-type $(p+1) \times (p+1)$ matrices where e in (10.4.14) is the first basic orth in $\mathbb{R}^{p+1}$, $f_\ell = [0; a_\ell(y)]$ and $G = I_{p+1}$.

Another example is the one arising in the chance constrained Truss Topology Design problem, see section 10.2.2.

The justification of Conjecture 10.1 in the arrow-type case is given by the following

Theorem 10.4.4. Let $m \times m$ matrices $A_1, ..., A_L$ of the form (10.4.14) along with a matrix $A \in \mathbf{S}^m$ satisfy the relation (10.1.13), and ζ_ℓ be independent with zero means and such that $\mathbf{E}\{\zeta_\ell^2\} \leq \sigma^2$, $\ell = 1, ..., L$ (under Assumption **A.III**, one can take $\sigma = \sqrt{\exp\{1\} - 1}$). Then, for every $\chi \in (0, 1)$, with $\Upsilon = \Upsilon(\chi) \equiv \frac{2\sqrt{2}\sigma}{\sqrt{1-\chi}}$ one has

$$\mathrm{Prob}\left\{ -\Upsilon A \preceq \sum_{\ell=1}^L \zeta_\ell A_\ell \preceq \Upsilon A \right\} \geq \chi \qquad (10.4.16)$$

(cf. (10.1.14)). When ζ satisfies Assumption **A.I**, or ζ satisfies Assumption **A.II** and $\chi \geq \frac{6}{7}$, relation (10.4.16) is satisfied with $\Upsilon = \Upsilon_{\mathrm{I}}(\chi) \equiv 2 + 4\sqrt{3 \ln \frac{4}{1-\chi}}$ and with $\Upsilon = \Upsilon_{\mathrm{II}}(\chi) \equiv \sqrt{3\left(1 + 3 \ln \frac{1}{1-\chi}\right)}$, respectively.

Proof. First of all, when ζ_ℓ, $\ell = 1, ..., L$, satisfy Assumption **A.III**, we indeed have $\mathbf{E}\{\zeta_\ell^2\} \leq \exp\{1\} - 1$ due to $t^2 \leq \exp\{t^2\} - 1$ for all t. Further, same as in the proof of Theorem 10.4.1, it suffices to consider the case when $A \succ 0$ and to prove the following statement:

> Let A_ℓ be of the form of (10.4.14) and such that the matrices $B_\ell = A^{-1/2} A_\ell A^{-1/2}$ satisfy $\sum_\ell B_\ell^2 \preceq I$. Let, further, ζ_ℓ satisfy the premise in Theorem 10.4.4. Then, for every $\chi \in (0,1)$, one has
>
> $$\mathrm{Prob}\{\|\sum_{\ell=1}^{L} \zeta_\ell B_\ell\| \leq \frac{2\sqrt{2}\sigma}{\sqrt{1-\chi}}\} \geq \chi. \qquad (10.4.17)$$

Observe that B_ℓ are also of the arrow-type form (10.4.14):

$$B_\ell = [gh_\ell^T + h_\ell g^T] + \lambda_\ell H \qquad [g = A^{-1/2}e, h_\ell = A^{-1/2}f_\ell, H = A^{-1/2}GA^{-1/2}]$$

Note that w.l.o.g. we can assume that $\|g\|_2 = 1$ and then rotate the coordinates to make g the first basic orth. In this situation, the matrices B_ℓ become

$$B_\ell = \left[\begin{array}{c|c} q_\ell & r_\ell^T \\ \hline r_\ell & \lambda_\ell Q \end{array} \right]; \qquad (10.4.18)$$

by appropriate scaling of λ_ℓ, we can ensure that $\|Q\| = 1$. We have

$$B_\ell^2 = \left[\begin{array}{c|c} q_\ell^2 + r_\ell^T r_\ell & q_\ell r_\ell^T + \lambda_\ell r_\ell^T Q \\ \hline q_\ell r_\ell + \lambda_\ell Q r_\ell & r_\ell r_\ell^T + \lambda_\ell^2 Q^2 \end{array} \right].$$

We conclude that $\sum_{\ell=1}^{L} B_\ell^2 \preceq I_m$ implies that $\sum_\ell(q_\ell^2 + r_\ell^T r_\ell) \leq 1$ and $[\sum_\ell \lambda_\ell^2]Q^2 \preceq I_{m-1}$; since $\|Q^2\| = 1$, we arrive at the relations

$$\begin{array}{ll} (a) & \sum_\ell \lambda_\ell^2 \leq 1, \\ (b) & \sum_\ell(q_\ell^2 + r_\ell^T r_\ell) \leq 1. \end{array} \qquad (10.4.19)$$

Now let $p_\ell = [0; r_\ell] \in \mathbb{R}^m$. We have

$$S[\zeta] \equiv \sum_\ell \zeta_\ell B_\ell = [g^T(\underbrace{\sum_\ell \zeta_\ell p_\ell}_{\xi}) + \xi^T g] + \mathrm{Diag}\{\underbrace{\sum_\ell \zeta_\ell q_\ell}_{\theta}, (\underbrace{\sum_\ell \zeta_\ell \lambda_\ell}_{\eta})Q\}$$

$$\Rightarrow \qquad \|S[\zeta]\| \leq \|g\xi^T + \xi g^T\| + \max[|\theta|, |\eta|\|Q\|] = \|\xi\|_2 + \max[|\theta|, |\eta|].$$

Setting

$$\alpha = \sum_\ell r_\ell^T r_\ell, \ \beta = \sum_\ell q_\ell^2,$$

we have $\alpha + \beta \le 1$ by (10.4.19.b). Besides this,

$$\mathbf{E}\{\xi^T\xi\} = \sum_{\ell,\ell'} \mathbf{E}\{\zeta_\ell\zeta_{\ell'}\}p_\ell^T p_{\ell'} = \sum_\ell \mathbf{E}\{\zeta_\ell^2\}r_\ell^T r_\ell \quad [\zeta_\ell \text{ are independent zero mean}]$$

$$\le \sigma^2 \sum_\ell r_\ell^T r_\ell = \sigma^2\alpha$$

$$\Rightarrow \quad \text{Prob}\{\|\xi\|_2 > t\} \le \frac{\sigma^2\alpha}{t^2} \forall t > 0 \qquad [\text{Tschebyshev Inequality}]$$

$$\mathbf{E}\{\eta^2\} = \sum_\ell \mathbf{E}\{\zeta_\ell^2\}\lambda_\ell^2 \le \sigma^2 \sum_\ell \lambda_\ell^2 \le \sigma^2 \qquad [\,(10.4.19.a)]$$

$$\Rightarrow \quad \text{Prob}\{|\eta| > t\} \le \frac{\sigma^2}{t^2} \forall t > 0 \qquad [\text{Tschebyshev Inequality}]$$

$$\mathbf{E}\{\theta^2\} = \sum_\ell \mathbf{E}\{\zeta_\ell^2\}q_\ell^2 \le \sigma^2\beta$$

$$\Rightarrow \quad \text{Prob}\{|\theta| > t\} \le \frac{\sigma^2\beta}{t^2} \forall t > 0 \qquad [\text{Tschebyshev Inequality}].$$

Thus, for every $\Upsilon > 0$ and all $\lambda \in (0,1)$ we have

$$\text{Prob}\{\|S[\zeta]\| > \Upsilon\} \quad \le \quad \text{Prob}\{\|\xi\|_2 + \max[|\theta|,|\eta|] > \Upsilon\} \le \text{Prob}\{\|\xi\|_2 > \lambda\Upsilon\}$$

$$+\text{Prob}\{|\theta| > (1-\lambda)\Upsilon\} + \text{Prob}\{|\eta| > (1-\lambda)\Upsilon\}$$

$$\le \quad \frac{\sigma^2}{\Upsilon^2}\left[\frac{\alpha}{\lambda^2} + \frac{\beta+1}{(1-\lambda)^2}\right],$$

whence, due to $\alpha + \beta \le 1$,

$$\text{Prob}\{\|S[\zeta]\| > \Upsilon\} \le \frac{\sigma^2}{\Upsilon^2} \max_{\alpha\in[0,1]} \min_{\lambda\in(0,1)} \left[\frac{\alpha}{\lambda^2} + \frac{2-\alpha}{(1-\lambda)^2}\right] = \frac{8\sigma^2}{\Upsilon^2};$$

with $\Upsilon = \Upsilon(\chi)$, this relation implies (10.4.16).

Assume now that ζ_ℓ satisfy Assumption **A.I**. We should prove that here the relation (10.4.16) holds true with $\Upsilon = \Upsilon_I(\chi)$, or, which is the same,

$$\text{Prob}\left\{\|S[\zeta]\| > \Upsilon\right\} \le 1 - \chi, \; S[\zeta] = \sum_\ell \zeta_\ell B_\ell = \left[\begin{array}{c|c} \sum_\ell \zeta_\ell q_\ell & \sum_\ell \zeta_\ell r_\ell^T \\ \hline \sum_\ell \zeta_\ell r_\ell & (\sum_\ell \zeta_\ell\lambda_\ell)Q \end{array}\right].$$

$$(10.4.20)$$

Observe that for a symmetric block-matrix $P = \left[\begin{array}{c|c} A & B^T \\ \hline B & C \end{array}\right]$ we have $\|P\| \le$ $\left\|\left[\begin{array}{c|c} \|A\| & \|B\| \\ \hline \|B\| & \|C\| \end{array}\right]\right\|$, and that the norm of a symmetric matrix does not exceed its Frobenius norm, whence

$$\|S[\zeta]\|^2 \le |\sum_\ell \zeta_\ell q_\ell|^2 + 2\|\sum_\ell \zeta_\ell r_\ell\|_2^2 + |\sum_\ell \zeta_\ell\lambda_\ell|^2 \equiv \alpha[\zeta] \qquad (10.4.21)$$

(recall that $\|Q\| = 1$). Let E_ρ be the ellipsoid $E_\rho = \{z : \alpha[z] \le \rho^2\}$. Observe that E_ρ contains the centered at the origin Euclidean ball of radius $\rho/\sqrt{3}$. Indeed, applying the Cauchy Inequality, we have

$$\alpha[z] \le \left(\sum_\ell z_\ell^2\right)\left[\sum_\ell q_\ell^2 + 2\sum_\ell \|r_\ell\|_2^2 + \sum_\ell \lambda_\ell^2\right] \le 3\sum_\ell z_\ell^2$$

(we have used (10.4.19)). Further, ζ_ℓ are independent with zero mean and $\mathbf{E}\{\zeta_\ell^2\} \leq 1$ for every ℓ; applying the same (10.4.19), we therefore get $\mathbf{E}\{\alpha[\zeta]\} \leq 3$. By the Tschebyshev Inequality, we have

$$\text{Prob}\{\zeta \in E_\rho\} \equiv \text{Prob}\{\alpha[\zeta] \leq \rho^2\} \geq 1 - \frac{3}{\rho^2}.$$

Invoking the Talagrand Inequality (see the proof of Lemma B.3.3 in section B.3), we have

$$\rho^2 > 3 \Rightarrow \mathbf{E}\left\{\exp\{\frac{\text{dist}_{\|\cdot\|_2}^2(\zeta, E_\rho)}{16}\}\right\} \leq \frac{1}{\text{Prob}\{\zeta \in E_\rho\}} \leq \frac{\rho^2}{\rho^2 - 3}.$$

On the other hand, if $r > \rho$ and $\alpha[\zeta] > r^2$, then $\zeta \notin (r/\rho)E_\rho$ and therefore $\text{dist}_{\|\cdot\|_2}(\zeta, E_\rho) \geq (r/\rho - 1)\rho/\sqrt{3} = (r - \rho)/\sqrt{3}$ (recall that E_ρ contains the centered at the origin $\|\cdot\|_2$-ball of radius $\rho/\sqrt{3}$). Applying the Tschebyshev Inequality, we get

$$r^2 > \rho^2 > 3 \Rightarrow \text{Prob}\{\alpha[\zeta] > r^2\} \leq \mathbf{E}\left\{\exp\{\frac{\text{dist}_{\|\cdot\|_2}^2(\zeta, E_\rho)}{16}\}\right\} \exp\{-\frac{(r-\rho)^2}{48}\}$$

$$\leq \frac{\rho^2 \exp\{-\frac{(r-\rho)^2}{48}\}}{\rho^2 - 3}.$$

With $\rho = 2$, $r = \Upsilon_{\mathrm{I}}(\chi) = 2 + 4\sqrt{3\ln\frac{4}{1-\chi}}$ this bound implies $\text{Prob}\{\alpha[\zeta] > r^2\} \leq 1 - \chi$; recalling that $\sqrt{\alpha[\zeta]}$ is an upper bound on $\|S[\zeta]\|$, we see that (10.4.16) indeed holds true with $\Upsilon = \Upsilon_{\mathrm{I}}(\chi)$.

Now consider the case when $\zeta \sim \mathcal{N}(0, I_L)$. Observe that $\alpha[\zeta]$ is a homogeneous quadratic form of ζ: $\alpha[\zeta] = \zeta^T A\zeta$, $A_{ij} = q_i q_j + 2r_i^T r_j + \lambda_i \lambda_j$. We see that the matrix A is positive semidefinite, and $\text{Tr}(A) = \sum_i(q_i^2 + \lambda_i^2 + 2\|r_i\|_2^2) \leq 3$. Denoting by μ_ℓ the eigenvalues of A, we have $\zeta^T A\zeta = \sum_{\ell=1}^L \mu_\ell \xi_\ell^2$, where $\xi \sim \mathcal{N}(0, I_L)$ is an appropriate rotation of ζ. Now we can use the Bernstein scheme to bound from above $\text{Prob}\{\alpha[\zeta] > \rho^2\}$:

$$\forall(\gamma \geq 0, \max_\ell \gamma\mu_\ell < 1/2):$$

$$\ln\left(\text{Prob}\{\alpha[\zeta] > \rho^2\}\right) \leq \ln\left(\mathbf{E}\left\{\exp\{\gamma\zeta^T A\zeta\}\right\}\exp\{-\gamma\rho^2\}\right)$$

$$= \ln\left(\mathbf{E}\left\{\exp\{\gamma\sum_\ell \mu_\ell \xi_\ell^2\}\right\}\right) - \gamma\rho^2 = \sum_\ell \ln\left(\mathbf{E}\left\{\exp\{\gamma\mu_\ell \xi_\ell^2\}\right\}\right) - \gamma\rho^2$$

$$= -\frac{1}{2}\sum_\ell \ln(1 - 2\mu_\ell\gamma) - \gamma\rho^2.$$

The concluding expression is a convex and monotone function of μ's running through the box $\{0 \leq \mu_\ell < \frac{1}{2\gamma}\}$. It follows that when $\gamma < 1/6$, the maximum of the expression over the set $\{\mu_1, ..., \mu_L \geq 0, \sum_\ell \mu_\ell \leq 3\}$ is $-\frac{1}{2}\ln(1 - 6\gamma) - \gamma\rho^2$. We get

$$0 \leq \gamma < \frac{1}{6} \Rightarrow \ln\left(\text{Prob}\{\alpha[\zeta] > \rho^2\}\right) \leq -\frac{1}{2}\ln(1 - 6\gamma) - \gamma\rho^2.$$

Optimizing this bound in γ and setting $\rho^2 = 3(1 + \Delta)$, $\Delta \geq 0$, we get $\text{Prob}\{\alpha[\zeta] > 3(1 + \Delta)\} \leq \exp\{-\frac{1}{2}[\Delta - \ln(1 + \Delta)]\}$. It follows that if $\chi \in (0, 1)$ and $\Delta = \Delta(\chi) \geq 0$

is such that $\Delta - \ln(1+\Delta) = 2\ln\frac{1}{1-\chi}$, then
$$\text{Prob}\{\|S[\zeta]\| > \sqrt{3(1+\Delta)}\} \le \text{Prob}\{\alpha[\zeta] > 3(1+\Delta)\} \le 1-\chi.$$
It is easily seen that when $1-\chi \le \frac{1}{7}$, one has $\Delta(\chi) \le 3\ln\frac{1}{1-\chi}$, that is, $\text{Prob}\{\|S[\zeta]\| > \sqrt{3\left(1+3\ln\frac{1}{1-\chi}\right)}\} \le 1-\chi$, which is exactly what was claimed in the case of Gaussian ζ. $\qquad\square$

10.4.3 Application: Recovering Signal from Indirect Noisy Observations

Consider the situation as follows (cf. section 6.6): we observe in noise a linear transformation
$$u = As + \rho\xi \tag{10.4.22}$$
of a random signal $s \in \mathbb{R}^n$; here A is a given $m \times n$ matrix, $\xi \sim \mathcal{N}(0, I_m)$ is the noise, (which is independent of s), and $\rho \ge 0$ is a (deterministic) noise level. Our goal is to find a linear estimator
$$\widehat{s}(u) = Gu \equiv GAs + \rho G\xi \tag{10.4.23}$$
such that
$$\text{Prob}\{\|\widehat{s}(u) - s\|_2 \le \tau_*\} \ge 1-\epsilon, \tag{10.4.24}$$
where $\tau_* > 0$ and $\epsilon \ll 1$ are given. Note that the probability in (10.4.24) is taken w.r.t. the joint distribution of s and ξ. We assume below that $s \sim \mathcal{N}(0, C)$ with known covariance matrix $C \succ 0$. Besides this, we assume that $m \ge n$ and A is of rank n. When there is no observation noise, we can recover s from u in a linear fashion without any error; it follows that when $\rho > 0$ is small enough, there exists G that makes (10.4.24) valid. Let us find the largest such ρ, that is, let us solve the optimization problem
$$\max_{G,\rho}\{\rho : \text{Prob}\{\|(GA - I_n)s + \rho G\xi\|_2 \le \tau_*\} \ge 1-\epsilon\}. \tag{10.4.25}$$
Setting $S = C^{1/2}$ and introducing a random vector $\theta \sim \mathcal{N}(0, I_n)$ independent of ξ (so that the random vector $[S^{-1}s; \xi]$ has exactly the same $\mathcal{N}(0, I_{n+m})$ distribution as the vector $\zeta = [\theta; \xi]$), we can rewrite our problem equivalently as
$$\max_{G,\rho}\{\rho : \text{Prob}\{\|H_\rho(G)\zeta\|_2 \le \tau_*\} \ge 1-\epsilon\}, \quad H_\rho(G) = [(GA-I_n)S, \rho G]. \tag{10.4.26}$$
Let $h_\rho^\ell(G)$ be the ℓ-th column in the matrix $H_\rho(G)$, $\ell = 1,...,L = m+n$. Invoking Lemma 6.3.3, our problem is nothing but the chance constrained program
$$\max_{G,\rho}\left\{\rho : \text{Prob}\left\{\sum_{\ell=1}^L \zeta_\ell \mathcal{A}_\rho^\ell(G) \preceq \tau_*\mathcal{A}^{\mathrm{n}} \equiv \tau_* I_{n+1}\right\} \ge 1-\epsilon\right\}$$
$$\mathcal{A}_\rho^\ell(G) = \left[\begin{array}{c|c} & [h_\rho^\ell(G)]^T \\ \hline h_\rho^\ell(G) & \end{array}\right]. \tag{10.4.27}$$
We intend to process the latter problem as follows:

A) We use our "Conjecture-related" approximation scheme to build a nonde-creasing continuous function $\Gamma(\rho) \to 0, \rho \to +0$, and matrix-valued function G_ρ (both functions are efficiently computable) such that

$$\text{Prob}\{\|(GA - I_n)s + \rho G\xi\|_2 > \tau_*\} = \text{Prob}\{\sum_{\ell=1}^{L} \zeta_\ell \mathcal{A}_\rho^\ell(G_\rho) \npreceq \tau_* I_{n+1}\} \leq \Gamma(\rho).$$

$$(10.4.28)$$

B) We then solve the approximating problem

$$\max_\rho \{\rho : \Gamma(\rho) \leq \epsilon\}. \qquad (10.4.29)$$

Clearly, a feasible solution ρ to the latter problem, along with the associated matrix G_ρ, form a feasible solution to the problem of interest (10.4.27). On the other hand, the approximating problem is efficiently solvable: $\Gamma(\rho)$ is nondecreasing, efficiently computable and $\Gamma(\rho) \to 0$ as $\rho \to +0$, so that the approximating problem can be solved efficiently by bisection. We find a feasible nearly optimal solution $\widehat{\rho}$ to the approximating problem and treat $(\widehat{\rho}, G_{\widehat{\rho}})$ as a suboptimal solution to the problem of interest. By our analysis, this solution is feasible for the latter problem.

Remark 10.4.5. In fact, the constraint in (10.4.26) is simpler than a general-type chance constrained conic quadratic inequality — it is a chance constrained Least Squares inequality (the right hand side is affected neither by the decision variables, nor by the noise), and as such it admits a Bernstein-type approximation described in section 4.5.5, see Corollary 4.5.11. Of course, in the outlined scheme one can use the Bernstein approximation as an alternative to the Conjecture-related approximation.

Now let us look at steps A, B in more details.

Step A). We solve the semidefinite program

$$\nu_*(\rho) = \min_{\nu, G} \left\{ \nu : \sum_{\ell=1}^{L} (\mathcal{A}_\rho^\ell(G))^2 \preceq \nu I_{n+1} \right\}; \qquad (10.4.30)$$

whenever $\rho > 0$, this problem clearly is solvable. Due to the fact that part of the matrices $\mathcal{A}_\rho^\ell(G)$ are independent of ρ, and the remaining ones are proportional to ρ, the optimal value is a positive continuous and nondecreasing function of $\rho > 0$. Finally, $\nu_*(\rho) \to +0$ as $\rho \to +0$ (look what happens at the point G satisfying the relation $GA = I_n$).

Let G_ρ be an optimal solution to (10.4.30). Setting $A_\ell = \mathcal{A}_\rho^\ell(G_\rho)\nu_*^{-\frac{1}{2}}(\rho)$, $A = I_{n+1}$, the arrow-type matrices $A, A_1, ..., A_L$ satisfy (10.1.13); invoking Theorem 10.4.4, we conclude that

$$\chi \in [\tfrac{6}{7}, 1) \Rightarrow \text{Prob}\{-\Upsilon(\chi)\nu_*^{\frac{1}{2}}(\rho)I_{n+1} \preceq \sum_{\ell=1}^{L} \zeta_\ell \mathcal{A}_\rho^\ell(G_\rho) \preceq \Upsilon(\chi)\nu_*^{\frac{1}{2}}(\rho)I_{n+1}\}$$

$$\geq \chi, \; \Upsilon(\chi) = \sqrt{3\left(1 + 3\ln\tfrac{1}{1-\chi}\right)}.$$

Now let χ and ρ be such that $\chi \in [6/7, 1)$ and $\Upsilon(\chi)\sqrt{\nu_*(\rho)} \leq \tau_*$. Setting

$$Q = \{z : \|\sum_{\ell=1}^{L} z_\ell \mathcal{A}_\rho^\ell(G_\rho)\| \leq \Upsilon(\chi)\sqrt{\nu_*(\rho)}\},$$

we get a closed convex set such that the random vector $\zeta \sim \mathcal{N}(0, I_{n+m})$ takes its values in Q with probability $\geq \chi > 1/2$. Invoking Theorem B.5.1 (where we set $\alpha = \tau_*/(\Upsilon(\chi)\sqrt{\nu_*(\rho)}))$, we get

$$\text{Prob}\left\{\sum_{\ell=1}^{L} \zeta_\ell \mathcal{A}_\rho^\ell(G_\rho) \npreceq \tau_* I_{n+1}\right\} \leq \text{Erf}\left(\frac{\tau_* \text{ErfInv}(1-\chi)}{\sqrt{\nu_*(\rho)}\Upsilon(\chi)}\right)$$

$$= \text{Erf}\left(\frac{\tau_* \text{ErfInv}(1-\chi)}{\sqrt{3\nu_*(\rho)\left[1 + 3\ln\frac{1}{1-\chi}\right]}}\right).$$

Setting

$$\Gamma(\rho) = \inf_\chi \left\{ \text{Erf}\left(\frac{\tau_* \text{ErfInv}(1-\chi)}{\sqrt{3\nu_*(\rho)\left[1 + 3\ln\frac{1}{1-\chi}\right]}}\right) : \begin{array}{l} \chi \in [6/7, 1), \\ 3\nu_*(\rho)\left[1 + 3\ln\frac{1}{1-\chi}\right] \leq \tau_*^2 \end{array} \right\}$$

(10.4.31)

(if the feasible set of the right hand side optimization problem is empty, then, by definition, $\Gamma(\rho) = 1$), we ensure (10.4.28). Taking into account that $\nu_*(\rho)$ is a nondecreasing continuous function of $\rho > 0$ that tends to 0 as $\rho \rightarrow +0$, it is immediately seen that $\Gamma(\rho)$ possesses these properties as well.

Solving (10.4.30). Good news is that problem (10.4.30) has a closed form solution. To see this, note that the matrices $\mathcal{A}_\rho^\ell(G)$ are pretty special arrow type matrices: their diagonal entries are zero, so that these $(n+1) \times (n+1)$ matrices are of the form $\left[\begin{array}{c|c} & [h_\rho^\ell(G)]^T \\ \hline h_\rho^\ell(G) & \end{array}\right]$ with n-dimensional vectors $h_\rho^\ell(G)$ affinely depending on G. Now let us make the following observation:

Lemma 10.4.6. Let $f_\ell \in \mathbb{R}^n$, $\ell = 1, ..., L$, and $\nu \geq 0$. Then

$$\sum_{\ell=1}^{L} \left[\begin{array}{c|c} & f_\ell^T \\ \hline f_\ell & \end{array}\right]^2 \preceq \nu I_{n+1} \qquad (*)$$

if and only if $\sum_\ell f_\ell^T f_\ell \leq \nu$.

Proof. Relation $(*)$ is nothing but

$$\sum_\ell \left[\begin{array}{c|c} f_\ell^T f_\ell & \\ \hline & f_\ell f_\ell^T \end{array}\right] \preceq \nu I_{n+1},$$

so it definitely implies that $\sum_{\ell} f_{\ell}^{T} f_{\ell} \leq \nu$. To prove the inverse implication, it suffices to verify that the relation $\sum_{\ell} f_{\ell}^{T} f_{\ell} \leq \nu$ implies that $\sum_{\ell} f_{\ell} f_{\ell}^{T} \preceq \nu I_{n}$. This is immediate due to $\text{Tr}(\sum_{\ell} f_{\ell} f_{\ell}^{T}) = \sum_{\ell} f_{\ell}^{T} f_{\ell} \leq \nu$, (note that the matrix $\sum_{\ell} f_{\ell} f_{\ell}^{T}$ is positive semidefinite, and therefore its maximal eigenvalue does not exceed its trace). $\qquad\square$

In view of Lemma 10.4.6, the optimal solution and the optimal value in (10.4.30) are exactly the same as their counterparts in the minimization problem

$$\nu = \min_{G} \sum_{\ell} [h_{\rho}^{\ell}(G)]^{T} h_{\rho}^{\ell}(G).$$

Thus, (10.4.30) is nothing but the problem

$$\nu_{*}(\rho) = \min_{G} \left\{ \text{Tr}((GA - I_{n})C(GA - I)^{T}) + \rho^{2}\text{Tr}(GG^{T}) \right\}. \qquad (10.4.32)$$

The objective in this unconstrained problem has a very transparent interpretation: it is the mean squared error of the linear estimator $\widehat{s} = Gu$, the noise intensity being ρ. The matrix G minimizing this objective is called the *Wiener filter*; a straightforward computation yields

$$\begin{aligned} G_{\rho} &= CA^{T}(ACA^{T} + \rho^{2}I_{m})^{-1}, \\ \nu_{*}(\rho) &= \text{Tr}\left((G_{\rho}A - I_{n})C(G_{\rho}A - I_{n})^{T} + \rho^{2}G_{\rho}G_{\rho}^{T}\right). \end{aligned} \qquad (10.4.33)$$

Remark 10.4.7. The Wiener filter is one of the oldest and the most basic tools in Signal Processing; it is good news that our approximation scheme recovers this tool, albeit from a different perspective: we were seeking a linear filter that ensures that with probability $1 - \epsilon$ the recovering error does not exceed a given threshold (a problem that seemingly does not admit a closed form solution); it turned out that the *sub*optimal solution yielded by our approximation scheme is the precise solution to a simple classical problem.

Refinements. The pair $(\widehat{\rho}, G_{\text{W}} = G_{\widehat{\rho}})$ ("W" stands for "Wiener") obtained via the outlined approximation scheme is feasible for the problem of interest (10.4.27). However, we have all reason to expect that our provably 100%-reliable approach is conservative — exactly because of its 100% reliability. In particular, it is very likely that $\widehat{\rho}$ is a too conservative lower bound on the actual feasibility radius $\rho_{*}(G_{\text{W}})$ — the largest ρ such that (ρ, G_{W}) is feasible for the chance constrained problem of interest. We can try to improve this lower bound by the Randomized r procedure, e.g., as follows:

Given a confidence parameter $\delta \in (0, 1)$, we run $\nu = 10$ steps of bisection on the segment $\Delta = [\widehat{\rho}, 100\widehat{\rho}]$. At a step t of this process, given the previous localizer Δ_{t-1} (a segment contained in Δ, with $\Delta_{0} = \Delta$), we take as the current trial value ρ_{t} of ρ the midpoint of Δ_{t-1} and apply the Randomized r procedure in order to check whether (ρ_{t}, G_{W}) is feasible for (10.4.27). Specifically, we

- compute the $L = m + n$ vectors $h^\ell_{\rho_t}(G_W)$ and the quantity $\mu_t = \sqrt{\sum\limits_{\ell=1}^{m+n} \|h^\ell_{\rho_t}(G_W)\|_2^2}$. By Lemma 10.4.6, we have

$$\sum_{\ell=1}^{L} \left[\mathcal{A}^\ell_{\rho_t}(G_W)\right]^2 \preceq \mu_t^2 I_{n+1},$$

so that the matrices $A = I_{n+1}$, $A_\ell = \mu_t^{-1}\mathcal{A}^\ell_{\rho_t}(G_W)$ satisfy (10.1.13);

- apply to the matrices $A, A_1, ..., A_L$ the Randomized r procedure with parameters $\epsilon, \delta/\nu$, thus ending up with a random quantity r_t such that "up to probability of bad sampling $\leq \delta/\nu$," one has

$$\text{Prob}\{\zeta : -I_{n+1} \preceq r_t \sum_{\ell=1}^{L} \zeta_\ell A_\ell \preceq I_{n+1}\} \geq 1 - \epsilon,$$

or, which is the same,

$$\text{Prob}\{\zeta : -\frac{\mu_t}{r_t}I_{n+1} \preceq \sum_{\ell=1}^{L} \zeta_\ell \mathcal{A}^\ell_\rho(G_W) \preceq \frac{\mu_t}{r_t}I_{n+1}\} \geq 1 - \epsilon. \qquad (10.4.34)$$

Note that when the latter relation is satisfied and $\frac{\mu_t}{r_t} \leq \tau_*$, the pair (ρ_t, G_W) is feasible for (10.4.27);

- finally, complete the bisection step, namely, check whether $\mu_t/r_t \leq \tau_*$. If it is the case, we take as our new localizer Δ_t the part of Δ_{t-1} to the right of ρ_t, otherwise Δ_t is the part of Δ_{t-1} to the left of ρ_t.

After ν bisection steps are completed, we claim that the left endpoint $\widetilde{\rho}$ of the last localizer Δ_ν is a lower bound on $\rho_*(G_W)$. Observe that this claim is valid, provided that all ν inequalities (10.4.34) take place, which happens with probability at least $1 - \delta$.

Illustration: Deconvolution. A rotating scanning head reads random signal s as shown in figure 10.2. The signal registered when the head observes bin i, $0 \leq i < n$, is

$$u_i = (As)_i + \rho\xi_i \equiv \sum_{j=-d}^{d} K_j s_{(i-j) \bmod n} + \rho\xi_i, \ 0 \leq i < n,$$

where $r = p \bmod n$, $0 \leq r < n$, is the remainder when dividing p by n. The signal s is assumed to be Gaussian with zero mean and known covariance $C_{ij} = \mathbf{E}\{s_i s_j\}$ depending on $(i - j) \bmod n$ only ("stationary periodic discrete-time Gaussian process"). The goal is to find a linear recovery $\widehat{s} = Gu$ and the largest ρ such that

$$\text{Prob}_{[s;\xi]}\{\|G(As + \rho\xi) - s\|_2 \leq \tau_*\} \geq 1 - \epsilon.$$

We intend to process this problem via the outlined approach using two safe approximations of the chance constraint of interest — the Conjecture-related and the Bernstein (see Remark 10.4.5). The recovery matrices and critical levels of noise

$$(K * s)_i = 0.2494s_{i-1} + 0.5012s_i + 0.2494s_{i+1}$$

Figure 10.2   A scanner.

as given by these two approximations will be denoted G_W, ρ_W ("W" for "Wiener") and G_B, ρ_B ("B" for "Bernstein"), respectively.

Note that in the case in question one can immediately verify that the matrices $A^T A$ and C commute. Whenever this is the case, the computational burden to compute G_W and G_B reduces dramatically. Indeed, after appropriate rotations of x and y we arrive at the situation where both A and C are diagonal, in which case in both our approximation schemes one loses nothing by restricting G to be diagonal. This significantly reduces the dimensions of the convex problems we need to solve.

In the experiment we use

$$n = 64, \, d = 1, \, \tau_* = 0.1\sqrt{n} = 0.8, \, \epsilon = 1.e\text{-}4;$$

C was set to the unit matrix, (meaning that $s \sim \mathcal{N}(0, I_{64})$), and the convolution kernel K is the one shown in figure 10.2. After (G_W, ρ_w) and (G_B, ρ_B) were computed, we used the Randomized r procedure with $\delta = 1.e\text{-}6$ to refine the critical values of noise for G_W and G_B; the refined values of ρ are denoted $\widehat{\rho}_W$ and $\widehat{\rho}_B$, respectively.

The results of the experiments are presented in table 10.3. While G_B and G_W turned out to be close, although not identical, the critical noise levels as yielded by the Conjecture-related and the Bernstein approximations differ by $\approx 30\%$. The refinement increases these critical levels by a factor ≈ 2 and makes them nearly equal. The resulting critical noise level 3.6e-4 is not too conservative: the simulation results shown in table 10.4 demonstrate that at a twice larger noise level, the probability for the chance constraint to be violated is by far larger than the required 1.e-4.

Admissible noise level	Bernstein approximation	Conjecture-related approximation
Before refinement	1.92e-4	1.50e-4
After refinement ($\delta = 1.e\text{-}6$)	3.56e-4	3.62e-4

Table 10.3 Results of deconvolution experiment.

Noise level	$\text{Prob}\{\|\widehat{s} - s\|_2 > \tau_*\}$	
	$G = G_B$	$G = G_W$
3.6e-4	0	0
7.2e-4	6.7e-3	6.7e-3
1.0e-3	7.4e-2	7.5e-2

Table 10.4 Empirical value of $\text{Prob}\{\|\widehat{s} - s\|_2 > 0.8\}$ based on 10,000 simulations.

10.4.3.1 Modifications

We have addressed the Signal Recovery problem (10.4.22), (10.4.23), (10.4.24) in the case when $s \sim \mathcal{N}(0, C)$ is random, the noise is independent of s and the probability in (10.4.24) is taken w.r.t. the joint distribution of ξ and s. Next we want to investigate two other versions of the problem.

Recovering a uniformly distributed signal. Assume that the signal s is

(a) uniformly distributed in the unit box $\{s \in \mathbb{R}^n : \|s\|_\infty \leq 1\}$,
or

(b) uniformly distributed on the vertices of the unit box
and is independent of ξ. Same as above, our goal is to ensure the validity of (10.4.24) with as large ρ as possible. To this end, let us use Gaussian Majorization. Specifically, in the case of (a), let $\widetilde{s} \sim \mathcal{N}(0, (2/\pi)I)$. As it was explained in section 10.3, the condition

$$\text{Prob}\{\|(GA - I)\widetilde{s} + \rho G\xi\|_2 \leq \tau_*\} \geq 1 - \epsilon$$

is sufficient for the validity of (10.4.24). Thus, we can use the Gaussian case procedure presented in section 10.4 with the matrix $(2/\pi)I$ in the role of C; an estimator that is good in this case will be at least as good in the case of the signal s.

In case of (b), we can act similarly, utilizing Theorem 10.3.3. Specifically, let $\widetilde{s} \sim \mathcal{N}(0, (\pi/2)I)$ be independent of ξ. Consider the parametric problem

$$\nu(\rho) \equiv \min_G \left\{ \frac{\pi}{2} \text{Tr} \left((GA - I)(GA - I)^T \right) + \rho^2 \text{Tr}(GG^T) \right\}, \qquad (10.4.35)$$

$\rho \geq 0$ being the parameter (cf. (10.4.32) and take into account that the latter problem is equivalent to (10.4.30)), and let G_ρ be an optimal solution to this problem.

The same reasoning as on p. 267 shows that

$$6/7 \leq \chi < 1 \Rightarrow \mathrm{Prob}\{(\widetilde{s}, \xi) : \|(G_\rho A - I)\widetilde{s} + \rho G_\rho \xi\|_2 \leq \Upsilon(\chi)\nu_*^{1/2}(\rho)\} \geq \chi,$$

$$\Upsilon(\chi) = \sqrt{3\left(1 + 3\ln\frac{1}{1-\chi}\right)}.$$

Applying Theorem 10.3.3 to the convex set $Q = \{(z, x) : \|(G_\rho A - I)z + \rho G_\rho x\|_2 \leq \Upsilon(\chi)\nu_*^{1/2}(\rho)\}$ and the random vectors $[s; \xi]$, $[\widetilde{s}; \xi]$, we conclude that

$$\forall\left(\begin{smallmatrix}\chi \in [6/7,1)\\ \gamma > 1\end{smallmatrix}\right) : \mathrm{Prob}\{(s, \xi) : \|(G_\rho A - I)s + \rho G_\rho \xi\|_2 > \gamma\Upsilon(\chi)\nu_*^{1/2}(\rho)\}$$

$$\leq \min_{\beta \in [1,\gamma]} \frac{1}{\gamma - \beta} \int_\beta^\infty \mathrm{Erf}(r\mathrm{ErfInv}(1 - \chi))dr.$$

We conclude that setting

$$\widetilde{\Gamma}(\rho) = \inf_{\chi,\gamma,\beta}\left\{\frac{1}{\gamma - \beta}\int_\beta^\infty \mathrm{Erf}(r\mathrm{ErfInv}(1 - \chi))dr : \begin{array}{l} 6/7 \leq \chi < 1, \gamma > 1 \\ 1 \leq \beta < \gamma \\ \gamma\Upsilon(\chi)\nu_*^{1/2}(\rho) \leq \tau_* \end{array}\right.$$
$$\left[\Upsilon(\chi) = \sqrt{3\left(1 + 3\ln\frac{1}{1-\chi}\right)}\right]$$

($\widetilde{\Gamma}(\rho) = 1$ when the right hand side problem is infeasible), one has

$$\mathrm{Prob}\{(s, \xi) : \|(G_\rho A - I)s + \rho G_\rho \xi\|_2 > \tau_*\} \leq \widetilde{\Gamma}(\rho)$$

(cf. p. 267). It is easily seen that $\widetilde{\Gamma}(\cdot)$ is a continuous nondecreasing function of $\rho > 0$ such that $\widetilde{\Gamma}(\rho) \to 0$ as $\rho \to +0$, and we end up with the following safe approximation of the Signal Recovery problem:

$$\max_\rho\left\{\rho : \widetilde{\Gamma}(\rho) \leq \epsilon\right\}$$

(cf. (10.4.29)).

Note that in the above "Gaussian majorization" scheme we could use the Bernstein approximation, based on Corollary 4.5.11, of the chance constraint $\mathrm{Prob}\{\|(GA - I)\widetilde{s} + \rho G\xi\|_2 \leq \tau_*\} \geq 1 - \epsilon$ instead of the Conjecture-related approximation.

The case of deterministic uncertain signal. Up to now, signal s was considered as random and independent of ξ, and the probability in (10.4.24) was taken w.r.t. the joint distribution of s and ξ; as a result, certain "rare" realizations of the signal can be recovered very poorly. Our current goal is to understand what happens when we replace the specification (10.4.24) with

$$\forall (s \in \mathcal{S}) :$$
$$\mathrm{Prob}\{\xi : \|Gu - s\|_2 \leq \tau_*\} \equiv \mathrm{Prob}\{\xi : \|(GA - I)s + \rho G\xi\|_2 \leq \tau_*\} \geq 1 - \epsilon, \tag{10.4.36}$$

where $\mathcal{S} \subset \mathbb{R}^n$ is a given compact set.

Our starting point is the following observation:

Lemma 10.4.8. Let G, $\rho \geq 0$ be such that

$$\Theta \equiv \frac{\tau_*^2}{\max_{s \in \mathcal{S}} s^T (GA - I)^T (GA - I) s + \rho^2 \mathrm{Tr}(G^T G)} \geq 1. \qquad (10.4.37)$$

Then for every $s \in \mathcal{S}$ one has

$$\mathrm{Prob}_{\zeta \sim \mathcal{N}(0,I)} \left\{ \|(GA - I)s + \rho G \zeta\|_2 > \tau_* \right\} \leq \exp \left\{ -\frac{(\Theta - 1)^2}{4(\Theta + 1)} \right\}. \qquad (10.4.38)$$

Proof. There is nothing to prove when $\Theta = 1$, so that let $\Theta > 1$. Let us fix $s \in \mathcal{S}$ and let $g = (GA - I)s$, $W = \rho^2 G^T G$, $w = \rho G^T g$. We have

$$\mathrm{Prob}\{\|(GA - I)s + \rho G \zeta\|_2 > \tau_*\} = \mathrm{Prob}\left\{\|g + \rho G \zeta\|_2^2 > \tau_*^2\right\}$$

$$= \mathrm{Prob}\left\{\zeta^T [\rho^2 G^T G]\zeta + 2\zeta^T \rho G^T g > \tau_*^2 - g^T g\right\} \qquad (10.4.39)$$

$$= \mathrm{Prob}\left\{\zeta^T W \zeta + 2\zeta^T w > \tau_*^2 - g^T g\right\}.$$

Denoting by λ the vector of eigenvalues of W, we can assume w.l.o.g. that $\lambda \neq 0$, since otherwise $W = 0$, $w = 0$ and thus the left hand side in (10.4.39) is 0 (note that $\tau_*^2 - g^T g > 0$ due to (10.4.37) and since $s \in \mathcal{S}$), and thus (10.4.38) is trivially true. Setting

$$\Omega = \frac{\tau_*^2 - g^T g}{\sqrt{\lambda^T \lambda + w^T w}}$$

and invoking Proposition 4.5.10, we arrive at

$$\mathrm{Prob}\{\|(GA - I)s + \rho G \zeta\|_2 > \tau_*\} \leq \exp \left\{ -\frac{\Omega^2 \sqrt{\lambda^T \lambda + w^T w}}{4 \left[2\sqrt{\lambda^T \lambda + w^T w} + \|\lambda\|_\infty \Omega \right]} \right\}$$

$$= \exp \left\{ -\frac{[\tau_*^2 - g^T g]^2}{4[2[\lambda^T \lambda + w^T w] + \|\lambda\|_\infty [\tau_*^2 - g^T g]]} \right\}$$

$$= \exp \left\{ -\frac{[\tau_*^2 - g^T g]^2}{4[2[\lambda^T \lambda + g^T [\rho^2 G G^T]g] + \|\lambda\|_\infty [\tau_*^2 - g^T g]]} \right\} \qquad (10.4.40)$$

$$\leq \exp \left\{ -\frac{[\tau_*^2 - g^T g]^2}{4\|\lambda\|_\infty [2[\|\lambda\|_1 + g^T g] + [\tau_*^2 - g^T g]]} \right\},$$

where the concluding inequality is due to $\rho^2 G G^T \preceq \|\lambda\|_\infty I$ and $\lambda^T \lambda \leq \|\lambda\|_\infty \|\lambda\|_1$. Further, setting $\alpha = g^T g$, $\beta = \mathrm{Tr}(\rho^2 G^T G)$ and $\gamma = \alpha + \beta$, observe that $\beta = \|\lambda\|_1 \geq \|\lambda\|_\infty$ and $\tau_*^2 \geq \Theta \gamma \geq \gamma$ by (10.4.37). It follows that

$$\frac{[\tau_*^2 - g^T g]^2}{4\|\lambda\|_\infty [2[\|\lambda\|_1 + g^T g] + [\tau_*^2 - g^T g]]} \geq \frac{(\tau_*^2 - \gamma + \beta)^2}{4\beta(\tau_*^2 + \gamma + \beta)} \geq \frac{(\tau_*^2 - \gamma)^2}{4\gamma(\tau_*^2 + \gamma)},$$

where the concluding inequality is readily given by the relations $\tau_*^2 \geq \gamma \geq \beta > 0$. Thus, (10.4.40) implies that

$$\mathrm{Prob}\{\|(GA - I)s + \rho G \zeta\|_2 > \tau_*\} \leq \exp \left\{ -\frac{(\tau_*^2 - \gamma)^2}{4\gamma(\tau_*^2 + \gamma)} \right\} \leq \exp \left\{ -\frac{(\Theta - 1)^2}{4(\Theta + 1)} \right\}. \quad \square$$

Lemma 10.4.8 suggests a safe approximation of the problem of interest as follows. Let $\Theta(\epsilon) > 1$ be given by

$$\exp\{-\frac{(\Theta - 1)^2}{4(\Theta + 1)}\} = \epsilon \qquad [\Rightarrow \Theta(\epsilon) = (4 + o(1)) \ln(1/\epsilon) \text{ as } \epsilon \to +0]$$

and let

$$\phi(G) = \max_{s \in \mathcal{S}} s^T (GA - I)^T (GA - I)s, \tag{10.4.41}$$

(this function clearly is convex). By Lemma 10.4.8, the optimization problem

$$\max_{\rho, G} \left\{ \rho : \phi(G) + \rho^2 \mathrm{Tr}(G^T G) \leq \gamma_* \equiv \Theta^{-1}(\epsilon)\tau_*^2 \right\} \tag{10.4.42}$$

is a safe approximation of the problem of interest. Applying bisection in ρ, we can reduce this problem to a "short series" of convex feasibility problems of the form

$$\text{find } G: \phi(G) + \rho^2 \mathrm{Tr}(G^T G) \leq \gamma_*. \tag{10.4.43}$$

Whether the latter problems are or are not computationally tractable depends on whether the function $\phi(G)$ is so, which happens if and only if we can efficiently optimize positive semidefinite quadratic forms $s^T Q s$ over $\mathcal{S}$.

Example 10.4.9. Let $\mathcal{S}$ be an ellipsoid centered at the origin:

$$\mathcal{S} = \{s = Hv : v^T v \leq 1\}$$

In this case, it is easy to compute $\phi(G)$ — this function is semidefinite representable:

$$\phi(G) \leq t \Leftrightarrow \max_{s \in \mathcal{S}} s^T (GA - I)^T (GA - I)s \leq t$$

$$\Leftrightarrow \max_{v : \|v\|_2 \leq 1} v^T (H^T (GA - I)^T (GA - I)H v \leq t$$

$$\Leftrightarrow \lambda_{\max}(H^T (GA - I)^T (GA - I)H) \leq t$$

$$\Leftrightarrow tI - H^T (GA - I)^T (GA - I)H \succeq 0 \Leftrightarrow \left[\begin{array}{c|c} tI & H^T (GA - I)^T \\ \hline (GA - I)H & I \end{array} \right] \succeq 0,$$

where the concluding $\Leftrightarrow$ is given by the Schur Complement Lemma. Consequently, (10.4.43) is the efficiently solvable convex feasibility problem

$$\text{Find } G, t: \quad t + \rho^2 \mathrm{Tr}(G^T G) \leq \gamma_*, \left[\begin{array}{c|c} tI & H^T (GA - I)^T \\ \hline (GA - I)H & I \end{array} \right] \succeq 0.$$

Example 10.4.9 allows us to see the dramatic difference between the case where we are interested in "highly reliable with high probability" recovery of a *random* signal and "highly reliable" recovery of *every* realization of uncertain signal. Specifically, assume that G, ρ are such that (10.4.24) is satisfied with $s \sim \mathcal{N}(0, I_n)$. Note that when n is large, s is nearly uniformly distributed over the sphere $\mathcal{S}$ of radius $\sqrt{n}$ (indeed, $s^T s = \sum_i s_i^2$, and by the Law of Large Numbers, for $\delta > 0$ the probability of the event $\{\|s\|_2 \notin [(1 - \delta)\sqrt{n}, (1 + \delta)\sqrt{n}]\}$ goes to 0 as $n \to \infty$, in fact exponentially fast. Also, the direction $s/\|s\|_2$ of s is uniformly distributed on the unit sphere). Thus, the recovery in question is, essentially, a highly reliable recovery of random signal uniformly distributed over the above sphere $\mathcal{S}$. Could we expect the recovery to "nearly satisfy" (10.4.36), that is, to be reasonably good in the worst case over the signals from $\mathcal{S}$? The answer is negative when n is large. Indeed, a *sufficient* condition for (10.4.24) to be satisfied is

$$\mathrm{Tr}((GA - I)^T (GA - I)) + \rho^2 \mathrm{Tr}(G^T G) \leq \frac{\tau_*^2}{O(1) \ln(1/\epsilon)} \tag{*}$$

with appropriately chosen absolute constant $O(1)$. A *necessary* condition for (10.4.36) to be satisfied is

$$n\lambda_{\max}((GA - I)^T(GA - I)) + \rho^2 \text{Tr}(G^T G) \leq O(1)\tau_*^2. \qquad (**)$$

Since the trace of the $n \times n$ matrix $Q = (GA - I)^T(GA - I)$ can be nearly n times less than $n\lambda_{\max}(Q)$, the validity of $(*)$ *by far* does not imply the validity of $(**)$. To be more rigorous, consider the case when $\rho = 0$ and $GA - I = \text{Diag}\{1, 0, ..., 0\}$. In this case, the $\|\cdot\|_2$-norm of the recovering error, in the case of $s \sim \mathcal{N}(0, I_n)$, is just $|s_1|$, and $\text{Prob}\{|s_1| > \tau_*\} \leq \epsilon$ provided that $\tau_* \geq \sqrt{2\ln(2/\epsilon)}$, in particular, when $\tau_* = \sqrt{2\ln(2/\epsilon)}$. At the same time, when $s = \sqrt{n}[1; 0; ...; 0] \in \mathcal{S}$, the norm of the recovering error is $\sqrt{n}$, which, for large n, is incomparably larger than the above τ_*.

Example 10.4.10. Here we consider the case where $\phi(G)$ cannot be computed efficiently, specifically, the case where $\mathcal{S}$ is the unit box $B_n = \{s \in \mathbb{R}^n : \|s\|_\infty \leq 1\}$ (or the set V_n of vertices of this box). Indeed, it is known that for a general-type positive definite quadratic form $s^T Q s$, computing its maximum over the unit box is NP-hard, even when instead of the precise value of the maximum its 4%-accurate approximation is sought. In situations like this we could replace $\phi(G)$ in the above scheme by its efficiently computable upper bound $\widehat{\phi}(G)$. To get such a bound in the case when $\mathcal{S}$ is the unit box, we can use the following wonderful result:

Nesterov's $\frac{\pi}{2}$ Theorem [88] *Let $A \in \mathbf{S}_+^n$. Then the efficiently computable quantity*

$$\text{SDP}(A) = \min_{\lambda \in \mathbb{R}^n} \left\{ \sum_i \lambda_i : \text{Diag}\{\lambda\} \succeq A \right\}$$

is an upper bound, tight within the factor $\frac{\pi}{2}$, on the quantity

$$\text{Opt}(A) = \max_{s \in B_n} s^T A s.$$

Assuming that $\mathcal{S}$ is B_n (or V_n), Nesterov's $\frac{\pi}{2}$ Theorem provides us with an efficiently computable and tight, within the factor $\frac{\pi}{2}$, upper bound

$$\widehat{\phi}(G) = \min_\lambda \left\{ \sum_i \lambda_i : \left[\begin{array}{c|c} \text{Diag}(\lambda) & (GA - I)^T \\ \hline GA - I & I \end{array} \right] \succeq 0 \right\}$$

on $\phi(G)$. Replacing $\phi(\cdot)$ by its upper bound, we pass from the intractable problems (10.4.43) to their tractable approximations

$$\text{find } G, \lambda: \quad \sum_i \lambda_i + \rho^2 \text{Tr}(G^T G) \leq \gamma_*, \left[\begin{array}{c|c} \text{Diag}(\lambda) & (GA - I)^T \\ \hline GA - I & I \end{array} \right] \succeq 0; \qquad (10.4.44)$$

we then apply bisection in ρ to rapidly approximate the largest $\rho = \rho_*$, along with the associated $G = G_*$, for which problems (10.4.44) are solvable, thus getting a feasible solution to the problem of interest.

10.5 NOTES AND REMARKS

NR 10.1. The celebrated Talagrand Inequality in the form we use to prove Theorem 10.1.1 can be found in [67]. Theorem B.5.1 underlying Theorem 10.1.2

was announced, in a slightly weaker form, in [82]; the proof, heavily exploiting the result of Borell [31], was published in [84].

We are grateful to A. Man-Cho So who brought to our attention the results of [78, 93, 35], which allow to justify easily the validity of Conjecture 10.1 with $\Upsilon = O(1)\sqrt{\ln m}$ in the general case.

The concept of convex majorization used in section 10.3 is, essentially, a symmetrized version of the well-studied notion of second order stochastic dominance [47, 60, 101, 102]. Proposition 10.3.2 and Theorem 10.3.3 originate from [84].

NR 10.2. For the basic results on Wiener filtering theory mentioned in section 10.4.3 see, e.g., [34].

NR 10.3. A surprising fact is that at the present level of our knowledge the chance constrained versions of "complicated" uncertain conic inequalities, like conic quadratic and especially semidefinite ones, seem to be better suited for tight tractable approximation than the RCs of these inequalities associated with deterministic uncertainty sets, even simple ones. Indeed, the RCs here typically are computationally intractable, and even building their tight tractable approximations requires severe restrictions on the structure of perturbations and/or on the geometry of uncertainty set. This is in sharp contrast with uncertain LO, where processing chance versions of uncertain constraints requires approximations, while processing the RCs of the constraints is easy.

Chapter Eleven
Globalized Robust Counterparts of Uncertain Conic Problems

In this chapter we study the Globalized Robust Counterparts of general-type uncertain conic problems and derive results on tractability of GRCs.

11.1 GLOBALIZED ROBUST COUNTERPARTS OF UNCERTAIN CONIC PROBLEMS: DEFINITION

Consider an uncertain conic problem (5.1.2), (5.1.3):

$$\min_x \left\{ c^T x + d : A_i x - b_i \in \mathbf{Q}_i, \ 1 \le i \le m \right\}, \tag{11.1.1}$$

where $\mathbf{Q}_i \subset \mathbb{R}^{k_i}$ are nonempty closed convex sets given by finite lists of conic inclusions:

$$\mathbf{Q}_i = \{ u \in \mathbb{R}^{k_i} : Q_{i\ell} u - q_{i\ell} \in \mathbf{K}_{i\ell}, \ \ell = 1, ..., L_i \}, \tag{11.1.2}$$

with closed convex pointed cones $\mathbf{K}_{i\ell}$, and let the data be affinely parameterized by the perturbation vector ζ:

$$(c, d, \{A_i, b_i\}_{i=1}^m) = (c^0, d^0, \{A_i^0, b_i^0\}_{i=1}^m) + \sum_{\ell=1}^L \zeta_\ell (c^\ell, d^\ell, \{A_i^\ell, b_i^\ell\}_{i=1}^m). \tag{11.1.3}$$

When extending the notion of Globalized Robust Counterparts (chapter 3) to this case, we need a small modification; when introducing the notion of GRCs in the LO case, we assumed that the set $\mathcal{Z}_+$ of all "physically possible" realizations of the perturbation vector ζ is of the form $\mathcal{Z}_+ = \mathcal{Z} + \mathcal{L}$, where $\mathcal{Z}$ is the closed convex normal range of ζ and $\mathcal{L}$ is a closed convex cone. We further said that a candidate solution $\bar{y}$ to uncertain scalar linear inequality

$$[a^0 + \sum_{\ell=1}^L \zeta_\ell a^\ell]^T y - [b^0 + \sum_{\ell=1}^L \zeta_\ell b^\ell] \le 0 \tag{$*$}$$

is robust feasible with global sensitivity α, if

$$[a^0 + \sum_{\ell=1}^L \zeta_\ell a^\ell]^T y - [b^0 + \sum_{\ell=1}^L \zeta_\ell b^\ell] \le \alpha \mathrm{dist}(\zeta, \mathcal{Z}|\mathcal{L}) \ \forall \zeta \in \mathcal{Z} + \mathcal{L}. \tag{$^*_*$}$$

Now we are in the situation when the left hand side of our uncertain constraints (11.1.1) are vectors rather than scalars, so that a straightforward analogy of $\binom{*}{*}$ does not make sense. Note, however, that when rewriting $(*)$ in our present "inclusion form"

$$[a^0 + \sum_{\ell=1}^{L} \zeta_\ell a^\ell]^T y - [b^0 + \sum_{\ell=1}^{L} \zeta_\ell b^\ell] \in \mathbf{Q} \equiv \mathbb{R}_-,$$

relation $\binom{*}{*}$ says exactly that the distance from the left hand side of $(*)$ to $\mathbf{Q}$ does not exceed $\alpha \mathrm{dist}(\zeta, \mathcal{Z}|\mathcal{L})$ for all $\zeta \in \mathcal{Z} + \mathcal{L}$. In this form, the notion of global sensitivity admits the following multi-dimensional extension:

Definition 11.1.1. Consider an uncertain convex constraint

$$[P_0 + \sum_{\ell=1}^{L} \zeta_\ell P_\ell] y - [p^0 + \sum_{\ell=1}^{L} \zeta_\ell p^\ell] \in \mathbf{Q}, \tag{11.1.4}$$

where $\mathbf{Q}$ is a nonempty closed convex subset in $\mathbb{R}^k$. Let $\|\cdot\|_Q$ be a norm on $\mathbb{R}^k$, $\|\cdot\|_Z$ be a norm on $\mathbb{R}^L$, $\mathcal{Z} \subset \mathbb{R}^L$ be a nonempty closed convex normal range of perturbation ζ, and $\mathcal{L} \subset\in \mathbb{R}^L$ be a closed convex cone. We say that a candidate solution y is robust feasible, with global sensitivity α, for (11.1.4), under the perturbation structure $(\|\cdot\|_Q, \|\cdot\|_Z, \mathcal{Z}, \mathcal{L})$, if

$$\mathrm{dist}([P_0 + \sum_{\ell=1}^{L} \zeta_\ell P_\ell]y - [p^0 + \sum_{\ell=1}^{L} \zeta_\ell p^\ell], \mathbf{Q}) \le \alpha \mathrm{dist}(\zeta, \mathcal{Z}|\mathcal{L})$$

$$\forall \zeta \in \mathcal{Z}_+ = \mathcal{Z} + \mathcal{L}$$

$$\left[\begin{array}{ccl} \mathrm{dist}(u, \mathbf{Q}) & = & \min\limits_{v}\{\|u - v\|_Q : v \in \mathbf{Q}\} \\ \mathrm{dist}(\zeta, \mathcal{Z}|\mathcal{L}) & = & \min\limits_{v}\{\|\zeta - v\|_Z : v \in \mathcal{Z}, \zeta - v \in \mathcal{L}\} \end{array} \right]. \tag{11.1.5}$$

Sometimes it is necessary to add some structure to the latter definition. Specifically, assume that the space $\mathbb{R}^L$ where ζ lives is given as a direct product:

$$\mathbb{R}^L = \mathbb{R}^{L_1} \times ... \times \mathbb{R}^{L_S}$$

and let $\mathcal{Z}^s \subset \mathbb{R}^{L_s}$, $\mathcal{L}^s \subset \mathbb{R}^{L_s}$, $\|\cdot\|_s$ be, respectively, closed nonempty convex set, closed convex cone and a norm on $\mathbb{R}^{L_s}$, $s = 1, ..., S$. For $\zeta \in \mathbb{R}^L$, let ζ^s, $s = 1, ..., S$, be the projections of ζ onto the direct factors $\mathbb{R}^{L_s}$ of $\mathbb{R}^L$. The "structured version" of Definition 11.1.1 is as follows:

Definition 11.1.2. A candidate solution y to the uncertain constraint (11.1.4) is said to be robust feasible with global sensitivities α_s, $s = 1, ..., S$, under the perturbation structure $(\|\cdot\|_Q, \{\mathcal{Z}^s, \mathcal{L}^s, \|\cdot\|_s\}_{s=1}^S)$, if

$$\mathrm{dist}([P_0 + \sum_{\ell=1}^{L} \zeta_\ell P_\ell]y - [p^0 + \sum_{\ell=1}^{L} \zeta_\ell p^\ell], \mathbf{Q}) \le \sum_{s=1}^{S} \alpha_s \mathrm{dist}(\zeta^s, \mathcal{Z}^s|\mathcal{L}^s)$$

$$\forall \zeta \in \mathcal{Z}_+ = \underbrace{(\mathcal{Z}^1 \times ... \times \mathcal{Z}^S)}_{\mathcal{Z}} + \underbrace{\mathcal{L}^1 \times ... \times \mathcal{L}^S}_{\mathcal{L}}$$

$$\left[\begin{array}{ccl} \mathrm{dist}(u, \mathbf{Q}) & = & \min\limits_{v}\{\|u - v\|_Q : v \in \mathbf{Q}\} \\ \mathrm{dist}(\zeta^s, \mathcal{Z}^s|\mathcal{L}^s) & = & \min\limits_{v^s}\{\|\zeta^s - v^s\|_s : v^s \in \mathcal{Z}^s, \zeta^s - v^s \in \mathcal{L}^s\}. \end{array} \right] \tag{11.1.6}$$

Note that Definition 11.1.1 can be obtained from Definition 11.1.2 by setting $S = 1$. We refer to the semi-infinite constraints (11.1.5), (11.1.6) as to *Globalized Robust Counterparts* of the uncertain constraint (11.1.4) w.r.t. the perturbations structure in question. When building the GRC of uncertain problem (11.1.1), (11.1.3), we first rewrite it as an uncertain problem

$$
\min_{y=(x,t)} \left\{ t : \begin{array}{c} \overbrace{[P_{00}+\sum_{\ell=1}^{L}\zeta_\ell P_{0\ell}]y-[p_0^0+\sum_{\ell=1}^{L}\zeta_\ell p_0^\ell]}^{} \\ c^T x + d - t \equiv [c^0 + \sum_{\ell=1}^{L}\zeta_\ell c^\ell]^T x + [d^0 + \sum_{\ell=1}^{L}\zeta_\ell d^\ell] - t \in \mathbf{Q}_0 \equiv \mathbb{R}_- \\ A_i x - b_i \equiv [A_i^0 + \sum_{\ell=1}^{L}\zeta_\ell A_i^\ell]x - [b_i^0 + \sum_{\ell=1}^{L}\zeta_\ell b_i^\ell] \in \mathbf{Q}_i,\ 1 \le i \le m \\ \underbrace{[P_{i0}+\sum_{\ell=1}^{L}\zeta_\ell P_{i\ell}]y-[p_i^0+\sum_{\ell=1}^{L}\zeta_\ell p_i^\ell]}_{} \end{array} \right\}
$$

with certain objective, and then replace the constraints with their Globalized RCs. The underlying perturbation structures and global sensitivities may vary from constraint to constraint.

11.2 SAFE TRACTABLE APPROXIMATIONS OF GRCS

A Globalized RC, the same as the plain one, can be computationally intractable, in which case we can look for the second best thing — a safe tractable approximation of the GRC. This notion is defined as follows (cf. Definition 5.3.1):

Definition 11.2.1. Consider the uncertain convex constraint (11.1.4) along with its GRC (11.1.6). We say that a system $\mathcal{S}$ of convex constraints in variables y, $\alpha = (\alpha_1, ..., \alpha_S) \ge 0$, and, perhaps, additional variables u, is a safe approximation of the GRC, if the projection of the feasible set of $\mathcal{S}$ on the space of (y, α) variables is contained in the feasible set of the GRC:

$$\forall(\alpha = (\alpha_1, ..., \alpha_S) \ge 0, y):$$
$$(\exists u :\ (y, \alpha, u) \text{ satisfies } \mathcal{S}) \Rightarrow (y, \alpha) \text{ satisfies (11.1.6)}.$$

This approximation is called tractable, provided that $\mathcal{S}$ is so, (e.g., $\mathcal{S}$ is an explicit system of CQIs/LMIs of, more general, the constraints in $\mathcal{S}$ are efficiently computable).

When quantifying the tightness of an approximation, we, as in the case of RC, assume that the normal range $\mathcal{Z} = \mathcal{Z}^1 \times ... \times \mathcal{Z}^S$ of the perturbations contains the origin and is included in the single-parametric family of normal ranges:

$$\mathcal{Z}_\rho = \rho \mathcal{Z},\ \rho > 0.$$

As a result, the GRC (11.1.6) of (11.1.4) becomes a member, corresponding to $\rho = 1$, of the single-parametric family of constraints

$$\text{dist}([P_0 + \sum_{\ell=1}^{L}\zeta_\ell P_\ell]y - [p^0 + \sum_{\ell=1}^{L}\zeta_\ell p^\ell], \mathbf{Q}) \le \sum_{s=1}^{S} \alpha_s \text{dist}(\zeta^s, \mathcal{Z}^s | \mathcal{L}^s)$$
$$\forall \zeta \in \mathcal{Z}_+^\rho = \underbrace{\rho(\mathcal{Z}^1 \times ... \times \mathcal{Z}^S)}_{\mathcal{Z}_\rho} + \underbrace{\mathcal{L}^1 \times ... \times \mathcal{L}^S}_{\mathcal{L}} \quad\quad (\text{GRC}_\rho)$$

in variables y, α. We define the tightness factor of a safe tractable approximation of the GRC as follows (cf. Definition 5.3.2):

Definition 11.2.2. Assume that we are given an approximation scheme that associates with (GRC$_\rho$) a finite system $\mathcal{S}_\rho$ of efficiently computable convex constraints on variables y, α and, perhaps, additional variables u, depending on $\rho > 0$ as a parameter. We say that this approximation scheme is a safe tractable approximation of the GRC tight, within tightness factor $\vartheta \geq 1$, if

(i) For every $\rho > 0$, $\mathcal{S}_\rho$ is a safe tractable approximation of (GRC$_\rho$): whenever $(y, \alpha \geq 0)$ can be extended to a feasible solution of $\mathcal{S}_\rho$, (y, α) satisfies (GRC$_\rho$);

(ii) Whenever $\rho > 0$ and $(y, \alpha \geq 0)$ are such that (y, α) can<u>not</u> be extended to a feasible solution of $\mathcal{S}_\rho$, the pair $(y, \vartheta^{-1}\alpha)$ is <u>not</u> feasible for (GRC$_{\vartheta\rho}$).

11.3 GRC OF UNCERTAIN CONSTRAINT: DECOMPOSITION

11.3.1 Preliminaries

Recall the notion of the *recessive cone* of a closed and nonempty convex set **Q**:

Definition 11.3.1. Let $\mathbf{Q} \subset \mathbb{R}^k$ be a nonempty closed convex set and $\bar{x} \in \mathbf{Q}$. The recessive cone $\text{Rec}(\mathbf{Q})$ of Q is comprised of all rays emanating from $\bar{x}$ and contained in **Q**:
$$\text{Rec}(\mathbf{Q}) = \{h \in \mathbb{R}^k : \bar{x} + th \in \mathbf{Q} \forall t \geq 0\}.$$

(Due to closedness and convexity of **Q**, the right hand side set in this formula is independent of the choice of $\bar{x} \in \mathbf{Q}$ and is a nonempty closed convex cone in $\mathbb{R}^k$.)

Example 11.3.2.

(i) The recessive cone of a nonempty bounded and closed convex set **Q** is trivial: $\text{Rec}(\mathbf{Q}) = \{0\}$;

(ii) The recessive cone of a closed convex cone **Q** is **Q** itself;

(iii) The recessive cone of the set $\mathbf{Q} = \{x : Ax - b \in \mathbf{K}\}$, where **K** is a closed convex cone, is the set $\{h : Ah \in \mathbf{K}\}$;

(iv.a) Let **Q** be a closed convex set and $e_i \to e$, $i \to \infty$, $t_i \geq 0$, $t_i \to \infty$, $i \to \infty$, be sequences of vectors and reals such that $t_i e_i \in \mathbf{Q}$ for all i. Then $e \in \text{Rec}(\mathbf{Q})$.

(iv.b) Vice versa: every $e \in \text{Rec}(\mathbf{Q})$ can be represented in the form of $e = \lim_{i \to \infty} e_i$ with vectors e_i such that $ie_i \in \mathbf{Q}$.

Proof. (iv.a): Let $\bar{x} \in \mathbf{Q}$. With our e_i and t_i, for every $t > 0$ we have $\bar{x} + te_i - t/t_i \bar{x} = (t/t_i)(t_i e_i) + (1 - t/t_i)\bar{x}$. For all but finitely many values of i, the right hand side in this equality is a convex combination of two vectors from **Q** and therefore belongs to **Q**; for $i \to \infty$, the left hand side converges to $\bar{x} + te$. Since **Q** is closed, we conclude that $\bar{x} + te \in \mathbf{Q}$; since $t > 0$ is arbitrary, we get $e \in \text{Rec}(\mathbf{Q})$.

(iv.b): Let $e \in \text{Rec}(\mathbf{Q})$ and $\bar{x} \in \mathbf{Q}$. Setting $e_i = i^{-1}(\bar{x} + ie)$, we have $ie_i \in \mathbf{Q}$ and $e_i \to e$ as $i \to \infty$. $\qquad\square$

11.3.2 The Main Result

The following statement is the "multi-dimensional" extension of Proposition 3.2.1:

Proposition 11.3.3. A candidate solution y is feasible for the GRC (11.1.6) of the uncertain constraint (11.1.4) if and only if x satisfies the following system of semi-infinite constraints:

$$(a) \quad \overbrace{[P_0 + \sum_{\ell=1}^{L} \zeta_\ell P_\ell]^T y - [p^0 + \sum_{\ell=1}^{L} \zeta_\ell p^\ell]}^{P(y,\zeta)} \in \mathbf{Q}$$

$$\forall \zeta \in \mathcal{Z} \equiv \mathcal{Z}^1 \times ... \times \mathcal{Z}^S$$

(11.3.1)

$$(b_s) \quad \text{dist}(\overbrace{\sum_{\ell=1}^{L} [P_\ell y - p^\ell](E_s \zeta^s)_\ell}^{\Phi(y)E_s \zeta^s}, \text{Rec}(\mathbf{Q})) \le \alpha_s$$

$$\forall \zeta^s \in \mathcal{L}_{\|\cdot\|_s}^s \equiv \{\zeta^s \in \mathcal{L}^s : \|\zeta^s\|_s \le 1\}, \quad s = 1, ..., S,$$

where E_s is the natural embedding of $\mathbb{R}^{L_s}$ into $\mathbb{R}^L = \mathbb{R}^{L_1} \times ... \times \mathbb{R}^{L_S}$ and $\text{dist}(u, \text{Rec}(\mathbf{Q})) = \min_{v \in \text{Rec}(\mathbf{Q})} \|u - v\|_Q$.

Proof. Assume that y satisfies (11.1.6), and let us verify that y satisfies (11.3.1). Relation (11.3.1.a) is evident. Let us fix $s \le S$ and verify that y satisfies (11.3.1.b_s). Indeed, let $\bar{\zeta} \in \mathcal{Z}$ and $\zeta^s \in \mathcal{L}_{\|\cdot\|_s}^s$. For $i = 1, 2, ...,$ let ζ_i be given by $\zeta_i^r = \bar{\zeta}^r$, $r \ne s$, and $\zeta_i^s = \bar{\zeta}^s + i\zeta^s$, so that $\text{dist}(\zeta_i^r, \mathcal{Z}^r|\mathcal{L}^r)$ is 0 for $r \ne s$ and is $\le i$ for $r = s$. Since y is feasible for (11.1.6), we have

$$\underbrace{\text{dist}([P_0 + \sum_{\ell=1}^{L} (\zeta_i)_\ell P_\ell]y - [p^0 + \sum_{\ell=1}^{L} (\zeta_i)_\ell p^\ell], \mathbf{Q})}_{P(y,\zeta_i) = P(y,\bar{\zeta}) + i\Phi(y)E_s \zeta^s} \le \alpha_s i,$$

that is, there exists $q_i \in \mathbf{Q}$ such that

$$\|P(y, \bar{\zeta}) + i\Phi(y)E_s \zeta^s - q_i\|_Q \le \alpha_s i.$$

From this inequality it follows that $\|q_i\|_Q/i$ remains bounded when $i \to \infty$; setting $q_i = ie_i$ and passing to a subsequence $\{i_\nu\}$ of indices i, we may assume that $e_{i_\nu} \to e$ as $\nu \to \infty$; by item (iv.a) of Example 11.3.2, we have $e \in \text{Rec}(\mathbf{Q})$. We further have

$$\begin{aligned}
\|\Phi(y)E_s \zeta^s - e_{i_\nu}\|_Q &= i_\nu^{-1}\|i_\nu \Phi(y)E_s \zeta - q_{i_\nu}\|_Q \\
&\le i_\nu^{-1}[\|P(y, \bar{\zeta}) + i_\nu \Phi(y)E_s \zeta^s - q_{i_\nu}\|_Q + i_\nu^{-1}\|P(y, \bar{\zeta})\|_Q] \\
&\le \alpha_s + i_\nu^{-1}\|P(y, \bar{\zeta})\|_Q,
\end{aligned}$$

whence, passing to limit as $\nu \to \infty$, $\|\Phi(y)E_s \zeta^s - e\|_Q \le \alpha_s$, whence, due to $e \in \text{Rec}(\mathbf{Q})$, we have $\text{dist}(\Phi(y)E_s \zeta^s, \text{Rec}(\mathbf{Q})) \le \alpha_s$. Since $\zeta^s \in \mathcal{L}_{\|\cdot\|_s}^s$ is arbitrary, (11.3.1.b_s) holds true.

Now assume that y satisfies (11.3.1), and let us prove that y satisfies (11.1.6). Indeed, given $\zeta \in \mathcal{Z} + \mathcal{L}$, we can find $\bar{\zeta}^s \in \mathcal{Z}^s$ and $\delta^s \in \mathcal{L}^s$ in such a way that $\zeta^s = \bar{\zeta}^s + \delta^s$ and $\|\delta^s\|_s = \text{dist}(\zeta^s, \mathcal{Z}^s|\mathcal{L}^s)$. Setting $\bar{\zeta} = (\bar{\zeta}^1, ..., \bar{\zeta}^S)$ and invoking (11.3.1.a), the vector $\bar{u} = P(y, \bar{\zeta})$ belongs to $\mathbf{Q}$. Further, for every s, by (11.3.1.b_s), there exists $\delta u^s \in \text{Rec}(\mathbf{Q})$ such that $\|\Phi(y)E_s \delta^s - \delta u^s\|_Q \le \alpha_s \|\delta^s\|_s = \alpha_s \text{dist}(\zeta^s, \mathcal{Z}^s|\mathcal{L}^s)$. Since

$P(y, \zeta) = P(y, \bar{\zeta}) + \sum_s \Phi(y) E_s \delta^s$, we have

$$\|P(y, \zeta) - \underbrace{[\bar{u} + \sum_s \delta u^s]}_{v}\|_Q \leq \underbrace{\|P(y, \bar{\zeta}) - \bar{u}\|_Q}_{=0} + \sum_s \underbrace{\|\Phi(y) E_s \delta^s - \delta u^s\|_Q}_{\leq \alpha_s \mathrm{dist}(\zeta^s, \mathcal{Z}^s | \mathcal{L}^s)};$$

since $\bar{u} \in \mathbf{Q}$ and $\delta u^s \in \mathrm{Rec}(\mathbf{Q})$ for all s, we have $v \in \mathbf{Q}$, so that the inequality implies that

$$\mathrm{dist}(P(y, \zeta), \mathbf{Q}) \leq \sum_s \alpha_s \mathrm{dist}(\zeta^s, \mathcal{Z}^s | \mathcal{L}^s).$$

Since $\zeta \in \mathcal{Z} + \mathcal{L}$ is arbitrary, y satisfies (11.1.6). $\qquad \square$

11.4 TRACTABILITY OF GRCS

11.4.1 Preliminaries

Proposition 11.3.3 demonstrates that the GRC of an uncertain constraint (11.1.4) is equivalent to the explicit system of semi-infinite constraints (11.3.1). We are well acquainted with the constraint (11.3.1.a) — it is nothing but the RC of the uncertain constraint (11.1.4) with the normal range $\mathcal{Z}$ of the perturbations in the role of the uncertainty set. As a result, we have certain knowledge of how to convert this semi-infinite constraint into a tractable form or how to build its tractable safe approximation. What is new is the constraint (11.3.1.b), which is of the following generic form:

We are given
• an Euclidean space E with inner product $\langle \cdot, \cdot \rangle_E$, a norm (not necessarily the Euclidean one) $\| \cdot \|_E$, and a closed convex cone K^E in E;
• an Euclidean space F with inner product $\langle \cdot, \cdot \rangle_F$, norm $\| \cdot \|_F$ and a closed convex cone K^F in F.
These data define a function on the space $\mathcal{L}(E, F)$ of linear mappings $\mathcal{M}$ from E to F, specifically, the function

$$\begin{aligned} \Psi(\mathcal{M}) &= \max_e \left\{ \mathrm{dist}_{\|\cdot\|_F}(\mathcal{M}e, K^F) : e \in K^E, \|e\|_E \leq 1 \right\}, \\ \mathrm{dist}_{\|\cdot\|_F}(f, K^F) &= \min_{g \in K^F} \|f - g\|_F. \end{aligned} \tag{11.4.1}$$

Note that $\Psi(\mathcal{M})$ is a kind of a norm: it is nonnegative, satisfies the requirement $\Psi(\lambda \mathcal{M}) = \lambda \Psi(\mathcal{M})$ when $\lambda \geq 0$, and satisfies the triangle inequality $\Psi(\mathcal{M} + \mathcal{N}) \leq \Psi(\mathcal{M}) + \Psi(\mathcal{N})$. The properties of a norm that are missing are symmetry (in general, $\Psi(-\mathcal{M}) \neq \Psi(\mathcal{M})$) and strict positivity (it may happen that $\Psi(\mathcal{M}) = 0$ for $\mathcal{M} \neq 0$). Note also that in the case when $K^F = \{0\}$, $K^E = E$, $\Psi(\mathcal{M}) = \max_{e: \|e\|_E \leq 1} \|\mathcal{M}e\|_F$ becomes the usual norm of a linear mapping induced by given norms in the origin and the destination spaces.

The above setting gives rise to a convex inequality

$$\Psi(\mathcal{M}) \leq \alpha \tag{11.4.2}$$

in variables $\mathcal{M}, \alpha$. Note that every one of the constraints (11.3.1.b) is obtained from a convex inequality of the form (11.4.2) by affine substitution

$$\mathcal{M} \leftarrow H_s(y), \ \alpha \leftarrow \alpha_s$$

where $H_s(y) \in \mathcal{L}(E_s, F_s)$ is affine in y. Indeed, (11.3.1.b_s) is obtained in this fashion when specifying

- $(E, \langle \cdot, \cdot \rangle_E)$ as the Euclidean space where $\mathcal{Z}^s, \mathcal{L}^s$ live, and $\| \cdot \|_E$ as $\| \cdot \|_s$;
- $(F, \langle \cdot, \cdot \rangle_F)$ as the Euclidean space where $\mathbf{Q}$ lives, and $\| \cdot \|_F$ as $\| \cdot \|_Q$;
- K^E as the cone $\mathcal{L}^s$, and K^F as the cone $\mathrm{Rec}(\mathbf{Q})$;
- $H(y)$ as the linear map $\zeta^s \mapsto \Phi(y) E_s \zeta^s$.

It follows that efficient processing of constraints (11.3.1.b) reduces to a similar task for the associated constraints

$$\Psi_s(\mathcal{M}_s) \leq \alpha_s \qquad\qquad (\mathcal{C}_s)$$

of the form (11.4.2). Assume, e.g., that we are smart enough to build, for certain $\vartheta \geq 1$,

(i) a ϑ-tight safe tractable approximation of the semi-infinite constraint (11.3.1.a) with $\mathcal{Z}_\rho = \rho \mathcal{Z}_1$ in the role of the perturbation set. Let this approximation be a system $\mathcal{S}_\rho^a$ of explicit convex constraints in variables y and additional variables u;

(ii) for every $s = 1, ..., S$ a ϑ-tight efficiently computable upper bound on the function $\Psi_s(\mathcal{M}_s)$, that is, a system $\mathcal{S}^s$ of efficiently computable convex constraints on matrix variable $\mathcal{M}_s$, real variable τ_s and, perhaps, additional variables u^s such that

(a) whenever $(\mathcal{M}_s, \tau_s)$ can be extended to a feasible solution of $\mathcal{S}^s$, we have $\Psi_s(\mathcal{M}_s) \leq \tau_s$,

(b) whenever $(\mathcal{M}_s, \tau_s)$ can<u>not</u> be extended to a feasible solution of $\mathcal{S}^s$, we have $\vartheta \Psi_s(\mathcal{M}_s) > \tau_s$.

In this situation, we can point out a safe tractable approximation, tight within the factor ϑ (see Definition 11.2.2), of the GRC in question. To this end, consider the system of constraints in variables $y, \alpha_1, ..., \alpha_S, u, u^1, ..., u^S$ as follows:

$$(y, u) \text{ satisfies } \mathcal{S}_\rho^a \text{ and } \{(H_s(y), \alpha_s, u^s) \text{ satisfies } \mathcal{S}^s, \ s = 1, ..., S\}, \qquad (\mathcal{S}_\rho)$$

and let us verify that this is a ϑ-tight safe computationally tractable approximation of the GRC. Indeed, $\mathcal{S}_\rho$ is an explicit system of efficiently computable convex constraints and as such is computationally tractable. Further, $\mathcal{S}_\rho$ is a safe approximation of the (GRC$_\rho$). Indeed, if (y, α) can be extended to a feasible solution of $\mathcal{S}_\rho$, then y satisfies (11.3.1.a) with $\mathcal{Z}_\rho$ in the role of $\mathcal{Z}$ (since (y, u) satisfies $\mathcal{S}_\rho^a$) and (y, α_s) satisfies (11.3.1.b_s) due to (ii.a) (recall that (11.3.1.b_s) is equivalent to $\Psi_s(H_s(y)) \leq \alpha_s$). Finally, assume that (y, α) can<u>not</u> be extended to a feasible solution of $(\mathcal{S}_\rho)$, and let us prove that then $(y, \vartheta^{-1}\alpha)$ is not feasible for (GRC$_{\vartheta\rho}$).

Indeed, if (y, α) cannot be extended to a feasible solution to $\mathcal{S}_\rho$, then either y cannot be extended to a feasible solution of $\mathcal{S}_\rho^a$, or for certain s (y, α_s) cannot be extended to a feasible solution of $\mathcal{S}^s$. In the first case, y does not satisfy $(11.3.1.a)$ with $\mathcal{Z}_{\vartheta\rho}$ in the role of $\mathcal{Z}$ by (i); in the second case, $\vartheta^{-1}\alpha_s < \Psi_s(H_s(y))$ by (ii.b), so that in both cases the pair $(y, \vartheta^{-1}\alpha)$ is not feasible for $(\mathrm{GRC}_{\vartheta\rho})$.

We have reduced the tractability issues related to Globalized RCs to similar issues for RCs (which we have already investigated in the CO case) and to the issue of efficient bounding of $\Psi(\cdot)$. The rest of this section is devoted to investigating this latter issue.

11.4.2 Efficient Bounding of $\Psi(\cdot)$

11.4.2.1 Symmetry

We start with observing that the problem of efficient computation of (a tight upper bound on) $\Psi(\cdot)$ possesses a kind of symmetry. Indeed, consider a setup

$$\Xi = (E, \langle \cdot, \cdot \rangle_E, \| \cdot \|_E, K^E; F, \langle \cdot, \cdot \rangle_F, \| \cdot \|_F, K^F)$$

specifying Ψ, and let us associate with Ξ its *dual* setup

$$\Xi_* = (F, \langle \cdot, \cdot \rangle_F, \| \cdot \|_F^*, K_*^F; E, \langle \cdot, \cdot \rangle_E, \| \cdot \|_E^*, K_*^E),$$

where

- for a norm $\| \cdot \|$ on a Euclidean space $(G, \langle \cdot, \cdot \rangle_G)$, its *conjugate* norm $\| \cdot \|^*$ is defined as
$$\|u\|^* = \max_v \{\langle u, v \rangle_G : \|v\| \leq 1\};$$

- For a closed convex cone K in a Euclidean space $(G, \langle \cdot, \cdot \rangle_G)$, its *dual* cone is defined as
$$K_* = \{y : \langle y, h \rangle_G \geq 0 \ \ \forall h \in K\}.$$

.

Recall that the *conjugate* to a linear map $\mathcal{M} \in \mathcal{L}(E, F)$ from Euclidean space E to Euclidean space F is the linear map $\mathcal{M}^* \in \mathcal{L}(F, E)$ uniquely defined by the identity

$$\langle \mathcal{M}e, f \rangle_F = \langle e, \mathcal{M}^* f \rangle_E \quad \forall (e \in E, f \in F);$$

representing linear maps by their matrices in a fixed pair of orthonormal bases in E, F, the matrix representing $\mathcal{M}^*$ is the transpose of the matrix representing $\mathcal{M}$. Note that twice taken dual/conjugate of an entity recovers the original entity: $(K_*)_* = K$, $(\| \cdot \|^*)^* = \| \cdot \|$, $(\mathcal{M}^*)^* = \mathcal{M}$, $(\Xi_*)_* = \Xi$.

Recall that the functions $\Psi(\cdot)$ are given by setups Ξ of the outlined type according to

$$\Psi(\mathcal{M}) \equiv \Psi_\Xi(\mathcal{M}) = \max_{e \in E} \left\{ \mathrm{dist}_{\| \cdot \|_F}(\mathcal{M}e, K^F) : e \in K^E, \|e\|_E \leq 1 \right\}.$$

The aforementioned symmetry is nothing but the following simple statement:

Proposition 11.4.1. For every setup $\Xi = (E, ..., K^F)$ and every $\mathcal{M} \in \mathcal{L}(E, F)$ one has

$$\Psi_\Xi(\mathcal{M}) = \Psi_{\Xi_*}(\mathcal{M}^*).$$

Proof. Let $H, \langle \cdot, \cdot \rangle_H$ be a Euclidean space. Recall that the *polar* of a closed convex set $X \subset H$, $0 \in X$, is the set $X^o = \{y \in H : \langle y, x \rangle_H \leq 1 \quad \forall x \in X\}$. We need the following facts:

(a) If $X \subset H$ is closed, convex and $0 \in X$, then so is X^o, and $(X^o)^0 = X$ [100];

(b) If $X \subset H$ is convex compact, $0 \in X$, and $K^H \subset H$ is closed convex cone, then $X + K^H$ is closed and

$$(X + K^H)^o = X^o \cap (-K_*^H).$$

Indeed, the arithmetic sum of a compact and a closed set is closed, so that $X + K^H$ is closed, convex, and contains 0. We have

$$f \in (X + K^H)^o \Leftrightarrow 1 \geq \sup_{x \in X, h \in K^H} \langle f, x + h \rangle_H = \sup_{x \in X} \langle f, x \rangle_H + \sup_{h \in K^H} \langle f, h \rangle_H;$$

since K^H is a cone, the concluding inequality is possible iff $f \in X^o$ and $f \in -K_*^H$.

(c) Let $\| \cdot \|$ be a norm in H. Then for every $\alpha > 0$ one has $(\{x : \|x\| \leq \alpha\})^o = \{x : \|x\|^* \leq 1/\alpha\}$ (evident).

When $\alpha > 0$, we have

$$\Psi_\Xi(\mathcal{M}) \leq \alpha$$

$$\Leftrightarrow \begin{cases} \forall e \in K^E \cap \{e : \|e\|_E \leq 1\} : \\ \mathcal{M}e \in \{f : \|f\|_F \leq \alpha\} + K^F \end{cases} \qquad \text{[by definition]}$$

$$\Leftrightarrow \begin{cases} \forall e \in K^E \cap \{e : \|e\|_E \leq 1\} : \\ \mathcal{M}e \in [[\{f : \|f\|_F \leq \alpha\} + K^F]^o]^o \end{cases} \qquad \text{[by (a)]}$$

$$\Leftrightarrow \begin{cases} \forall e \in K^E \cap \{e : \|e\|_E \leq 1\} : \\ \langle \mathcal{M}e, f \rangle_F \leq 1 \; \forall f \in \underbrace{[\{f : \|f\|_F \leq \alpha\} + K^F]^o}_{=\{f : \|f\|_F^* \leq \alpha^{-1}\} \cap (-K_*^F)} \end{cases} \qquad \text{[by (b), (c)]}$$

$$\Leftrightarrow \begin{cases} \forall e \in K^E \cap \{e : \|e\|_E \leq 1\} : \\ \langle e, \mathcal{M}^*f \rangle_E \leq 1 \; \forall f \in \{f : \|f\|_F^* \leq \alpha^{-1}\} \cap (-K_*^F) \end{cases}$$

$$\Leftrightarrow \begin{cases} \forall e \in K^E \cap \{e : \|e\|_E \leq \alpha^{-1}\} : \\ \langle e, \mathcal{M}^*f \rangle_E \leq 1 \; \forall f \in \{f : \|f\|_F^* \leq 1\} \cap (-K_*^F) \end{cases} \qquad \text{[evident]}$$

$$\Leftrightarrow \begin{cases} \forall e \in [-(-K_*^E)_*] \cap [\{e : \|e\|_E^* \leq \alpha\}^o] : \\ \langle e, \mathcal{M}^*f \rangle_E \leq 1 \; \forall f \in \{f : \|f\|_F^* \leq 1\} \cap (-K_*^F) \end{cases} \qquad \text{[by (c)]}$$

$$\Leftrightarrow \begin{cases} \forall e \in [(-K_*^E) + \{e : \|e\|_E^* \leq \alpha\}]^o : \\ \langle \mathcal{M}^*f, e \rangle_E \leq 1 \; \forall f \in \{f : \|f\|_F^* \leq 1\} \cap (-K_*^F) \end{cases} \qquad \text{[by (b)]}$$

$$\Leftrightarrow \begin{cases} \forall f \in \{f : \|f\|_F^* \leq 1\} \cap (-K_*^F) : \\ \langle \mathcal{M}^*f, e \rangle_E \leq 1 \; \forall e \in [(-K_*^E) + \{e : \|e\|_E^* \leq \alpha\}]^o \end{cases}$$

$$\Leftrightarrow \begin{cases} \forall f \in \{f : \|f\|_F^* \leq 1\} \cap (-K_*^F) : \\ \mathcal{M}^*f \in (-K_*^E) + \{e : \|e\|_E^* \leq \alpha\} \end{cases} \qquad \text{[by (a)]}$$

$$\Leftrightarrow \begin{cases} \forall f \in K_F^* \cap \{f : \|f\|_F^* \leq 1\} : \\ \mathcal{M}^*f \in K_*^E + \{e : \|e\|_E^* \leq \alpha\} \end{cases}$$

$$\Leftrightarrow \Psi_{\Xi_*}(\mathcal{M}^*) \leq \alpha. \qquad \qquad \square$$

11.4.2.2 Good GRC setups

Proposition 11.4.1 says that "good" setups Ξ — those for which $\Psi_\Xi(\cdot)$ is efficiently computable or admits a tight, within certain factor ϑ, efficiently computable upper bound — always come in symmetric pairs: if Ξ is good, so is Ξ_*, and vice versa. In what follows, we refer to members of such a symmetric pair as to *counterparts* of each other. We are about to list a number of good pairs. From now on, we assume that all components of a setup in question are "computationally tractable," specifically, that the cones K^E, K^F and the epigraphs of the norms $\|\cdot\|_E$, $\|\cdot\|_F$ are given by LMI representations (or, more general, by systems of efficiently computable convex constraints). Below, we denote by B_E and B_F the unit balls of the norms $\|\cdot\|_E$, $\|\cdot\|_F$, respectively.

Here are several good GRC setups:

A: $K^E = \{0\}$. The counterpart is
A*: $K^F = F$.

These cases are trivial: $\Psi_\Xi(\mathcal{M}) \equiv 0$.

B: $K^E = E$, $B_E = \mathrm{Conv}\{e^1, ..., e^N\}$, *the list* $\{e^i\}_{i=1}^N$ *is available. The counterpart is the case*
B*: $K^F = \{0\}$, $B_F = \{f : \langle f^i, f\rangle_F \leq 1, i = 1, ..., N\}$, *the list* $\{f^i\}_{i=1}^N$ *is available.*

Standard example for **B** is $E = \mathbb{R}^n$ with the standard inner product, $K^E = E$, $\|e\| = \|e\|_1 \equiv \sum_j |e_j|$. Standard example for **B*** is $F = \mathbb{R}^m$ with the standard inner product, $\|f\|_F = \|f\|_\infty = \max_j |f_j|$.

The cases in question are easy. Indeed, in the case of **B** we clearly have

$$\Psi(\mathcal{M}) = \max_{1 \leq j \leq N} \mathrm{dist}_{\|\cdot\|_F}(\mathcal{M}e_i, K^F),$$

and thus $\Psi(\mathcal{M})$ is efficiently computable (as the maximum of a finite family of efficiently computable quantities $\mathrm{dist}_{\|\cdot\|_F}(\mathcal{M}e_i, K^F)$). Assuming, e.g., that E, F are, respectively, $\mathbb{R}^m$ and $\mathbb{R}^n$ with the standard inner products, and that K^F, $\|\cdot\|_F$ are given by strictly feasible conic representations:

$$K^F = \{f : \exists u : Pf + Qu \in \mathbf{K}^1\},$$
$$\{t \geq \|f\|_F\} \Leftrightarrow \{\exists v : Rf + tr + Sv \in \mathbf{K}^2\}$$

the relation

$$\Psi(\mathcal{M}) \leq \alpha$$

can be represented equivalently by the following explicit system of conic constraints

$$(a) \quad Pf^i + Qu^i \in \mathbf{K}^1, \, i = 1, ..., N$$
$$(b) \quad R(\mathcal{M}e^i - f^i) + \alpha r + Sv^i \in \mathbf{K}^2, \, i = 1, ..., N$$

in variables $\mathcal{M}, \alpha, u^i, f^i, v^i$. Indeed, relations (a) equivalently express the requirement $f^i \in K^F$, while relations (b) say that $\|\mathcal{M}e^i - f^i\|_F \leq \alpha$.

C: $K^E = E$, $K^F = \{0\}$. The counterpart case is exactly the same.

In the case of **C**, $\Psi(\cdot)$ is the norm of a linear map from E to F induced by given norms on the origin and the destination spaces:

$$\Psi(\mathcal{M}) = \max_e \{\|\mathcal{M}e\|_F : \|e\|_E \leq 1\}.$$

Aside of situations covered by **B**, **B***, there is only one generic situation where computing the norm of a linear map is easy — this is the situation where both $\|\cdot\|_E$ and $\|\cdot\|_F$ are Euclidean norms. In this case, we lose nothing by assuming that $E = \ell_2^n$ (that is, E is $\mathbb{R}^n$ with the standard inner product and the standard norm $\|e\|_2 = \sqrt{\sum_i e_i^2}$), $F = \ell_2^m$, and let M be the $m \times n$ matrix representing the map $\mathcal{M}$ in the standard bases of E and F. In this case, $\Psi(\mathcal{M}) = \|M\|_{2,2}$ is the maximal singular value of M and as such is efficiently computable. A semidefinite representation of the constraint $\|M\|_{2,2} \leq \alpha$ is

$$\left[\begin{array}{c|c} \alpha I_n & M^T \\ \hline M & \alpha I_m \end{array}\right] \succeq 0.$$

Now consider the case when $E = \ell_p^n$ (that is, E is $\mathbb{R}^n$ with the standard inner product and the norm

$$\|e\|_p = \begin{cases} \left(\sum_j |e_j|^p\right)^{1/p} & , 1 \leq p < \infty \\ \max_j |e_j| & , p = \infty \end{cases},$$

and $F = \ell_r^m$, $1 \leq r, p \leq \infty$. Here again we can naturally identify $\mathcal{L}(E, F)$ with the space $\mathbb{R}^{m \times n}$ of real $m \times n$ matrices, and the problem of interest is to compute

$$\|M\|_{p,r} = \max_e \{\|Me\|_r : \|e\|_p \leq 1\}.$$

The case of $p = r = 2$ is the just considered "purely Euclidean" situation; the cases of $p = 1$ and of $r = \infty$ are covered by **B**, **B***. These are the only 3 cases when computing $\|\cdot\|_{p,r}$ is known to be easy. It is also known that it is NP-hard to compute the matrix norm in question when $p > r$. However, in the case of $p \geq 2 \geq r$ there exists a tight efficiently computable upper bound on $\|M\|_{p,r}$ due to Nesterov [115, Theorem 13.2.4]. Specifically, Nesterov shows that when $\infty \geq p \geq 2 \geq r \geq 1$, the explicitly computable quantity

$$\Psi_{p,r}(M) = \frac{1}{2} \min_{\substack{\mu \in \mathbb{R}^n \\ \nu \in \mathbb{R}^m}} \left\{ \|\mu\|_{\frac{p}{p-2}} + \|\nu\|_{\frac{r}{2-r}} : \left[\begin{array}{c|c} \mathrm{Diag}\{\mu\} & M^T \\ \hline M & \mathrm{Diag}\{\nu\} \end{array}\right] \succeq 0 \right\}$$

is an upper bound on $\|M\|_{p,r}$, and this bound is tight within the factor $\vartheta = \left[\frac{2\sqrt{3}}{\pi} - \frac{2}{3}\right]^{-1} \approx 2.2936$:

$$\|M\|_{p,r} \leq \Psi_{p,r}(M) \leq \left[\frac{2\sqrt{3}}{\pi} - \frac{2}{3}\right]^{-1} \|M\|_{p,r}.$$

It follows that the explicit system of efficiently computable convex constraints

$$\left[\begin{array}{c|c} \mathrm{Diag}\{\mu\} & M^T \\ \hline M & \mathrm{Diag}\{\nu\} \end{array}\right] \succeq 0, \quad \frac{1}{2}\left[\|\mu\|_{\frac{p}{p-2}} + \|\nu\|_{\frac{r}{2-r}}\right] \leq \alpha \qquad (11.4.3)$$

in variables M, α, μ, ν is a safe tractable approximation of the constraint

$$\|M\|_{p,r} \leq \alpha,$$

which is tight within the factor ϑ. In some cases the value of the tightness factor can be improved; e.g., when $p = \infty$, $r = 2$ and when $p = 2$, $r = 1$, the tightness factor does not exceed $\sqrt{\pi/2}$.

Most of the tractable (or nearly so) cases considered so far deal with the case when $K^F = \{0\}$ (the only exception is the case **B*** that, however, imposes severe restrictions on $\|\cdot\|_E$). In the GRC context, that means that we know nearly nothing about what to do when the recessive cone of the right hand side set $\mathbf{Q}$ in (11.1.1) is nontrivial, or, which is the same, $\mathbf{Q}$ is unbounded. This is not that disastrous — in many cases, boundedness of the right hand side set is not a severe restriction. However, it is highly desirable, at least from the academic viewpoint, to know something about the case when K^F is nontrivial, in particular, when K^F is a nonnegative orthant, or a Lorentz, or a Semidefinite cone (the two latter cases mean that (11.1.1) is an uncertain CQI, respectively, uncertain LMI). We are about to consider several such cases.

D: $F = \ell_\infty^m$, K^F is a "sign" cone, meaning that $K^F = \{u \in \ell_\infty^m : u_i \geq 0, i \in I_+, u_i \leq 0, i \in I_-, u_i = 0, i \in I_0\}$, where I_+, I_-, I_0 are given non-intersecting subsets of the index set $i = \{1, ..., m\}$.

The counterpart is

D*: $E = \ell_1^m$, $K^E = \{v \in \ell_1^m : v_j \geq 0, j \in J_+, v_j \leq 0, j \in J_-, v_j = 0, j \in J_0\}$, where J_+, J_-, J_0 are given non-overlapping subsets of the index set $\{1, ..., m\}$.

In the case of **D***, assuming, for the sake of notational convenience, that $J_+ = \{1, ..., p\}$, $J_- = \{p+1, ..., q\}$, $J_0 = \{r+1, ..., m\}$ and denoting by e^j the standard basic orths in ℓ_1, we have

$$\begin{aligned} B &\equiv \{v \in K^E : \|v\|_E \leq 1\} = \mathrm{Conv}\{e^1, ..., e^p, -e^{p+1}, ..., -e^q, \pm e^{q+1}, ..., \pm e^r\} \\ &\equiv \mathrm{Conv}\{g^1, ..., g^s\}, s = 2r - q. \end{aligned}$$

Consequently,

$$\Psi(\mathcal{M}) = \max_{1 \leq j \leq s} \mathrm{dist}_{\|\cdot\|_F}(\mathcal{M}g^j, K^F)$$

is efficiently computable (cf. case **B**).

E: $F = \ell_2^m$, $K^F = \mathbf{L}^m \equiv \{f \in \ell_2^m : f_m \geq \sqrt{\sum_{i=1}^{m-1} f_i^2}\}$, $E = \ell_2^n$, $K^E = E$.

The counterpart is

E*: $F = \ell_2^n$, $K^F = \{0\}$, $E = \ell_2^m$, $K^E = \mathbf{L}^m$.

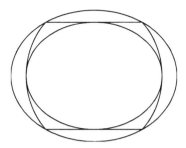

Figure 11.1 2-D cross-sections of the solids B, $\sqrt{3/2}B$ (ellipses) and D_s by a 2-D plane passing through the common symmetry axis $e_1 = ... = e_{m-1} = 0$ of the solids.

In the case of $\mathbf{E}^*$, let $D = \{e \in K^E : \|e\|_2 \le 1\}$, and let

$$B = \{e \in E : e_1^2 + ... + e_{m-1}^2 + 2e_m^2 \le 1\}.$$

Let us represent a linear map $\mathcal{M} : \ell_2^m \to \ell_2^n$ by its matrix M in the standard bases of the origin and the destination spaces. Observe that

$$B \subset D_s \equiv \mathrm{Conv}\{D \cup (-D)\} \subset \sqrt{3/2}B \qquad (11.4.4)$$

(see figure 11.1). Now, let B_F be the unit Euclidean ball, centered at the origin, in $F = \ell_2^m$. By definition of $\Psi(\cdot)$ and due to $K^F = \{0\}$, we have

$$\Psi(\mathcal{M}) \le \alpha \Leftrightarrow \mathcal{M}D \subset \alpha B_F \Leftrightarrow (\mathcal{M}D \cup (-\mathcal{M}D)) \subset \alpha B_F \Leftrightarrow \mathcal{M}D_s \subset \alpha B_F.$$

Since $D_s \subset \sqrt{3/2}B$, the inclusion $\mathcal{M}(\sqrt{3/2}B) \subset \alpha B_F$ is a sufficient condition for the validity of the inequality $\Psi(\mathcal{M}) \le \alpha$, and since $B \subset D_s$, this condition is tight within the factor $\sqrt{3/2}$. (Indeed, if $\mathcal{M}(\sqrt{3/2}B) \not\subset \alpha B_F$, then $MB \not\subset \sqrt{2/3}\alpha B_F$, meaning that $\Psi(\mathcal{M}) > \sqrt{2/3}\alpha$.) Noting that $\mathcal{M}(\sqrt{3/2}B) \le \alpha$ if and only if $\|M\Delta\|_{2,2} \le \alpha$, where $\Delta = \mathrm{Diag}\{\sqrt{3/2}, ..., \sqrt{3/2}, \sqrt{3/4}\}$, we conclude that the efficiently verifiable convex inequality

$$\|M\Delta\|_{2,2} \le \alpha$$

is a safe tractable approximation, tight within the factor $\sqrt{3/2}$, of the constraint $\Psi(\mathcal{M}) \le \alpha$.

F: $F = \mathbf{S}^m$, $\|\cdot\|_F = \|\cdot\|_{2,2}$, $K^F = \mathbf{S}_+^m$, $E = \ell_\infty^n$, $K^E = E$.

The counterpart is

F*: $F = \ell_1^n$, $K^F = \{0\}$, $E = \mathbf{S}^m$, $\|e\|_E = \sum_{i=1}^m |\lambda_i(e)|$, where $\lambda_1(e) \ge \lambda_2(e) \ge$ ... $\ge \lambda_m(e)$ are the eigenvalues of e, $K^E = \mathbf{S}_+^m$.

In the case of $\mathbf{F}$, given $\mathcal{M} \in \mathcal{L}(\ell_\infty^n, \mathbf{S}^m)$, let $e^1, ..., e^n$ be the standard basic orths of ℓ_∞^n, and let $B_E = \{v \in \ell_\infty^n : \|u\|_\infty \le 1\}$. We have

$$\{\Psi(\mathcal{M}) \le \alpha\} \Leftrightarrow \left\{\forall v \in B_E \, \exists V \succeq 0 : \max_i |\lambda_i(\mathcal{M}v - V)| \le \alpha\right\}$$
$$\Leftrightarrow \{\forall v \in B_E : \mathcal{M}v + \alpha I_m \succeq 0\}.$$

Thus, the constraint
$$\Psi(\mathcal{M}) \le \alpha \qquad\qquad (*)$$
is equivalent to
$$\alpha I + \sum_{i=1}^{n} v_i(\mathcal{M}e^i) \succeq 0 \ \forall(v : \|v\|_\infty \le 1).$$
It follows that the explicit system of LMIs
$$Y_i \succeq \pm \mathcal{M}e^i, \ i = 1, ..., n$$
$$\alpha I_m \succeq \sum_{i=1}^{n} Y_i \qquad\qquad (11.4.5)$$
in variables $\mathcal{M}$, α, $Y_1, ..., Y_n$ is a safe tractable approximation of the constraint $(*)$. Now let
$$\Theta(\mathcal{M}) = \vartheta(\mu(\mathcal{M})), \ \mu(\mathcal{M}) = \max_{1 \le i \le n} \text{Rank}(\mathcal{M}e^i),$$
where $\vartheta(\mu)$ is the function defined in the Real Case Matrix Cube Theorem, so that $\vartheta(1) = 1$, $\vartheta(2) = \pi/2$, $\vartheta(4) = 2$, and $\vartheta(\mu) \le \pi\sqrt{\mu/2}$ for $\mu \ge 1$. Invoking this Theorem (see the proof of Theorem 7.1.2), we conclude that the *local* tightness factor of our approximation does not exceed $\Theta(\mathcal{M})$, meaning that if $(\mathcal{M}, \alpha)$ cannot be extended to a feasible solution of (11.4.5), then
$$\Theta(\mathcal{M})\Psi(\mathcal{M}) > \alpha.$$

11.5 ILLUSTRATION: ROBUST ANALYSIS OF NONEXPANSIVE DYNAMICAL SYSTEMS

We are about to illustrate the techniques we have developed by applying them to the problem of *robust nonexpansiveness analysis* coming from Robust Control; in many aspects, this problem resembles the Robust Lyapunov Stability Analysis problem we have considered in sections 8.2.3 and 9.1.2.

11.5.1 Preliminaries: Nonexpansive Linear Dynamical Systems

Consider an uncertain time-varying linear dynamical system (cf. (8.2.18)):
$$\begin{array}{rcl} \dot{x}(t) &=& A_t x(t) + B_t u(t) \\ y(t) &=& C_t x(t) + D_t u(t) \end{array} \qquad\qquad (11.5.1)$$
where $x \in \mathbb{R}^n$ is the state, $y \in \mathbb{R}^p$ is the output and $u \in \mathbb{R}^q$ is the control. The system is assumed to be uncertain, meaning that all we know about the matrix $\Sigma_t = \left[\begin{array}{c|c} A_t & B_t \\ \hline C_t & D_t \end{array}\right]$ is that at every time instant t it belongs to a given *uncertainty set* $\mathcal{U}$.

System (11.5.1) is called *nonexpansive* (more precisely, *robustly nonexpansive w.r.t. uncertainty set $\mathcal{U}$*), if

$$\int_0^t y^T(s)y(s)ds \leq \int_0^t u^T(s)u(s)ds$$

for all $t \geq 0$ and for all trajectories of (all realizations of) the system such that $z(0) = 0$. In what follows, we focus on the simplest case of a system with $y(t) \equiv x(t)$, that is, on the case of $C_t \equiv I$, $D_t \equiv 0$. Thus, from now on the system of interest is

$$\begin{aligned}
\dot{x}(t) &= A_t x(t) + B_t u(t) \\
&\quad [A_t, B_t] \in \mathcal{AB} \subset \mathbb{R}^{n \times m} \; \forall t, \\
&\quad m = n + q = \dim x + \dim u.
\end{aligned} \tag{11.5.2}$$

Robust nonexpansiveness now reads

$$\int_0^t x^T(s)x(s)ds \leq \int_0^t u^T(s)u(s)ds \tag{11.5.3}$$

for all $t \geq 0$ and all trajectories $x(\cdot)$, $x(0) = 0$, of all realizations of (11.5.2).

Similarly to robust stability, robust nonexpansiveness admits a *certificate* that is a matrix $X \in \mathbf{S}^n_+$. Specifically, such a certificate is a solution of the following system of LMIs in matrix variable $X \in \mathbf{S}^m$:

(a) $X \succeq 0$

(b) $\forall [A, B] \in \mathcal{AB}$:

$$\mathcal{A}(A, B; X) \equiv \left[\begin{array}{c|c} -I_n - A^T X - XA & -XB \\ \hline -B^T X & I_q \end{array} \right] \succeq 0. \tag{11.5.4}$$

The fact that solvability of (11.5.4) is a *sufficient* condition for robust nonexpansiveness of (11.5.2) is immediate: if X solves (11.5.4), $x(\cdot)$, $u(\cdot)$ satisfy (11.5.2) and $x(0) = 0$, then

$$\begin{aligned}
u^T(s)u(s) - x^T(s)x(s) - \tfrac{d}{ds}\left[x^T(s)Xx(s)\right] &= u^T(s)u(s) - x^T(s)x(s) \\
-[\dot{x}^T(s)Xx(s) + x^T(s)X\dot{x}(s)] &= u^T(s)u(s) - x^T(s)x(s) \\
-[A_s x(s) + B_s u(s)]^T Xx(s) - x^T(s)X[A_s x(s) + B_s u(s)] & \\
= \left[x^T(s), u^T(s)\right] \mathcal{A}(A_s, B_s; X) \left[\begin{array}{c} x(s) \\ u(s) \end{array} \right] &\geq 0,
\end{aligned}$$

whence

$$\begin{aligned}
t > 0 \Rightarrow \int_0^t [u^T(s)u(s) - x^T(s)x(s)]ds &\geq x^T(t)Xx(t) - x^T(0)Xx(0) \\
= x^T(t)Xx(t) &\geq 0.
\end{aligned}$$

It should be added that when (11.5.2) is time-invariant, (i.e., $\mathcal{AB}$ is a singleton) and satisfies mild regularity conditions, the existence of the outlined certificate, (i.e., the solvability of (11.5.4)), is sufficient *and necessary* for nonexpansiveness.

Now, (11.5.4) is nothing but the RC of the system of LMIs in matrix variable $X \in \mathbf{S}^n$:

$$
\begin{array}{ll}
(a) & X \succeq 0 \\
(b) & \mathcal{A}(A, B; X) \in \mathbf{S}_+^m,
\end{array}
\tag{11.5.5}
$$

the uncertain data being $[A, B]$ and the uncertainty set being $\mathcal{AB}$. From now on we focus on the *interval uncertainty*, where the uncertain data $[A, B]$ in (11.5.5) is parameterized by perturbation $\zeta \in \mathbb{R}^L$ according to

$$
[A, B] = [A_\zeta, B_\zeta] \equiv [A^{\mathrm{n}}, B^{\mathrm{n}}] + \sum_{\ell=1}^L \zeta_\ell e_\ell f_\ell^T;
\tag{11.5.6}
$$

here $[A^{\mathrm{n}}, B^{\mathrm{n}}]$ is the nominal data and $e_\ell \in \mathbb{R}^n$, $f_\ell \in \mathbb{R}^m$ are given vectors.

Imagine, e.g., that the entries in the uncertain matrix $[A, B]$ drift, independently of each other, around their nominal values. This is a particular case of (11.5.6) where $L = nm$, $\ell = (i, j)$, $1 \le i \le n$, $1 \le j \le m$, and the vectors e_ℓ and f_ℓ associated with $\ell = (i, j)$ are, respectively, the i-th standard basic orth in $\mathbb{R}^n$ multiplied by a given deterministic real δ_ℓ ("typical variability" of the data entry in question) and the j-th standard basic orth in $\mathbb{R}^m$.

11.5.2 Robust Nonexpansiveness: Analysis via GRC

11.5.2.1 The GRC setup and its interpretation

We are about to consider the GRC of the uncertain system of LMIs (11.5.5) affected by interval uncertainty (11.5.6). Our "GRC setup" will be as follows:

i) We equip the space $\mathbb{R}^L$ where the perturbation ζ lives with the uniform norm $\|\zeta\|_\infty = \max_\ell |\zeta_\ell|$, and specify the normal range of ζ as the box

$$
\mathcal{Z} = \{\zeta \in \mathbb{R}^L : \|\zeta\|_\infty \le r\}
\tag{11.5.7}
$$

with a given $r > 0$.

ii) We specify the cone $\mathcal{L}$ as the entire $E = \mathbb{R}^L$, so that all perturbations are "physically possible."

iii) The only uncertainty-affected LMI in our situation is (11.5.5.b); the right hand side in this LMI is the positive semidefinite cone $\mathbf{S}_+^{n+m}$ that lives in the space $\mathbf{S}^m$ of symmetric $m \times m$ matrices equipped with the Frobenius Euclidean structure. We equip this space with the standard spectral norm $\|\cdot\| = \|\cdot\|_{2,2}$.

Note that our setup belongs to what was called "case $\mathbf{F}$" on p. 291.

Before processing the GRC of (11.5.5), it makes sense to understand what does it actually mean that X is a feasible solution to the GRC with global sensitivity α. By definition, this means three things:

A. $X \succeq 0$;

B. X is a robust feasible solution to (11.5.5.b), the uncertainty set being

$$\mathcal{AB}_r \equiv \{[A_\zeta, b_\zeta] : \|\zeta\|_\infty \leq r\},$$

see (11.5.6); this combines with **A** to imply that if the perturbation $\zeta = \zeta^t$ underlying $[A_t, B_t]$ all the time remains in its normal range $\mathcal{Z} = \{\zeta : \|\zeta\|_\infty \leq r\}$, the uncertain dynamical system (11.5.2) is robustly nonexpansive.

C. When $\rho > r$, we have

$$\forall(\zeta, \|\zeta\|_\infty \leq \rho) : \mathrm{dist}(\mathcal{A}(A_\zeta, B_\zeta; X), \mathbf{S}^m_+) \leq \alpha\mathrm{dist}(\zeta, \mathcal{Z}|\mathcal{L}) = \alpha(\rho - r),$$

or, recalling what is the norm on $\mathbf{S}^m$,

$$\forall(\zeta, \|\zeta\|_\infty \leq \rho) : \mathcal{A}(A_\zeta, B_\zeta; X) \succeq -\alpha(\rho - r)I_m. \tag{11.5.8}$$

Now, repeating word for word the reasoning we used to demonstrate that (11.5.4) is sufficient for robust nonexpansiveness of (11.5.2), one can extract from (11.5.8) the following conclusion:

> (!) *Whenever in uncertain dynamical system* (11.5.2) *one has* $[A_t, B_t] = [A_{\zeta^t}, B_{\zeta^t}]$ *and the perturbation* ζ^t *remains all the time in the range* $\|\zeta^t\|_\infty \leq \rho$, *one has*
>
> $$(1 - \alpha(\rho - r)) \int\limits_0^t x^T(s)x(s)ds \leq (1 + \alpha(\rho - r)) \int\limits_0^t u^T(s)u(s)ds \tag{11.5.9}$$
>
> *for all* $t \geq 0$ *and all trajectories of the dynamical system such that* $x(0) = 0$.

We see that global sensitivity α indeed controls "deterioration of nonexpansiveness" as the perturbations run out of their normal range $\mathcal{Z}$: when the $\|\cdot\|_\infty$ distance from ζ^t to $\mathcal{Z}$ all the time remains bounded by $\rho - r \in [0, \frac{1}{\alpha})$, relation (11.5.9) guarantees that the L_2 norm of the state trajectory on every time horizon can be bounded by constant times the L_2 norm of the control on the this time horizon. The corresponding constant $\left(\frac{1+\alpha(\rho-r)}{1-\alpha(\rho-r)}\right)^{1/2}$ is equal to 1 when $\rho = r$ and grows with ρ, blowing up to $+\infty$ as $\rho - r$ approaches the critical value α^{-1}, and the larger α, the smaller is this critical value.

11.5.2.2 Processing the GRC

Observe that (11.5.4) and (11.5.6) imply that

$$\mathcal{A}(A_\zeta, B_\zeta; X) = \mathcal{A}(A^{\mathrm{n}}, B^{\mathrm{n}}; X) - \sum_{\ell=1}^{L} \zeta_\ell \left[L_\ell^T(X)R_\ell + R_\ell^T L_\ell(X)\right],$$
$$L_\ell^T(X) = [Xe_\ell; 0_{m-n,1}], \; R_\ell^T = f_\ell. \tag{11.5.10}$$

Invoking Proposition 11.3.3, the GRC in question is equivalent to the following system of LMIs in variables X and α:

(a) $X \succeq 0$

(b) $\forall (\zeta, \|\zeta\|_\infty \le r):$

$$\mathcal{A}(A^{\mathrm{n}}, B^{\mathrm{n}}; X) + \sum_{\ell=1}^L \zeta_\ell \left[L_\ell^T(X) R_\ell + R_\ell^T L_\ell(X) \right] \succeq 0 \qquad (11.5.11)$$

(c) $\forall (\zeta, \|\zeta\|_\infty \le 1) : \sum_{\ell=1}^L \zeta_\ell \left[L_\ell^T(X) R_\ell + R_\ell^T L_\ell(X) \right] \succeq -\alpha I_m.$

Note that the semi-infinite LMIs $(11.5.11.b,c)$ are affected by structured norm-bounded uncertainty with 1×1 scalar perturbation blocks (see section 9.1.1). Invoking Theorem 9.1.2, the system of LMIs

$$(a) \quad X \succeq 0$$

$$(b.1) \quad Y_\ell \succeq \pm \left[L_\ell^T(X) R_\ell + R_\ell^T L_\ell(X) \right], 1 \le \ell \le L$$

$$(b.2) \quad \mathcal{A}(A^{\mathrm{n}}, B^{\mathrm{n}}; X) - r \sum_{\ell=1}^L Y_\ell \succeq 0$$

$$(c.1) \quad Z_\ell \succeq \pm \left[L_\ell^T(X) R_\ell + R_\ell^T L_\ell(X) \right], 1 \le \ell \le L$$

$$(c.2) \quad \alpha I_m - \sum_{\ell=1}^L Z_\ell \succeq 0$$

in matrix variables $X, \{Y_\ell, Z_\ell\}_{\ell=1}^L$ and in scalar variable α is a safe tractable approximation of the GRC, tight within the factor $\frac{\pi}{2}$. Invoking the result of Exercise 9.1, we can reduce the design dimension of this approximation; the equivalent reformulation of the approximation is the SDO program

$$\min \alpha$$

s.t.

$$X \succeq 0$$

$$\left[\begin{array}{c|ccc} \mathcal{A}(A^{\mathrm{n}}, B^n; X) - r \sum_{\ell=1}^L \lambda_\ell R_\ell^T R_\ell & L_1^T(X) & \cdots & L_L^T(X) \\ \hline L_1(X) & \lambda_1/r & & \\ \vdots & & \ddots & \\ L_L(X) & & & \lambda_L/r \end{array} \right] \succeq 0$$

$$\left[\begin{array}{c|ccc} \alpha I_m - \sum_{\ell=1}^L \mu_\ell R_\ell^T R_\ell & L_1^T(X) & \cdots & L_L^T(X) \\ \hline L_1(X) & \mu_1 & & \\ \vdots & & \ddots & \\ L_L(X) & & & \mu_L \end{array} \right] \succeq 0$$

$$(11.5.12)$$

in variable $X \in \mathbf{S}^m$ and scalar variables $\alpha, \{\lambda_\ell, \mu_\ell\}_{\ell=1}^L$. Note that we have equipped our (approximate) GRC with the objective to minimize the global sensitivity of X; of course, other choices of the objective are possible as well.

11.5.2.3 Numerical illustration

The data. In the illustration we are about to present, the state dimension is $n = 5$, and the control dimension is $q = 2$, so that $m = \dim x + \dim u = 7$. The nominal data (chosen at random) are as follows:

$$[A^{\mathrm{n}}, B^{\mathrm{n}}]$$

$$= M := \begin{bmatrix} -1.089 & -0.079 & -0.031 & -0.575 & -0.387 & 0.145 & 0.241 \\ -0.124 & -2.362 & -2.637 & 0.428 & 1.454 & -0.311 & 0.150 \\ -0.627 & 1.157 & -1.910 & -0.425 & -0.967 & 0.022 & 0.183 \\ -0.325 & 0.206 & 0.500 & -1.475 & 0.192 & 0.209 & -0.282 \\ 0.238 & -0.680 & -0.955 & -0.558 & -1.809 & 0.079 & 0.132 \end{bmatrix}.$$

The interval uncertainty (11.5.6) is specified as

$$[A_\zeta, b_\zeta] = M + \sum_{i=1}^{5} \sum_{j=1}^{7} \zeta_{ij} \underbrace{|M_{ij}| g_i}_{e_i} f_j^T,$$

where g_i, f_j are the standard basic orths in $\mathbb{R}^5$ and $\mathbb{R}^7$, respectively; in other words, every entry in $[A, B]$ is affected by its own perturbation, and the variability of an entry is the magnitude of its nominal value.

Normal range of perturbations. Next we should decide how to specify the normal range $\mathcal{Z}$ of the perturbations, i.e., the quantity r in (11.5.7). "In reality" this choice could come from the nature of the dynamical system in question and the nature of its environment. In our illustration there is no "nature and environment," and we specify r as follows. Let r_* be the largest r for which the robust nonexpansiveness of the system at the perturbation level r, (i.e., the perturbation set being the box $B_r = \{\zeta : \|\zeta\|_\infty \leq r\}$) admits a certificate. It would be quite reasonable to choose, as the normal range of perturbations $\mathcal{Z}$, the box B_{r_*}, so that the normal range of perturbations is the largest one where the robust nonexpansiveness still can be certified. Unfortunately, precise checking the existence of a certificate for a given box in the role of the perturbation set means to check the feasibility status of the system of LMIs

$$(a) \quad X \succeq 0$$
$$(b) \quad \forall(\zeta, \|\zeta\|_\infty \leq r) : \mathcal{A}(A_\zeta, B_\zeta; X) \succeq 0$$

in matrix variable X, with $\mathcal{A}(\cdot, \cdot; \cdot)$ given in (11.5.4). This task seems to be intractable, so that we are forced to replace this system with its safe tractable approximation, tight within the factor $\pi/2$, specifically, with the system

$$X \succeq 0$$

$$\begin{bmatrix} \mathcal{A}(A^{\mathrm{n}}, B^n; X) - r \sum_{\ell=1}^{L} \lambda_\ell R_\ell^T R_\ell & L_1^T(X) & \cdots & L_L^T(X) \\ \hline L_1(X) & \lambda_1/r & & \\ \vdots & & \ddots & \\ L_L(X) & & & \lambda_L/r \end{bmatrix} \succeq 0 \quad (11.5.13)$$

in matrix variable X and scalar variables λ_ℓ (cf. (11.5.12)), with $R_\ell(X)$ and L_ℓ given by (11.5.10). The largest value r_1 of r for which the latter system is solvable (this quantity can be easily found by bisection) is a lower bound, tight within the factor $pi/2$, on r_*, and this is the quantity we use in the role of r when specifying the normal range of perturbations according to (11.5.7).

Applying this approach to the outlined data, we end up with

$$r = r_1 = 0.0346.$$

The results. With the outlined nominal and perturbation data and r, the optimal value in (11.5.12) turns out to be

$$\alpha_{\text{GRC}} = 27.231.$$

It is instructive to compare this quantity with the global sensitivity of the RC-certificate X_{RC} of robust nonexpansiveness; by definition, X_{RC} is the X component of a feasible solution to (11.5.13) where r is set to r_1. This X clearly can be extended to a feasible solution to our safe tractable approximation (11.5.12) of the GRC; the smallest, over all these extensions, value of the global sensitivity α is

$$\alpha_{\text{RC}} = 49.636,$$

which is by a factor 1.82 larger than α_{GRC}. It follows that the GRC-based analysis of the robust nonexpansiveness properties of the uncertain dynamical system in question provides us with essentially more optimistic results than the RC-based analysis. Indeed, a feasible solution $(\alpha, ...)$ to (11.5.12) provides us with the *upper bound*

$$C_*(\rho) \le C_\alpha(\rho) \equiv \begin{cases} 1, & 0 \le \rho \le r \\ \frac{1+\alpha(\rho-r)}{1-\alpha(\rho-r)}, & r \le \rho < r + \alpha^{-1} \end{cases} \tag{11.5.14}$$

(cf. (11.5.9)) on the "existing in the nature, but difficult to compute" quantity

$$C_*(\rho) = \inf \left\{ C : \int_0^t x^T(s)x(s)ds \le C \int_0^t u^T(s)u(s)ds \, \forall (t \ge 0, x(\cdot), u(\cdot)) : \right.$$

$$\left. x(0) = 0, \dot{x}(s) = A_{\zeta^s}x(s) + B_{\zeta^s}u(s), \|\zeta^s\|_\infty \le \rho \, \forall s \right\}$$

responsible for the robust nonexpansiveness properties of the dynamical system. The upper bounds (11.5.14) corresponding to α_{RC} and α_{GRC} are depicted on the left plot in figure 11.2 where we see that the GRC-based bound is much better than the RC-based bound.

Of course, both the bounds in question are conservative, and their "level of conservatism" is difficult to access theoretically: while we do understand how conservative our tractable approximations to intractable RC/GRC are, we have no idea how conservative the sufficient condition (11.5.4) for robust nonexpansiveness is (in this respect, the situation is completely similar to the one in Lyapunov Stability Analysis, see section 9.1.2). We can, however, run a brute force simulation to bound $C_*(\rho)$ from below. Specifically, generating a sample of perturbations of a given magnitude and checking the associated matrices $[A_\zeta, B_\zeta]$ for nonexpansiveness, we can build an upper bound $\bar{\rho}_1$ on the largest ρ for which every matrix $[A_\zeta, B_\zeta]$ with $\|\zeta\|_\infty \le \rho$ generates a nonexpansive time-invariant

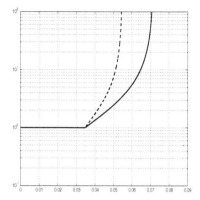

Figure 11.2 RC/GRC-based analysis: bounds (11.5.14) vs. ρ for $\alpha = \alpha_{\mathrm{GRC}}$ (solid) and $\alpha = \alpha_{\mathrm{RC}}$ (dashed).

dynamical system; $\overline{\rho}_1$ is, of course, greater than or equal to the largest $\rho = \rho_1$ for which $C_*(\rho) \leq 1$. Similarly, testing matrices A_ζ for stability, we can build an upper bound $\overline{\rho}_\infty$ on the largest $\rho = \rho_\infty$ for which all matrices A_ζ, $\|\zeta\|_\infty \leq \rho$, have all their eigenvalues in the closed left hand side plane; it is immediately seen that $C_*(\rho) = \infty$ when $\rho > \rho_\infty$. For our nominal and perturbation data, simulation yields

$$\overline{\rho}_1 = 0.310, \quad \overline{\rho}_\infty = 0.7854.$$

These quantities should be compared, respectively, to $r_1 = 0.0346$, (which clearly is a *lower* bound on the range ρ_1 of ρ's where $C_*(\rho) \leq 1$) and $r_\infty = r_1 + \alpha_{\mathrm{GRC}}^{-1}$ (this is the range of values of ρ where the GRC-originating upper bound (11.5.14) on $C_*(\rho)$ is finite; as such, r_∞ is a *lower* bound on ρ_∞). We see that in our numerical example the conservatism of our approach is "within one order of magnitude": $\overline{\rho}_1 / r_1 \approx 8.95$ and $\overline{\rho}_\infty / r_\infty \approx 11.01$.

Chapter Twelve
Robust Classification and Estimation

In this chapter, we present some applications of Robust Optimization in the context of Machine Learning and Linear Regression.

12.1 ROBUST SUPPORT VECTOR MACHINES

We begin our development with an overview, the focus of which is the specific example of Support Vector Machines for binary classification.

12.1.1 Support Vector Machines

BINARY LINEAR CLASSIFICATION.

Let X denote the $n \times m$ matrix of data points (they are columns in X), each one belonging to one of two classes. Let $y \in \{-1, 1\}^m$ be the corresponding label vector, so that $y_i = 1$ when i-th data point is in the first class, and $y_i = -1$ when i-th data point is in the second class. We refer to the pair (X, y) as the *training data*.

In linear classification, we seek to separate, if possible, the two classes by a hyperplane $\mathcal{H}(w, b) = \{x : w^T x + b = 0\}$, where $w \in \mathbb{R}^n$ and $b \in \mathbb{R}$ are the hyperplane's parameters. To any candidate hyperplane $\mathcal{H}(w, b)$ corresponds a decision rule of the form $z = \text{sign}(w^T x + b)$, which can be used to predict the label z of a new point x.

MAXIMALLY ROBUST SEPARATION FOR SEPARABLE DATA.

Perfect linear separation occurs when the decision rule makes no errors on the data set. This translates as a set of linear inequalities in (w, b):

$$y_i(w^T x_i + b) > 0, \quad i = 1, \ldots, m. \tag{12.1.1}$$

Let us assume that the data is separable, in the sense that the above conditions are feasible. Assume now that the data is uncertain, specifically, for every i, i-th "true" data point is only known to belong to the interior of an Euclidian ball of radius ρ centered at the "nominal" data point x_i^n. In the following, we refer to this as to *spherical uncertainty*. Insisting that the above inequalities be valid for all choices of the data points within their respective balls leads to the Robust

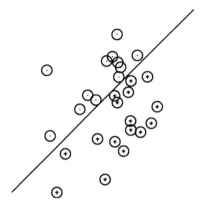

Figure 12.1 Maximally robust classifier for separable data, with spherical uncertainties around each data point.

Counterpart to the above inequalities:

$$y_i(w^T x_i^{\mathrm{n}} + b) \geq \rho \|w\|_2, \quad i = 1, \ldots, m. \tag{12.1.2}$$

The maximally robust classifier is the one that maximizes ρ subject to the conditions (12.1.2). By homogeneity of the above in (w, b), we can always enforce $\rho \|w\|_2 = 1$, so that maximizing ρ leads to minimizing $\|w\|_2$, via the quadratic optimization problem:

$$\min_{w,b} \left\{ \|w\|_2 : y_i(w^T x_i^{\mathrm{n}} + b) \geq 1, \, 1 \leq i \leq m \right\}. \tag{12.1.3}$$

The above problem and its optimal solution are illustrated in figure 12.1. The maximally robust classifier corresponds to the largest radius such that the corresponding balls around each data point are still perfectly separated. In Machine Learning literature, the optimal quantity ρ is referred to as the margin of the classifier, and the corresponding classifier as the maximum margin classifier.

NON-SEPARABLE CASE: THE HINGE LOSS FUNCTION.

In general, perfect separation may not be possible. To cope with this, we modify the "separation constraints" $y_i(w^T x_i + b) \geq 1$ as

$$y_i(w^T x_i + b) \geq 1 - v_i, \quad v_i \geq 0, \quad i = 1, \ldots, m,$$

where the number of nonzero entries in the slack vector v is the number of errors the classifier makes on the training data. We could look for a classifier that minimizes this number, but this would be a computationally intractable problem. Instead, we can seek to minimize the more tractable sum of the elements of v. This leads to

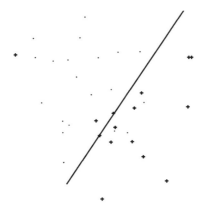

Figure 12.2 The classical SVM separator for non-separable data, as defined in (12.1.6), with regularization parameter $\lambda = 0.1$.

the linear optimization problem

$$\min_{w,b} \left\{ \sum_{i=1}^{m} v_i : y_i(w^T x_i + b) \geq 1 - v_i, \, v_i \geq 0, \, 1 \leq i \leq m \right\}. \qquad (12.1.4)$$

This can be written in the equivalent form of minimizing the so-called *realized hinge loss* function

$$\mathcal{R}_{\text{svm}}(w,b) := \sum_{i=1}^{m} [1 - y_i(w^T x_i + b)]_{+}, \qquad (12.1.5)$$

where we used the term "realized" to emphasize that the function above depends on a particular realization of the data.

The above function is based on replacing the indicator function, which would arise if we were to minimize the actual number of errors, with a convex upper bound.

THE CLASSICAL SVM FORMULATION.

In practice, we need to trade-off the number of training set errors (or its proxy, which is the loss function above), and the amount of robustness with respect to spherical perturbations of the data points. One way to formulate this trade-off is via the classical Support Vector Machine (SVM) formulation:

$$\min_{w,b,v} \left\{ \lambda \|w\|_2^2 + \sum_{i=1}^{m} v_i : y_i(w^T x_i + b) \geq 1 - v_i, \, 1 \leq i \leq m, \, v \geq 0 \right\}$$
$$\Updownarrow \qquad\qquad (12.1.6)$$
$$\min_{w,b} \left\{ \mathcal{R}_{\text{svm}}(w,b) + \lambda \|w\|_2^2 \right\},$$

where $\lambda > 0$ is a regularization parameter. An example is shown in figure 12.2. A less classical approach to accomplish the above trade-off is

$$\min_{w,b} \{\mathcal{R}_{\text{svm}}(w,b) + \lambda\|w\|_2\}. \tag{12.1.7}$$

The above, which we call the *norm-penalized SVM*, is equivalent to the classical formulation (12.1.6), in the sense that the set of solutions obtained when λ spans the positive real line is the same for both problems.

12.1.2 Minimizing Worst-Case Realized Loss

An alternate (and perhaps more versatile) approach to the classical SVM is to consider the minimization of the realized loss function $\mathcal{R}_{\text{svm}}$, and *then* apply a robust optimization procedure, in order to minimize its worst-case value under perturbations of the data points. The corresponding problem has the form of minimizing (over (w,b)) the *worst-case realized loss function*

$$\max_{X \in \mathcal{X}} \mathcal{R}_{\text{svm}}(w,b), \tag{12.1.8}$$

where the set $\mathcal{X}$ describes our uncertainty model about the data matrix X. (Our notation here is somewhat loose, as the dependence of the realized loss function on the data X is implicit.)

12.1.3 Measurement-Wise Uncertainty Models

We examine the worst-case hinge loss minimization problem in the case when perturbations affect each measurement independently.

SPHERICAL UNCERTAINTY.

Return to our spherical uncertainty model, for which the set $\mathcal{X}$ is $\mathcal{X}_{\text{sph}}$, where

$$\mathcal{X}_{\text{sph}} := \{X^{\text{n}} + \Delta \; : \; \Delta = [\delta_1, \ldots, \delta_m], \; \|\delta_i\|_2 \leq \rho, \; i = 1, \ldots, m\}.$$

Here, the matrix X^{n} contains the nominal data, and its columns x_i^{n} are the nominal data points (nominal *feature vectors*, in the SVM terminology). We obtain an explicit expression for the worst-case realized loss:

$$\max_{X \in \mathcal{X}_{\text{bll}}} \mathcal{R}_{\text{svm}}(w,b) = \sum_{i=1}^{m} [1 - y_i(w^T x_i^{\text{n}} + b) + \rho\|w\|_2]_+.$$

The worst-case realized loss function above can be minimized via second-order cone optimization:

$$\min_{w,b} \left\{ \sum_{i=1}^{m} [1 - y_i(w^T x_i^{\text{n}} + b) + \rho\|w\|_2]_+ \right\}. \tag{12.1.9}$$

Note that the robust version of worst-case loss minimization is not the same as the classical Support Vector Machine (12.1.6): in the former case, the penalty

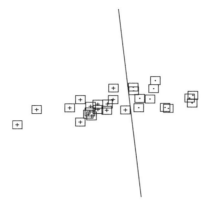

Figure 12.3 Maximally robust classifier for separable data, with box-type uncertainties.

term is "inside" the loss function, while it is outside in the latter. In the worst-case loss minimization, we are trying to separate balls drawn around data points, as we did in the separable case. If the interior of one of these balls intersects the separating hyperplane, then the procedure counts this as an error. In contrast, the classical SVM procedure only considers errors corresponding to the centers of the balls. It turns out that we can bound one approach relative to the other. An upper bound on the worst-case loss is readily given by

$$\sum_{i=1}^{m}[1 - y_i(w^T x_i^{\mathrm{n}} + b) + \rho\|w\|_2]_+ \leq \sum_{i=1}^{m}[1 - y_i(w^T x_i^{\mathrm{n}} + b)]_+ + m\rho\|w\|_2.$$

Minimizing the upper bound above is of the same form as the norm-penalized SVM (12.1.7), itself closely linked to the classical Vector Machine solution, as noted before.

INTERVAL UNCERTAINTY MODELS

We can modify our assumptions about the uncertainty affecting the data, which leads to different classification algorithms. Of particular interest is the case when the uncertainty affects each element of the data matrix independently, in a component-wise fashion.

For example, consider the *box uncertainty* model, where each data point i is only known to belong to the $\|\cdot\|_\infty$-ball of radius ρ centered at the nominal data. The associated version of (12.1.2) is

$$y_i(w^T x_i^{\mathrm{n}} + b) \geq \rho\|w\|_1, \quad i = 1, \ldots, m.$$

The corresponding maximally robust classifier is obtained via the Linear Optimization problem

$$\min_{w,b}\left\{\|w\|_1 : y_i(w^T x_i^{\mathrm{n}} + b) \geq 1, \, 1 \leq i \leq m\right\}. \tag{12.1.10}$$

An example is illustrated in figure 12.3. Observe the contrast with the case when uncertainties are spherical (figure(12.1)): with box uncertainties, the classifier's coefficient vector tends to be sparser. (This effect becomes more pronounced as n/m becomes larger.)

Likewise, minimizing the worst-case realized loss function (12.1.8) where the set $\mathcal{X}$ is described as a box:

$$\mathcal{X}_{\text{box}} = \{X^{\text{n}} + \Delta \ : \ \Delta = [\delta_1, \ldots, \delta_m], \ \|\delta_i\|_\infty \leq \rho, \ i = 1, \ldots, m\}, \qquad (12.1.11)$$

is solved via the Linear Optimization problem

$$\min_{w,b} \left\{ \sum_{i=1}^{m} [1 - y_i(w^T x_i^{\text{n}} + b) + \rho\|w\|_1]_+ \right\}.$$

We can extend some of the above results to more general interval uncertainty models, of the form

$$\mathcal{X}_{\text{int}} = \{X^{\text{n}} + \Delta \ : \ |\Delta_{pq}| \leq \rho R_{pq}, \ 1 \leq p \leq n, \ 1 \leq q \leq m\}, \qquad (12.1.12)$$

where $R \in \mathbb{R}_+^{n \times m}$ is a matrix with nonnegative entries that specifies the relative ranges of the uncertainties around each component of X^{n}. The corresponding worst-case realized loss takes the form

$$\max_{X \in \mathcal{X}_{\text{int}}} \mathcal{R}_{\text{svm}}(w, b) = \sum_{i=1}^{m} [1 - y_i(w^T x_i^{\text{n}} + b) + \rho\sigma_i^T |w|]_+,$$

where σ_i is the i-th column of R, $1 \leq i \leq m$, and $|w| = [|w_1|; \ldots; |w_n|]$.

Note that, for general uncertainty models, the approach we followed to devise maximally robust classifiers becomes a little more complicated. Indeed, the condition for robust separability writes now

$$y_i(w^T x_i^{\text{n}} + b) \geq \rho\sigma_i^T |w|, \ i = 1, \ldots, m.$$

Unless the vectors σ_i, $i = 1, \ldots, m$, are all equal, there is no way to formulate the problem of maximizing ρ subject to the above conditions, as a convex optimization problem, as we did before. Of course, the problem is quasi-convex, and can be solved as a sequence of convex ones via Bisection in ρ.

12.1.4 Coupled Uncertainty Models

In the previous models, uncertainty independently affects each measurement (each column of X). In some cases, it makes sense to assume instead a global bound on the perturbation matrix, which couples uncertainties affecting different measurements. These models are part of a family referred to as coupled uncertainty models.

A NORM-BOUND UNCERTAINTY MODEL

Perhaps the simplest of coupled uncertainty models corresponds to the uncertainty set

$$\mathcal{X}_{LSV} = \left\{ X^{\mathrm{n}} + \Delta \; : \; \Delta \in \mathbb{R}^{n \times m}, \; \|\Delta\| \leq \rho \right\}, \qquad (12.1.13)$$

where X^{n} is the nominal data and $\| \cdot \|$ denotes the largest singular value (LSV) norm.

For separable data, the maximally separable classifier is based on the robust separability condition

$$\forall \, i = 1, \ldots, m, \; \; \forall \Delta = [\delta_1, \ldots, \delta_m], \; \; \|\Delta\| \leq \rho \; : \; y_i(w^T(x_i^{\mathrm{n}} + \delta_i) + b) \geq 1.$$

It turns out that the above condition is exactly the same as the robust separability condition encountered for spherical uncertainties, (12.1.2). This comes from the fact that the conditions above only involve the projection of the unit ball (for the matrix norm $\| \cdot \|$) on the subspaces generated by the columns of Δ. The maximally robust separating classifier is the same in the present norm-bound model, as it was in the case of spherical uncertainties.

In contrast, when we look at minimizing the worst-case realized loss function (12.1.8), the situation is different, since the norm induces a coupling between terms corresponding to different measurements. The robust problem of interest now is

$$\min_{w,b} \left\{ \max_{X \in \mathcal{X}} \sum_{i=1}^{m} [1 - y_i(w^T x_i + b)]_+ \right\}. \qquad (12.1.14)$$

In the case of $\mathcal{X} = \mathcal{X}_{LSV}$, as it will be seen later, the worst-case realized cost function, which is the objective in (12.1.14), can be expressed as

$$\max_{k \in \{0, \ldots, m\}} \left\{ \min_{\mu} \left\{ \rho \sqrt{k} \|w\|_2 + k\mu + \sum_{i=1}^{m} [1 - y_i(w^T x_i^{\mathrm{n}} + b) - \mu]_+ \right\} \right\}.$$

The problem of minimizing this function writes as a second-order cone optimization problem:

$$\min_{w,b,t,\{\mu_k\}} \left\{ t \; : \; \begin{array}{l} t \geq \rho \sqrt{k} \|w\|_2 + k\mu_k + \sum_{i=1}^{m} [1 - y_i(w^T x_i^{\mathrm{n}} + b) - \mu_k]_+, \\ \qquad\qquad\qquad\qquad\qquad\qquad\qquad\qquad 0 \leq k \leq m \end{array} \right\}. \qquad (12.1.15)$$

Setting $\mu_k = 0$ for every k in (12.1.15), we obtain an upper bound of the form

$$\min_{w,b} \left\{ \sum_{i=1}^{m} [1 - y_i(w^T x_i^{\mathrm{n}} + b)]_+ + \rho \sqrt{m} \|w\|_2 \right\},$$

which is similar (up to a scaling of the penalty term) to the norm-penalized SVM, (12.1.7). Also, setting $\mu_k = -k^{-1/2} \rho \|w\|_2$ for every k in (12.1.15), we obtain the problem (12.1.9) (with $\rho \sqrt{m}$ in the role of ρ) which we encountered with spherical uncertainties.

12.1.5 Worst-Case Loss and Adjustable Variables

As noted before, the norm-bound uncertainty model couples the uncertainties affecting different measurements. Our previous development shows that the problem of minimizing the worst-case loss function under norm-bounded uncertainties is not equivalent to the same problem under spherical (measurement-wise) uncertainties.

This discrepancy might be understood geometrically. In the worst-case loss problem (12.1.8) with the LSV model, we can replace the set $\mathcal{X}$ by a bigger set of the form $\mathcal{X}_1 \times \ldots \times \mathcal{X}_m$, where $\mathcal{X}_i$'s are the projections of $\mathcal{X}$ on the δ_i variables, where δ_i's are the columns of Δ. Exploiting the fact that the loss function decomposes as a sum of terms that depend on δ_i only, we obtain the spherical model's SVM (12.1.9).

Another interpretation calls into play Robust Optimization with *adjustable variables*. Start from the Linear Optimization representation of the problem of minimizing the realized loss function, (12.1.4), and then apply a Robust Optimization procedure, requiring that the constraints be satisfied irrespective of the choice of the data matrix in the set $\mathcal{X}$. This "naive" approach would lead us to replace the constraints by their robust counterpart:

$$\forall \Delta = [\delta_1, \ldots, \delta_m], \ \|\Delta\| \le \rho \ : \quad y_i(w^T(x_i + \delta_i) + b) \ge 1 - v_i, \quad i = 1, \ldots, m,$$
$$v_i \ge 0,$$

which is the same as

$$y_i(w^T x_i + b) \ge 1 - v_i + \rho \|w\|_2, \ v_i \ge 0, \ i = 1, \ldots, m,$$

Minimizing $\sum_{i=1}^m v_i$ subject to the above constraints is precisely the same as the problem corresponding to spherical (measurement-wise) uncertainties, (12.1.9).

Contrarily to what happens with maximally robust classifiers for separable data, a naive approach to robustifying the problem fails to produce an accurate answer. Indeed, in the naive approach above, the slack variable v is assumed to be independent of the perturbation. In reality, this variable should be considered a function of the perturbation, as an adjustable variable, in order to accurately model our problem of minimizing the worst-case realized loss function.[1] The naive approach does work when imposing $v = 0$, as we do with maximally robust classifiers for separable data.

Our discussion motivates us to study the problem of computing, and optimizing, the worst-case loss function in more detail. The next section is devoted to building a framework of specific models for which exact answers are possible.

---

[1] For in-depth study of adjustability and adjustable Robust Counterparts, see chapter 14.

12.2 ROBUST CLASSIFICATION AND REGRESSION

12.2.1 Nominal Problem and Robust Counterpart

LOSS FUNCTION MINIMIZATION

We start from the following "nominal" problem, which is

$$\min_{\theta \in \Theta} \mathcal{L}(Z^T \theta), \qquad (12.2.1)$$

where the function $\mathcal{L} : \mathbb{R}^m \to \mathbb{R}$ is convex; the variable θ contains the regressor or classifier coefficients, and is constrained to a given convex set $\Theta \subseteq \mathbb{R}^n$; the matrix $Z := [z_1, \ldots, z_m] \in \mathbb{R}^{n \times m}$ contains the data of the problem. We assume that the set Θ is computationally tractable, and that the nominal problem is so as well.

In the following, we refer to the columns z_i's of the data matrix Z as *measurements*, and to its rows as *features*. We call the vector θ the *parameter vector*. For a given parameter vector $\theta \in \Theta$, we define $r = Z^T \theta$ to be the corresponding *residual vector*. In our setup, we refer to $r \to \mathcal{L}(r)$ as the loss function (a function of the residual vector), and to $\theta \to \mathcal{L}(Z^T \theta)$ as the realized loss function (a function of the parameter vector). With this definition, the loss function is data-independent, whereas the realized loss is not.

ROBUST COUNTERPART

We address the Robust Counterpart to the nominal problem (12.2.1), which is to minimize the worst-case realized loss function:

$$\min_{\theta \in \Theta} \max_{Z \in \mathcal{Z}} \mathcal{L}(Z^T \theta), \qquad (12.2.2)$$

where $\mathcal{Z} \subseteq \mathbb{R}^{m \times n}$ is a given subset of matrices that describes the uncertainty on the data matrix Z. We assume that

$$\mathcal{Z} = Z^{\mathrm{n}} + \rho U \mathcal{D}, \qquad (12.2.3)$$

where the matrix $Z^{\mathrm{n}} \in \mathbb{R}^{n \times m}$ contains the nominal data, the uncertainty level $\rho \geq 0$ allows one to control the size of the perturbation, and the given matrix $U \in \mathbb{R}^{n \times l}$ allows for modeling some structural information about the perturbation, as when some rows of the matrix Z are not affected by uncertainty. Here, the triple $(Z^{\mathrm{n}}, \rho, U)$ encodes the robust problem's data. The set $\mathcal{D} \subseteq \mathbb{R}^{l \times m}$ (convex, compact, and containing the origin) is reserved for describing structural information about the perturbation, such as norm bounds. For now, the only assumption we make about the set $\mathcal{D}$ is that it is computationally tractable, meaning that its support function

$$\phi_{\mathcal{D}}(Y) := \max_{\Delta \in \mathcal{D}} \langle Y, \Delta \rangle$$

is so; from now on, for $n \times m$ matrices Y, Δ, $\langle Y, \Delta \rangle = \mathrm{Tr}(Y^T \Delta)$ is the Frobenius inner product of the matrices.

Our goal is to obtain further conditions that ensure that a tractable representation of the semi-infinite inequality

$$\forall \, Z \in \mathcal{Z} \ : \ \mathcal{L}(Z^T \theta) \leq \tau, \tag{12.2.4}$$

where $\tau \in \mathbb{R}$ is given, exists. The Robust Counterpart (12.2.2) of (12.2.1) then also writes in tractable form, as

$$\min_{\theta \in \Theta, \tau} \left\{ \tau : (\theta, \tau) \text{ satisfies (12.2.4)} \right\}.$$

We are about to present two motivating examples.

Example 12.2.1. Robust Linear Regression. Assume that we have observed m inputs ("regressors") $x_i \in \mathbb{R}^{n-2}$, $i = 1, \ldots, m$, to certain system along with the corresponding outputs $y_i \in \mathbb{R}$, $1 \leq i \leq m$, of the system, and seek for a linear regression model

$$y_i \approx x_i^T w + b$$

that best fits the data, in the sense that a given norm $\| \cdot \|$ of the residual vector $[y_1 - x_1^T w - b; \ldots; y_m - x_m^T w - b]$ is as small as possible (perhaps, under certain additional restrictions on the coefficients w, b of the regressions model). Setting

$$Z = \begin{bmatrix} x_1 & \cdots & x_m \\ y_1 & \cdots & y_m \\ 1 & \cdots & 1 \end{bmatrix} \in \mathbb{R}^{n \times m}, \ \theta = [-w; 1; -b] \in \mathbb{R}^n, \ \mathcal{L}(r) = \|r\| : \mathbb{R}^m \to \mathbb{R}, \tag{12.2.5}$$

we can write down the problem of finding the best linear regression model for our data in the form of (12.2.1), where $\Theta \subset \mathbb{R}^n$ is contained in the plane $\theta_{n-1} = 1$ (and can be a proper subset of this plane, provided we intend to impose some restrictions on w and b). If we assume now that the regressors x_i and the outputs y_i are not measured exactly, so that all we know about the "true" data matrix Z is that it belongs to a given uncertainty set $\mathcal{Z}$, a natural course of action is to seek the linear regression model that guarantees the best possible worst-case, over the data matrices $Z \in \mathcal{Z}$, fit, thus arriving at the Robust Counterpart of (12.2.1).

Example 12.2.2. Robust SVM. Consider the situation that in the previous section was called "worst-case realized loss function minimization", specifically, the situation where we are given m data points, assembled into the data matrix $X = [x_1, \ldots, x_m]$, along with labels $y_1, \ldots, y_m \in \{-1, 1\}$ of these points, and seek a linear classifier $\text{sign}(w^T x + b)$ capable of minimizing the classification error. As was explained in section 12.1, in the case of an uncertain data matrix X known to belong to a given uncertainty set $\mathcal{X}$, minimization of the worst-case realized loss in this case reduces to the semi-infinite problem (12.1.14). Setting

$$Z = \begin{bmatrix} y_1 x_1 & \cdots & y_m x_m \\ y_1 & \cdots & y_m \\ 1 & \cdots & 1 \end{bmatrix}, \ \theta = [-\omega; -b; 1], \ \mathcal{L}(r) = \sum_{i=1}^m \max[r_i, 0], \tag{12.2.6}$$

and straightforwardly converting the uncertainty set $\mathcal{X}$ for X into an uncertainty set $\mathcal{Z}$ for Z, we represent the Robust Counterpart (12.1.14) in the form of (12.2.2).

12.2.2 Some Simple Cases

SCENARIO UNCERTAINTY

Perhaps the simplest case involves an uncertainty set given as a (convex hull) of a finite number of given matrices. Namely:

$$\mathcal{Z} = \mathrm{Conv}\left\{ Z^{(1)}, \ldots, Z^{(K)} \right\},$$

where $Z^{(k)} \in \mathbb{R}^{m \times n}$, $k = 1, \ldots, K$, are given. Then the semi-infinite inequality (12.2.4) writes in the convex, tractable form

$$\max_{1 \le k \le K} \mathcal{L}((Z^{(k)})^T \theta) \le \tau$$

(cf. section 6.1).

ASSUMPTION ABOUT THE LOSS FUNCTION

Another set of results obtains when making specific assumptions about the loss function.

> **Assumption L:** *The loss function* $\mathcal{L} : \mathbb{R}^m \to \mathbb{R}$ *is of the form*
>
> $$\mathcal{L}(r) = \pi(\mathrm{abs}(P(r))),$$
>
> *where* $\mathrm{abs}(\cdot)$ *acts componentwise,* $\pi : \mathbb{R}^m_+ \to \mathbb{R}$ *is a computationally tractable convex, monotone function on the non-negative orthant, and* $P : \mathbb{R}^m \to \mathbb{R}^m$ *is the vector-valued function*
>
> $$P(r) = \begin{cases} r & (\text{``symmetric case''}) \\ r_+ & (\text{``asymmetric case''}) \end{cases}$$
>
> *where* r_+ *is the vector with components* $\max[r_i, 0]$, $i = 1, \ldots, m$.

In the following, Assumption **L** is always in force, unless the opposite is stated explicitly.

EXAMPLES

The following specific loss functions satisfy assumption **L**.

- The case of the hinge loss function, (12.1.5), which arises in Support Vector Machines, is recovered upon choosing $P(r) = r_+$, and $\pi(r) = \sum_{i=1}^m r_i$, see Example 12.2.2.

- *Least-squares regression.* This is the problem from Example 12.2.1 with $\|\cdot\| = \|\cdot\|_2$; here the loss function satisfies **L** with $P(r) = r$, $\pi(r) = \|r\|_2$.

- ℓ_p-*norm regression.* This is the problem from Example 12.2.1 with $\|\cdot\| = \|\cdot\|_p$, with a $p \in [1, \infty]$. Here the loss function satisfies **L** with $P(r) = r$, $\pi(r) = \|r\|_p$.

- *Huber penalty regression,* which is useful for outlier rejection. This is the problem from Example 12.2.1 with

$$\mathcal{L}(r) = \sum_{i=1}^{m} H(r_i, 1)),$$

where $H : \mathbb{R}_+ \times \mathbb{R}_{++} \to \mathbb{R}$ is the (generalized) *Huber function*

$$H(t, \mu) = \max_{|\xi| \leq 1} \left\{ t\xi - \frac{\mu \xi^2}{2} \right\} = \begin{cases} \frac{t^2}{2\mu}, & |t| \leq \mu, \\ |t| - \frac{\mu}{2}, & |t| \geq \mu. \end{cases} \tag{12.2.7}$$

Here assumption **L** is satisfied with $P(r) = r$ and $\pi(r) = \sum_{i=1}^{m} H(r_i, 1)$.

Without more assumptions on the problem, we can find a tractable representation of the semi-infinite inequality (12.2.4) in some "simple" cases. We examine two of these simple cases now.

WEIGHTED ℓ_∞-NORM LOSS

Assume that the loss function is such that $\pi(u) = \max_i \alpha_i u_i$, where the weighting vector $\alpha \in \mathbb{R}_+^m$ is given. Thus, the loss assumes the form

$$\mathcal{L}(r) = \begin{cases} \max_{1 \leq i \leq m} \alpha_i |r_i|, & \text{(symmetric case)}, \\ \max_{1 \leq i \leq m} \alpha_i [r_i]_+, & \text{(asymmetric case)}. \end{cases} \tag{12.2.8}$$

We can represent the constraint $\pi(u) \leq \tau$ as a system of at most $2m$ inequalities of the form $u \in \mathcal{P}$, where

$$\mathcal{P} := \{ u \in \mathbb{R}^m \ : \ -\gamma\tau \leq \alpha_i u_i \leq \tau, \ i = 1, \ldots, m \},$$

with $\gamma = 1$ in the symmetric case, and $\gamma = 0$ in the asymmetric case. The condition (12.2.4) thus writes

$$\forall \Delta \in \mathcal{D} \ : \ -\gamma\tau \leq \alpha_i \left[[z_i^{\text{n}}]^T \theta + \rho e_i^T \Delta^T U^T \theta \right] \leq \tau, \ i = 1, \ldots, m,$$

where e_i are the standard basic orths in $\mathbb{R}^m$. This translates as the set of tractable constraints:

$$-\gamma\tau + \rho\alpha_i \phi_{\mathcal{D}}(-U^T \theta e_i^T) \leq \alpha_i [z_i^{\text{n}}]^T \theta \leq \tau - \rho\alpha_i \phi_{\mathcal{D}}(U^T \theta e_i^T), \ i = 1, \ldots, m. \tag{12.2.9}$$

Theorem 12.2.3. [Weighted ℓ_∞-norm loss] If $\mathcal{L}$ is given by (12.2.8) for some weighting vector $\alpha \in \mathbb{R}_+^m$, then the semi-infinite inequality (12.2.4) can be represented as the system of explicit convex constraints (12.2.9), with $\gamma = 1$ in the symmetric case, and $\gamma = 0$ in the asymmetric case.

MEASUREMENT-WISE UNCERTAINTY

Here, we assume that the columns of Z (each one of which, as the reader recalls, corresponds to a specific measurement) are independently perturbed. Specifically, we assume that

$$\mathcal{D} = \mathcal{D}_1 \times \ldots \times \mathcal{D}_m,$$

where each $\mathcal{D}_i$ describes the uncertainty about a specific column i. Let us denote by ϕ_i the support function of $\mathcal{D}_i$.

We observe that when Δ runs through $\mathcal{D}$, the vector $u = P((Z^n + \rho U \Delta)^T \theta)$ covers the box $\{u : 0 \le u \le u^{\mathrm{up}}(\theta)\}$, where

$$u^{\mathrm{up}}(\theta) = \begin{cases} \max\left[-[z_i^n]^T \theta + \rho\phi_i(-U^T\theta), [z_i^n]^T\theta + \rho\phi_i(U^T\theta)\right], \\ \hspace{5.5cm} \text{(symmetric case)}, \\ \max\left[0, [z_i^n]^T\theta + \rho\phi_i(U^T\theta)\right], \quad \text{(asymmetric case)}. \end{cases} \quad (12.2.10)$$

Exploiting the monotonicity of $\pi(\cdot)$, we obtain that the bound (12.2.4) on the worst-case loss holds if and only if

$$\pi(u^{\mathrm{up}}(\theta)) \le \tau.$$

We have obtained the following result.

Theorem 12.2.4. [Measurement-wise uncertainty] If $\mathcal{L}$ satisfies assumption **L**, and the uncertainty set $\mathcal{D}$ is measurement-wise, that is, it is given as a product $\mathcal{D}_1 \times \ldots \times \mathcal{D}_m$, where each subset $\mathcal{D}_i$ describes the uncertainty on the i-th column (measurement) of the matrix Z, then the semi-infinite constraint (12.2.4) can be represented as the tractable convex constraint

$$\pi(u^{\mathrm{up}}(\theta)) \le \tau,$$

where u^{up} is given by (12.2.10).

12.2.3 Generalized Bounded Additive Uncertainty

A second set of results derives from making further assumptions, this time mostly on the uncertainty model (the set $\mathcal{D}$, see (12.2.3)) considered in the Robust Counterpart (12.2.2).

ASSUMPTION ON THE UNCERTAINTY MODEL

As a convenient starting point, observe that since the function $\mathcal{L}$ is convex and finite-valued on $\mathbb{R}^m$ (by assumption **L**), it is the bi-conjugate of itself:

$$\mathcal{L}(r) = \sup_v \left[v^T r - \mathcal{L}^*(v)\right] \quad (12.2.11)$$

where $\mathcal{L}^*$ is a convex lower semicontinuous function on $\mathbb{R}^m$ taking values in $\mathbb{R} \cup \{+\infty\}$.

This is how (12.2.11) looks for loss functions we are especially interested in:

- [p-norm] $\mathcal{L}(r) = \|r\|_p$:

$$\mathcal{L}(r) = \max_{v:\|v\|_{p_*} \le 1} v^T r, \quad \frac{1}{p} + \frac{1}{p_*} = 1$$

  that is, $\mathcal{L} * (v) = 0$ when $\|v\|_{p_*} \le 1$ and $\mathcal{L}^*(v) = +\infty$ otherwise;

- [hinge loss function] $\mathcal{L}(r) = \sum_{i=1}^m \max[r_i, 0]$:

$$\mathcal{L}(r) = \max_{0 \le u \le 1} v^T r,$$

  where $\mathbf{1}$ is the all-ones vector. In other words, $\mathcal{L}^*(v) = 0$ when $0 \le v \le \mathbf{1}$ and $\mathcal{L}^*(v) = +\infty$ otherwise;

- [Huber loss] $\mathcal{L}(r) = \sum_{i=1}^m H(r_i, 1)$, see (12.2.7):

$$\mathcal{L}(r) = \max_{-1 \le v \le 1} \left[v^T r - \frac{1}{2}\|v\|_2^2 \right],$$

  i.e., $\mathcal{L}^*(v) = \frac{1}{2}\|v\|_2^2$ when $-1 \le v \le \mathbf{1}$ and $\mathcal{L}^*(v) = +\infty$ otherwise.

We continue to focus on the additive uncertainty model (12.2.3), and make a further assumption about the set $\mathcal{D}$. To motivate our assumption, we observe that in view of (12.2.11), the objective of the robust problem (12.2.2) reads

$$\begin{aligned}\max_{Z \in \mathcal{Z}} \mathcal{L}(Z^T \theta) &= \max_{\Delta \in \mathcal{D}, v} \left\{ v^T (Z^{\mathrm{n}} + \rho U \Delta)^T \theta - \mathcal{L}^*(v) \right\} \\ &= \max_v \left\{ v^T [Z^{\mathrm{n}}]^T \theta - \mathcal{L}^*(v) + \rho \max_{\Delta \in \mathcal{D}} \left[\theta^T U \Delta v \right] \right\}.\end{aligned}$$
(12.2.12)

Hence, the function from $\mathbb{R}^l \times \mathbb{R}^m$ to $\mathbb{R}$ defined as

$$(u, v) \to \max_{\Delta \in \mathcal{D}} u^T \Delta v$$
(12.2.13)

plays a crucial role, as it fully encapsulates the way in which the perturbation structure, that is, the set $\mathcal{D}$, enters the robust problem. Note that, as is common in Robust Optimization, the convexity of the set $\mathcal{D}$ plays no role in the robust counterpart: $\mathcal{D}$ can be safely replaced by its convex hull there.

We now make a fundamental assumption about the set $\mathcal{D}$ in regards of the function defined in (12.2.13). Recall that a *Minkowski function* $\phi(\cdot)$ is a (finite everywhere) *nonnegative* convex function ϕ that is positively homogeneous of degree 1: $\phi(tv) = t\phi(v)$ whenever $t \ge 0$.

Our assumption on the set $\mathcal{D}$ is as follows:

Assumption A: *The set $\mathcal{D}$ is such that there exists a Minkowski function ϕ on $\mathbb{R}^l$ and a norm ψ on $\mathbb{R}^m$ such that*

$$\forall u \in \mathbb{R}^l, \ \forall v \in \mathbb{R}^m \ : \ \max_{\Delta \in \mathcal{D}} u^T \Delta v = \phi(u)\psi(v).$$

A way to interpret assumption **A** is that it provides an expression for the support function of the matrix set $\mathcal{D}$, but only for rank-one matrices. As such, the assumption does not fully characterize $\mathcal{D}$.

Here are a few examples of sets $\mathcal{D}$ that satisfy assumption **A**. We identify each case on our list with an acronym that will allow us to easily refer to a specific uncertainty model. For example, model LSV refers to the first model detailed below.

[LSV] *Largest singular value model:* With $\mathcal{D} = \{\Delta \in \mathbb{R}^{l \times m} : \|\Delta\| \leq 1\}$, where $\|\cdot\|$ is the largest singular value of Δ, one can capture possible dependencies between uncertainties affecting different data points. This set satisfies assumption **A**, with ϕ, ψ being the Euclidean norms in $\mathbb{R}^l$ and $\mathbb{R}^m$, respectively.

[FRO] The *Frobenius norm model* is the same as above, with the Frobenius norm instead of the largest singular value norm. This set satisfies assumption **A**, with the same ϕ, ψ.

[IND] *Induced norm model:* Consider, as an extension of the LSV model, the set $\mathcal{D} = \{\Delta \in \mathbb{R}^{l \times m} : \|\Delta v\|_{p_*} \leq \|v\|_q \ \forall v \in \mathbb{R}^m\}$, where $p, q \in [1, \infty]$ and $\frac{1}{p_*} + \frac{1}{p} = 1$. This set satisfies assumption **A** with $\phi(\cdot) = \|\cdot\|_p$, $\psi(\cdot) = \|\cdot\|_q$.

[MWU] *Measurement-wise uncertainty models,* already seen in section 12.1.3, correspond to the following choice of the set $\mathcal{D}$:

$$\mathcal{D} = \left\{\Delta = [\delta_1, \ldots, \delta_m] \in \mathbb{R}^{l \times m} \ : \ \|\delta_i\| \leq 1, \ i = 1, \ldots, m\right\},$$

where $\|\cdot\|$ is a norm on $\mathbb{R}^l$ (the case of box uncertainty (12.1.11) corresponds to $U = I$, $\|\cdot\| \equiv \|\cdot\|_\infty$). Such sets satisfy assumption **A** with $\psi(\cdot) = \|\cdot\|_1$ and $\phi(\cdot) = \|\cdot\|_*$, where $\|\cdot\|_*$ is the norm *conjugate* to a norm $\|\cdot\|$:

$$\|\eta\|_* = \max_h \{h^T \eta : \|h\| \leq 1\} \quad [\Leftrightarrow \|h\| = \max_\eta \{\eta^T h : \|\eta\|_* \leq 1\}].$$

[COM] *Composite norm models:* This is a variation of the previous case as follows

$$\mathcal{D} = \left\{\Delta = [\delta_1, \ldots, \delta_m] \in \mathbb{R}^{l \times m} \ : \ \|[\|\delta_1\|_p; \ldots; \|\delta_m\|_p]\|_q \leq 1\right\},$$

with $p, q \in [1, \infty]$. Here assumption **A** is satisfied with $\phi(\cdot) = \|\cdot\|_{p_*}$ and $\psi(\cdot) = \|\cdot\|_{q_*}$, where for an $s \in [1, \infty]$ s_* is given by $\frac{1}{s_*} + \frac{1}{s} = 1$ (see Exercise 12.2). The MWU models are obtained with $q = \infty$. When $q < \infty$, the above allows one to capture dependencies across perturbations affecting different measurements.

[KER] *K-error models:* For $p \in [1, \infty]$ and $K \in \{1, \ldots, m\}$, the set

$$\mathcal{D} = \ \text{Conv}\left\{[\lambda_1 \delta_1, \ldots, \lambda_m \delta_m] \in \mathbb{R}^{l \times m} : \|\delta_i\|_p \leq 1, 1 \leq i \leq m, \\ \textstyle\sum_{i=1}^m \lambda_i \leq K, \ \lambda \in \{0, 1\}^m\right\},$$

allows to model the fact that there are at most K (norm-bounded) errors affecting the measurements, which again couples them. Here, assumption **A** is satisfied with $\phi(\cdot) = \|\cdot\|_{p_*}$ and with the norm ψ defined as

$$\psi(v) = \sum_{i=1}^K |v|_{[i]}, \tag{12.2.14}$$

where $|v|_{[i]}$ is the i-th largest component of the vector $[|v_1|; \ldots; |v_m|]$. Note that this norm has both the ℓ_1 and ℓ_∞ norms as special cases, obtained with $K = m$ and $K = 1$, respectively.

THE WORST-CASE LOSS FUNCTION

Under assumption A, invoking (12.2.12), we obtain the worst-case realized loss function

$$\max_{Z \in \mathcal{Z}} \mathcal{L}(Z^T \theta) = \max_v \left\{ v^T [Z^n]^T \theta - \mathcal{L}^*(v) + \rho \phi(U^T \theta) \psi(v) \right\}. \tag{12.2.15}$$

Introducing convex function $\mathcal{L}_{\mathrm{wc}}(r, \kappa) : \mathbb{R}^m \times \mathbb{R}_+ \to \mathbb{R}$ given by

$$\mathcal{L}_{\mathrm{wc}}(r, \kappa) := \max_v \left[v^T r - \mathcal{L}^*(v) + \kappa \psi(v) \right], \tag{12.2.16}$$

the semi-infinite inequality (12.2.4) becomes the convex inequality

$$\mathcal{L}_{\mathrm{wc}}([Z^n]^T \theta, \rho \phi(U^T \theta)) \equiv \min_\kappa \left\{ \mathcal{L}_{\mathrm{wc}}([Z^n]^T \theta, \kappa) : \kappa \geq \rho \phi(U^T \theta) \right\} \leq \tau. \tag{12.2.17}$$

Note that $\equiv$ in the latter relation is due to the evident fact that $\mathcal{L}_{\mathrm{wc}}$ is nondecreasing with respect to its second argument.

We will refer to the function (12.2.16) as the *worst-case loss function* associated with our robust problem. The worst-case loss function is indeed a loss function in the classical sense, since it is convex and independent of the problem's data, and depends only on problem structure. Note that $\mathcal{L}_{\mathrm{wc}}(\cdot, 0) = \mathcal{L}(\cdot)$, so the worst-case loss function is really an extension of the original loss function.

We can alternatively express the worst-case loss function as

$$\mathcal{L}_{\mathrm{wc}}(r, \kappa) = \max_\xi \left\{ \mathcal{L}(r + \kappa \xi) : \psi_*(\xi) \leq 1 \right\}, \tag{12.2.18}$$

where $\psi_*(\cdot)$ stands for the norm conjugate to $\psi(\cdot)$. In the above, $\psi_*(\cdot)$ defines the shape of allowable additive perturbations to the residual vector r, while κ defines the size of this set. The worst-case function fully describes the effect of such allowable perturbations on the original loss function, for arbitrary residual vectors r and perturbation sizes κ.

When $\psi(\cdot) = \| \cdot \|_2$, the function $\mathcal{L}_{\mathrm{wc}}(\cdot, 1)$ as given by (12.2.18) is the robust regularization of the original loss function $\mathcal{L}$, in the sense of Lewis [76].

With our perturbation model, the robust problem (12.2.17) becomes

$$\max_{\xi \, : \, \psi_*(\xi) \leq 1} \mathcal{L}([Z^n]^T \theta + [\rho \phi(U^T \theta)] \xi) \leq \tau,$$

exactly as if the realized loss function were subject to additive perturbations on the residual $r = Z^T \theta$, with the amplitude of the perturbation depending on the parameter vector θ.

Our abilities to process efficiently the semi-infinite inequality (12.2.4), or, which is the same under our structural assumptions on the perturbation model, the inequality (12.2.17), hinges on our ability to efficiently compute, and find sub-gradients of, the worst-case loss function (12.2.16). We now examine the situation for specific choices of the norm ψ.

THE CASE OF $\psi(\cdot) = \|\cdot\|_\infty$

This case includes for example a composite norm model (labelled as COM in our list). Denoting by e_i the standard basic orths in $\mathbb{R}^m$, we have

$$
\begin{aligned}
\mathcal{L}_{\mathrm{wc}}(r, \kappa) &= \max_v \left\{ v^T r - \mathcal{L}^*(v) + \kappa \max_{1 \le i \le m} |v_i| \right\} \\
&= \max_{1 \le i \le m} \left\{ \max_v \left[v^T r - \mathcal{L}^*(v) + \kappa |v_i| \right] \right\} \\
&= \max_{1 \le i \le m} \left\{ \max_v \max \left[v^T r - \mathcal{L}^*(v) - \kappa v_i, v^T r - \mathcal{L}^*(v) + \kappa v_i \right] \right\} \\
&= \max_{1 \le i \le m} \max \left[\max_v \left[v^T [r - \kappa e_i] - \mathcal{L}^*(v) \right], \max_v \left[v^T [r + \kappa e_i] - \mathcal{L}^*(v) \right] \right] \\
&= \max_{1 \le i \le m} \max \left(\mathcal{L}(r + \kappa e_i), \mathcal{L}(r - \kappa e_i) \right).
\end{aligned}
$$

We have arrived at the following result:

Theorem 12.2.5. If the loss function satisfies assumption **L**, and the uncertainty set satisfies assumption **A**, with $\psi(\cdot) = \|\cdot\|_\infty$, then the semi-infinite inequality (12.2.4) can be represented by the explicit system of efficiently computable convex constraints

$$
\rho \phi(U^T \theta) \le \kappa, \ \mathcal{L}([Z^{\mathrm{n}}]^T \theta \pm \kappa e_i) \le \tau, \ 1 \le i \le m
$$

in variables θ, κ.

THE CASE OF $\psi(\cdot) = \|\cdot\|_1$

This case includes as a special case the MWU models. In particular, this recovers the situation we have encountered in Support Vector Machines with box uncertainty (section 12.1.3).

This time, we start with the expression (12.2.18). Invoking assumption **L**, we obtain

$$
\mathcal{L}_{\mathrm{wc}}(r, \kappa) = \max_{\xi, \|\xi\|_\infty \le 1} \mathcal{L}(r + \kappa \xi) = \pi(u(r, \kappa)),
$$

where

$$
(u(r, \kappa))_i := \begin{cases} |r_i| + \kappa, & \text{(symmetric case)}, \\ (r_i + \kappa)_+, & \text{(asymmetric case)}. \end{cases} \tag{12.2.19}
$$

Theorem 12.2.6. If the loss function satisfies assumption **L** and the uncertainty set satisfies assumption **A** with $\psi(\cdot) = \|\cdot\|_1$, then the semi-infinite inequality (12.2.4) can be represented by the explicit efficiently computable convex constraint

$$
\pi(u([Z^{\mathrm{n}}]^T \theta, \rho \phi(U^T \theta))) \le \tau,
$$

with $u(\cdot, \cdot)$ given by (12.2.19).

THE CASE OF $\psi = \| \cdot \|_2$

This case includes in particular the LSV and FRO models. The worst-case loss function now expresses as (12.2.18), with the constraint involving the Euclidean norm:

$$\mathcal{L}_{\mathrm{wc}}(r, \kappa) = \max_v \left\{ v^T r - \mathcal{L}^*(v) + \kappa \|v\|_2 \right\} \qquad (12.2.20)$$

$$= \max_\xi \left\{ \mathcal{L}(r + \kappa\xi) : \|\xi\|_2 \le 1 \right\}.$$

We can process the problem above in a computationally tractable fashion when $\mathcal{L}$ is separable. Indeed, as the parameter κ spans the positive real line, the set of solutions to problem (12.2.20) is the same as that obtained upon replacing the Euclidean norm by its square. The problem is then separable and a (unique) optimal solution

$$v^{\mathrm{opt}}(\kappa) := \mathrm{argmax}_v \left[v^T r - \mathcal{L}^*(v) + \kappa \|v\|_2^2 \right]$$

can be efficiently computed. The solution to the original problem (12.2.20) corresponds to the value of κ for which $\kappa = \|v^{\mathrm{opt}}(\kappa)\|_2$. It is easily shown that this fixed point equation has a unique solution.

For general loss functions, the problem of computing the worst-case loss function, is apparently intractable. For some specific ones, such as the least-squares loss function $\mathcal{L}(r) = \|r\|_2$, in the case of $\psi(\cdot) = \| \cdot \|_2$ the problem has a trivial solution. In addition, the problem is tractable for Support Vector Machine classification or Huber regression. We consider these cases next.

12.2.4 Examples

SUPPORT VECTOR MACHINES

Consider the Robust SVM problem described in Example 12.2.2. Recall that the data matrix Z in this case is built upon the matrix $X = [x_1, \ldots, x_m] \in \mathbb{R}^{n \times m}$ of measured "feature vectors" (this matrix can be uncertain) and a sequence $y_1, \ldots, y_m \in \{-1, 1\}$ of labels assumed to be certain. We assume that X is subject to an additive bounded uncertainty, the corresponding uncertainty set being

$$\mathcal{X} := \{ X + \Delta \; : \; \Delta \in \rho \mathcal{D}_0 \},$$

with $\mathcal{D}_0 \subset \mathbb{R}^{n \times m}$ satisfying assumption **A**, the corresponding norm and Minkowski function being, respectively, ψ and ϕ. We further assume that ψ is a symmetric gauge (a norm that is invariant under permutation and sign changes in the argument).

The corresponding set $\mathcal{Z}$ then is

$$\mathcal{Z} = \{ Z^{\mathrm{n}} + U \Delta \mathrm{Diag}\{y\} : \Delta \in \rho \mathcal{D}_0 \}, \qquad (12.2.21)$$

where $U = [I_n; 0_{2 \times n}] \in \mathbb{R}^{(n+2) \times n}$. Exploiting the fact that the norm ψ is sign-invariant, the set $\mathcal{D} = \{ \Delta \mathrm{Diag}\{y\} : \Delta \in \mathcal{D}_0 \}$ also satisfies assumption **A**, with the same functions ϕ, ψ. Thus, the corresponding set $\mathcal{Z}$ is as (12.2.3) with $\mathcal{D}$ satisfying assumption **A**.

Since the norm ψ is permutation-invariant, we have, for every $k \in \{0, \ldots, m\}$, and every $v \in \{0,1\}^m$ such that $v^T \mathbf{1} = k$:

$$\psi(v) = \psi(\sum_{i=1}^{k} e_i) := c_k,$$

where e_i is the i-th basis vector in $\mathbb{R}^m$. For example, if $\psi(\cdot) = \|\cdot\|_p$, then $c_k = k^{1/p}$ for every k.

The worst-case loss function reads

$$\mathcal{L}_{\mathrm{wc}}(r, \kappa) = \max_{0 \leq v \leq 1} \left[v^T r + \kappa \psi(v)\right] = \max_{v \in \{0,1\}^m} \left[v^T r + \kappa \psi(v)\right],$$

where we have exploited the convexity of the objective in the most left maximization problem.

Now observe that, for every scalar $\kappa \geq 0$, and vector $r \in \mathbb{R}^m$, denoting by $r_{[i]}$ the i-th largest component of r, we have

$$\begin{aligned}
\mathcal{L}_{\mathrm{wc}}(r, \kappa) &= \max_{v \in \{0,1\}^m} \left[\kappa \psi(v) + v^T r\right] \\
&= \max_{k \in \{0,\ldots,m\}} \max_{v \in \{0,1\}^m, v^T \mathbf{1} = k} \left[\kappa \psi(v) + v^T r\right] \\
&= \max_{k \in \{0,\ldots,m\}} \max_{v \in \{0,1\}^m, v^T \mathbf{1} = k} \left[\kappa c_k + v^T r\right] \\
&= \max_{k \in \{0,\ldots,m\}} \left[\kappa c_k + \sum_{i=1}^{k} r_{[i]}\right] \\
&= \max_{k \in \{0,\ldots,m\}} \min_{\mu} \left[\kappa c_k + k\mu + \sum_{i=1}^{m} [1 - r_i - \mu]_+\right].
\end{aligned}$$

The resulting equality shows that the worst-case loss function can be computed via Linear Optimization.

The semi-infinite inequality (12.2.4) therefore can be represented as

$$\exists \{\mu_k\}_{k=0}^{m} : \rho c_k \phi(U^T \theta) + k\mu_k + \sum_{i=1}^{m} \left[1 - [z_i^{\mathrm{n}}]^T \theta - \mu_k\right]_+ \leq \tau, \ 0 \leq k \leq n, \quad (12.2.22)$$

where z_i^{n} are the columns of the matrix Z^{n}, see (12.2.6). With $\theta = [-w; -b; 1]$, and using the original problem's notation (see Example 12.2.2), we represent the semi-infinite inequality (12.2.4) by the system of explicit convex constraints

$$\rho c_k \phi(w) + k\mu_k + \sum_{i=1}^{m} \left[1 - y_i(w^T x_i^{\mathrm{n}} + b) - \mu_k\right]_+ \leq \tau, \ 0 \leq k \leq n$$

in variables $\tau, w, b, \{\mu_k\}$; here x_i^{n} are the measured feature vectors.

Let us consider the following more specific examples. The case referred to as LSV or FRO in our list of models, where the set $\mathcal{D}$ is the set of matrices Δ with

$\|\Delta\| \le \rho$, where $\| \cdot \|$ either the largest singular value or the Frobenius norm, has $c_k = \sqrt{k}$. Setting $\phi(\cdot) = \| \cdot \|_2$, this proves the claims made in the earlier discussion of SVMs with norm-bounded uncertainty (section 12.1.4).

As another specific example, consider the model referred to as KER, which allows one to control the number of perturbations affecting the data. In this case, we have $\phi(\cdot) = \| \cdot \|_{p_*}$, while $\psi(\cdot)$ is defined by (12.2.14). Hence, $c_k = \min(k, K)$ for every $k \in \{0, \ldots, m\}$.

In the case of measurement-wise uncertainty (MWU models) we have $\psi(\cdot) = \| \cdot \|_1$, which yields $c_k = k$. Here (12.2.4) can be represented by the system of explicit convex constraints

$$k(\rho\phi(w) + \mu_k) + \sum_{i=1}^{m} \left[1 - y_i(w^T x_i^{\mathrm{n}} + b) - \mu_k \right]_+ \le \tau, \ 0 \le k \le n$$

in variables $\tau, w, b, \{\mu_k\}$.

We easily recover the problems encountered in section 12.1.3 upon introducing new variables $\widetilde{\mu}_k = \mu_k + \rho\phi(w)$.

The notion of maximally robust separation can be extended to a general bounded additive perturbation structure. Assume that the data is separable, that is, inequalities (12.1.1) are feasible. Also, assume that ϕ is a norm. The problem is to maximize ρ such that the inequalities

$$\forall \, \Delta \in \rho\mathcal{D} \ : \ \theta^T(Z^{\mathrm{n}} + U\Delta)e_i \ge 0, \ \ i = 1, \ldots, m$$

hold, where e_i is the i-th basis orth. The above can be written as

$$\forall \, i = 1, \ldots, m \ : \ [z_i^{\mathrm{n}}]^T\theta \ge \rho \max_{\Delta \in \mathcal{D}} \theta^T U^T \Delta(-e_i).$$

Exploiting assumption **A** and still assuming that ψ is a symmetric gauge, the latter condition can be rewritten as

$$\forall \, i = 1, \ldots, m \ : \ [z_i^{\mathrm{n}}]^T\theta \ge \rho\phi(U^T\theta) \cdot \psi(e_1).$$

Using homogeneity together with the fact that ϕ is a norm, and the original problem's notation, we obtain that maximizing ρ subject to the above conditions can be written as

$$\min_{w,b} \left\{ \phi(w) : y_i([x_i^{\mathrm{n}}]^T w + b) \ge 1, \ 1 \le i \le m \right\}.$$

The actual maximally robust classifier does not depend on the norm ψ, only on the Minkowski function ϕ. However, the optimal margin $\rho^{\mathrm{opt}} = 1/(\phi(w^{\mathrm{opt}})\psi(e_1))$ depends on both norms.

This generalizes the results we have obtained for the spherical and box uncertainty models, in (12.1.3) and (12.1.10) respectively. It also confirms our prior observation that the maximally robust separating classifier is the same, wether we choose an LSV or spherical uncertainty model.

Maximum hinge loss, interval data

As an example illustrating Theorem 12.2.3, consider a Support Vector Machine problem with the "maximum hinge" loss

$$\max_{1 \le i \le m} \left[1 - y_i (w^T x_i + b) \right]_+ .$$

Assume that the data matrix $X = [x_1, \ldots, x_m]$ is only known to belong to the interval matrix set $\mathcal{X}_{\text{int}}$, as given in (12.1.12). Defining Z, θ as in (12.2.6) and applying Theorem 12.2.3, after straightforward computations, we arrive the following tractable representation of the semi-infinite inequality (12.2.4):

$$\max_{1 \le i \le m} \left[1 - y_i (w^T x_i + b) + \rho \sigma_i^T |w| \right]_+ \le \tau,$$

where σ_i^T is i-th row of the matrix R participating in the description (12.1.12) of the interval matrix set $\mathcal{X}_{\text{int}}$.

Least-squares regression

We now turn to the robust least-squares regression problem, that is, the problem of Example 12.2.1 with $\| \cdot \| = \| \cdot \|_2$. Here the semi-infinite constraint (12.2.4) reads

$$\max_{Z \in \{Z^n + \rho U \Delta : \Delta \in \mathcal{D}\}} \| Z^T \theta \|_2 \le \tau,$$

$$Z = \begin{bmatrix} x_1 & \cdots & x_m \\ y_1 & \cdots & y_m \\ 1 & \cdots & 1 \end{bmatrix}, \ \theta = [-w; 1; -b], \ U = \left[I_{n+1}; 0_{1 \times (n+1)} \right]. \tag{12.2.23}$$

where $\dim x_i = n$, $y_i \in \mathbb{R}$ and $\mathcal{D}$ is an uncertainty set in the space $\mathbb{R}^{(n+1) \times m}$. We assume that this set satisfies assumption **A** with certain ϕ and ψ.

A. Assume first that $\psi(\cdot) = \| \cdot \|_2$. We are in the situation where the function $\mathcal{L}^*$ in (12.2.11) is the indicator of the unit Euclidean ball in $\mathbb{R}^m$:

$$\mathcal{L}(r) \equiv \|r\|_2 = \max_{v : \|v\|_2 \le 1} r^T v,$$

so that the worst-case loss (12.2.16) is

$$\max_{\|v\|_2 \le 1} \left\{ v^T r + \kappa \|v\|_2 \right\} = \|r\|_2 + \kappa.$$

Consequently, (12.2.23) is equivalent to the explicit convex inequality

$$\| y^n - [X^n]^T w - b\mathbf{1} \|_2 + \rho \phi([w; b]) \le \tau,$$

where $\begin{bmatrix} X^n \\ [y^n]^T \end{bmatrix} = \begin{bmatrix} x_1^n & \cdots & x_m^n \\ y_1^n & \cdots & y_m^n \end{bmatrix}$ is the nominal data.

We can specialize the result to obtain some well-known penalties for least-squares regression. For example, assume that the uncertainty model is based on the largest singular value norm (LSV model). Then ϕ is the Euclidean norm, and we recover the result derived in [50]. Alternatively, assume that the uncertainty model is in the class IND with $p = 2$ and $q = \infty$, which corresponds to $\mathcal{D} = \{\Delta : \|\Delta\|_{2,\infty} \leq 1\}$, where $\|\cdot\|_{2,\infty}$ is the induced norm defined as

$$
\begin{aligned}
\|\Delta\|_{2,\infty} &= \max_v \left\{\|\Delta v\|_\infty : \|v\|_2 \leq 1\right\} \\
&= \max_{u,v} \left\{u^T \Delta v : \|u\|_1 \leq 1, \ \|v\|_2 \leq 1\right\} = \max_{1 \leq i \leq n} \sqrt{\sum_{j=1}^m \Delta_{ij}^2}.
\end{aligned}
$$

In this case (12.2.23) bears the form

$$
\|y^n - [X^n]^T w - b\mathbf{1}\|_2 + \rho[\|w\|_1 + |b|] \leq \tau.
$$

The problem of minimizing τ under this constraint, with b set to 0, reads

$$
\min_w \left\{\|y^n - [X^n]^T w\|_2 + \rho\|w\|_1\right\},
$$

which is essentially the same (up to a squared first term) as LASSO regression [113]. Note that the induced norm above couples the uncertainties across different measurements, but allows the features (rows of data matrix Z) to be independently perturbed.

B. We can extend the results to other symmetric gauges ψ in lieu of the Euclidean norm. For example, consider the case of KER models, where the norm ψ is defined in (12.2.14). The worst-case loss function is

$$
\mathcal{L}_{\text{wc}}(r,\kappa) = \max_{v \,:\, \|v\|_2 \leq 1} \left\{v^T r + \kappa \sum_{i=1}^K |v|_{[i]}\right\} = \sqrt{\sum_{i=1}^K (|r|_{[i]} + \kappa)^2 + \sum_{i=K+1}^m |r|_{[i]}^2},
$$

so that (12.2.23) is equivalent to

$$
\sqrt{\sum_{i=1}^K (|[Z^n[-w;1;-b]|_{[i]} + \rho\phi([w;b]))^2 + \sum_{i=K+1}^m |[Z^n]^T][-w;1;-b]|_{[i]}^2} \leq \tau.
$$

ℓ_1 REGRESSION

Now consider the semi-infinite inequality (12.2.4) in the case of ℓ_1-regression, where the inequality reads

$$
\max_{Z \in \{Z^n + \rho U\Delta : \Delta \in \mathcal{D}\}} \|Z^T \theta\|_1 \leq \tau,
$$

$$
Z = \begin{bmatrix} x_1 & \cdots & x_m \\ y_1 & \cdots & y_m \\ 1 & \cdots & 1 \end{bmatrix}, \ \theta = [-w;1;-b], \ U = \begin{bmatrix} I_{n+1} ; 0_{1 \times (n+1)} \end{bmatrix}. \tag{12.2.24}
$$

We assume that the set $\mathcal{D}$ satisfies assumption **A**, with arbitrary norm ϕ and symmetric gauge function ψ. Here, the loss function is

$$
\mathcal{L}(r) = \|r\|_1 = \max_{v, \|v\|_\infty \leq 1} v^T r,
$$

and the worst-case loss function now reads

$$\mathcal{L}_{\mathrm{wc}}(r, \kappa) = \max_{v, \|v\|_\infty \leq 1} \left[v^T r + \kappa \psi(v) \right] = \max_{v: v_i = \pm 1, 1 \leq i \leq m} \left[v^T r + \kappa \psi(v) \right]$$
$$= \|r\|_1 + \kappa \psi(\mathbf{1}),$$

where we used the property of sign invariance of ψ in the last line. Thus, (12.2.24) is equivalent to the explicit convex constraint

$$\|y^{\mathrm{n}} - [X^{\mathrm{n}}]^T w - b\mathbf{1}\|_1 + \rho \phi([w; b]) \psi(\mathbf{1}) \leq \tau.$$

In the special cases of the LSV and FRO models, this constraint reads

$$\|y^{\mathrm{n}} - [X^{\mathrm{n}}]^T w - b\mathbf{1}\|_1 + \rho \sqrt{m} \| [w; b] \|_2 \leq \tau.$$

ℓ_∞ REGRESSION

Consider now the same problem as above, with the same assumption about the set $\mathcal{D}$ (in the latter assumption, ψ can be an arbitrary norm), but with the loss function

$$\mathcal{L}(r) = \|r\|_\infty = \max_{v: \|v\|_1 \leq 1} v^T r.$$

Denoting by e_i the i-th unit vector in $\mathbb{R}^m$, the worst-case loss function reads

$$\mathcal{L}_{\mathrm{wc}}(r, \kappa) = \max_{v: \|v\|_1 \leq 1} \left\{ v^T r + \kappa \psi(v) \right\} = \max_{v \in \{e_1, -e_1\}_{i=1}^m} v^T r + \kappa \psi(v)$$
$$= \max_{v \in \{e_1, -e_1, e_1, -e_2, \ldots, e_m, -e_m\}} \left\{ v^T r + \kappa \psi(v) \right\}$$
$$= \max_{1 \leq i \leq m} \left[|r_i| + \kappa \psi(e_i) \right].$$

Consequently, the semi-infinite inequality (12.2.4) can be represented by the system of explicit convex inequalities

$$|y_i^{\mathrm{n}} - [x_i^{\mathrm{n}}]^T w - b| + \rho \phi([w; b]) \psi(e_i) \leq \tau, \ 1 \leq i \leq m.$$

HUBER PENALTY REGRESSION

Consider the variant of the regression problem from Example 12.2.1 with a separable loss function with Huber-type components (12.2.7). Here the semi-infinite inequality (12.2.4) reads

$$\max_{Z \in \{Z^{\mathrm{n}} + \rho U \Delta : \Delta \in \mathcal{D}\}} \mathcal{L}(Z^T \theta) \leq \tau,$$

$$Z = \begin{bmatrix} x_1 & \ldots & x_m \\ y_1 & \ldots & y_m \\ 1 & \ldots & 1 \end{bmatrix}, \ \theta = [-w; 1; -b], \ U = \left[I_{n+1}; 0_{1 \times (n+1)} \right],$$

$$\mathcal{L}(r) = \sum_{i=1}^m H(r_i), \ H(t) = \max_s [ts - h(s)], \ h(s) = \begin{cases} s^2/2, & |s| \leq 1, \\ +\infty, & |s| \geq 1. \end{cases}$$

$$(12.2.25)$$

Assume that the perturbation set $\mathcal{D} \in \mathbb{R}^{(n+1) \times m}$ satisfies assumption **A** with $\psi(\cdot) = \|\cdot\|_2$ (this corresponds, for example, to the LSV or FRO models.) Here the worst-case loss function reads

$$\mathcal{L}_{\mathrm{wc}}(r, \kappa) = \max_{v, \|v\|_\infty \leq 1} \left[v^T r - v^T v/2 + \kappa \|v\|_2 \right]$$
$$= \max_{v, 0 \leq v \leq 1} \left[v^T |r| - v^T v/2 + \kappa \|v\|_2 \right].$$

We proceed with a change of variables $\nu_i = \sqrt{v_i}$, $1 \leq i \leq m$, which yields a concave maximization problem:

$$\mathcal{L}_{\mathrm{wc}}(r, \kappa) = \max_{\nu: 0 \leq \nu \leq 1} \left[\sum_{i=1}^m [|r_i| \sqrt{\nu_i} - \nu_i/2] + \kappa \sqrt{\sum_{i=1}^m \nu_i} \right].$$

Expressing the second term as

$$\kappa \sqrt{\sum_{i=1}^m \nu_i} = \min_{\lambda \geq 0} \left\{ \frac{\kappa^2}{2\lambda} + \frac{\lambda}{2} \sum_{i=1}^m \nu_i \right\},$$

and applying duality, we obtain:

$$\mathcal{L}_{\mathrm{wc}}(r, \kappa) = \max_{0 \leq \nu \leq 1} \min_{\lambda \geq 0} \left[\frac{\kappa^2}{2\lambda} + \sum_{i=1}^m \left[|r_i| \sqrt{\nu_i} + \frac{\nu_i(\lambda - 1)}{2} \right] \right\}$$
$$= \min_{\lambda \geq 0} \max_{0 \leq \nu \leq 1} \left\{ \frac{\kappa^2}{2\lambda} + \sum_{i=1}^m \left[|r_i| \sqrt{\nu_i} + \frac{\nu_i(\lambda - 1)}{2} \right] \right\}$$
$$= \min_{\lambda \geq 0} \left\{ \frac{\kappa^2}{2\lambda} + \sum_{i=1}^m \max_{0 \leq \tau \leq 1} \left[|r_i| \sqrt{t} - \frac{t(1 - \lambda)}{2} \right] \right\}$$
$$= \min_{\lambda \geq 0} \left\{ \frac{\kappa^2}{2\lambda} + \sum_{i=1}^m \widetilde{H}(r_i, 1 - \lambda) \right\},$$

where

$$\widetilde{H}(t, \mu) = \max_{0 \leq \xi \leq 1} \left[|t|\xi - \mu \frac{\xi^2}{2} \right] = \begin{cases} \frac{t^2}{2\mu} & |t| \leq \mu, \\ |t| - \frac{\mu}{2} & |t| \geq \mu. \end{cases}$$

Note that $\widetilde{H}(t, \mu)$ is a convex in t, μ extension of the Huber function $H(t, \mu)$ (originally defined only for $\mu > 0$) to the entire space of variables t, μ.

We conclude that in the case in question (12.2.25) can be represented by the system

$$\sum_{i=1}^m \widetilde{H}(|y_i^{\mathrm{n}} - [x_i^{\mathrm{n}}]^T w - b|, 1 - \lambda) + \frac{\rho^2 \phi^2([w; b])}{2\lambda} \leq \tau, \ \lambda \geq 0$$

of explicit convex constraints in variables τ, w, b, λ.

It is interesting to compare the above formulation of the semi-infinite inequality (12.2.25) with that of the nominal inequality (that is, the one with $\rho = 0$); in the latter, the second (the penalty) term is dropped, and then λ is set to zero. Also, in practical applications, the nominal inequality is often modified as follows:

$$\sum_{i=1}^{m} H(y_i^{\mathrm{n}} - [x_i^{\mathrm{n}}]^T w - b, M) + \alpha \|w\|_2^2 / 2 \leq \tau,$$

where parameters $M > 0$ and $\alpha \geq 0$ are chosen by the user, or via cross-validation. The robust formulation perhaps provides guidance about the choice of these parameters, as well as for the norm used in the penalty.

12.3 AFFINE UNCERTAINTY MODELS

Up to now, we have considered robust counterparts to classification and regression problems, using a certain class of models of perturbations. Our specific modeling assumptions allowed us to end up with tractable robust counterparts.

For more general perturbations models, such exact answers are difficult to obtain, and we must settle for upper bounds. In this section, we consider a class of models where the uncertainty enters affinely in the problem's data.

12.3.1 Norm-Bounded Affine Uncertainty Models

We assume that the set $\mathcal{Z}$ appearing in the Robust Counterpart (12.2.2) is of the form

$$\mathcal{Z} = \left\{ Z(\zeta) := Z^{\mathrm{n}} + \sum_{\ell=1}^{L} \zeta_\ell Z_\ell \ : \ \|\zeta\|_p \leq \rho \right\}, \tag{12.3.1}$$

where Z_ℓ are given $n \times m$ matrices, and $p \in \{1, 2, \infty\}$. For simplicity, we assume that the matrices Z_ℓ are all rank-one, and let $Z_\ell = u_\ell v_\ell^T$, with given vectors $u_\ell \in \mathbb{R}^n$, $v_\ell \in \mathbb{R}^m$. We define the matrices $U := [u_1, \ldots, u_L] \in \mathbb{R}^{n \times L}$, $V := [v_1, \ldots, v_L] \in \mathbb{R}^{m \times L}$.

Note that when V is the identity, the set $\mathcal{Z}$ has exactly the form we assumed in (12.2.3), with $\mathcal{D} = \{\mathrm{Diag}\{\zeta\} \ : \ \|\zeta\|_p \leq 1\}$. However, this set does not satisfy assumption **A**. Hence, the previous theory cannot be directly applied, even in the case when V is the identity matrix.

Note that there are two cases of loss functions $\mathcal{L}(\cdot)$ where we already know how to handle the semi-infinite inequality (12.2.4), that is, the RC of the uncertain constraint $\mathcal{L}(Z^T \theta) \leq \tau$, the uncertainty set being (12.3.1). These cases are as follows:

POLYHEDRAL LOSS FUNCTION $\mathcal{L}(r) = \max_{1 \leq \mu \leq M} [a_\mu^T r + b_\mu]$

This case (in particular, the one with $\mathcal{L}(r) = \|r\|_\infty$) is nothing but the case of uncertain system of affinely perturbed scalar linear inequalities, and here all results of Part I are applicable. In particular, here (12.2.4) is computationally tractable (since the uncertainty set (12.3.1) is so).

THE CASE OF $\mathcal{L}(r) = \|r\|_2$

In this case, the uncertain constraint $\{\mathcal{L}(Z^T\theta) \leq \tau\}_{Z \in \mathcal{Z}}$ is nothing but an uncertain affinely perturbed conic quadratic inequality with a certain right hand side, so that the results of chapters 6, 7 are readily applicable. In particular, with the specific perturbation model (12.3.1), the RC (12.2.4) of our uncertain inequality admits tractable reformulation when $p = 1$ (scenario uncertainty, section 6.1) and $p = 2$ (simple ellipsoidal uncertainty, section 6.3) and admits a safe tractable approximation, tight within the factor $O(\ln(L))$, when $p = \infty$ (section 7.2).

In the following, we intend to consider several other cases when the semi-infinite inequality (12.2.4), the uncertainty set being (12.3.1), is tractable or admits safe tractable approximations.

12.3.2 Pseudo Worst-Case Loss Function

With perturbation model (12.3.1), the worst-case realized loss function now reads

$$\max_{Z \in \mathcal{Z}} \mathcal{L}(Z^T\theta) = \max_v \left\{ v^T Z^T \theta - \mathcal{L}^*(v) + \rho \max_{\zeta : \|\zeta\|_p \leq 1} \left[\sum_{\ell=1}^{L} \zeta_\ell (u_\ell^T \theta)(v_\ell^T v) \right] \right\}$$

$$= \max_v \left\{ v^T Z^T \theta - \mathcal{L}^*(v) + \rho \left(\sum_{\ell=1}^{L} |u_\ell^T \theta|^q \cdot |v_\ell^T v|^q \right)^{1/q} \right\}.$$

where $1/q + 1/p = 1$.

The semi-infinite constraint (12.2.4) now reads

$$\max_{Z \in \mathcal{Z}} \mathcal{L}(Z^T\theta) \leq \tau$$

$$\Updownarrow \tag{12.3.2}$$

$$\min \left\{ \mathcal{L}_{\mathrm{pwc}}([Z^n]^T\theta, \kappa) : \kappa_\ell \geq \rho |u_\ell^T \theta|, \, 1 \leq \ell \leq L \right\} \leq \tau,$$

where $\mathcal{L}_{\mathrm{pwc}}(r, \kappa) : \mathbb{R}^m \times \mathbb{R}_+^L \to \mathbb{R}$ is the convex function

$$\mathcal{L}_{\mathrm{pwc}}(r, \kappa) = \max_v \left\{ v^T r - \mathcal{L}^*(v) + \left[\sum_{\ell=1}^{L} \kappa_\ell^q |v_\ell^T v|^q \right]^{1/q} \right\}$$

$$= \max_v \left\{ v^T r - \mathcal{L}^*(v) + \|V^T(\kappa)v\|_q \right\}$$

$$= \max_\xi \left\{ \mathcal{L}(r + V(\kappa)\xi) : \|\xi\|_p \leq 1 \right\},$$

where

$$V(\kappa) := [\kappa_1 v_1, \ldots, \kappa_L v_L] \in \mathbb{R}^{m \times L}.$$

We refer to $\mathcal{L}_{\mathrm{pwc}}$ as the pseudo worst-case loss function, as it depends on problem data through vectors v_ℓ, $\ell = 1, \ldots, l$. Observe how the above expression is an extension extends that one found for norm-bound additive models (12.2.18). Note that same as in the previous section, all we need in order to robustify an uncertain classification/regression problem is a tractable representation (or at least a safe tractable approximation) of the associated semi-infinite inequality (12.3.2) (or, which is the same, efficient computability of $\mathcal{L}_{\mathrm{pwc}}$, or at least an efficiently computable convex upper bound on this function).

12.3.3 Main Results

THE CASE OF $p = 1$

In this case (12.3.2) admits a tractable reformulation, cf. section 6.1. Indeed, when $p = 1$, we have $q = \infty$, and we can further express function $\mathcal{L}_{\mathrm{pwc}}$ as

$$
\begin{aligned}
\mathcal{L}_{\mathrm{pwc}}(r, \kappa) &= \max_v \left\{ v^T r - \mathcal{L}^*(v) + \max_{1 \le \ell \le L} \kappa_\ell |v_\ell^T v| \right\} \\
&= \max_{1 \le \ell \le L} \max_v \left\{ v^T r - \mathcal{L}^*(v) + \kappa_\ell |v_\ell^T v| \right\} \\
&= \max_{1 \le \ell \le L} \max_{|t| \le 1} \max_v \left\{ v^T r - \mathcal{L}^*(v) + t\kappa_\ell v_\ell^T v \right\} \\
&= \max_{1 \le \ell \le L} \max_{|t| \le 1} \mathcal{L}(r + t\kappa_\ell v_\ell) \\
&= \max_{1 \le \ell \le L} \max \left[\mathcal{L}(r - \kappa_\ell v_\ell), \mathcal{L}(r + \kappa_\ell v_\ell) \right].
\end{aligned}
$$

The semi-infinite inequality (12.3.2) now becomes an explicit convex inequality

$$
\max_{1 \le \ell \le L} \max \left[\mathcal{L}([Z^{\mathrm{n}}]^T \theta + \rho |u_\ell^T \theta| v_\ell), \mathcal{L}([Z^{\mathrm{n}}]^T \theta - \rho |u_\ell^T \theta| v_\ell) \right] \le \tau.
$$

THE CASE OF $p = 2$, HINGE LOSS

In contrast with the case of $p = 1$, the case $p = 2$ is computationally hard in general. Instead of proceeding in full generality, let us now specialize our problem to have the SVM (hinge) loss.

For the hinge loss, the function $\mathcal{L}_{\mathrm{pwc}}$ defined above reads

$$
\mathcal{L}_{\mathrm{pwc}}(r, \kappa) = \max_{0 \le v \le 1} \left\{ v^T r + \|V(\kappa)^T v\|_2 \right\}.
$$

Computing the above quantity is NP-hard. Writing

$$
\mathcal{L}_{\mathrm{pwc}}(r, \kappa) \le \inf_{\lambda > 0} \max_{0 \le v \le 1} \left[v^T r + \frac{\lambda}{2} + \frac{1}{2\lambda} \|V^T(\kappa)v\|_2^2 \right], \tag{12.3.3}
$$

we can now produce a safe tractable approximation, based on semidefinite relaxation, of (12.3.2), specifically, as follows. In (12.3.3), the domain of maximization in v can be represented by a system of quadratic inequalities $f_\ell(v) := v_\ell^2 - v_\ell \le 0$, $\ell = 1, \ldots, m$. With $\lambda > 0$ in (12.3.3) fixed, let nonnegative $\mu_1, \ldots, \mu_m$ be such that

$$
\forall v \in \mathbb{R}^m : v^T r + \frac{\lambda}{2} + \frac{1}{2\lambda} \|V(\kappa)^T v\|_2^2 \le \sum_{\ell=1}^m \mu_\ell f_\ell(v) + \tau; \tag{12.3.4}
$$

since the right hand side in this inequality on the box $0 \le v \le 1$ is $\le \tau$, (12.3.4) implies that $\mathcal{L}_{\mathrm{pwc}}(r, \kappa) \le \tau$. On the other hand, the condition (12.3.4) says merely that a certain quadratic form $v^T A v + 2 b^T v + c$ is everywhere nonnegative, which is the case if and only if $\begin{bmatrix} A & b \\ b^T & c \end{bmatrix} \succeq 0$. The latter condition, with A, b, c coming from (12.3.4), reads

$$\left[\begin{array}{c|c} \mathrm{Diag}\{\mu\} - \frac{1}{2\lambda} V(\kappa) V^T(\kappa) & -\frac{1}{2}[r+1] \\ \hline -\frac{1}{2}[r+1]^T & \tau - \frac{\lambda}{2} \end{array} \right] \succeq 0,$$

which, by the Schur Complement Lemma, is equivalent to

$$\left[\begin{array}{c|c|c} \mathrm{Diag}\{\mu\} & -\frac{1}{2}[r+1] & V(\kappa) \\ \hline -\frac{1}{2}[r+1]^T & \tau - \frac{\lambda}{2} & \\ \hline V^T(\kappa) & & 2\lambda I_L \end{array} \right] \ge 0.$$

Invoking (12.3.2), we arrive at the following result:

Proposition 12.3.1. The system of explicit convex constraints

$$\left[\begin{array}{c|c|c} \mathrm{Diag}\{\mu_1, \ldots, \mu_m\} & -\frac{1}{2}\left[[Z^\mathrm{n}]^T \theta + 1\right] & V(\kappa) \\ \hline -\frac{1}{2}\left[[Z^\mathrm{n}]^T \theta + 1\right]^T & \tau - \frac{\lambda}{2} & \\ \hline V^T(\kappa) & & 2\lambda I_L \end{array} \right] \succeq 0$$

$$\rho |u_\ell^T \theta| \le \kappa_\ell, \ 1 \le \ell \le L \tag{12.3.5}$$

in variables $\theta, \kappa, \mu, \lambda$ is a safe tractable approximation of the semi-infinite constraint (12.3.2) in the case of affine uncertainty (12.3.1) with $p = 2$.

THE CASE OF $p = \infty$, HINGE LOSS

For the hinge loss, the function $\mathcal{L}_{\mathrm{pwc}}$ defined above reads

$$\mathcal{L}_{\mathrm{pwc}}(r, \kappa) = \max_{0 \le v \le 1} \left\{ v^T r + \|V^T(\kappa) v\|_1 \right\}.$$

Again, computing this quantity is NP-hard, but we can bound it via the same scheme as in the case of $p = 2$. Specifically, given a positive vector $\lambda \in \mathbb{R}^L$, we have

$$\mathcal{L}_{\mathrm{pwc}}(r, \kappa) \le \max_{0 \le v \le 1} \left\{ v^T r + \sum_{\ell=1}^{L} \left[\frac{\lambda_\ell}{2} + \frac{\kappa_\ell^2 (v_\ell^T v)^2}{2\lambda_\ell} \right] \right\}.$$

Applying semidefinite relaxation in exactly the same fashion as when deriving Proposition 12.3.1, we arrive at the following result:

Proposition 12.3.2. The system of explicit convex constraints

$$\left[\begin{array}{c|c|c} \mathrm{Diag}\{\mu_1, \ldots, \mu_m\} & -\frac{1}{2}\left[[Z^\mathrm{n}]^T \theta + 1\right] & V(\kappa) \\ \hline -\frac{1}{2}\left[[Z^\mathrm{n}]^T \theta + 1\right]^T & \tau - \frac{1}{2}\sum_{\ell=1}^{L} \lambda_\ell & \\ \hline V^T(\kappa) & & 2\mathrm{Diag}\{\lambda\} \end{array} \right] \succeq 0$$

$$\rho |u_\ell^T \theta| \le \kappa_\ell, \ 1 \le \ell \le L \tag{12.3.6}$$

in variables $\theta, \kappa, \mu, \lambda$ is a safe tractable approximation of the semi-infinite constraint (12.3.2) in the case of affine uncertainty (12.3.1) with $p = \infty$.

12.3.4 Globalized Robust Counterparts

PROBLEM DEFINITION

In this section, we consider a variation on the approach taken up to now, based on the notion of Globalized Robust Counterparts developed in chapters 3, 11. Instead of the semi-infinite inequality (12.2.4), which is the Robust Counterpart of the uncertain constraint $\{\mathcal{L}(Z^T\theta) \leq \tau\}_{Z \in \mathcal{Z}}$, we address the GRC of this uncertain constraint, that is, the semi-infinite constraint

$$\forall Z : \mathcal{L}(Z^T\theta) \leq \tau + \alpha \mathrm{dist}(Z, \mathcal{Z}), \tag{12.3.7}$$

where $\alpha > 0$ is given. The interpretation of the constraint in the above is as follows. Our model now allows for perturbed matrices Z to take values outside their normal range $\mathcal{Z}$. However, we seek to control the resulting degradation in the loss function: the further away Z is from the set $\mathcal{Z}$, the more degradation we tolerate. The parameter α controls the "rate" of the degradation in the value of the loss function.

To illustrate this approach, we consider two examples of loss functions; in both these examples, $\mathcal{Z}$ is just the singleton of the nominal data.

EXAMPLE: $\mathcal{L}(r) = \|r\|_s$

Let the loss function be $\mathcal{L}(r) = \|r\|_s$. We assume also that $\mathcal{Z} = \{Z^\mathrm{n}\}$, and that the norm in the space $\mathbb{R}^{n \times m} \ni Z$ that underlies the distance in the right hand side of (12.3.7) is the largest singular value norm $\|\cdot\|$. With these assumptions, (12.3.7) becomes the semi-infinite inequality

$$\forall \Delta : \|(Z^\mathrm{n} + \Delta)^T\theta\|_s \leq \tau + \alpha\|\Delta\|.$$

Setting $\mathbf{Q} = \{r : \|r\|_s \leq \tau\}$, the latter relation is nothing but the semi-infinite constraint

$$\forall \Delta : \mathrm{dist}_{\|\cdot\|_s}((Z^\mathrm{n} + \Delta)^T\theta, \mathbf{Q}) \leq \alpha\|\Delta\|, \tag{12.3.8}$$

which, according to Definition 11.1.5, is indeed the GRC of the uncertain constraint

$$(Z^\mathrm{n} + \Delta)^T\theta \in \mathbf{Q},$$

the data perturbation being Δ, in the case when the normal range of the perturbation is the origin in $\mathbb{R}^{n \times m}$, the cone participating in the perturbation structure is the entire $\mathbb{R}^{n \times m}$, and the norms in the spaces where $\mathbf{Q}$ and Δ live are specified as $\|\cdot\|_s$ and the LSV norm $\|\cdot\|$, respectively. Invoking Proposition 11.3.3, the GRC (12.3.8) can be represented by the constraints

$$\begin{array}{ll} (a) & \|[Z^\mathrm{n}]^T\theta\|_s \leq \tau \\ (b) & \|\Delta^T\theta\|_s \leq \alpha \; \forall(\Delta : \|\Delta\| \leq 1). \end{array} \tag{12.3.9}$$

The semi-infinite constraint (b) is easy to process. Indeed, the image of the LSV ball $\{\|\Delta\| \leq 1\}$ under the mapping $\Delta \mapsto \Delta^T\theta$ is exactly the Euclidean ball $\{w \in$

$\mathbb{R}^m : \|w\|_2 \leq \|\theta\|_2\}$, and the maximum of the $\|\cdot\|_s$-norm on the latter ball is $\chi\|\theta\|_2$, where

$$\chi = \chi(m,s) = \begin{cases} m^{\frac{2-s}{2s}} & 1 \leq s \leq 2, \\ 1 & s \geq 2. \end{cases} \tag{12.3.10}$$

We have arrived at the following

Proposition 12.3.3. When $\mathcal{L}(\cdot) = \|\cdot\|_s$, the GRC (12.3.7) of the uncertain inequality $\mathcal{L}(Z^T\theta) \leq \tau$ can be represented by the system of explicit convex constraints

$$\begin{array}{ll} (a) & \|[Z^n]^T\theta\|_s \leq \tau \\ (b) & \chi(m,s)\|\theta\|_2 \leq \alpha \end{array} \tag{12.3.11}$$

in variables θ, with $\chi(m,s)$ given by (12.3.10).

EXAMPLE: THE HINGE LOSS FUNCTION

Now let $\mathcal{L}(r) = \sum_{i=1}^m [r_i]_+$. As above, we assume that $\mathcal{Z} = \{Z^n\}$ and that the norm underlying the distance in the right hand side of (12.3.7) is the LSV norm $\|\cdot\|$. Here (12.3.7) reads

$$\forall \Delta \in \mathbb{R}^{n \times m} : \sum_{i=1}^m \left[\left([Z^n]^T\theta + \Delta^T\theta\right)_i\right]_+ \leq \tau + \alpha\|\Delta\|,$$

or, which is the same

$$\forall \Delta \in \mathbb{R}^{n \times m} : \max \left[\sum_{i=1}^m \left[\left([Z^n]^T\theta + \Delta^T\theta\right)_i\right]_+ - \tau, 0\right] \leq \alpha\|\Delta\|. \tag{12.3.12}$$

Observing that $\max[\sum_i [r_i]_+ - \tau, 0]$ is nothing but the distance, induced by the $\|\cdot\|_1$-norm, from r to the closed convex set $\mathbf{Q} = \{r \in \mathbb{R}^m : \sum_i [r_i]_+ \leq \tau\}$, (12.3.12) is nothing but the semi-infinite constraint

$$\forall \Delta \in \mathbb{R}^{n \times m} : \text{dist}_{\|\cdot\|_1}([Z^n + \Delta]^T\theta, \mathbf{Q}) \leq \alpha\|\Delta\|;$$

same as in the previous example, this is nothing but the GRC, as defined in section 11.1, of the uncertain inclusion $[Z^n + \Delta]^T\theta \in \mathbf{Q}$, the normal range of the perturbation Δ being the origin in the space $\mathbb{R}^{n \times m}$, the cone in the perturbation space being the entire $\mathbb{R}^{n \times m}$, and the norms in the space where $\mathbf{Q}$ and Δ live being $\|\cdot\|_1$ and the LSV norm $\|\cdot\|$, respectively. Invoking Proposition 11.3.3, the GRC can be represented by the constraints

$$\begin{array}{ll} (a) & \sum_i \left[([Z^n]^T\theta)_i\right]_+ \leq \tau \\ (b) & \text{dist}_{\|\cdot\|_1}(\Delta^T\theta, \mathbb{R}_-^m) \leq \alpha \ \forall(\Delta : \|\Delta\| \leq 1) \end{array} \tag{12.3.13}$$

(we have taken into account that the recessive cone of $\mathbf{Q}$ is the nonpositive orthant $\mathbb{R}_-^m$). Applying exactly the same argument as in the previous Example, we can rewrite (b) equivalently as

$$\sqrt{m}\|\theta\|_2 \leq \alpha.$$

We have arrived at the following result:

Proposition 12.3.4. When $\mathcal{L}(r) = \sum_i [r_i]_+$, the GRC (12.3.7) of the uncertain inequality $\mathcal{L}(Z^T\theta) \leq \tau$ can be represented by the system of explicit convex constraints

$$
\begin{array}{ll}
(a) & \sum_{i=1}^{m} \left[([Z^{\mathrm{n}}]^T\theta)_i \right]_+ \leq \tau \\
(b) & \sqrt{m}\|\theta\|_2 \leq \alpha
\end{array}
\qquad (12.3.14)
$$

in variables θ.

We observe that the GRC approach leads directly to a loss function minimization problem with constraint on the size of the variable θ.

12.4 RANDOM AFFINE UNCERTAINTY MODELS

We now examine a variation on robust classification and regression problems, where the perturbation affecting the data is random.

12.4.1 Problem Formulations

RANDOM AFFINE UNCERTAINTY

As in the previous section, we assume that the perturbation enters affinely in the data. Precisely, the data matrix Z is assumed to be an affine function of a random vector $\zeta \in \mathbb{R}^l$:

$$
Z_\zeta = Z^{\mathrm{n}} + \sum_{\ell=1}^{L} \zeta_\ell u_\ell v_\ell^T, \qquad (12.4.1)
$$

where $u_\ell \in \mathbb{R}^n$, $v_\ell \in \mathbb{R}^m$ are given vectors. Here, we assume that its distribution of the random vector ζ is only known to belong to a given class Π of distributions on $\mathbb{R}^L$. (We will be more specific about our modeling assumptions shortly.)

ROBUST COUNTERPARTS

There are two kinds of robust counterparts that naturally arise in the context of loss function minimization with random perturbations. One processes the worst-case (over the class Π) expected loss, and the other handles the (worst-case) probability of the loss being larger than a target.

A first formulation, which we call *worst-case expected loss minimization,* focuses on robust, w.r.t. Π, upper bounding of the expected loss function, i.e., on the constraint

$$
\max_{\pi \in \Pi} \mathbf{E}_\pi \{\mathcal{L}(Z_\zeta^T\theta)\} \leq \tau, \qquad (12.4.2)
$$

where $\mathbf{E}_\pi$ denotes the expected value with respect to distribution $\pi \in \Pi$ of the random variable ζ.

The above approach has no concern on the "spread" of the values of the realized loss around its mean. This motivates us to study robust bounding, called *guaranteed loss-at-risk bounding,* of the loss function posed as the constraint

$$\max_{\pi \in \Pi} \text{Prob}_\pi \left\{ \mathcal{L}(Z_\xi^T \theta) \geq \tau \right\} \leq \epsilon, \tag{12.4.3}$$

where the "risk level" $\epsilon \in (0, 1)$ is given, and Prob_π is the probability taken with respect to π. If ϵ is set to be very small in the above for a given value of τ, then with high probability the loss is smaller than τ, irrespective of which distribution $\pi \in \Pi$ the random perturbation obeys. Of course, there is a trade-off between how small we can guarantee the loss to be (measured via τ), and the level of certainty as set by ϵ. Note that (12.4.3) is what was called an ambiguous chance constraint in chapters 2, 4, 10. The new, as compared with what we did in these chapters, aspect of the situation is that we are not speaking directly about linear/conic chance constraints: the randomly perturbed data are now inside a nonlinear loss function. However, we can still use the techniques from chapters 2, 4, 10 for straightforward processing of (12.4.3) in at least the following two cases (where the rank 1 matrices $u_\ell v_\ell^T$ can be replaced with arbitrary given matrices Z_ℓ):

• $\mathcal{L}(r) = \max_{1 \leq i \leq I} [a_i^T r + b_i]$ is a piecewise linear convex function given by the list of its linear pieces; what is applicable in this case, are the results of chapters 2, 4 on chance constrained scalar linear inequalities, and the results of section 10.4.1 on chance constrained systems of linear inequalities;

• $\mathcal{L}(r) = \|r\|_2$. This case is covered by the results of chapters 10 on chance constrained conic quadratic inequalities.

In contrast to guaranteed loss-at-risk bounding, the problem of worst-case expected loss minimization is completely new for us, and this is the problem we intend to focus on in the rest of this section.

12.4.2 Moment Constraints

THREE CLASSES OF DISTRIBUTIONS

We consider three specific sets of allowable distributions Π.

The first class, denoted by Π_2, is the set of distributions with given first- and second-order moments. Without loss of generality, we may assume that the mean is zero, and the covariance matrix is the identity.

The second class, denoted by Π_∞, is the set of distributions with given first moment, and given variances. Again, without loss of generality we assume that the mean is zero, and the variances are all equal to one. A norm-bound counterpart to this model is given by (12.3.1), with $p = \infty$.

The last class is defined as the set of distributions with zero mean and total variance equal to one. This class, denoted by Π_{tot} in the following, can be described as the stochastic counterpart to the norm-bound model (12.3.1), with $p = 2$.

To the class Π, we associate a subspace $\mathcal{Q}$ of matrices $Q \in \mathbf{S}^L$ such that for every $\pi \in \Pi$, and every $Q \in \mathcal{Q}$, we have

$$\mathbf{E}_\pi \{\zeta^T Q \zeta\} = \mathrm{Tr}(Q).$$

When $\Pi = \Pi_2$, the corresponding set $\mathcal{Q}$ is simply the entire space of $L \times L$ symmetric matrices; when $\Pi = \Pi_\infty$, it is the set of $L \times L$ diagonal matrices; and when $\Pi = \Pi_{\mathrm{tot}}$, it reduces to the set of scaled versions of the $L \times L$ identity matrix.

<small>WORST-CASE EXPECTED LOSS MINIMIZATION</small>

From the definition of the set $\mathcal{Q}$, for every $Q \in \mathcal{Q}$, $q \in \mathbb{R}^L$ and $t \in \mathbb{R}$, the condition

$$\forall \zeta \in \mathbb{R}^L \ : \ \begin{bmatrix} \zeta \\ 1 \end{bmatrix}^T \begin{bmatrix} Q & q \\ q^T & t \end{bmatrix} \begin{bmatrix} \zeta \\ 1 \end{bmatrix} \geq \mathcal{L}(Z_\zeta^T \theta) \tag{12.4.4}$$

implies that $\mathrm{Tr}(Q) + t$ is an upper bound on the worst-case expected loss. (This is readily seen by taking expectations on both sides of the above.) Thus, we may compute an upper bound on the worst-case expected loss by minimizing $\mathrm{Tr}(Q) + t$ subject to the condition above, with $Q \in \mathcal{Q}$, q, t the variables.

Standard results from duality theory for moment problems imply that the bound we compute this way is actually tight. Thus:

$$\max_{\pi \in \Pi} \mathbf{E}_\pi \left\{ \mathcal{L}(Z_\zeta^T \theta) \right\} = \min_{Q \in \mathcal{Q}, q, t} \{ \mathrm{Tr}(Q) + t : \ t, q, Q \text{ satisfy } (12.4.4) \}$$

$$= \min_{Q \in \mathcal{Q}, q, t} \left\{ \mathrm{Tr}(Q) + t : \forall (\zeta \in \mathbb{R}^L, v \in \mathbb{R}^m) : \right.$$

$$\left. \begin{bmatrix} \zeta \\ 1 \end{bmatrix}^T \begin{bmatrix} Q & q \\ q^T & t \end{bmatrix}^T \begin{bmatrix} \zeta \\ 1 \end{bmatrix} \geq v^T [Z^{\mathrm{n}}]^T \theta - \mathcal{L}^*(v) + \sum_{\ell=1}^L \zeta_\ell (u_\ell^T \theta)(v_\ell^T v) \right\}.$$

Eliminating the variable ζ from the quadratic constraint above leads to

$$\max_{\pi \in \Pi} \mathbf{E}_\pi \{ \mathcal{L}(Z_\zeta^T \theta) \} = \mathcal{L}_{\mathrm{pwc}}([Z^{\mathrm{n}}]^T \theta, U^T \theta),$$

where $U = [u_1, \ldots, u_L]$ and $\mathcal{L}_{\mathrm{pwc}} : \mathbb{R}^m \times \mathbb{R}^L \to \mathbb{R}$ is the pseudo worst-case loss function associated with our problem:

$$\mathcal{L}_{\mathrm{pwc}}(r, \kappa)$$
$$= \min_{Q \in \mathcal{Q}, q, t} \left\{ \mathrm{Tr}(Q) + t : \left[\begin{array}{c|c} Q & q - \frac{1}{2} V^T(\kappa) v \\ \hline q^T - \frac{1}{2} v^T V(\kappa) & t - v^T r + \mathcal{L}^*(v) \end{array} \right] \succeq 0 \ \forall v \right\}$$
$$= \min_{Q \in \mathcal{Q}, q} \left\{ \mathrm{Tr}(Q) \right.$$
$$\left. + \max_v \left[v^T r - \mathcal{L}^*(v) + (q - \tfrac{1}{2} V^T(\kappa) v)^T Q^{-1} (q - \tfrac{1}{2} V^T(\kappa) v) \right] : Q \succ 0 \right\}$$

,

with $V(\kappa) := [\kappa_1 v_1, \ldots, \kappa_L v_L]$, as before. Consequently, the constraint of interest (12.4.2) writes

$$\mathcal{L}_{\mathrm{pwc}}([Z^{\mathrm{n}}]^T \theta, U^T \theta) \leq \tau, \tag{12.4.5}$$

and all we need in order to handle it (or its safe approximation) efficiently is the ability to compute efficiently the convex function $\mathcal{L}_{\mathrm{pwc}}(\cdot, \cdot)$ (or a convex upper bound on this function). While computing $\mathcal{L}_{\text{pwc}}$ can be NP-hard in general, this task is tractable in a variety of cases, ranging from Huber regression to ℓ_1-regression [39].

EXAMPLE: HINGE LOSS

To illustrate the point, we focus on the case of SVMs (hinge loss). The pseudo worst-case loss function here reads

$$\mathcal{L}_{\text{pwc}}(r, \kappa) = \inf_{Q,q} \{\text{Tr}(Q)$$
$$+ \max_{v:0 \leq v \leq 1} \{v^T r + (q - \tfrac{1}{2} V^T(\kappa)v)^T Q^{-1}(q - \tfrac{1}{2} V^T(\kappa)v)\} : Q \in \mathcal{Q}, Q \succ 0\}.$$

The inner maximum is NP-hard to compute in general. However, we can build its efficiently computable upper bound via semidefinite relaxation, completely similar to what we did in section 12.3.3 when deriving (12.3.5) and (12.3.6). The result now is as follows:

Proposition 12.4.1. The system of explicit convex constraints

$$\left[\begin{array}{c|c|c} \text{Diag}\{\mu_1, \ldots, \mu_m\} & -\tfrac{1}{2}[Z^n]^T\theta & \tfrac{1}{2}V(\kappa) \\ \hline -\tfrac{1}{2}\theta^T Z^n & \tau & -q^T \\ \hline \tfrac{1}{2}V^T(\kappa) & -q & Q \end{array}\right] \succeq 0$$
$$\kappa_\ell = u_\ell^T \theta, \ \ell = 1, \ldots, L, \ Q \in \mathcal{Q}$$
(12.4.6)

in variables $\tau, \theta, Q, q, \kappa, \mu$ is a safe tractable approximation of the constraint of interest (12.4.2).

12.4.3 Bernstein Approximation for Independent Perturbations

We now illustrate how the Bernstein bound from chapter 4 can be used in the case when the random perturbations $\zeta_1, \ldots, \zeta_L$ in (12.4.1) are independent. In fact, we use below a slightly more general perturbation model

$$Z_\zeta = Z^n + \sum_{\ell=1}^{L} \zeta_\ell Z_\ell;$$
(12.4.7)

the difference with (12.4.1) is that now we do not require Z_ℓ to be of rank 1. We assume that the loss function satisfies assumption **L**, see p. 311. Moreover, we assume that the class of allowable distributions Π of the perturbation vector $\zeta \in \mathbb{R}^L$ is of the form $\Pi = \mathcal{P}_1 \times \ldots \times \mathcal{P}_L$, where $\mathcal{P}_\ell$ is a given set of distributions on the real axis, $1 \leq \ell \leq L$.

12.4.3.1 Worst-case expected hinge loss

Consider the case when $\mathcal{L}(r) = \sum_{i=1}^{m}[r_i]_+$ and we want to bound the corresponding expected loss from above, in a fashion that is robust with respect to the distribution

$\pi \in \Pi$. Thus, we are interested in a safe tractable approximation of the inequality

$$\sup_{\pi \in \mathcal{P}} \mathbf{E} \left\{ \sum_{i=1}^{m} \left[e_i^T [Z^{\mathrm{n}}]^T \theta + \sum_{\ell=1}^{L} \zeta_\ell e_i^T Z_\ell^T \theta \right]_+ \right\} \leq \tau, \tag{12.4.8}$$

Let $a_{i0}(\theta) := e_i^T [Z^{\mathrm{n}}]^T \theta$, $a_{i\ell}(\theta) := e_i^T Z_\ell^T \theta$, $i = 1, \ldots, m$, $\ell = 1, \ldots, L$, and let

$$\xi_{i,\theta} := e_i^T [Z^{\mathrm{n}} + \sum_{\ell=1}^{L} \zeta_\ell Z^\ell]^T \theta = a_{i0}(\theta) + \sum_{\ell=1}^{L} \zeta_\ell a_{i\ell}(\theta), \quad i = 1, \ldots, m.$$

so that (12.4.8) reads

$$\sup_{\pi \in \mathcal{P}} \mathbf{E} \left\{ \sum_{i=1}^{m} [\xi_{i,\theta}]_+ \right\} \leq \tau.$$

We now apply a sort of Bernstein approximation. Specifically, we have $\exp\{s\} \geq$ e $\max[s, 0]$ for all s. Thus, whenever $\beta > 0$, we have $\beta \exp\{s/\beta\} \geq$ e $\max[s, 0]$ for all s. Thus, whenever $\alpha = (\alpha_1, \ldots, \alpha_m) > 0$, we have $\max[\xi_{i,\theta}, 0] \leq e^{-1} \alpha_i \exp\{\xi_{i,\theta}/\alpha_i\}$. Exploiting the fact that a distribution $\pi \in \Pi$ is a product $\pi = \pi_1 \times \ldots \times \pi_L$, we obtain that for every $i = 1, \ldots, m$:

$$\mathbf{E}_{\zeta \sim \pi} \{\max[\xi_{i,\theta}, 0]\} \leq G_\pi^i(\alpha_i, \theta) := e^{-1} \alpha_i \mathbf{E}_{\zeta \sim \pi} \{\exp\{\xi_{i,\theta}/\alpha_i\}\}$$
$$= e^{-1} \alpha_i \exp\{a_{i0}(\theta)/\alpha_i\} \prod_{\ell=1}^{L} G_\pi^{i\ell}(\alpha_i, \theta),$$

where

$$G_\pi^{i\ell}(\alpha_i, \theta) := \mathbf{E}_{\zeta_\ell \sim P_\ell} \{\exp\{a_{i\ell}(\theta)\zeta_\ell/\alpha_i\}\} .$$

The function $F_\pi(w) = \mathbf{E}_{\zeta \sim \pi} \{\exp\{w_0 + \sum_{\ell=1}^{L} w_\ell \zeta_\ell\}\}$ is convex in w, whence the function $H_\pi(\alpha, w) = \alpha F_\pi(w/\alpha)$ is convex in $(\alpha > 0, w)$. It follows that $G_\pi^i(\alpha_i, \theta)$ is convex in $(\alpha_i > 0, \theta)$, since this function is obtained from H_π by affine substitution of argument (look what $\xi_{i,\theta}$ is and note that $a_{i\ell}(\theta)$, $0 \leq \ell \leq L$, are linear in θ). It follows that the function

$$G^i(\alpha_i, \theta) = \sup_{\pi \in \Pi} G_\pi^i(\alpha_i, \theta) = e^{-1} \alpha_i \exp\{a_{i0}(\theta)/\alpha_i\} \prod_{\ell=1}^{L} G_{i\ell}(\alpha_i, \theta),$$
$$G_{i\ell}(\alpha_i, \theta) := \sup_{\pi_\ell \in \mathcal{P}_\ell} \mathbf{E}_{\zeta_\ell \sim P_\ell} \{\exp\{a_{i\ell}(\theta)\zeta_\ell/\alpha_i\}\}$$

is convex in (α_i, θ) when $\alpha_i > 0$ as well.

The approach leads to a tractable approximation in the case when we can compute the functions $G_{i\ell}(\alpha_i, \theta)$ (as is the case in examples presented in section 2.4.2), since then $G^i(\alpha_i, \theta)$ is efficiently computable. Whenever this is the case, the tractable convex constraint

$$\inf_{\alpha > 0} \sum_{i=1}^{m} G^i(\alpha_i, \theta) \leq \tau \tag{12.4.9}$$

is a safe tractable approximation of (12.4.8). Note that this is a "double conservative" bounding, one source of conservatism is the Bernstein approximation per se, and another source of conservatism is that this approximation is applied termwise, we just sum up the optimal Bernstein bounds for the quantities $\sup_{\pi \in \Pi} \mathbf{E}_{\zeta \sim \pi} \{[\xi_{i,\theta}]_+\}$, $1 \le i \le m$.

12.5 EXERCISES

Exercise 12.1. [Implementation errors] Consider the separability condition (12.1.1). Now assume that the classifier vector w is not implemented exactly, but with some relative error δw, which we assume to be such that $\|\delta w\|_\infty \le \rho \|w\|_2$, where $\rho \ge 0$. Formulate the corresponding robust separability condition. How would you find the classifier that is maximally robust with respect to such implementation errors?

Exercise 12.2. Show that the sets defined as COM satisfy condition **A**, for appropriate choice of the functions ϕ and ψ.

Exercise 12.3. [Label uncertainty] We start with the minimization problem involving the hinge loss function (see (12.1.4), (12.1.5)):

$$\min_{w,b} \sum_{i=1}^{m} [1 - y_i(w^T x_i + b)]_+,$$

where the notation is the same as in Section 12.1.1. We consider the case when the data points x_i, $i = 1, \ldots, m$, are exactly known, but the label vector $y \in \{-1, 1\}^m$ is only partially known.

i) First we consider the case when a subset of the labels is completely unknown. (This situation is sometimes referred to as *semi-supervised* learning.) To model this situation, we assume that there is a partition of $\{1, \ldots, m\}$ into two disjoint subsets $\mathcal{I}, \mathcal{J}$, with $\mathcal{I}$ (resp. $\mathcal{J}$) the set of indices corresponding to known (resp. unknown) labels. Formulate the corresponding robust counterpart as a linear optimization problem.

ii) In some situations, it is possible that some labels are given the wrong sign. We assume that a subset $\mathcal{J} \subseteq \{1, \ldots, m\}$ of cardinality k has the corresponding labels switched in sign. Again, formulate the corresponding robust counterpart as a linear optimization problem.

Exercise 12.4. [Robust SVM with boolean data] Many classification problems, such as those involving co-occurrence data in text documents, involve boolean data. The classical SVM implicitly assumes (say) spherical uncertainty, which may not be consistent with the fact that the process generating the data is such that it is boolean. In this exercise, we explore the idea of robust SVM with uncertainty models that preserve the boolean nature of the perturbed data. We thus consider the problem described in Section 12.1, where the data matrix $X = [x_1, \ldots, x_m]$ is boolean. Throughout, we assume that the perturbation affecting the data points is measurement-wise.

i) In a first approach, we assume that each measurement is subject to an additive perturbation: $x_i \to x_i + \delta_i$, where $\delta_i \in \{-1, 0, 1\}^n$, and $\|\delta_i\|_1 \le k$, where k is given. Hence the number of changes in the data is constrained by k. Note that our model allows for changes from 0 to -1, or 1 to 2, which is not consistent with our boolean assumption, and might lead to conservative results. Form the robust counterpart for this uncertainty model.

ii) A more realistic model, which does preserve the boolean nature of the perturbed data, involves allowing for "flips" in the data, but constraining the total number of flips per measurement. Thus, we impose $\delta_i \in \{-x_i, -x_i + \mathbf{1}\}$ for each i. Indeed, this means that if $x_i(j) = 0$, then $\delta_i(j)$ can only take the value 0 or 1; if $x_i(j) = 1$, then $\delta_i(j)$ can only take the value 0 or -1. We still constrain the total number of flips per measurement, with the constraint $\|\delta_i\|_1 \leq k$, where k is given. Again, form the robust counterpart for this uncertainty model.

Exercise 12.5. [Uncertainty in a future data point] In this problem, we consider a classification problem with a hinge loss function, involving m data points $x_1, \ldots, x_m$ and their associated label $y_1, \ldots, y_m$. We now add a new data point x_{m+1} and label y_{m+1} to the training set, which are not completely known. Precisely, all we know about the new data pair (x_{m+1}, y_{m+1}) is that x_{m+1} will be close to *one* of the previous points x_i, $i = 1, \ldots, m$, and will have the same label. We assume further that $\|x - x_i\|_2 \leq \rho$ for some $i = 1, \ldots, m$. Our goal is to design a classifier that is robust with respect to uncertainty in the new data point. Express the corresponding robust counterpart as a second-order cone optimization problem.

12.6 NOTES AND REMARKS

NR 12.1. A number of authors have studied machine learning problems from the point of view of robust optimization, mostly with a focus on supervised learning. Early work focused on least-squares regression [49].

On the topic of robust classification, prior work has focused mostly on measurement-wise uncertainty models. An approach to binary classification based on modeling each class as partially known distributions was introduced in [54]. Support vector machines with interval uncertainty (and their connections to sparse classification) are studied in [51], while the case with ellipsoidal uncertainty on the data points has been introduced and applied in a biology context in [29]. The approach has been further developed in [105]. Related work includes [114].

To our knowledge, the results in this chapter pertaining to uncertainties that couple measurements are new. Caramanis and co-authors [116] and Bertsimas and Fertis [28] both independently developed a theory that recovers some of these results.

NR 12.2. The term "robust statistics" is generally used to refer to methods that nicely handle (reject) outliers in data. A standard reference on the topic is Huber's book [65]. As said in the Preface, a precise and rigorous connection with robust optimization remains to be made. It is our belief that the two approaches are radically different, even contradictory, in nature: rejecting outliers is akin to discarding data points that yield large values of the loss function, while robust optimization takes into account all the data points and focuses on those that do result in large losses.

To make this discussion a little more precise, consider the case of a classification problem with hinge loss, with data points $x_i \in \mathbb{R}^n$, and label $y_i \in \{-1, 1\}$, $i = 1, \ldots, m$. In a robust statistics approach, we would look for a classifier that only takes into account the "best" k ($k \leq m$) points, from the standpoint of the considered hinge loss. The problem can be formulated as

$$\min_{w,b} \min_{\delta \in \mathcal{D}} \sum_{i=1}^{m} \delta_i [1 - y_i(w^T x_i + b)]_+,$$

where $\mathcal{D} = \{\delta \in \{0, 1\}^m : \sum_{i=1}^{m} \delta_i = k\}$. The above amounts to find a classifier that is optimal for the best k points. This is a non-convex problem.

In contrast, a robust optimization approach would seek a classifier that is optimal for the *worst* k points:

$$\min_{w,b} \max_{\delta \in \mathcal{D}} \sum_{i=1}^{m} \delta_i [1 - y_i(w^T x_i + b)]_+,$$

which is a convex problem.

Part III

Robust Multi-Stage Optimization

Chapter Thirteen
Robust Markov Decision Processes

This chapter is devoted to a robust dynamical decision making problem involving a finite-state, finite-action stochastic system. The system's dynamics is described by state transition probability distributions, which we assume to be uncertain and varying in a given uncertainty set. At each time period, nature is playing against the decision-maker, by picking at will transition distributions within their ranges. The goal of the robust decision making is to minimize the worst-case expected value of a given cost function, where "worst-case" is with respect to the considered class of policies of nature. We show that when the cardinalities of the (finite!) state and action spaces are moderate, the problem can be solved in a computationally tractable fashion using an extension to Bellman's famous Dynamic Programming algorithm, which requires the solution of a convex optimization problem at each step. We illustrate the approach on a path planning problem arising in aircraft routing through random weather conditions.

13.1 MARKOV DECISION PROCESSES

13.1.1 The Nominal Control Problems

Markov decision processes (MDPs) are used to model the random behavior of a dynamical system, based on the assumption that the state of the system, as well as the possible control actions, belong to given finite collections. Due to the great versatility of these models, MDPs are increasingly ubiquitous in applications, including finance, system biology, communications engineering, and so on.

MDP models are described in terms of state transition probabilities, which inform us on the probabilities of transition from a given state to another, conditional on a particular control action. The goal of the corresponding decision making problem, which will be our nominal problem in this chapter, is to minimize the expected value of a given cost function, which is itself described by a finite set of values assigned to each state-control pair. The nominal problem comes in two flavors, depending on whether the horizon (decision span) of the problem is finite or infinite.

Let us define the finite-horizon nominal problem more precisely. We consider a Markov decision process with finite state and finite action sets, and finite decision horizon $\mathcal{T} = \{0, 1, 2, \ldots, N-1\}$. At each time period, the system occupies a state $i \in \mathcal{X}$, where $n = |\mathcal{X}|$ is finite, and a decision maker is allowed to choose an action a deterministically from a finite set of allowable actions $\mathcal{A} = \{a_1, \ldots, a_m\}$ (for notational simplicity we assume that $\mathcal{A}$ is not state-dependent). The states make random, Markovian transitions according to a collection of (possibly time-dependent) transition probability distributions $\tau := \{p_{ti}^a : a \in \mathcal{A}, t \in \mathcal{T}, i \in \mathcal{X}\}$, where for every $a \in \mathcal{A}$, $t \in \mathcal{T}$, $i \in \mathcal{X}$ the vector $p_{ti}^a = [p_{ti}^a(1); \ldots; p_{ti}^a(n)] \in \mathbb{R}^n$ contains the probabilities $p_{ti}^a(j)$ of transition under control action a at stage t from state $i \in \mathcal{X}$ to state $j \in \mathcal{X}$. We refer to a collection τ of the outlined structure as to *nature's policy*. In addition, we assume that the probability distribution q_0 of the states at time $t = 0$ is given. We denote by $u = (u_0(\cdot), \ldots, u_{N-1}(\cdot))$ a generic *control policy*, where $u_t(\cdot) : \mathcal{X} \to \mathcal{A}$ is the decision rule at time $t \in \mathcal{T}$, so that the control action at this time, the state of the system being $i \in \mathcal{X}$, is $u_t(i)$. We denote by $\Pi = \mathcal{A}^{nN}$ the corresponding strategy space.

We denote by $c_t(i, a)$ the cost corresponding to state $i \in \mathcal{X}$ and action $a \in \mathcal{A}$ at time $t \in \mathcal{T}$, and by $c_N(i)$ the cost function at the terminal stage. We assume that $c_t(i, a)$ is finite for every $t \in \mathcal{T}$, $i \in \mathcal{X}$ and $a \in \mathcal{A}$.

We are ready to define $C_N(u, \tau)$, the expected total cost under control policy u and nature's policy $\tau = \{p_{ti}^a : a \in \mathcal{A}, t \in \mathcal{T}, i \in \mathcal{X}\}$:

$$C_N(u, \tau) := \mathbf{E}\left(\sum_{t=0}^{N-1} c_t(i_t, u_t(i_t)) + c_N(i_N)\right), \qquad (13.1.1)$$

with i_t the (random) state at time t corresponding to u, τ.

For a given collection τ, and a given initial state distribution vector, q_0, we define the *finite-horizon nominal problem* by

$$\phi_N(\Pi, \tau) := \min_{u \in \Pi} C_N(u, \tau). \qquad (13.1.2)$$

A special case of interest is when the expected total cost function bears the form (13.1.1), where the terminal cost is zero, and $c_t(i, a) = \nu^t c(i, a)$, with $c(i, a)$ now a time-invariant cost function, which we assume finite everywhere, and $\nu \in (0, 1)$ is a discount factor. We refer to this function as the *discounted cost function*. With such a function, one can pose the corresponding *infinite-horizon nominal problem*, where $N \to \infty$ in (13.1.1).

Example 13.1.1. (Aircraft path planning problem) We consider the problem of routing an aircraft whose path is obstructed by stochastic obstacles, representing storms or other severe weather disturbances. The goal of the stochastic decision-making problem is to route (say, one) aircraft from a given originating city, in order to minimize the expected value of the fuel consumption required to reach a given destination. To model this problem as an MDP, we first discretize the entire airspace, using a simple two-dimensional grid (ignoring the third dimension to simplify). Some nodes in the grid correspond to positions of likely storms or obstacles. The state vector comprises the current position of the aircraft on the grid, as well as the current states (from severe, mild, to absent) of each storm. The action in the MDP corresponds to the choice of nodes to fly towards, from any given node. There are k obstacles, associated to a Markov chain with a $3^k \times 3^k$

transition matrix, according to the number of possible states of an obstacle. The size of the transition matrix data for the routing problem is thus of the order $3^k N$, where N is the number of nodes in the grid.

13.1.2 Solving the Nominal Problems

According to the above setup, for a given control policy encoded in $u = (u_0(\cdot), \ldots, u_{N-1}(\cdot)) \in \Pi$ and a given nature policy $\tau = \{p_{ti}^a : a \in \mathcal{A}, t \in \mathcal{T}, i \in \mathcal{X}\}$, the system's random behavior is described by the deterministic system

$$q_{t+1}(j) = \sum_{i \in \mathcal{X}} p_{ti}^{u_t(i)}(j)q_t(i), \quad j \in \mathcal{X}, t \in \mathcal{T}, \tag{13.1.3}$$

where $q_t = [q_t(1); \ldots; q_t(n)] \in \mathbb{R}^n$ is the distribution, associated with u, τ, of the states at time $t \in \mathcal{T}$, and q_0 is the initial distribution of states. The total expected cost then is

$$C_N(u, \tau) = \sum_{t \in \mathcal{T}} \sum_{i \in \mathcal{X}} q_t(i)c_t(i, u_t(i)) + q_N^T c_N. \tag{13.1.4}$$

The following theorem shows how to compute the expected cost (13.1.1), for a given control policy $u = (u_0(\cdot), \ldots, u_{N-1}(\cdot))$.

Theorem 13.1.2. [LO representation of finite-horizon expected cost] The expected cost (13.1.1), the control policy being $u = \{u_t(\cdot)\}_{t \in T}$ and the nature policy being $\tau = \{p_{ti}^a : a \in \mathcal{A}, t \in \mathcal{T}, i \in \mathcal{X}\}$, is the optimal value in the Linear Optimization problem

$$\phi_N(u, \tau) = \max_{v_0, \ldots, v_{N-1}} \left\{ q_0^T v_0 : v_t(i) \le c_t(i, u_t(i)) + \sum_j p_{ti}^{u_t(i)}(j)v_{t+1}(j), \right. \\ \left. i \in \mathcal{X}, t \in \mathcal{T} \right\}, \tag{13.1.5}$$

where $v_N = c_N$.

In the above, the vectors v_t^* optimal for the LO problem (13.1.5) represent the expected costs-to-go from a particular state and time.

The result can be leveraged to a LO solution to the finite-horizon nominal problem.

Theorem 13.1.3. [LO representation of finite-horizon nominal problem] For a fixed nature policy $\tau = \{p_{ti}^a : a \in \mathcal{A}, t \in \mathcal{T}, i \in \mathcal{X}\}$, the finite-horizon nominal problem (13.1.2) can be solved as the linear optimization problem

$$\phi_N(\Pi, \tau) = \max_{v_0, \ldots, v_{N-1}} \left\{ q_0^T v_0 : v_t(i) \le \min_{a \in \mathcal{A}} \left[c_t(i, a) + \sum_j p_{ti}^{u_t(i)}(j)v_{t+1}(j) \right] \right. \\ \left. \forall i \in \mathcal{X}, t \in \mathcal{T} \right\}, \tag{13.1.6}$$

where $v_t(i)$ is i-th coordinate of v_t. A corresponding optimal control policy $u^* = (u_0^*(\cdot), \ldots, u_{N-1}^*(\cdot))$ is obtained by setting

$$u_t^*(i) \in \underset{a \in \mathcal{A}}{\operatorname{argmin}} \left\{ c_t(i,a) + \sum_j p_{ti}^a(j) v_{t+1}(j) \right\}, i \in \mathcal{X}, t \in \mathcal{T}, \qquad (13.1.7)$$

where the vectors $v_0, \ldots, v_{N-1}$ are optimal for the LO (13.1.6).

Here, the entries in the vectors v_t^* optimal for the LO (13.1.6) can be interpreted as optimal expected costs-to-go from a particular state and time, and are also referred to collectively as the value function. The celebrated Dynamic Programming algorithm, due to Bellman (1953), is based on a recursion providing a solution to the LO (13.1.6) .

Theorem 13.1.4. [Dynamic Programming algorithm] The nominal problem can be solved via the backward recursion

$$v_t(i) = \min_{a \in \mathcal{A}} \left\{ c_t(i,a) + \sum_j p_{ti}^a(j) v_{t+1}(j) \right\}, i \in \mathcal{X}, t = N-1, N-2, \ldots, 0, \quad (13.1.8)$$

initiated with $v_N = c_N$. Here $v_t(i)$ is the optimal value function in state i at stage t. The corresponding optimal control policy is obtained via (13.1.7).

Bellman's Dynamic Programming algorithm complexity is $O(nmN)$ arithmetic operations in the finite-horizon case.

13.1.3 The Curse of Uncertainty

For systems with moderate number of state and controls, the Bellman recursion provides an attractive and elegant solution, which has earned the algorithm a place in the pantheon of the top algorithms of the twentieth century. The application of the Dynamic Programming algorithm remains challenging for larger-scale systems, due to the famous "curse of dimensionality": in many applications, such as illustrated by Example 13.1.1, the states correspond to a discretization of several real-valued variables, representing position for example, and their number grows exponentially with the number of such real-valued variables. This curse is well-studied and makes the MDP field a very active area of research.

In this chapter, we explore a different "curse" associated with MDP models: the curse of uncertainty. As we will see, this curse may be present, but, in contrast with its earlier cousin the curse of dimensionality, it can be cured in a computationally tractable fashion.

The curse of uncertainty refers to the fact that optimal solutions to Markov decision problems may be very sensitive with respect to the state transition probabilities. In many practical problems, the estimation of these probabilities is far from being accurate, and represents a tremendous challenge, often complicated by the time-varying (non-stationary) nature of the system. Hence, estimation errors arc limiting factors in applying Markov decision processes to real-world problems. This motivates us to examine Robust Counterparts to problem (13.1.2), and gives us an example of a problem originally formulated as a stochastic control problem, with an added layer of uncertainty, in the state distributions.

13.2 THE ROBUST MDP PROBLEMS

In this section, we address the curse of uncertainty by assuming that the second player, which we refer to as nature, is allowed to change the transition probability distributions within prescribed bounds, and seek a control policy that is robust against the action of nature.

13.2.1 Uncertainty Models

We assume that for each action a, time t, and state i the corresponding transition probability distribution p_{ti}^a chosen by nature is only known to lie in some given subset $\mathcal{P}_{ti}^a$ of the set of probability distributions on $\mathcal{X}$; the latter is nothing but the standard simplex $\Delta_n = \{p = [p(1); \dots; p(n)] \in \mathbb{R}_+^n : \sum_j p(j) = 1\}$. Loosely speaking, we can think of the sets $\mathcal{P}_{ti}^a$ as *sets of confidence* for the transition probability distributions. Let us provide some specific examples of uncertainty sets $\mathcal{P}_{ti}^a$; when presenting these examples, we skip the indices a, t, i.

(a) The *scenario model* involves a finite collection of distributions:

$$\mathcal{P} = \mathrm{Conv}\{p^1, \dots, p^k\},$$

where $p^s \in \Delta_n$, $s = 1, \dots, k$, are given. This is the case when $\mathcal{P} \subset \Delta_n$ is a polytope given by its vertices.

(b) The *interval model* assumes

$$\mathcal{P} = \left\{ p : \underline{p} \le p \le \overline{p}, \sum_j p(j) = 1 \right\},$$

where $\overline{p}, \underline{p}$ are given non-negative vectors in $\mathbb{R}^n$ (whose elements do not necessarily sum to one), with $\overline{p} \ge \underline{p}$.

(c) The *likelihood model* has the form

$$\mathcal{P} = \mathcal{P}(\rho) := \left\{ p \in \Delta_n : L(p) := \sum_{i=1}^n q(i) \ln(q(i)/p(i)) \le \rho \right\}, \qquad (13.2.1)$$

where $q \in \Delta_n$ is a fixed reference distribution (for instance, the maximum likelihood estimate of the "true" transition distribution, the current time, state, and control action being given) and $\rho \ge 0$ is the "uncertainty level." Note that when $\rho = 0$, the set $\mathcal{P}(\rho)$ becomes the singleton $\{q\}$. A slightly more general model in this category is the Maximum A Posteriori (MAP) model, with $L(p)$ replaced by $L(p) - \ln(g_{\mathrm{prior}}(p))$, where g_{prior} refers to an a priori density function of the parameter vector p. It is customary to choose the prior to be a Dirichlet distribution, the density of which is of the form

$$g_{\text{prior}}(p) = K \cdot \prod_j [p(j)]^{\alpha_j - 1},$$

where $\alpha_j \geq 1$ are given, and K is a normalizing constant. Choosing $\alpha_j = 1$, for all j, we recover the "non-informative prior," which is the uniform distribution on the n-dimensional simplex, in which case the MAP model, up to a shift in ρ, reduces to the likelihood model.

(d) The *entropy model* is

$$\mathcal{P}(\rho) := \left\{ p \in \Delta_n : D(p\|q) := \sum_{j=1}^n p(j) \ln(p(j)/q(j)) \leq \rho \right\},$$

so that $D(p\|q)$ is the Kullback-Leibler divergence between distribution p and a reference distribution $q \in \Delta_n$. Here again $\rho \geq 0$ is the uncertainty level. This model mirrors the likelihood model; the latter is obtained from the former by exchanging the roles of p and q in the expression for the divergence.

(e) The *ellipsoidal model* has the form

$$\mathcal{P}(\rho) = \left\{ p \in \Delta_n : (p-q)^T H (p-q) \leq \rho^2 \right\},$$

where $q \in \Delta_n$ and $H \succ 0$ are given.

Example 13.2.1. [**Building uncertainty models**] Uncertainty models can be derived from a controlled experiment starting from a state $i \in \mathcal{X}$, in which we record the number of transitions between state pairs. This way, we obtain vectors of empirical transition frequencies q_{ti}^a for each $a \in \mathcal{A}$. It turns out that these vectors are the maximum-likelihood estimates of the true transition probability distributions. The corresponding likelihood model is (13.2.1), with ρ being the parameter controlling the uncertainty set's size. Ellipsoidal models can then be derived by a second-order approximation to the log-likelihood function, while interval models can be obtained by projections of the likelihood uncertainty set on the coordinate axes.

The above models describe the bounds that are imposed on the transition probability distributions. To fully describe the uncertainty model, we need to further specify the ways the second player, nature, can change these distributions, dynamically over time, or otherwise. In this regard, two uncertainty models are possible, leading to two possible forms of finite-horizon robust control problems.

In a first model, referred to as the *stationary uncertainty* model, the transition distributions p_{ti}^a chosen by nature are independent of t, and thus are represented by a two-index collection $\{p_i^a : a \in \mathcal{A}, i \in \mathcal{X}\}$ chosen by the nature from a given uncertainty set. In the second model, which we refer to as the *time-varying uncertainty* model, we are given a collection $\{\mathcal{P}_{ti}^a : a \in \mathcal{A}, t \in \mathcal{T}, i \in \mathcal{X}\}$ of subsets in Δ_n, and nature can "at will" choose, for every time instant t, current state i and current action a, the transition probability distribution p_{ti}^a from the set $\mathcal{P}_{ti}^a$. For technical reasons, we assume the sets $\mathcal{P}_{ti}^a$ nonempty and closed. Each problem leads to a game between the decision maker and nature, where the decision maker seeks to minimize the maximum expected cost, and with nature being the maximizing player.

13.2.2 Robust Counterparts

Equipped with uncertainty models, we are ready to define our robust control problems more formally.

As above, a *policy of nature* is a specific collection $\tau = \{p_{ti}^a : a \in \mathcal{A}, t \in \mathcal{T}, i \in \mathcal{X}\}$ of time-dependent transition probability distributions chosen by nature. In the non-stationary model, the set of admissible policies of nature is the entire direct product $\prod_{a \in \mathcal{A}, t \in \mathcal{T}, i \in \mathcal{X}} \mathcal{P}_{ti}^a$ of given sets $\mathcal{P}_{ti}^a \subset \Delta_n$. In the case of the stationary model, these policies are further restricted to have p_{ti}^a independent of t (in which case it makes sense to assume that the sets $\mathcal{P}_{ti}^a$ also are independent of t). The stationary uncertainty model leads to the Robust Counterpart

$$\phi_N(\Pi) := \min_{u \in \Pi} \max_{\{p_i^a \in \mathcal{P}_i^a : a \in \mathcal{A}, i \in \mathcal{X}\}} C_N(p, \{p_i^a\}). \tag{13.2.2}$$

In contrast, the time-varying uncertainty model leads to a relaxed version of the above:

$$\phi_N(\Pi) \leq \psi_N(\Pi) := \min_{u \in \Pi} \max_{\{p_{ti}^a \in \mathcal{P}_{ti}^a : a \in \mathcal{A}, t \in \mathcal{T}, i \in \mathcal{X}\}} C_N(u, \{p_{ti}^a\}). \tag{13.2.3}$$

The first model is attractive for statistical reasons, as it is much easier to develop statistically accurate sets of confidence when the underlying process is time-invariant. Unfortunately, the resulting game (13.2.2) seems to be hard to solve. The second model is attractive as one can solve the corresponding game (13.2.3) using a variant of the Dynamic Programming algorithm seen later, but we are left with a difficult task, that of estimating a meaningful set of confidence for the time-varying distributions p_{ti}^a.

In the finite-horizon case, we would like to use the first model of uncertainty, where the transition probability distributions are time-invariant. This would allow us to describe uncertainty in a statistically accurate way using likelihood or entropy bounds. However, the associated Robust Counterpart problem "as it is" seems to be too difficult computationally, and we pass to its safe approximation that is common in Control, specifically, extend the time-invariant uncertainty to a time-varying one. This means that we solve the second problem (13.2.3) as a safe approximation of the problem of actual interest (13.2.2), using uncertainty sets $\mathcal{P}_{ti}^a \equiv \mathcal{P}_i^a$ derived from the time-invariance assumption on the transition probabilities.

13.3 THE ROBUST BELLMAN RECURSION ON FINITE HORIZON

We consider the finite-horizon robust control problem defined in (13.2.3).

The following theorem extends Bellman's Dynamic Programming to the Robust Counterpart (13.2.3).

Theorem 13.3.1. [Robust Dynamic Programming] The Robust Counterpart problem (13.2.3) can be solved via the backward recursion

$$v_t(i) = \min_{a \in \mathcal{A}} \left\{ c_t(i,a) + \max_{p \in \mathcal{P}_{ti}^a} \sum_j p(j) v_{t+1}(j) \right\}, i \in \mathcal{X}, t = N-1, N-2, \ldots \quad (13.3.1)$$

initialized by setting $v_N = c_N$. Here $v_t(i)$ is the minimum, w.r.t. control policies over the horizon $t, t+1, \ldots, N-1$, of the maximal, over the nature policies over the same horizon, expected control cost, provided that at time t the controlled system is at state i.

A corresponding optimal control policy $u^* = (u_0^*(\cdot), \ldots, u_{N-1}^*(\cdot))$ is obtained by setting

$$u_t^*(i) \in \operatorname*{argmin}_{a \in \mathcal{A}} \left\{ c_t(i,a) + \max_{p \in \mathcal{P}_{ti}^a} \sum_j p(j) v_{t+1}(j) \right\}, \quad i \in \mathcal{X}, \, t \in \mathcal{T}, \quad (13.3.2)$$

and the corresponding worst-case nature policy is obtained by setting

$$p_{ti}^a \in \operatorname*{argmax}_{p \in \mathcal{P}_{ti}^a} \left\{ p^T v_{t+1} : p \in \mathcal{P}_{ti}^a \right\}, \quad i \in \mathcal{X}, \, a \in \mathcal{A}, \, t \in \mathcal{T}. \quad (13.3.3)$$

The optimal value in (13.2.3) is

$$\psi_N(\Pi) = q_0^T v_0,$$

q_0 being the initial distribution of states. Finally, the effect of uncertainty on a *given* strategy $u = (u_0(\cdot), \ldots, u_{N-1}(\cdot))$ can be evaluated by the following backward recursion

$$v_t^u(i) = c_t(i, u_t(i)) + \max_{p \in \mathcal{P}_{ti}^{u_t(i)}} p^T v_{t+1}^u, \quad i \in \mathcal{X}, t = N-1, N-2, \ldots, 0,$$
$$(13.3.4)$$

initialized with $v_N = c_N$; this recursion provides the worst-case value function v^u for the strategy u.

Proof. The proof of (13.3.1) is given by the standard Dynamic Programming reasoning. Let us *define* $v_t(i)$ as $c_N(i)$ for $t = N$ and as the minimum, over control policies on the time horizon $t, t+1, \ldots, N-1$, of the worst-case expected control cost over this time horizon, where the worst case is taken w.r.t. the policies of nature over this horizon. Recalling the definition of q_0, all we need to prove is that the quantities $v_t(i)$ obey the recurrence in (13.3.1). The latter is readily given by "backward induction" in t. Indeed, the base $t = N$ is evident. To carry out the induction step, assume that our recurrence holds true for $t \geq \tau + 1$ and all states, and let us verify that it holds true for every state i at time τ as well. Indeed, denoting by a a candidate control action at this time and state, note that nature can choose as the transition probability distribution from this state at this time an arbitrary vector $p \in \mathcal{P}_{\tau i}^a$. With this choice of nature, taking into account the Markovian property of the system, our expected losses over the time horizon $\tau, \tau + 1, \ldots, N - 1$, according to the inductive hypothesis, will be

$$c_\tau(i, a) + \sum_j p(j) v_{\tau+1}(j).$$

The worst value of this quantity, over nature's choice at time t in the state i, is

$$\max_{p \in \mathcal{P}_{\tau i}^a} \left\{ c_\tau(i, a) + \sum_j p(j) v_{\tau+1}(j) \right\} = c_\tau(i, a) + \max_{p \in \mathcal{P}_{\tau i}^a} \sum_j p(j) v_{\tau+1}(j).$$

Consequently, our minimal worst-case expected loss $v_\tau(i)$ over the time horizon $\tau, \tau + 1, \ldots, N - 1$, the state at time τ being i, is nothing but

$$\min_{a \in \mathcal{A}} \left\{ c_\tau(i, a) + \max_{p \in \mathcal{P}_{\tau i}^a} \sum_j p(j) v_{\tau+1}(j) \right\},$$

as claimed in (13.3.1). Induction is completed.

The remaining statements of Theorem 13.3.1 are evident.

13.3.1 Tractability Issues

Assume that $\mathcal{P}_{ti}^a$ are computationally tractable convex sets, e.g., they are given by explicit semidefinite representations. Then every time step of the backward recursion in (13.3.1) requires solving $\mathrm{Card}(\mathcal{A})\mathrm{Card}(\mathcal{X}) = mn$ problems of maximizing a given linear function over a set of the form $\mathcal{P}_{ti}^a$. It follows that the overall complexity of solving the Robust Counterpart is bounded by $mnN\mathcal{C}$, where $\mathcal{C}$ is the (maximal over a, t, i) complexity of maximizing a linear form over the computationally tractable convex set $\mathcal{P}_{ti}^a$. We conclude that the robust Bellman recursion is computationally tractable, provided m, n, N are moderate.

Note also that in the proof of Theorem 13.3.1 we never used the convexity of the sets $\mathcal{P}_{ti}^a \subset \Delta_n$, only the fact that they are nonempty and closed (the latter makes all the required maxima attainable). Besides this, from the structure of (13.3.1) we see that this recursion remains intact when the sets $\mathcal{P}_{ti}^a$ are extended to their convex hulls. As a result, the robust Bellman recursion is tractable when the sets $\mathcal{P}_{ti}^a$ are not necessary convex, but we are smart enough to represent their convex hulls in a computationally tractable fashion (and, in addition, m, n, N are moderate).

We are about to illustrate the above constructions and results numerically.

EXAMPLE 13.1.1 CONTINUED

Figure 13.1 represents a hexagonal grid with 127 vertices; a plane should fly from the origin O to the destination D, moving along the edges of the grid; flying along an edge takes a unit of time. The hexagonal region W in the middle of the grid (the corresponding nodes are marked by asterisks, and the edges are solid) represents an area that can be affected by bad weather; at every time period $[t, t+1)$, $t \in \mathbb{Z}$, this weather can be either in state "g" (good), or "b" (bad). When the plane is flying along an edge in W, the fuel consumption depends on the state of the weather, specifically, it is equal to $\ell > 0$ when the weather is good, and to $u > \ell$ when the weather is bad. Fuel consumption when flying along an edge outside of W is always equal to ℓ.

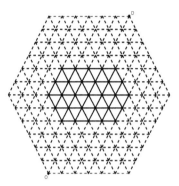

Figure 13.1 "Flying grid" with origin O and destination D. Asterisks and solid lines represent the area of potentially bad weather.

Now, the weather "lives its own life" and is described by a Markov chain, with the transition probabilities

$$\left[\begin{array}{c|c} p_{g2g} & p_{g2b} := 1 - p_{g2g} \\ \hline p_{b2g} := 1 - p_{b2b} & p_{b2b} \end{array}\right],$$

where p_{g2g} is the probability to remain good, (i.e., to pass from the state "g" to itself during a single time period), and p_{b2b} is the probability to stay bad. We assume that these probabilities are not known exactly and can, independently of each other, run through the "uncertainty box"

$$\mathcal{U} = \{[p_{g2g}; p_{b2b}] : |p_{g2g} - p_{g2g}^n| \le \delta_g, |p_{b2b} - p_{b2b}^n| \le \delta_b\},$$

where $p_{\cdot}^n$ are the corresponding nominal probabilities, and

$$\delta_g \le \min[p_{g2g}^n, 1 - p_{g2g}^n], \quad \delta_b \le \min[p_{b2b}^n, 1 - p_{b2b}^n]$$

specify the maximal magnitudes of the uncertainties. We assume that the weather transition probabilities are time varying, meaning that the probability $p_{s_{t-1}2s_t}^t$ of the weather to be at a state $s_t \in \{\text{"g", "b"}\}$ in the period $[t, t+1)$, conditioned on being in the state s_{t-1} in the period $[t-1, t)$, is chosen at time t by "nature" and can be an arbitrary point of the corresponding uncertainty interval.

Our goal is to find a strategy for the plane that minimizes the worst, over the actions of nature, expected cost of the total fuel consumption when flying from the origin to the destination.

Modeling the situation as an uncertain Dynamic Programming problem

In order to apply the outlined machinery, we proceed as follows.

- We model the state of the system at a time $t \in \mathbb{Z}_+$ as the pair (p_t, s_t), where p_t is the grid point where the plane is at time t, and s_t is the state of the weather in the period $[t, t+1)$. Thus, there is a total of $127 \times 2 = 254$ states.

- A control action a_t at time t is just the decision of the plane to move along which one of the 6 edges emanating from the current position p_t in the period $[t, t+1)$. Thus, a_t, in general, takes values $1, 2, 3, 4, 5, 6$, where, say, 1 means the bearing 0°, 2 means the bearing 60°, and so on. At a boundary point of the grid, some of these actions (those which would lead the plane outside of the grid) are forbidden. In addition, we allow for the control action $a_t = 0$ ("not to move at all"), but only in the state where the position of the plane is the destination D; this action is interpreted as staying at the destination after arriving there. Thus, there is a total of 7 control actions.

- The "on-line" costs $c_t("\text{state}", "\text{action}")$ represent the fuel consumption in the period $[t, t+1)$: when the plane at time t is at a point grid $p = p_t$, the weather at this time, i.e., in the period $[t, t+1)$, is at a state $s = s_t$, and the control action at time t is $a = a_t \neq 0$, the cost $c_t((p, s), a)$ is, in general, the fuel consumption when flying along the edge emanating from p in the direction a, the state of the weather being s. To account for the fact that some control actions $a \neq 0$ are forbidden when p is a boundary point of the grid distinct from the destination D, we set the corresponding costs to a large value M. Similarly, the only grid point where the action $a = 0$ is allowed, is the destination D, and we set $c_t((p, s), 0) = M$ when $p \neq$ D. In contrast to this, the only control action allowed at the destination is $a = 0$, and we set $c_t((D, s), a)$ to be equal to M when $a \neq 0$ and to be equal to 0 when $a = 0$.

As for the terminal costs, we set them to M for all states (p, s) where $P \neq$ D, and to 0 otherwise.

With the outlined modeling, the situation falls in the realm of Robust Markov Decision processes and can be processed accordingly.

Numerical results

We have specified the nominal weather transition probabilities by setting $p^{\mathrm{n}}_{\mathrm{g2g}} = p^{\mathrm{n}}_{\mathrm{b2b}} = 0.9$ and $\delta_{\mathrm{b}} = \delta_{\mathrm{g}} = \delta := 0.075$, meaning that there is a strong tendency for the weather to stay as it is, with relatively low probabilities to change from bad to good and vice versa. To make the phenomenon of data uncertainty more "pronounced," we used $\ell = 1$ and $u = 5$, meaning that bad weather significantly increases the fuel consumption; since our example is used for illustration purposes only, we do not care how realistic this assumption is. We further have computed two routing policies: the *nominal* one, where the transition probabilities are at their nominal values all the time, and the *robust* policy, where nature indeed can choose these probabilities, in a time-dependent fashion, within our uncertainty set. We then simulated every one of these two policies in the situa-

	Optimal value	
Policy	Good weather at departure	Bad weather at departure
nominal	14.595	15.342
robust	15.705	15.916

Table 13.1   Optimal values in the nominal and in the robust routing problems

tions where (a) the weather transition probabilities stay at their nominal values, and (b) nature "does it best" to reduce the probability of the weather to be good, that is, the actual weather transition probabilities are given by $p_{\text{g2g}} = p^{\text{n}}_{\text{g2g}} - \delta = 0.825$, $p_{\text{b2b}} = p^{\text{n}}_{\text{b2b}} + \delta = 0.975$. We ran simulations separately, 100 at a time, for the cases of good and bad weather at the departure. The results are displayed in tables 13.1, 13.2 and in figure 13.2. From table 13.2 we conclude that with an "aggressive" behavior of nature, the robust routing policy has non-negligible advantages, in terms of our objective the (worst-case) expected total fuel consumption, as compared to the nominal one. It is even more instructive to pay attention to the "structural differences" in routing for the two policies in question. As is seen from figure 13.2.(a,c), with the nominal routing policy, the plane can go "deep inside" the region that can be affected by bad weather, while with the robust policy, it never happens: the routes can only go along the boundary edges of this region. The explanation is as follows: since with our model, when choosing action a_t at instant t, we already know the weather on the time period $[t, t + 1)$, it is not costly to move along a "potentially dangerous" edge when the weather in the period $[t, t + 1)$ is good; all we need in order to avoid high fuel consumption, is the possibility to escape from the dangerous region as soon as the weather in this region changes from good to bad. This possibility does exist when the route does not go inside the dangerous region, as in figure 13.2.(b,d), where we also clearly see the "escapes" caused by changing the weather from good to bad. As a result, with the robust policy, the fuel consumption in fact remains low all the time, the price being a potential increase in travelling time because of longer routes. With the nominal routing policy (which relies on more "optimistic," as compared to the robust policy, assumptions about the probabilities for the weather to stay good or to change from bad to good), the tradeoff between the lengths of the routes and the fuel consumption along the edges is resolved differently; we see in figure 13.2.(a,c) that the corresponding routes can go deep inside the potentially dangerous region, meaning that with this policy, high fuel consumption is indeed possible. Note that this structural difference is caused by a rather subtle uncertainty in the weather transition probabilities.

13.4 NOTES AND REMARKS

NR 13.1. The results in this Chapter originate from [90].

Policy	Weather at departure	Nominal weather transition probabilities	Worst case weather transition probabilities
nominal	good	14.594	15.808
nominal	bad	15.342	16.000
robust	good	15.425	15.705
robust	bad	15.638	15.916

Table 13.2 Worst-case, over nature's strategies, expected fuel consumption for the nominal and the robust routing policies.

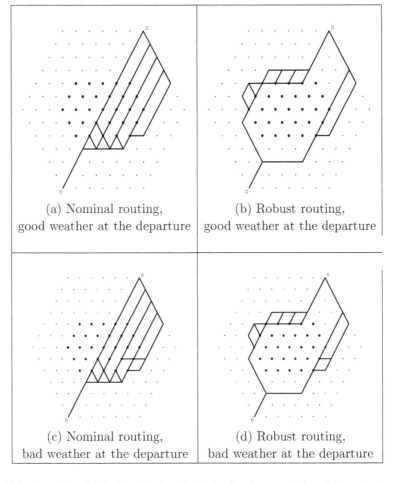

Figure 13.2 Samples of 100 simulated trajectories for the nominal and the robust routing. When simulating the routing policies, the weather transition probabilities at every step were chosen as $p_{g2g} = p_{g2g}^n - \xi\delta$, $p_{b2b} = p_{b2b}^n + \xi\delta$ with ξ uniformly distributed on $[0, 1]$ ("nature is aggressive, but not overly so").

The actual number of routes we see in every picture is much less than the number (100) of routes sampled, since many such routes are identical to each other.

Chapter Fourteen
Robust Adjustable Multistage Optimization

In this chapter we continue investigating robust multi-stage decision making processes started in chapter 13. Note that in the context of chapter 13, computational tractability of the robust counterparts stems primarily from the fact that both the state and the action spaces associated with the decision making process under consideration are finite with moderate cardinalities. These assumptions combine with the Markovian nature of the process to allow for solving the robust counterpart in a computationally efficient way by properly adapted Dynamic Programming techniques. In what follows we intend to consider multi-stage decision making in the situations where Dynamic Programming hardly is applicable, primarily because of the "curse of dimensionality" discussed in chapter 13.

14.1 ADJUSTABLE ROBUST OPTIMIZATION: MOTIVATION

Consider a general-type uncertain optimization problem — a collection

$$\mathcal{P} = \left\{ \min_x \left\{ f(x,\zeta) : F(x,\zeta) \in \mathbf{K} \right\} : \zeta \in \mathcal{Z} \right\} \qquad (14.1.1)$$

of *instances* — optimization problems of the form

$$\min_x \left\{ f(x,\zeta) : F(x,\zeta) \in \mathbf{K} \right\},$$

where $x \in \mathbb{R}^n$ is the decision vector, $\zeta \in \mathbb{R}^L$ represents the uncertain data or data perturbation, the real-valued function $f(x,\zeta)$ is the objective, and the vector-valued function $F(x,\zeta)$ taking values in $\mathbb{R}^m$ along with a set $\mathbf{K} \subset \mathbb{R}^m$ specify the constraints; finally, $\mathcal{Z} \subset \mathbb{R}^L$ is the uncertainty set where the uncertain data is restricted to reside.

> Format (14.1.1) covers all uncertain optimization problems considered in Parts I and II; moreover, in these latter problems the objective f and the right hand side F of the constraints always were *bi-affine* in x, ζ, (that is, affine in x when ζ is fixed, and affine in ζ, x being fixed), and $\mathbf{K}$ was a "simple" convex cone (a direct product of nonnegative rays/Lorentz cones/Semidefinite cones, depending on whether we were speaking about uncertain Linear, Conic Quadratic or Semidefinite Optimization). We shall come back to this "well-structured" case later; for our immediate purposes the specific conic structure of instances plays no role, and we can focus on "general" uncertain problems in the form of (14.1.1).

The Robust Counterpart of uncertain problem (14.1.1) is defined as the semi-infinite optimization problem

$$\min_{x,t} \{t : \forall \zeta \in \mathcal{Z} : f(x,\zeta) \le t, F(x,\zeta) \in \mathbf{K}\} ; \qquad (14.1.2)$$

this is exactly what was called the RC of an uncertain problem in the situations considered in Parts I and II.

Recall that our interpretation of the RC (14.1.2) as the natural source of robust/robust optimal solutions to the uncertain problem (14.1.1) is not self-evident, and its "informal justification" relies upon the specific assumptions A.1–3 on our "decision environment," see page 9. We have already relaxed somehow the last of these assumptions, thus arriving at the notion of Globalized Robust Counterpart. What is on our agenda now is to revise the first assumption, which reads

> A.1. All decision variables in (14.1.1) represent "here and now" decisions; they should get specific numerical values as a result of solving the problem *before* the actual data "reveals itself" and as such should be independent of the actual values of the data.

In Parts I, II we have considered numerous examples of situations where this assumption is valid. At the same time, there are situations when it is too restrictive, since "in reality" some of the decision variables can adjust themselves, to some extent, to the actual values of the data. One can point out at least two sources of such adjustability: presence of *analysis variables* and *wait-and-see decisions*.

Analysis variables. Not always all decision variables x_j in (14.1.1) represent actual decisions; in many cases, some of x_j are slack, or analysis, variables introduced in order to convert the instances into a desired form, e.g., the one of Linear Optimization programs. It is very natural to allow for the analysis variables to depend on the true values of the data — why not?

Example 14.1.1. [cf. Example 1.2.7] Consider an "ℓ_1 constraint"

$$\sum_{k=1}^{K} |a_k^T x - b_k| \le \tau; \qquad (14.1.3)$$

you may think, e.g., about the Antenna Design problem (section 3.3) where the "fit" between the actual diagram of the would-be antenna array and the target diagram is quantified by the $\|\cdot\|_1$ distance. Assuming that the data and x are real, (14.1.3) can be represented equivalently by the system of linear inequalities

$$-y_k \le a_k^T x - b_k \le y_k, \quad \sum_k y_k \le \tau$$

in variables x, y, τ. Now, when the data a_k, b_k are uncertain and the components of x do represent "here and now" decisions and should be independent of the actual values of the data, there is absolutely no reason to impose the latter requirement on the slack variables y_k as well: they do not represent decisions at all and just certify the fact that the actual decisions x, τ meet the requirement (14.1.3). While we can, of course, impose this requirement "by force," this perhaps will lead to a too conservative model. It seems to be completely natural to allow for the certificates y_k to depend on actual values of the data — it may well happen that then we shall be able to certify robust feasibility for (14.1.3) for a larger set of pairs (x, τ).

Wait-and-see decisions. This source of adjustability comes from the fact that some of the variables x_j represent decisions that are not "here and now" decisions, i.e., those that should be made before the true data "reveals itself." In multi-stage decision making processes, like those considered in chapter 13, some x_j can represent "wait and see" decisions, which could be made after the controlled system "starts to live," at time instants when part (or all) of the true data is revealed. It is fully legitimate to allow for these decisions to depend on the part of the data that indeed "reveals itself" before the decision should be made.

    Example 14.1.2. Consider a multi-stage inventory system affected by uncertain demand. The most interesting of the associated decisions — the *replenishment orders* — are made one at a time, and the replenishment order of "day" t is made when we already know the actual demands in the preceding days. It is completely natural to allow for the orders of day t to depend on the preceding demands.

14.2 ADJUSTABLE ROBUST COUNTERPART

A natural way to model adjustability of variables is as follows: for every $j \leq n$, we allow for x_j to depend on a prescribed "portion" $P_j\zeta$ of the true data ζ:

$$x_j = X_j(P_j\zeta), \qquad (14.2.1)$$

where $P_1, ..., P_n$ are given in advance matrices specifying the "information base" of the decisions x_j, and $X_j(\cdot)$ are *decision rules* to be chosen; these rules can in principle be arbitrary functions on the corresponding vector spaces. For a given j, specifying P_j as the zero matrix, we force x_j to be completely independent of ζ, that is, to be a "here and now" decision; specifying P_j as the unit matrix, we allow for x_j to depend on the entire data (this is how we would like to describe the analysis variables). And the "in-between" situations, choosing P_j with $1 \leq \text{Rank}(P_j) < L$ enables one to model the situation where x_j is allowed to depend on a "proper portion" of the true data.

    We can now replace in the usual RC (14.1.2) of the uncertain problem (14.1.1) the independent of ζ decision variables x_j with functions $X_j(P_j\zeta)$, thus arriving at the problem

$$\min_{t, \{X_j(\cdot)\}_{j=1}^n} \{t : \forall \zeta \in \mathcal{Z} : f(X(\zeta), \zeta) \leq t, F(X(\zeta), \zeta) \in \mathbf{K}\},$$
$$X(\zeta) = [X_1(P_1\zeta); ...; X_n(P_n\zeta)]. \qquad (14.2.2)$$

The resulting optimization problem is called the *Adjustable Robust Counterpart* (ARC) of the uncertain problem (14.1.1), and the (collections of) decision rules $X(\zeta)$, which along with certain t are feasible for the ARC, are called *robust feasible decision rules*. The ARC is then the problem of specifying a collection of decision rules with prescribed information base that is feasible for as small t as possible. The *robust optimal decision rules* now replace the *constant* (non-adjustable, data-independent) robust optimal decisions that are yielded by the usual Robust Counterpart (14.1.2) of our uncertain problem. Note that the ARC is an extension of the RC; the latter is a "trivial" particular case of the former corresponding to the case of trivial information base in which all matrices P_j are zero.

14.2.1 Examples

We are about to present two instructive examples of uncertain optimization programs with adjustable variables.

Information base induced by time precedences. In many cases, decisions are made subsequently in time; whenever this is the case, a natural information base of the decision to be made at instant t ($t = 1, ..., N$) is the part of the true data that becomes known at time t. As an instructive example, consider a simple Multi-Period Inventory model mentioned in Example 14.1.2:

Example 14.1.2 continued. Consider an inventory system where d products share common warehouse capacity, the time horizon is comprised of N periods, and the goal is to minimize the total inventory management cost. Allowing for backlogged demand, the simplest model of such an inventory looks as follows:

$$
\begin{array}{lll}
\text{minimize} & C & \text{[inventory management cost]} \\
\text{s.t.} & & \\
(a) & C \geq \sum_{t=1}^{N} \left[c_{\mathrm{h},t}^T y_t + c_{\mathrm{b},t}^T z_t + c_{\mathrm{o},t}^T w_t \right] & \text{[cost description]} \\
(b) & x_t = x_{t-1} + w_t - \zeta_t,\ 1 \leq t \leq N & \text{[state equations]} \\
(c) & y_t \geq 0, y_t \geq x_t,\ 1 \leq t \leq N & \\
(d) & z_t \geq 0, z_t \geq -x_t,\ 1 \leq t \leq N & \\
(e) & \underline{w}_t \leq w_t \leq \overline{w}_t,\ 1 \leq t \leq N & \\
(f) & q^T y_t \leq r &
\end{array}
$$

$$(14.2.3)$$

The variables in this problem are:

- $C \in \mathbb{R}$ — (upper bound on the total inventory management cost;
- $x_t \in \mathbb{R}^d$, $t = 1, ..., N$ — states. i-th coordinate x_t^i of vector x_t is the amount of product of type i that is present in the inventory at the time instant t (end of time interval # t). This amount can be nonnegative, meaning that the inventory at this time has x_t^i units of free product # i; it may be also negative, meaning that the inventory at the moment in question owes the customers $|x_t^i|$ units of the product i ("backlogged demand"). The initial state x_0 of the inventory is part of the data, and not part of the decision vector;
- $y_t \in \mathbb{R}^d$ are upper bounds on the positive parts of the states x_t, that is, (upper bounds on) the "physical" amounts of products stored in the inventory at time t, and the quantity $c_{\mathrm{h},t}^T y_t$ is the (upper bound on the) holding cost in the period t; here $c_{\mathrm{h},t} \in \mathbb{R}_+^d$ is a given vector of the holding costs per unit of the product. Similarly, the quantity $q^T y_t$ is (an upper bound on) the warehouse capacity used by the products that are "physically present" in the inventory at time t, $q \in \mathbb{R}_+^d$ being a given vector of the warehouse capacities per units of the products;

- $z_t \in \mathbb{R}^d$ are (upper bounds on) the backlogged demands at time t, and the quantities $c_{\mathrm{b},t}^T z_t$ are (upper bounds on) the penalties for these backlogged demands. Here $c_{\mathrm{b,t}} \in \mathbb{R}_+^d$ are given vectors of the penalties per units of the backlogged demands;

- $w_t \in \mathbb{R}^d$ is the vector of replenishment orders executed in period t, and the quantities $c_{\mathrm{o},t}^T w_t$ are the costs of executing these orders. Here $c_{\mathrm{o},t} \in \mathbb{R}_+^d$ are given vectors of per unit ordering costs.

With these explanations, the constraints become self-evident:

- (a) is the "cost description": it says that the total inventory management cost is comprised of total holding and ordering costs and of the total penalty for the backlogged demand;

- (b) are state equations: "what will be in the inventory at the end of period t (x_t) is what was there at the end of preceding period (x_{t-1}) plus the replenishment orders of the period (w_t) minus the demand of the period (ζ_t);

- (c), (d) are self-evident;

- (e) represents the upper and lower bounds on replenishment orders, and (f) expresses the requirement that (an upper bound on) the total warehouse capacity $q^T y_t$ utilized by products that are "physically present" in the inventory at time t should not be greater than the warehouse capacity r.

In our simple example, we assume that out of model's parameters

$$x_0, \{c_{\mathrm{h},t}, c_{\mathrm{b},t}, c_{\mathrm{o},t}, \underline{w}_t, \overline{w}_t\}_{t=1}^N, q, r, \{\zeta_t\}_{t=1}^N$$

the only uncertain element is the *demand trajectory* $\zeta = [\zeta_1; ...; \zeta_N] \in \mathbb{R}^{dN}$, and that this trajectory is known to belong to a given uncertainty set $\mathcal{Z}$. The resulting uncertain Linear Optimization problem is comprised of instances (14.2.3) parameterized by the uncertain data — demand trajectory ζ — running through a given set $\mathcal{Z}$.

As far as the adjustability is concerned, all variables in our problem, except for the replenishment orders w_t, are analysis variables. As for the orders, the simplest assumption is that w_t should get numerical value at time t, and that at this time we already know the past demands $\zeta^{t-1} = [\zeta_1; ...; \zeta_{t-1}]$. Thus, the information base for w_t is $\zeta^{t-1} = P_t \zeta$ (with the convention that $\zeta^s = 0$ when $s < 0$). For the remaining analysis variables the information base is the entire demand trajectory ζ. Note that we can easily adjust this model to the case when there are lags in demand acquisition, so that w_t should depend on a prescribed initial segment $\zeta^{\tau(t)-1}$, $\tau(t) \leq t$, of ζ^{t-1} rather than on the entire ζ^{t-1}. We can equally easily account for the possibility, if any, to observe the demand "on line," by allowing w_t to depend on ζ^t rather than on ζ^{t-1}. Note that in all these cases the information base of the decisions is readily given by the natural time precedences between the "actual decisions" augmented by a specific demand acquisition protocol.

Example 14.2.1. Project management. Figure 14.1 is a simple *PERT diagram* — a graph representing a Project Management problem. This is an acyclic directed graph with nodes corresponding to *events*, and arcs corresponding to *activities*. Among the nodes there is a *start node* S with no incoming arcs and an *end node* F with no outgoing arcs, interpreted as "start of the project" and "completion of the project," respectively. The remaining nodes correspond to the events "a specific stage of the project is completed, and one can pass to another stage". For example, the diagram could represent creating a factory, with A, B, C being, respectively, the events "equipment to be installed is acquired and delivered," "facility #1 is built and equipped," "facility # 2 is built and equipped." The activities are jobs comprising the project. In our example, these jobs could be as follows:

a:	acquiring and delivering the equipment for facilities ## 1,2
b:	building facility # 1
c:	building facility # 2
d:	installing equipment in facility # 1
e:	installing equipment in facility # 2
f:	training personnel and preparing production at facility # 1
g:	training personnel and preparing production at facility # 2

The topology of a PERT diagram represents *logical precedences* between the activities and events: a particular activity, say g, can start only after the event C occurs, and the latter event happens when both activities c and e are completed.

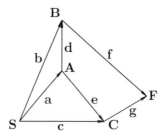

Figure 14.1   A PERT diagram.

In PERT models it is assumed that activities γ have nonnegative durations τ_γ (perhaps depending on control parameters), and are executed without interruptions, with possible idle periods between the moment when the start of an activity is allowed by the logical precedences and the moment when it is actually started. With these assumptions, one can write down a system of constraints on the time instants t_ν when events ν can take place. Denoting by $\Gamma = \{\gamma = (\mu_\gamma, \nu_\gamma)\}$ the set of arcs in a PERT diagram (μ_γ is the start- and ν_γ is the end-node of an arc γ), this system reads

$$t_{\mu_\gamma} - t_{\nu_\gamma} \geq \tau_\gamma \ \forall \gamma \in \Gamma. \tag{14.2.4}$$

"Normalizing" this system by the requirement

$$t_S = 0,$$

the values of t_F, which can be obtained from feasible solutions to the system, are achievable durations of the entire project. In a typical Project Management problem, one imposes an upper bound on t_F and minimizes, under this restriction, coupled with the system of constraints (14.2.4), some objective function.

As an example, consider the situation where the "normal" durations τ_γ of activities can be reduced at certain price ("in reality" this can correspond to investing into an activity extra manpower, machines, etc.). The corresponding model becomes

$$\tau_\gamma = \zeta_\gamma - x_\gamma, \ c_\gamma = f_\gamma(x_\gamma),$$

where ζ_γ is the "normal duration" of the activity, x_γ ("crush") is a nonnegative decision variable, and $c_\gamma = f_\gamma(x_\gamma)$ is the cost of the crush; here $f_\gamma(\cdot)$ is a given function. The associated optimization model might be, e.g., the problem of minimizing the total cost of the crushes under a given upper bound T on project's duration:

$$\min_{\substack{x=\{x_\gamma:\gamma\in\Gamma\} \\ \{t_\nu\}}} \left\{ \sum_\gamma f_\gamma(x_\gamma) : \begin{array}{c} t_{\mu_\gamma} - t_{\nu_\gamma} \geq \zeta_\gamma - x_\gamma \\ 0 \leq x_\gamma \leq \overline{x}_\gamma \end{array} \right\} \forall \gamma \in \Gamma, t_S = 0, t_F \leq T \right\},$$

$$(14.2.5)$$

where $\overline{x}_\gamma$ are given upper bounds on crushes. Note that when $f_\gamma(\cdot)$ are convex functions, (14.2.5) is an explicit convex problem, and when, in addition to convexity, $f_\gamma(\cdot)$ are piecewise linear, (which is usually the case in reality and which we assume from now on), (14.2.5) can be straightforwardly converted to a Linear Optimization program.

Usually part of the data of a PERT problem are uncertain. Consider the simplest case when the only uncertain elements of the data in (14.2.5) are the normal durations ζ_γ of the activities (their uncertainty may come from varying weather conditions, inaccuracies in estimating the forthcoming effort, etc.). Let us assume that these durations are random variables, say, independent of each other, distributed in given segments $\Delta_\gamma = [\underline{\zeta}_\gamma, \overline{\zeta}_\gamma]$. To avoid pathologies, assume also that $\underline{\zeta}_\gamma \geq \overline{x}_\gamma$ for every γ ("you cannot make the duration negative"). Now (14.2.5) becomes an uncertain LO program with uncertainties affecting only the right hand sides of the constraints. A natural way to "immunize" the solutions to the problem against data uncertainty is to pass to the usual RC of the problem — to think of both t_γ and x_γ as of variables with values to be chosen in advance in such a way that the constraints in (14.2.4) are satisfied for all values of the data ζ_γ from the uncertainty set. With our model of the latter set the RC is nothing but the "worst instance" of our uncertain problem, the one where ζ_γ are set to their maximum possible values $\overline{\zeta}_\gamma$. For large PERT graphs, such an approach is very conservative: why should we care about the highly improbable case where all the normal durations — independent random variables! — are simultaneously at their worst-case values? Note that even taking into account that the normal durations are random and replacing the uncertain constraints in (14.2.5) by their chance constrained versions, we essentially do not reduce the conservatism. Indeed, every one of randomly perturbed constraints in (14.2.5) contains a *single* random perturbation, so that we cannot hope that random perturbations of a constraint will to some extent cancel each other. As a result, to require the validity of every uncertain constraint with probability 0.9 or 0.99 is the same as to require its validity "in the worst case" with just slightly reduced maximal normal durations of the activities.

A much more promising approach is to try to adjust our decisions "on line." Indeed, we are speaking about a process that evolves in time, with "actual decisions" represented by variables x_γ and t_ν's being the analysis variables. Assuming that the decision on x_γ can be postponed till the event μ_γ (the earliest time when the activity γ can be started) takes place, at that time we already know the actual durations of the activities terminated before the event μ_γ, we could then adjust our decision on x_γ in accordance with this information. The difficulty is that *we do not know in advance what will be the actual time precedences between the events — these precedences depend on our decisions and on the actual values of the uncertain data.* For example, in the situation described by figure 14.1, we, in general, cannot know in advance which one of the events B, C will precede the other one in time. As a result, in our present situation, in sharp contrast to the situation of Example 14.1.2, an attempt to fully utilize the possibilities to adjust the decisions to the actual values of the data results in an extremely complicated problem, where not only the decisions themselves, but the very information base of the decisions become dependent on the uncertain data and our policy. However, we could stick to something in-between "no adjustability at all" and "as much adjustability as possible." Specifically, we definitely know that if a pair of activities γ', γ are linked by a logical precedence, so that there exists an oriented route in the graph that starts with γ' and ends with γ, then the actual duration of γ' will be known before γ can start. Consequently, we can take, as the information base of an activity γ, the collection $\zeta^\gamma = \{\zeta_{\gamma'} : \gamma' \in \Gamma_-(\gamma)\}$, where $\Gamma_-(\gamma)$ is the set of all activities that logically precede the activity γ. In favorable circumstances, such an approach could reduce significantly the price of robustness as compared to the non-adjustable RC. Indeed, when plugging into the randomly perturbed constraints of (14.2.5) instead of constants x_γ functions $X_\gamma(\zeta^\gamma)$, and requiring from the resulting inequalities to be valid with probability $1 - \epsilon$, we end up with a system of chance constraints such that some of them (in good cases, even most of them) involve many independent random perturbations each. When the functions $X_\gamma(\zeta^\gamma)$ are regular enough, (e.g., are affine), we can hope that the numerous independent perturbations affecting a chance constraint will to some extent cancel each other, and consequently, the resulting system of chance constraints will be significantly less conservative than the one corresponding to non-adjustable decisions.

14.2.2 Good News on the ARC

Passing from a trivial information base to a nontrivial one — passing from robust optimal *data-independent decisions* to robust optimal *data-based decision rules* can indeed dramatically reduce the associated robust optimal value.

Example 14.2.2. Consider the toy uncertain LO problem

$$\left\{ \min_x \left\{ x_1 : \begin{array}{ll} x_2 \geq \frac{1}{2}\zeta x_1 + 1 & (a_\zeta) \\ x_1 \geq (2 - \zeta)x_2 & (b_\zeta) \\ x_1, x_2 \geq 0 & (c_\zeta) \end{array} \right\} : 0 \leq \zeta \leq \rho \right\},$$

where $\rho \in (0, 1)$ is a parameter (uncertainty level). Let us compare the optimal value of its non-adjustable RC (where both x_1 and x_2 must be independent of ζ) with the optimal value of the ARC where x_1 still is assumed to be independent of ζ ($P_1\zeta \equiv 0$) but x_2 is allowed to depend on ζ ($P_2\zeta \equiv \zeta$).

A feasible solution (x_1, x_2) of the RC should remain feasible for the constraint (a_ζ) when $\zeta = \rho$, meaning that $x_2 \geq \frac{\rho}{2}x_1 + 1$, and should remain feasible for the constraint (b_ζ) when $\zeta = 0$, meaning that $x_1 \geq 2x_2$. The two resulting inequalities imply that $x_1 \geq \rho x_1 + 2$, whence $x_1 \geq \frac{2}{1-\rho}$. Thus, Opt(RC)$\geq \frac{2}{1-\rho}$, whence Opt(RC)$\rightarrow \infty$ as $\rho \rightarrow 1 - 0$.

Now let us solve the ARC. Given $x_1 \geq 0$ and $\zeta \in [0, \rho]$, it is immediately seen that x_1 can be extended, by properly chosen x_2, to a feasible solution of (a_ζ) through (c_ζ) if and only if the pair $(x_1, x_2 = \frac{1}{2}\zeta x_1 + 1)$ is feasible for (a_ζ) through (c_ζ), that is, if and only if $x_1 \geq (2 - \zeta)\left[\frac{1}{2}\zeta x_1 + 1\right]$ whenever $1 \leq \zeta \leq \rho$. The latter relation holds true when $x_1 = 4$ and $\rho \leq 1$ (since $(2-\zeta)\zeta \leq 1$ for $0 \leq \zeta \leq 2$). Thus, Opt(ARC)≤ 4, and the difference between Opt(RC) and Opt(ARC) and the ratio Opt(RC)/Opt(ARC) go to ∞ as $\rho \rightarrow 1 - 0$.

14.2.3 Bad News on the ARC

Unfortunately, from the computational viewpoint the ARC of an uncertain problem more often than not is wishful thinking rather than an actual tool. The reason comes from the fact that *the ARC is typically severely computationally intractable.* Indeed, (14.2.2) is an *infinite-dimensional problem*, where one wants to optimize over *functions* — decision rules — rather than vectors, and these functions, in general, depend on many real variables. It is unclear even how to *represent* a general-type candidate decision rule — a general-type multivariate function — in a computer. Seemingly the only option here is sticking to a chosen in advance *parametric family* of decision rules, like piece-wise constant/linear/quadratic functions of $P_j\zeta$ with simple domains of the pieces (say, boxes). With this approach, a candidate decision rule is identified by the vector of values of the associated parameters, and the ARC becomes a finite-dimensional problem, the parameters being our new decision variables. This approach is indeed possible and in fact will be the focus of what follows. However, it should be clear from the very beginning that if the parametric family in question is "rich enough" to allow for good approximation of "truly optimal" decision rules (think of polynomial splines of high degree as approximations to "not too rapidly varying" general-type multivariate functions), the number of parameters involved should be astronomically large, unless the dimension of ζ is really small, like $1 - 3$ (think of how many coefficients there are in a single algebraic polynomial of degree 10 with 20 variables). Thus, aside of "really low dimensional" cases, "rich" general-purpose parametric families of decision rules are for all practical purposes as intractable as non-parametric families. In other words, when the dimension L of ζ is not too small, tractability of parametric families of decision rules is something opposite to their "approximation abilities,"

and sticking to tractable parametric families, we lose control of how far the optimal value of the "parametric" ARC is away from the optimal value of the "true"
infinite-dimensional ARC. The only exception here seems to be the case when we
are smart enough to utilize our knowledge of the structure of instances of the uncertain problem in question in order to identify the optimal decision rules up to a
moderate number of parameters. *If we indeed are that smart and if the parameters in question can be further identified numerically in a computationally efficient
fashion, we indeed can end up with an optimal solution to the "true" ARC.* Unfortunately, the two "if's" in the previous sentence are big if's indeed — to the best of
our knowledge, the only *generic* situation when these conditions are satisfied is the
"environment" of Markov Decision Processes considered in chapter 13 and the Dynamic Programming techniques that can be used in this environment. It seems that
these techniques form the only component in the existing "optimization toolbox"
that could be used to process the ARC numerically, at least when approximations
of a provably high quality are sought. Unfortunately, the Dynamic Programming
techniques are very "fragile" — they require instances of a very specific structure,
suffer from "curse of dimensionality," etc., cf. chapter 13. The bottom line, in
our opinion, is that *aside of situations,* like those considered in chapter 13, *where
Dynamic Programming is computationally efficient,* (which is an exception rather
than a rule), *the only hopefully computationally tractable approach to optimizing
over decision rules is to stick to their simple parametric families,* even at the price
of giving up full control over the losses in optimality that can be incurred by such
a simplification.

Before moving to an in-depth investigation of (a version of) the just outlined
"simple approximation" approach to adjustable robust decision-making, it is worth
pointing out two situations when no simple approximations are necessary, since the
situations in question are very simple from the very beginning.

14.2.3.1 Simple case I: fixed recourse and scenario-generated uncertainty set

Consider an uncertain *conic* problem

$$\mathcal{P} = \left\{ \min_x \left\{ c_\zeta^T x + d_\zeta : A_\zeta x + b_\zeta \in \mathbf{K} \right\} : \zeta \in \mathcal{Z} \right\} \tag{14.2.6}$$

($A_\zeta, b_\zeta, c_\zeta, d_\zeta$ are affine in ζ, $\mathbf{K}$ is a computationally tractable convex cone) and
assume that

   i) $\mathcal{Z}$ is a scenario-generated uncertainty set, that is, a set given as a convex hull
     of finitely many "scenarios" ζ^s, $1 \leq s \leq S$;

  ii) The information base ensures that every variable x_j either is non-adjustable
     ($P_j = 0$), or is fully adjustable ($P_j = I$);

 iii) We are in the situation of *fixed recourse,* that is, for every adjustable variable
     x_j (one with $P_j \neq 0$), all its coefficients in the objective and the left hand
     side of the constraint are certain, (i.e., are independent of ζ).

W.l.o.g. we can assume that $x = [u; v]$, where the u variables are non-adjustable, and the v variables are fully adjustable; under fixed recourse, our uncertain problem can be written down as

$$\mathcal{P} = \left\{ \min_{u,v} \left\{ p_\zeta^T u + q^T v + d_\zeta : P_\zeta u + Qv + r_\zeta \in \mathbf{K} \right\} : \zeta \in \mathrm{Conv}\{\zeta^1, ..., \zeta^S\} \right\}$$

($p_\zeta, d_\zeta, P_\zeta, r_\zeta$ are affine in ζ). An immediate observation is that:

Theorem 14.2.3. Under assumptions $1-3$, the ARC of the uncertain problem $\mathcal{P}$ is equivalent to the computationally tractable conic problem

$$\mathrm{Opt} = \min_{t,u,\{v^s\}_{s=1}^S} \left\{ t : p_{\zeta^s}^T u + q^T v^s + d_{\zeta^s} \le t,\ P_{\zeta^s} u + Qv^s + r_{\zeta^s} \in \mathbf{K} \right\}. \quad (14.2.7)$$

Specifically, the optimal values in the latter problem and in the ARC of $\mathcal{P}$ are equal. Moreover, if $\bar{t}, \bar{u}, \{\bar{v}^s\}_{s=1}^S$ is a feasible solution to (14.2.7), then the pair $\bar{t}, \bar{u}$ augmented by the decision rule for the adjustable variables:

$$v = \bar{V}(\zeta) = \sum_{s=1}^S \lambda_s(\zeta) \bar{v}^s$$

form a feasible solution to the ARC. Here $\lambda(\zeta)$ is an arbitrary nonnegative vector with the unit sum of entries such that

$$\zeta = \sum_{s=1}^S \lambda_s(\zeta) \zeta^S. \quad (14.2.8)$$

Proof. Observe first that $\lambda(\zeta)$ is well-defined for every $\zeta \in \mathcal{Z}$ due to $\mathcal{Z} = \mathrm{Conv}\{\zeta^1, ..., \zeta^S\}$. Further, if $\bar{t}, \bar{u}, \{\bar{v}^s\}$ is a feasible solution of (14.2.7) and $\bar{V}(\zeta)$ is as defined above, then for every $\zeta \in \mathcal{Z}$ the following implications hold true:

$$\begin{aligned}
&\bar{t} \ge p_{\zeta^s} \bar{u} + q^T \bar{v}^s + d_{\zeta^s}\ \forall s \Rightarrow \bar{t} \ge \textstyle\sum_s \lambda_s(\zeta) \left[p_{\zeta^s}^T \bar{u} + q^T \bar{v}^s + d_{\zeta^s} \right] \\
&= p_\zeta^T \bar{u} + q^T \bar{V}(\zeta) + d_\zeta, \\
&\mathbf{K} \ni P_{\zeta^s} \bar{u} + Q\bar{v}^s + r_{\zeta^s}\ \forall s \Rightarrow \mathbf{K} \ni \textstyle\sum_s \lambda_s(\zeta) \left[P_{\zeta^s} \bar{u} + Q\bar{v}^s + r_{\zeta^s} \right] \\
&= P_\zeta \bar{u} + Q\bar{V}(\zeta) + r_\zeta
\end{aligned}$$

(recall that $p_\zeta, ..., r_\zeta$ are affine in ζ). We see that $(\bar{t}, \bar{u}, \bar{V}(\cdot))$ is indeed a feasible solution to the ARC

$$\min_{t,u,V(\cdot)} \left\{ t : p_\zeta^T u + q^T V(\zeta) + d_\zeta \le t, P_\zeta u + QV(\zeta) + r_\zeta \in \mathbf{K}\ \forall \zeta \in \mathcal{Z} \right\}$$

of $\mathcal{P}$. As a result, the optimal value of the latter problem is $\le \mathrm{Opt}$. It remains to verify that the optimal value of the ARC and Opt are equal. We already know that the first quantity is $\le$ the second one. To prove the opposite inequality, note that if $(t, u, V(\cdot))$ is feasible for the ARC, then clearly $(t, u, \{v^s = V(\zeta^s)\})$ is feasible for (14.2.7). $\qquad \square$

The outlined result shares the same shortcoming as Theorem 6.1.2 from section 6.1: scenario-generated uncertainty sets are usually too "small" to be of much interest, unless the number L of scenarios is impractically large. It is also worth noticing that the assumption of fixed recourse is essential: it is easy to show (see [13]) that without it, the ARC may become intractable.

14.2.3.2 Simple case II: uncertain LO with constraint-wise uncertainty

Consider an uncertain LO problem

$$\mathcal{P} = \left\{ \min_x \left\{ c_\zeta^T x + d_\zeta : a_{i\zeta}^T x \le b_{i\zeta}, \ i = 1, ..., m \right\} : \zeta \in \mathcal{Z} \right\}, \tag{14.2.9}$$

where, as always, $c_\zeta, d_\zeta, a_{i\zeta}, b_{i\zeta}$ are affine in ζ. Assume that

i) The uncertainty is constraint-wise: ζ can be split into blocks $\zeta = [\zeta^0; ...; \zeta^m]$ in such a way that the data of the objective depend solely on ζ^0, the data of the i-th constraint depend solely on ζ^i, and the uncertainty set $\mathcal{Z}$ is the direct product of convex compact sets $\mathcal{Z}_0, \mathcal{Z}_1, ..., \mathcal{Z}_m$ in the spaces of $\zeta^0, ..., \zeta^m$;

ii) One can point out a convex compact set $\mathcal{X}$ in the space of x variables such that whenever $\zeta \in \mathcal{Z}$ and x is feasible for the instance of $\mathcal{P}$ with the data ζ, one has $x \in \mathcal{X}$.

The validity of the latter, purely technical, assumption can be guaranteed, e.g., when the constraints of the uncertain problem contain (certain) finite upper and lower bounds on every one of the decision variables. The latter assumption, for all practical purposes, is non-restrictive.

Our goal is to prove the following

Theorem 14.2.4. Under the just outlined assumptions *i*) and *ii*), the ARC of (14.2.9) is equivalent to its usual RC (no adjustable variables): both ARC and RC have equal optimal values.

Proof. All we need is to prove that the optimal value in the ARC is $\ge$ the one of the RC. When achieving this goal, we can assume w.l.o.g. that all decision variables are *fully adjustable* — are allowed to depend on the entire vector ζ. The "fully adjustable" ARC of (14.2.9) reads

$$\begin{aligned}
\text{Opt(ARC)} &= \min_{X(\cdot), t} \left\{ t : \begin{array}{l} c_{\zeta^0}^T X(\zeta) + d_{\zeta^0} - t \le 0 \\ a_{i\zeta^i}^T X(\zeta) - b_{i\zeta^i} \le 0, 1 \le i \le m \end{array} \right. \\
&\qquad\qquad\qquad \forall (\zeta \in \mathcal{Z}_0 \times ... \times \mathcal{Z}_m) \Bigg\} \\
&= \inf \left\{ t : \forall (\zeta \in \mathcal{Z}_0 \times ... \times \mathcal{Z}_m) \exists x \in \mathcal{X} : \right. \\
&\qquad\qquad\qquad \left. \alpha_{i,\zeta^i}^T x - \beta_i t + \gamma_{i,\zeta^i} \le 0, 0 \le i \le m \right\},
\end{aligned} \tag{14.2.10}$$

(the restriction $x \in \mathcal{X}$ can be added due to assumption *i*)), while the RC is the problem

$$\text{Opt(RC)} = \inf \left\{ t : \exists x \in \mathcal{X} : \alpha_{i\zeta^i}^T x - \beta_i t + \gamma_{i\zeta^i} \le 0 \forall (\zeta \in \mathcal{Z}_0 \times ... \times \mathcal{Z}_m) \right\}; \tag{14.2.11}$$

here $\alpha_{i\zeta^i}$, $\gamma_{i\zeta^i}$ are affine in ζ^i and $\beta_i \geq 0$.

In order to prove that $\mathrm{Opt(ARC)} \geq \mathrm{Opt(RC)}$, it suffices to consider the case when $\mathrm{Opt(ARC)} < \infty$ and to show that whenever a real $\bar{t}$ is $> \mathrm{Opt(ARC)}$, we have $\bar{t} \geq \mathrm{Opt(RC)}$. Looking at (14.2.11), we see that to this end it suffices to lead to a contradiction the statement that for some $\bar{t} > \mathrm{Opt(ARC)}$ one has

$$\forall x \in \mathcal{X} \exists (i = i_x \in \{0, 1, ..., m\}, \zeta^i = \zeta^{i_x}_x \in \mathcal{Z}_{i_x}) : \alpha^T_{i_x \zeta^{i_x}_x} x - \beta_{i_x} \bar{t} + \gamma_{i_x \zeta^{i_x}_x} > 0.$$
(14.2.12)

Assume that $\bar{t} > \mathrm{Opt(ARC)}$ and that (14.2.12) holds. For every $x \in \mathcal{X}$, the inequality

$$\alpha^T_{i_x \zeta^{i_x}_x} y - \beta_{i_x} \bar{t} + \gamma_{i_x \zeta^{i_x}_x} > 0$$

is valid when $y = x$; therefore, for every $x \in \mathcal{X}$ there exist $\epsilon_x > 0$ and a neighborhood U_x of x such that

$$\forall y \in U_x : \alpha^T_{i_x \zeta^{i_x}_x} y - \beta_{i_x} \bar{t} + \gamma_{i_x \zeta^{i_x}_x} \geq \epsilon_x.$$

Since $\mathcal{X}$ is a compact set, we can find finitely many points $x^1, ..., x^N$ such that $\mathcal{X} \subset \bigcup_{j=1}^{N} U_{x^j}$. Setting $\epsilon = \min_j \epsilon_{x^j}$, $i[j] = i_{x^j}$, $\zeta[j] = \zeta^{i_{x^j}}_{x^j} \in \mathcal{Z}_{i[j]}$, and

$$f_j(y) = \alpha^T_{i[j], \zeta[j]} y - \beta_{i[j]} \bar{t} + \gamma_{i[j], \zeta[j]},$$

we end up with N affine functions of y such that

$$\max_{1 \leq j \leq N} f_j(y) \geq \epsilon > 0 \; \forall y \in \mathcal{X}.$$

Since $\mathcal{X}$ is a convex compact set and $f_j(\cdot)$ are affine (and thus convex and continuous) functions, the latter relation, by well-known facts from Convex Analysis (namely, the von Neumann Lemma), implies that there exists a collection of nonnegative weights λ_j with $\sum_j \lambda_j = 1$ such that

$$f(y) \equiv \sum_{j=1}^{N} \lambda_j f_j(y) \geq \epsilon \; \forall y \in \mathcal{X}.$$
(14.2.13)

Now let

$$\begin{aligned}
\omega_i &= \sum_{j:i[j]=i} \lambda_j, \; i = 0, 1, ..., m; \\
\bar{\zeta}^i &= \begin{cases} \sum_{j:i[j]=i} \frac{\lambda_j}{\omega_i} \zeta[j], & \omega_i > 0 \\ \text{a point from } \mathcal{Z}_i, & \omega_i = 0 \end{cases}, \\
\bar{\zeta} &= [\bar{\zeta}^0; ...; \bar{\zeta}^m].
\end{aligned}$$

Due to its origin, every one of the vectors $\bar{\zeta}^i$ is a convex combination of points from $\mathcal{Z}_i$ and as such belongs to $\mathcal{Z}_i$, since the latter set is convex. Since the uncertainty is constraint-wise, we conclude that $\bar{\zeta} \in \mathcal{Z}$. Since $\bar{t} > \mathrm{Opt(ARC)}$, we conclude from (14.2.10) that there exists $\bar{x} \in \mathcal{X}$ such that the inequalities

$$\alpha^T_{i\bar{\zeta}^i} \bar{x} - \beta_i \bar{t} + \gamma_{i\bar{\zeta}^i} \leq 0$$

hold true for every i, $0 \leq i \leq m$. Taking a weighted sum of these inequalities, the weights being ω_i, we get

$$\sum_{i:\omega_i>0} \omega_i[\alpha_{i\bar{\zeta}^i}^T\bar{x} - \beta_i\bar{t} + \gamma_{i\bar{\zeta}^i}] \leq 0. \tag{14.2.14}$$

At the same time, by construction of $\bar{\zeta}^i$ and due to the fact that $\alpha_{i\zeta^i}$, $\gamma_{i\zeta^i}$ are affine in ζ^i, for every i with $\omega_i > 0$ we have

$$[\alpha_{i\bar{\zeta}^i}^T\bar{x} - \beta_i\bar{t} + \gamma_{i\bar{\zeta}^i}] = \sum_{j:i[j]=i} \frac{\lambda_j}{\omega_i} f_j(\bar{x}),$$

so that (14.2.14) reads

$$\sum_{j=1}^{N} \lambda_j f_j(\bar{x}) \leq 0,$$

which is impossible due to (14.2.13) and to $\bar{x} \in \mathcal{X}$. We have arrived at the desired contradiction.   □

14.3 AFFINELY ADJUSTABLE ROBUST COUNTERPARTS

We are about to investigate in-depth a specific version of the "parametric decision rules" approach we have outlined previously. At this point, we prefer to come back from general-type uncertain problem (14.1.1) to affinely perturbed uncertain conic problem

$$\mathcal{C} = \left\{ \min_{x\in\mathbb{R}^n} \left\{ c_\zeta^T x + d_\zeta : A_\zeta x + b_\zeta \in \mathbf{K} \right\} : \zeta \in \mathcal{Z} \right\}, \tag{14.3.1}$$

where $c_\zeta, d_\zeta, A_\zeta, b_\zeta$ are affine in ζ, $\mathbf{K}$ is a "nice" cone (direct product of nonnegative rays/Lorentz cones/semidefinite cones, corresponding to uncertain LP/CQP/SDP, respectively), and $\mathcal{Z}$ is a convex compact uncertainty set given by a strictly feasible SDP representation

$$\mathcal{Z} = \left\{ \zeta \in \mathbb{R}^L : \exists u : \mathcal{P}(\zeta, u) \succeq 0 \right\},$$

where $\mathcal{P}$ is affine in $[\zeta; u]$. Assume that along with the problem, we are given an information base $\{P_j\}_{j=1}^n$ for it; here P_j are $m_j \times n$ matrices. To save words (and without risk of ambiguity), we shall call such a pair "uncertain problem $\mathcal{C}$, information base" merely an *uncertain conic problem*. Our course of action is to restrict the ARC of the problem to a specific parametric family of decision rules, namely, the *affine* ones:

$$x_j = X_j(P_j\zeta) = p_j + q_j^T P_j \zeta, \ j = 1,...,n. \tag{14.3.2}$$

The resulting restricted version of the ARC of (14.3.1), which we call the *Affinely Adjustable Robust Counterpart* (AARC), is the semi-infinite optimization program

$$\min_{t,\{p_j,q_j\}_{j=1}^n} \left\{ t : \begin{array}{l} \sum_{j=1}^n c_\zeta^j[p_j + q_j^T P_j \zeta] + d_\zeta - t \leq 0 \\ \sum_{j=1}^n A_\zeta^j[p_j + q_j^T P_j \zeta] + b_\zeta \in \mathbf{K} \end{array} \right\} \forall \zeta \in \mathcal{Z}, \tag{14.3.3}$$

where c_ζ^j is j-th entry in c_ζ, and A_ζ^j is j-th column of A_ζ. Note that the variables in this problem are t and the coefficients p_j, q_j of the affine decision rules (14.3.2). As such, these variables do *not* specify uniquely the actual decisions x_j; these decisions are uniquely defined by these coefficients *and* the corresponding portions $P_j\zeta$ of the true data once the latter become known.

14.3.1 Tractability of the AARC

The rationale for focusing on affine decision rules rather than on other parametric families is that *there exists at least one important case when the AARC of an uncertain conic problem is, essentially, as tractable as the RC of the problem.* The "important case" in question is the one of *fixed recourse* and is defined as follows:

Definition 14.3.1. Consider an uncertain conic problem (14.3.1) augmented by an information base $\{P_j\}_{j=1}^n$. We say that this pair is with fixed recourse, if the coefficients of every adjustable, (i.e., with $P_j \neq 0$), variable x_j are certain:

$$\forall(j : P_j \neq 0): \text{ both } c_\zeta^j \text{ and } A_\zeta^j \text{ are independent of } \zeta.$$

For example, both Examples 14.1.1 (Inventory) and 14.1.2 (Project Management) are uncertain problems with fixed recourse.

An immediate observation is as follows:

(!) *In the case of fixed recourse, the AARC, similarly to the RC, is a semi-infinite conic problem — it is the problem*

$$\min_{t,y=\{p_j,q_j\}} \left\{ t : \begin{array}{l} \widehat{c}_\zeta^T y + d_\zeta \leq t \\ \widehat{A}_\zeta y + b_\zeta \in \mathbf{K} \end{array} \right\} \forall \zeta \in \mathcal{Z} \right\}, \qquad (14.3.4)$$

with $\widehat{c}_\zeta, d_\zeta, \widehat{A}_\zeta, b_\zeta$ affine in ζ:

$$\begin{array}{rl} \widehat{c}_\zeta^T y &= \sum_j c_\zeta^j [p_j + q_j^T P_j \zeta] \\ \widehat{A}_\zeta y &= \sum_j A_\zeta^j [p_j + q_j^T P_j \zeta]. \end{array} \qquad [y = \{[p_j, q_j]\}_{j=1}^n]$$

Note that it is exactly fixed recourse that makes $\widehat{c}_\zeta, \widehat{A}_\zeta$ affine in ζ; without this assumption, these entities are quadratic in ζ.

As far as the tractability issues are concerned, observation (!) is *the* main argument in favor of affine decision rules, *provided we are in the situation of fixed recourse.* Indeed, in the latter situation the AARC is a semi-infinite conic problem, and we can apply to it all the results of Parts I and II related to tractable reformulations/tight safe tractable approximations of semi-infinite conic problems. Note that many of these results, while imposing certain restrictions on the geometries of the uncertainty set and the cone $\mathbf{K}$, require from the objective (if it is uncertain) and the left hand sides of the uncertain constraints nothing more than bi-affinity in the decision variables and in the uncertain data. *Whenever this is the case, the "tractability status" of the AARC is not worse than the one of the usual RC.* In particular, *in the case of fixed recourse we can:*

i) Convert the AARC of an uncertain LO problem into an explicit efficiently solvable "well-structured" convex program (see Theorem 1.3.4).

ii) Process efficiently the AARC of an uncertain conic quadratic problem with (common to all uncertain constraints) simple ellipsoidal uncertainty (see section 6.5).

iii) Use a tight safe tractable approximation of an uncertain problem with linear objective and convex quadratic constraints with (common for all uncertain constraints) ∩-ellipsoidal uncertainty (see section 7.2.3): whenever $\mathcal{Z}$ is the intersection of M ellipsoids centered at the origin, the problem admits a safe tractable approximation tight within the factor $O(1)\sqrt{\ln(M)}$ (see Theorem 7.2.3).

The reader should be aware, however, that the AARC, in contrast to the usual RC, is *not* a constraint-wise construction, since when passing to the coefficients of affine decision rules as our new decision variables, the portion of the uncertain data affecting a particular constraint can change when allowing the original decision variables entering the constraint to depend on the uncertain data not affecting the constraint directly. This is where the words "common" in the second and the third of the above statements comes from. For example, the RC of an uncertain conic quadratic problem with the constraints of the form

$$\|A_\zeta^i x + b_\zeta^i\|_2 \leq x^T c_\zeta^i + d_\zeta^i, \ i = 1, ..., m,$$

is computationally tractable, provided that the projection $\mathcal{Z}_i$ of the overall uncertainty set $\mathcal{Z}$ onto the subspace of data perturbations of i-th constraint is an ellipsoid (section 6.5). To get a similar result for the AARC, we need *the overall uncertainty set $\mathcal{Z}$ itself* to be an ellipsoid, since otherwise the projection of $\mathcal{Z}$ on the data of the "AARC counterparts" of original uncertain constraints can be different from ellipsoids. The bottom line is that the claim that with fixed recourse, the AARC of an uncertain problem is "as tractable" as its RC should be understood with some caution. This, however, is not a big deal, since the "recipe" is already here: *Under the assumption of fixed recourse, the AARC is a semi-infinite conic problem, and in order to process it computationally, we can use all the machinery developed in Parts I and II. If this machinery allows for tractable reformulation/tight safe tractable approximation of the problem, fine, otherwise too bad for us.* " Recall that there exists at least one really important case when everything is fine — this is the case of uncertain LO problem with fixed recourse.

It should be added that when processing the AARC in the case of fixed recourse, we can enjoy all the results on safe tractable approximations of chance constrained affinely perturbed scalar, conic quadratic and linear matrix inequalities developed in Parts I and II. Recall that these results imposed certain restrictions on the distribution of ζ (like independence of $\zeta_1, ..., \zeta_L$), but never required more than affinity of the bodies of the constraints w.r.t. ζ, so that these results work equally well in the cases of RC and AARC.

Last, but not least, the concept of an Affinely Adjustable Robust Counterpart can be straightforwardly "upgraded" to the one of Affinely Adjustable *Globalized* Robust Counterpart. We have no doubts that a reader can carry out such an "upgrade" on his/her own and understands that in the case of fixed recourse, the above "recipe" is equally applicable to the AARC and the AAGRC.

14.3.2 Is Affinity an Actual Restriction?

Passing from *arbitrary* decision rules to *affine* ones seems to be a dramatic simplification. On a closer inspection, the simplification is not as severe as it looks, or, better said, the "dramatics" is not exactly where it is seen at first glance. Indeed, assume that we would like to use decision rules that are quadratic in $P_j\zeta$ rather than linear. Are we supposed to introduce a special notion of a "Quadratically Adjustable Robust Counterpart"? The answer is negative. All we need is to augment the data vector $\zeta = [\zeta_1; ...; \zeta_L]$ by extra entries — the pairwise products $\zeta_i\zeta_j$ of the original entries — and to treat the resulting "extended" vector $\widehat{\zeta} = \widehat{\zeta}[\zeta]$ as our new uncertain data. With this, the decision rules that are quadratic in $P_j\zeta$ become *affine* in $\widehat{P}_j\widehat{\zeta}[\zeta]$, where $\widehat{P}_j$ is a matrix readily given by P_j. More generally, assume that we want to use decision rules of the form

$$X_j(\zeta) = p_j + q_j^T \widehat{P}_j \widehat{\zeta}[\zeta], \tag{14.3.5}$$

where $p_j \in \mathbb{R}, q_j \in \mathbb{R}^{m_j}$ are "free parameters," (which can be restricted to reside in a given convex set), $\widehat{P}_j$ are given $m_j \times D$ matrices and

$$\zeta \mapsto \widehat{\zeta}[\zeta] : \mathbb{R}^L \to \mathbb{R}^D$$

is a given, possibly nonlinear, mapping. Here again we can pass from the original data vector ζ to the data vector $\widehat{\zeta}[\zeta]$, thus making the desired decision rules (14.3.5) merely affine in the "portions" $\widehat{P}_j\widehat{\zeta}$ of the new data vector. We see that when allowing for a seemingly harmless redefinition of the data vector, affine decision rules become as powerful as arbitrary affinely parameterized parametric families of decision rules. This latter class is really huge and, for all practical purposes, is as rich as the class of *all* decision rules. Does it mean that the concept of AARC is basically as flexible as the one of ARC? Unfortunately, the answer is negative, and the reason for the negative answer comes not from potential difficulties with extremely complicated nonlinear transformations $\zeta \mapsto \widehat{\zeta}[\zeta]$ and/or "astronomically large" dimension D of the transformed data vector. The difficulty arises already when the transformation is pretty simple, as is the case, e.g., when the coordinates in $\widehat{\zeta}[\zeta]$ are just the entries of ζ and the pairwise products of these entries. Here is where the difficulty arises. Assume that we are speaking about a single uncertain affinely perturbed scalar linear constraint, allow for quadratic dependence of the original decision variables on the data and pass to the associated adjustable robust counterpart of the constraint. As it was just explained, this counterpart is nothing but a semi-infinite scalar inequality

$$\forall(\widehat{\zeta} \in \mathcal{U}) : a_{0,\widehat{\zeta}} + \sum_{j=1}^{J} a_{j,\widehat{\zeta}} y_j \leq 0$$

where $a_{j,\widehat{\zeta}}$ are affine in $\widehat{\zeta}$, the entries in $\widehat{\zeta} = \widehat{\zeta}[\zeta]$ are the entries in ζ and their pairwise products, $\mathcal{U}$ is the image of the "true" uncertainty set $\mathcal{Z}$ under the *non-linear* mapping $\zeta \to \widehat{\zeta}[\zeta]$, and y_j are our new decision variables (the coefficients of the quadratic decision rules). While the body of the constraint in question is bi-affine in y and in $\widehat{\zeta}$, this semi-infinite constraint can well be intractable, since the *uncertainty set $\mathcal{U}$ may happen to be intractable, even when $\mathcal{Z}$ is tractable.* Indeed, the tractability of a semi-infinite bi-affine scalar constraint

$$\forall(u \in \mathcal{U}) : f(y, u) \leq 0$$

heavily depends on whether the underlying uncertainty set $\mathcal{U}$ is convex and computationally tractable. When it is the case, we can, modulo minor technical assumptions, solve efficiently the *Analysis problem* of checking whether a given candidate solution y is feasible for the constraint — to this end, it suffices to maximize the affine function $f(y, \cdot)$ over the computationally tractable convex set $\mathcal{U}$. This, under minor technical assumptions, can be done efficiently. The latter fact, in turn, implies (again modulo minor technical assumptions) that we can optimize efficiently linear/convex objectives under the constraints with the above features, and this is basically all we need. The situation changes dramatically when the uncertainty set $\mathcal{U}$ is *not* a convex computationally tractable set. By itself, the convexity of $\mathcal{U}$ costs nothing: since f is bi-affine, the feasible set of the semi-infinite constraint in question remains intact when we replace $\mathcal{U}$ with its convex hull $\widehat{\mathcal{Z}}$. The actual difficulty is that the convex hull $\widehat{\mathcal{Z}}$ of the set $\mathcal{U}$ can be computationally intractable. In the situation we are interested in — the one where $\widehat{\mathcal{Z}} = \mathrm{Conv}\mathcal{U}$ and $\mathcal{U}$ is the image of a computationally tractable convex set $\mathcal{Z}$ under a *nonlinear* transformation $\zeta \mapsto \widehat{\zeta}[\zeta]$, $\widehat{\mathcal{Z}}$ can be computationally intractable already for pretty simple $\mathcal{Z}$ and nonlinear mappings $\zeta \mapsto \widehat{\zeta}[\zeta]$. It happens, e.g., when $\mathcal{Z}$ is the unit box $\|\zeta\|_\infty \leq 1$ and $\widehat{\zeta}[\zeta]$ is comprised of the entries in ζ and their pairwise products. In other words, the "Quadratically Adjustable Robust Counterpart" of an uncertain linear inequality with interval uncertainty is, in general, computationally intractable.

In spite of the just explained fact that "global linearization" of nonlinear decision rules via nonlinear transformation of the data vector not necessarily leads to tractable adjustable RCs, one should keep in mind this option, since it is important methodologically. Indeed, "global linearization" allows one to "split" the problem of processing the ARC, restricted to decision rules (14.3.5), into two subproblems:

(a) Building a tractable representation (or a tight tractable approximation) of the convex hull $\widehat{\mathcal{Z}}$ of the image $\mathcal{U}$ of the original uncertainty set $\mathcal{Z}$ under the nonlinear mapping $\zeta \mapsto \widehat{\zeta}[\zeta]$ associated with (14.3.5). Note that this problem by itself has nothing to do with adjustable robust counterparts and the like;

(b) Developing a tractable reformulation (or a tight safe tractable approximation) of the *Affinely Adjustable* Robust Counterpart of the uncertain problem in question, with $\widehat{\zeta}$ in the role of the data vector, the tractable convex set, yielded by (a), in the role of the uncertainty set, and the information base given by the matrices $\widehat{P}_j$.

Of course, the resulting two problems are not completely independent: the tractable convex set $\widehat{\mathcal{Z}}$ with which we, upon success, end up when solving (a) should be simple enough to allow for successful processing of (b). Note, however, that this "coupling of problems (a) and (b)" is of no importance when the uncertain problem in question is an LO problem with fixed recourse. Indeed, in this case the AARC of the problem is computationally tractable whatever the uncertainty set as long as it is tractable, therefore every tractable set $\widehat{\mathcal{Z}}$ yielded by processing of problem (a) will do.

Example 14.3.2. Assume that we want to process an uncertain LO problem

$$
\mathcal{C} = \left\{ \min_x \left\{ c_\zeta^T x + d_\zeta : A_\zeta x \geq b_\zeta \right\} : \zeta \in \mathcal{Z} \right\}
$$
$$
[c_\zeta, d_\zeta, A_\zeta, b_\zeta : \text{ affine in } \zeta] \tag{14.3.6}
$$

with fixed recourse and a tractable convex compact uncertainty set $\mathcal{Z}$, and consider a number of affinely parameterized families of decision rules.

A. *"Genuine" affine decision rules: x_j is affine in $P_j\zeta$.* As we have already seen, the associated ARC — the usual AARC of $\mathcal{C}$ — is computationally tractable.

B. *Piece-wise linear decision rules with fixed breakpoints.* Assume that the mapping $\zeta \mapsto \widehat{\zeta}[\zeta]$ augments the entries of ζ with finitely many entries of the form $\phi_i(\zeta) = \max\left[r_i, s_i^T\zeta\right]$, and the decision rules we intend to use should be affine in $\widehat{P}_j\widehat{\zeta}$, where $\widehat{P}_j$ are given matrices. In order to process the associated ARC in a computationally efficient fashion, all we need is to build a tractable representation of the set $\widehat{\mathcal{Z}} = \text{Conv}\{\widehat{\zeta}[\zeta] : \zeta \in \mathcal{Z}\}$. While this could be difficult in general, there are useful cases when the problem is easy, e.g., the case where

$$
\mathcal{Z} = \{\zeta \in \mathbb{R}^L : f_k(\zeta) \leq 1, 1 \leq k \leq K\},
$$
$$
\widehat{\zeta}[\zeta] = [\zeta; (\zeta)_+; (\zeta)_-], \text{ with } (\zeta)_- = \max[\zeta, 0_{L\times 1}], (\zeta)_+ = \max[-\zeta, 0_{L\times 1}].
$$

Here, for vectors u, v, $\max[u, v]$ is taken coordinate-wise, and $f_k(\cdot)$ are lower semicontinuous and *absolutely symmetric* convex functions on $\mathbb{R}^L$, absolute symmetry meaning that $f_k(\zeta) \equiv f_k(\text{abs}(\zeta))$ (abs acts coordinate-wise). (Think about the case when $f_k(\zeta) = \|[\alpha_{k1}\zeta_1; ...; \alpha_{kL}\zeta_L]\|_{p_k}$ with $p_k \in [1, \infty]$.) It is easily seen that if $\mathcal{Z}$ is bounded, then

$$
\widehat{\mathcal{Z}} = \left\{ \widehat{\zeta} = [\zeta; \zeta^+; \zeta^-] : \begin{array}{ll} (a) & f_k(\zeta^+ + \zeta^-) \leq 1, 1 \leq k \leq K \\ (b) & \zeta = \zeta^+ - \zeta^- \\ (c) & \zeta^\pm \geq 0 \end{array} \right\}.
$$

Indeed, (a) through (c) is a system of convex constraints on vector $\widehat{\zeta} = [\zeta; \zeta^+; \zeta^-]$, and since f_k are lower semicontinuous, the feasible set C of this system is convex and closed; besides, for $[\zeta; \zeta^+; \zeta^-] \in C$ we have $\zeta^+ + \zeta^- \in \mathcal{Z}$; since the latter set is bounded by assumption, the sum $\zeta^+ + \zeta^-$ is bounded uniformly in $\widehat{\zeta} \in C$, whence, by (a) through (c),

C is bounded. Thus, C is a closed and bounded convex set. The image $\mathcal{U}$ of the set $\mathcal{Z}$ under the mapping $\zeta \mapsto [\zeta; (\zeta)_+; (\zeta)_-]$ clearly is contained in C, so that the convex hull $\widehat{\mathcal{Z}}$ of $\mathcal{U}$ is contained in C as well. To prove the inverse inclusion, note that since C is a (nonempty) convex compact set, it is the convex hull of the set of its extreme points, and therefore in order to prove that $\widehat{\mathcal{Z}} \supset C$ it suffices to verify that every extreme point $[\zeta; \zeta^+, \zeta^-]$ of C belongs to $\mathcal{U}$. But this is immediate: in an extreme point of C we should have $\min[\zeta_\ell^+, \zeta_\ell^-] = 0$ for every ℓ, since if the opposite were true for some $\ell = \bar{\ell}$, then C would contain a nontrivial segment centered at the point, namely, points obtained from the given one by the "3-entry perturbation" $\zeta_{\bar{\ell}}^+ \mapsto \zeta_{\bar{\ell}}^+ + \delta$, $\zeta_{\bar{\ell}}^- \mapsto \zeta_{\bar{\ell}}^- - \delta$, $\zeta_{\bar{\ell}} \mapsto \zeta_{\bar{\ell}} + 2\delta$ with small enough $|\delta|$. Thus, every extreme point of C has $\min[\zeta^+, \zeta^-] = 0$, $\zeta = \zeta^+ - \zeta^-$, and a point of this type satisfying (a) clearly belongs to $\mathcal{U}$. $\square$

C. *Separable decision rules.* Assume that $\mathcal{Z}$ is a box: $\mathcal{Z} = \{\zeta : \underline{a} \leq \zeta \leq \overline{a}\}$, and we are seeking for *separable* decision rules with a prescribed "information base," that is, for the decision rules of the form

$$x_j = \xi_j + \sum_{\ell \in I_j} f_\ell^j(\zeta_\ell), \ j = 1, ..., n, \tag{14.3.7}$$

where the only restriction on functions f_ℓ^j is to belong to given finite-dimensional linear spaces $\mathcal{F}_\ell$ of univariate functions. The sets I_j specify the information base of our decision rules. Some of these sets may be empty, meaning that the associated x_j are non-adjustable decision variables, in full accordance with the standard convention that a sum over an empty set of indices is 0. We consider two specific choices of the spaces $\mathcal{F}_\ell$:

C.1: $\mathcal{F}_\ell$ is comprised of all piecewise linear functions on the real axis with fixed breakpoints $a_{\ell 1} < ... < a_{\ell m}$ (w.l.o.g., assume that $\underline{a}_\ell < a_{\ell 1}$, $a_{\ell m} < \overline{a}_\ell$);

C.2: $\mathcal{F}_\ell$ is comprised of all algebraic polynomials on the axis of degree $\leq \kappa$.

Note that what follows works when m in **C.1** and κ in **C.2** depend on ℓ; in order to simplify notation, we do not consider this case explicitly.

C.1: Let us augment every entry ζ_ℓ of ζ with the reals $\zeta_{\ell i}[\zeta_\ell] = \max[\zeta_\ell, a_{\ell i}]$, $i = 1, ..., m$, and let us set $\zeta_{\ell 0}[\zeta_\ell] = \zeta_\ell$. In the case of **C.1**, decision rules (14.3.7) are exactly the rules where x_j is affine in $\{\zeta_{\ell i}[\zeta] : \ell \in I_j\}$; thus, all we need in order to process efficiently the ARC of (14.3.6) restricted to the decision rules in question is a tractable representation of the convex hull of the image $\mathcal{U}$ of $\mathcal{Z}$ under the mapping $\zeta \mapsto \{\zeta_{\ell i}[\zeta]\}_{\ell, i}$. Due to the direct product structure of $\mathcal{Z}$, the set $\mathcal{U}$ is the direct product, over $\ell = 1, ..., d$, of the sets

$$\mathcal{U}_\ell = \{[\zeta_{\ell 0}[\zeta_\ell]; \zeta_{\ell 1}[\zeta_\ell]; ...; \zeta_{\ell m}[\zeta_\ell]] : \underline{a}_\ell \leq \zeta_\ell \leq \overline{a}_\ell\},$$

so that all we need are tractable representations of the convex hulls of the sets $\mathcal{U}_\ell$. The bottom line is, that all we need is a tractable description of a set C of the form

$$C_m = \text{Conv} S_m, \ S_m = \{[s_0; \max[s_0, a_1]; ...; \max[s_0, a_m]] : a_0 \leq s_0 \leq a_{m+1}\},$$

where $a_0 < a_1 < a_2 < ... < a_m < a_{m+1}$ are given.

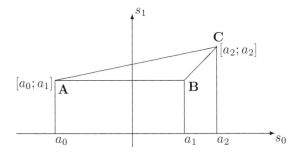

Figure 14.2 S_1 (union of segments AB and BC) and $C_1 = \text{Conv} S_1$ (triangle ABC)

Let us first consider the case of $m = 1$. Here

$$S_1 = \{[s_0; s_1] = [s_0; \max[s_0, a_1]] : a_0 \le s_0 \le a_2\}.$$

This set and its convex hull $C_1 = \text{Conv} S_1$ are shown in figure 14.2. The set C_1 is given by the following inequalities:

$$a_0 \le s_0 \le a_2, \; s_1 \ge \max[s_0, a_1], \; s_1 \le \frac{a_2 - a_1}{a_2 - a_0}(s_0 - a_0) + a_1.$$

After rearranging these three inequalities, a representation of C_1 is given equivalently as

$$C_1 = \text{Conv} S_1 = \left\{ [s_0; s_1] : \begin{array}{c} 0 \le \frac{s_1 - s_0}{a_1 - a_0} \le \frac{a_2 - s_1}{a_2 - a_1} \le 1 \\ a_0 \le s_0 \le a_2 \end{array} \right\}.$$

The result can be generalized for $m > 1$ as follows:

Lemma 14.3.3. The convex hull C_m of the set S_m is

$$C_m = \left\{ [s_0; s_1; ...; s_m] : \left\{ \begin{array}{l} a_0 \le s_0 \le a_{m+1} \\ 0 \le \frac{s_1 - s_0}{a_1 - a_0} \le \frac{s_2 - s_1}{a_2 - a_1} \le ... \le \frac{s_{m+1} - s_m}{a_{m+1} - a_m} \le 1 \end{array} \right. \right\}, \quad (14.3.8)$$

where $s_{m+1} = a_{m+1}$.

Proof. It is convenient to make affine substitution of variables as follows:

$$\mathcal{P} : [s_0; s_1; ...; s_m] \mapsto \begin{array}{l} [\delta_0 = s_0 - a_0; \delta_1 = s_1 - s_0; \delta_2 = s_2 - s_1; \\ ...; \delta_m = s_m - s_{m-1}; \delta_{m+1} = a_{m+1} - s_m]. \end{array}$$

$\mathcal{P}$ is an affine embedding that maps, in a one-to-one fashion, the $(m+1)$-dimensional space of s variables onto the hyperplane $\delta_0 + \delta_1 + ... + \delta_{m+1} = a_{m+1} - a_0$ in the $(m+2)$-dimensional space of δ variables. The image of the right hand side of (14.3.8) under this mapping is the set

$$P = \left\{ \delta = [\delta_0; ...; \delta_{m+1}] : \left\{ \begin{array}{ll} 0 \le \delta_0 \le d \equiv a_{m+1} - a_0 & (a) \\ 0 \le \frac{\delta_1}{d_1} \le \frac{\delta_2}{d_2} \le ... & \\ \quad \le \frac{\delta_{m+1}}{d_{m+1}} \le 1, \; d_i = a_i - a_{i-1} & (b) \\ \delta_0 + \delta_1 + ... + \delta_{m+1} = d & (c) \end{array} \right. \right\}; \quad (14.3.9)$$

the image of S_m under the same mapping we denote by S_+. Since $\mathcal{P}$ is an affine embedding, to prove (14.3.8) is exactly the same as to prove that $P = \text{Conv} S_+$, and this is what we intend to do.

Let us first prove that $P \supset \mathrm{Conv}S_+$. Since P clearly is convex, it suffices to verify that if

$$\begin{aligned}
\delta &= \mathcal{P}([s_0; s_1 = \max[s_0, a_1]; ...; s_{m+1} = \max[s_0, a_{m+1}]]) \\
&\equiv [s_0 - a_0; s_1 - s_0; s_2 - s_1; s_3 - s_2; ...; s_{m+1} - s_m]
\end{aligned}$$

with $a_0 \le s_0 \le a_{m+1} \equiv s_{m+1}$, then $\delta \in P$. The fact that δ satisfies (a) and (c) (from now on, (a) through (c) refer to the respective relations in (14.3.9)) is evident. To verify (b), observe, first, that $a_0 \le s_0 \le s_1 \le ... \le s_m \le a_{m+1}$, whence all δ_i are ≥ 0. Let j be such that $s_0 \in [a_{j-1}, a_j]$. Then $s_i = s_0$ for $i < j$ and $s_i = a_i$ for $i \ge j$, whence $\delta_i/d_i = 0$ for $1 \le i \le j - 1$, $\delta_j/d_j = \frac{a_j - s_0}{d_j}$ and $\delta_i/d_i = 1$ for $i > j$. Since $0 \le a_j - s_0 \le d_j$, (b) follows. Thus, $P \supset \mathrm{Conv}S_+$. It remains to prove the opposite inclusion. Since P clearly is a nonempty convex compact set, in order to prove that $P \subset \mathrm{Conv}S_+$ it suffices to verify that if $\delta = [\delta_0; ...; \delta_{m+1}]$ is an extreme point of P, then $\delta = \mathcal{P}(s)$ for certain $s \in S_m$, and this is what we are about to do. By (a), we have $0 \le \delta_0 \le d$, and by (b) we have $0 \le \delta_i \le d_i$, $1 \le i \le m+1$. Since $d = d_1 + ... + d_{m+1}$, in the case of $\delta_0 = 0$ (c) implies that $\delta_i = d_i$ for $i = 1, ..., m + 1$, so that δ is the point $[0; d_1; ...; d_{m+1}]$, and this point indeed is $\mathcal{P}(s)$ with $s = [a_0; a_1; ...; a_m] \in S_m$. When $\delta_0 = d$, (c) implies that $\delta_1 = ... = \delta_{m+1} = 0$ (note that by (a), (b) $\delta \ge 0$ for all $\delta \in P$). Thus, here $\delta = [a_{m+1}; 0; ...; 0]$, and this point is $\mathcal{P}(s)$ with $s = [a_{m+1}; a_{m+1}; ...; a_{m+1}] \in S_m$. It remains to consider the case when δ is an extreme point of P and $0 < \delta_0 < d$. We claim that the fractions in (b) take at most two values, namely, 0 and 1. We shall justify this claim later, and meanwhile let us derive from it that δ indeed is $\mathcal{P}(s)$ with $s \in S_m$. Given the claim, just three options are possible:

— all fractions in (b) are equal to 0. In this case $\delta_1 = ... = \delta_{m+1} = 0$, and thus $\delta_0 = d$ by (c), which is not the case;

— all fractions in (b) are equal to 1. In this case $\delta_i = d_i$, $i = 1, ..., m+1$, and $\delta_0 = 0$ by (c), which again is not the case;

— for certain j, $1 \le j \le m$, the fractions $\frac{\delta_i}{d_i}$ are equal to 0 when $i \le j$ and are equal to 1 when $i > j$, or, which is the same, $\delta_i = 0$ for $1 \le i \le j$ and $\delta_i = d_i$ for $i > j$. Invoking (c), we see that $\delta_0 = d - d_{j+1} - d_{j+2} - ... - d_{m+1} = d_1 + ... + d_j = a_j - a_0$, so that $\delta = [a_j - a_0; 0, ..., 0; a_{j+1} - a_j; a_{j+2} - a_{j+1}; ...; a_{m+1} - a_m]$. But this point indeed is $\mathcal{P}(s)$ for $s = [a_j; a_j; ...; a_j; a_{j+1}; a_{j+2}; a_{j+3}; ...; a_m] \in S_m$.

It remains to justify our claim. Assume, on the contrary to what should be proved, that among the fractions $\frac{\delta_i}{d_i}$, $i = 1, ..., m + 1$, there is a fraction taking a value $\theta \in (0, 1)$, and let I be the set of indices of all fractions that are equal to θ; note that by (b) I is a segment of consecutive indices from the sequence $1, 2, ..., m + 1$. Setting $q = \sum_{i \in I} d_i$, consider the following perturbation of vector δ:

$$\delta \mapsto \delta[t] = [\delta_0[t]; ...; \delta_{m+1}[t]], \quad \delta_i[t] = \begin{cases} \delta_0 + t, & i = 0 \\ \delta_i, & i \ge 1, i \notin I \\ \delta_i - \frac{d_i}{q} t, & i \in I \end{cases}.$$

We claim that when $|t|$ is small enough, we have $\delta[t] \in P$. Indeed, for small $|t|$ the vector $\delta[t]$

— satisfies (a) due to $0 < \delta_0 < d$;

— satisfies (b), since the fractions $\frac{\delta_i[t]}{d_i}$ for $i \in I$ are equal to each other and close to θ, and the remaining fractions stay intact;

— satisfies (c), since $\sum_{i=0}^{m+1} \delta_i[t]$ is independent of t due to the origin of q.

We see that δ is the midpoint of a nontrivial segment $\{\delta[t] : -\epsilon \le t \le \epsilon\}$ which for small enough $\epsilon > 0$ is contained in P; but this is impossible, since δ is an extreme point of P. $\square$

C.2: Similar to the case of **C.1**, in the case of **C.2** all we need in order to process efficiently the ARC of (14.3.6), restricted to decision rules (14.3.7), is a tractable representation of the set

$$C = \text{Conv} S, \quad S = \{\widehat{s} = [s; s^2; ...; s^\kappa] : |s| \le 1\}.$$

(We have assumed w.l.o.g. that $\underline{a}_\ell = -1$, $\overline{a}_\ell = 1$.) Here is the description (originating from [87]):

Lemma 14.3.4. The set $C = \text{Conv} S$ admits the explicit semidefinite representation

$$C = \left\{ \widehat{s} \in \mathbb{R}^\kappa : \exists \lambda = [\lambda_0; ...; \lambda_{2\kappa}] \in \mathbb{R}^{2\kappa+1} : [1; \widehat{s}] = Q^T \lambda, [\lambda_{i+j}]_{i,j=0}^\kappa \succeq 0 \right\}, \tag{14.3.10}$$

where the $(2\kappa+1) \times (\kappa+1)$ matrix Q is defined as follows: take a polynomial $p(t) = p_0 + p_1 t + ... + p_\kappa t^\kappa$ and convert it into the polynomial $\widehat{p}(t) = (1+t^2)^\kappa p(2t/(1+t^2))$. The vector of coefficients of $\widehat{p}$ clearly depends linearly on the vector of coefficients of p, and Q is exactly the matrix of this linear transformation.

Proof. 1^0. Let $P \subset \mathbb{R}^{\kappa+1}$ be the cone of vectors p of coefficients of polynomials $p(t) = p_0 + p_1 t + p_2 t^2 + ... + p_\kappa t^\kappa$ that are nonnegative on $[-1, 1]$, and P_* be the cone dual to P. We claim that

$$C = \{\widehat{s} \in \mathbb{R}^\kappa : [1; \widehat{s}] \in P_*\}. \tag{14.3.11}$$

Indeed, let C' be the right hand side set in (14.3.11). If $\widehat{s} = [s; s^2; ...; s^\kappa] \in S$, then $|s| \le 1$, so that for every $p \in P$ we have $p^T [1; \widehat{s}] = p(s) \ge 0$. Thus, $[1; \widehat{s}] \in P_*$ and therefore $\widehat{s} \in C'$. Since C' is convex, we arrive at $C \equiv \text{Conv} S \subset C'$. To prove the inverse inclusion, assume that there exists $\widehat{s} \notin C$ such that $z = [1; \widehat{s}] \in P_*$, and let us lead this assumption to a contradiction. Since $\widehat{s}$ is not in C and C is a closed convex set and clearly contains the origin, we can find a vector $q \in \mathbb{R}^\kappa$ such that $q^T \widehat{s} = 1$ and $\max_{r \in C} q^T r \equiv \alpha < 1$, or, which is the same due to $C = \text{Conv} S$, $q^T [s; s^2; ...; s^\kappa] \le \alpha < 1$ whenever $|s| \le 1$. Setting $p = [\alpha; -q]$, we see that $p(s) \ge 0$ whenever $|s| \le 1$, so that $p \in P$ and therefore $\alpha - q^T \widehat{s} = p^T [1; \widehat{s}] \ge 0$, whence $1 = q^T \widehat{s} \le \alpha < 1$, which is a desired contradiction.

2^0. It remains to verify that the right hand side in (14.3.11) indeed admits representation (14.3.10). We start by deriving a semidefinite representation of the cone P_+ of (vectors of coefficients of) all polynomials $p(s)$ of degree not exceeding 2κ that are nonnegative on the entire axis. The representation is as follows. A $(\kappa + 1) \times (\kappa + 1)$ symmetric matrix W can be associated with the polynomial of degree $\le 2\kappa$ given by $p_W(t) = [1; t; t^2; ...; t^\kappa]^T W [1; t; t^2; ...; t^\kappa]$, and the mapping $\mathcal{A} : W \mapsto p_W$ clearly is linear: $\left(\mathcal{A}[w_{ij}]_{i,j=0}^\kappa \right)_\nu = \sum_{0 \le i \le \nu} w_{i,\nu-i}$, $0 \le \nu \le 2\kappa$. A dyadic matrix $W = ee^T$ "produces" in this way a polynomial that is the square of another polynomial: $\mathcal{A}ee^T = e^2(t)$ and as such is ≥ 0 on the entire axis. Since every matrix $W \succeq 0$ is a sum of dyadic matrices, we conclude that $\mathcal{A}W \in P_+$ whenever $W \succeq 0$. Vice versa, it is well known that every polynomial $p \in P_+$ is the sum of squares of polynomials of degrees $\le \kappa$, meaning that every $p \in P_+$ is $\mathcal{A}W$ for certain W that is the sum of dyadic matrices and as such is $\succeq 0$. Thus,

$$P_+ = \{p = \mathcal{A}W : W \in \mathbf{S}_+^{\kappa+1}\}.$$

Now, the mapping $t \mapsto 2t/(1 + t^2) : \mathbb{R} \to \mathbb{R}$ maps $\mathbb{R}$ onto the segment $[-1, 1]$. It follows that a *polynomial p of degree $\le \kappa$ is ≥ 0 on $[-1, 1]$ if and only if the polynomial $\widehat{p}(t) = (1 + t^2)^\kappa p(2t/(1 + t^2))$ of degree $\le 2\kappa$ is ≥ 0 on the entire axis*, or, which is the same, $p \in P$ if and only if $Qp \in P_+$. Thus,

$$P = \{p \in \mathbb{R}^{\kappa+1} : \exists W \in \mathbf{S}^{\kappa+1} : W \succeq 0, \mathcal{A}W = Qp\}.$$

Given this semidefinite representation of P, we can immediately obtain a semidefinite representation of P_*. Indeed,

$$q \in P_* \Leftrightarrow 0 \leq \min_{p \in P}\{q^T p\} \Leftrightarrow 0 \leq \min_{p \in \mathbb{R}^\kappa}\{q^T p : \exists W \succeq 0 : Qp = \mathcal{A}W\}$$
$$\Leftrightarrow 0 \leq \min_{p, W}\{q^T p : Qp - \mathcal{A}W = 0, W \succeq 0\}$$
$$\Leftrightarrow \{q = Q^T \lambda : \lambda \in \mathbb{R}^{2\kappa+1}, \mathcal{A}^* \lambda \succeq 0\},$$

where the concluding $\Leftrightarrow$ is due to semidefinite duality. Computing $\mathcal{A}^* \lambda$, we arrive at (14.3.10). $\qquad\square$

Remark 14.3.5. Note that **C.2** admits a straightforward modification where the spaces $\mathcal{F}_\ell$ are comprised of trigonometric polynomials $\sum_{i=0}^{\kappa}[p_i \cos(i\omega_\ell s) + q_i \sin(i\omega_\ell s)]$ rather than of algebraic polynomials $\sum_{i=0}^{\kappa} p_i s^i$. Here all we need is a tractable description of the convex hull of the curve

$$\{[s; \cos(\omega_\ell s); \sin(\omega_\ell s); ...; \cos(\kappa\omega_\ell s); \sin(\kappa\omega_\ell s)] : -1 \leq s \leq 1\}$$

which can be easily extracted from the semidefinite representation of the cone P_+.

Discussion. There are items to note on the results stated in **C**. The bad news is that *understood literally, these results have no direct consequences in our context — when $\mathcal{Z}$ is a box, decision rules* (14.3.7) *never outperform "genuine" affine decision rules with the same information base* (that is, the decision rules (14.3.7) with the spaces of affine functions on the axis in the role of $\mathcal{F}_\ell$).

The explanation is as follows. Consider, instead of (14.3.6), a more general problem, specifically, the uncertain problem

$$\mathcal{C} = \left\{\min_x \left\{c_\zeta^T x + d_\zeta : A_\zeta x - b_\zeta \in \mathbf{K}\right\} : \zeta \in \mathcal{Z}\right\} \qquad (14.3.12)$$
$$[c_\zeta, d_\zeta, A_\zeta, b_\zeta : \text{ affine in } \zeta]$$

where $\mathbf{K}$ is a convex set. Assume that $\mathcal{Z}$ is a direct product of simplexes: $\mathcal{Z} = \Delta_1 \times ... \times \Delta_L$, where Δ_ℓ is a k_ℓ-dimensional simplex (the convex hull of $k_\ell + 1$ affinely independent points in $\mathbb{R}^{k_\ell}$). Assume we want to process the ARC of this problem restricted to the decision rules of the form

$$x_j = \xi_j + \sum_{\ell \in I_j} f_\ell^j(\zeta_\ell), \qquad (14.3.13)$$

where ζ_ℓ is the projection of $\zeta \in \mathcal{Z}$ on Δ_ℓ, and the only restriction on the functions f_ℓ^j is that they belong to given families $\mathcal{F}_\ell$ of functions on $\mathbb{R}^{k_\ell}$. We still assume fixed recourse: the columns of A_ζ and the entries in c_ζ associated with adjustable, (i.e., with $I_j \neq \emptyset$) decision variables x_j are independent of ζ.

The above claim that "genuinely affine" decision rules are not inferior as compared to the rules (14.3.7) is nothing but the following simple observation:

Lemma 14.3.6. Whenever certain $t \in \mathbb{R}$ is an achievable value of the objective in the ARC of (14.3.12) restricted to the decision rules (14.3.13), that is, there exist decision rules of the latter form such that

$$\left. \begin{array}{l} \sum_{j=1}^{n} \left[\xi_j + \sum_{\ell \in I_j} f_\ell^j(\zeta_\ell) \right] (c_\zeta)_j + d_\zeta \leq t \\ \sum_{j=1}^{n} \left[\xi_j + \sum_{\ell \in I_j} f_\ell^j(\zeta_\ell) \right] A_\zeta^j - b_\zeta \in \mathbf{K} \end{array} \right\} \begin{array}{l} \forall \zeta \ \in \ [\zeta_1; ...; \zeta_L] \in \mathcal{Z} \\ \quad = \ \Delta_1 \times ... \times \Delta_L, \end{array}$$

(14.3.14)

t is also an achievable value of the objective in the ARC of the uncertain problem restricted to affine decision rules with the same information base: there exist affine in ζ_ℓ functions $\phi_\ell^j(\zeta_\ell)$ such that (14.3.14) remains valid with ϕ_ℓ^j in the role of f_ℓ^j.

Proof is immediate: since every collection of $k_\ell + 1$ reals can be obtained as the collection of values of an affine function at the vertices of k_ℓ-dimensional simplex, we can find affine functions $\phi_\ell^j(\zeta_\ell)$ such that $\phi_\ell^j(\zeta_\ell) = f_\ell^j(\zeta_\ell)$ whenever ζ_ℓ is a vertex of the simplex Δ_ℓ. When plugging into the left hand sides of the constraints in (14.3.14) the functions $\phi_\ell^j(\zeta_\ell)$ instead of $f_\ell^j(\zeta_\ell)$, these left hand sides become affine functions of ζ (recall that we are in the case of fixed recourse). Due to this affinity and to the fact that $\mathcal{Z}$ is a convex compact set, in order for the resulting constraints to be valid for all $\zeta \in \mathcal{Z}$, it suffices for them to be valid at every one of the extreme points of $\mathcal{Z}$. The components $\zeta_1, ..., \zeta_L$ of such an extreme point ζ are vertices of $\Delta_1, ..., \Delta_L$, and therefore the validity of "ϕ constraints" at ζ is readily given by the validity of the "f constraints" at this point — by construction, at such a point the left hand sides of the "ϕ" band the "f" constraints coincide with each other. $\square$

Does the bad news mean that our effort in **C.1–2** was just wasted? The good news is that this effort still can be utilized. Consider again the case where ζ_ℓ are scalars, assume that $\mathcal{Z}$ is not a box, in which case Lemma 14.3.6 is not applicable. Thus, we have hope that the ARC of (14.3.6) restricted to the decision rules (14.3.7) is indeed less conservative (has a strictly less optimal value) than the ARC restricted to the affine decision rules. What we need in order to process the former, "more promising," ARC, is a tractable description of the convex hull $\widehat{\mathcal{Z}}$ of the image $\mathcal{U}$ of $\mathcal{Z}$ under the mapping

$$\zeta \mapsto \widehat{\zeta}[\zeta] = \{\zeta_{\ell i}[\zeta_\ell]\}_{\substack{0 \leq i \leq m, \\ 1 \leq \ell \leq L}}$$

where $\zeta_{\ell 0} = \zeta_\ell$, $\zeta_{\ell i}[\zeta_\ell] = f_{i\ell}(\zeta_\ell)$, $1 \leq i \leq m$, and the functions $f_{i\ell} \in \mathcal{F}_\ell$, $i = 1, ..., m$, span $\mathcal{F}_\ell$. The difficulty is that *with $\mathcal{F}_\ell$ as those considered in **C.1–2*** (these families are "rich enough" for most of applications), *we, as a matter of fact, do not know how to get a tractable representation of $\widehat{\mathcal{Z}}$, unless $\mathcal{Z}$ is a box.* Thus, $\mathcal{Z}$ more complicated than a box seems to be too complex, and when $\mathcal{Z}$ is a box, we gain nothing from allowing for "complex" $\mathcal{F}_\ell$. Nevertheless, we can proceed as follows. Let us include $\mathcal{Z}$, (which is not a box), into a box $\mathcal{Z}^+$, and let us apply the outlined approach to $\mathcal{Z}^+$ in the role of $\mathcal{Z}$, that is, let us try to build a tractable description of the convex hull $\widehat{\mathcal{Z}}^+$ of the image $\mathcal{U}^+$ of $\mathcal{Z}^+$ under the mapping $\zeta \mapsto \widehat{\zeta}[\zeta]$. With luck, (e.g., in situations **C.1–2**), we will succeed, thus getting a tractable representation of $\widehat{\mathcal{Z}}^+$; the latter set is, of course, larger than the "true" set $\widehat{\mathcal{Z}}$ we want to describe. There is another "easy to describe" set that contains $\widehat{\mathcal{Z}}$, namely, the inverse image $\widehat{\mathcal{Z}}^0$ of $\mathcal{Z}$ under the natural projection $\Pi : \{\zeta_{\ell i}\}_{\substack{0 \leq i \leq m, \\ 1 \leq \ell \leq L}} \mapsto \{\zeta_{\ell 0}\}_{1 \leq \ell \leq L}$ that recovers ζ from

$\widehat{\zeta}[\zeta]$. And perhaps we are smart enough to find other easy to describe convex sets $\widehat{\mathcal{Z}}^1,...,\widehat{\mathcal{Z}}^k$ that contain $\widehat{\mathcal{Z}}$.

Assume, e.g., that $\mathcal{Z}$ is the Euclidean ball $\{\|\zeta\|_2 \leq r\}$, and let us take as $\mathcal{Z}^+$ the embedding box $\{\|\zeta\|_\infty \leq r\}$.

In the case of **C.1** we have for $i \geq 1$: $\zeta_{\ell i}[\zeta_\ell] = \max[\zeta_\ell, a_{\ell i}]$, whence $|\zeta_{\ell i}[\zeta_\ell]| \leq \max[|\zeta_\ell|, |a_{\ell i}|]$. It follows that when $\zeta \in \mathcal{Z}$, we have $\sum_\ell \zeta_{\ell i}^2[\zeta_\ell] \leq \sum_\ell \max[\zeta_\ell^2, a_{\ell i}^2] \leq \sum_\ell [\zeta_\ell^2 + a_{\ell i}^2] \leq r^2 + \sum_\ell a_{\ell i}^2$, and we can take as $\widehat{\mathcal{Z}}^p$, $p = 1,...,m$, the elliptic cylinders $\{\{\zeta_{\ell i}\}_{\ell,i} : \sum_\ell \zeta_{\ell p}^2 \leq r^2 + \sum_\ell a_{\ell p}^2\}$. In the case of **C.2**, we have $\zeta_{\ell i}[\zeta_\ell] = \zeta_\ell^{i+1}$, $1 \leq i \leq \kappa - 1$, so that $\sum_\ell |\zeta_{\ell i}[\zeta_\ell]| \leq \max_{z \in \mathbb{R}^L}\{\sum_\ell |z_\ell|^{i+1} : \|z\|_2 \leq r\} = r^{i+1}$. Thus, we can take $\widehat{\mathcal{Z}}^p = \{\{\zeta_{\ell i}\}_{\ell,i} : \sum_\ell |\zeta_{\ell p}| \leq r^{p+1}\}$, $1 \leq p \leq \kappa - 1$.

Since all the easy to describe convex sets $\widehat{\mathcal{Z}}^+$, $\widehat{\mathcal{Z}}^0,...,\widehat{\mathcal{Z}}^k$ contain $\widehat{\mathcal{Z}}$, the same is true for the easy to describe convex set

$$\widetilde{\mathcal{Z}} = \widehat{\mathcal{Z}}^+ \cap \widehat{\mathcal{Z}}^0 \cap \widehat{\mathcal{Z}}^1 \cap ... \cap \widehat{\mathcal{Z}}^k,$$

so that the (tractable along with $\widetilde{\mathcal{Z}}$) semi-infinite LO problem

$$\min_{\substack{t, \\ \{X_j(\cdot) \in \mathcal{X}_j\}_{j=1}^n}} \left\{ t : \left. \begin{array}{l} d_{\Pi(\widehat{\zeta})} + \sum_{j=1}^n X_j(\widehat{\zeta})(c_{\Pi(\widehat{\zeta})})_j \leq t \\ \sum_{j=1}^n X_j(\widehat{\zeta})A_{\Pi(\widehat{\zeta})}^j - b_{\Pi(\widehat{\zeta})} \geq 0 \end{array} \right\} \forall \widehat{\zeta} = \{\zeta_{\ell i}\} \in \widetilde{\mathcal{Z}} \right\} \quad (S)$$

$$\left[\Pi\left(\{\zeta_{\ell i}\}_{\substack{0 \leq i \leq m \\ 1 \leq \ell \leq L}} \right) = \{\zeta_{\ell 0}\}_{1 \leq \ell \leq L}, \mathcal{X}_j = \{X_j(\widehat{\zeta}) = \xi_j + \sum_{\substack{\ell \in I_j, \\ 0 \leq i \leq m}} \eta_{\ell i}\zeta_{\ell i}\} \right]$$

is a safe tractable approximation of the ARC of (14.3.6) restricted to decision rules (14.3.7). Note that this approximation is *at least* as flexible as the ARC of (14.3.6) restricted to genuine affine decision rules. Indeed, a rule $X(\cdot) = \{X_j(\cdot)\}_{j=1}^n$ of the latter type is "cut off" the family of all decision rules participating in (S) by the requirement "X_j depend solely on $\zeta_{\ell 0}$, $\ell \in I_j$," or, which is the same, by the requirement $\eta_{\ell i} = 0$ whenever $i > 0$. Since by construction the projection of $\widetilde{\mathcal{Z}}$ on the space of variables $\zeta_{\ell 0}$, $1 \leq \ell \leq L$, is exactly $\mathcal{Z}$, a pair $(t, X(\cdot))$ is feasible for (S) if and only if it is feasible for the AARC of (14.3.6), the information base being given by $I_1, ..., I_n$. The bottom line is, that when $\mathcal{Z}$ is not a box, the tractable problem (S), while still producing robust feasible decisions, is at least as flexible as the AARC. Whether this "at least as flexible" is or is not "more flexible," depends on the application in question, and since both (S) and AARC are tractable, it is easy to figure out what the true answer is.

Here is a toy example. Let $L = 2$, $n = 2$, and let (14.3.6) be the uncertain problem

$$\left\{ \min_x \left\{ x_2 : \begin{array}{l} x_1 \geq \zeta_1 \\ x_1 \geq -\zeta_1 \\ x_2 \geq x_1 + 3\zeta_1/5 + 4\zeta_2/5 \\ x_2 \geq x_1 - 3\zeta_1/5 - 4\zeta_2/5 \end{array} \right\}, \|\zeta\|_2 \leq 1 \right\},$$

with fully adjustable variable x_1 and non-adjustable variable x_2. Due to the extreme simplicity of our problem, we can immediately point out an optimal solution to the *unrestricted* ARC, namely,

$$X_1(\zeta) = |\zeta_1|, \ x_2 \equiv \mathrm{Opt(ARC)} = \max_{\|\zeta\|_2 \leq 1}[|\zeta_1| + |3\zeta_1 + 4\zeta_2|/5] = \frac{4\sqrt{5}}{5} \approx 1.7889.$$

Now let us compare Opt(ARC) with the optimal value Opt(AARC) of the AARC and with the optimal value Opt(RARC) of the restricted ARC where the decision rules are allowed to be affine in $[\zeta_\ell]_\pm$, $\ell = 1, 2$ (as always, $[a]_+ = \max[a, 0]$ and $[a]_- = \max[-a, 0]$). The situation fits **B**, so that we can process the RARC as it is. Noting that $a = [a]_+ - [a]_-$, the decision rules that are affine in $[\zeta_\ell]_\pm$, $\ell = 1, 2$, are exactly the same as the decision rules (14.3.7), where $\mathcal{F}_\ell$, $\ell = 1, 2$, are the spaces of piecewise linear functions on the axis with the only breakpoint 0. We see that *up to the fact that $\mathcal{Z}$ is a circle rather than a square*, the situation fits **C.1** as well, and we can process RARC via its safe tractable approximation (S). Let us look what are the optimal values yielded by these 3 schemes.

- The AARC of our toy problem is

$$\mathrm{Opt(AARC)} \ = \ \min_{x_2, \xi, \eta} \left\{ x_2 : \begin{array}{ll} \overbrace{\xi + \eta^T\zeta}^{X_1(\zeta)} \geq |\zeta_1| & (a) \\ x_2 \geq X_1(\zeta) + |3\zeta_1 + 4\zeta_2|/5 & (b) \end{array} \right.$$
$$\left. \forall(\zeta : \|\zeta\|_2 \leq 1) \right\}$$

This problem can be immediately solved. Indeed, (a) should be valid for $\zeta = \zeta^1 \equiv [1; 0]$ and for $\zeta = \zeta^2 \equiv -\zeta^1$, meaning that $X_1(\pm\zeta^1) \geq 1$, whence $\xi \geq 1$. Further, (b) should be valid for $\zeta = \zeta^3 \equiv [3; 4]/5$ and for $\zeta = \zeta^4 \equiv -\zeta^3$, meaning that $x_2 \geq X_1(\pm\zeta^3) + 1$, whence $x_2 \geq \xi + 1 \geq 2$. We see that the optimal value is ≥ 2, and this bound is achievable (we can take $X_1(\cdot) \equiv 1$ and $x_2 = 2$). As a byproduct, in our toy problem the AARC is as conservative as the RC.

- The RARC of our problem as given by **B** is

$$\mathrm{Opt(RARC)} \ = \ \min_{x_2, \xi, \eta, \eta_\pm} \left\{ x_2 : \begin{array}{l} \overbrace{\xi + \eta^T\zeta + \eta_+^T\zeta^+ + \eta_-^T\zeta^-}^{X_1(\widehat{\zeta})} \geq |\zeta_1| \\ x_2 \geq X_1(\widehat{\zeta}) + |3\zeta_1 + 4\zeta_2|/5 \end{array} \right.$$
$$\left. \forall(\widehat{\zeta} = [\underbrace{\zeta_1; \zeta_2}_{\zeta}; \underbrace{\zeta_1^+; \zeta_2^+}_{\zeta^+}; \underbrace{\zeta_1^-; \zeta_2^-}_{\zeta^-}] \in \widehat{\mathcal{Z}}) \right\},$$
$$\widehat{\mathcal{Z}} = \left\{ \widehat{\zeta} : \zeta = \zeta^+ - \zeta^-, \zeta^\pm \geq 0, \|\zeta^+ + \zeta^-\|_2 \leq 1 \right\}.$$

We can say in advance what are the optimal value and the optimal solution to the RARC — they should be the same as those of the ARC, since the latter, as a matter of fact, admits optimal decision rules that are affine in $|\zeta_1|$, and thus in $[\zeta_\ell]_\pm$. Nevertheless, we have carried out numerical optimization which yielded another optimal solution to the RARC (and thus - to ARC):

$$\mathrm{Opt(RARC)} = x_2 = 1.7889,$$
$$\xi = 1.0625, \eta = [0; 0], \eta_+ = \eta_- = [0.0498; -0.4754],$$

which corresponds to $X_1(\zeta) = 1.0625 + 0.0498|\zeta_1| - 0.4754|\zeta_2|$.

• The safe tractable approximation of the RARC looks as follows. The mapping $\zeta \mapsto \widehat{\zeta}[\zeta]$ in our case is

$$[\zeta_1; \zeta_2] \mapsto [\zeta_{1,0} = \zeta_1; \zeta_{1,1} = \max[\zeta_1, 0]; \zeta_{2,0} = \zeta_2; \zeta_{2,1} = \max[\zeta_2, 0]],$$

the tractable description of $\widehat{\mathcal{Z}}^+$ as given by **C.1** is

$$\widehat{\mathcal{Z}}^+ = \left\{ \{\zeta_{\ell i}\}_{\substack{i=0,1 \\ \ell=1,2}} : \begin{array}{c} -1 \leq \zeta_{\ell 0} \leq 1 \\ 0 \leq \frac{\zeta_{\ell 1} - \zeta_{\ell 0}}{1} \leq \frac{1 - \zeta_{\ell 1}}{1} \leq 1 \end{array} \right\}, \ell = 1, 2 \right\}$$

and the sets $\widehat{\mathcal{Z}}^0$, $\widehat{\mathcal{Z}}^1$ are given by

$$\widehat{\mathcal{Z}}^i = \left\{ \{\zeta_{\ell i}\}_{\substack{i=0,1 \\ \ell=1,2}} : \zeta_{1i}^2 + \zeta_{2i}^2 \leq 1 \right\}, i = 0, 1.$$

Consequently, (S) becomes the semi-infinite LO problem

$$\text{Opt}(S) = \min_{x_2, \xi, \{\eta_{\ell i}\}} \left\{ x_2 : \begin{array}{l} X_1(\widehat{\zeta}) \equiv \xi + \sum_{\substack{\ell=1,2 \\ i=0,1}} \eta_{\ell i} \zeta_{\ell i} \geq \zeta_{1,0} \\ X_1(\widehat{\zeta}) \equiv \xi + \sum_{\substack{\ell=1,2 \\ i=0,1}} \eta_{\ell i} \zeta_{\ell i} \geq -\zeta_{1,0} \\ x_2 \geq \xi + \sum_{\substack{\ell=1,2 \\ i=0,1}} \eta_{\ell i} \zeta_{\ell i} + [3\zeta_{1,0} + 4\zeta_{2,0}]/5 \\ x_2 \geq \xi + \sum_{\substack{\ell=1,2 \\ i=0,1}} \eta_{\ell i} \zeta_{\ell i} - [3\zeta_{1,0} + 4\zeta_{2,0}]/5 \\ -1 \leq \zeta_{\ell 0} \leq 1, \ell = 1, 2 \\ \forall \widehat{\zeta} = \{\zeta_{\ell i}\} : \ 0 \leq \zeta_{\ell 1} - \zeta_{\ell 0} \leq 1 - \zeta_{\ell 1} \leq 1, \ell = 1, 2 \\ \zeta_{1i}^2 + \zeta_{2i}^2 \leq 1, i = 0, 1 \end{array} \right\}.$$

Computation results in

$$\text{Opt}(S) = x_2 = \frac{25 + \sqrt{8209}}{60} \approx 1.9267,$$
$$X_1(\zeta) = \frac{5}{12} - \frac{3}{5}\zeta_{1.0}[\zeta_1] + \frac{6}{5}\zeta_{1,1}[\zeta_1] + \frac{7}{60}\zeta_{2,0}[\zeta_2] = \frac{5}{12} + \frac{3}{5}|\zeta_1| + \frac{7}{60}\zeta_2.$$

As it could be expected, we get $2 = \text{Opt}(\text{AARC}) > 1.9267 = \text{Opt}(S) > 1.7889 = \text{Opt}(\text{RARC}) = \text{Opt}(\text{ARC})$. Note that in order to get $\text{Opt}(S) < \text{Opt}(\text{AARC})$, taking into account $\widehat{\mathcal{Z}}^1$ is a must: in the case of **C.1**, whatever be $\mathcal{Z}$ and a box $\mathcal{Z}^+ \supset \mathcal{Z}$, with $\widetilde{\mathcal{Z}} = \widehat{\mathcal{Z}}^+ \cap \widehat{\mathcal{Z}}^0$ we gain nothing as compared to the genuine affine decision rules.

D. *Quadratic decision rules, ellipsoidal uncertainty set.* In this case,

$$\widehat{\zeta}[\zeta] = \left[\begin{array}{c|c} & \zeta^T \\ \hline \zeta & \zeta\zeta^T \end{array} \right]$$

is comprised of the entries of ζ and their pairwise products (so that the associated decision rules (14.3.5) are quadratic in ζ), and $\mathcal{Z}$ is the ellipsoid $\{\zeta \in \mathbb{R}^L : \|Q\zeta\|_2 \leq 1\}$, where Q has a trivial kernel. The convex hull of the image of $\mathcal{Z}$ under the quadratic mapping $\zeta \to \widehat{\zeta}[\zeta]$ is easy to describe:

Lemma 14.3.7. In the above notation, the set $\widehat{\mathcal{Z}} = \text{Conv}\{\widehat{\zeta}[\zeta] : \|Q\zeta\|_2 \leq 1\}$ is a convex compact set given by the semidefinite representation as follows:

$$\widehat{\mathcal{Z}} = \left\{ \widehat{\zeta} = \left[\begin{array}{c|c} & v^T \\ \hline v & W \end{array} \right] \in \mathbf{S}^{L+1} : \widehat{\zeta} + \left[\begin{array}{c|c} 1 & \\ \hline & \end{array} \right] \succeq 0, \text{Tr}(QWQ^T) \leq 1 \right\}.$$

Proof. It is immediately seen that it suffices to prove the statement when $Q = I$, which we assume from now on. Besides this, when we add to the mapping $\widehat{\zeta}[\zeta]$ the constant matrix $\left[\begin{array}{c|c} 1 & \\ \hline & \end{array} \right]$, the convex hull of the image of $\mathcal{Z}$ is translated by the same matrix. It follows that all we need is to prove that the convex hull $\mathcal{Q}$ of the image of the unit Euclidean ball under the mapping $\zeta \mapsto \widetilde{\zeta}[\zeta] = \left[\begin{array}{c|c} 1 & \zeta^T \\ \hline \zeta & \zeta\zeta^T \end{array}\right]$ can be represented as

$$\mathcal{Q} = \left\{\widehat{\zeta} = \left[\begin{array}{c|c} 1 & v^T \\ \hline v & W \end{array}\right] \in \mathbf{S}^{L+1} : \widehat{\zeta} \succeq 0, \mathrm{Tr}(QWQ^T) \leq 1 \right\}. \tag{14.3.15}$$

Denoting the right hand side in (14.3.15) by $\widehat{\mathcal{Q}}$, both $\mathcal{Q}$ and $\widehat{\mathcal{Q}}$ are nonempty convex compact sets. Therefore they coincide if and only if their support functions are identical.[1] We are in the situation where $\mathcal{Q}$ is the convex hull of the set $\left\{\left[\begin{array}{c|c} 1 & \zeta^T \\ \hline \zeta & \zeta\zeta^T \end{array}\right] : \zeta^T\zeta \leq 1 \right\}$, so that the support function of $\mathcal{Q}$ is

$$\phi(P) = \max_Z \left\{\mathrm{Tr}(PZ) : Z = \left[\begin{array}{c|c} 1 & \zeta^T \\ \hline \zeta & \zeta\zeta^T \end{array}\right] : \zeta^T\zeta \leq 1 \right\} \qquad \left[P = \left[\begin{array}{c|c} p & q^T \\ \hline q & R \end{array}\right] \in \mathbf{S}^{L+1}\right].$$

We have

$$\phi(P) = \max_Z \left\{\mathrm{Tr}(PZ) : Z = \left[\begin{array}{c|c} 1 & \zeta^T \\ \hline \zeta & \zeta\zeta^T \end{array}\right] \text{ with } \zeta^T\zeta \leq 1 \right\}$$
$$= \max_\zeta \left\{\zeta^T R\zeta + 2q^T\zeta + p : \zeta^T\zeta \leq 1 \right\}$$
$$= \min_\tau \left\{\tau : \tau \geq \zeta^T R\zeta + 2q^T\zeta + p \,\forall(\zeta : \zeta^T\zeta \leq 1)\right\}$$
$$= \min_\tau \left\{\tau : (\tau - p)t^2 - \zeta^T R\zeta - 2tq^T\zeta \geq 0 \,\forall((\zeta,t) : \zeta^T\zeta \leq t^2)\right\}$$
$$= \min_\tau \left\{\tau : \exists\lambda \geq 0 : (\tau - p)t^2 - \zeta^T R\zeta - 2tq^T\zeta - \lambda(t^2 - \zeta^T\zeta) \geq 0 \,\forall(\zeta,t)\right\} \ [\mathcal{S}\text{-Lemma}]$$
$$= \min_{\tau,\lambda} \left\{\tau : \left[\begin{array}{c|c} \tau - p - \lambda & -q^T \\ \hline -q & \lambda I - R \end{array}\right] \succeq 0, \lambda \geq 0\right\}$$
$$= \max_{u,v,W,r} \left\{up + 2v^T q + \mathrm{Tr}(RW) : \mathrm{Tr}\left(\left[\begin{array}{c|c} \tau - \lambda & \\ \hline & \lambda I \end{array}\right]\left[\begin{array}{c|c} u & v^T \\ \hline v & W \end{array}\right]\right) + r\lambda \right.$$
$$\left. \equiv \tau \forall(\tau,\lambda), \left[\begin{array}{c|c} u & v^T \\ \hline v & W \end{array}\right] \succeq 0, r \geq 0\right\} \ [\text{semidefinite duality}]$$
$$= \max_{v,W} \left\{p + 2v^T q + \mathrm{Tr}(RW) : \left[\begin{array}{c|c} 1 & v^T \\ \hline v & W \end{array}\right] \succeq 0, \mathrm{Tr}(W) \leq 1\right\}$$
$$= \max_{v,W} \left\{\mathrm{Tr}\left(P\left[\begin{array}{c|c} 1 & v^T \\ \hline v & W \end{array}\right]\right) : \left[\begin{array}{c|c} 1 & v^T \\ \hline v & W \end{array}\right] \in \widehat{\mathcal{Q}}\right\}.$$

Thus, the support function of $\mathcal{Q}$ indeed is identical to the one of $\widehat{\mathcal{Q}}$. $\qquad \square$

Corollary 14.3.8. Consider a fixed recourse uncertain LO problem (14.3.6) with an ellipsoid as an uncertainty set, where the adjustable decision variables are allowed to be quadratic functions of prescribed portions $P_j\zeta$ of the data. The associated ARC of the problem is computationally tractable and is given by an explicit semidefinite program of the sizes polynomial in those of instances and in the dimension L of the data vector.

---

[1]The support function of a nonempty convex set $X \subset \mathbb{R}^n$ is the function $f(\xi) = \sup_{x \in X} \xi^T x : \mathbb{R}^n \to \mathbb{R} \cup \{+\infty\}$. The fact that two closed nonempty convex sets in $\mathbb{R}^n$ are identical, if and only if their support functions are so, is readily given by the Separation Theorem.

E. *Quadratic decision rules and an uncertainty set that is an intersection of concentric ellipsoids.* Here the uncertainty set $\mathcal{Z}$ is $\cap$-ellipsoidal:

$$\mathcal{Z} = \mathcal{Z}_\rho \equiv \{\zeta \in \mathbb{R}^L : \zeta^T Q_j \zeta \leq \rho^2, 1 \leq j \leq J\}$$
$$\left[Q_j \succeq 0, \sum_j Q_j \succ 0\right] \tag{14.3.16}$$

(cf. section 7.2), where $\rho > 0$ is an uncertainty level, and, as above, $\widehat{\zeta}[\zeta] = \left[\begin{array}{c|c} & \zeta^T \\ \hline \zeta & \zeta\zeta^T \end{array}\right]$, so that our intention is to process the ARC of an uncertain problem corresponding to quadratic decision rules. As above, all we need is to get a tractable representation of the convex hull of the image of $\mathcal{Z}_\rho$ under the nonlinear mapping $\zeta \mapsto \widehat{\zeta}[\zeta]$. This is essentially the same as to find a similar representation of the convex hull $\widehat{\mathcal{Z}}_\rho$ of the image of $\mathcal{Z}_\rho$ under the nonlinear mapping

$$\zeta \mapsto \widehat{\zeta}_\rho[\zeta] = \left[\begin{array}{c|c} & \zeta^T \\ \hline \zeta & \frac{1}{\rho}\zeta\zeta^T \end{array}\right];$$

indeed, both convex hulls in question can be obtained from each other by simple linear transformations. The advantage of our normalization is that now $\mathcal{Z}_\rho = \rho\mathcal{Z}_1$ and $\widehat{\mathcal{Z}}_\rho = \rho\widehat{\mathcal{Z}}_1$, as it should be for respectable perturbation sets.

While the set $\widehat{\mathcal{Z}}_\rho$ is, in general, computationally intractable, we are about to demonstrate that this set admits a tight tractable approximation, and that the latter induces a tight tractable approximation of the "quadratically adjustable" RC of the Linear Optimization problem in question. The main ingredient we need is as follows:

Lemma 14.3.9. Consider the semidefinite representable set

$$\mathcal{W}_\rho = \rho\mathcal{W}_1, \ \mathcal{W}_1 = \left\{\widehat{\zeta} = \left[\begin{array}{c|c} & v^T \\ \hline v & W \end{array}\right] : \left[\begin{array}{c|c} 1 & v^T \\ \hline v & W \end{array}\right] \succeq 0, \mathrm{Tr}(WQ_j) \leq 1, 1 \leq j \leq J\right\}. \tag{14.3.17}$$

Then

$$\forall \rho > 0 : \widehat{\mathcal{Z}}_\rho \subset \mathcal{W}_\rho \subset \widehat{\mathcal{Z}}_{\vartheta\rho}, \tag{14.3.18}$$

where $\vartheta = O(1)\ln(J+1)$ and J is the number of ellipsoids in the description of $\mathcal{Z}_\rho$.

Proof. Since both $\widehat{\mathcal{Z}}_\rho$ and $\widehat{\mathcal{W}}_\rho$ are nonempty convex compact sets containing the origin and belonging to the subspace $\mathbf{S}_0^{L+1}$ of $\mathbf{S}^{L+1}$ comprised of matrices with the first diagonal entry being zero, to prove (14.3.18) is the same as to verify that the corresponding support functions

$$\phi_{\mathcal{W}_\rho}(P) = \max_{\widehat{\zeta}\in\mathcal{W}_\rho} \mathrm{Tr}(P\widehat{\zeta}), \ \phi_{\widehat{\mathcal{Z}}_\rho}(P) = \max_{\widehat{\zeta}\in\widehat{\mathcal{Z}}_\rho} \mathrm{Tr}(P\widehat{\zeta}),$$

considered as functions of $P \in \mathbf{S}_0^{L+1}$, satisfy the relation

$$\phi_{\widehat{\mathcal{Z}}_\rho}(\cdot) \leq \phi_{\mathcal{W}_\rho}(\cdot) \leq \phi_{\widehat{\mathcal{Z}}_{\vartheta\rho}}(\cdot).$$

Taking into account that $\widehat{\mathcal{Z}}_s = s\widehat{\mathcal{Z}}_1$, $s > 0$, this task reduces to verifying that

$$\phi_{\widehat{\mathcal{Z}}_\rho}(\cdot) \leq \phi_{\mathcal{W}_\rho}(\cdot) \leq \vartheta\phi_{\widehat{\mathcal{Z}}_\rho}(\cdot).$$

Thus, all we should prove is that whenever $P = \left[\begin{array}{c|c} & p^T \\ \hline p & R \end{array}\right] \in \mathbf{S}_0^{L+1}$, one has

$$\max_{\widehat{\zeta} \in \widehat{\mathcal{Z}}_\rho} \operatorname{Tr}(P\widehat{\zeta}) \leq \max_{\widehat{\zeta} \in \mathcal{W}_\rho} \operatorname{Tr}(P\widehat{\zeta}) \leq \vartheta \max_{\widehat{\zeta} \in \widehat{\mathcal{Z}}_\rho} \operatorname{Tr}(P\widehat{\zeta}).$$

Recalling the origin of $\widehat{\mathcal{Z}}_\rho$, the latter relation reads

$$\begin{aligned}
\forall P = \left[\begin{array}{c|c} & p^T \\ \hline p & R \end{array}\right] : \operatorname{Opt}_P(\rho) &\equiv \max_\zeta \left\{ 2p^T\zeta + \tfrac{1}{\rho}\zeta^T R\zeta : \zeta^T Q_j \zeta \leq \rho^2, 1 \leq j \leq J \right\} \\
&\leq \operatorname{SDP}_P(\rho) \equiv \max_{\widehat{\zeta} \in \mathcal{W}_\rho} \operatorname{Tr}(P\widehat{\zeta}) \leq \vartheta \operatorname{Opt}_P(\rho) \equiv \operatorname{Opt}_P(\vartheta\rho).
\end{aligned}$$

$$(14.3.19)$$

Observe that the three quantities in the latter relation are of the same homogeneity degree w.r.t. $\rho > 0$, so that it suffices to verify this relation when $\rho = 1$, which we assume from now on.

We are about to derive (14.3.19) from the Approximate $\mathcal{S}$-Lemma (Theorem B.3.1 in the Appendix). To this end, let us specify the entities participating in the latter statement as follows:

- $x = [t; \zeta] \in \mathbb{R}_t^1 \times \mathbb{R}_\zeta^L$;
- $A = P$, that is, $x^T A x = 2tp^T\zeta + \zeta^T R\zeta$;
- $B = \left[\begin{array}{c|c} 1 & \\ \hline & \end{array}\right]$, that is, $x^T B x = t^2$;
- $B_j = \left[\begin{array}{c|c} & \\ \hline & Q_j \end{array}\right], 1 \leq j \leq J$, that is, $x^T B_j x = \zeta^T Q_j \zeta$;
- $\rho = 1$.

With this setup, the quantity $\operatorname{Opt}(\rho)$ from (B.3.1) becomes nothing but $\operatorname{Opt}_P(1)$, while the quantity $\operatorname{SDP}(\rho)$ from (B.3.2) is

$$\begin{aligned}
\operatorname{SDP}(1) &= \max_X \left\{ \operatorname{Tr}(AX) : \operatorname{Tr}(BX) \leq 1, \operatorname{Tr}(B_j X) \leq 1, 1 \leq j \leq J, X \succeq 0 \right\} \\
&= \max_X \left\{ 2p^T v + \operatorname{Tr}(RW) : \begin{array}{l} u \leq 1 \\ \operatorname{Tr}(WQ_j) \leq 1, 1 \leq j \leq J \\ X = \left[\begin{array}{c|c} u & v^T \\ \hline v & W \end{array}\right] \succeq 0 \end{array} \right\} \\
&= \max_{v,W} \left\{ 2p^T v + \operatorname{Tr}(RW) : \begin{array}{l} \operatorname{Tr}(WQ_j) \leq 1, 1 \leq j \leq J \\ \left[\begin{array}{c|c} 1 & v^T \\ \hline v & W \end{array}\right] \succeq 0 \end{array} \right\} \\
&= \max_{\widehat{\zeta}} \left\{ \operatorname{Tr}(P\widehat{\zeta}) : \widehat{\zeta} = \left[\begin{array}{c|c} & v^T \\ \hline v & W \end{array}\right] : \begin{array}{l} \left[\begin{array}{c|c} 1 & v^T \\ \hline v & W \end{array}\right] \succeq 0 \\ \operatorname{Tr}(WQ_j) \leq 1, 1 \leq j \leq J \end{array} \right\} \\
&= \operatorname{SDP}_P(1).
\end{aligned}$$

With these observations, the conclusion (B.3.4) of the Approximate $\mathcal{S}$-Lemma reads

$$\operatorname{Opt}_P(1) \leq \operatorname{SDP}_P(1) \leq \operatorname{Opt}(\Omega(J)), \quad \Omega(J) = 9.19\sqrt{\ln(J+1)} \tag{14.3.20}$$

where for $\Omega \geq 1$

$$\text{Opt}(\Omega) = \max_x \left\{ x^T A x : x^T B x \leq 1, x^T B_j x \leq \Omega^2 \right\}$$
$$= \max_{t,\zeta} \left\{ 2t p^T \zeta + \zeta^T R \zeta : t^2 \leq 1, \zeta^T Q_j \zeta \leq \Omega^2, 1 \leq j \leq J \right\}$$
$$= \max_\zeta \left\{ 2p^T \zeta + \zeta^T R \zeta : \zeta^T Q_j \zeta \leq \Omega^2, 1 \leq j \leq J \right\}$$
$$= \max_{\eta = \Omega^{-1}\zeta} \left\{ \Omega(2p^T \eta) + \Omega^2 \eta^T R \eta : \eta^T Q_j \eta \leq 1, 1 \leq j \leq J \right\}$$
$$\leq \Omega^2 \max_\eta \left\{ 2p^T \eta + \eta^T R \eta : \eta^T Q_j \eta \leq 1, 1 \leq j \leq J \right\}$$
$$= \Omega^2 \text{Opt}(1).$$

Setting $\vartheta = \Omega^2(J)$, we see that (14.3.20) implies (14.3.19). $\qquad \square$

Corollary 14.3.10. Consider a fixed recourse uncertain LO problem (14.3.6) with $\cap$-ellipsoidal uncertainty set $\mathcal{Z}_\rho$ (see (14.3.16)) where one seeks robust optimal quadratic decision rules:

$$
\left[\begin{array}{l}
\quad\quad\quad x_j = p_j + q_j^T \widehat{P}_j \left(\widehat{\zeta}_\rho[\zeta] \right) \\
\bullet \quad \widehat{\zeta}_\rho[\zeta] = \left[\begin{array}{c|c} & \zeta^T \\ \hline \zeta & \frac{1}{\rho}\zeta\zeta^T \end{array} \right] \\
\bullet \quad \widehat{P}_j : \text{ linear mappings from } \mathbf{S}^{L+1} \text{ to } \mathbb{R}^{m_j} \\
\bullet \quad p_j \in \mathbb{R}, q_j \in \mathbb{R}^{m_j} : \text{ parameters to be specified}
\end{array} \right].
\tag{14.3.21}
$$

The associated Adjustable Robust Counterpart of the problem admits a safe tractable approximation that is tight within the factor ϑ given by Lemma 14.3.9.

Here is how the safe approximation of the Robust Counterpart mentioned in Corollary 14.3.10 can be built:

i) We write down the optimization problem

$$
\min_{t,x} \left\{ t : \begin{array}{l} a_{0\zeta}^T[t;x] + b_{0\zeta} \equiv t - c_\zeta^T x - d_\zeta \geq 0 \\ a_{i\zeta}^T[t;x] + b_{i,\zeta} \equiv A_{i\zeta}^T x - b_{i\zeta} \geq 0, \ i = 1, ..., m \end{array} \right\}
\tag{P}
$$

where $A_{i\zeta}^T$ is i-th row in A_ζ and $b_{i\zeta}$ is i-th entry in b_ζ;

ii) We plug into the $m+1$ constraints of (P), instead of the original decision variables x_j, the expressions $p_j + q_j^T \widehat{P}_j \left(\widehat{\zeta}_\rho[\zeta] \right)$, thus arriving at the optimization problem of the form

$$
\min_{[t;y]} \left\{ t : \alpha_{i\widehat{\zeta}}^T[t;y] + \beta_{i\widehat{\zeta}} \geq 0, \ 0 \leq i \leq m \right\},
\tag{P'}
$$

where y is the collection of coefficients p_j, q_j of the quadratic decision rules, $\widehat{\zeta}$ is our new uncertain data — a matrix from $\mathbf{S}_0^{L+1}$ (see p. 384), and $\alpha_{i\widehat{\zeta}}, \beta_{i\widehat{\zeta}}$ are affine in $\widehat{\zeta}$, the affinity being ensured by the assumption of fixed recourse. The "true" quadratically adjustable RC of the problem of interest is the semi-infinite problem

$$
\min_{[t;y]} \left\{ t : \forall \widehat{\zeta} \in \widehat{\mathcal{Z}}_\rho : \alpha_{i\widehat{\zeta}}^T[t;y] + \beta_{i\widehat{\zeta}} \geq 0, \ 0 \leq i \leq m \right\}
\tag{R}
$$

obtained from (P') by requiring the constraints to remain valid for all $\widehat{\zeta} \in \widehat{\mathcal{Z}}_\rho$, the latter set being the convex hull of the image of $\mathcal{Z}_\rho$ under the mapping $\zeta \mapsto \widehat{\zeta}_\rho[\zeta]$. The semi-infinite problem (R) in general is intractable, and we replace it with its safe tractable approximation

$$\min_{[t;y]} \left\{ t : \forall \widehat{\zeta} \in \mathcal{W}_\rho : \alpha_{i\widehat{\zeta}}^T[t;y] + \beta_{i\widehat{\zeta}} \geq 0, 0 \leq i \leq m \right\}, \qquad (R')$$

where $\mathcal{W}_\rho$ is the semidefinite representable convex compact set defined in Lemma 14.3.9. By Theorem 1.3.4, (R') is tractable and can be straightforwardly converted into a semidefinite program of sizes polynomial in $n = \dim x$, m and $L = \dim \zeta$. Here is the conversion: recalling the structure of $\widehat{\zeta}$ and setting $z = [t; x]$, we can rewrite the body of i-th constraint in (R') as

$$\alpha_{i\widehat{\zeta}}^T z + \beta_{i\widehat{\zeta}} \equiv a_i[z] + \mathrm{Tr}\Big(\underbrace{\left[\begin{array}{c|c} & v^T \\ \hline v & W \end{array} \right]}_{\widehat{\zeta}} \left[\begin{array}{c|c} & p_i^T[z] \\ \hline p_i[z] & P_i[z] \end{array} \right] \Big),$$

where $a_i[z]$, $p_i[z]$ and $P_i[z] = P_i^T[z]$ are affine in z. Therefore, invoking the definition of $\mathcal{W}_\rho = \rho \mathcal{W}_1$ (see Lemma 14.3.9), the RC of the i-th semi-infinite constraint in (R') is the first predicate in the following chain of equivalences:

$$\min_{v,W} \Big\{ a_i[z] + 2\rho v^T p_i[z] + \rho \mathrm{Tr}(W P_i[z]) :$$
$$\left[\begin{array}{c|c} 1 & v^T \\ \hline v & W \end{array} \right] \succeq 0, \mathrm{Tr}(W Q_j) \leq 1, 1 \leq j \leq J \Big\} \geq 0 \qquad (a_i)$$

$$\Updownarrow$$

$$\exists \lambda^i = [\lambda_1^i; ...; \lambda_J^i] : \left\{ \begin{array}{l} \lambda^i \geq 0 \\ \left[\begin{array}{c|c} a_i[z] - \sum_j \lambda_j^i & \rho p_i^T[z] \\ \hline \rho p_i[z] & \rho P_i[z] + \sum_j \lambda_j^i Q_j \end{array} \right] \succeq 0 \end{array} \right. \qquad (b_i)$$

where $\Updownarrow$ is given by Semidefinite Duality. Consequently, we can reformulate (R') equivalently as the semidefinite program

$$\min_{\substack{z=[t;y] \\ \{\lambda_j^i\}}} \left\{ t : \begin{array}{l} \left[\begin{array}{c|c} a_i[z] - \sum_j \lambda_j^i & \rho p_i^T[z] \\ \hline \rho p_i[z] & \rho P_i[z] + \sum_j \lambda_j^i Q_j \end{array} \right] \succeq 0 \\ \lambda_j^i \geq 0, 0 \leq i \leq m, 1 \leq j \leq J \end{array} \right\}.$$

The latter SDP is a ϑ-tight safe tractable approximation of the quadratically adjustable RC with ϑ given by Lemma 14.3.9.

14.3.3 The AARC of Uncertain Linear Optimization Problem Without Fixed Recourse

We have seen that the AARC of an uncertain LO problem

$$\mathcal{C} = \Big\{ \min_x \big\{ c_\zeta^T x + d_\zeta : A_\zeta x \geq b_\zeta \big\} : \zeta \in \mathcal{Z} \Big\}$$
$$[c_\zeta, d_\zeta, A_\zeta, b_\zeta : \text{ affine in } \zeta] \qquad (14.3.22)$$

with computationally tractable convex compact uncertainty set $\mathcal{Z}$ *and with fixed recourse* is computationally tractable. What happens when the assumption of fixed recourse is removed? The answer is that in general the AARC can become intractable (see [13]). However, we are about to demonstrate that for an ellipsoidal uncertainty set $\mathcal{Z} = \mathcal{Z}_\rho = \{\zeta : \|Q\zeta\|_2 \leq \rho\}$, $\mathrm{Ker}Q = \{0\}$, the AARC is computationally tractable, and for the $\cap$-ellipsoidal uncertainty set $\mathcal{Z} = \mathcal{Z}_\rho$ given by (14.3.16), the AARC admits a tight safe tractable approximation. Indeed, for affine decision rules

$$x_j = X_j(P_j\zeta) \equiv p_j + q_j^T P_j\zeta$$

the AARC of (14.3.22) is the semi-infinite problem of the form

$$\min_{z=[t;y]} \left\{ t : \forall \zeta \in \mathcal{Z}_\rho : a_{i\zeta}[z] + b_{i\zeta} \geq 0, \, 0 \leq i \leq m \right\}, \tag{14.3.23}$$

where $y = \{p_j, q_j\}_{j=1}^n$, $a_{i\zeta}[z]$ is affine in z and quadratic in ζ, and $b_{i\zeta}$ is quadratic in ζ (in fact just affine). Introducing, as we already have on several occasions, the nonlinear mapping

$$\zeta \mapsto \widehat{\zeta}_\rho[\zeta] = \left[\begin{array}{c|c} & \zeta^T \\ \hline \zeta & \frac{1}{\rho}\zeta\zeta^T \end{array} \right]$$

and denoting by $\widehat{\mathcal{Z}}_\rho$ the convex hull of the image of $\mathcal{Z}_\rho$ under this mapping (so that $\widehat{\mathcal{Z}}_\rho = \rho\widehat{\mathcal{Z}}_1$), we can rewrite the AARC equivalently as the semi-infinite problem

$$\min_{z=[t;y]} \left\{ t : \forall \widehat{\zeta} \in \widehat{\mathcal{Z}}_\rho : \alpha_{i\widehat{\zeta}}^T z + \beta_{i\widehat{\zeta}} \geq 0, \, 0 \leq i \leq m \right\} \tag{14.3.24}$$

with $\alpha_{i\widehat{\zeta}}$, $\beta_{i\widehat{\zeta}}$ affine in $\widehat{\zeta} = \left[\begin{array}{c|c} & v^T \\ \hline v & W \end{array} \right]$. In view of Theorem 1.3.4, all we need in order to process (14.3.24) efficiently is a computationally tractable representation of convex compact set $\widehat{\mathcal{Z}}_\rho = \rho\widehat{\mathcal{Z}}_1$, which we do have when $\mathcal{Z}_\rho$ is an ellipsoid (see Lemma 14.3.7). When $\mathcal{Z}_\rho$ is the $\cap$-ellipsoidal uncertainty (14.3.16), Lemma 14.3.9 provides us with a computationally tractable outer approximation $\mathcal{W}_\rho$ of the set $\widehat{\mathcal{Z}}_\rho$ tight within factor $\vartheta = O(1)\ln(J + 1)$. Replacing in (14.3.24) the "difficult" set $\widehat{\mathcal{Z}}_\rho$ with the "easy one" $\mathcal{W}_\rho$, we end up with an efficiently solvable problem (completely similar to the one in Corollary 14.3.10), and this problem is a ϑ-tight safe approximation of (14.3.24).

In fact the above approach can be extended even slightly beyond just affine decision rules. Specifically, in the case of an uncertain LO we could allow for the adjustable "fixed recourse" variables x_j — those for which all the coefficients in the objective and the constraints of instances are certain — to be *quadratic* in $P_j\zeta$, and for the remaining "non-fixed recourse" adjustable variables to be affine in $P_j\zeta$. This modification does not alter the structure of (14.3.23) (that is, quadratic dependence of $\alpha_{i\zeta}$, $\beta_{i\zeta}$ on ζ), and we could process (14.3.24) in exactly the same manner as before.

14.3.4 Illustration: the AARC of Multi-Period Inventory Affected by Uncertain Demand

We are about to illustrate the AARC methodology by its application to the simple multi-product multi-period inventory model presented in Example 14.1.1 (see also p. 358).

Building the AARC of (14.2.3). We first decide on the information base of the "actual decisions" — vectors w_t of replenishment orders of instants $t = 1, ..., N$. Assuming that the part of the uncertain data, (i.e., of the demand trajectory $\zeta = \zeta^N = [\zeta_1; ...; \zeta_N]$) that becomes known when the decision on w_t should be made is the vector $\zeta^{t-1} = [\zeta_1; ...; \zeta_{t-1}]$ of the demands in periods preceding time t, we introduce affine decision rules

$$w_t = \omega_t + \Omega_t \zeta^{t-1} \qquad (14.3.25)$$

for the orders; here ω_t, Ω_t form the coefficients of the decision rules we are seeking.

The remaining variables in (14.2.3), with a single exception, are analysis variables, and we allow them to be arbitrary affine functions of the entire demand trajectory ζ^N:

$$
\begin{array}{ll}
x_t = \xi_t + \Xi_t \zeta^N,\ t = 2, ..., N+1 & \text{[states]} \\
y_t = \eta_t + H_t \zeta^N,\ t = 1, ..., N & \text{[upper bounds on } [x_t]_+] \\
z_t = \pi_t + \Pi_t \zeta^N,\ t = 1, ..., N & \text{[upper bounds on } [x_t]_-].
\end{array}
\qquad (14.3.26)
$$

The only remaining variable C — the upper bound on the inventory management cost we intend to minimize — is considered as non-adjustable.

We now plug the affine decision rules in the objective and the constraints of (14.2.3), and require the resulting relations to be satisfied for all realizations of the uncertain data ζ^N from a given uncertainty set $\mathcal{Z}$, thus arriving at the AARC of our inventory model:

$$
\begin{aligned}
& \text{minimize} \quad C \\
& \text{s.t.}\ \forall \zeta^N \in \mathcal{Z}: \\
& C \geq \sum_{t=1}^{N} \left[c_{\text{h},t}^T [\eta_t + H_t \zeta^N] + c_{\text{b},t}^T [\pi_t + \Pi_t \zeta^N] + c_{\text{o},t}^T [\omega_t + \Omega_t \zeta^{t-1}] \right] \\
& \xi_t + \Xi_t \zeta^N = \begin{cases} \xi_{t-1} + \Xi_{t-1} \zeta^N + [\omega_t + \Omega_t \zeta^{t-1}] - \zeta_t,\ 2 \leq t \leq N \\ x_0 + \omega_1 - \zeta_1,\ t = 1 \end{cases} \\
& \eta_t + H_t \zeta^N \geq 0,\ \eta_t + H_t \zeta^N \geq \xi_t + \Xi_t \zeta^N,\ 1 \leq t \leq N \\
& \pi_t + \Pi_t \zeta^N \geq 0,\ \pi_t + \Pi_t \zeta^N \geq -\xi_t - \Xi_t \zeta^N,\ 1 \leq t \leq N \\
& \underline{w}_t \leq \omega_t + \Omega_t \zeta^{t-1} \leq \overline{w}_t,\ 1 \leq t \leq N \\
& q^T \left[\eta_t + H_t \zeta^N \right] \leq r
\end{aligned}
\qquad (14.3.27)
$$

the variables being C and the coefficients $\omega_t, \Omega_t, ..., \pi_t, \Pi_t$ of the affine decision rules.

We see that the problem in question has fixed recourse (it always is so when the uncertainty affects just the constant terms in conic constraints) and is nothing but an explicit semi-infinite LO program. Assuming the uncertainty set $\mathcal{Z}$ to be computationally tractable, we can invoke Theorem 1.3.4 and reformulate this semi-infinite problem as a computationally tractable one. For example, with *box uncertainty*:

$$\mathcal{Z} = \{\zeta^N \in \mathbb{R}_+^{N \times d} : \underline{\zeta}_t \le \zeta_t \le \overline{\zeta}_t, \; 1 \le t \le N\},$$

the semi-infinite LO program (14.3.27) can be immediately rewritten as an explicit "certain" LO program. Indeed, after replacing the semi-infinite coordinate-wise vector inequalities/equations appearing in (14.3.27) by equivalent systems of scalar semi-infinite inequalities/equations and representing the semi-infinite linear equations by pairs of opposite semi-infinite linear inequalities, we end up with a semi-infinite optimization program with a certain linear objective and finitely many constraints of the form

$$\forall \left(\zeta_t^i \in [\underline{\zeta}_t^i, \overline{\zeta}_t^i], t \le N, i \le d \right) : p^\ell[y] + \sum_{i,t} \zeta_t^i p_{ti}^\ell[y] \le 0$$

(ℓ is the serial number of the constraint, y is the vector comprised of the decision variables in (14.3.27), and $p^\ell[y]$, $p_{ti}^\ell[y]$ are given affine functions of y). The above semi-infinite constraint can be represented by a system of linear inequalities

$$\underline{\zeta}_t^i p_{ti}^\ell[y] \le u_{ti}^\ell$$
$$\overline{\zeta}_t^i p_{ti}^\ell[y] \le u_{ti}^\ell$$
$$p^\ell[y] + \sum_{t,i} u_{ti}^\ell \le 0,$$

in variables y and additional variables u_{ti}^ℓ. Putting all these systems of inequalities together and augmenting the resulting system of linear constraints with our original objective to be minimized, we end up with an explicit LO program that is equivalent to (14.3.27).

Some remarks are in order:

i) We could act similarly when building the AARC of any uncertain LO problem with fixed recourse and "well-structured" uncertainty set, e.g., one given by an explicit polyhedral/conic quadratic/semidefinite representation. In the latter case, the resulting tractable reformulation of the AARC would be an explicit linear/conic quadratic/semidefinite program of sizes that are polynomial in the sizes of the instances and in the size of conic description of the uncertainty set. Moreover, the "tractable reformulation" of the AARC can be built automatically, by a kind of compilation.

ii) Note how flexible the AARC approach is: we could easily incorporate additional constraints, (e.g., those forbidding backlogged demand, expressing lags in acquiring information on past demands and/or lags in executing the replenishment orders, etc.). Essentially, the only thing that matters is that we are dealing with an uncertain LO problem with fixed recourse. This is in sharp contrast with the ARC. As we have already mentioned, there is, essentially, only one optimization technique — Dynamic Programming — that with luck can be used to process the (general-type) ARC numerically. To do so, one needs indeed a lot of luck — to be "computationally tractable," Dynamic Programming imposes many highly "fragile" limitations on the structure and the sizes of instances. For example, the effort to solve the "true" ARC of our toy Inventory problem by Dynamic Programming blows up exponentially with the number of products d (we can say that $d = 4$ is already "too big"); in contrast to this, the AARC does not suffer of "curse of dimensionality" and scales reasonably well with problem's sizes.

iii) Note that we have no difficulties processing uncertainty-affected *equality constraints* (such as state equations above) — this is something that we cannot afford with the usual — non-adjustable — RC (how could an equation remain valid when the variables are kept constant, and the coefficients are perturbed?).

iv) Above, we "immunized" affine decision rules against uncertainty in the worst-case-oriented fashion — by requiring the constraints to be satisfied for *all* realizations of uncertain data from $\mathcal{Z}$. Assuming ζ to be random, we could replace the worst-case interpretation of the uncertain constraints with their chance constrained interpretation. To process the "chance constrained" AARC, we could use all the "chance constraint machinery" we have developed so far for the RC, exploiting the fact that for fixed recourse there is no essential difference between the structure of the RC and that of the AARC.

Of course, all the nice properties of the AARC we have just mentioned have their price — in general, as in our toy inventory example, we have no idea of how much we lose in terms of optimality when passing from general decision rules to affine rules. At present, we are not aware of any theoretical tools for evaluating such a loss. Moreover, it is easy to build examples showing that sticking to affine decision rules can indeed be costly; it even may happen that the AARC is infeasible, while the ARC is not. Much more surprising is the fact that there are meaningful situations where the AARC is unexpectedly good. Here we present a single simple example (a much more advanced one is presented in section 15.2).

Consider our inventory problem in the single-product case with added constraints that no backlogged demand is allowed and that the amount of product in the inventory should remain between two given positive bounds. Assuming box uncertainty in the demand, the "true" ARC of the uncertain problem is well within the grasp of Dynamic Programming, and thus we can measure the "non-optimality" of affine decision rules experimentally — by comparing the optimal values of the true ARC with those of the AARC as well as of the non-adjustable RC. To this end, we generated at random several hundreds of data sets for the problem with time horizon $N = 10$ and filtered out all data sets that led to infeasible ARC (it indeed can be infeasible due to the presence of upper and lower bounds on the inventory level and the fact that we forbid backlogged demand). We did our best to get as rich a family of examples as possible — those with time-independent and with time-dependent costs, various levels of demand uncertainty (from 10% to 50%), etc. We then solved ARCs, AARCs and RCs of the remaining "well-posed" problems — the ARCs by Dynamic Programming, the AARCs and RCs — by reduction to

Range of $\dfrac{\mathrm{Opt(RC)}}{\mathrm{Opt(AARC)}}$	1	$(1, 2]$	$(2, 10]$	$(10, 1000]$	∞
Frequency in the sample	38%	23%	14%	11%	15%

Table 14.1 Experiments with ARCs, AARCs and RCs of randomly generated single-product inventory problems affected by uncertain demand.

explicit LO programs. The number of "well-posed" problems we processed was 768, and the results were as follows:

i) To our great surprise, *in every one of the 768 cases we have analyzed, the computed optimal values of the "true" ARC and the AARC were identical.*

Thus, there is an "experimental evidence" that in the case of our single-product inventory problem, the affine decision rules allow one to reach "true optimality." It should be added that the phenomenon in question seems to be closely related to our intention to optimize the *guaranteed*, (i.e., the worst-case, w.r.t. demand trajectories from the uncertainty set), inventory management cost. When optimizing the "average" cost, the ARC frequently becomes significantly less expensive than the AARC.[2]

ii) The (equal to each other) optimal values of the ARC and the AARC in many cases were much better than the optimal value of the RC, as it is seen from table 14.1. In particular, in 40% of the cases the RC was at least twice as bad in terms of the (worst-case) inventory management cost as the ARC/AARC, and in 15% of the cases the RC was in fact infeasible.

The bottom line is twofold. First, we see that in multi-stage decision making there exist meaningful situations where the AARC, while "not less computationally tractable" than the RC, is much more flexible and much less conservative. Second, the AARC is not necessarily "significantly inferior" as compared to the ARC.

14.4 ADJUSTABLE ROBUST OPTIMIZATION AND SYNTHESIS OF LINEAR CONTROLLERS

While the usefulness of affine decision rules seems to be heavily underestimated in the "OR-style multi-stage decision making," they play one of the central roles in Control. Our next goal is to demonstrate that the use of AARC can render important Control implications.

---

[2]On this occasion, it is worthy of mention that affine decision rules were proposed many years ago, in the context of Multi-Stage Stochastic Programming, by A. Charnes. In Stochastic Programming, people are indeed interested in optimizing the expected value of the objective, and soon it became clear that in this respect, the affine decision rules can be pretty far from being optimal. As a result, the simple — and extremely useful from the computational perspective — concept of affine decision rules remained completely forgotten for many years.

14.4.1 Robust Affine Control over Finite Time Horizon

Consider a discrete time linear dynamical system

$$
\begin{aligned}
x_0 &= z \\
x_{t+1} &= A_t x_t + B_t u_t + R_t d_t \ , \ t = 0, 1, ... \\
y_t &= C_t x_t + D_t d_t
\end{aligned}
\tag{14.4.1}
$$

where $x_t \in \mathbb{R}^{n_x}$, $u_t \in \mathbb{R}^{n_u}$, $y_t \in \mathbb{R}^{n_y}$ and $d_t \in \mathbb{R}^{n_d}$ are the state, the control, the output and the exogenous input (disturbance) at time t, and A_t, B_t, C_t, D_t, R_t are known matrices of appropriate dimension.

Notational convention. Below, given a sequence of vectors $e_0, e_1, ...$ and an integer $t \geq 0$, we denote by e^t the initial fragment of the sequence: $e^t = [e_0; ...; e_t]$. When t is negative, e^t, by definition, is the zero vector.

Affine control laws. A typical problem of (finite-horizon) Linear Control associated with the "open loop" system (14.4.1) is to "close" the system by a non-anticipative affine output-based control law

$$
u_t = g_t + \sum_{\tau=0}^{t} G_{t\tau} y_\tau
\tag{14.4.2}
$$

(here the vectors g_t and matrices $G_{t\tau}$ are the parameters of the control law). The closed loop system (14.4.1), (14.4.2) is required to meet prescribed design specifications. We assume that these specifications are represented by a system of linear inequalities

$$
Aw^N \leq b
\tag{14.4.3}
$$

on the *state-control trajectory* $w^N = [x_0; ...; x_{N+1}; u_0; ...; u_N]$ over a given finite time horizon $t = 0, 1, ..., N$.

An immediate observation is that for a given control law (14.4.2) the dynamics (14.4.1) specifies the trajectory as an affine function of the initial state z and the sequence of disturbances $d^N = (d_0, ..., d_N)$:

$$
w^N = w_0^N[\gamma] + W^N[\gamma]\zeta, \ \ \zeta = (z, d^N),
$$

where $\gamma = \{g_t, G_{t\tau}, 0 \leq \tau \leq t \leq N\}$, is the "parameter" of the underlying control law (14.4.2). Substituting this expression for w^N into (14.4.3), we get the following system of constraints on the decision vector γ:

$$
A\left[w_0^N[\gamma] + W^N[\gamma]\zeta\right] \leq b.
\tag{14.4.4}
$$

If the disturbances d^N and the initial state z are certain, (14.4.4) is "easy" — it is a system of constraints on γ with certain data. Moreover, in the case in question we lose nothing by restricting ourselves with "off-line" control laws (14.4.2) — those with $G_{t\tau} \equiv 0$; when restricted onto this subspace, let it be called Γ, in the γ space, the function $w_0^N[\gamma] + W^N[\gamma]\zeta$ turns out to be bi-affine in γ and in ζ, so that (14.4.4) reduces to a system of explicit linear inequalities on $\gamma \in \Gamma$. Now, when the disturbances and/or the initial state are *not* known in advance, (which is the only case of interest in Robust Control), (14.4.4) becomes an uncertainty-affected system of constraints, and we could try to solve the system in a robust fashion, e.g., to seek a solution γ that makes the constraints feasible for all realizations of $\zeta = (z, d^N)$ from a given uncertainty set $\mathcal{Z}\mathcal{D}^N$, thus arriving at the system of semi-infinite scalar constraints

$$A\left[w_0^N[\gamma] + W^N[\gamma]\zeta\right] \leq b \ \forall \zeta \in \mathcal{Z}\mathcal{D}^N. \tag{14.4.5}$$

Unfortunately, the semi-infinite constraints in this system are *not* bi-affine, since the dependence of w_0^N, W^N on γ is highly nonlinear, unless γ is restricted to vary in Γ. Thus, when seeking "on-line" control laws (those where $G_{t\tau}$ can be nonzero), (14.4.5) becomes a system of highly nonlinear semi-infinite constraints and as such seems to be severely computationally intractable (the feasible set corresponding to (14.4.4) can be in fact nonconvex). One possibility to circumvent this difficulty would be to switch from control laws that are affine in the outputs y_t to those affine in disturbances and the initial state (cf. approach of [55]). This, however, could be problematic in the situations when we do not observe z and d_t directly. The good news is that we can overcome this difficulty without requiring d_t and z to be observable, the remedy being a suitable re-parameterization of affine control laws.

14.4.2 Purified-Output-Based Representation of Affine Control Laws and Efficient Design of Finite-Horizon Linear Controllers

Imagine that in parallel with controlling (14.4.1) with the aid of a non-anticipating output-based control law $u_t = U_t(y_0, ..., y_t)$, we run the *model* of (14.4.1) as follows:

$$
\begin{aligned}
\widehat{x}_0 &= 0 \\
\widehat{x}_{t+1} &= A_t\widehat{x}_t + B_t u_t \\
\widehat{y}_t &= C_t\widehat{x}_t \\
v_t &= y_t - \widehat{y}_t.
\end{aligned}
\tag{14.4.6}
$$

Since we know past controls, we can run this system in an "on-line" fashion, so that the *purified output* v_t becomes known when the decision on u_t should be made. An immediate observation is that *the purified outputs are completely independent of the control law in question — they are affine functions of the initial state and the disturbances* $d_0, ..., d_t$, *and these functions are readily given by the dynamics of* (14.4.1).

> Indeed, from the descriptions of the open-loop system and the model, it follows that the differences $\delta_t = x_t - \widehat{x}_t$ evolve with time according to the equations
>
> $$
> \begin{aligned}
> \delta_0 &= z \\
> \delta_{t+1} &= A_t + R_t d_t, \ t = 0, 1, ...
> \end{aligned}
> $$
>
> while
>
> $$v_t = C_t\delta_t + D_t d_t.$$
>
> From these relations it follows that
>
> $$v_t = \mathcal{V}_t^d d^t + \mathcal{V}_t^z z \tag{14.4.7}$$

with matrices $\mathcal{V}_t^d$, $\mathcal{V}_t^z$ depending solely on the matrices $A_\tau, B_\tau, ..., 0 \leq \tau \leq t$, and readily given by these matrices.

Now, it was mentioned that $v_0, ..., v_t$ are known when the decision on u_t should be made, so that we can consider *purified-output-based* (POB) affine control laws

$$u_t = h_t + \sum_{\tau=0}^{t} H_{t\tau} v_\tau.$$

The complete description of the dynamical system "closed" by this control is

plant:	
$(a):$ $\begin{cases} x_0 &= z \\ x_{t+1} &= A_t x_t + B_t u_t + R_t d_t \\ y_t &= C_t x_t + D_t d_t \end{cases}$	
model:	
$(b):$ $\begin{cases} \widehat{x}_0 &= 0 \\ \widehat{x}_{t+1} &= A_t \widehat{x}_t + B_t u_t \\ \widehat{y}_t &= C_t \widehat{x}_t \end{cases}$	(14.4.8)
purified outputs:	
$(c):$ $v_t = y_t - \widehat{y}_t$	
control law:	
$(d):$ $u_t = h_t + \sum_{\tau=0}^{t} H_{t\tau} v_\tau$	

The main result. We are about to prove the following simple and fundamental fact:

Theorem 14.4.1.

(i) For every affine control law in the form of (14.4.2), there exists a control law in the form of (14.4.8.*d*) that, whatever be the initial state and a sequence of inputs, results in exactly the same state-control trajectories of the closed loop system;

(ii) Vice versa, for every affine control law in the form of (14.4.8.*d*), there exists a control law in the form of (14.4.2) that, whatever be the initial state and a sequence of inputs, results in exactly the same state-control trajectories of the closed loop system;

(iii) [bi-affinity] The state-control trajectory w^N of closed loop system (14.4.8) is affine in z, d^N when the parameters $\eta = \{h_t, H_{t\tau}\}_{0 \leq \tau \leq t \leq N}$ of the underlying control law are fixed, and is affine in η when z, d^N are fixed:

$$w^N = \omega[\eta] + \Omega_z[\eta]z + \Omega_d[\eta]d^N \tag{14.4.9}$$

for some vectors $\omega[\eta]$ and matrices $\Omega_z[\eta]$, $\Omega_d[\eta]$ depending affinely on η.

Proof. (i): Let us fix an affine control law in the form of (14.4.2), and let $x_t = X_t(z, d^{t-1})$, $u_t = U_t(z, d^t)$, $y_t = Y_t(z, d^t)$, $v_t = V_t(z, d^t)$ be the corresponding states, controls, outputs, and purified outputs. To prove (i) it suffices to show that for every $t \geq 0$ with properly chosen vectors q_t and matrices $Q_{t\tau}$ one has

$$\forall(z, d^t) : Y_t(z, d^t) = q_t + \sum_{\tau=0}^{t} Q_{t\tau} V_\tau(z, d^\tau). \tag{I_t}$$

Indeed, given the validity of these relations and taking into account (14.4.2), we would have

$$U_t(z, d^t) \equiv g_t + \sum_{\tau=0}^{t} G_{t\tau} Y_\tau(z, d^\tau) \equiv h_t + \sum_{\tau=0}^{t} H_{t\tau} V(z, d^\tau) \tag{II_t}$$

with properly chosen h_t, $H_{t\tau}$, so that the control law in question can indeed be represented as a linear control law via purified outputs.

We shall prove (I_t) by induction in t. The base $t = 0$ is evident, since by (14.4.8.a–c) we merely have $Y_0(z, d^0) \equiv V_0(z, d^0)$. Now let $s \geq 1$ and assume that relations (I_t) are valid for $0 \leq t < s$. Let us prove the validity of (I_s). From the validity of (I_t), $t < s$, it follows that the relations (II_t), $t < s$, take place, whence, by the description of the model system, $\widehat{x}_s = \widehat{X}_s(z, d^{s-1})$ is affine in the purified outputs, and consequently the same is true for the model outputs $\widehat{y}_s = \widehat{Y}_s(z, d^{s-1})$:

$$\widehat{Y}_s(z, d^{s-1}) = p_s + \sum_{\tau=0}^{s-1} P_{s\tau} V_\tau(z, d^\tau).$$

We conclude that with properly chosen p_s, $P_{s\tau}$ we have

$$Y_s(z, d^s) \equiv \widehat{Y}_s(z, d^{s-1}) + V_s(z, d^s) = p_s + \sum_{\tau=0}^{s-1} P_{s\tau} V_\tau(z, d^\tau) + V_s(z, d^s),$$

as required in (I_s). Induction is completed, and (i) is proved.

(ii): Let us fix a linear control law in the form of (14.4.8.d), and let $x_t = X_t(z, d^{t-1})$, $\widehat{x}_t = \widehat{X}_t(z, d^{t-1})$, $u_t = U_t(z, d^t)$, $y_t = Y_t(z, d^t)$, $v_t = V_t(z, d^t)$ be the corresponding actual and model states, controls, and actual and purified outputs. We should verify that the state-control dynamics in question can be obtained from an appropriate control law in the form of (14.4.2). To this end, similarly to the proof of (i), it suffices to show that for every $t \geq 0$ one has

$$V_t(z, d^t) \equiv q_t + \sum_{\tau=0}^{t} Q_{t\tau} Y_\tau(z, d^\tau) \tag{III_t}$$

with properly chosen q_t, $Q_{t\tau}$. We again apply induction in t. The base $t = 0$ is again trivially true due to $V_0(z, d^0) \equiv Y_0(z, d^0)$. Now let $s \geq 1$, and assume that relations (III_t) are valid for $0 \leq t < s$, and let us prove that (III_s) is valid as well. From the validity of (III_t), $t < s$, and from (14.4.8.d) it follows that

$$t < s \Rightarrow U_t(z, d^t) = c_t + \sum_{\tau=0}^{t} C_{t\tau} Y_\tau(z, d^\tau)$$

with properly chosen c_t and $C_{t\tau}$. From these relations and the description of the model system it follows that its state $\widehat{X}_s(z, d^{s-1})$ at time s, and therefore the model output $\widehat{Y}_s(z, d^{s-1})$, are affine functions of $Y_0(z, d^0), ..., Y_{s-1}(z, d^{s-1})$:

$$\widehat{Y}_s(z, d^{s-1}) = p_s + \sum_{\tau=0}^{s-1} P_{s\tau} Y_\tau(z, d^\tau)$$

with properly chosen p_s, $P_{s\tau}$. It follows that

$$V_s(z, d^s) \equiv Y_s(z, d^s) - \widehat{Y}_s(z, d^{s-1}) = Y_s(z, d^s) - p_s - \sum_{\tau=0}^{s-1} P_{s\tau} Y_\tau(z, d^\tau),$$

as required in (III$_s$). Induction is completed, and (ii) is proved.

(iii): For $0 \le s \le t$ let

$$A_s^t = \begin{cases} \prod_{r=s}^{t-1} A_r, & s < t \\ I, & s = t \end{cases}$$

Setting $\delta_t = x_t - \widehat{x}_t$, we have by (14.4.8.$a$–$b$)

$$\delta_{t+1} = A_t \delta_t + R_t d_t, \ \delta_0 = z \Rightarrow \delta_t = A_0^t z + \sum_{s=0}^{t-1} A_{s+1}^t R_s d_s$$

(from now on, sums over empty index sets are zero), whence

$$v_\tau = C_\tau \delta_\tau + D_\tau d_\tau = C_\tau A_0^\tau z + \sum_{s=0}^{\tau-1} C_\tau A_{s+1}^\tau R_s d_s + D_\tau d_\tau. \qquad (14.4.10)$$

Therefore control law (14.4.8.d) implies that

$$\begin{aligned} u_t &= h_t + \sum_{\tau=0}^{t} H_{t\tau} v_\tau = \underbrace{h_t}_{\nu_t[\eta]} + \underbrace{\left[\sum_{\tau=0}^{t} H_{t\tau} C_\tau A_0^\tau\right]}_{N_t[\eta]} z \\ &\quad + \sum_{s=0}^{t-1} \underbrace{\left[H_{ts} D_s + \sum_{\tau=s+1}^{t} H_{t\tau} C_\tau A_{s+1}^\tau R_s\right]}_{N_{ts}[\eta]} d_s + \underbrace{H_{tt} D_t}_{N_{tt}[\eta]} d_t \\ &= \nu_t[\eta] + N_t[\eta] z + \sum_{s=0}^{t} N_{ts}[\eta] d_s, \end{aligned}$$

$$(14.4.11)$$

whence, invoking (14.4.8.a),

$$
\begin{aligned}
x_t &= A_0^t z + \sum_{\tau=0}^{t-1} A_{\tau+1}^t [B_\tau u_\tau + R_\tau d_\tau] = \underbrace{\left[\sum_{\tau=0}^{t-1} A_{\tau+1}^t B_\tau h_t \right]}_{\mu_t[\eta]} \\
&+ \underbrace{\left[A_0^t + \sum_{\tau=0}^{t-1} A_{\tau+1}^t B_\tau N_\tau[\eta] \right]}_{M_t[\eta]} z \\
&+ \sum_{s=0}^{t-1} \underbrace{\left[\sum_{\tau=s}^{t-1} A_{\tau+1}^t B_\tau N_{\tau s}[\eta] + A_{s+1}^t B_s R_s \right]}_{M_{ts}[\eta]} d_s \\
&= \mu_t[\eta] + M_t[\eta] z + \sum_{s=0}^{t-1} M_{ts}[\eta] d_s.
\end{aligned}
\tag{14.4.12}
$$

We see that the states x_t, $0 \le t \le N+1$, and the controls u_t, $0 \le t \le N$, of the closed loop system (14.4.8) are affine functions of z, d^N, and the corresponding "coefficients" $\mu_t[\eta],...,N_{ts}[\eta]$ are affine vector- and matrix-valued functions of the parameters $\eta = \{h_t, H_{t\tau}\}_{0 \le \tau \le t \le N}$ of the underlying control law (14.4.8.d).   □

The consequences. The representation (14.4.8.d) of affine control laws is incomparably better suited for design purposes than the representation (14.4.2), since, as we know from Theorem 14.4.1.(iii), *with controller (14.4.8.d), the state-control trajectory w^N becomes bi-affine in $\zeta = (z, d^N)$ and in the parameters $\eta = \{h_t, H_{t\tau}, 0 \le \tau \le t \le N\}$ of the controller:*

$$
w^N = \omega^N[\eta] + \Omega^N[\eta]\zeta
\tag{14.4.13}
$$

with vector- and matrix-valued functions $\omega^N[\eta]$, $\Omega^N[\eta]$ affinely depending on η and readily given by the dynamics (14.4.1). Substituting (14.4.13) into (14.4.3), we arrive at the system of semi-infinite *bi-affine* scalar inequalities

$$
A \left[\omega^N[\eta] + \Omega^N[\eta]\zeta \right] \le b
\tag{14.4.14}
$$

in variables η, and can use the tractability results from chapters 1, 3, 11 in order to solve efficiently the RC/GRC of this uncertain system of scalar linear constraints. For example, we can process efficiently the GRC setting of the semi-infinite constraints (14.4.13)

$$
a_i^T \left[\omega^N[\eta] + \Omega^N[\eta][z; d^N] \right] - b_i \le \alpha_i^z \mathrm{dist}(z, \mathcal{Z}) + \alpha_d^i \mathrm{dist}(d^N, \mathcal{D}^N)
\\
\forall [z; d^N] \; \forall i = 1, ..., I
\tag{14.4.15}
$$

where $\mathcal{Z}$, $\mathcal{D}^N$ are "good," (e.g., given by strictly feasible semidefinite representations), closed convex normal ranges of z, d^N, respectively, and the distances are defined via the $\|\cdot\|_\infty$ norms (this setting corresponds to the "structured" GRC, see Definition 11.1.2). By the results of section 11.3, system (14.4.15) is equivalent to the system of constraints

$\forall (i, 1 \leq i \leq I):$

$(a) \quad a_i^T \left[\omega^N[\eta] + \Omega^N[\eta][z; d^N] \right] - b_i \leq 0 \ \forall [z; d^N] \in \mathcal{Z} \times \mathcal{D}^N$ $\qquad$ (14.4.16)

$(b) \quad \|a_i^T \Omega_z^N[\eta]\|_1 \leq \alpha_z^i \qquad (c) \quad \|a_i^T \Omega_d^N[\eta]\|_1 \leq \alpha_d^i,$

where $\Omega^N[\eta] = \left[\Omega_z^N[\eta], \Omega_d^N[\eta] \right]$ is the partition of the matrix $\Omega^N[\eta]$ corresponding to the partition $\zeta = [z; d^N]$. Note that in (14.4.16), the semi-infinite constraints (a) admit explicit semidefinite representations (Theorem 1.3.4), while constraints $(b-c)$ are, essentially, just linear constraints on η and on α_z^i, α_d^i. As a result, (14.4.16) can be thought of as a computationally tractable system of convex constraints on η and on the sensitivities α_z^i, α_d^i, and we can minimize under these constraints a "nice," (e.g., convex), function of η and the sensitivities. Thus, after passing to the POB representation of affine control laws, we can process efficiently specifications expressed by systems of linear inequalities, to be satisfied in a robust fashion, on the (finite-horizon) state-control trajectory.

> The just summarized nice consequences of passing to the POB control laws are closely related to the tractability of AARCs of uncertain LO problems with fixed recourse, specifically, as follows. Let us treat the state equations (14.4.1) coupled with the design specifications (14.4.3) as a system of uncertainty-affected linear constraints on the state-control trajectory w, the uncertain data being $\zeta = [z; d^N]$. Relations (14.4.10) say that the purified outputs v_t are known in advance, completely independent of what the control law in use is, *linear* functions of ζ. With this interpretation, a POB control law becomes a collection of affine decision rules that specify the decision variables u_t as affine functions of $P_t\zeta \equiv [v_0; v_1; ...; v_t]$ and simultaneously, via the state equations, specify the states x_t as affine functions of $P_{t-1}\zeta$. Thus, when looking for a POB control law that meets our design specifications in a robust fashion, we are doing nothing but solving the RC (or the GRC) of an uncertain LO problem in affine decision rules possessing a prescribed "information base." On closest inspection, this uncertain LO problem is with fixed recourse, and therefore its robust counterparts are computationally tractable.

Remark 14.4.2. It should be stressed that the re-parameterization of affine control laws underlying Theorem 14.4.1 (and via this Theorem — the nice tractability results we have just mentioned) is nonlinear. As a result, it can be of not much use when we are optimizing over affine control laws satisfying additional restrictions rather than over all affine control laws.

> Assume, e.g., that we are seeking control in the form of a simple output-based linear feedback:
>
> $$u_t = G_t y_t.$$
>
> This requirement is just a system of simple linear constraints on the parameters of the control law in the form of (14.4.2), which, however, does not help

much, since, as we have already explained, optimization over control laws in this form is by itself difficult. And when passing to affine control laws in the form of (14.4.8.d), the requirement that our would-be control should be a linear output-based feedback becomes a system of highly nonlinear constraints on our new design parameters η, and the synthesis again turns out to be difficult.

Example: Controlling finite-horizon gains. Natural design specification pertaining to finite-horizon Robust Linear Control are in the form of bounds on finite-horizon *gains* z2x^N, z2u^N, d2x^N, d2u^N defined as follows: with a linear, (i.e., with $h_t \equiv 0$) control law (14.4.8.d), the states x_t and the controls u_t are linear functions of z and d^N:

$$x_t = X_t^z[\eta]z + X_t^d[\eta]d^N, \; u_t = U_t^z[\eta]z + U_t^d[\eta]d^N$$

with matrices $X_t^z[\eta],...,U_t^d[\eta]$ affinely depending on the parameters η of the control law. Given t, we can define the z to x_t *gains* and the *finite-horizon* z to x gain as z2x$_t(\eta) = \max_z\{\|X_t^z[\eta]z\|_\infty : \|z\|_\infty \leq 1\}$ and z2x$^N(\eta) = \max_{0\leq t\leq N}$ z2x$_t(\eta)$. The definitions of the z to u gains z2u$_t(\eta)$, z2u$^N(\eta)$ and the "disturbance to x/u" gains d2x$_t(\eta)$, d2x$^N(\eta)$, d2u$_t(\eta)$, d2u$^N(\eta)$ are completely similar, e.g., d2u$_t(\eta) = \max_{d^N}\{\|U_t^d[\eta]d^N\|_\infty : \|d^N\|_\infty \leq 1\}$ and d2u$^N(\eta) = \max_{0\leq t\leq N}$ d2u$_t(\eta)$. The finite-horizon gains clearly are nonincreasing functions of the time horizon N and have a transparent Control interpretation; e.g., d2x$^N(\eta)$ ("peak to peak d to x gain") is the largest possible perturbation in the states x_t, $t = 0, 1, ..., N$ caused by a unit perturbation of the sequence of disturbances d^N, both perturbations being measured in the $\|\cdot\|_\infty$ norms on the respective spaces. Upper bounds on ^{N}-gains (and on *global* gains like d2x$^\infty(\eta) = \sup_{N\geq 0}$ d2x$^N(\eta)$) are natural Control specifications. With our purified-output-based representation of linear control laws, the finite-horizon specifications of this type result in explicit systems of linear constraints on η and thus can be processed routinely via LO. For example, an upper bound d2x$^N(\eta) \leq \lambda$ on d2x^N gain is equivalent to the requirement $\sum_j|(X_t^d[\eta])_{ij}| \leq \lambda$ for all i and all $t \leq N$; since X_t^d is affine in η, this is just a system of linear constraints on η and on appropriate slack variables. Note that imposing bounds on the gains can be interpreted as passing to the GRC (14.4.15) in the case where the "desired behavior" merely requires $w^N = 0$, and the normal ranges of the initial state and the disturbances are the origins in the corresponding spaces: $\mathcal{Z} = \{0\}$, $\mathcal{D}^N = \{0\}$.

14.4.2.1 Non-affine control laws

So far, we focused on synthesis of finite-horizon *affine* POB controllers. Acting in the spirit of section 14.3.2, we can handle also synthesis of *quadratic* POB control laws — those where every entry of u_t, instead of being affine in the purified outputs $v^t = [v_0; ...; v_t]$, is allowed to be a quadratic function of v^t. Specifically, assume that we want to "close" the open loop system (14.4.1) by a non-anticipating control law in order to ensure that the state-control trajectory w^N of the closed loop system satisfies a given system S of linear constraints in a robust fashion, that is, for all realizations of the "uncertain data" $\zeta = [z; d^N]$ from a given uncertainty set $\mathcal{Z}_\rho^N = \rho \mathcal{Z}^N$ ($\rho > 0$ is, as always, the uncertainty level, and $\mathcal{Z} \ni 0$ is a closed convex set of "uncertain data of magnitude ≤ 1"). Let us use a *quadratic POB control law* in the form of

$$u_t^i = h_{it}^0 + h_{i,t}^T v^t + \frac{1}{\rho}[v^t]^T H_{i,t} v^t, \qquad (14.4.17)$$

where u_t^i is i-th coordinate of the vector of controls at instant t, and h_{it}^0, h_{it} and H_{it} are, respectively, real, vector, and matrix parameters of the control law.[3] On a finite time horizon $0 \leq t \leq N$, such a quadratic control law is specified by ρ and the finite-dimensional vector $\eta = \{h_{it}^0, h_{it}, H_{it}\}_{\substack{1 \leq i \leq \dim u \\ 0 \leq t \leq N}}$. Now note that the purified outputs are well defined for any non-anticipating control law, not necessary affine, and they are *independent of the control law linear functions of $\zeta^t \equiv [z; d^t]$. The coefficients of these linear functions are readily given by the data $A_\tau, ..., D_\tau$, $0 \leq \tau \leq t$ (see (14.4.7)). With this in mind, we see that the controls, as given by (14.4.17), are quadratic functions of the initial state and the disturbances, the coefficients of these quadratic functions being affine in the vector η of parameters of our quadratic control law:*

$$u_t^i = \mathcal{U}_{it}^{(0)}[\eta] + [z; d^t]^T \mathcal{U}_{it}^{(1)}[\eta] + \frac{1}{\rho}[z; d^t]^T \mathcal{U}_{it}^{(2)}[\eta][z; d^t] \qquad (14.4.18)$$

with affine in η reals/vectors/matrices $\mathcal{U}_{it}^{(\kappa)}[\eta]$, $\kappa = 0, 1, 2$. Plugging these representations of the controls into the state equations of the open loop system (14.4.1), we conclude that the states x_t^j of the closed loop system obtained by "closing" (14.4.1) by the quadratic control law (14.4.17), have the same "affine in η, quadratic in $[z; d^t]$" structure as the controls:

$$x_t^i = \mathcal{X}_{jt}^{(0)}[\eta] + [z; d^{t-1}]^T \mathcal{X}_{jt}^{(1)}[\eta] + \frac{1}{\rho}[z; d^{t-1}]^T \mathcal{X}_{jt}^{(2)}[\eta][z; d^{t-1}] \qquad (14.4.19)$$

with affine in η reals/vectors/matrices $\mathcal{X}_{jt}^{(\kappa)}$, $\kappa = 0, 1, 2$.

Plugging representations (14.4.18), (14.4.19) into the system S of our target constraints, we end up with a system of semi-infinite constraints on the parameters η of the control law, specifically, the system

$$a_k[\eta] + 2\zeta^T p_k[\eta] + \frac{1}{\rho}\zeta^T R_k[\eta]\zeta \leq 0 \,\forall \zeta = [z; d^N] \in \mathcal{Z}_\rho^N = \rho \mathcal{Z}^N, \; k = 1, ..., K,$$

$$(14.4.20)$$

where $a_k[\eta]$, $p_k[\eta]$ and $R_k[\zeta]$ are affine in η. Setting $P_k[\eta] = \left[\begin{array}{c|c} & p_k^T[\eta] \\ \hline p_k[\eta] & R_k[\eta] \end{array}\right]$, $\widehat{\zeta}_\rho[\zeta] = \left[\begin{array}{c|c} & \zeta^T \\ \hline \zeta & \zeta\zeta^T \end{array}\right]$ and denoting by $\widehat{\mathcal{Z}}_\rho^N$ the convex hull of the image of the set $\mathcal{Z}_\rho^N$ under the mapping $\zeta \mapsto \widehat{\zeta}_\rho[\zeta]$, system (14.4.20) can be rewritten equivalently

---

[3]The specific way in which the uncertainty level ρ affects the controls is convenient technically and is of no practical importance, since "in reality" the uncertainty level is a known constant.

as

$$a_k[\eta] + \mathrm{Tr}(P_k[\eta]\widehat{\zeta}) \le 0 \ \forall(\widehat{\zeta} \in \widehat{\mathcal{Z}}_\rho^N \equiv \rho\widehat{\mathcal{Z}}_1^N, k = 1, ..., K) \qquad (14.4.21)$$

and we end up with a system of semi-infinite bi-affine scalar inequalities. From the results of section 14.3.2 it follows that this semi-infinite system:

- is computationally tractable, provided that $\mathcal{Z}^N$ is an ellipsoid $\{\zeta : \zeta^T Q\zeta \le 1\}$, $Q \succ 0$. Indeed, here $\widehat{\mathcal{Z}}_1^N$ is the semidefinite representable set

$$\left\{ \left[\begin{array}{c|c} & \omega^T \\ \hline \omega & \Omega \end{array} \right] : \left[\begin{array}{c|c} 1 & \omega^T \\ \hline \omega & \Omega \end{array} \right] \succeq 0, \mathrm{Tr}(\Omega Q) \le 1 \right\};$$

- admits a safe tractable approximation tight within the factor $\vartheta = O(1)\ln(J+1)$, provided that $\mathcal{Z}^N$ is the $\cap$-ellipsoidal uncertainty set $\{\zeta : \zeta^T Q_j\zeta \le 1, 1 \le j \le J\}$, where $Q_j \succeq 0$ and $\sum_j Q_j \succ 0$. This approximation is obtained when replacing the "true" uncertainty set $\widehat{\mathcal{Z}}_\rho^N$ with the semidefinite representable set

$$\mathcal{W}_\rho = \rho \left\{ \left[\begin{array}{c|c} & \omega^T \\ \hline \omega & \Omega \end{array} \right] : \left[\begin{array}{c|c} 1 & \omega^T \\ \hline \omega & \Omega \end{array} \right] \succeq 0, \mathrm{Tr}(\Omega Q_j) \le 1, 1 \le j \le J \right\}$$

(recall that $\widehat{\mathcal{Z}}_\rho^N \subset \mathcal{W}_\rho \subset \widehat{\mathcal{Z}}_{\vartheta\rho}^N$).

14.4.3 Handling Infinite-Horizon Design Specifications

One might think that the outlined reduction of (discrete time) Robust Linear Control problems to Convex Programming, based on passing to the POB representation of affine control laws and deriving tractable reformulations of the resulting semi-infinite bi-affine scalar inequalities is intrinsically restricted to the case of finite-horizon control specifications. In fact our approach is well suited for handling infinite-horizon specifications — those imposing restrictions on the asymptotic behavior of the closed loop system. Specifications of the latter type usually have to do with the *time-invariant* open loop system (14.4.1):

$$\begin{array}{rcl} x_0 & = & z \\ x_{t+1} & = & Ax_t + Bu_t + Rd_t \ , \ t = 0, 1, ... \\ y_t & = & Cx_t + Dd_t \end{array} \qquad (14.4.22)$$

From now on we assume that *the open loop system* (14.4.22) *is stable*, that is, the spectral radius of A is < 1 (in fact this restriction can be somehow circumvented, see below). Imagine that we "close" (14.4.22) by a *nearly time-invariant* POB control law *of order* k, that is, a law of the form

$$u_t = h_t + \sum_{s=0}^{k-1} H_s^t v_{t-s}, \qquad (14.4.23)$$

where $h_t = 0$ for $t \ge N_*$ and $H_\tau^t = H_\tau$ for $t \ge N_*$ for a certain *stabilization time* N_*. From now on, all entities with negative indices are set to 0. While the "time-varying" part $\{h_t, H_\tau^t, 0 \le t < N_*\}$ of the control law can be used to adjust the finite-horizon behavior of the closed loop system, its asymptotic behavior is as if the law were time-invariant: $h_t \equiv 0$ and $H_\tau^t \equiv H_\tau$ for all $t \geq 0$. Setting $\delta_t = x_t - \widehat{x}_t$, $H^t = [H_0^t, ..., H_{k-1}^t]$, $H = [H_0, ..., H_{k-1}]$, the dynamics (14.4.22), (14.4.6), (14.4.23) is given by

$$
\underbrace{\begin{bmatrix} x_{t+1} \\ \delta_{t+1} \\ \delta_t \\ \vdots \\ \delta_{t-k+2} \end{bmatrix}}_{\omega_{t+1}} = \underbrace{\left[\begin{array}{c|cccc} A & BH_0^t C & BH_1^t C & \dots & BH_{k-1}^t C \\ \hline & A & & & \\ & & A & & \\ & & & \ddots & \\ & & & & A \end{array}\right]}_{A_+[H^t]} \omega_t
$$

$$
+ \underbrace{\left[\begin{array}{c|cccc} R & BH_0^t D & BH_1^t D & \dots & BH_{k-1}^t D \\ \hline & R & & & \\ & & R & & \\ & & & \ddots & \\ & & & & R \end{array}\right]}_{R_+[H^t]} \begin{bmatrix} d_t \\ d_t \\ d_{t-1} \\ \vdots \\ d_{t-k+1} \end{bmatrix} \quad (14.4.24)
$$

$$
+ \begin{bmatrix} Bh_t \\ 0 \\ \vdots \\ 0 \end{bmatrix}, \; t = 0, 1, 2, ...,
$$

$$
u_t = h_t + \sum_{\nu=0}^{k-1} H_\nu^t [C\delta_{t-\nu} + Dd_{t-\nu}].
$$

We see that starting with time N_*, dynamics (14.4.24) is exactly as if the underlying control law were the time invariant POB law with the parameters $h_t \equiv 0$, $H^t \equiv H$. Moreover, since A is stable, we see that *system (14.4.24) is stable independently of the parameter H of the control law, and the resolvent $\mathcal{R}_H(s) := (sI - A_+[H])^{-1}$ of $A_+[H]$ is the affine in H matrix*

$$
\left[\begin{array}{c|cccc} \mathcal{R}_A(s) & \mathcal{R}_A(s)BH_0C\mathcal{R}_A(s) & \mathcal{R}_A(s)BH_1C\mathcal{R}_A(s) & \dots & \mathcal{R}_A(s)BH_{k-1}C\mathcal{R}_A(s) \\ \hline & \mathcal{R}_A(s) & & & \\ & & \mathcal{R}_A(s) & & \\ & & & \ddots & \\ & & & & \mathcal{R}_A(s) \end{array}\right],
$$
$$(14.4.25)$$

where $\mathcal{R}_A(s) = (sI - A)^{-1}$ is the resolvent of A.

Now imagine that the sequence of disturbances d_t is of the form $d_t = s^t d$, where $s \in \mathbb{C}$ differs from 0 and from the eigenvalues of A. From the stability of (14.4.24) it follows that as $t \to \infty$, the solution ω_t of the system, independently of the initial state, approaches the "steady-state" solution $\widehat{\omega}_t = s^t \mathcal{H}(s)d$, where $\mathcal{H}(s)$ is certain matrix. In particular, the state-control vector $w_t = \begin{bmatrix} x_t \\ u_t \end{bmatrix}$ approaches, as $t \to \infty$, the trajectory $\widehat{w}_t = s^t \mathcal{H}_{xu}(s)d$. The associated *disturbance-to-state/control transfer matrix $\mathcal{H}_{xu}(s)$* is easily computable:

$$
\mathcal{H}_{xu}(s) = \begin{bmatrix} \overbrace{\mathcal{R}_A(s)\left[R + \sum_{\nu=0}^{k-1} s^{-\nu}BH_\nu\left[D + C\mathcal{R}_A(s)R\right]\right]}^{\mathcal{H}_x(s)} \\ \underbrace{\left[\sum_{\nu=0}^{k-1} s^{-\nu}H_\nu\right]\left[D + C\mathcal{R}_A(s)R\right]}_{\mathcal{H}_u(s)} \end{bmatrix}. \quad (14.4.26)
$$

The crucial fact is *that the transfer matrix $\mathcal{H}_{xu}(s)$ is affine in the parameters $H = [H_0, ..., H_{k-1}]$ of the nearly time invariant control law* (14.4.23). As a result, *design specifications representable as explicit convex constraints on the transfer matrix $\mathcal{H}_{xu}(s)$* (these are typical specifications in infinite-horizon design of linear controllers) *are equivalent to explicit convex constraints on the parameters H of the underlying POB control law and therefore can be processed efficiently via Convex Optimization.*

Example: Discrete time H_∞ control. Discrete time H_∞ design specifications impose constraints on the behavior of the transfer matrix along the unit circumference $s = \exp\{\imath\omega\}$, $0 \leq \omega \leq 2\pi$, that is, on the steady state response of the closed loop system to a disturbance in the form of a harmonic oscillation.[4]. A rather general form of these specifications is a system of constraints

$$\|Q_i(s) - M_i(s)\mathcal{H}_{xu}(s)N_i(s)\| \leq \tau_i \ \forall(s = \exp\{\imath\omega\} : \omega \in \Delta_i), \qquad (14.4.27)$$

where $Q_i(s)$, $M_i(s)$, $N_i(s)$ are given rational matrix-valued functions with no singularities on the unit circumference $\{s : |s| = 1\}$, $\Delta_i \subset [0, 2\pi]$ are given segments, and $\|\cdot\|$ is the standard matrix norm (the largest singular value).

We are about to demonstrate that constraints (14.4.27) can be represented by an explicit finite system of LMIs; as a result, specifications (14.4.27) can be efficiently processed numerically. Here is the derivation. Both "transfer functions" $\mathcal{H}_x(s)$, $\mathcal{H}_u(s)$ are of the form $q^{-1}(s)Q(s, H)$, where $q(s)$ is a scalar polynomial independent of H, and $Q(s, H)$ is a matrix-valued polynomial of s with coefficients *affinely depending on H*. With this in mind, we see that the constraints are of the generic form

$$\|p^{-1}(s)P(s, H)\| \leq \tau \forall(s = \exp\{\imath\omega\} : \omega \in \Delta), \qquad (14.4.28)$$

where $p(\cdot)$ is a scalar polynomial independent of H and $P(s, H)$ is a polynomial in s with $m \times n$ matrix coefficients affinely depending on H. Constraint (14.4.28) can be expressed equivalently by the semi-infinite matrix inequality

$$\begin{bmatrix} \tau I_m & P(z, H)/p(z) \\ (P(z, H))^*/(p(z))^* & \tau I_n \end{bmatrix} \succeq 0 \forall(z = \exp\{\imath\omega\} : \omega \in \Delta)$$

(* stands for the Hermitian conjugate, $\Delta \subset [0, 2\pi]$ is a segment) or, which is the same,

$$\begin{aligned} S_{H,\tau}(\omega) &\equiv \begin{bmatrix} \tau p(\exp\{\imath\omega\})(p(\exp\{\imath\omega\}))^* I_m & (p(\exp\{\imath\omega\}))^* P(\exp\{\imath\omega\}, H) \\ p(\exp\{\imath\omega\})(P(\exp\{\imath\omega\}, H))^* & \tau p(\exp\{\imath\omega\})(p(\exp\{\imath\omega\}))^* I_n \end{bmatrix} \\ &\succeq 0 \forall\omega \in \Delta. \end{aligned}$$

---

[4]The entries of $\mathcal{H}_x(s)$ and $\mathcal{H}_u(s)$, restricted onto the unit circumference $s = \exp\{\imath\omega\}$, have very transparent interpretation. Assume that the only nonzero entry in the disturbances is the j-th one, and it varies in time as a harmonic oscillation of unit amplitude and frequency ω. The steady-state behavior of i-th state then will be a harmonic oscillation of the same frequency, but with another amplitude, namely, $|(\mathcal{H}_x(\exp\{\imath\omega\})_{ij}|$ and phase shifted by $\arg((\mathcal{H}_x(\exp\{\imath\omega\})_{ij})$. Thus, the *state-to-input frequency responses* $(\mathcal{H}_x(\exp\{\imath\omega\}))_{ij}$ explain the steady-state behavior of states when the input is comprised of harmonic oscillations. The interpretation of the *control-to-input frequency responses* $(\mathcal{H}_u(\exp\{\imath\omega\}))_{ij}$ is completely similar.

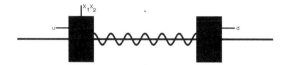

Figure 14.3 Double pendulum: two masses linked by a spring sliding without friction along a rod. Position and velocity of the first mass are observed.

Observe that $S_{H,\tau}(\omega)$ is a trigonometric polynomial taking values in the space of Hermitian matrices of appropriate size, the coefficients of the polynomial being affine in H, τ. It is known [53] that the cone $\mathcal{P}_m$ of (coefficients of) all Hermitian matrix-valued trigonometric polynomials $S(\omega)$ of degree $\leq m$, which are $\succeq 0$ for all $\omega \in \Delta$, is semidefinite representable, i.e., there exists an explicit LMI

$$\mathcal{A}(S, u) \succeq 0$$

in variables S (the coefficients of a polynomial $S(\cdot)$) and additional variables u such that $S(\cdot) \in \mathcal{P}_m$ if and only if S can be extended by appropriate u to a solution of the LMI. Consequently, the relation

$$\mathcal{A}(S_{H,\tau}, u) \succeq 0, \tag{$*$}$$

which is an LMI in H, τ, u, is a semidefinite representation of (14.4.28): H, τ solve (14.4.28) if and only if there exists u such that H, τ, u solve $(*)$.

14.4.4 Putting Things Together: Infinite- and Finite-Horizon Design Specifications

For the time being, we have considered optimization over purified-output-based affine control laws in two different settings, finite- and infinite-horizon design specifications. In fact we can to some extent combine both settings, thus seeking affine purified-output-based controls ensuring both a good steady-state behavior of the closed loop system and a "good transition" to this steady-state behavior. The proposed methodology will become clear from the example that follows.

Consider the open-loop time-invariant system representing the discretized double-pendulum depicted on figure 14.3. The dynamics of the continuous time prototype plant is given by

$$\begin{aligned} \dot{x} &= A_c x + B_c u + R_c d \\ y &= C x, \end{aligned}$$

where

$$A_c = \begin{bmatrix} 0 & 1 & 0 & 0 \\ -1 & 0 & 1 & 0 \\ 0 & 0 & 0 & 1 \\ 1 & 0 & -1 & 0 \end{bmatrix}, \; B_c = \begin{bmatrix} 0 \\ 1 \\ 0 \\ 0 \end{bmatrix}, \; R_c = \begin{bmatrix} 0 \\ 0 \\ 0 \\ -1 \end{bmatrix}, \; C = \begin{bmatrix} 1 & 0 & 0 & 0 \\ 0 & 1 & 0 & 0 \end{bmatrix}$$

$(x_1, x_2$ are the position and the velocity of the first mass, and x_3, x_4 those of the second mass). The discrete time plant we will actually work with is

$$
\begin{aligned}
x_{t+1} &= A_0 x_t + B u_t + R d_t \\
y_t &= C x_t
\end{aligned}
\tag{14.4.29}
$$

where $A_0 = \exp\{\Delta \cdot A_c\}$, $B \overset{\Delta}{=} \int_0^\Delta \exp\{s A_c\} B_c ds$, $R \overset{\Delta}{=} \int_0^\Delta \exp\{s A_c\} R_c ds$. System (14.4.29) is not stable (absolute values of all eigenvalues of A_0 are equal to 1), which seemingly prevents us from addressing infinite-horizon design specifications via the techniques developed in section 14.4.3. The simplest way to circumvent the difficulty is to augment the original plant by a stabilizing time-invariant linear feedback; upon success, we then apply the purified-output-based synthesis to the augmented, already stable, plant. Specifically, let us look for a controller of the form

$$
u_t = K y_t + w_t.
\tag{14.4.30}
$$

With such a controller, (14.4.29) becomes

$$
\begin{aligned}
x_{t+1} &= A x_t + B w_t + R d_t, \quad A = A_0 + BKC \\
y_t &= C x_t.
\end{aligned}
\tag{14.4.31}
$$

If K is chosen in such a way that the matrix $A = A_0 + BKC$ is stable, we can apply all our purified-output-based machinery to the plant (14.4.31), with w_t in the role of u_t, however keeping in mind that the "true" controls u_t will be $K y_t + w_t$.

For our toy plant, a stabilizing feedback K can be found by "brute force" — by generating a random sample of matrices of the required size and selecting from this sample a matrix, if any, which indeed makes (14.4.31) stable. Our search yielded feedback matrix $K = [-0.6950, -1.7831]$, with the spectral radius of the matrix $A = A_0 + BKC$ equal to 0.87. From now on, we focus on the resulting plant (14.4.31), which we intend to "close" by a control law from $\mathcal{C}_{8,0}$, where $\mathcal{C}_{k,0}$ is the family of all time invariant control laws of the form

$$
w_t = \sum_{\tau=0}^t H_{t-\tau} v_\tau \quad
\left[
\begin{array}{l}
v_t = y_t - C\hat{x}_t, \\
\hat{x}_{t+1} = A\hat{x}_t + B w_t, \ \hat{x}_0 = 0
\end{array}
\right]
\tag{14.4.32}
$$

where $H_s = 0$ when $s \geq k$. Our goal is to pick in $\mathcal{C}_{8,0}$ a control law with desired properties (to be precisely specified below) expressed in terms of the following 6 criteria:

- the four peak to peak gains z2x, z2u, d2x, d2u defined on p. 400;

- the two H_∞ gains

$$
H_{\infty,x} = \max_{|s|=1, i, j} |(\mathcal{H}_x(s))|_{ij}, \quad H_{\infty,u} = \max_{|s|=1, i, j} |(\mathcal{H}_u(s))|_{ij},
$$

where $\mathcal{H}_x$ and $\mathcal{H}_u$ are the transfer functions from the disturbances to the states and the controls, respectively.

Optimized criterion	Resulting values of the criteria					
	$z2x^{40}$	$z2u^{40}$	$d2x^{40}$	$d2u^{40}$	$H_{\infty,x}$	$H_{\infty,u}$
$z2x^{40}$	<u>25.8</u>	205.8	1.90	3.75	10.52	5.87
$z2u^{40}$	58.90	<u>161.3</u>	1.90	3.74	39.87	20.50
$d2x^{40}$	5773.1	13718.2	<u>1.77</u>	6.83	1.72	4.60
$d2u^{40}$	1211.1	4903.7	1.90	<u>2.46</u>	66.86	33.67
$H_{\infty,x}$	121.1	501.6	1.90	5.21	<u>1.64</u>	5.14
$H_{\infty,u}$	112.8	460.4	1.90	4.14	8.13	<u>1.48</u>
	z2x	z2u	d2x	d2u	$H_{\infty,x}$	$H_{\infty,u}$
(14.4.34)	31.59	197.75	1.91	4.09	1.82	2.04
(14.4.35)	2.58	0.90	1.91	4.17	1.77	1.63

Table 14.2 Gains for time invariant control laws of order 8 yielded by optimizing, one at a time, the criteria z2x^{40},...,$H_{\infty,u}$ over control laws from $\mathcal{F} = \{\eta \in \mathcal{C}_{8,0} :$ d2x$^{40}[\eta] \leq 1.90\}$ (first six lines), and by solving programs (14.4.34), (14.4.35) (last two lines).

Note that while the purified-output-based control w_t we are seeking is defined in terms of the stabilized plant (14.4.31), the criteria z2u,d2u, $H_{\infty,u}$ are defined in terms of the original controls $u_t = Ky_t + w_t = KCx_t + w_t$ affecting the actual plant (14.4.29).

In the synthesis we are about to describe our primary goal is to minimize the global disturbance to state gain d2x, while the secondary goal is to avoid too large values of the remaining criteria. We achieve this goal as follows.

Step 1: Optimizing d2x. As it was explained on p. 400, the optimization problem

$$\text{Opt}_{d2x}(k, 0; N_+) = \min_{\eta \in \mathcal{C}_{k,0}} \max_{0 \leq t \leq N_+} d2x_t[\eta] \qquad (14.4.33)$$

is an explicit convex program (in fact, just an LO), and its optimal value is a lower bound on the best possible global gain d2x achievable with control laws from $\mathcal{C}_{k,0}$. In our experiment, we solve (14.4.33) for $k = 8$ and $N_+ = 40$, arriving at $\text{Opt}_{d2x}(8, 0; 40) = 1.773$. The global d2x gain of the resulting time-invariant control law is 1.836 — just 3.5% larger than the outlined lower bound. We conclude that the control yielded by the solution to (14.4.33) is nearly the best one, in terms of the global d2x gain, among time-invariant controls of order 8. At the same time, part of the other gains associated with this control are far from being good, see line "d2x^{40}" in table 14.2.

Step 2: Improving the remaining gains. To improve the "bad" gains yielded by the nearly d2x-optimal control law we have built, we act as follows: we look at the family $\mathcal{F}$ of all time invariant control laws of order 8 with the finite-horizon d2x gain d2x$^{40}[\eta] = \max_{0 \leq t \leq 40} d2x_t[\eta]$ not exceeding 1.90 (that is, look at the controls from $\mathcal{C}_{8,0}$ that are within 7.1% of the optimum in terms of their d2x^{40} gain) and act as follows:

A. We optimize over $\mathcal{F}$, one at a time, every one of the remaining criteria z2x$^{40}[\eta]$ = $\max_{0 \leq t \leq 40}$ z2x$_t[\eta]$, z2u$^{40}[\eta]$ = $\max_{0 \leq t \leq 40}$ z2u$_t[\eta]$, d2u$^{40}[\eta]$ = $\max_{0 \leq t \leq 40}$ d2u$_t[\eta]$, $H_{\infty,x}[\eta]$, $H_{\infty,u}[\eta]$, thus obtaining "reference values" of these criteria; these are lower bounds on the optimal values of the corresponding global gains, optimization being carried out over the set $\mathcal{F}$. These lower bounds are the underlined data in table 14.2.

B. We then minimize over $\mathcal{F}$ the "aggregated gain"

$$\frac{\text{z2x}^{40}[\eta]}{25.8} + \frac{\text{z2u}^{40}[\eta]}{161.3} + \frac{\text{d2u}^{40}[\eta]}{2.46} + \frac{H_{\infty,x}[\eta]}{1.64} + \frac{H_{\infty,u}[\eta]}{1.48} \tag{14.4.34}$$

(the denominators are exactly the aforementioned reference values of the corresponding gains). The global gains of the resulting time-invariant control law of order 8 are presented in the "(14.4.34)" line of table 14.2.

Step 3: Finite-horizon adjustments. Our last step is to improve the z2x and z2u gains by passing from a time invariant affine control law of order 8 to a nearly time invariant law of order 8 with stabilization time $N_* = 20$. To this end, we solve the convex optimization problem

$$\min_{\eta \in \mathcal{C}_{8,20}} \left\{ \text{z2x}^{50}[\eta] + \text{z2u}^{50}[\eta] : \begin{array}{rcl} \text{d2x}^{50}[\eta] & \leq & 1.90 \\ \text{d2u}^{50}[\eta] & \leq & 4.20 \\ H_{\infty,x}[\eta] & \leq & 1.87 \\ H_{\infty,u}[\eta] & \leq & 2.09 \end{array} \right\} \tag{14.4.35}$$

(the right hand sides in the constraints for d2u$^{50}[\cdot]$, $H_{\infty,x}[\cdot]$, $H_{\infty,u}[\cdot]$ are the slightly increased (by 2.5%) gains of the time invariant control law obtained in Step 2). The global gains of the resulting control law are presented in the last line of table 14.2, see also figure 14.4. We see that finite-horizon adjustments allow us to reduce by orders of magnitude the global z2x and z2u gains and, as an additional bonus, result in a substantial reduction of H_∞-gains.

Simple as this control problem may be, it serves well to demonstrate the importance of purified-output-based representation of affine control laws and the associated possibility to express various control specifications as explicit convex constraints on the parameters of such laws.

14.5 EXERCISES

Exercise 14.1. Consider a discrete time linear dynamical system

$$\begin{aligned} x_0 &= z \\ x_{t+1} &= A_t x_t + B_t u_t + R_t d_t, \ t = 0, 1, \dots \end{aligned} \tag{14.5.1}$$

where $x_t \in \mathbb{R}^n$ are the states, $u_t \in \mathbb{R}^m$ are the controls, and $d_t \in \mathbb{R}^k$ are the exogenous disturbances. We are interested in the behavior of the system on the

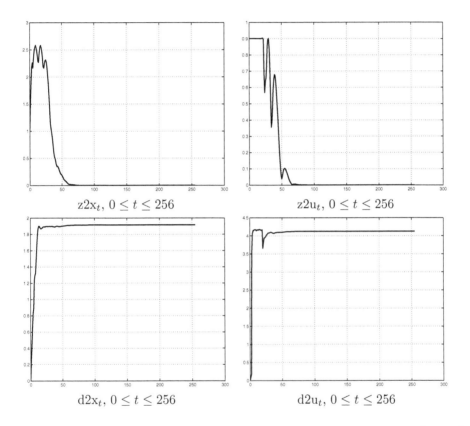

Figure 14.4 Frequency responses and gains of control law given by solution to (14.4.35).

finite time horizon $t = 0, 1, ..., N$. A "desired behavior" is given by the requirement

$$\|Pw - q\|_\infty \leq R \tag{14.5.2}$$

on the state-control trajectory $w = [x_0; ...; x_{N+1}; u_0; ...; u_N]$.

Let us treat $\zeta = [z; d_0; ...; d_N]$ as an uncertain perturbation with perturbation structure $(\mathcal{Z}, \mathcal{L}, \| \cdot \|_r)$, where

$$\mathcal{Z} = \{\zeta : \|\zeta - \bar{\zeta}\|_s \leq R\}, \ \mathcal{L} = \mathbb{R}^L \qquad [L = \dim \zeta]$$

and $r, s \in [1, \infty]$, so that (14.5.1), (14.5.2) become a system of uncertainty-affected linear constraints on w^N. We want to process the Affinely Adjustable GRC of the system, where u_t are allowed to be affine functions of the initial state z and the vector of disturbances $d^t = [d_0; ...; d_t]$ up to time t, and the states x_t are allowed to be affine functions of z and d^{t-1}. We wish to minimize the corresponding global sensitivity.

> In control terms: we want to "close" the open-loop system (14.5.1) with a non-anticipative affine control law
>
> $$u_t = U_t^z z + U_t^d d^t + u_t^0 \tag{14.5.3}$$
>
> based on observations of initial states and disturbances up to time t in such a way that the "closed loop system" (14.5.1), (14.5.3) exhibits the desired behavior in a robust w.r.t. the initial state and the disturbances fashion.

Write down the AAGRC of our uncertain problem as an explicit convex program with efficiently computable constraints.

Exercise 14.2. Consider the modification of Exercise 14.1 where the cone $\mathcal{L} = \mathbb{R}^L$ is replaced with

$$\mathcal{L} = \{[0; d_0; ...; d_N] : d_t \geq 0, \ 0 \leq t \leq N\},$$

and solve the corresponding version of the Exercise.

Exercise 14.3. Consider the simplest version of Exercise 14.1, where (14.5.1) reads

$$\begin{aligned} x_0 &= z \in \mathbb{R} \\ x_{t+1} &= x_t + u_t - d_t, \ t = 0, 1, ..., 15, \end{aligned}$$

(14.5.2) reads

$$|\theta x_t| = 0, \ t = 1, 2, ..., 16, \ |u_t| = 0, \ t = 0, 1, ..., 15$$

and the perturbation structure is

$$\mathcal{Z} = \{[z; d_0; ...; d_{15}] = 0\} \subset \mathbb{R}^{17}, \ \mathcal{L} = \{[0; d_0; d_1; ...; d_{15}]\}, \|\zeta\| \equiv \|\zeta\|_2.$$

Assuming the same "adjustability status" of u_t and x_t as in Exercise 14.1,

  i) Represent the AAGRC of (the outlined specializations of) (14.5.1), (14.5.2), where the goal is to minimize the global sensitivity, as an explicit convex program;

ii) Interpret the AAGRC in Control terms;

iii) Solve the AAGRC for the values of θ equal to 1.e6, 10, 2, 1.

Exercise 14.4. Consider a communication network — an oriented graph G with the set of nodes $V = \{1, ..., n\}$ and the set of arcs Γ. Several ordered pairs of nodes (i, j) are marked as "source-sink" nodes and are assigned traffic d_{ij} — the amount of information to be transmitted from node i to node j per unit time; the set of all source-sink pairs is denoted by $\mathcal{J}$. Arcs $\gamma \in \Gamma$ of a communication network are assigned with capacities — upper bounds on the total amount of information that can be sent through the arc per unit time. We assume that the arcs already possess certain capacities p_γ, which can be further increased; the cost of a unit increase of the capacity of arc γ is a given constant c_γ.

1) Assuming the demands d_{ij} certain, formulate the problem of finding the cheapest extension of the existing network capable to ensure the required source-sink traffic as an LO program.

2) Now assume that the vector of traffic $d = \{d_{ij} : (i, j) \in \mathcal{J}\}$ is uncertain and is known to run through a given semidefinite representable compact uncertainty set $\mathcal{Z}$. Allowing the amounts x_γ^{ij} of information with origin i and destination j traveling through the arc γ to depend affinely on traffic, build the AARC of the (uncertain version of the) problem from 1). Consider two cases: (a) for every $(i, j) \in \mathcal{J}$, x_γ^{ij} can depend affinely solely on d_{ij}, and (b) x_γ^{ij} can depend affinely on the entire vector d. Are the resulting problems computationally tractable?

3) Assume that the vector d is random, and its components are independent random variables uniformly distributed in given segments Δ_{ij} of positive lengths. Build the chance constrained versions of the problems from 2).

14.6 NOTES AND REMARKS

NR 14.1. Multi-stage decision making problems, including those where the decisions should be made in an uncertain environment, are of extreme applied importance and therefore were on the "optimization agenda" for several decades, essentially, since the birth of Mathematical Programming in late 1940s. Unfortunately, because of the immense intrinsic complexity of these problems, there still is a dramatic gap between what we would like to do and what we indeed can do. Here is the opinion of George Dantzig, the founder of Mathematical Programming: *"In retrospect it is interesting to note that the original problem that started my research* [on Linear and Mathematical Programming] *is still outstanding — namely, the problem of planning or scheduling dynamically over time, particularly planning dynamically under uncertainty. If such a problem could be successfully solved it could eventually through better planning contribute to the well-being and stability of the world."* [43, p. 30]. We strongly believe that this opinion reflects equally well the situations today and in 1991, when it was expressed.

We think that the only "well defined" existing optimization technique for uncertainty-affected multi-stage optimization problems is Dynamic Programming (DP). When applicable, DP allows to solve the "true" ARC of the problem, which is the huge advantage of the technique. Unfortunately, DP suffers of the "curse of dimensionality" and becomes computationally impractical (except for rare cases of problems with very specific structure) when the state dimension of the underlying Markov decision process becomes something like 4–5 or more. Aside of DP, the main traditional approach to multi-stage decision making under uncertainty is offered by *Multi-Stage Stochastic Programming* that, typically, assumes the uncertain data to be random with known distribution and offers to solve the uncertain problem in general-type decision rules with a prescribed information base (typically, the decisions of stage $t = 1, ..., N$ are allowed to depend on the portion $\zeta^t = [\zeta_1; ...; \zeta_{t-1}]$ of the complete data ζ^N). These decision rules should satisfy the constraints (exactly or with a given close to 1 probability) and to minimize under these restrictions the expected value of a given objective. While the model in question seems to be adequate for what we actually want in multi-stage optimization, the difficulty comes from the fact that as a rule, the multi-stage stochastic programming models are severely computationally intractable, aside of problems of really rare and very "fragile" structure. Specifically, the best known to us complexity bounds for N-stage Linear Stochastic Programming [104] are $O(\epsilon^{-2(N-1)})$, ϵ being the required accuracy. Practically speaking, it means that at the present level of our knowledge, problems with $N = 3$ "most probably," and problems with $N \geq 4$ "surely" are far beyond the reach of computational methods capable of producing solutions of reasonable accuracy in reasonable time. In light of these disastrous complexity results, the reader could ask to which extent the Multi-Stage SP can be considered as a practical tool for processing "really multi-stage" ($N > 2$) decision making problems, and how should one interpret frequent claims of successful processing of complicated problems with 5, 10, or even more stages. Well, this is how this processing typically looks: first people discretize possible values of ζ_t and build "scenario trees," in the simplest case something like "ζ_1 can take values from such and such set of low cardinality. Every one of these values can be augmented by the values of ζ_2 from such and such low cardinality set (perhaps depending on ζ_1); the resulting pairs $[\zeta_1; \zeta_2]$ can be augmented by the values of ζ_3 from such and such low cardinality set, perhaps depending on the pair, etc. The actual set of possible realizations of ζ is then replaced with the tree, the routes in the tree are somehow assigned probabilities, and the "true" multi-stage problem is approximated with the problem where the decision rules of step t are functions of the values of ζ^{t-1} coming from scenarios, i.e., they are functions on a finite set and thus can be represented by vectors. When, as is usually the case, the original uncertain problem is an LO program, the resulting "restricted to the tree of scenarios" multi-stage decision-making problem is just the usual large LO program that can be solved by the LO machinery, perhaps adjusted to the specific "staircase structure" of the LO in question.

With all due respect to practical results that can be obtained with this approach, it has a severe methodological drawback: it is absolutely unclear what the resulting solution has to do with the problem we intend to solve. Strictly speaking, we even cannot treat this solution as a candidate solution, bad or good alike, to the original problem — the decision rules we end up with simply do not say what our decisions should be when the actual realizations of the uncertain data differ from the scenario realizations (this will happen with probability 1, provided that the true distribution of uncertain data has no atoms). The standard answer to this question is as follows: all we need are the first stage decisions, and they are independent of uncertain data and thus are indeed yielded by the scenario approximation. In "real life" we shall implement these decisions; after arriving at the second stage, we apply the same scenario approximation to the problem with the number of stages reduced by one, implement the resulting "here and now" decisions, etc. This answer still is far from being satisfactory. First, there is no guarantee that with this approach at the second, third, etc., stage we shall not meet an infeasible scenario approximation — and this well can happen even when the "true" multi-stage problem is perfectly feasible. The standard way to avoid this unpleasant possibility is to postulate "complete recourse" — whatever be our "here and now" decisions that satisfy the "here and now" constraints, the problem of the next stage will be feasible.[5] However, even under complete recourse and with all numerous tricks of Multi-Stage SP aimed at reducing the number of scenarios, the question "how far from optimum, in terms of the criterion we intend to minimize, are the decision rules we get with the scenario approximation" remains unanswered; to the best of our knowledge, in typical situations meaningful optimality guarantees become possible only with an astronomically large, completely impractical, number of scenarios in the tree.

The dramatic theoretical gap between what Multi-Stage Stochastic Programming intends to achieve and what, if any, it provably achieves disappears when passing from the general-type decision rules to affine ones. Here we indeed achieve what we intend to achieve, at least in the case of multi-stage uncertain LO with a fixed recourse and tractable uncertainty set. Needless to say, the gap is closed "from the bad end" — by replacing our actual (unreachable at the present state of our knowledge) goal with an incomparably more modest one, and not by inventing "Wunderwaffen" capable of solving a multi-stage problem to "true optimality." On a good side of the AARC approach, when the AARC is tractable *and feasible*, we are indeed able to guarantee the validity of the constraints whatever be the realization of the uncertain data from the uncertainty set — a feature that is not shared by the scenario approximation of a multi-stage problem with incomplete recourse. The bottom line is, that both the scenario and the AARC approach are very far from being "ideal" tools for solving multi-stage decision making problems; what is

---

[5] "In reality," complete recourse means that when running out of money and other resources, we can lend/buy what is absent, perhaps at a high price. More often than not this assumption is as relevant as the famous advice given by the Queen of France Marie Antoinette (1755–1793) to the peasants coming to her gate begging for food: "qu'ils mangent de la brioche." ([If they have no bread,] "let them eat cake.")

better in a particular situation depends on the situation and should be decided on a "case by case" basis; thus both approaches seem to have the "right to exist."

NR 14.2. The idea of affine decision rules is too old and too simple to be easily attributable to a particular person/paper (especially taking into account that linear controllers are commonplace in Control, a "close scientific relative" of Optimization). To the best of our knowledge, (which in this particular case is not that much of a guarantee), in the optimization literature first mention of this approach should be attributed to Charnes. The major bulk of the AARC methodology and results as presented in the main body of this chapter originate from [13].

NR 14.3. The main results of Section 14.4 originate from [15] (the finite horizon results, including Theorem 14.4.1) and from [16] (infinite horizon results). These results are close to (although not completely covered by) the well known in Control results on Youla parameterization [117]; the "common roots" of the results in question lie in the simple fact that the purified outputs, as defined in (14.4.6), are affine functions of the initial state and disturbances, and these functions remain the same whatever non-anticipating control law we use. In hindsight, our results are somehow connected to those in [80]; we are grateful to M. Campi for making us aware of this connection.

Part IV

Selected Applications

Chapter Fifteen
Selected Applications

We have considered already numerous examples illustrating applications of the Robust Optimization methodology, but these were, essentially, toy examples aimed primarily at clarifying particular RO techniques. In this chapter, we present a number of additional examples, with emphasis on potential and actual "real-life" aspects of Robust Optimization models in question. Many more examples can be found in the literature, see, e.g., [9, 16, 110, 89] and references therein.

15.1 ROBUST LINEAR REGRESSION AND MANUFACTURING OF TV TUBES

The application of RO to follow is from E. Stinstra and D. den Hertog [108], to whom we are greatly indebted for the permission to reproduce here part of their results.

The problem we want to solve is

$$\min_x \left\{ f_0(x) : f_i(x) \leq 0, \ i = 1, ..., r, x \in X \right\}, \tag{15.1.1}$$

where

$$f_i(x) = \alpha_i^T g(x), \ 0 \leq i \leq m, \tag{15.1.2}$$

$g(x) = [g_1(x); ...; g_t(x)] : \mathbb{R}^n \to \mathbb{R}^t$ is comprised of basic functions given in advance, and $X \subset \mathbb{R}^n$ is a given computationally tractable closed convex set.

Data uncertainty comes from the fact that the coefficients $\alpha_i \in \mathbb{R}^t$ in (15.1.2) are not known in advance. All we know are *inexact* measurements

$$y_{ik}^s \approx y_{ik}^r := f_i(\chi_k), \ 0 \leq i \leq r, 1 \leq k \leq p$$

of the values of f_i ("responses") along a given set $\chi_1, ..., \chi_p$ of the values of the design vector.

> In the situation of [108], y_{ik}^s are the responses of a *simulation model* of the true physical system, while y_{ik}^r are the responses of the system itself; thus, the inexactness of the measurements reflects the simulation errors.

We assume that the relation between the true and the measured responses is given by

$$y_{ik}^r = (1 + \zeta_{ik}^m)y_{ik}^s + \zeta_{ik}^a, \tag{15.1.3}$$

where ζ_{ik}^{m} and ζ_{ik}^{a} are the multiplicative and the additive errors, respectively; all we know about these errors is that their collection $\zeta = \{\zeta_{ik}^{m}, \zeta_{ik}^{a}\}_{\substack{0 \le i \le m, \\ 1 \le k \le p}}$ belongs to a given convex and closed perturbation set $\mathcal{Z}$.

The robust counterparts. Assume that the "design matrix"

$$D = \begin{bmatrix} g_1(\chi_1) & \cdots & g_t(\chi_1) \\ \vdots & \vdots & \vdots \\ g_1(\chi_p) & \cdots & g_t(\chi_p) \end{bmatrix}$$

is of rank t, so that from the relations $y_{ik}^{r} = \alpha_i^T g(\chi_k)$, $k = 1, ..., p$, it follows that

$$\alpha_i = G y_i^{r}, \; G = (D^T D)^{-1} D^T, \; y_i^{r} = [y_{i1}^{r}; ...; y_{ip}^{r}], \; 0 \le i \le m, \qquad (15.1.4)$$

and, in addition, that

$$y_i^{r} \in \operatorname{Im} D. \qquad (15.1.5)$$

The latter information allows one to reduce, given y^s, the perturbation set $\mathcal{Z}$ to the set

$$\mathcal{Z}(y^s) = \{\zeta \in \mathcal{Z} : y_i^s + Y_i^s \zeta_i^m + \zeta_i^a \in \operatorname{Im} D, \; 0 \le i \le m\},$$

$$\left[y_i^s = [y_{i1}^s; ...; y_{ip}^s], \; Y_i^s = \operatorname{Diag}\{y_i^s\}, \; \zeta_i^m = [\zeta_{i1}^m; ...; \zeta_{ip}^m], \zeta_i^a = [\zeta_{i1}^a; ...; \zeta_{ip}^a] \right].$$

Note that all we know about y_i^{r} given y^s is that

$$y_i^{r} = y_i^s + Y_i^s \zeta_i^m + \zeta_i^a \text{ for some } \zeta \in \mathcal{Z}(y^s).$$

Consequently, all we know about α_i given y^s is that

$$\alpha_i \in \mathcal{U}_i = \{a = G[y_i^s + Y_i^s \zeta_i^m + \zeta_i^a], \zeta \in \mathcal{Z}(y^s)\}.$$

Therefore the robust version of (15.1.1), where we require the validity of the constraints for all collections $\alpha_0, ...\alpha_m$ compatible with our measurements y^s and minimize under this restriction the guaranteed value of the objective, is the optimization problem

$$\min_{z,x} \left\{ z : \begin{array}{l} a_0^T g(x) \le z \, \forall a_0 \in \mathcal{U}_0 \\ a_i^T g(x) \le 0 \, \forall a_i \in \mathcal{U}_i, \\ \qquad\qquad 1 \le i \le m \\ x \in X \end{array} \right\}. \qquad (15.1.6)$$

Along with this "true" RC of the uncertain problem in question, one can consider its simplified, somehow more conservative, version where we ignore the information contained in (15.1.5). The simplified RC reads

$$\min_{z,x} \left\{ z : \begin{array}{l} a_0^T g(x) \le z \, \forall a_0 \in \widetilde{\mathcal{U}}_0 \\ a_i^T g(x) \le 0 \, a_i \in \widetilde{\mathcal{U}}_i, \\ \qquad\qquad 1 \le i \le m \\ x \in X \end{array} \right\}, \qquad (15.1.7)$$

$$\widetilde{\mathcal{U}}_i = \{a = G[y_i^s + Y_i^s \zeta_i^m + \zeta_i^a], \zeta \in \mathcal{Z}\}, \; 0 \le i \le m.$$

Note that what actually is used in [108] is the simplified RC (15.1.7).

Tractability of the RCs. Assume that the perturbation set $\mathcal{Z}$ is a computationally tractable convex set, e.g., a set given by a polyhedral, or conic quadratic, or semidefinite representation. Then so is $\mathcal{Z}(y^s)$ (as a set cut off $\mathcal{Z}$ by finitely many linear equations on ζ expressing the fact that $y_i^s + Y_i^s \zeta_i^m + \zeta_i^a \in \operatorname{Im} D$). Invoking Theorem 1.3.4, we conclude that *in the case of linear regression models,* (i.e., when all the basic functions $g_j(x)$, $1 \leq j \leq t$, are affine), both (15.1.6) and (15.1.7) are computationally tractable.

The assumption that $g_j(x)$ are affine is essential, otherwise the RCs in question can lose convexity even when no measurement errors are allowed ($\mathcal{Z} = \{0\}$). What we indeed can do efficiently in the case of general regression models (those where $g_j(x)$ non necessarily are affine) is to solve the Analysis problem, i.e., to check whether a given x is feasible for the RCs, since such a verification reduces to maximizing the linear form $a^T g(x)$ of a over the computationally tractable convex sets $\mathcal{U}_i$ (in the case of (15.1.6)) or $\widetilde{\mathcal{U}}_i$ (in the case of (15.1.6)). When there are reasons to conclude that (15.1.6), (15.1.7) by themselves are convex programs, this possibility to solve efficiently the Analysis problem implies, modulo minor technical assumptions, the possibility to solve efficiently the Synthesis problems (15.1.6), (15.1.7) (this is a well known fact of Convex Programming complexity theory, see, e.g., [56]).

> Assume, e.g., that all non-affine functions $g_j(x)$ are convex and that the corresponding coefficients $(\alpha_i)_j$ in (15.1.2) are known to be nonnegative, so that the "true" problem (15.1.1) indeed is convex. Adding to the description of $\mathcal{U}_i$, $\widetilde{\mathcal{U}}_i$ the (valid for the true coefficients) requirements that the entries in a with indices $j \in J = \{j : g_j \text{ is not affine}\}$ are nonnegative, we end up with reduced, still computationally tractable, uncertainty sets $\mathcal{U}_i^+$, $\widetilde{\mathcal{U}}_i^+$. At the same time, the semi-infinite constraints $a_i^T g(x) \leq \ldots \forall a_i \in \mathcal{U}_i^+$ in the resulting modification of (15.1.6) ('$\ldots$' is either z, or 0) can be written down as $\bar{f}_i(x) := \max_{a \in \mathcal{U}_i^+} a^T g(x) \leq \ldots$. Observing that the functions $a^T g(x)$, $a \in \mathcal{U}_i^+$, are convex, we conclude that $\bar{f}_i(x)$ is convex as well. Moreover, given x, we can find efficiently $a_x \in \mathcal{U}_i^+$ such that $a_x^T g(x) = \bar{f}_i(x)$, (since $\mathcal{U}_i^+$ is computationally tractable). Note that the vector $\sum_{j=1}^t (a_x)_j g_j'(x)$ is a subgradient of $\bar{f}_i(x)$. Thus, (15.1.6) in our case is a convex problem with efficiently computable objective and constraints and as such is computationally tractable, and similarly for (15.1.7).

Numerical illustration. In [108], the outlined methodology was applied to optimizing the temperature profile in enameling of TV tubes. In this process, a tube is heated in a designated oven. The resulting thermal stresses in the tube depend on the "temperature profile" x — the collection of temperatures along certain grid on the surface of the tube, see figure 15.1. A bad profile can result in stresses that are too large and therefore in much of scrap. The designer's goal is to choose a temperature profile in a way that ensures the temperature values are between given bounds and the differences in temperature at nearby locations are not too big, and to minimize under these restrictions the maximal thermal stress in a specified area.

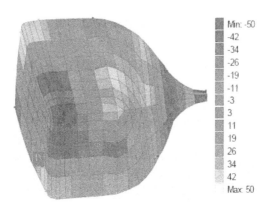

Figure 15.1   A temperature profile

The mathematical model of the problem is

$$
\min_{s_{\max}, x} \left\{ s_{\max} : \begin{array}{ll} s_i(x) \leq s_{\max},\, 1 \leq i \leq m & (a) \\ \ell \leq x \leq u & (b) \\ -\Delta \leq p_j^T x + q_j \leq \Delta,\, j \in J & (c) \end{array} \right\}, \tag{15.1.8}
$$

where $x \in \mathbb{R}^{23}$ stands for the temperature profile, $s_i(x)$ are the thermal stresses at $m = 210$ control points, and constraints (c) impose bounds on the absolute values of differences in temperature at neighboring points. It is assumed that $s_i(x)$ are given by linear regression model:

$$
s_i(x) = \alpha_i^T \underbrace{[1; x]}_{g(x)}. \tag{15.1.9}
$$

The simulated responses are the stresses at control points yielded by a finite element model, with a typical simulation, (i.e., running the model for a given x) taking several hours.

As for the uncertain perturbations, (i.e., the simulation errors), it is assumed that the only nonzero components in the perturbation ζ are the multiplicative errors ζ_{ik}^{m}, and two models of the perturbation set $\mathcal{Z}$ are considered:

box uncertainty:
$$
\mathcal{Z} = \left\{ \zeta = \{\zeta_{ik}^{\mathrm{m}}, \zeta_{ik}^{\mathrm{a}} = 0\}_{\substack{1 \leq i \leq m, \\ 1 \leq k \leq p}} : -\sigma_i^{\mathrm{b}} \leq \zeta_{ik}^{\mathrm{m}} \leq \sigma_i^{\mathrm{b}} \,\forall i, k \right\},
$$
ellipsoidal uncertainty:
$$
\mathcal{Z} = \left\{ \zeta = \{\zeta_{ik}^{\mathrm{m}}, \zeta_{ik}^{\mathrm{a}} = 0\}_{\substack{1 \leq i \leq m, \\ 1 \leq k \leq p}} : \sum_{k=1}^{p} [\zeta_{ik}^{\mathrm{m}}]^2 \leq [\sigma_i^{\mathrm{e}}]^2 \,\forall i \right\}.
$$

The corresponding tractable reformulations of the simplified robust counterpart (15.1.7) (this is what was used in [108] to get a robust solution) are

case of box uncertainty:

$$\min_{x, s_{\max}} \left\{ s_{\max} : \begin{array}{r} [y_i^s]^T G^T [1; x] + \sigma_i^b \| Y_i^s G^T [1; x] \|_1 \leq s_{\max}, \\ 1 \leq i \leq m \\ \ell \leq x \leq u, \, -\Delta \leq p_j^T x + q_j \leq \Delta, \, 1 \leq j \leq J \end{array} \right\},$$

case of ellipsoidal uncertainty:

$$\min_{x, s_{\max}} \left\{ s_{\max} : \begin{array}{r} [y_i^s]^T G^T [1; x] + \sigma_i^e \| Y_i^s G^T [1; x] \|_2 \leq s_{\max}, \\ 1 \leq i \leq m \\ \ell \leq x \leq u, \, -\Delta \leq p_j^T x + q_j \leq \Delta, \, 1 \leq j \leq J \end{array} \right\}.$$

(15.1.10)

The experiments reported in [108] were conducted as follows. After y^s was generated,

• a sample of 100 independent realizations $\zeta^1, ..., \zeta^{100}$ was built. When generating ζ^μ, the additive errors were set to 0, and the multiplicative errors were drawn, independently of each other, either from the uniform, or from normal distribution, depending on whether the box or the ellipsoidal model of uncertainty was explored (for details, see [108]);

• a realization ζ^i of ζ along with y^s according to (15.1.3) yields a realization $y^{r,\mu}$, of "true" responses;[1] the latter, according to (15.1.4), yields a collection α_i^μ, $i = 1, ..., m$ of the "true" coefficients α_i in (15.1.9), thus allowing one to compute the corresponding "true" stresses and their maximum $s_{\max}^\mu(x)$ for any given temperature profile x.

The goal of the outlined simulation was to compare the "true" values of the maximal stresses associated with the robust optimal and the nominally optimal temperature profiles. The robust optimal profile is the optimal solution to the robust problem (15.1.10) associated with the uncertainty model in question, while the nominally optimal profile is the optimal solution to the same problem with σ_i^b and σ_i^e set to 0. The results of the experiment are presented on figure 15.2. It is clearly seen that the robust temperature profile significantly outperforms the nominal one in terms of both the expectation and the variance of the resulting maximal thermal stress.

15.2 INVENTORY MANAGEMENT WITH FLEXIBLE COMMITMENT CONTRACTS

The content of this section originates from [14].

---

[1]Here and below, quotation marks in "true" express the fact that what is considered as true responses in the reported experiments comes from the simulated responses y^s and the perturbation model (15.1.3), rather than from measuring the actual responses of the physical system.

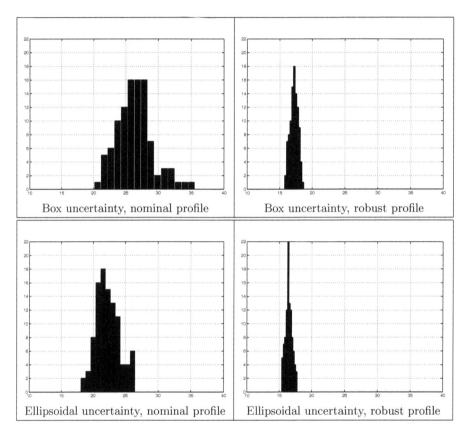

Figure 15.2 Distributions of maximal stresses for nominal and robust temperature profiles (x-axis: values of the stress, y-axis: frequency in a 100-element sample).

15.2.1 The Problem

Consider a single product inventory functioning at a finite time horizon. The state of the inventory at time $t = 1, 2, ..., T$ is specified by the amount x_t of product in the inventory at the beginning of period t. During the period, the inventory management ("the retailer") orders q_t units of product from the supplier, that we assume arrive immediately, and satisfies external demand for d_t units of the product. Thus, the state equations of the inventory are

$$
\begin{array}{rcl}
x_1 & = & z \\
x_{t+1} & = & x_t + q_t - d_t,\ 1 \le t \le T
\end{array}
\tag{15.2.1}
$$

where z is a given initial state. We assume that backlogged demand is allowed, so that the states x_t can be nonpositive. Our additional constraints include:
- *lower and upper bounds on the orders* $L_t \le q_t \le U_t,\ 1 \le t \le T$, and
- *lower and upper bounds on cumulative orders* $\widehat{L}_t \le \sum_{\tau=1}^{t} q_\tau \le \widehat{U}_t,\ 1 \le t \le$
T, where $L_t \le U_t$, $\widehat{L}_t \le \widehat{U}_t$ are given bounds and $L_t, \widehat{L}_t \ge 0$.

Our goal is to minimize the overall inventory management cost that includes the following components:

i) *Holding cost* $\sum_{t=1}^{T} h_t \max[x_{t+1}, 0]$, where $h_t \ge 0$ is the cost of storing a unit of the product in period t.

ii) *Shortage cost* $\sum_{t=1}^{T} p_t \max[0, -x_{t+1}]$, where $p_t \ge 0$ is the per unit penalty for backlogged demand in period t.

iii) *Ordering cost* $\sum_{t=1}^{T} c_t q_t$, where $c_t \ge 0$ is the per unit cost of replenishing the inventory in period t.

iv) *Salvage term* $-s \max[x_{T+1}, 0]$, where $s \ge 0$ is the salvage coefficient. In other words, we assume that after T periods the product remaining in the inventory can be sold at the per unit price s.

At this point, our model is pretty similar to the one of Example 14.1.2. We, however, are about to enrich this simple model by an important additional component — *commitments*. In this model "as it is," the only restrictions on the replenishment policy are given by bounds on the instant and the cumulative orders, and the retailer has complete freedom in choosing the orders within these bounds, with no care of how this freedom affects the supplier. The latter therefore is supposed to work "on very short notice," with very limited options of predicting what will be required from him in the future. In other words, in the model we have presented so far the retailer and the supplier are in different positions as far as the risk from the inevitable demand uncertainty is concerned: the retailer can vary, within certain bounds, the replenishment orders, thus adjusting himself, to some extent, to the actual demand; while the supplier must blindly execute the orders, with no compensation for their unpredictable variations. One of the real life mechanisms aimed at a more fair distribution of the risk along a supply chain is offered by *flexible commitments contracts* as follows. "At time 0," before the inventory starts to function, the supplier and the retailer make an agreement about *commitments* $w_t, 1 \le t \le T$ — the "projected" future orders for the entire horizon $1 \le t \le T$. The retailer is not required to "fully respect" the commitments — to make the future orders q_t exactly equal to w_t — but is supposed to pay to the supplier penalties for deviations of the actual orders from the commitments. As a result, the inventory management cost gets an extra component

$$\sum_{t=1}^{T} \left[\alpha_t^+ \max[q_t - w_t, 0] + \alpha_t^- \max[w_t - q_t, 0] \right]$$
$$+ \sum_{t=2}^{T} \left[\beta_t^+ \max[w_t - w_{t-1}, 0] + \beta_t^- \max[w_{t-1} - w_t, 0] \right],$$

where $\alpha_t^\pm \ge 0$ are given penalties for per unit excess/recess of the actual orders as compared to commitments, and $\beta_t^\pm \ge 0$ are given penalties for variations in the commitments. The resulting inventory management model becomes the optimization problem:

> minimize
> $$C = \quad \sum_{t=1}^{T} h_t \max[x_{t+1}, 0] \quad + \sum_{t=1}^{T} p_t \max[0, -x_{t+1}] \quad + \sum_{t=1}^{T} c_t q_t$$
> [holding cost] $\qquad\qquad$ [shortage cost] $\qquad\qquad$ [ordering cost]
> $$+ \sum_{t=1}^{T} \left[\alpha_t^+ \max[q_t - w_t, 0] + \alpha_t^- \max[w_t - q_t, 0] \right]$$
> [penalty for deviations from commitments]
> $$+ \sum_{t=2}^{T} \left[\beta_t^+ \max[w_t - w_{t-1}, 0] + \beta_t^- \max[w_{t-1} - w_t, 0] \right]$$
> [penalty for commitments variability]
> $$- s \max[x_{T+1}, 0]$$
> [salvage term]

> subject to
> $$x_1 = z$$
> $$x_{t+1} = x_t + q_t - d_t, \ 1 \le t \le T \quad \text{[state equations]}$$
> $$L_t \le q_t \le U_t, \ 1 \le t \le T \quad \text{[bounds on orders]}$$
> $$\widehat{L}_t \le \sum_{\tau=1}^{t} q_\tau \le \widehat{U}_t, \ 1 \le t \le T \quad \text{[bounds on accumulated orders]}$$
>
> $$(15.2.2)$$

Introducing analysis variables, the problem reduces to the LO program as follows:

$$\min \left\{ C : \begin{cases} C \ge \sum_{t=1}^{T} [c_t q_t + y_t + u_t] + \sum_{t=2}^{T} z_t & (a) \\ x_{t+1} = x_t + q_t - d_t, \ 1 \le t \le T & (b) \\ x_1 = z & (c) \\ L_t \le q_t \le U_t, \ 1 \le t \le T & (d) \\ \widehat{L}_t \le \sum_{\tau=1}^{t} q_\tau \le \widehat{U}_t, \ 1 \le t \le T & (e) \\ y_t \ge \overline{h}_t x_{t+1}, y_t \ge -p_t x_{t+1}, \ 1 \le t \le T & (f) \\ u_t \ge \alpha_t^+ (q_t - w_t), u_t \ge \alpha_t^- (w_t - q_t), \ 1 \le t \le T & (g) \\ z_t \ge \beta_t^+ (w_t - w_{t-1}) + \beta_t^- (w_{t-1} - w_t), \ 2 \le t \le T & (h) \end{cases} \right\} \quad (15.2.3)$$

in variables $C, \{x_t\}_{t=1}^{T+1}, \{q_t, w_t, y_t, u_t\}_{t=1}^{T}, \{z_t\}_{t=2}^{T}$; here $\overline{h}_t = h_t$ for $1 \le t \le T-1$ and $\overline{h}_T = h_T - s$. Note that (15.2.3) is indeed equivalent to (15.2.2) under the additional restriction that

$$h_T + p_T \ge s, \qquad (15.2.4)$$

which we assume from now on.

The role of assumption (15.2.4) can be explained as follows. It indeed is true, without any assumptions, that every feasible solution to (15.2.2) can clearly be extended to a feasible solution to (15.2.3) by setting

$$y_t = \max[h_t x_{t+1}, -p_t x_{t+1}], \quad u_t = \max[\alpha_t^+(q_t - w_t), \alpha_t^-(w_t - q_t)],$$
$$z_t = \max[\beta_t^+(w_t - w_{t-1}), \beta_t^-(w_{t-1} - w_t)].$$

In order to conclude that (15.2.2) and (15.2.3) are equivalent, we also need the inverse to be true — that is, every feasible solution to (15.2.3) should induce a feasible solution to (15.2.2) with the same or better value of the objective. On a closest inspection, it turns out that in order for the latter statement to be true, the quantity $\max[\bar{h}_T s, -p_T s]$ should be equal to $\bar{h}_T s$ or to $-p_T s$ depending on whether $s \geq 0$ or $s < 0$, which is the case if and only if (15.2.4) takes place.

15.2.2 Specifying Uncertainty and Adjustability

"In reality," the definitely uncertain element of the data in (15.2.3) is the demand trajectory $d = [d_1; ...; d_T]$. In our model, we consider the demands as the *only* uncertain component of the data, thus treating all the cost coefficients, bounds on the orders, and the initial state z as known in advance. We further should decide on the "adjustability status" of our decision variables, and this is easy: the actual decisions in our problem are the commitments w_t, that by their origin are non-adjustable, and the replenishment orders q_t, which it makes sense to consider as adjustable: according to the "covering story," a decision on the actual value of q_t should be made at the beginning of the period t and as such can depend on the part $d^{t-1} = [d_1; ...; d_{t-1}]$ of the demand trajectory that "reveals itself" at this moment.

> Of course, there is no necessity to take our "covering story" completely at its "face value." It may happen, e.g., that there are delays in register-ing the demands and/or in executing the orders, so that what actually will be delivered in period t should be determined according to the de-mands d_τ of "remote past." On the other hand, it may happen that not only the past demands, but also the current demand d_t can be used when making the decision on q_t. To cover all these possibilities, we as-sume that we are given in advance certain sets $I_t \subset \{1, ..., t\}$ of indices of the demands d_τ, which are known when the decision on q_t is being made. Note that some (or even all) of I_t can be empty, meaning that the corresponding orders q_t are non-adjustable.

The remaining variables in (15.2.3) are the analysis variables; all of them, except for the (upper bound on the) management cost C are, in principle, fully adjustable. However, it is clear that it makes no sense to make adjustable the variables z_t that neither appear in uncertainty-affected constraints, nor are linked to other variables that do appear in such constraints. Thus, the analysis variables that indeed make sense to treat as fully adjustable — depending on the entire demand trajectory — are y_t and u_t. With this convention, specifying somehow the (closed, convex, and bounded) uncertainty set $\mathcal{D}$ for the uncertain demand trajectory, we end up with an uncertain LO problem with *fixed recourse* and are in a good position to process this problem via the AARC methodology as presented in chapter 14.

15.2.3 Building an Affinely Adjustable Robust Counterpart of (15.2.3)

To build the AARC of (15.2.3), we

i) Keep the non-adjustable decision variables $w_1, ..., w_t$, (which represent the commitments), and the non-adjustable analysis variables $C, z_2, ..., z_T$ "as they are" and introduce affine decision rules for the "actual decisions" $q_1,...,q_t$, respecting the given "information base" of these decisions:

$$q_t = q_t^0 + \sum_{\tau \in I_t} q_t^\tau d_\tau, \ 1 \leq t \leq T. \tag{15.2.5}$$

ii) Introduce "fully adjustable" affine decision rules for the remaining analysis variables:

$$\left. \begin{array}{rcl} y_t &=& y_t^0 + \sum_{\tau=1}^{T} y_t^\tau d_\tau \\ u_t &=& u_t^0 + \sum_{\tau=1}^{T} u_t^\tau d_\tau \end{array} \right\}, 1 \leq t \leq T. \tag{15.2.6}$$

iii) Replace in (15.2.3) the adjustable variables q_t, y_t, u_t with the just introduced affine decision rules, thus arriving at a semi-infinite LO problem in variables $\xi = \{C, \{w_t\}, \{z_t\}, \{q_t^\tau, y_t^\tau, u_t^\tau\}\}$. In this problem, the objective and part of the constraints are certain, while the remaining constraints are affinely perturbed by the demand trajectory d.

iv) Finally, impose on all the constraints the requirement to be satisfied for all $d \in \mathcal{D}$, and minimize the (upper bound on the) inventory management cost under the resulting constraints.

Of course, in order to implement the latter recommendation efficiently, we need to reformulate in tractable form the semi-infinite constraints we end up with. As we know, this is possible, provided that $\mathcal{D}$ is a computationally tractable convex set; however, what exactly are the "tractable reformulations," depends on the description of $\mathcal{D}$. In what follows, we restrict ourselves to the simplest case of box uncertainty:

$$\mathcal{D} = \{d \in \mathbb{R}^T : \underline{d}_t \leq d_t \leq \bar{d}_t, \ 1 \leq t \leq T\}.$$

In this case, the tractable reformulation of a semi-infinite constraint

$$\forall (d \in \mathcal{D}) : \sum_{\tau=1}^{T} d_\tau a_\tau(\xi) \leq b(\xi)$$

(here $a_\tau(\xi)$, $b(\xi)$ are known in advance *affine* functions of the decision vector ξ of the AARC; note that all semi-infinite constraints we end up with are of this generic form) is really simple: this is the convex constraint

$$\sum_{\tau=1}^{T} d_\tau^* A_\tau(\xi) + \sum_{\tau=1}^{T} \delta_\tau |a_\tau(\xi)| \leq b(\xi)$$
$$\left[d_\tau^* = \tfrac{1}{2}[\underline{d}_\tau + \overline{d}_\tau], \, \delta_\tau = \tfrac{1}{2}[\overline{d}_\tau - \underline{d}_\tau] \right],$$

which further can be converted into a system of linear constraints by introducing slack variables. The bottom line is that the AARC of (15.2.3), the uncertainty set being a box, is just an explicit LO program.

While the outlined construction is correct, it is not as "economical" as it could be. Indeed,

• We can use the state equations (15.2.3.b,c) to eliminate the state variables x_t, $1 \leq t \leq T+1$; with affine decision rules (15.2.5), the resulting affine decision rule for x_t expresses this variable as an affine function of $d^{t-1} = [d_1; ...; d_{t-1}]$, the coefficients being known linear combinations of the variables u_t^τ. Thus, we do not need variables x_t^τ at all.

• The latter observation taken along with the direct product structure of the uncertainty set $\mathcal{D}$ allows one to "save" on the decision rules for y_t and u_t — while we allowed these variables to be affine functions of the *entire* demand trajectory d, we in fact lose nothing when restricting these affine functions to depend on appropriate parts of this trajectory. Indeed, let us look at the constraints (15.2.3.f) for a particular value of t. There are two of them, both of the form

$$y_t \geq a_t x_{t+1}.$$

We are interested in the case when x_{t+1} is substituted by an affine function $X_{t+1}(d^t)$ of d^t with coefficients depending solely on q_t^τ, and y_t is substituted by an affine function $Y_t(d) \equiv y_t^0 + \sum_{\tau=1}^{T} y_t^\tau d_\tau$ of d such that the constraints in question are satisfied for all d from the box $\mathcal{D}$. Clearly this is the case if and only if

$$\overline{Y}_t(d^t) \equiv \left[y_t^0 + \min_{d_{t+1},...,d_T} \left\{ \sum_{\tau=t+1}^{T} y_t^\tau d_\tau : \begin{array}{c} \underline{d}_\tau \leq d_\tau \leq \overline{d}_\tau, \\ t < \tau \leq T \end{array} \right\} \right] + \sum_{\tau=1}^{t} y_t^\tau d_\tau \geq a_t X_{t+1}(d^t)$$

for all d^t from the "truncated box" $\mathcal{D}_t = \{ d^t : \underline{d}_\tau \leq d_\tau \leq \overline{d}_\tau, 1 \leq \tau \leq t \}$. We see that as far as the constraints (15.2.3.f) are concerned, we lose nothing when replacing the affine decision rule Y_t for y_t with the affine decision rule $\overline{Y}_t$. Since by construction $\overline{Y}_t(d^t) \leq Y_t(d)$ for all $d \in \mathcal{D}$, this updating clearly preserves robust validity of the only other constraint involving our particular variable y_t, namely, the constraint (15.2.3.a). The bottom line is that *we lose nothing when restricting y_t to be an affine function of d^t rather than of the entire d*. Am similar argument is applicable to the variables u_t — the corresponding decision rules can without any harm be restricted to be affine functions of the part d^t of the demand trajectory. Thus, the actual number of y_t^τ and u_t^τ

variables in AARC can be made essentially smaller than in the above "straightforward" construction.

15.2.4 Numerical Results

The outlined model was proposed in [14]; this paper also reports on intensive numerical study of the model, and we are about to reproduce here part of the results. In all reported experiments, the uncertainty set $\mathcal{D}$ is a box of the form $\{d \in \mathbb{R}^T : |d_t - d_t^*| \leq \rho d_t^*, 1 \leq t \leq T\}$ with positive "nominal demands" d_t^*, "uncertainty level" ρ varying in the range from 10% to 70%, and time horizon T set to 12.

15.2.4.1 AARC vs. optimal decision rules

The most interesting question is, of course, how much we lose when restricting ourselves with affine decision rules — that is, how far is the optimal value Opt(AARC) of the AARC from the optimal value Opt(ARC) of the Adjustable Robust Counterpart of the uncertain problem (15.2.2). While in general the latter quantity is difficult to compute, the extreme simplicity of the box uncertainty set we consider allows us to compute it, provided T is not too large, specifically, as follows. We start with the following simple fact from [14]:

Lemma 15.2.1. Consider a multi-stage problem

$$
\min_{\{S_t(d^{t-1})\}_{t=1}^{T+1}} E :
\left\{
\begin{array}{l}
\forall d^T \in F_0 \times \dots \times F_T : \\
\left\{
\begin{array}{l}
E \geq \sum\limits_{t=1}^{T+1} f_t(S_t(d^{t-1})), \\
A_1 S_1(d^0) \geq b_1, \\
A_{t+1} S_{t+1}(d^t) \geq B_{t+1} d^t + C_{t+1} S_t(d^{t-1}) \\
\qquad\qquad\qquad + b_{t+1}, 1 \leq t \leq T \\
\|S_t(d^{t-1})\|_\infty \leq R, 1 \leq t \leq T
\end{array}
\right.
\end{array}
\right\}
\qquad (P)
$$

where $\emptyset \neq F_t \subset D_t$, D_t are polytopes in $\mathbb{R}^{n_t}$, and f_t are lower semicontinuous convex functions with polyhedral domains. Then the optimal value in the problem corresponding to $F_t = D_t$, $0 \leq t \leq T$, is equal to the optimal value in the problem corresponding to $F_t = \text{Ext}(D_t)$, $0 \leq t \leq T$, where $\text{Ext}(D)$ is the set of extreme points of a polytope D. The ARC of uncertain problem (15.2.2) clearly satisfies the assumptions of Lemma, and here D_t are the segments $[\underline{d}_t, \overline{d}_t]$, and the extreme points of D_t are $\underline{d}_t$ and $\overline{d}_t$. By Lemma, the optimal value of the ARC of (15.2.2) remains intact when passing from the box uncertainty set $\mathcal{D}$ to the finite uncertainty set $\mathcal{D}'$ comprised of 2^T "extreme" demand trajectories — those where the demand at every time instant t is either $\underline{d}_t$, or $\overline{d}_t$. Now, the ARC of a multi-stage uncertain LO problem with a *finite* uncertainty set $\mathcal{D}'$ is just a large LO problem. Indeed, we can assign every pair comprised of a time instant $t \in \{1, ..., T\}$ and a possible "scenario" $s \in \mathcal{D}'$ with a set of decision variables representing the decisions that should be made at the instant t when the realization of the uncertain data is s. We impose on these variables the *non-anticipativity restriction* "the decisions associated with (t, s) and (t, s') should be identical to each other, provided that the information on uncertain data, available at time t, is the same for both scenarios s and s'," and then optimize the objective over the resulting set of decision variables under the non-anticipativity constraint coupled with the requirement that the original constraints are valid for every one of the scenarios $s \in \mathcal{D}'$.

Here is an illustrative example: there are two assets, and we want to invest \$1 in these assets at the beginning of time period 1, to sell the assets at the end of this period, and to reinvest the resulting capital in the same two assets at the beginning of time period 2 in order to maximize the guaranteed value of the resulting portfolio at the end of the period 2. Denoting $p_{t,i} \geq 0$ the (uncertain) return of asset i, $i = 1, 2$, in period t, $t = 1, 2$, the uncertain problem is

$$\max_{[x_{t,i}]_{1 \leq i, t \leq 2}} \left\{ p_{2,1}x_{2,1} + p_{2,2}x_{2,2} : \begin{array}{l} x_{2,1} + x_{2,2} \\ \qquad \leq p_{1,1}x_{1,1} + p_{1,2}x_{1,2} \\ x_{1,1} + x_{1,2} \leq 1 \\ x_{t,i} \geq 0 \end{array} \right\},$$

where $x_{t,i}$ is the capital invested in asset i at the beginning of time period t. Assume that all we know at the end of time period t are the returns of the assets in this and in the preceding periods, and that there are just 5 possible scenarios represented in the following table

Scenario	p_{ti}	
#	$[p_{1,1}, p_{1,2}]$	$[p_{2,1}, p_{2,2}]$
1	$[1, 1]$	$[1, 1]$
2	$[1, 1]$	$[1, 2]$
3	$[2, 1]$	$[2, 1]$
4	$[1, 2]$	$[1, 1]$
5	$[3, 1]$	$[2, 2]$

The LO representing the ARC of this uncertain problem reads

$$\max_{C, x_{t,i}^k} \left\{ C : \begin{array}{l} C \leq p_{2,1}^k x_{2,1}^k + p_{2,2}^k x_{2,2}^k, 1 \leq k \leq 5 \\ x_{2,1}^k + x_{2,2}^k \leq p_{1,1}^k x_{1,1}^k + p_{1,2}^k x_{1,2}^k, 1 \leq k \leq 5 \\ x_{1,1}^k + x_{1,2}^k \leq 1, 1 \leq i \leq k \\ x_{ti}^k \geq 0, 1 \leq k \leq 5, q \leq t, i \leq 2 \\ \hline x_{1,i}^k = x_{1,i}^1, 1 \leq k \leq 5, i = 1, 2 \\ x_{2,i}^1 = x_{2,i}^2, i = 1, 2 \\ x_{2,i}^3 = x_{2,i}^4, i = 1, 2 \end{array} \right\}$$

where $x_{t,i}^k$ is the investment in asset i at the beginning of period t in the scenario k, $p_{t,i}^k$ are the corresponding returns, and the concluding lines in the list of constraints represent the non-anticipativity restrictions, specifically, say that
• all decisions made at the beginning of period 1 (at this time we do not know what the scenario is) should be the same for all scenarios;
• with the information available when the decisions $x_{2,i}^k$ should be made, the scenarios 1 and 2 are undistinguishable, so that the corresponding decisions should be the same, and similarly for the scenarios 3 and 4.

$\rho,\%$	Opt(ARC)	Opt(AARC)	Opt(RC)
10	13531.8	13531.8 (+0.0%)	15033.4 (+11.1%)
20	15063.5	15063.5 (+0.0%)	18066.7 (+19.9%)
30	16595.3	16595.3 (+0.0%)	21100.0 (+27.1%)
40	18127.0	18127.0 (+0.0%)	24300.0 (+34.1%)
50	19658.7	19658.7 (+0.0%)	27500.0 (+39.9%)
60	21190.5	21190.5 (+0.0%)	30700.0 (+44.9%)
70	22722.2	22722.2 (+0.0%)	33960.0 (+49.5%)

Table 15.1   Optimal values of ARC, AARC, and RC of the Inventory Management problems with Flexible Commitments Contracts, data W12 (for the description of the data, see [14]). In parentheses: the excess of the optimal value as compared to Opt(ARC).

We see that the ARC of our toy uncertain problem is an explicit LO program.

Of course, the sizes of the LO representing the ARC of problem (15.2.2) associated with the 2^T point uncertainty set $\mathcal{D}'$ grow exponentially with T, which makes this naive approach intractable unless T is small. However, with $T = 12$ (this is the time horizon used in the experiments) the sizes of the LO representing the ARC of (15.2.2) (45,072 inequalities with 24,597 variables) are still amenable for the state-of-the-art LP solvers, which makes it possible to compare Opt(AARC) and Opt(ARC).

Our experiment was organized as follows. We generated several hundreds of data sets for (15.2.2), picking at random the (both varying in time and time invariant) cost coefficients, the bounds on instant and cumulative orders, the nominal demand trajectory $\{d_t^*\}_{t=1}^{12}$, and the uncertainty level ρ, and filtered out all data sets that either result in problems with infeasible ARCs, or are such that our LO solver (the state-of-the-art commercial LO solver mosekopt) reported numerical difficulties when processing either ARC, or the AARC of the problem. For the remaining data sets (there were 300 of them) we computed Opt(ARC), Opt(AARC), the information base being $I_t = \{1, ..., t-1\}$ ("when decision on q_t is made, the past demands are known, while the current and the future ones are unknown"), and, finally, the optimal value of the RC (no adjustable variables at all). The results of the experiment were striking: among the 300 processed data sets, there were just two (!) where the computed Opt(AARC) was > Opt(ARC), and the difference of the optimal values in both these cases was less than 4% of Opt(ARC). Note that this surprisingly good performance of affine decision rules is in full accordance with the experimental results on a simpler inventory problem reported in section 14.3.4.

It should be stressed that the RC in our experiments sometimes was essentially inferior as compared to AARC, see table 15.1.

ρ, %	Opt(AARC), $I_t = [1:t-1]$	Opt(AARC), $I_t = [t-3:t-1]$	Opt(AARC), $I_t = [1:t-3]$	Opt(AARC), $I_t = \emptyset$
30	16595	16595 (+0.0%)	17894 (+8.4%)	21100 (+27.1%)
70	22722	22722 (+0.0%)	26044 (+14.6%)	33960 (+49.5%)

Table 15.2   The role of information base, data W12. $[a:b]$ stands for the set $\{a, a + 1, ..., b\}$. In parentheses: excess over Opt(AARC) for the case of complete information base $[1:t-1]$.

15.2.4.2 The role of information base

The results reported so far correspond to the case when the decisions on replenishment orders q_t can depend on all the preceding demands $d_1, ..., d_{t-1}$. Table 15.2 illustrates possible consequences of changes in the information base. We see that we lose nothing when making decisions solely on the basis of last three demands, suffer from non-negligible losses when the last two demands are unavailable when making decisions on the replenishment orders, and lose a lot when demands are not observed at all (that is, our AARC reduces to the RC).

15.2.4.3 Folding horizon

For the time being, we have been using a straightforward interpretation of the AARC of (15.2.3): this is a problem that we solve when the agreement on commitments is being made and before the inventory starts to function. The "here and now" components of the resulting solution — that is, the commitments and the initial order q_1 — are executed immediately; as for the remaining orders, we get decision rules that never will be revised in the future. When the time comes to specify a future replenishment order q_t, we shall just plug into the corresponding decision rule the actual demands d_τ, $\tau \in I_t$. On a close inspection, there is a smarter way to implement our methodology, specifically, the *folding horizon* scheme as follows. Just as with our present approach, we solve the AARC of (15.2.3) before the inventory starts to function and implement the resulting "here and now" decisions (commitments and the first replenishment order q_1). At the beginning of period 2, we resolve the AARC for the reduced time horizon $2, ..., T$, treating the current state of the inventory as the initial state, and the already computed commitment-related quantities $w_1, .., w_T$, $z_2, ..., z_T$ as known constants rather than variables. Solving the AARC of the problem on the reduced time horizon, we get the value of the new "here and now" decision — the replenishment order q_2, and implement this order. We proceed in the same fashion, solving at a every step $t = 1, 2, ..., T$ the AARC of the uncertain problem associated with the remaining part of the original time horizon and using the solution to specify the decisions that should be made at time t.

It is clear that the "folding horizon" strategy, (which can be applied to every multi-stage decision-making problem, not necessarily to our Inventory one), can only outperform our initial strategy, where we never revise the decision rules yielded by the AARC of the "full-horizon" uncertain problem. Indeed, let the optimal value of the latter AARC be C_*, meaning that with the original decision rules our total expenses will never exceed C_*, whatever the demand trajectory from the uncertainty set in question. When resolving the problem at the beginning of the second period, with "already implemented" variables $w_1, ..., w_t$, $z_2, ..., z_T$, q_1 set to their values computed at the beginning of the first period and the original uncertainty set replaced with its cross-section with the plane "d_1 equals to its already observed value," the optimal value of the new AARC cannot be greater than C_* (since this value is guaranteed already by the original decision rules), and can happen to be less than C_*, so that it definitely makes sense to switch to the decision rules given by the AARC of the problem on the reduced time horizon.

Note that the above reasoning is applicable to any kind of "worst-case-oriented" multi-stage decision-making in an uncertainty-affected environment, and not only to the AARC-based decision making. If we were smart enough to solve the ARCs, and thus were able to build decision rules with the best possible worst-case guarantees, it still would make sense to implement the folding horizon strategy, since it preserves the worst-case performance guarantees yielded by the full-horizon ARC and is capable, to some extent, of utilizing "the luck." Attractiveness of the folding horizon strategy only increases when passing from the situation where we can solve ARCs (and thus achieve the optimal worst-case performance guarantees) to the situation where we use restricted versions of ARCs, (e.g., use AARCs) and thus achieve only suboptimal worst-case performance guarantees. In this case it well may happen that the folding horizon strategy results in *better* worst-case performance guarantees than the "full horizon" AARC-based strategy (in this respect, note that when folding horizon is combined with AARC-based decision rules, the actual decision rules we end up with are not necessarily affine.)

All this being said, the numerical experimentation suggests that in the case of the particular uncertain problem we are considering the folding horizon strategy yields only marginal savings, see table 15.3.

15.3 CONTROLLING A MULTI-ECHELON MULTI-PERIOD SUPPLY CHAIN

In this section we describe an application of the AARC methodology of section 14.3, combined with the Globalized Robust Optimization model studied in chapter 3, to derive optimal policies for controlling a multi-echelon supply chain. We treat the problem as synthesizing a discrete time dynamical system, using the "purified outputs" scheme developed in section 14.4. The presentation to follow is based on [20].

15.3.1 The Problem

Consider a multi-echelon serial supply chain as in figure 15.3.

ρ, %	Opt(AARC), full horison	Full horizon AARC policy	Folding horizon AARC policy
10	13532	13375 (-1.2%): 41	13373 (-1.2%): 41
20	15064	14745 (-2.1%): 86	14743 (-2.1%): 86
30	16595	16122 (-2.8%): 124	16115 (-2.9%): 127
40	18127	17477 (-3.6%): 170	17464 (-3.7%): 174
50	19659	18858 (-4.1%): 207	18848 (-4.1%): 209
60	21191	20267 (-4.4%): 236	20261 (-4.4%): 229
70	22722	21642 (-4.8%): 287	21633 (-4.8%): 280

Table 15.3 Full horizon AARC policy vs. the Folding horizon strategy, data W12. Numbers $a(b\%) : c$ in the second and the third columns stand for the average (a) and the empirical standard deviation (c), as computed over 100 simulated demand trajectories, of the actual inventory management cost; b is the excess of a as compared to the optimal value of the full horizon AARC given in the second column. At every uncertainty level ρ, full and folding horizon policies were tested at the same 100 randomly selected demand trajectories.

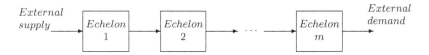

Figure 15.3 A serial supply chain.

Let us denote by $j = 1, 2, \ldots, m$, the index of an echelon, with echelon j being the predecessor of echelon $j + 1$, $j = 1, 2, \ldots, m - 1$. There is an external demand d_t faced by echelon m in period t $(t = 1, 2, \ldots, n)$, where n is the planning horizon.

Let $x_t^j \geq 0$ denote the amount of product echelon j orders from echelon $j + 1$ at the beginning of period t and y_t^j denote the inventory level in echelon j at the end of time period t. The initial inventory level in echelon j is denoted by z^j. Delays between the time that an order is placed and the time it is supplied can occur. There are 3 types of delays: (1) *information delay*: the time it takes the information about the order to reach the preceding echelon, (2) *manufacturing delay*: the time it takes to manufacture or assemble the order (measured from the time the order is received), and (3) *lead time*: the time it takes the replenishment to travel from its origin to its destination. The 3 delays are nonnegative integers for each echelon j, which are denoted by $I(j), M(j)$ and $L(j)$, respectively. $I(m + 1)$ denotes the information delay between the external demand and echelon m. The dynamics of the system is given by:

$$
\begin{aligned}
y_t^j &= y_{t-1}^j + x_{t-(I(j)+M(j-1)+L(j))}^j - x_{t-(I(j+1)+M(j))}^{j+1}, \ 1 \leq j \leq m - 1 \\
y_0^j &= z^j, \ 1 \leq j \leq m \\
y_t^m &= y_{t-1}^m + x_{t-(I(m)+M(m-1)+L(m))}^m - d_{t-(I(m+1)+M(m))},
\end{aligned}
$$

$$(15.3.1)$$

which simply says that the change in inventory level from one period to the next is equal to the quantities received minus the requirements. Negative levels of inventory, which may occur, represent unsatisfied requirements or *backlogging*.

The objective is to minimize the total cost, that is comprised of three components: (1) buying or manufacturing costs, (2) inventory holding costs, and (3) backlogging cost. Let c_t^j be the buying/manufacturing cost per item at echelon j and time period t, h_t^j the holding cost per item per unit of time in echelon j at time t, and p_t^j the backlogging (or shortage) cost per item per unit of time in echelon j at time t. The index t of the various costs allows us to consider capitalization (for instance $c_t^j = c^j(1+r)^{t-1}$), which can greatly impact the cost when the planning horizon is long.

Instead of minimizing the cost elements above, one may opt to control supply chains so as to minimize or even eliminate the "bullwhip effect" — the amplification of demand variability from a downstream site to an upstream site (see [73]). Reducing this effect has implications beyond cost minimization since bullwhip peaks and ebbs often cause disruptions that are difficult to quantify, e.g., loss of reputation and goodwill among customers and suppliers. Bullwhip effects may be caused by the use of heuristics [77, 57]; by irrational behavior of the supply chain members, as illustrated in the "Beer Distribution Game" in [107], or as a result of the strategic interactions among rational supply chain members in [74]. There are many real-world evidences of the occurrence of the bullwhip effect. Examples include diapers [75], TV sets [64], food products [74, 59], pharmaceutical products [36], and more. It is noted in [112] that the semiconductor equipment industry is more volatile than the personal computer industry, and [30] shows evidence of bullwhip existence through an empirical study conducted in the automotive industry.

There were many studies attempted to construct strategies aimed at minimizing the bullwhip effect. The objective of most of the studies is to minimize the bullwhip effect by minimizing either the ratio or the difference between the order variance and the demand variance [41, 36, 118]. In contrast, in this section we follow the approach of [20] and apply an economic rationale to the control problem; thus, we want to control the chain not merely for the sake of stabilizing the system for operational reasons but first and foremost for minimizing the cost. Such an optimal controller is likely to generate a small bullwhip effect.

The problem is posed as the following optimization program:

$$\min_{y,x} \sum_{j,t}[c_t^j x_t^j + \max(h^j y_t^j, -p^j y_t^j)]$$

s.t.

$$
\left.
\begin{array}{rcl}
y_t^j &=& y_{t-1}^j + x_{t-(I(j)+M(j-1)+L(j))}^j \\
&& -x_{t-(I(j+1)+M(j))}^{j+1}, \ 1 \le j \le m-1 \\
y_t^m &=& y_{t-1}^m + x_{t-(I(m)+M(m-1)+L(m))}^m \\
&& -d_{t-(I(m+1)+M(m))}
\end{array}
\right\} \ 1 \le t \le n \quad (15.3.2)
$$

$$
\left.
\begin{array}{rcl}
x_t^j &\ge& 0 \\
\bar{a}^j \ge y_t^j &\ge& \underline{a}^j \\
y_0^j &=& z^j
\end{array}
\right\} \ \forall j \in \{1, \dots, m\}
$$

To simplify the notation, let $T^L(j) = I(j)+M(j-1)+L(j)$ and $T^M(j) = I(j+1)+M(j)$. Introducing slack variables for the max-terms in the objective, we transform (15.3.2) into the LO program as follows:

$$\min_{y,x} \sum_{j,t}[c_t^j x_t^j + w_t^j]$$

s.t.

$$
\left.
\begin{array}{l}
\left.
\begin{array}{rcl}
y_t^j &=& y_{t-1}^j + x_{t-T^L(j)}^j - x_{t-T^M(j)}^{j+1} \ 1 \le j \le m-1 \\
y_t^m &=& y_{t-1}^m + x_{t-T^L(m)}^m - d_{t-T^M(m)}
\end{array}
\right\} \\
\left.
\begin{array}{l}
w_t^j \ge h_t^j y_t^j, \ w_t^j \ge -p_t^j y_t^j, \ w_t^j \ge 0 \\
\bar{a}^j \ge y_t^j \ge \underline{a}^j, \ b^j \ge x_t^j \ge 0, \ y_0^j = z^j
\end{array}
\right\}, \ 1 \le j \le m
\end{array}
\right\}, \ 1 \le t \le n. \quad (15.3.3)
$$

Using the equality constraints to eliminate the y variables, we arrive at the final LO formulation of the nominal problem:

$$\min_{\sigma,w,x} \sigma$$

s.t.

$$
\left.
\begin{array}{l}
\sigma \ge \sum_{j,t}[c_t^j x_t^j + w_t^j] \\[2mm]
\left.
\begin{array}{rcl}
w_t^j &\ge& h_t^j(z^j + \sum_{t'=1}^{t}(x_{t'-T^L(j)}^j - x_{t'-T^M(j)}^{j+1})), \\[2mm]
w_t^j &\ge& -p_t^j((z^j + \sum_{t'=1}^{t}(x_{t'-T^L(j)}^j - x_{t'-T^M(j)}^{j+1})), \\[2mm]
\underline{a}^j &\le& z^j + \sum_{t'=1}^{t}(x_{t'-T^L(j)}^j - x_{t'-T^M(j)}^{j+1}), \\[2mm]
\bar{a}^j &\ge& z^j + \sum_{t'=1}^{t}(x_{t'-T^L(j)}^j - x_{t'-T^M(j)}^{j+1}), \\[2mm]
&& 1 \le j \le m-1
\end{array}
\right\} \\[10mm]
\begin{array}{rcl}
w_t^m &\ge& h_t^m(z^m + \sum_{t'=1}^{t}(x_{t'-T^L(m)}^m - d_{t'-T^M(m)})) \\[2mm]
w_t^m &\ge& -p_t^m((z^m + \sum_{t'=1}^{t}(x_{t'-T^L(m)}^m - d_{t'-T^M(m)})) \\[2mm]
\underline{a}^m &\le& z^m + \sum_{t'=1}^{t}(x_{t'-T^L(m)}^m - d_{t'-T^M(m)}) \\[2mm]
\bar{a}^m &\ge& z^m + \sum_{t'=1}^{t}(x_{t'-T^L(m)}^m - d_{t'-T^M(m)}) \\[2mm]
b^j \ge x_t^j \ge 0, \ w_t^j \ge 0, \ 1 \le j \le m
\end{array}
\end{array}
\right\}, \ 1 \le t \le n. \quad (15.3.4)
$$

We assume that the demand $d = \{d_t\}_{t=1}^n$ and the initial inventory levels $z = \{z^j\}_{j=1}^m$ are uncertain; all we know is that they belong to some uncertainty sets: $d_t \in D_t$, $z^j \in Z^j$. Thus, formulation (15.3.4) in fact represents a family of LPs — one for each possible realization of the uncertain data.

15.3.2 Illustrating the Bullwhip Effect

To illustrate the bullwhip effect, we use an example based on Love [77].

The example uses a fluctuating demand that is shown in table 15.4. The planning horizon consists of $n = 20$ time periods and there are $m = 3$ echelons. Furthermore, we assume that there is 2-unit delay in executing replenishment orders $(T^L(j) = 2)$, while $T^M(j) = 0$, $1 \le j \le m$. The initial inventory level is assumed to be 12 for every echelon ($z^j = 12$ for all j).

t	1	2	3	4	5	6	7	8	9	10
d_t	6	6	6	6	6	6	6	6	7	8
t	11	12	13	14	15	16	17	18	19	20
d_t	9	10	9	8	7	6	5	4	5	6

Table 15.4 Demand for Love's data.

Love [77] uses the following simple control law to solve (the deterministic) problem (15.3.4):

$$x_t^j = x_{t-1}^{j+1} + \frac{1}{2}(\Upsilon^j - y_{t-1}^j) \quad \begin{matrix} \forall j \in 1, \dots, m \\ \forall t \in 1, \dots, n. \end{matrix} \tag{15.3.5}$$

Here $x_t^m = d_t$, and Υ^j is the "target" inventory for echelon j (equal to 12, for all the echelons, in the example) and acts as insurance against unforeseen production or supply disruptions.

Figure 15.4 shows the inventory level of the 3 echelons resulting from implementing the heuristic control law (15.3.5). The bullwhip effect here is evident; small demand fluctuations (only between 4 and 10) cause huge fluctuations in the inventory levels.

15.3.3 Building the Affinely Adjustable Globalized Robust Counterpart (AA-GRC) of the Supply Chain Problem

For our supply chain problem (15.3.3), the discrete time dynamic system is:

$$\begin{aligned}
y_t^j &= y_{t-1}^j + x_{t-T^L(j)}^j - x_{t-T^M(j)}^{j+1}, \; 1 \le j \le m-1 \\
y_0^j &= z^j, \; 1 \le j \le m \\
y_t^m &= y_{t-1}^m + x_{t-T^L(m)}^m - d_{t-T^M(m)}.
\end{aligned}$$

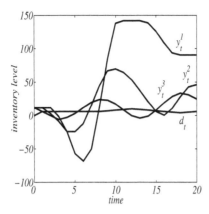

Figure 15.4 Inventory levels in each of the 3 echelons

Here the purified outputs (see section 14.4) are given by:

$$v_t^j = y_t^j - \widehat{y}_t^j = \begin{cases} z^m - \sum_{\tau=1}^{t-T^M(m)} d_\tau & \text{if } j = m \\ z^j & \text{otherwise.} \end{cases} \tag{15.3.6}$$

After eliminating the equalities in (15.3.3) we arrived at the LO problem (15.3.4), which we showed to be of a form amenable to treatment by the RO methodology. We use a purified output-based linear control law and also make the associated auxiliary variables affinely dependent on the uncertain data, specifically:

$$\begin{aligned} x_t^j \equiv x_t^j(d, z) &= \overline{\eta}_0^{x,t,j} + \sum_{j'=1}^{m} \sum_{\tau=1}^{n} \eta_{\tau,j'}^{x,t,j} z^{j'} - \sum_{\tau=1}^{n} \sum_{\tau'=1}^{\tau-M(m)} \eta_{\tau,m}^{x,t,j} d_{\tau'} \\ w_t^j &= \overline{\eta}_0^{w,t,j} + \sum_{j'=1}^{m} \widetilde{\eta}_{j'}^{w,t,j} z^{j'} + \sum_{\tau=1}^{n} \eta_\tau^{w,t,j} d_\tau \end{aligned} \tag{15.3.7}$$

Of course we impose the constraints $\eta_{\tau,j'}^{x,t,j} = 0 \quad \forall \tau \geq t$ and set $\eta_\tau^{w,t,j} = 0 \quad \forall \tau \geq t$ to make the affine decision rules non-anticipative.

What we arrived at is, essentially, the AARC of the uncertain problem (15.3.4), see section 14.3. As it should be, the AARC is bi-affine in the decision variables and the uncertain data, and thus is amenable to processing via the GRC methodology (see chapter 3).

Let us illustrate our approach by processing the second constraint in (15.3.4). The original constraint is of the form

$$w_t^j \geq h_t^j\Big(z^j + \sum_{t'=1}^{t} (x_{t'-T^L(j)}^j - x_{t'-T^M(j)}^{j+1})\Big). \tag{15.3.8}$$

Implementing the decision rules given by (15.3.7), we arrive at

$$
\begin{aligned}
0 \;\geq\; & h_t^j \sum_{t'=1}^{t} (\overline{\eta}_0^{x,t'-T^L(j),j} - \overline{\eta}_0^{x,t'-T^M(j),j+1}) - \overline{\eta}_0^{w,t,j} + h_t^j z^j \\
& + \sum_{j'=1}^{m} z^{j'} [h_t^j \sum_{t'=1}^{t} \sum_{\tau=1}^{n} (\eta_{\tau,j'}^{x,t'-T^L(j),j} - \eta_{\tau,j'}^{x,t'-T^M(j),j+1}) - \widetilde{\eta}_{j'}^{w,t,j}] \\
& + \sum_{\tau'=1}^{n} d_{\tau'} [-h_t^j \sum_{t'=1}^{t} \sum_{\tau=\tau'+M(m)}^{n} (\eta_{\tau',m}^{x,t'-T^L(j),j} - \eta_{\tau,m}^{x,t'-T^M(j),j+1}) - \eta_{\tau'}^{w,t,j}],
\end{aligned}
$$
$$
1 \leq j \leq m.
$$
$$(15.3.9)$$

We now implement the GRC assuming that the normal ranges of both d_t and z^j are the intervals $[\underline{d}_t, \overline{d}_t]$ and $[\underline{z}^j, \overline{z}^j]$ respectively and the norm defining the distance function (see Definition 3.1.1 in chapter 3) is the ℓ_1 norm. Now, by Example 3.2.3.(a) the GRC of (15.3.9) is the linear system

$$
0 \geq h_t^j \sum_{t'=1}^{t} (\overline{\eta}_0^{x,t'-T^L(j),j} - \overline{\eta}_0^{x,t'-T^M(j),j+1}) - \overline{\eta}_0^{w,t,j} + \sum_{j'=1}^{m} v_{j'}^{2,t,j} + \sum_{\tau'=1}^{n} \vartheta_{\tau'}^{2,t,j},
$$
$$
1 \leq j \leq m
$$

$$
v_{j'}^{2,t,j} \geq \underline{z}^{j'} [h_t^j \sum_{t'=1}^{t} \sum_{\tau=1}^{n} (\eta_{\tau,j'}^{x,t'-T^L(j),j} - \eta_{\tau,j'}^{x,t'-T^M(j),j+1}) - \widetilde{\eta}_{j'}^{w,t,j} + h_t^j \delta_j^{j'}],
$$
$$
1 \leq j' \leq m
$$

$$
v_{j'}^{2,t,j} \geq \overline{z}^{j'} [h_t^j \sum_{t'=1}^{t} \sum_{\tau=1}^{n} (\eta_{\tau,j'}^{x,t'-T^L(j),j} - \eta_{\tau,j'}^{x,t'-T^M(j),j+1}) - \widetilde{\eta}_{j'}^{w,t,j} + h_t^j \delta_j^{j'}],
$$
$$
1 \leq j' \leq m
$$

$$
\vartheta_{\tau'}^{2,t,j} \geq \underline{d}_{\tau'} [-h_t^j \sum_{t'=1}^{t} \sum_{\tau=\tau'+M(m)}^{n} (\eta_{\tau,m}^{x,t'-T^L(j),j} - \eta_{\tau,m}^{x,t'-T^M(j),j+1}) - \eta_{\tau'}^{w,t,j}],
$$
$$
1 \leq \tau' \leq n
$$

$$
\vartheta_{\tau'}^{2,t,j} \geq \overline{d}_{\tau'} [-h_t^j \sum_{t'=1}^{t} \sum_{\tau=\tau'+M(m)}^{n} (\eta_{\tau,m}^{x,t'-T^L(j),j} - \eta_{\tau,m}^{x,t'-T^M(j),j+1}) - \eta_{\tau'}^{w,t,j}],
$$
$$
1 \leq \tau' \leq n
$$

$$
\mu_{j'}^{2,t,j} \geq [h_t^j \sum_{t'=1}^{t} \sum_{\tau=1}^{n} (\eta_{\tau,j'}^{x,t'-T^L(j),j} - \eta_{\tau,j'}^{x,t'-T^M(j),j+1}) - \widetilde{\eta}_{j'}^{w,t,j} + h_t^j I_{\{j'=j\}}],
$$
$$
1 \leq j' \leq m
$$

$$
\mu_{j'}^{2,t,j} \geq -[h_t^j \sum_{t'=1}^{t} \sum_{\tau=1}^{n} (\eta_{\tau,j'}^{x,t'-T^L(j),j} - \eta_{\tau,j'}^{x,t'-T^M(j),j+1}) - \widetilde{\eta}_{j'}^{w,t,j} + h_t^j I_{\{j'=j\}}],
$$
$$
1 \leq j' \leq m
$$

$$
\widetilde{\mu}_{\tau'}^{2,t,j} \geq [-h_t^j \sum_{t'=1}^{t} \sum_{\tau=\tau'+M(m)}^{n} (\eta_{\tau,m}^{x,t'-T^L(j),j} - \eta_{\tau,m}^{x,t'-T^M(j),j+1}) - \eta_{\tau'}^{w,t,j}],
$$
$$
1 \leq \tau' \leq n
$$

$$
\widetilde{\mu}_{\tau'}^{2,t,j} \geq -[-h_t^j \sum_{t'=1}^{t} \sum_{\tau=\tau'+M(m)}^{n} (\eta_{\tau,m}^{x,t'-T^L(j),j} - \eta_{\tau,m}^{x,t'-T^M(j),j+1}) - \eta_{\tau'}^{w,t,j}],
$$
$$
1 \leq \tau' \leq n
$$

Here $\mu^{2,t,j}$ and $\widetilde{\mu}^{2,t,j}$ are the "sensitivity parameters" (denoted by α in chapter 3), and $\delta_j^{j'} = \begin{cases} 1, & j = j' \\ 0, & j \neq j' \end{cases}$ are the Kronecker symbols. Rather than choosing $\mu^{2,t,j}$,

Data type	Notation	Data set I	Data set II
number of periods	n	20	20
number of echelons	m	3	3
delays (all j)	$L(j)$	2	2
	$M(j)$	0	0
	$I(j)$	0	0
costs (all j and t)	c_j^t	2	2
	h_j^t	1	1
	p_j^t	3	3
normal ranges of uncertainty set (all j and t)	D_t	[4,10]	[90,110] if $t \leq 10$ [135,165] if $t > 10$
	Z_j	[10,14]	[130,220]

Table 15.5 Data sets for the supply chain problem.

$\tilde{\mu}^{2,t,j}$ in advance, we treat them here as variables, but limit their variability by adding the following constraints:

$$\sum_{t=1}^{n} \sum_{j=1}^{m} \mu_{j'}^{2,t,j} \leq \alpha_{2,Z_{j'}}, \ 1 \leq j' \leq m,$$

$$\sum_{t=1}^{n} \sum_{j=1}^{m} \tilde{\mu}_{\tau'}^{2,t,j} \leq \alpha_{2,D_{\tau'}}, \ 1 \leq \tau' \leq n.$$

The complete AAGRC formulation of the supply chain problem (15.3.4) is an LP with $O(m^2 n + mn^2)$ constraints and $O(m^2 n^2)$ variables (more precisely $1 + 27m + 27n + mn(8 + 28m + 28n)$ constraints and $1 + 8m + 8n + 2mn + mn(2 + 15m + 15n + mn)$ variables). As an example for a problem with 3 echelons and 20 time periods, and where all the linear decision rules (LDRs) use the entire demand history, the LP has 39,742 constraints and 24,605 variables. Such a problem is solved in about 10 minutes using a state-of-the-art LP solver on a PC with an AMD 1.8 GHz processor and 1GB of memory.

15.3.4 Computational Results

We have tested the outlined AAGRC approach as applied to the supply chain problem (15.3.4) in an intensive simulation study. We used two different data sets given in table 15.5.

15.3.4.1 Bullwhip effect results

Here we used data set I, which complies with the data used in the example of section 15.3.2. We solved problem (15.3.3) by the RC, the AARC, and the AAGRC methods. The results are shown in figure 15.5.

All three robust methods resulted in a dramatic reduction of the bullwhip effect compared to the heuristic method employed in section 15.3.2 (See figure

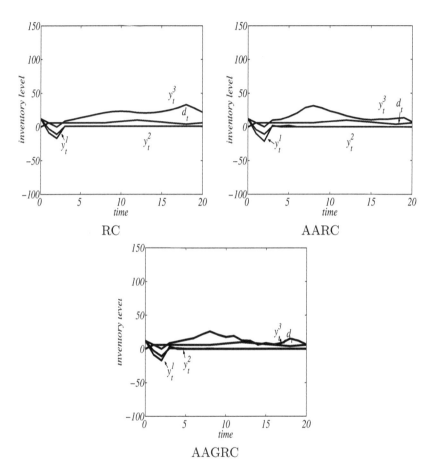

Figure 15.5   Bullwhip effects.

15.4). Of the three, the AAGRC method exhibited the smallest fluctuation range in the inventory level.

15.3.4.2 Optimal cost results

Here we used data set II, in which the demand is of a step-type: the demand starts at a constant value and after some time jumps to a new, higher value, and remains at this new level until the end of the planning horizon. This type of demand is used in [103] since, according to [69], it invokes all resonance frequencies of the dynamical system and is therefore very useful when studying the dynamics of the system. We chose the demand average over the first 10 periods to be 100 units, and the average over the remaining 10 periods to be 150 units; the actual demand fluctuated within a given margin around these averages.

In our simulations we used 4 different demand distributions as given in table 15.6. In two of them, the fluctuation margin was 10%, and in the remaining two it was 20%.

Distribution	Input number	Demand (D_t)			Initial inventory (Z^j)	
		LB	UB	Relevance	LB	UB
Uniform	1(a)	90	110	$t \leq 10$	180	220
		135	165	$t > 10$		
	1(b)	80	120	$t \leq 10$	160	240
		120	180	$t > 10$		
		Mean	Std	Relevance	Mean	Std
Normal	2(a)	100	$3\frac{1}{3}$	$t \leq 10$	200	$6\frac{2}{3}$
		150	2.5	$t > 10$		
	2(b)	100	$6\frac{2}{3}$	$t \leq 10$	200	$13\frac{1}{3}$
		150	5	$t > 10$		

Table 15.6 Step demand — input distributions.

In order to estimate the quality of our solution, we have used the *utopian solution* as a benchmark. For a given simulated demand/initial inventory, the utopian solution is the optimal solution of the corresponding deterministic LP in (15.3.4). The average (over all simulated trajectories) of the optimal utopian cost is denoted by OPT. Let C_A be the average optimal cost for a solution method A. We use the *deviation ratio*:

$$R^A = \frac{C_A}{\text{OPT}} - 1$$

as our measure of the effectiveness of method A.

We can see in table 15.7 that the average cost tends to decrease as we move from the RC to the AAGRC. The methods are comparatively close to the utopian solution, and for both the AARC and AAGRC the cost is, on average, just 15% higher than the "utopian" cost. Note that the difference with the utopian cost is an *upper bound* on the deviation from the "true optimal" solution, that is, the solution of the (untractable) ARC.

Input	Measure	Average		
		RC	AARC	AAGRC
1(a)	Cost	14,910	14,330	14,330
	R	0.19	0.14	0.14
1(b)	Cost	15,598	14,386	14,386
	R	0.24	0.15	0.15
2(a)	Cost	14,701	14,271	14,113
	R	0.18	0.14	0.13
2(b)	Cost	15,112	14,335	14,229
	R	0.21	0.15	0.14

Table 15.7 Step type demand — method comparison — averages of the cost and of deviation ratio R.

15.3.4.3 The effect of the sensitivity parameter

The main feature of the AAGRC approach as compared to the AARC is the possibility to control to some extent the behavior of the system when uncertain data run outside of their normal range by playing with the *sensitivity parameter* α (see Definition 3.1.1 in chapter 3). The choice of $\alpha = 0$ results in the most conservative attitude: the constraint must be satisfied not just for parameters in the normal range, but for *all* physically possible values, which typically results in infeasible AAGRC. The choice $\alpha = \infty$ corresponds to focusing solely on the normal range of uncertain data and makes the AAGRC identical to the AARC. An intermediate choice $\alpha \in (0, \infty)$ balances these two extreme attitudes. With such an α, the AAGRC is "more constrained" than the AARC and thus leads to a larger (or the same) optimal value. This comparison, however, has to do with the worst-case realization of the data *in the normal range*; when the data can run out of their normal range, the AAGRC-based decision rules can outperform the AARC-based rules. We are about to present the related numerical results as obtained for our supply chain problem.

We focus on the constraint in (15.3.4) that requires nonnegativity of the vectors x and w and impose, when solving the AAGRC, an upper bound $\bar{\alpha}$ on the related sensitivity parameters. Note that in the AAGRC, the sensitivity parameters are treated as variables rather than given constants. The possible values of $\bar{\alpha}$ in our experiments were values $\infty = \bar{\alpha}_0 > \bar{\alpha}_1 > \bar{\alpha}_2 \geq \bar{\alpha}_3 \geq \bar{\alpha}_4$. Our goal was to investigate the influence of $\bar{\alpha}$ on the cumulative distribution function (cdf) of the deviation ratio R. To this end we used the data set I from table 15.5 and tested 6 different cases of the probability distribution of the demands and initial inventories (see table 15.7). For each input distribution we simulated 50 realizations to build the empirical cdf of the deviation ratio R. The results are plotted in figure 15.6.

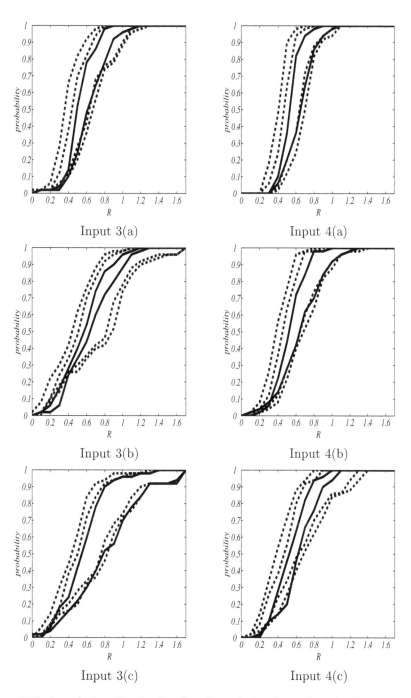

Figure 15.6 Cumulative distribution function of the demand ratio R vs. α. From left to right: solid: GRC(α_0), RC; dashed: AAGRC(α_2), AAGRC(α_1), AAGRC(α_3), AAGRC(α_4)

It is clearly seen that for all input distributions we examined, the results are consistent. First, the AARC, as it should be, is better than the RC. Second, as $\overline{\alpha}$ decreases, starting from ∞, which is the AARC case, the optimal value of the AAGRC increases (since the problem becomes more constrained). Specifically, it is Opt(AARC) = 1410 for $\overline{\alpha} \in [\overline{\alpha}_1, \overline{\alpha}_0 = \infty]$, 1628 for $\overline{\alpha} = \overline{\alpha}_2$, 2055.5 for $\overline{\alpha} = \overline{\alpha}_3$, and 2137 for $\overline{\alpha} = \overline{\alpha}_4$. However, the simulation results (reflecting *average* performance of the corresponding solutions) are of a different flavor: we get better results when using the AAGRC with $\overline{\alpha} = \overline{\alpha}_1$ and $\overline{\alpha} = \overline{\alpha}_2$ than when using the AAGRC with $\overline{\alpha} = \overline{\alpha}_0$, which is the AARC. We also see that too strong bounds on the sensitivity parameters could be dangerous: when setting $\overline{\alpha}$ to $\overline{\alpha}_3$ and $\overline{\alpha}_4$, the results become worse than even those for the RC solution. The bottom line is, that with our distributions of initial states and demands, the intermediate choice $\overline{\alpha} = \overline{\alpha}_2$ in the AAGRC results in a control policy that outperforms those given by the AARC and RC. For example, figure 15.6 shows that with the AAGRC-based policy corresponding to $\overline{\alpha} = \overline{\alpha}_2$, the deviation ratio is $\leq 40\%$ with probability 0.7, while for the AARC-based policy similar probability is < 0.2.

15.3.4.4 The effect of the normal range

In the previous section we discussed how changes in the upper bound $\overline{\alpha}$ on the sensitivity parameters affect the performance of the associated AAGRC-based control of our supply chain. Note, however, that the decision on $\overline{\alpha}$ depends on the chosen normal range of the uncertain data. Reducing this range with $\overline{\alpha}$ being fixed, we reduce both Opt(AAGRC) and Opt($AARC$) (since now we have to protect the system against a smaller set of data perturbations), and at the same time spoil the guarantees on the system's behavior in the original range of uncertain data (since what used to be in the normal range, now is outside of it). We are about to report on an experiment that gives an impression of the associated tradeoffs. In this experiment, we use data set I (see table 15.5); in particular, when evaluating the performance of a control policy by simulation, we used a fixed demand distribution P with the support $[4, 10]$. In contrast to this, the AARC and the AAGRC (with $\overline{\alpha} = \overline{\alpha}_2$) were solved for two different normal data ranges shown in table 15.8, which also presents the corresponding costs. We see that the optimal values of both AARC and AAGRC indeed decrease as the normal data range shrinks, and that with the same normal range, the optimal value of the AARC is indeed better than that of the AAGRC. At the same time we see that in terms of the "empirical worst case cost" (defined as the largest cost observed when processing a sample of 50 demand trajectories drawn from the distribution P) the winner is the AAGRC-based policy associated with the "small" normal range $[5, 9]$. This means that using a smaller normal range and protecting against infeasibility outside this range via the AAGRC methodology, we can get better results than those yielded by the AARC associated with a wider normal range.

| Assumed | AARC | | AAGRC | |
| normal | Optimal | Empirical worst | Optimal | Empirical worst |
range	value	case	value	case
[4,10]	1410	1240	1628	1124
[5,9]	1149	1100	1311	1068

Table 15.8 AARC and AAGRC costs for different normal ranges of the uncertain data.

How much protection against infeasibility does the AAGRC provide? To test this we again used simulations with demand trajectories drawn from P. The AARC was run with three uncertainty sets [4,10], (which is the actual range of the demand distributed according to P), [5,9] and [6,8], while the AAGRC was run with the same sets as the normal ranges of the uncertain data, $\bar{\alpha}$ being set to $\bar{\alpha}_2$. Since the second and the third sets of uncertain data underestimate the true range of the uncertain demand, the resulting control policies from time to time can generate infeasible controls (negative replenishment orders). Table 15.9 reports on the frequency of these failures for the control policies in question, specifically, it presents empirical probabilities, (built upon a sample of 50 demand trajectories drawn from P) to issue at least one and at least two infeasible order(s).

| Assumed | AARC | | AAGRC | |
normal range	$N \geq 1$	$N \geq 2$	$N \geq 1$	$N \geq 2$
[4,10]	0%	0%	0%	0%
[5,9]	24%	22%	28%	0%
[6,8]	24%	24%	0%	0%

Table 15.9 Percent of replications with $N > 0$ infeasible orders.

Here again we see how advantageous the AAGRC approach is: in every one of the experiments with time horizon 20, the AAGRCs associated with "underestimated" uncertainty result in at most one infeasible order. In contrast to this, AARCs associated with "underestimated" uncertainty result in more than one infeasible order in about 20% of experiments (and in some of them produce as much as 6 infeasible orders).

Appendix A
Notation and Prerequisites

A.1 NOTATION

- $\mathbb{Z}$, $\mathbb{R}$, $\mathbb{C}$ stand for the sets of all integers, reals, and complex numbers, respectively.

- $\mathbb{C}^{m \times n}$, $\mathbb{R}^{m \times n}$ stand for the spaces of complex, respectively, real $m \times n$ matrices. We write $\mathbb{C}^n$ and $\mathbb{R}^n$ as shorthands for $\mathbb{C}^{n \times 1}$, $\mathbb{R}^{n \times 1}$, respectively.

For $A \in \mathbb{C}^{m \times n}$, A^T stands for the transpose, and A^H for the conjugate transpose of A:
$$(A^H)_{rs} = \overline{A_{sr}},$$
where $\bar{z}$ is the conjugate of $z \in \mathbb{C}$.

- Both $\mathbb{C}^{m \times n}$, $\mathbb{R}^{m \times n}$ are equipped with the inner product
$$\langle A, B \rangle = \mathrm{Tr}(AB^H) = \sum_{r,s} A_{rs} \overline{B_{rs}}.$$

The norm associated with this inner product is denoted by $\| \cdot \|_2$.

- For $p \in [1, \infty]$, we define the p-norms $\| \cdot \|_p$ on $\mathbb{C}^n$ and $\mathbb{R}^n$ by the relation
$$\|x\|_p = \begin{cases} \left(\sum_i |x_i|^p \right)^{1/p}, & 1 \leq p < \infty \\ \lim_{p \to \infty} \|x\|_p = \max_i |x_i|, & p = \infty \end{cases}, \ 1 \leq p \leq \infty.$$

Note that when $p, q \in [1, \infty]$ and $\frac{1}{p} + \frac{1}{q} = 1$, then the norms $\| \cdot \|_p$ and $\| \cdot \|_q$ are conjugates of each other:
$$\|x\|_p = \max_{y:\|y\|_q \leq 1} |\langle x, y \rangle|.$$

In particular, $|\langle x, y \rangle| \leq \|x\|_p \|y\|_q$ (Hölder inequality).

- We use the notation I_m, $0_{m \times n}$ for the unit $m \times m$, respectively, the zero $m \times n$ matrices.

- $\mathbf{H}^m$, $\mathbf{S}^m$ are real vector spaces of $m \times m$ Hermitian, respectively, real symmetric matrices. Both are Euclidean spaces w.r.t. the inner product $\langle \cdot, \cdot \rangle$.

- We use "MATLAB notation": when $A_1, ..., A_k$ are matrices with the same number of rows, $[A_1, ..., A_k]$ denotes the matrix with the same number of rows obtained by writing, from left to right, first the columns of A_1, then the columns of A_2, and so on. When $A_1, ..., A_k$ are matrices with the same number of columns, $[A_1; A_2; ...; A_k]$ stands for the matrix with the same number of columns obtained by writing, from top to bottom, first the rows of A_1, then the rows of A_2, and so on.

• For a Hermitian/real symmetric $m \times m$ matrix A, $\lambda(A)$ is the vector of eigenvalues $\lambda_r(A)$ of A taken with their multiplicities in the non-ascending order:

$$\lambda_1(A) \geq \lambda_2(A) \geq ... \geq \lambda_m(A).$$

• For an $m \times n$ matrix A, $\sigma(A) = (\sigma_1(A), ..., \sigma_n(A))^T$ is the vector of singular values of A:

$$\sigma_r(A) = \lambda_r^{1/2}(A^H A),$$

and

$$\|A\|_{2,2} = \|A\| = \sigma_1(A) = \max\{\|Ax\|_2 : x \in \mathbb{C}^n, \|x\|_2 \leq 1\}$$

(by evident reasons, when A is real, one can replace $\mathbb{C}^n$ in the right hand side with $\mathbb{R}^n$).

• For Hermitian/real symmetric matrices A, B, we write $A \succeq B$ ($A \succ B$) to express that $A - B$ is positive semidefinite (resp., positive definite).

A.2 CONIC PROGRAMMING

A.2.1 Euclidean Spaces, Cones, Duality

A.2.1.1 Euclidean spaces

A *Euclidean space* is a finite dimensional linear space over reals equipped with an *inner product* $\langle x, y \rangle_E$ — a bilinear and symmetric real-valued function of $x, y \in E$ such that $\langle x, x \rangle_E > 0$ whenever $x \neq 0$.

Example: The standard Euclidean space $\mathbb{R}^n$. This space is comprised of n-dimensional real column vectors with the standard coordinate-wise linear operations and the inner product $\langle x, y \rangle_{\mathbb{R}^n} = x^T y$. $\mathbb{R}^n$ is a universal example of an Euclidean space: for every Euclidean n-dimensional space $(E, \langle \cdot, \cdot \rangle_E)$ there exists a one-to-one linear mapping $x \mapsto Ax : \mathbb{R}^n \to E$ such that $x^T y \equiv \langle Ax, Ay \rangle_E$. All we need in order to build such a mapping, is to find an *orthonormal basis* $e_1, ..., e_n$, $n = \dim E$, in E, that is, a basis such that $\langle e_i, e_j \rangle_E = \delta_{ij} \equiv \begin{cases} 1, & i = j \\ 0, & i \neq j \end{cases}$; such a basis always exists. Given an orthonormal basis $\{e_i\}_{i=1}^n$, a one-to-one mapping $A : \mathbb{R}^n \to E$ preserving the inner product is given by $Ax = \sum_{i=1}^n x_i e_i$.

Example: The space $\mathbb{R}^{m \times n}$ of $m \times n$ real matrices with the Frobenius inner product. The elements of this space are $m \times n$ real matrices with the standard linear operations and the inner product $\langle A, B \rangle_F = \mathrm{Tr}(AB^T) = \sum_{i,j} A_{ij} B_{ij}$.

Example: The space $\mathbf{S}^n$ of $n \times n$ real symmetric matrices with the Frobenius inner product. This is the subspace of $\mathbb{R}^{n \times n}$ comprised of all symmetric $n \times n$ matrices; the inner product is inherited from the embedding space. Of course, for symmetric matrices, this product can be written down without transposition:

$$A, B \in \mathbf{S}^n \Rightarrow \langle A, B \rangle_F = \mathrm{Tr}(AB) = \sum_{i,j} A_{ij} B_{ij}.$$

Example: The space $\mathbf{H}^n$ of $n \times n$ Hermitian matrices with the Frobenius inner product. This is the real linear space comprised of $n \times n$ Hermitian matrices; the inner product is

$$\langle A, B \rangle = \operatorname{Tr}(AB^H) = \operatorname{Tr}(AB) = \sum_{i,j=1}^{n} A_{ij} \overline{B_{ij}}.$$

A.2.1.2 Linear forms on Euclidean spaces

Every homogeneous linear form $f(x)$ on a Euclidean space $(E, \langle \cdot, \cdot \rangle_E)$ can be represented in the form $f(x) = \langle e_f, x \rangle_E$ for certain vector $e_f \in E$ uniquely defined by $f(\cdot)$. The mapping $f \mapsto e_f$ is a one-to-one linear mapping of the space of linear forms on E onto E.

A.2.1.3 Conjugate mapping

Let $(E, \langle \cdot, \cdot \rangle_E)$ and $(F, \langle \cdots \rangle_F)$ be Euclidean spaces. For a linear mapping $A : E \to F$ and every $f \in F$, the function $\langle Ae, f \rangle_F$ is a linear function of $e \in E$ and as such it is representable as $\langle e, A^*f \rangle_E$ for certain uniquely defined vector $A^*f \in E$. It is immediately seen that the mapping $f \mapsto A^*f$ is a linear mapping of F into E; the characteristic identity specifying this mapping is

$$\langle Ae, f \rangle_F = \langle e, A^*f \rangle \; \forall (e \in E, f \in F).$$

The mapping A^* is called *conjugate* to A. It is immediately seen that the conjugation is a linear operation with the properties $(A^*)^* = A$, $(AB)^* = B^*A^*$. If $\{e_j\}_{j=1}^{m}$ and $\{f_i\}_{i=1}^{n}$ are orthonormal bases in E, F, then every linear mapping $A : E \to F$ can be associated with the matrix $[a_{ij}]$ ("matrix of the mapping in the pair of bases in question") according to the identity

$$A \sum_{j=1}^{m} x_j e_j = \sum_i \left[\sum_j a_{ij} x_j \right] f_i$$

(in other words, a_{ij} is the i-th coordinate of the vector Ae_j in the basis $f_1, ..., f_n$). With this representation of linear mappings by matrices, the matrix representing A^* in the pair of bases $\{f_i\}$ in the argument and $\{e_j\}$ in the image spaces of A^* is the transpose of the matrix representing A in the pair of bases $\{e_j\}, \{f_i\}$.

A.2.1.4 Cones in Euclidean space

A nonempty subset $\mathbf{K}$ of a Euclidean space $(E, \langle \cdot, \cdot \rangle_E)$ is called a cone, if it is a convex set comprised of rays emanating from the origin, or, equivalently, whenever $t_1, t_2 \geq 0$ and $x_1, x_2 \in \mathbf{K}$, we have $t_1 x_1 + t_2 x_2 \in \mathbf{K}$.

A cone $\mathbf{K}$ is called *regular*, if it is closed, possesses a nonempty interior and is *pointed* — does not contain lines, or, which is the same, is such that $a \in \mathbf{K}$, $-a \in \mathbf{K}$ implies that $a = 0$.

Dual cone. If $\mathbf{K}$ is a cone in a Euclidean space $(E, \langle \cdot, \cdot \rangle_E)$, then the set

$$\mathbf{K}^* = \{ e \in E : \langle e, h \rangle_E \geq 0 \, \forall h \in \mathbf{K} \}$$

also is a cone called the cone *dual* to $\mathbf{K}$. The dual cone always is closed. The cone dual to dual is the closure of the original cone: $(\mathbf{K}^*)^* = \mathrm{cl}\, \mathbf{K}$; in particular, $(\mathbf{K}^*)^* = \mathbf{K}$ for every closed cone $\mathbf{K}$. The cone $\mathbf{K}^*$ possesses a nonempty interior if and only if $\mathbf{K}$ is pointed, and $\mathbf{K}^*$ is pointed if and only if $\mathbf{K}$ possesses a nonempty interior; in particular, $\mathbf{K}$ is regular if and only if $\mathbf{K}^*$ is so.

Example: Nonnegative ray and nonnegative orthants. The simplest one-dimensional cone is the nonnegative ray $\mathbb{R}_+ = \{ t \geq 0 \}$ on the real line $\mathbb{R}^1$. The simplest cone in $\mathbb{R}^n$ is the *nonnegative orthant* $\mathbb{R}^n_+ = \{ x \in \mathbb{R}^n : x_i \geq 0, 1 \leq i \leq n \}$. This cone is regular and self-dual: $(\mathbb{R}^n_+)^* = \mathbb{R}^n_+$.

Example: Lorentz cone $\mathbf{L}^n$. The cone $\mathbf{L}^n$ "lives" in $\mathbb{R}^n$ and is comprised of all vectors $x = [x_1; ...; x_n] \in \mathbb{R}^n$ such that $x_n \geq \sqrt{\sum_{j=1}^{n-1} x_j^2}$; same as $\mathbb{R}^n_+$, the Lorentz cone is regular and self-dual.

By definition, $\mathbf{L}^1 = \mathbb{R}_+$ is the nonnegative orthant; this is in full accordance with the "general" definition of a Lorentz cone combined with the standard convention "a sum over an empty set of indices is 0."

Example: Semidefinite cone $\mathbf{S}^n_+$. The cone $\mathbf{S}^n_+$ "lives" in the Euclidean space $\mathbf{S}^n$ of $n \times n$ symmetric matrices equipped with the Frobenius inner product. The cone is comprised of all $n \times n$ symmetric *positive semidefinite* matrices A, i.e., matrices $A \in \mathbf{S}^n$ such that $x^T A x \geq 0$ for all $x \in \mathbb{R}^n$, or, equivalently, such that all eigenvalues of A are nonnegative. Same as $\mathbb{R}^n_+$ and $\mathbf{L}^n$, the cone $\mathbf{S}^n_+$ is regular and self-dual.

Example: Hermitian semidefinite cone $\mathbf{H}^n_+$. This cone "lives" in the space $\mathbf{H}^n$ of $n \times n$ Hermitian matrices and is comprised of all positive semidefinite Hermitian $n \times n$ matrices; it is regular and self-dual.

A.2.2 Conic Problems and Conic Duality

A.2.2.1 Conic problem

A *conic problem* is an optimization problem of the form

$$\mathrm{Opt}(P) = \min_x \left\{ \langle c, x \rangle_E : \begin{array}{l} A_i x - b_i \in \mathbf{K}_i, \, i = 1, ..., m, \\ A x = b \end{array} \right\} \tag{P}$$

where

- $(E, \langle \cdot, \cdot \rangle_E)$ is a Euclidean space of *decision vectors* x and $c \in E$ is the *objective*;

- A_i, $1 \leq i \leq m$, are linear maps from E into Euclidean spaces $(F_i, \langle \cdot, \cdot \rangle_{F_i})$, $b_i \in F_i$ and $\mathbf{K}_i \subset F_i$ are regular cones;

- A is a linear mapping from E into a Euclidean space $(F, \langle \cdot, \cdot \rangle_F)$ and $b \in F$.

Examples: Linear, Conic Quadratic and Semidefinite Optimization.
We will be especially interested in the three generic conic problems as follows:

- *Linear Optimization*, or *Linear Programming*: this is the family of all conic problems associated with nonnegative orthants $\mathbb{R}_+^m$, that is, the family of all usual LPs $\min_x \{c^T x : Ax - b \geq 0\}$;

- *Conic Quadratic Optimization*, or *Conic Quadratic Programming*, or *Second Order Cone Programming*: this is the family of all conic problems associated with the cones that are *finite direct products* of Lorentz cones, that is, the conic programs of the form

$$\min_x \left\{ c^T x : [A_1; ...; A_m]x - [b_1; ...; b_m] \in \mathbf{L}^{k_1} \times ... \times \mathbf{L}^{k_m} \right\}$$

  where A_i are $k_i \times \dim x$ matrices and $b_i \in \mathbb{R}^{k_i}$. The "Mathematical Programming" form of such a program is

$$\min_x \left\{ c^T x : \|\bar{A}_i x - \bar{b}_i\|_2 \leq \alpha_i^T x - \beta_i, \, 1 \leq i \leq m \right\},$$

  where $A_i = [\bar{A}_i; \alpha_i^T]$ and $b_i = [\bar{b}_i; \beta_i]$, so that α_i is the last row of A_i, and β_i is the last entry of b_i;

- *Semidefinite Optimization*, or *Semidefinite Programming*: this is the family of all conic problems associated with the cones that are *finite direct products* of Semidefinite cones, that is, the conic programs of the form

$$\min_x \left\{ c^T x : A_i^0 + \sum_{j=1}^{\dim x} x_j A_i^j \succeq 0, \, 1 \leq i \leq m \right\},$$

  where A_i^j are symmetric matrices of appropriate sizes.

A.2.3 Conic Duality

A.2.3.1 Conic duality — derivation

The origin of conic duality is the desire to find a systematic way to bound from below the optimal value in a conic problem (P). This way is based on *linear aggregation* of the constraints of (P), namely, as follows. Let $y_i \in \mathbf{K}_i^*$ and $z \in F$. By the definition of the dual cone, for every x feasible for (P) we have

$$\langle A_i^* y_i, x \rangle_E - \langle y_i, b_i \rangle_{F_i} \equiv \langle y_i, Ax_i - b_i \rangle_{F_i} \geq 0, \, 1 \leq i \leq m,$$

and of course

$$\langle A^* z, x \rangle_E - \langle z, b \rangle_F = \langle z, Ax - b \rangle_F = 0.$$

Summing up the resulting inequalities, we get

$$\langle A^* z + \sum_i A_i^* y_i, x \rangle_E \geq \langle z, b \rangle_F + \sum_i \langle y_i, b_i \rangle_{F_i}. \tag{C}$$

By its origin, this scalar linear inequality on x is a consequence of the constraints of (P), that is, it is valid for all feasible solutions x to (P). It may happen that the left hand side in this inequality is, identically in $x \in E$, equal to the objective $\langle c, x \rangle_E$; this happens if and only if

$$A^* z + \sum_i A_i^* y_i = c.$$

Whenever it is the case, the right hand side of (C) is a valid lower bound on the optimal value in (P). The dual problem is nothing but the problem

$$\mathrm{Opt}(D) = \max_{z, \{y_i\}} \left\{ \langle z, b \rangle_F + \sum_i \langle y_i, b_i \rangle_{F_i} : \begin{array}{l} y_i \in \mathbf{K}_i^*, 1 \le i \le m, \\ A^* z + \sum_i A_i^* y_i = c \end{array} \right\} \qquad (D)$$

of maximizing this lower bound.

By the origin of the dual problem, we have

Weak Duality: *One has* $\mathrm{Opt}(D) \le \mathrm{Opt}(P)$.

We see that (D) is a conic problem. A nice and important fact is that *conic duality is symmetric*.

Symmetry of Duality: *The conic dual to* (D) *is (equivalent to)* (P).

Proof. In order to apply to (D) the outlined recipe for building the conic dual, we should rewrite (D) as a *minimization* problem

$$-\mathrm{Opt}(D) = \min_{z, \{y_i\}} \left\{ \langle z, -b \rangle_F + \sum_i \langle y_i, -b_i \rangle_{F_i} : \begin{array}{l} y_i \in \mathbf{K}_i^*, 1 \le i \le m \\ A^* z + \sum_i A_i^* y_i = c \end{array} \right\}; \qquad (D')$$

the corresponding space of decision vectors is the direct product $F \times F_1 \times ... \times F_m$ of Euclidean spaces equipped with the inner product

$$\langle [z; y_1, ..., y_m], [z'; y_1', ..., y_m'] \rangle = \langle z, z' \rangle_F + \sum_i \langle y_i, y_i' \rangle_{F_i}.$$

The above "duality recipe" as applied to (D') reads as follows: pick weights $\eta_i \in (\mathbf{K}_i^*)^* = \mathbf{K}_i$ and $\zeta \in E$, so that the scalar inequality

$$\underbrace{\langle \zeta, A^* z + \sum_i A_i^* y_i \rangle_E + \sum_i \langle \eta_i, y_i \rangle_{F_i}}_{= \langle A\zeta, z \rangle_F + \sum_i \langle A_i \zeta + \eta_i, y_i \rangle_{F_i}} \ge \langle \zeta, c \rangle_E \qquad (C')$$

in variables z, $\{y_i\}$ is a consequence of the constraints of (D'), and impose on the "aggregation weights" $\zeta, \{\eta_i \in \mathbf{K}_i\}$ an additional restriction that the left hand side in this inequality is, identically in $z, \{y_i\}$, equal to the objective of (D'), that is, the restriction that

$$A\zeta = -b, \ A_i \zeta + \eta_i = -b_i, \ 1 \le i \le m,$$

and maximize under this restriction the right hand side in (C'), thus arriving at the problem

$$\max_{\zeta, \{\eta_i\}} \left\{ \langle c, \zeta \rangle_E : \begin{array}{l} \mathbf{K}_i \ni \eta_i = A_i[-\zeta] - b_i, 1 \le i \le m \\ A[-\zeta] = b \end{array} \right\}.$$

Substituting $x = -\zeta$, the resulting problem, after eliminating η_i variables, is nothing but

$$\max_x \left\{ -\langle c, x \rangle_E : \begin{array}{l} A_i x - b_i \in \mathbf{K}_i, 1 \le i \le m \\ Ax = b \end{array} \right\},$$

which is equivalent to (P). $\qquad \square$

A.2.3.2 Conic Duality Theorem

A conic program (P) is called *strictly feasible*, if it admits a feasible solution $\bar{x}$ such that $A_i \bar{x} = -b_i \in \text{int} K_i$, $i = 1, ..., m$.

Conic Duality Theorem is the following statement resembling very much the standard Linear Programming Duality Theorem:

Theorem A.2.1. [Conic Duality Theorem] Consider a primal-dual pair of conic problems (P), (D). Then

(i) [Weak Duality] One has $\text{Opt}(D) \le \text{Opt}(P)$.

(ii) [Symmetry] The duality is symmetric: (D) is a conic problem, and the problem dual to (D) is (equivalent to) (P).

(iii) [Strong Duality] If one of the problems (P), (D) is strictly feasible and bounded, then the other problem is solvable, and $\text{Opt}(P) = \text{Opt}(D)$.

If both the problems are strictly feasible, then both are solvable with equal optimal values.

Proof. We have already verified Weak Duality and Symmetry. Let us prove the first claim in Strong Duality. By Symmetry, we can restrict ourselves to the case when the strictly feasible and bounded problem is (P).

Consider the following two sets in the Euclidean space $G = \mathbb{R} \times F \times F_1 \times ... \times F_m$:

$$\begin{aligned} T &= \{[t; z; y_1; ...; y_m] : \exists x : t = \langle c, x \rangle_E; y_i = A_i x - b_i, 1 \le i \le m; \\ &\quad z = Ax - b\}, \\ S &= \{[t; z; y_1; ...; y_m] : t < \text{Opt}(P), y_1 \in \mathbf{K}_1, ..., y_m \in \mathbf{K}_m, z = 0\}. \end{aligned}$$

The sets T and S clearly are convex and nonempty; observe that they do not intersect. Indeed, assuming that $[t; z; y_1; ...; y_m] \in S \cap T$, we should have $t < \text{Opt}(P)$, and $y_i \in \mathbf{K}_i$, $z = 0$ (since the point is in S), and at the same time for certain $x \in E$ we should have $t = \langle c, x \rangle_E$ and $A_i x - b_i = y_i \in \mathbf{K}_i$, $Ax - b = z = 0$, meaning that there exists a feasible solution to (P) with the value of the objective $< \text{Opt}(P)$, which is impossible. Since the convex and nonempty sets S and T do not intersect, they can be separated by a linear form: there exists $[\tau; \zeta; \eta_1; ...; \eta_m] \in$

$G = \mathbb{R} \times F \times F_1 \times ... \times F_m$ such that

(a)
$$\sup_{[t;z;y_1;...;y_m]\in S} \langle [\tau; \zeta; \eta_1; ...; \eta_m], [t; z; y_1; ...; y_m] \rangle_G$$
$$\leq \inf_{[t;z;y_1;...;y_m]\in T} \langle [\tau; \zeta; \eta_1; ...; \eta_m], [t; z; y_1; ...; y_m] \rangle_G,$$

(b)
$$\inf_{[t;z;y_1;...;y_m]\in S} \langle [\tau; \zeta; \eta_1; ...; \eta_m], [t; z; y_1; ...; y_m] \rangle_G$$
$$< \sup_{[t;z;y_1;...;y_m]\in T} \langle [\tau; \zeta; \eta_1; ...; \eta_m], [t; z; y_1; ...; y_m] \rangle_G,$$

or, which is the same,

(a)
$$\sup_{t<\mathrm{Opt}(P), y_i\in \mathbf{K}_i} [\tau t + \sum_i \langle \eta_i, y_i \rangle_{F_i}]$$
$$\leq \inf_{x\in E} [\tau \langle c, x \rangle_E + \langle \zeta, Ax - b \rangle_F + \sum_i \langle \eta_i, A_i x - b_i \rangle_{F_i}],$$

(b)
$$\inf_{t<\mathrm{Opt}(P), y_i\in \mathbf{K}_i} [\tau t + \sum_i \langle \eta_i, y_i \rangle_{F_i}]$$
$$< \sup_{x\in E} [\tau \langle c, x \rangle + \langle \zeta, Ax - b \rangle_F + \sum_i \langle \eta_i, Ax - b_i \rangle_{F_i}].$$

$$(\text{A.2.1})$$

Since the left hand side in (A.2.1.a) is finite, we have

$$\tau \geq 0, \ -\eta_i \in \mathbf{K}_i^*, 1 \leq i \leq m, \qquad (\text{A.2.2})$$

whence the left hand side in (A.2.1.a) is equal to $\tau\mathrm{Opt}(P)$. Since the right hand side in (A.2.1.a) is finite and $\tau \geq 0$, we have

$$A^*\zeta + \sum_i A_i^*\eta_i + \tau c = 0 \qquad (\text{A.2.3})$$

and the right hand side in (a) is $\langle -\zeta, b \rangle_F - \sum_i \langle \eta_i, b_i \rangle_{F_i}$, so that (A.2.1.$a$) reads

$$\tau\mathrm{Opt}(P) \leq \langle -\zeta, b \rangle_F - \sum_i \langle \eta_i, b_i \rangle_{F_i}. \qquad (\text{A.2.4})$$

We claim that $\tau > 0$. Believing in our claim, let us extract from it Strong Duality. Indeed, setting $y_i = -\eta_i/\tau$, $z = -\zeta/\tau$, (A.2.2), (A.2.3) say that $z, \{y_i\}$ is a feasible solution for (D), and by (A.2.4) the value of the dual objective at this dual feasible solution is $\geq \mathrm{Opt}(P)$. By Weak Duality, this value cannot be larger than $\mathrm{Opt}(P)$, and we conclude that our solution to the dual is in fact an optimal one, and that $\mathrm{Opt}(P) = \mathrm{Opt}(D)$, as claimed.

It remains to prove that $\tau > 0$. Assume this is not the case; then $\tau = 0$ by (A.2.2). Now let $\bar{x}$ be a strictly feasible solution to (P). Taking inner product of both sides in (A.2.3) with $\bar{x}$, we have

$$\langle \zeta, A\bar{x} \rangle_F + \sum_i \langle \eta_i, A_i\bar{x} \rangle_{F_i} = 0,$$

while (A.2.4) reads

$$-\langle \zeta, b \rangle_F - \sum_i \langle \eta_i, b_i \rangle_{F_i} \geq 0.$$

Summing up the resulting inequalities and taking into account that $\bar{x}$ is feasible for (P), we get

$$\sum_i \langle \eta_i, A_i\bar{x} - b_i \rangle \geq 0.$$

Since $A_i\bar{x} - b_i \in \text{int}\mathbf{K}_i$ and $\eta_i \in -\mathbf{K}_i^*$, the inner products in the left hand side of the latter inequality are nonpositive, and i-th of them is zero if and only if $\eta_i = 0$; thus, the inequality says that $\eta_i = 0$ for all i. Adding this observation to $\tau = 0$ and looking at (A.2.3), we see that $A^*\zeta = 0$, whence $\langle \zeta, Ax \rangle_F = 0$ for all x and, in particular, $\langle \zeta, b \rangle_F = 0$ due to $b = A\bar{x}$. The bottom line is that $\langle \zeta, Ax - b \rangle_F = 0$ for all x. Now let us look at (A.2.1.b). Since $\tau = 0$, $\eta_i = 0$ for all i and $\langle \zeta, Ax - b \rangle_F = 0$ for all x, both sides in this inequality are equal to 0, which is impossible. We arrive at a desired contradiction.

We have proved the first claim in Strong Duality. The second claim there is immediate: if both (P), (D) are strictly feasible, then both problems are bounded as well by Weak Duality, and thus are solvable with equal optimal values by the already proved part of Strong Duality. $\qquad\square$

A.2.3.3 Optimality conditions in Conic Programming

Optimality conditions in Conic Programming are given by the following statement:

Theorem A.2.2. Consider a primal-dual pair (P), (D) of conic problems, and let both problems be strictly feasible. A pair $(x, \xi \equiv [z; y_1; ...; y_m])$ of feasible solutions to (P) and (D) is comprised of optimal solutions to the respective problems if and only if

(i) [Zero duality gap] One has

$$\begin{aligned} \text{DualityGap}(x;\xi) \quad &:= \quad \langle c, x \rangle_E - [\langle z, b \rangle_F + \textstyle\sum_i \langle b_i, y_i \rangle_{F_i}] \\ &= \quad 0, \end{aligned}$$

same as if and only if

(ii) [Complementary slackness]

$$\forall i : \langle y_i, A_i x_i - b_i \rangle_{F_i} = 0.$$

Proof. By Conic Duality Theorem, we are in the situation when $\text{Opt}(P) = \text{Opt}(D)$. Therefore

$$\begin{aligned} \text{DualityGap}(x;\xi) = \quad &\underbrace{[\langle c, x \rangle_E - \text{Opt}(P)]}_{a} \\ &+ \underbrace{\left[\text{Opt}(D) - \left[\langle z, b \rangle_F + \sum_i \langle b_i, y_i \rangle_{F_i} \right] \right]}_{b} \end{aligned}$$

Since x and ξ are feasible for the respective problems, the duality gap is nonnegative and it can vanish if and only if $a = b = 0$, that is, if and only if x and ξ are optimal solutions to the respective problems, as claimed in (i). To prove (ii), note that since x is feasible, we have

$$Ax = b, \ A_i x - b_i \in \mathbf{K}_i, \ c = A^* z + \sum_i A_i^* y_i, y_i \in \mathbf{K}_i^*,$$

whence

$$\begin{aligned}
\text{DualityGap}(x; \xi) &= \langle c, x \rangle_E - [\langle z, b \rangle_F + \sum_i \langle b_i, y_i \rangle_{F_i}] \\
&= \langle A^* z + \sum_i A_i^* y_i, x \rangle_E - [\langle z, b \rangle_F + \sum_i \langle b_i, y_i \rangle_{F_i}] \\
&= \underbrace{\langle z, Ax - b \rangle_F}_{=0} + \sum_i \underbrace{\langle y_i, A_i x - b_i \rangle_{F_i}}_{\geq 0},
\end{aligned}$$

where the nonnegativity of the terms in the last $\sum_i$ follows from $y_i \in \mathbf{K}_i^*$, $A_i x_i - b_i \in \mathbf{K}_i$. We see that the duality gap, as evaluated at a pair of primal-dual feasible solutions, vanishes if and only if the complementary slackness holds true, and thus (ii) is readily given by (i).   □

A.2.4 Conic Representations of Sets and Functions

A.2.4.1 Conic representations of sets

When asked whether the optimization programs

$$\min_y \sum_{i=1}^m |a_i^T y - b_i| \tag{A.2.5}$$

and

$$\min_y \max_{1 \leq i \leq m} |a_i^T y - b_i| \tag{A.2.6}$$

are Linear Optimization programs, the answer definitely will be "yes", in spite of the fact that an LO program is defined as

$$\min_x \left\{ c^T x : Ax \geq b, Px = p \right\} \tag{A.2.7}$$

and neither (A.2.5), nor (A.2.6) are in this form. What the "yes" answer actually means, is that both (A.2.5) and (A.2.6) can be straightforwardly *reduced to*, or, which is the same, *represented by* LO programs, e.g., the LO program

$$\min_{y,u} \left\{ \sum_{i=1}^m u_i : -u_i \leq a_i^T y - b_i \leq u_i, 1 \leq i \leq m \right\} \tag{A.2.8}$$

in the case of (A.2.5), and the LO program

$$\min_{y,t} \left\{ t : -t \leq a_i^T y - b_i \leq t, 1 \leq i \leq m \right\} \tag{A.2.9}$$

in the case of (A.2.6).

An "in-depth" explanation of what actually takes place in these and similar examples is as follows.

i) The "initial form" of a typical Mathematical Programming problem is $\min_{v \in V} f(v)$, where $f(v) : V \to \mathbb{R}$ is the objective, and $V \subset \mathbb{R}^n$ is the feasible set of the problem. It is technically convenient to assume that the objective is "as simple as possible" — just linear: $f(v) = e^T v$; this assumption does not restrict generality, since we can always pass from the original problem, given in the form $\min_{v \in V} \phi(v)$, to the equivalent problem

$$\min_{y=[v;s]} \left\{ c^T y \equiv s : y \in Y = \{[v;s] : v \in V, s \geq \phi(v)\} \right\}.$$

Thus, from now on we assume w.l.o.g. that the original problem is

$$\min_y \left\{ d^T y : y \in Y \right\}. \tag{A.2.10}$$

ii) All we need in order to reduce (A.2.10) to an LO program is what is called a *polyhedral representation* of Y, that is, a representation of the form

$$U = \{y \in \mathbb{R}^n : \exists u : Ay + Bu - b \in \mathbb{R}_+^N\}.$$

Indeed, given such a representation, we can reformulate (A.2.10) as the LO program

$$\min_{x=[y;u]} \left\{ c^T x := d^T y : \mathcal{A}(x) := Ay + Bu - b \geq 0 \right\}.$$

For example, passing from (A.2.5) to (A.2.8), we first rewrite the original problem as

$$\min_{t,y} \left\{ t : \sum_i |a_i^T y - b_i| \leq t \right\}$$

and then point out a polyhedral representation

$$
\{[y;t] : \sum_i |a_i^T y - b_i| \leq t\}
$$
$$
= \{[y;t] : \exists u : \underbrace{\left\{ \begin{array}{l} u_i - a_i^T y + b_i \geq 0, \\ u_i + a_i^T y - b_i \geq 0, \\ t - \sum_i u_i \geq 0 \end{array} \right\}}_{A[y;t]+Bu-b\geq 0} \}
$$

of the feasible set of the latter problem, thus ending up with reformulating the problem of interest as an LO program in variables y, t, u. The course of actions for (A.2.6) is completely similar, up to the fact that after "linearizing the objective" we get the optimization problem

$$\min_{y,t} \left\{ t : -t \leq a_i^T y - b_i \leq t, \ 1 \leq i \leq m \right\}$$

where the feasible set is polyhedral "as it is" (i.e., with polyhedral representation not requiring u-variables).

The notion of polyhedral representation naturally extends to conic problems, specifically, as follows. Let $\mathcal{K}$ be a family of regular cones, every one "living" in its own Euclidean space. A set $Y \subset \mathbb{R}^n$ is called $\mathcal{K}$-*representable*, if it can be represented in the form

$$Y = \{y \in \mathbb{R}^n : \exists u \in \mathbb{R}^m : Ay + Bu - b \in \mathbf{K}\}, \tag{A.2.11}$$

where $\mathbf{K} \in \mathcal{K}$ and A, B, b are matrices and vectors of appropriate dimensions. A representation of Y of the form (A.2.11), (i.e., the corresponding collection $A, B, b, \mathbf{K}$), is called a $\mathcal{K}$-*representation* ($\mathcal{K}$-r. for short) of Y.

Geometrically, a $\mathcal{K}$-r. of Y is the representation of Y as the *projection* on the space of y variables of the set $Y_+ = \{[y;u] : Ax + Bu - b \in \mathbf{K}\}$, which, in turn, is given as the inverse image of a cone $\mathbf{K} \in \mathcal{K}$ under the affine mapping $[y; u] \mapsto Ay + Bu - b$.

The role of the notion of a conic representation stems from the fact that given a $\mathcal{K}$-r. of the feasible domain Y of (A.2.10), we can immediately rewrite this optimization program as a conic program involving a cone from the family $\mathcal{K}$, specifically, as the program

$$\min_{x=[y;u]} \left\{ c^T x := d^T y : \mathcal{A}(x) := Ay + Bu - b \in \mathbf{K} \right\}. \qquad (A.2.12)$$

In particular,

- When $\mathcal{K} = \mathcal{LO}$ is the family of all nonnegative orthants (or, which is the same, the family of all finite direct products of nonnegative rays), a $\mathcal{K}$-representation of Y allows one to rewrite (A.2.10) as a Linear program;

- When $\mathcal{K} = \mathcal{CQO}$ is the family of all finite direct products of Lorentz cones, a $\mathcal{K}$-representation of Y allows one to rewrite (A.2.10) as a Conic Quadratic program;

- When $\mathcal{K} = \mathcal{SDO}$ is the family of all finite direct products of positive semidefinite cones, a $\mathcal{K}$-representation of Y allows one to rewrite (A.2.10) as a Semidefinite program.

Note that a $\mathcal{K}$-representable set is always convex.

A.2.4.2 Elementary calculus of $\mathcal{K}$-representations

It turns out that when the family of cones $\mathcal{K}$ is "rich enough," $\mathcal{K}$-representations admit a kind of simple "calculus" that allows to convert $\mathcal{K}$-r.'s of operands participating in a standard convexity-preserving operation, like taking intersection, into a $\mathcal{K}$-r. of the result of this operation. "Richness" here means that $\mathcal{K}$

- contains a nonnegative ray $\mathbb{R}_+$;

- is closed w.r.t. taking finite direct products: whenever $\mathbf{K}_i \in \mathcal{K}$, $1 \leq i \leq m < \infty$, one has $\mathbf{K}_1 \times ... \times \mathbf{K}_m \in \mathcal{K}$;

- is closed w.r.t. passing from a cone to its dual: whenever $\mathbf{K} \in \mathcal{K}$, one has $\mathbf{K}^* \in \mathcal{K}$.

In particular, every one of the three aforementioned families of cones $\mathcal{LO}$, $\mathcal{CQO}$, $\mathcal{SDO}$ is rich.

We present here the most basic and most frequently used "calculus rules" (for more rules and for instructive examples of $\mathcal{LO}$-, $\mathcal{CQO}$-, and $\mathcal{SDO}$-representable sets, see [8]). Let $\mathcal{K}$ be a rich family of cones. Then *i*) [taking finite intersections] If the sets $Y_i \subset \mathbb{R}^n$ are $\mathcal{K}$-representable, $1 \leq i \leq m$, then so is their intersection $Y = \bigcap\limits_{i=1}^{m} Y_i$.

Indeed, if $Y_i = \{y \in \mathbb{R}^n : \exists u_i : A_i x + B_i u - b_i \in \mathbf{K}_i$ with $\mathbf{K}_i \in \mathcal{K}\}$, then

$$
\begin{aligned}
Y = \quad &\{y \in \mathbb{R}^n : \exists u = [u_1; ...; u_m] : \\
&[A_1; ...; A_m]y + \mathrm{Diag}\{B_1, ..., B_m\}[u_1; ...; u_m] - [b_1; ...; b_m] \\
&\in \mathbf{K} := \mathbf{K}_1 \times ... \times \mathbf{K}_m\},
\end{aligned}
$$

and $\mathbf{K} \in \mathcal{K}$, since $\mathcal{K}$ is closed w.r.t. taking finite direct products.

ii) [taking finite direct products] If the sets $Y_i \subset \mathbb{R}^{n_i}$ are $\mathcal{K}$-representable, $1 \leq i \leq m$, then so is their direct product $Y = Y_1 \times ... \times Y_m$.

Indeed, if $Y_i = \{y \in \mathbb{R}^n : \exists u_i : A_i x + B_i u - b_i \in \mathbf{K}_i$ with $\mathbf{K}_i \in \mathcal{K}\}$, then

$$
\begin{aligned}
Y = \quad &\{y = [y_1; ...; y_m] \in \mathbb{R}^{n_1 + ... + n_m} : \exists u = [u_1; ...; u_m] : \\
&\mathrm{Diag}\{A_1, ..., A_m\}y + \mathrm{Diag}\{B_1, ..., B_m\}[u_1; ...; u_m] - [b_1; ...; b_m] \\
&\in \mathbf{K} := \mathbf{K}_1 \times ... \times \mathbf{K}_m\},
\end{aligned}
$$

and, as above, $\mathbf{K} \in \mathcal{K}$.

iii) [taking inverse affine images] Let $Y \subset \mathbb{R}^n$ be $\mathcal{K}$-representable, let $z \mapsto Pz + p : \mathbb{R}^N \to \mathbb{R}^n$ be an affine mapping. Then the inverse affine image $Z = \{z : Pz + p \in Y\}$ of Y under this mapping is $\mathcal{K}$-representable.

Indeed, if $Y = \{y \in \mathbb{R}^n : \exists u : Ay + Bu - b \in \mathbf{K}\}$ with $\mathbf{K} \in \mathcal{K}$, then

$$
Z = \{z \in \mathbb{R}^N : \exists u : \underbrace{A[Pz + p] + Bu - b}_{\equiv \tilde{A}z + Bu - \tilde{b}} \in \mathbf{K}\}.
$$

iv) [taking affine images] If a set $Y \subset \mathbb{R}^n$ is $\mathcal{K}$-representable and $y \mapsto z = Py + p : \mathbb{R}^n \to \mathbb{R}^m$ is an affine mapping, then the image $Z = \{z = Py + p : y \in Y\}$ of Y under the mapping is $\mathcal{K}$-representable.

Indeed, if $Y = \{y \in \mathbb{R}^n : \exists u : Au + Bu - b \in \mathbf{K}\}$, then

$$
Z = \{z \in \mathbb{R}^m : \exists [y; u] : \underbrace{\begin{bmatrix} Py + p - z \\ -Py - p + z \\ Ay + Bu - b \end{bmatrix}}_{\equiv \tilde{A}z + \tilde{B}[y;u] - \tilde{b}} \in \mathbf{K}_+ := \mathbb{R}_+^m \times \mathbb{R}_+^m \times \mathbf{K}\},
$$

and the cone $\mathbf{K}_+$ belongs to $\mathcal{K}$ as the direct product of several nonnegative rays (every one of them belongs to $\mathcal{K}$) and the cone $\mathbf{K} \in \mathcal{K}$.

Note that the above "calculus rules" are "completely algorithmic" — a $\mathcal{K}$-r. of the result of an operation is readily given by $\mathcal{K}$-r.'s of the operands.

A.2.4.3 Conic representation of functions

By definition, the *epigraph* of a function $f(y) : \mathbb{R}^n \to \mathbb{R} \cup \{+\infty\}$ is the set

$$
\mathrm{Epi}\{f\} = \{[y; t] \in \mathbb{R}^n \times \mathbb{R} : t \geq f(y)\}.
$$

Note that a function is convex if and only if its epigraph is so.

Let $\mathcal{K}$ be a family of regular cones. A function f is called $\mathcal{K}$-representable, if its epigraph is so:

$$\text{Epi}\{f\} := \{[y,t] : \exists u : Ay + ta + Bu - b \in \mathbf{K}\} \qquad (\text{A.2.13})$$

with $\mathbf{K} \in \mathcal{K}$. A $\mathcal{K}$-representation ($\mathcal{K}$-r. for short) of a function is, by definition, a $\mathcal{K}$-r. of its epigraph. Since $\mathcal{K}$-representable sets always are convex, so are $\mathcal{K}$-representable functions.

Examples of $\mathcal{K}$-r.'s of functions:

- the function $f(y) = |y| : \mathbb{R} \to \mathbb{R}$ is $\mathcal{LO}$-representable:
$$\{[y;t] : t \geq |y|\} = \{[y;t] : A[y;t] := [t - y; t + y] \in \mathbb{R}_+^2\};$$

- the function $f(y) = \|y\|_2 : \mathbb{R}^n \to \mathbb{R}$ is $\mathcal{CQO}$-representable:
$$\{[y;t] \in \mathbb{R}^{n+1} : t \geq \|y\|_2\} = \{[y;t] \in \mathbf{L}^{n+1}\};$$

- the function $f(y) = \lambda_{\max}(y) : \mathbf{S}^n \to \mathbb{R}$ (the maximal eigenvalue of a symmetric matrix y) is $\mathcal{SDO}$-representable:
$$\{[y;t] \in \mathbf{S}^n \times \mathbb{R} : t \geq \lambda_{\max}(y)\} = \{[y;t] : \mathcal{A}[y;t] := tI_n - y \in \mathbf{S}_+^n\}.$$

Observe that a $\mathcal{K}$-r. (A.2.13) of a function f induces $\mathcal{K}$-r.'s of its *level sets* $\{y : f(y) \leq c\}$:

$$\{y : f(y) \leq c\} = \{y : \exists u : Ay + Bu - [b - ca] \in \mathbf{K}\}.$$

This explains the importance of $\mathcal{K}$-representations of functions: usually, the feasible set Y of a convex problem (A.2.10) is given by a system of convex constraints:

$$Y = \{y : f_i(y) \leq 0, \, 1 \leq i \leq m\}.$$

If now all functions f_i are $\mathcal{K}$-representable, then, by the above observation and by the "calculus rule" related to intersections, Y is $\mathcal{K}$-representable as well, and a $\mathcal{K}$-r. of Y is readily given by $\mathcal{K}$-r.'s of f_i.

$\mathcal{K}$-representable functions admit simple calculus, which is similar to the one of $\mathcal{K}$-representable sets, and is equally algorithmic; for details and instructive examples, see [8].

A.3 EFFICIENT SOLVABILITY OF CONVEX PROGRAMMING

The goal of this section is to explain the precise meaning of the informal (and in fact slightly exaggerated) claim,

An optimization problem with convex efficiently computable objective and constraints is efficiently solvable.

that on many different occasions was reiterated in the main body of the book. Our exposition follows the one from [8, chapter 5].

A.3.1 Generic Convex Programs and Efficient Solution Algorithms

In what follows, it is convenient to represent optimization programs as

$$(p): \quad \mathrm{Opt}(p) = \min_x \left\{ p_0(x) : x \in X(p) \subset \mathbb{R}^{n(p)} \right\},$$

where $p_0(\cdot)$ and $X(p)$ are the objective, which we assume to be a real-valued function on $\mathbb{R}^{n(p)}$, and the feasible set of program (p), respectively, and $n(p)$ is the dimension of the decision vector.

A.3.1.1 A generic optimization problem

A *generic optimization program* $\mathcal{P}$ is a collection of optimization programs (p) ("instances of $\mathcal{P}$") such that every instance of $\mathcal{P}$ is identified by a finite-dimensional *data vector* $\mathrm{data}(p)$; the dimension of this vector is called the *size* $\mathrm{Size}(p)$ of the instance:

$$\mathrm{Size}(p) = \dim \mathrm{data}(p).$$

For example, *Linear Optimization* is a generic optimization problem $\mathcal{LO}$ with instances of the form

$$(p): \quad \min_x \left\{ c_p^T x : x \in X(p) := \{x : A_p x - b_p \geq 0\} \right\}$$
$$[A_p : m(p) \times n(p)],$$

where $m(p), n(p), c_p, A_p, b_p$ can be arbitrary. The data of an instance can be identified with the vector

$$\mathrm{data}(p) = [m(p); n(p); c_p; b_p; A_p^1; ...; A_p^{n(p)}],$$

where A_p^i is i-th column in A_p.

Similarly, *Conic Quadratic Optimization* is a generic optimization problem $\mathcal{CQO}$ with instances

$$(p): \quad \min_x \left\{ c_p^T x : x \in X(p) \right\},$$
$$X(p) := \{x : \|A_{pi} x - b_{pi}\|_2 \leq e_{pi}^T x - d_{pi}, \ 1 \leq i \leq m(p)\}$$
$$[A_{pi} : k_i(p) \times n(p)].$$

The data of an instance can be defined as the vector obtained by listing, in a fixed order, the dimensions $m(p)$, $n(p)$, $\{k_i(p)\}_{i=1}^{m(p)}$ and the entries of the reals d_{pi}, vectors c_p, b_{pi}, e_{pi} and the matrices A_{pi}^ℓ.

Finally, *Semidefinite Optimization* is a generic optimization problem $\mathcal{SDO}$ with instances of the form

$$(p): \quad \min_x \left\{ c_p^T x : x \in X(p) := \{x : A_p^i(x) \succeq 0, 1 \leq i \leq m(p)\} \right\}$$
$$A_p^i(x) = A_{pi}^0 + x_1 A_{pi}^1 + ... + x_{n(p)} A_{pi}^{n(p)},$$

where A_{pi}^ℓ are symmetric matrices of size $k_i(p)$. The data of an instance can be defined in the same fashion as in the case of $\mathcal{CQO}$.

A.3.1.2 Approximate solutions

In order to quantify the quality of a candidate solution of an instance (p) of a generic problem $\mathcal{P}$, we assume that $\mathcal{P}$ is equipped with an *infeasibility measure* $\text{Infeas}_{\mathcal{P}}(p, x)$ — a real-valued nonnegative function of an instance $(p) \in \mathcal{P}$ and a candidate solution $x \in \mathbb{R}^{n(p)}$ to the instance such that $x \in X(p)$ if and only if $\text{Infeas}_{\mathcal{P}}(p, x) = 0$.

Given an infeasibility measure and a tolerance $\epsilon > 0$, we define an ϵ *solution* to an instance $(p) \in \mathcal{P}$ as a point $x_{\epsilon} \in \mathbb{R}^{n(p)}$ such that

$$p_0(x_{\epsilon}) - \text{Opt}(p) \leq \epsilon \ \& \ \text{Infeas}_{\mathcal{P}}(p, x_{\epsilon}) \leq \epsilon.$$

For example, a natural infeasibility measure for a generic optimization problem $\mathcal{P}$ with instances of the form

$$(p): \quad \min_{x} \left\{ p_0(x) : x \in X(p) := \{x : p_i(x) \leq 0, \ 1 \leq i \leq m(p)\} \right\} \quad (\text{A.3.1})$$

is

$$\text{Infeas}_{\mathcal{P}}(p, x) = \max \left[0, p_1(x), p_2(x), ..., p_{m(p)}(x) \right]; \quad (\text{A.3.2})$$

this recipe, in particular, can be applied to the generic problems $\mathcal{LO}$ and $\mathcal{CQO}$. A natural infeasibility measure for $\mathcal{SDO}$ is

$$\text{Infeas}_{\mathcal{SDO}}(p, x) = \min \left\{ t \geq 0 : A_p^i(x) + t I_{k_i(p)} \succeq 0, \ 1 \leq i \leq m(p) \right\}.$$

A.3.1.3 Convex generic optimization problems

A generic problem $\mathcal{P}$ is called *convex*, if for every instance (p) of the problem, $p_0(x)$ and $\text{Infeas}_{\mathcal{P}}(p, x)$ are convex functions of $x \in \mathbb{R}^{n(p)}$. Note that then $X(p) = \{x \in \mathbb{R}^{n(p)} : \text{Infeas}_{\mathcal{P}}(p, x) \leq 0\}$ is a convex set for every $(p) \in \mathcal{P}$.

For example, $\mathcal{LO}$, $\mathcal{CQO}$ and $\mathcal{SDO}$ with the just defined infeasibility measures are generic convex programs. The same is true for generic problems with instances (A.3.1) and infeasibility measure (A.3.2), provided that all instances are convex programs, i.e., $p_0(x), p_1(x), ..., p_{m(p)}(x)$ are restricted to be real-valued *convex* functions on $\mathbb{R}^{n(p)}$.

A.3.1.4 A solution algorithm

A solution algorithm $\mathcal{B}$ for a generic problem $\mathcal{P}$ is a code for the Real Arithmetic Computer — an idealized computer capable to store real numbers and to carry out the operations of Real Arithmetics (the four arithmetic operations, comparisons and computing elementary functions like $\sqrt{\cdot}$, $\exp\{\cdot\}$, $\sin(\cdot)$) with real arguments. Given on input the data vectors $\text{data}(p)$ of an instance $(p) \in \mathcal{P}$ and a tolerance $\epsilon > 0$ and executing on this input the code $\mathcal{B}$, the computer should eventually stop and output

— either a vector $x_{\epsilon} \in \mathbb{R}^{n(p)}$ that must be an ϵ solution to (p),

— or a correct statement "(p) is infeasible"/"(p) is not below bounded."

The *complexity* of the generic problem $\mathcal{P}$ with respect to a solution algorithm $\mathcal{B}$ is quantified by the function $\mathrm{Compl}_{\mathcal{P}}(p, \epsilon)$; the value of this function at a pair $(p) \in \mathcal{P}$, $\epsilon > 0$ is exactly the number of elementary operations of the Real Arithmetic Computer in the course of executing the code $\mathcal{B}$ on the input $(\mathrm{data}(p), \epsilon)$.

A.3.1.5 Polynomial time solution algorithms

A solution algorithm for a generic problem $\mathcal{P}$ is called *polynomial time* ("efficient"), if the complexity of solving instances of $\mathcal{P}$ within (an arbitrary) accuracy $\epsilon > 0$ is bounded by a polynomial in the size of the instance and the *number of accuracy digits* $\mathrm{Digits}(p, \epsilon)$ in an ϵ solution:

$$\mathrm{Compl}_{\mathcal{P}}(p, \epsilon) \leq \chi \left(\mathrm{Size}(p) \mathrm{Digits}(p, \epsilon) \right)^{\chi},$$
$$\mathrm{Size}(p) = \dim \mathrm{data}(p), \ \mathrm{Digits}(p, \epsilon) = \ln \left(\frac{\mathrm{Size}(p) + \|\mathrm{data}(p)\|_1 + \epsilon^2}{\epsilon} \right);$$

from now on, χ stands for various "characteristic constants" (not necessarily identical to each other) of the generic problem in question, i.e., for positive quantities depending on $\mathcal{P}$ and independent of $(p) \in \mathcal{P}$ and $\epsilon > 0$. Note also that while the "strange" numerator in the fraction participating in the definition of Digits arises by technical reasons, the number of accuracy digits for small $\epsilon > 0$ becomes independent of this numerator and close to $\ln(1/\epsilon)$.

A generic problem $\mathcal{P}$ is called *polynomially solvable* ("computationally tractable"), if it admits a polynomial time solution algorithm.

A.3.2 Polynomial Solvability of Generic Convex Programming Problems

The main fact about generic convex problems that underlies the remarkable role played by these problems in Optimization is that *under minor non-restrictive technical assumptions, a generic convex problem*, in contrast to typical generic non-convex problems, *is computationally tractable.*

The just mentioned "minor non-restrictive technical assumptions" are those of *polynomial computability*, *polynomial growth*, and *polynomial boundedness of feasible sets.*

A.3.2.1 Polynomial computability

A generic convex optimization problem $\mathcal{P}$ is called *polynomially computable*, if it can be equipped with two codes, $\mathcal{O}$ and $\mathcal{C}$, for the Real Arithmetic Computer, such that:

• for every instance $(p) \in \mathcal{P}$ and any candidate solution $x \in \mathbb{R}^{n(p)}$ to the instance, executing $\mathcal{O}$ on the input $(\mathrm{data}(p), x)$ takes a polynomial in $\mathrm{Size}(p)$ number of elementary operations and produces a value and a subgradient of the objective $p_0(\cdot)$ at the point x;

  • for every instance $(p) \in \mathcal{P}$, any candidate solution $x \in \mathbb{R}^{n(p)}$ to the instance and any $\epsilon > 0$, executing $\mathcal{C}$ on the input $(\text{data}(p), x, \epsilon)$ takes a polynomial in Size(p) and Digits(p, ϵ) number of elementary operations and results
— either in a correct claim that Infeas$_\mathcal{P}(p, x) \leq \epsilon$,
— or in a correct claim that Infeas$_\mathcal{P}(p, x) > \epsilon$ and in computing a linear form $e \in \mathbb{R}^{n(p)}$ that separates x and the set $\{y : \text{Infeas}_\mathcal{P}(p, y) \leq \epsilon\}$, so that

$$\forall(y, \text{Infeas}_\mathcal{P}(p, y) \leq \epsilon) : e^T y < e^T x.$$

Consider, for example, a generic convex program $\mathcal{P}$ with instances of the form (A.3.1) and the infeasibility measure (A.3.2) and assume that the functions $p_0(\cdot), p_1(\cdot), ..., p_{m(p)}(\cdot)$ are real-valued and convex for all instances of $\mathcal{P}$. Assume, moreover, that the objective and the constraints of instances are efficiently computable, meaning that there exists a code $\mathcal{CO}$ for the Real Arithmetic Computer, which being executed on an input of the form $(\text{data}(p), x \in \mathbb{R}^{n(p)})$ computes in a polynomial in Size(p) number of elementary operations the values and subgradients of $p_0(\cdot), p_1(\cdot), ..., p_{m(p)}(\cdot)$ at x. In this case, $\mathcal{P}$ is polynomially computable. Indeed, the code $\mathcal{O}$ allowing to compute in polynomial time the value and a subgradient of the objective at a given candidate solution is readily given by $\mathcal{CO}$. In order to build $\mathcal{C}$, let us execute $\mathcal{CO}$ on an input $(\text{data}(p), x)$ and compare the quantities $p_i(x)$, $1 \leq i \leq m(p)$, with ϵ. If $p_i(x) \leq \epsilon$, $1 \leq i \leq m(p)$, we output the correct claim that Infeas$_\mathcal{P}(p, x) \leq \epsilon$, otherwise we output a correct claim that Infeas$_\mathcal{P}(p, x) > \epsilon$ and return, as e, a subgradient, taken at x, of a constraint $p_{i(x)}(\cdot)$, where $i(x) \in \{1, 2, ..., m(p)\}$ is such that $p_{i(x)}(x) > \epsilon$.

By the reasons outlined above, the generic problems $\mathcal{LO}$ and $\mathcal{CQO}$ of Linear and Conic Quadratic Optimization are polynomially computable. The same is true for Semidefinite Optimization, see [8, chapter 5].

A.3.2.2 Polynomial growth

We say that $\mathcal{P}$ is of *polynomial growth*, if for properly chosen $\chi > 0$ one has

$$\forall((p) \in \mathcal{P}, x \in \mathbb{R}^{n(p)}) :$$
$$\max\left[|p_0(x)|, \text{Infeas}_\mathcal{P}(p, x)\right] \leq \chi \left(\text{Size}(p) + \|\text{data}(p)\|_1\right)^{\chi \text{Size}^\chi(p)}.$$

For example, the generic problems of Linear, Conic Quadratic and Semidefinite Optimization clearly are with polynomial growth.

A.3.2.3 Polynomial boundedness of feasible sets

We say that $\mathcal{P}$ is with polynomially bounded feasible sets, if for properly chosen $\chi > 0$ one has

$$\forall((p) \in \mathcal{P}) : x \in X(p) \Rightarrow \|x\|_\infty \leq \chi \left(\text{Size}(p) + \|\text{data}(p)\|_1\right)^{\chi \text{Size}^\chi(p)}.$$

While the generic convex problems $\mathcal{LO}$, $\mathcal{CQO}$, and $\mathcal{SDO}$ are polynomially computable and with polynomial growth, neither one of these problems (same as neither one of other natural generic convex problems) "as it is" possesses polynomially bounded feasible sets. We, however, can enforce the latter property by passing from a generic problem $\mathcal{P}$ to its "bounded version" $\mathcal{P}_b$ as follows: the instances of $\mathcal{P}_b$ are the instances (p) of $\mathcal{P}$ augmented by bounds on the variables; thus, an instance $(p_+) = (p, R)$ of $\mathcal{P}_b$ is of the form

$$(p, R): \quad \min_x \left\{ p_0(x) : x \in X(p, R) = X(p) \cap \{x \in \mathbb{R}^{n(p)} : \|x\|_\infty \leq R\} \right\}$$

where (p) is an instance of $\mathcal{P}$ and $R > 0$. The data of (p, R) is the data of (p) augmented by R, and

$$\text{Infeas}_{\mathcal{P}_b}((p, R), x) = \text{Infeas}_{\mathcal{P}}(p, x) + \max[\|x\|_\infty - R, 0].$$

Note that $\mathcal{P}_b$ inherits from $\mathcal{P}$ the properties of polynomial computability and/or polynomial growth, if any, and always is with polynomially bounded feasible sets. Note also that R can be really large, like $R = 10^{100}$, which makes the "expressive abilities" of $\mathcal{P}_b$, for all practical purposes, as strong as those of $\mathcal{P}$. Finally, we remark that the "bounded versions" of $\mathcal{LO}$, $\mathcal{CQO}$, and $\mathcal{SDO}$ are sub-problems of the original generic problems.

A.3.2.4 Main result

The main result on computational tractability of Convex Programming is the following:

Theorem A.3.1. Let $\mathcal{P}$ be a polynomially computable generic convex program with a polynomial growth that possesses polynomially bounded feasible sets. Then $\mathcal{P}$ is polynomially solvable.

As a matter of fact, "in real life" the only restrictive assumption in Theorem A.3.1 is the one of polynomial computability. This is the assumption that is usually violated when speaking about *semi-infinite* convex programs like the RCs of uncertain conic problems

$$\min_x \left\{ c_p^T x : x \in X(p) = \{x \in \mathbb{R}^{n(p)} : A_{p\zeta} x + a_{p\zeta} \in \mathbf{K} \, \forall (\zeta \in \mathcal{Z})\} \right\}.$$

associated with simple *non-polyhedral* cones $\mathbf{K}$. Indeed, when $\mathbf{K}$ is, say, a Lorentz cone, so that

$$X(p) = \{x : \|B_{p\zeta} x + b_{p\zeta}\|_2 \leq c_{p\zeta}^T x + d_{p\zeta} \, \forall (\zeta \in \mathcal{Z})\},$$

to compute the natural infeasibility measure

$$\min \left\{ t \geq 0 : \|B_{p\zeta} x + b_{p\zeta}\|_2 \leq c_{p\zeta}^T x + d_{p\zeta} + t \, \forall (\zeta \in \mathcal{Z}) \right\}$$

at a given candidate solution x means to *maximize* the function $f_x(\zeta) = \|B_{p\zeta} x + b_{p\zeta}\|_2 - c_{p\zeta}^T x - d_{p\zeta}$ over the uncertainty set $\mathcal{Z}$. When the uncertain data are affinely parameterized by ζ, this requires a *maximization of a nonlinear convex function $f_x(\zeta)$ over $\zeta \in \mathcal{Z}$*, and this problem can be (and generically is) computationally intractable, even when $\mathcal{Z}$ is a simple convex set. It becomes also clear why the outlined difficulty does not occur in uncertain LO with the data affinely parameterized by ζ: here $f_x(\zeta)$ is an affine function of ζ, and as such can be efficiently maximized over $\mathcal{Z}$, provided the latter set is convex and "not too complicated."

A.3.3 "What is Inside": Efficient Black-Box-Oriented Algorithms in Convex Optimization

Theorem A.3.1 is a direct consequence of a fact that is instructive in its own right and has to do with "black-box-oriented" Convex Optimization, specifically, with solving an optimization problem

$$\min_{x \in X} f(x), \qquad (A.3.3)$$

where

- $X \subset \mathbb{R}^n$ is a solid (a convex compact set with a nonempty interior) known to belong to a given Euclidean ball $E_0 = \{x : \|x\|_2 \le R\}$ and represented by a *Separation oracle* — a routine that, given on input a point $x \in \mathbb{R}^n$, reports whether $x \in X$, and if it is not the case, returns a vector $e \ne 0$ such that

$$e^T x \ge \max_{y \in X} e^T y;$$

- f is a convex real-valued function on $\mathbb{R}^n$ represented by a *First Order oracle* that, given on input a point $x \in \mathbb{R}^n$, returns the value and a subgradient of f at x.

In addition, we assume that we know in advance an $r > 0$ such that X contains a Euclidean ball of the radius r (the center of this ball can be unknown).

Theorem A.3.1 is a straightforward consequence of the following important fact:

Theorem A.3.2. [8, Theorem 5.2.1] There exists a Real Arithmetic algorithm (the Ellipsoid method) that, as applied to (A.3.3), the required accuracy being $\epsilon > 0$, finds a feasible ϵ solution x_ϵ to the problem (i.e., $x_\epsilon \in X$ and $f(x_\epsilon) - \min_X f \le \epsilon$) after at most

$$N(\epsilon) = \mathrm{Ceil}\left(2n^2\left[\ln\left(\tfrac{R}{r}\right) + \ln\left(\tfrac{\epsilon + \mathrm{Var}_R(f)}{\epsilon}\right)\right]\right) + 1$$
$$\mathrm{Var}_R(f) = \max_{\|x\|_2 \le R} f(x) - \min_{\|x\|_2 \le R} f(x)$$

steps, with a step reducing to a single call to the Separation and to the First Order oracles accompanied by $O(1)n^2$ additional arithmetic operations to process the answers of the oracles. Here $O(1)$ is an absolute constant.

Recently, the Ellipsoid method was equipped with "on line" accuracy certificates, which yield a slightly strengthened version of the above theorem, namely, as follows:

Theorem A.3.3. [86] Consider problem (A.3.3) and assume that
- $X \in \mathbb{R}^n$ is a solid contained in the centered at the origin Euclidean ball E_0

of a known in advance radius R and given by a Separation oracle that, given on input a point $x \in \mathbb{R}^n$, reports whether $x \in \text{int}X$, and if it is not the case, returns a nonzero e such that $e^T x \geq \max_{y \in X} e^T y$;

• $f : \text{int}X \to \mathbb{R}$ is a convex function represented by a First Order oracle that, given on input a point $x \in \text{int}X$, reports the value $f(x)$ and a subgradient $f'(x)$ of f at x. In addition, assume that f is semibounded on X, meaning that $V_X(f) \equiv \sup_{x,y \in \text{int}X} (y - x)^T f'(x) < \infty$.

There exists an explicit Real Arithmetic algorithm that, given on input a desired accuracy $\epsilon > 0$, terminates with a strictly feasible ϵ-solution x_ϵ to the problem $(x_\epsilon \in \text{int}X, \ f(x_\epsilon) - \inf_{x \in \text{int}X} f(x) \leq \epsilon)$ after at most

$$N(\epsilon) = O(1) \left(n^2 \left[\ln \left(\frac{nR}{r} \right) + \ln \left(\frac{\epsilon + V_X(f)}{\epsilon} \right) \right] \right)$$

steps, with a step reducing to a single call to the Separation and to the First Order oracles accompanied by $O(1)n^2$ additional arithmetic operations to process the answers of the oracles. Here r is the supremum of the radii of Euclidean balls contained in X, and $O(1)$'s are absolute constants.

The progress, as compared to Theorem A.3.1, is that now we do not need a priori knowledge of $r > 0$ such that X contains a Euclidean ball of radius r, f is allowed to be undefined outside of $\text{int}X$ and the role of $\text{Var}_R(f)$ (the quantity that now can be $+\infty$) is played by $V_X(f) \leq \sup_{\text{int}X} f - \inf_{\text{int}X} f$.

Appendix B

Some Auxiliary Proofs

B.1 PROOFS FOR CHAPTER 4

B.1.1 Proposition 4.2.2

1^0. Let us first verify that $Z_\epsilon \subset Z_*$, where Z_* is the feasible set of (4.0.1). Observe, first, that $Z_\epsilon^o \subset Z_*$. Indeed, let $z = [z_0; w] \in Z_\epsilon^o$, and let P be the distribution of ζ. Since $z \in Z_\epsilon^o$, there exists $\alpha > 0$ such that $\alpha z_0 + \Phi(\alpha w) \leq \ln(\epsilon)$. We have

$$\text{Prob}\{\zeta : z_0 + w^T \zeta > 0\} \leq \mathbf{E}\{\exp\{\alpha z_0 + \alpha w^T \zeta\}\} \leq \exp\{\alpha z_0 + \Phi(\alpha w)\} \leq \epsilon$$

(the second $\leq$ is given by (4.2.3)), as claimed. Since Z_* clearly is closed, we conclude that $Z_\epsilon = \text{cl}\, Z_\epsilon^o \subset Z_*$ as well.

2^0. Now let us prove that Z_ϵ is exactly the solution set of the convex inequality (4.2.6). We need the following

Lemma B.1.1. Let $H(z) : \mathbb{R}^N \to \mathbb{R} \cup \{+\infty\}$ be a lower semicontinuous convex function and a be a real. Assume that $H(0) > a$, $0 \in \text{int}\,\text{Dom}\,H$ and the set $\{z : H(z) < a\}$ is nonempty. Consider the sets

$$\mathcal{H}^o = \{z : \exists \beta > 0 : H(\beta^{-1}z) \leq a\}, \ \mathcal{H} = \text{cl}\, \mathcal{H}^o.$$

Then the function $G(z) = \inf_{\beta > 0} \left[\beta H(\beta^{-1}z) - \beta a\right]$ is convex and finite everywhere,

$$\mathcal{H} = \{z : G(z) \leq 0\} \tag{B.1.1}$$

and $\mathcal{H}$ is a nonempty closed convex cone.

Lemma B.1.1 $\Rightarrow$ Proposition 4.2.2: Setting $H(z_0, z_1, ..., z_L) = z_0 + \Phi(z_1, ..., z_L)$, $a = \ln(\epsilon)$, we clearly satisfy the premise in Lemma B.1.1; with this setup, the conclusion of Lemma clearly completes the proof of Proposition 4.2.2.

Proof of Lemma B.1.1: 0^0. H is convex, whence the function $\beta H(\beta^{-1}z)$ is convex in $(\beta > 0, z)$. It follows that $G(z)$ is convex, provided that it is finite everywhere. The latter indeed is the case. To see it, note that since $0 \in \text{int}\,\text{Dom}\,H$, $\beta H(\beta^{-1}z)$ is finite whenever β is large enough, so that $G(z) < \infty$ for every z. Due to the same inclusion $0 \in \text{int}\,H$, we have $H(u) \geq H(0) + g^T u$ for certain g and all u, whence $\beta H(\beta^{-1}z) - \beta a \geq \beta(H(0) - a) + g^T z$, so that $G(z) > -\infty$ due to $H(0) > a$. Thus, $G(z)$ is a real-valued convex function, as claimed.

1^0. Let $\mathcal{G} = \{z : H(z) \leq a\}$. Then $\mathcal{G}$ is a nonempty closed and convex set, and $\mathcal{H}^o = \{z : \exists \alpha > 0 : \alpha z \in \mathcal{G}\}$, so that the set $\mathcal{H}^o$ is convex, nonempty and satisfies the relation $\alpha \mathcal{H}^o = \mathcal{H}^o$ for all $\alpha > 0$. It follows that $\mathcal{H} = \mathrm{cl}\, \mathcal{H}^o$ is a nonempty closed convex cone. All we need to prove is that $\mathcal{H}$ admits the representation (B.1.1). Let $\bar{\mathcal{H}}$ be the right hand side set in (B.1.1); note that this set clearly contains $\mathcal{H}^o$.

2^0. We first prove that $\bar{\mathcal{H}}$ contains $\mathcal{H}$. To this end it suffices to verify that if $\beta_i > 0$ and z_i are such that $H(\beta_i^{-1}z_i) \le a$ for all i and $z_i \to \bar{z}$ as $i \to \infty$, then $\bar{z} \in \bar{\mathcal{H}}$. Indeed, passing to a subsequence, we may assume that as $i \to \infty$, one of the following 3 cases takes place:

1) $\beta_i \to \bar{\beta} \in (0, \infty)$, 2) $\beta_i \to +\infty$, 3) $\beta_i \to +0$.

In case 1) we have $\beta_i^{-1}z_i \to \bar{\beta}^{-1}\bar{z}$ and $H(\beta_i^{-1}z_i) \le a$; since H is lower semicontinuous, it follows that $H(\bar{\beta}^{-1}\bar{z}) \le a$, and since $\bar{\beta} > 0$, we get $\bar{z} \in \mathcal{H}^o \subset \bar{\mathcal{H}}$, as required.

Case 2) is impossible, since here $a \ge H(\beta_i^{-1}z_i) \to H(0)$ as $i \to \infty$ due to the continuity of H at $0 \in \mathrm{int}\,\mathrm{Dom}\, H$, while $H(0) > a$ by assumption.

In case 3), $\beta_i^{-1}z_i \in \mathcal{G}$, $\beta_i^{-1} \to +\infty$ and $z_i \to \bar{z}$ as $i \to \infty$, whence $\bar{z}$ is a recessive direction of the nonempty closed convex set $\mathcal{G}$. Let $z_0 \in \mathcal{G}$. Since $H(0) > a$, we have $z_0 \ne 0$, and since $0 \in \mathrm{int}\,\mathrm{Dom}\, H$, we can find $\lambda \in (0, 1)$ and $w \in \mathrm{Dom}\, H$ such that $\lambda z_0 + (1 - \lambda)w = 0$. Since $z_0 + \mathbb{R}_+\bar{z} \in \mathcal{G}$ and H is convex, H is bounded above on the convex hull of the ray $z_0 + \mathbb{R}_+\bar{z}$ and w, and this convex hull, by construction, contains the ray $\mathbb{R}_+\bar{z}$. We conclude that $H(\beta^{-1}\bar{z})$ is a bounded above function of $\beta > 0$, whence $\lim_{\beta \to +0}\left[\beta H(\beta^{-1}\bar{z}) - \beta a\right] \le 0$, and $\bar{z} \in \bar{\mathcal{H}}$, as claimed.

3^0. It remains to prove that $\bar{\mathcal{H}} \subset \mathcal{H}$. Let $z \in \bar{\mathcal{H}}$, so that
$$\lim_{i \to \infty} \beta_i[H(\beta_i^{-1}z) - a] \le 0$$
for certain sequence $\{\beta_i > 0\}$; we should prove that $z \in \mathrm{cl}\,\mathcal{H}^o$. Passing to a subsequence, we can assume that as $i \to \infty$, one of the above 3 cases 1), 2), 3) takes place.

In case 1), we, same as above, have $H(\bar{\beta}^{-1}z) \le a$, whence $z \in \mathcal{H}^o$, as required. Case 2 is impossible by the same reasons as above. In case 3) H clearly is bounded above on the ray $\mathbb{R}_+z$: $H(\alpha z) \le \bar{a} < \infty$ for certain $\bar{a}$. Now let z_0 be such that $H(z_0) < a$; then with properly chosen $\lambda \in (0, 1)$ we have $H(\lambda z_0 + (1 - \lambda)\alpha z) \le \lambda H(z_0) + (1 - \lambda)\bar{a} \le a$ for all $\alpha \ge 0$, whence the points $z_i = [(1 - \lambda)i]^{-1}[\lambda z_0 + (1 - \lambda)iz]$ are in $\mathcal{H}^o$ due to $H([(1 - \lambda)i]z_i) \le a$. As $i \to \infty$, we have $z_i \to z$, meaning that $z \in \mathrm{cl}\,\mathcal{H}^o$. $\qquad\qquad\square$

B.1.2 Proposition 4.2.3

Assume, on the contrary to what should be proved, that there exists $c \in R$ and a sequence u^i, $\|u^i\| \to \infty$, $i \to \infty$, such that $\phi(u^i) \le c$ $\forall i$. Since A has trivial kernel, the sequence $Au^i + a$ is unbounded, so that we can find w such that the sequence of reals $w^T(Au^i + a)$ is above unbounded. On the other hand, from (4.2.7) it follows that $w^T(Au + a) \le \Phi(w) + \phi(u)$ for all w, u, whence $w^T(Au^i + a) \le \Phi(w) + \phi(u^i) \le$

$\Phi(w) + c$, that is, the sequence $w^T(Au^i + a)$ is above bounded, which is the desired contradiction.   □

B.1.3 Theorem 4.2.5

Theorem 4.2.5 is an immediate corollary of the following statement:

Theorem B.1.2. Let $\Psi(z) : \mathbb{R}^n \to \mathbb{R}$ be a convex function and $\psi(u) : \mathbb{R}^m \to \mathbb{R} \cup \{+\infty\}$ be a lower semicontinuous convex function with bounded level sets such that

$$\Psi(z) = \sup_u \left\{ z^T(Bu + b) - \psi(u) \right\}. \tag{B.1.2}$$

Let, further, real ρ and a direction $e \in \mathbb{R}^L$ be such that

$$\rho < \Psi(0) \tag{B.1.3}$$

and

$$\lim_{t \to \infty} \Psi(z + te) < \rho \ \forall z \in \mathbb{R}^n. \tag{B.1.4}$$

Let us set $Z_o^\rho = \{z : \exists \alpha > 0 : \Psi(\alpha z) \leq \rho\}$, $Z^\rho = \text{cl}\, Z_o^\rho$. Then the set $\mathcal{U}^\rho = \{u : \psi(u) \leq -\rho\}$ is a nonempty convex compact set and

$$z \in Z^\rho \Leftrightarrow z^T(Bu + b) \leq 0 \ \forall u \in \mathcal{U}^\rho. \tag{B.1.5}$$

Theorem B.1.2 ⇒ Theorem 4.2.5. Let us set $\Psi(z_0, z_1, ..., z_L) = z_0 + \Phi([z_1; ...; z_L])$, $\psi(\cdot) \equiv \phi(\cdot)$, $Bu + b = [1; Au + a]$, $\rho = \ln(\epsilon)$, $e = [-1; 0; ...; 0]$. These data clearly satisfy the premise in Theorem B.1.2. It remains to note that $Z_o^\rho = Z_\epsilon^o$ (so that $Z^\rho = Z_\epsilon$), $\mathcal{U}^\rho = \mathcal{U}_\epsilon$ and $z^T(Bu + b) \equiv z_0 + [z_1; ...; z_L]^T(Au + a)$.

Proof of Theorem B.1.2.

1^0. First, let us verify that

$$\inf_u \psi(u) = -\Psi(0) \tag{B.1.6}$$

and extract from this relation that $\mathcal{U}^\rho$ is a nonempty compact convex set.

Indeed, by (B.1.2) we have

$$\Psi(0) = \sup_u \left\{ 0^T(Bu + b) - \phi(u) \right\} = -\inf_u \psi(u),$$

and (B.1.6) follows. Now, since $\rho < \Phi(0)$, (B.1.6) says that $-\rho > \inf_u \psi(u)$, so that the set $\mathcal{U}^\rho$ is nonempty. This set is convex, closed and bounded due to the fact that ψ is a convex lower semicontinuous function with bounded level sets.

2^0. The result of 1^0 states that $\mathcal{U}^\rho$ is a nonempty convex compact set, that is the first statement of Theorem B.1.2. To complete the proof, we need to justify the equivalence in (B.1.5), which is the goal of items 3^0 and 4^0 to follow.

3^0. We claim that whenever $z \in Z^\rho$, one has

$$z^T(Bu + b) \leq 0 \ \forall u \in \mathcal{U}^\rho. \tag{B.1.7}$$

Indeed, assuming, on the contrary, that $z^T(B\bar{u} + b) > 0$ for certain $\bar{u}$ with $\psi(\bar{u}) \leq -\rho$, observe that there exists a neighborhood U_z of z and $\delta > 0$ such that $[z']^T(B\bar{u} + b) > \delta$ whenever $z' \in U_z$. Consequently, for every $\alpha > 0$ and every $z' \in U_z$ we have $\Psi(\alpha z') \geq \alpha[z']^T(B\bar{u} + b) - \psi(\bar{u}) \geq \alpha\delta + \rho > \rho$, so that U_z does not intersect Z_o^ρ and therefore $z \notin Z^\rho$, which is a desired contradiction.

4^0. To complete the proof of Theorem B.1.2, it suffices to justify the following statement:

   (!) *If z satisfies (B.1.7), then $z \in Z^\rho$.*

To this end, let us fix z satisfying (B.1.7).

   $4^0.1.$ We claim that $e^T(Bu + b) \leq 0$ for all $u \in \mathrm{Dom}\,\psi$.
Indeed, assuming the opposite, there exists $\bar{u} \in \mathrm{Dom}\,\psi$ such that $e^T(B\bar{u} + b) > 0$, whence $\Psi(te) \geq te^T(B\bar{u} + b) - \psi(\bar{u}) \to \infty$ as $t \to \infty$, which is impossible due to (B.1.4).

   $4^0.2.$ Consider the case when z is such that $z^T(Bu + b) \leq 0$ for all $u \in \mathrm{Dom}\,\psi$. We claim that in this case $z + \delta e \in Z_o^\rho$ for all $\delta > 0$, whence, of course, $z \in Z^\rho$. Let us fix $\delta > 0$.

   $4^0.2.1)$ Let us first prove that $(z + \delta e)^T(Bu + b) < 0$ for every $u \in \mathcal{U}^\rho$. Indeed, assuming the opposite, there exists $\bar{u} \in \mathcal{U}^\rho$ with $(z + \delta e)^T(B\bar{u} + b) \geq 0$. Since $z^T(Bu + b) \leq 0$ and $e^T(Bu + b) \leq 0$ for all $u \in \mathrm{Dom}\,\psi$ (by assumption in $4^0.2$ and by $4^0.1$, respectively), we conclude that $z^T(B\bar{u} + b) = e^T(B\bar{u} + b) = 0$, whence $(z + te)^T(B\bar{u} + b) \geq 0$ for all $t \geq 0$, and therefore

$$\forall t > 0 : \Psi(z + te) \geq (z + te)^T(B\bar{u} + b) - \psi(\bar{u}) \geq 0 - (-\rho) = \rho,$$

which contradicts (B.1.4).

   $4^0.2.2)$ Since $\mathcal{U}^\rho$ is a compact set, $4^0.2.1)$ implies that there exists $\gamma > 0$ such that

$$(z + \delta e)^T(Bu + b) \leq -\gamma \ \forall u \in \mathcal{U}^\rho.$$

Now let $\alpha > 0$. We have

$$\begin{aligned}
\Psi(\alpha(z + \delta e)) &= \sup_{u \in \mathrm{Dom}\,\psi} \{\alpha(z + \delta e)^T(Bu + b) - \psi(u)\} \\
&= \max\Big[\sup_{u \in \mathcal{U}^\rho} \{\alpha(z + \delta e)^T(Bu + b) - \psi(u)\}, \\
&\qquad\quad \sup_{u \in (\mathrm{Dom}\,\psi)\setminus\mathcal{U}^\rho} \{\alpha(z + \delta e)^T(Bu + b) - \psi(u)\}\Big].
\end{aligned}$$

When $u \in \mathcal{U}^\rho$, we have $\alpha(z + \delta e)^T(Bu + b) - \psi(u) \leq -\alpha\gamma + \Psi(0)$; this quantity is $\leq \rho$ for all large enough $\alpha \geq 0$. When $u \in (\mathrm{Dom}\,\psi)\setminus\mathcal{U}^\rho$, we have $\alpha(z + \delta e)^T(Bu + b) - \psi(u) \leq 0 + \rho = \rho$ due to $\psi(u) > -\rho$ and $(z + \delta e)^T(Bu + b) \leq 0$ for all $u \in \mathrm{Dom}\,\psi$. We see that $\Psi(\alpha(z + \delta e)) \leq \rho$ for all large enough values of α, whence $z + \delta e \in Z_o^\rho$, as claimed in $4^0.2$.

   $4^0.3.$ We have seen that the inclusion $z \in \mathcal{Z}^\rho$ takes place in the case of $4^0.2$. It remains to verify that this inclusion takes place also when $z^T(Bu + b) > 0$ for certain $u \in \operatorname{Dom} \psi$. Assume that the latter is the case, and let us set

$$
\begin{aligned}
S &= \{[p;q] \in \mathbb{R}^2 : \exists u : p \geq \psi(u), q \geq c(u) \equiv -z^T(Bu+b)\} \\
T &= \{[p;q] \in \mathbb{R}^2 : p \leq -\rho, q < 0\}.
\end{aligned}
$$

The sets S, T clearly are convex and nonempty; let us prove that S, T do not intersect. Indeed, assuming that $[\bar{p};\bar{q}] \in S \cap T$, we would have $\bar{p} \leq -\rho$, $\bar{q} < 0$ and for certain $\bar{u}$ it holds $\bar{p} \geq \psi(\bar{u})$, $\bar{q} \geq c(\bar{u})$, that is, $z^T(B\bar{u}+b) > 0$, while $\psi(\bar{u}) \leq -\rho$; this is the desired contradiction, since z satisfies (B.1.7).

Since S, T are nonempty non-intersecting convex sets, they can be separated: there exists $[\mu;\nu] \neq 0$ such that

$$
\inf_{[p;q] \in S}[\mu p + \nu q] \geq \sup_{[p;q] \in T}[\mu p + \nu q].
$$

Due to the structure of S, T, this relation implies that $\mu, \nu \geq 0$ and that

$$
\inf_{u \in \operatorname{Dom} \psi}[\mu\psi(u) + \nu c(u)] \geq -\mu\rho. \tag{B.1.8}
$$

We claim that $\nu > 0$. Indeed, otherwise $\mu > 0$, and (B.1.8) would imply that $\psi(u) \geq -\rho$ for all $u \in \operatorname{Dom} \psi$, which is not the case (indeed, $\inf_u \psi(u) = -\Psi(0) < -\rho$ according to (B.1.6) and (B.1.3)). Thus, $\nu > 0$. We claim that $\mu > 0$ as well. Indeed, otherwise (B.1.8) would imply that $\inf_{u \in \operatorname{Dom} \psi} c(u) \geq 0$, that is, $z^T(Bu+b) \leq 0$ for all $u \in \operatorname{Dom} \psi$, which contradicts the premise in $4^0.3$. Thus, $\mu > 0, \nu > 0$ and therefore (B.1.8) implies that with $\alpha = \nu/\mu > 0$ one has $\inf_{u \in \operatorname{Dom} \psi}[\psi(u) + \alpha c(u)] \geq -\rho$, that is,

$$
\Psi(\alpha z) = \sup_u \{\alpha z^T(Bu+b) - \psi(u)\} \leq \rho,
$$

whence $z \in Z_o^\rho$. The proof of (!) is completed. $\qquad\square$

B.1.4 Proposition 4.3.1

The set Γ_ϵ^0 is contained in the feasible set Z_* of the chance constraint (4.0.1) by (4.3.3), and since Z_* is closed, it contains $\Gamma_\epsilon \equiv \operatorname{cl}\Gamma_\epsilon^o$ as well. All remaining assertions are readily given by Lemma B.1.1 (where one should set $H(z) = \Psi(z)$, $a = \epsilon$ and use (4.3.2) to verify the validity of Lemma's premise) combined with Remark 4.1.2. $\square$

B.1.5 Propositions 4.4.2, 4.4.4

B.1.5.1 Proof of Proposition 4.4.2

Let $\pi, \theta \in \Pi$. We first prove that $\theta \succeq_m \pi$ is equivalent to every one of the relations (4.4.1), (4.4.2). Let p, q be the densities, and let μ, ν be the probability distributions of π, θ. Let, further, $m_\pi = \operatorname{Prob}\{\pi = 0\}$, $m_\theta = \operatorname{Prob}\{\theta = 0\}$, $P(a) = \int_a^\infty p(s)ds$,

$Q(a) = \int_a^\infty q(s)ds$. For $f \in \mathcal{M}_b$ we have

$$\int f(s)d\mu(s) = m_\pi f(0) + 2\int_0^\infty f(s)p(s)ds = m_\pi f(0) + 2\int_0^\infty f(s)[-\tfrac{d}{ds}P(s)]ds$$
$$= m_\pi f(0) + 2f(0)P(0) + 2\int_0^\infty P(s)f'(s)ds = f(0) + 2\int_0^\infty P(s)f'(s)ds,$$

and similarly $\int f(s)d\nu(s) = f(0) + 2\int\limits_0^\infty f'(s)Q(s)ds$. Thus, $\int f(s)d\nu(s) \geq$ $\int f(s)d\mu(s)$ for all $f \in \mathcal{M}_b$ if and only if $\int_0^\infty f'(s)(Q(s) - P(s))ds \geq 0$ for every $f \in \mathcal{M}_b$, and the latter clearly is equivalent to $Q(s) \geq P(s)$ for all $s \geq 0$ (since P, Q are continuous, and the image of $\mathcal{M}_b$ under the mapping $f \mapsto f'|_{s\geq0}$ is exactly the set of all nonnegative continuous functions g on $\mathbb{R}_+$ such that $g(0) = 0$ and $\int_0^\infty g(s)ds < \infty$). Thus, (4.4.1) is equivalent to $Q(s) \geq P(s)$ for all $s \geq 0$, which is nothing but $\theta \succeq_m \pi$. Observing that $2\int_0^\infty f'(s)P(s)ds = \int f(s)p(s)ds$, $2\int_0^\infty f'(s)Q(s)ds = \int f(s)q(s)ds$, the same argument proves that (4.4.2) is equivalent to $\theta \succeq_m \pi$.

The fact that (4.4.1) implies the inequalities

$$(a) \quad \int f(s)d\nu(s) \geq \int f(s)d\mu(s), \qquad (b) \quad \int f(s)q(s)ds \geq \int f(s)p(s)ds \quad \text{(B.1.9)}$$

for every even function f that is monotone on $\mathbb{R}_+$ is due to the standard approximation argument. Indeed, let (4.4.1) takes place. Every bounded f with the outlined properties is the pointwise limit of a uniformly bounded sequence $f_i \in \mathcal{M}_b$. Passing to limit, as $i \to \infty$, in the relation $\int f_i(s)d\nu(s) \geq \int f_i(s)d\mu(s)$, we conclude that (B.1.9.a) takes place for every bounded even function on the axis that is monotone on $\mathbb{R}_+$. Passing to limit, as $i \to \infty$, in the relation $\int \min[f(s), i]d\nu(s) \geq \int \min[f(s), i]d\mu(s)$, we further conclude that the target relation holds true for every even function that is monotone on $\mathbb{R}_+$. By completely similar argument, the relation (4.4.2) which, as we have already verified, is equivalent to (4.4.1)), implies (B.1.9.b). $\qquad\qquad\square$

B.1.5.2 Proof of Proposition 4.4.4

Item (i) is evident. Let us prove item (ii). We claim that

(a) If $\xi, \xi' \in \Pi$ are independent, then $\xi + \xi' \in \Pi$.

(b) If $p \in \mathcal{P}$ and $f \in \mathcal{M}_b$, then $f_+ := f * p$ belongs to $\mathcal{M}_b$, where $*$ stands for the convolution: $(f * g)(s) = \int f(t)g(s - t)dt$.

(c) Let $\zeta, \widetilde{\zeta} \in \Pi$, $\zeta \succeq_m \widetilde{\zeta}$, and let $\delta, \widetilde{\delta}$ be uniformly distributed in $[-d, d]$. Assume also that $\zeta, \widetilde{\zeta}, \delta, \widetilde{\delta}$ are independent. Then $\zeta + \delta \in \Pi$, $\widetilde{\zeta} + \widetilde{\delta} \in \Pi$, both these random variables are regular, and $\zeta + \delta \succeq_m \widetilde{\zeta} + \widetilde{\delta}$.

Let us prove (a). Denoting by p, r the densities of ξ, ξ', and setting $m = \text{Prob}\{\xi = 0\}$, $m' = \text{Prob}\{\xi' = 0\}$, the density of $\xi + \xi'$ is $mr + m'p + p * r$. We should prove that this density is even (this is evident) and is nondecreasing when $s < 0$. To this end, it clearly suffices to verify that the density $p * r$ is nondecreasing when $s < 0$. By the standard approximation argument, it suffices to establish the latter fact when $p, r \in \mathcal{P}$ are smooth. We have

$$
\begin{aligned}
(p * r)'(s) &= \int p'(s-t) r(t) dt = \int p(s-t) r'(t) ds \\
&= \int_{-\infty}^{0} (p(s-t) - p(s+t)) r'(t) dt.
\end{aligned}
\tag{B.1.10}
$$

Since $s < 0$, with $t < 0$ we have $|s - t| \le |s + t| = |s| + |t|$; and since p is even and nonincreasing on $\mathbb{R}_+$, we conclude that $p(s - t) = p(|s - t|) \ge p(|s + t|) = p(s + t)$, so that $p(s - t) - p(s + t) \ge 0$ when $s, t \le 0$. Since, in addition, $r'(t) \ge 0$ when $t \le 0$, the concluding quantity in (B.1.10) is nonnegative. (a) is proved.

To prove (b), observe that the facts that f_+ is even, continuously differentiable and bounded are evident. All we should prove is that f_+ is nondecreasing on $\mathbb{R}_+$; by the standard approximation argument, it suffices to verify this fact when $p \in \mathcal{P}$ is smooth. In the latter case we have $f'_+(s) = \int f(s-t) p'(t) ds = \int_{-\infty}^{0} (f(s-t) - f(s+t)) p'(t) dt$. Assuming $s \ge 0$, $t \le 0$, and taking into account that f is even and is nondecreasing on $\mathbb{R}_+$, we have $f(s-t) = f(|s-t|) = f(|s|+|t|) \ge f(|s+t|) = f(s+t)$; since $p'(t) \ge 0$ when $t \le 0$, we conclude that $\int_{-\infty}^{0} (f(s - t) - f(s + t)) p'(t) dt \ge 0$ when $s \ge 0$. (b) is proved.

To prove (c), note that the inclusions $\zeta + \delta \in \Pi$, $\widetilde{\zeta} + \widetilde{\delta} \in \Pi$ are given by (a), and the fact that both these random variables are regular is evident. It remains to verify that $\zeta + \delta \succeq_m \widetilde{\zeta} + \widetilde{\delta}$. Given $f \in \mathcal{M}_b$, let $f_+(s) = \frac{1}{2d} \int_{s-d}^{s+d} f(r) dr$, so that $f_+ \in \mathcal{M}_b$ by (b). We have

$$
\mathbf{E}\{f(\zeta + \delta)\} = \mathbf{E}\{f_+(\zeta)\} \ge \mathbf{E}\{f_+(\widetilde{\zeta})\} = \mathbf{E}\{f(\widetilde{\zeta} + \widetilde{\delta})\},
$$

where $\ge$ is due to Proposition 4.4.2 combined with $\zeta \succeq_m \widetilde{\zeta}$ and $f_+ \in \mathcal{M}_b$. The resulting inequality, by Proposition 4.4.2, implies that $\zeta + \delta \succeq_m \widetilde{\zeta} + \widetilde{\delta}$. (c) is proved.

Now we can complete the proof of (ii). Let the premise of (ii) hold true. Observe that from (a) it follows that $\xi + \bar{\xi} \in \Pi$, $\eta + \bar{\eta} \in \Pi$, so that all we need is to verify that $\eta + \bar{\eta} \succeq_m \xi + \bar{\xi}$. Let us first verify this conclusion in the case when all four random variables $\xi, ..., \bar{\eta}$ are regular. Denoting the density of a regular random variable $\omega \in \Pi$ by p_ω and taking into account that such a density is even, for $f \in \mathcal{M}_b$ we have

$$
\mathbf{E}\{f(\xi + \bar{\xi})\} = \int f(s)(p_\xi * p_{\bar{\xi}})(s) ds = \int p_{\bar{\xi}}(s) \underbrace{(f * p_\xi)(s)}_{f_+(s)} ds,
$$

$$
\mathbf{E}\{f(\xi + \bar{\eta})\} = \int f(s)(p_\xi * p_{\bar{\eta}})(s) ds = \int p_{\bar{\eta}}(s) \underbrace{(f * p_\xi)(s)}_{f_+(s)} ds,
$$

and $f_+ \in \mathcal{M}_b$ by (b). Since $\bar{\eta} \succeq_m \bar{\xi}$ and $f_+ \in \mathcal{M}_b$, Proposition 4.4.2 implies that

$$
\mathbf{E}\{f(\xi + \bar{\xi})\} \le \mathbf{E}\{f(\xi + \bar{\eta})\}.
\tag{B.1.11}
$$

Completely similar reasoning demonstrates that

$$
\mathbf{E}\{f(\xi + \bar{\eta})\} \le \mathbf{E}\{f(\eta + \bar{\eta})\}.
\tag{B.1.12}
$$

Combining (B.1.11) and (B.1.12), we get

$$\mathbf{E}\{f(\xi + \bar{\xi})\} \leq \mathbf{E}\{f(\eta + \bar{\eta})\}.$$

This inequality holds true for all $f \in \mathcal{M}_b$; applying Proposition 4.4.2, we get $\xi + \bar{\xi} \preceq_m \eta + \bar{\eta}$, as claimed.

We have proved (ii) in the special case when, in addition to the premise of (ii), the four random variables $\xi, .., \bar{\eta}$ are regular. To justify the validity of (ii) in the general case, let δ_i^κ, $\kappa = 1, 2, 3, 4$, $i = 1, 2, ...$, be independent of each other and of the variables $\xi, ..., \bar{\eta}$ and uniformly distributed in $[-1/i, 1/i]$ random variables. For a fixed i, setting $\xi_i = \xi + \delta_i^1$, $\bar{\xi}_i = \bar{\xi} + \delta_i^2$, $\eta_i = \eta + \delta_i^3$, $\bar{\eta}_i = \bar{\eta} + \delta_i^4$, we get four independent random variables. By (c), these variables belong to Π, are regular and satisfy

$$\eta_i \succeq_m \xi_i, \ \bar{\eta}_i \succeq \bar{\xi}_i,$$

whence, by the already proved version of (ii),

$$\eta_i + \bar{\eta}_i \succeq_m \xi_i + \bar{\xi}_i,$$

and thus, by Proposition 4.4.2, for every $f \in \mathcal{M}_b$ one has

$$\mathbf{E}\{f(\eta_i + \bar{\eta}_i)\} \geq \mathbf{E}\{f(\xi_i + \bar{\xi}_i)\}.$$

This combines with the evident relations

$$\lim_{i \to \infty} \mathbf{E}\{f(\eta_i + \bar{\eta}_i) = \mathbf{E}\{f(\eta + \bar{\eta})\}, \ \lim_{i \to \infty} \mathbf{E}\{f(\xi_i + \bar{\xi}_i) = \mathbf{E}\{f(\xi + \bar{\xi})\}$$

to imply that $\mathbf{E}\{f(\eta + \bar{\eta})\} \geq \mathbf{E}\{f(\xi + \bar{\xi})\}$. Since $f \in \mathcal{M}_b$ is arbitrary, Proposition 4.4.2 implies that $\eta + \bar{\eta} \succeq_m \xi + \bar{\xi}$. $\qquad \square$

B.1.6 Theorem 4.4.6

The case of $L = 1$ is evident, so that in the sequel we assume that $L > 1$.

1^0. We start with the following

Lemma B.1.3. Let $p_1, ..., p_L$, $q_1, ..., q_L$ be unimodal and symmetric w.r.t. 0 probability densities such that $p_\ell \preceq_m q_\ell$, $1 \leq \ell \leq L$, and T be a symmetric w.r.t. the origin convex compact set in $\mathbb{R}^L$. Then

$$\int_T p_1(x_1)...p_L(x_L)dx \geq \int_T q_1(x_1)...q_L(x_L)dx. \tag{B.1.13}$$

Proof. By evident reasons it suffices to prove the lemma in the particular case when $p_\ell(\cdot) = q_\ell(\cdot)$ when $1 \leq \ell \leq L - 1$. Thus, we want to prove that if $p_1, ..., p_L, q_L$ are symmetric and unimodal w.r.t. 0 probability densities and $p_\ell \preceq_m q_\ell$, then

$$\int_T p_1(x_1)...p_{L-1}(x_{L-1})p_L(x_L)dx \geq \int_T p_1(x_1)...p_{L-1}(x_{L-1})q_L(x_L)dx. \tag{B.1.14}$$

Observe that if p is an unimodal and symmetric w.r.t. 0 probability density on the axis, then there exists a sequence $\{p^t\}_{t=1}^\infty$ of probability densities on the axis such that

— every p^t is a convex combination of the densities of uniform symmetric w.r.t. 0 distributions;

— the sequence $\{p^t\}$ converges to p in the sense that

$$\int f(s)p^t(s)ds \to \int f(s)p(s)ds \text{ as } t \to \infty$$

for every bounded piecewise continuous function f on the axis.

Approximating $p_1,...,p_{L-1}$ in this fashion, we see that it suffices to verify (B.1.14) under the assumption that $p_1, ..., p_{L-1}$ are densities of uniform distributions supported on symmetric w.r.t. 0 segments $\Sigma_1, ..., \Sigma_{L-1}$.

To proceed, we need the following fundamental fact:

Symmetrization Principle [Brunn-Minkowski] *Let $S \subset \mathbb{R}^n$, $n > 1$, be a nonempty convex compact set, $e \in \mathbb{R}^n$ be a unit vector, and Δ be the projection of S onto the axis $\mathbb{R}e$: $\Delta = [\min_{x \in S} e^T x, \max_{x \in S} e^T x]$. Then the function*

$$f(s) = \left(\text{mes}_{n-1}\{x \in S : e^T x = s\}\right)^{\frac{1}{n-1}}$$

is concave and continuous on Δ.

Now let $\Sigma = \Sigma_1 \times ... \times \Sigma_{L-1} \times \mathbb{R}$, $\widehat{T} = T \cap \Sigma$, so that $\widehat{T}$ is a convex compact set in $\mathbb{R}^L$, and let

$$f(s) = \text{mes}_{n-1}\{x \in \widehat{T} : x_L = s\}.$$

The function $f(s)$ is even; denoting by Δ the projection of $\widehat{T}$ onto the x_L axis and applying the Symmetrization Principle, we conclude that $f^{\frac{1}{L-1}}(s)$ is concave, even and continuous on Δ, whence, of course, $f^{\frac{1}{L-1}}(s)$ is nonincreasing on $\Delta \cap \mathbb{R}_+$. We see that the function $f(s)$ is even and nonnegative and is nonincreasing on $\mathbb{R}_+$, whence

$$\int f(s)p_L(s)ds \geq \int f(s)q_L(s)ds \tag{B.1.15}$$

due to $p_L \preceq_m q_L$ and Proposition 4.4.2. It remains to note that with p_ℓ being the uniform densities on Σ_ℓ, $1 \leq \ell \leq L - 1$, the left and the right hand sides in (B.1.14) are proportional, with a common positive proportionality coefficient, to the respective sides in (B.1.15). Lemma is proved. □

2^0. Lemma B.1.3 says that the Majorization Theorem is valid under the additional assumption that all random variables ζ_ℓ, η_ℓ are regular, and all we need is to get rid of this extra assumption. This we do next.

$2^0.1$. Let $f(x)$ be an even continuous and nonnegative function on $\mathbb{R}^L$ that has a bounded support and is quasiconcave, so that for all a, $0 < a \leq f(0) = \max f$ the sets $\{x : f(x) \geq a\}$ are convex compacts symmetric w.r.t. the origin. We claim that under the premise of the Majorization Theorem we have

$$\mathbf{E}\{f([\xi_1; ...; \xi_L])\} \geq \mathbf{E}\{f([\eta_1; ...; \eta_L])\}. \tag{B.1.16}$$

Indeed, for $i = 1, 2, ...,$ let $\delta_\ell^i, \zeta_\ell^i,$ $1 \leq \ell \leq L$, be random variables uniformly distributed in $[-1/i, 1/i]$ and independent of each other and of ξ and η variables. For a fixed i, setting $\xi_\ell^i = \xi_\ell + \delta_\ell^i$, $\eta_\ell^i = \eta_\ell + \zeta_\ell^i$, we get a collection of $2L$ independent regular random variables from Π, and $\xi_\ell^i \preceq_m \eta_\ell^i$ by Proposition 4.4.4.(ii). Observe that since f is with a bounded support and continuous, we clearly have

$$\mathbf{E}\{f([\xi_1^i; ...; \xi_L^i])\} \to \mathbf{E}\{f([\xi_1; ...; \xi_L])\}, \quad \mathbf{E}\{f([\eta_1^i; ...; \eta_L^i])\} \to \mathbf{E}\{f([\eta_1; ...; \eta_L])\}$$
$$\tag{B.1.17}$$

as $i \to \infty$. At the same time, we have

$$\mathbf{E}\{f([\xi_1^i; ...; \xi_L^i])\} \geq \mathbf{E}\{f([\eta_1^i; ...; \eta_L^i])\} \ \forall i. \tag{B.1.18}$$

Indeed, f can be represented as the limit of a uniformly converging sequence $\{f^t\}_{t=1}^\infty$ of functions that are weighted sums, with nonnegative weights, of the characteristic functions $\chi_a(\cdot)$ of the sets $\{x : f(x) \geq a\}$ associated with positive values of a. Since these sets are convex compacts symmetric w.r.t. the origin, Lemma B.1.3 as applied to regular random variables ξ_ℓ^i, η_ℓ^i implies that

$$\mathbf{E}\{\chi_a([\xi_1^i; ...; \xi_L^i])\} \geq \mathbf{E}\{\chi_a([\eta_1^i; ...; \eta_L^i])\}$$

whence

$$\mathbf{E}\{f^t([\xi_1^i; ...; \xi_L^i])\} \geq \mathbf{E}\{f^t([\eta_1^i; ...; \eta_L^i])\} \ \forall t.$$

As $t \to \infty$, the left and the right hand sides in this inequality converge to the respective sides in (B.1.18), so that the latter inequality indeed takes place.

Combining (B.1.18) and (B.1.17), we arrive at (B.1.16).

$2^0.2.$ Now we can complete the proof of the Majorization Theorem. The characteristic function χ of S clearly is the pointwise limit of a uniformly bounded sequence $\{\chi^t\}_{t=1}^\infty$ of functions with the properties imposed on f in item $2^0.1$, whence, by the latter item,

$$\mathbf{E}\{\chi^t([\xi_1; ...; \xi_L])\} \geq \mathbf{E}\{\chi^t([\eta_1; ...; \eta_L])\}.$$

As $t \to \infty$, the left and the right hand sides in the latter inequality converge to the respective sides in the target inequality (4.4.7), thus implying its validity. $\qquad\square$

B.1.7 Proposition 4.5.4

As it is explained immediately after formulating Proposition 4.5.4, the situation can be reduced to the case when $z_0 = 0$ and $z_1 = ... = z_L = 1$, which we assume from now on. Besides this, we assume that $\zeta_1, ..., \zeta_L$ are the random variables (4.5.17).

$1^0.$ We start with the following conditional statement:

(!) *If* (4.5.16) *holds true for all affine functions f and all functions f of the form*

$$f(s) = \max[0, a + s],$$

then (4.5.16) holds true for all piecewise linear convex functions f.

Indeed, every piecewise linear convex function $f(\cdot)$ is a linear combination, with nonnegative coefficients λ_i, of an affine function f_0 and functions f_i, $i = 1, ..., I = I(f)$, of the form $\max[0, a + s]$. Under the premise of (!), we have

$$\Phi[f_i, z] = \mathbf{E}\{f_i(\sum_\ell \zeta_\ell)\}, \, i = 0, 1, ..., I(f), \tag{B.1.19}$$

whence by the results of Proposition 4.5.3

$$\Phi[f, z] = \Phi[\sum_{i=0}^{I} \lambda_i f_i, z] \le \sum_{i=0}^{I} \lambda_i \Phi[f_i, z] = \mathbf{E}\{f(\sum_\ell \zeta_\ell)\},$$

where the concluding equality is given by (B.1.19) combined with the fact that $\sum_{i=0}^{I} \lambda_i f_i = f$. Thus, $\Phi[f, z] \le \mathbf{E}\{f(\sum_\ell \zeta_\ell)\}$; since the opposite inequality always is true, we get $\Phi[f, z] = \mathbf{E}\{f(\sum_\ell \zeta_\ell)\}$, as claimed in (!).

2^0. In view of (!), all we need in order to prove Proposition 4.5.4 is to verify that the relation (4.5.16) indeed takes place when f is linear or $f(s) = \max[0, a + s]$.

When f is linear, relation (4.5.16) holds true independently of whether $\zeta_1, ..., \zeta_L$ are linked to each other by (4.5.17) or are arbitrary random variables with given distributions. Indeed, when $f(s) = a + bs$, then, setting

$$\gamma_\ell(u_\ell) = \frac{1}{L}a + bu_\ell,$$

we clearly ensure that

$$\sum_\ell \gamma_\ell(u_\ell) = f(\sum_\ell u_\ell) \, \forall u \in \mathbb{R}^L,$$

which, by (4.5.12), implies that $\Phi[f, z] \le \sum_\ell \mathbf{E}\{\gamma_\ell(\zeta_\ell)\}$; but the latter quantity is exactly $\mathbf{E}\{f(\sum_\ell \zeta_\ell)\} \le \Phi[f, z]$, so that $\Phi[f, z] = \mathbf{E}\{f(\sum_\ell \zeta_\ell)\}$.

Now let us prove that (4.5.16) takes place when $f(s) = \max[0, a+s]$. To save notation, let $a = 0$ (the case of an arbitrary a is completely similar). As above, all we need is to verify that

$$\Phi[f, z] \le \mathbf{E}\{\max[0, \sum_\ell \zeta_\ell]\}. \tag{B.1.20}$$

Let $\phi(t) = \sum_\ell \phi_\ell(t)$, $t \in (0, 1)$. There are three possibilities:

    a) $\phi(t) \le 0$ for all $t \in (0, 1)$;

    b) $\phi(t) \ge 0$ for all $t \in (0, 1)$;

    c) $\phi(t_-) < 0$ and $\phi(t_+) > 0$ for appropriately chosen $t_\pm$, $0 < t_- < t_+ < 1$.

In the case of a), the nondecreasing functions $\phi_\ell(t)$ are above bounded, and $0 \ge \lim_{t \to 1-0} \phi(t) = \sum_\ell d_\ell$, $d_\ell = \lim_{t \to 1-0} \phi_\ell(t)$. Setting $\gamma_\ell(u_\ell) = \max[0, u_\ell - d_\ell]$, we clearly get

$$\sum_\ell \gamma_\ell(u_\ell) \geq \max[0, \sum_\ell (u_\ell - d_\ell)] \geq \max[0, \sum_\ell u_\ell],$$

where the concluding inequality is given by $\sum_\ell d_\ell \leq 0$. Invoking (4.5.12), we conclude that

$$\Phi[f,z] \leq \sum_\ell \mathbf{E}\{\max[0, \zeta_\ell - d_\ell]\} = \sum_\ell \int_0^1 \max[0, \phi_\ell(t) - d_\ell]dt$$
$$= 0 = \int_0^1 \max[0, \sum_\ell \phi_\ell(t)]dt = \mathbf{E}\{f(\sum_\ell \zeta_\ell)\},$$

where the second and the third equalities follow from the fact that $\phi_\ell(t) \leq d_\ell$ and $\phi(t) \leq 0$ when $t \in (0,1)$. The resulting inequality is exactly the relation (B.1.20) we need.

In the case of b), the nondecreasing functions ϕ_ℓ are below bounded on $(0,1)$, and $0 \leq \lim_{t \to +0} \phi(t) = \sum_\ell d_\ell$, $d_\ell = \lim_{t \to +0} \phi_\ell(t)$. Setting $\gamma_\ell(u_\ell) = \max[d_\ell, u_\ell]$, we ensure that

$$\sum_\ell \gamma_\ell(u_\ell) \geq \max[0, \sum_\ell u_\ell] \,\forall u$$

due to $\sum_\ell d_\ell \geq 0$, whence, invoking (4.5.12),

$$\Phi[f,z] \leq \sum_\ell \mathbf{E}\{\max[d_\ell, \zeta_\ell]\} = \sum_\ell \int_0^1 \max[d_\ell, \phi_\ell(t)]dt$$
$$= \int_0^1 [\sum_\ell \phi_\ell(t)]dt = \int_0^1 \max[0, \sum_\ell \phi_\ell(t)]dt = \mathbf{E}\{f(\sum_\ell \zeta_\ell)\},$$

where the second and the third equalities follow from the fact that $\phi_\ell(t) \geq d_\ell$ and $\phi(t) \geq 0$ when $t \in (0,1)$. The resulting inequality is exactly (B.1.20).

In the case of c), the quantity $t_* = \sup\{t \in (0,1) : \phi(t) \leq 0\}$ is well defined and belongs to $(0,1)$; since ϕ_ℓ are continuous from the left at t_*, we have

$$0 \geq \phi(t_*) = \sum_\ell d_\ell^-, \, d_\ell^- = \phi_\ell(t_*),$$

and since $\phi(t) > 0$ when $t > t_*$, we have

$$0 \leq \lim_{t \to t_*+0} \phi(t) = \sum_\ell d_\ell^+, \, d_\ell^+ = \lim_{t \to t_*+0} \phi_\ell(t).$$

Since ϕ_ℓ are nondecreasing, we have $d_\ell^+ \geq d_\ell^-$; since $\sum_\ell d_\ell^- \leq 0 \leq \sum_\ell d_\ell^+$, we can find reals $d_\ell \in [d_\ell^-, d_\ell^+]$ in such a way that $\sum_\ell d_\ell = 0$. Setting $\gamma_\ell(u_\ell) = \max[0, u_\ell - d_\ell]$, we clearly get

$$\sum_\ell \gamma_\ell(u_\ell) \geq \max[0, \sum_\ell u_\ell - \sum_\ell d_\ell] = \max[0, \sum_\ell u_\ell],$$

whence, invoking (4.5.12),

$$\Phi[f,z] \leq \sum_\ell \mathbf{E}\{\max[0, \zeta_\ell - d_\ell]\} = \sum_\ell \int_0^1 \max[0, \phi_\ell(t) - d_\ell]dt$$
$$= \sum_\ell \int_{t_*}^1 [\phi_\ell(t) - d_\ell]dt = \int_{t_*}^1 \phi(t)dt \leq \int_{t_*}^1 \max[0, \phi(t)]dt = \mathbf{E}\{f(\sum_\ell \zeta_\ell)\},$$

where the second equality follows from the fact that $\phi_\ell(t) - d_\ell \leq d_\ell^- - d_\ell \leq 0$ when $t \leq t_*$ and $\phi_\ell(t) - d_\ell \geq d_\ell^+ - d_\ell \geq 0$ when $t > t_*$, and the third equality is given by $\sum_\ell d_\ell = 0$. The resulting inequality is exactly (B.1.20). $\qquad\square$

B.1.8 Theorem 4.5.9

In view of Lemma 4.5.8, the only reason for Theorem 4.5.9 to need a proof rather than to be qualified as an immediate corollary of Proposition 4.2.2 is that the function F (that now plays the role of function $z_0 + \Phi(z_1, ..., z_L)$ in Proposition 4.2.2) is not finite-valued, which was assumed in the Proposition. However, the fact that the domain of F is not the entire space clearly does not affect the conclusion that Γ_ϵ^o (and therefore Γ_ϵ) is contained in the feasible set of the chance constraint (4.5.24). All remaining assertions are readily given by Lemma B.1.1 (where one should set $z = (W, w)$, $H(z) = F(W, w)$ and $a = \ln(\epsilon)$) combined with Remark 4.1.2. $\hspace{2cm}\square$

B.2 $\mathcal{S}$-LEMMA

Theorem B.2.1. [$\mathcal{S}$-Lemma] (i) [homogeneous version] Let A, B be symmetric matrices of the same size such that $\bar{x}^T A \bar{x} > 0$ for certain $\bar{x}$. Then the implication

$$x^T A x \geq 0 \Rightarrow x^T B x \geq 0$$

holds true if and only if

$$\exists \lambda \geq 0 : B \succeq \lambda A.$$

(ii) [inhomogeneous version] Let A, B be symmetric matrices of the same size, and let the quadratic form $x^T A x + 2a^T x + \alpha$ be strictly positive at certain point $\bar{x}$. Then the implication

$$x^T A x + 2a^T x + \alpha \geq 0 \Rightarrow x^T B x + 2b^T x + \beta \geq 0 \tag{B.2.1}$$

holds true if and only if

$$\exists \lambda \geq 0 : \left[\begin{array}{c|c} B - \lambda A & b^T - \lambda a^T \\ \hline b - \lambda a & \beta - \lambda \alpha \end{array} \right] \succeq 0.$$

Proof. (i): In one direction the statement is evident: if $B \succeq \lambda A$ with $\lambda \geq 0$, then $x^T B x \geq \lambda x^T A x$ for all x and therefore $x^T A x \geq 0$ implies $x^T B x \geq 0$.

Now assume that $x^T A x \geq 0$ implies that $x^T B x \geq 0$, and let us prove that $B \succeq \lambda A$ for certain $\lambda \geq 0$. Consider the optimization problem

$$\text{Opt} = \min_X \left\{ \text{Tr}(BX) : \text{Tr}(AX) \geq 0, \text{Tr}(X) = 1, X \succeq 0 \right\}. \tag{B.2.2}$$

This problem clearly is strictly feasible. Indeed, by assumption there exists $\bar{X} = \bar{x}\bar{x}^T \succeq 0$ such that $\text{Tr}(A\bar{X}) > 0$; adding to $\bar{X}$ a small positive definite matrix and normalizing the result to have unit trace, we get a strictly feasible solution to (B.2.2). Moreover, the problem is below bounded, since its feasible set is compact. Applying Semidefinite Duality, we conclude that there exists $\lambda \geq 0$ such that $B - \lambda A \succeq \text{Opt} \cdot I$. We see that it suffices for us to prove that $\text{Opt} \geq 0$.

Problem (B.2.2) is clearly solvable. Let X_* be its optimal solution, and let $\bar{A} = X_*^{1/2} A X_*^{1/2}$, $\bar{B} = X_*^{1/2} B X_*^{1/2}$. Then

$$\text{Tr}(\bar{A}) = \text{Tr}(AX_*) \geq 0, \ \text{Tr}(\bar{B}) = \text{Tr}(BX_*) = \text{Opt}, x^T \bar{A} x \geq 0 \Rightarrow x^T \bar{B} x \geq 0.$$

Now let $\bar{A} = U \Lambda U^T$ be the eigenvalue decomposition of $\bar{A}$, so that U is orthogonal and Λ is diagonal, and let ζ be a random vector with independent coordinates taking values ± 1 with probability $1/2$, and let $\xi = U\zeta$. For all realizations of ζ, we have

$$\xi^T \bar{A} \xi = \zeta^T U^T (U \Lambda U^T) U \zeta = \zeta^T \Lambda \zeta = \text{Tr}(\Lambda) = \text{Tr}(\bar{A}) \geq 0,$$

whence $\xi^T \bar{B} \xi \geq 0$. Taking expectation, we have

$$0 \leq \mathbf{E}\{\xi^T \bar{B} \xi\} = \mathbf{E}\{\zeta^T (U^T \bar{B} U) \zeta\} = \mathbf{E}\{\text{Tr}([U^T \bar{B} U][\zeta \zeta^T])\}$$
$$= \text{Tr}([U^T \bar{B} U] \underbrace{\mathbf{E}\{\zeta \zeta^T\}}_{=I}) = \text{Tr}([U^T \bar{B} U]) = \text{Tr}(\bar{B}) = \text{Opt}.$$

Thus, $\text{Opt} \geq 0$, as claimed.

(ii): Let us pass from original inhomogeneous quadratic forms on $\mathbb{R}^n$ to their homogenizations:

$$f_A(x) \equiv x^T A x + 2a^T x + \alpha$$
$$\mapsto \widehat{f}_A([x;t]) = [x;t]^T \widehat{A}[x;t] \equiv x^T A x + 2ta^T x + \alpha t^2,$$
$$f_B(x) \equiv x^T B x + 2b^T x + \beta$$
$$\mapsto \widehat{f}_B([x;t]) = [x;t]^T \widehat{B}[x;t] \equiv x^T B x + 2tb^T x + \beta t^2.$$

Claim: *In the situation of* (ii), $\exists \bar{y} : \bar{y}^T \widehat{A} \bar{y} > 0$, *and implication* (B.2.1) *is equivalent to the implication*

$$y^T \widehat{A} y \geq 0 \Rightarrow y^T \widehat{B} y \geq 0 \qquad\qquad (*)$$

Claim $\Rightarrow$ Inhomogeneous $\mathcal{S}$-Lemma: Combining Claim and Homogeneous $\mathcal{S}$-Lemma as applied to matrices $\widehat{A}$, $\widehat{B}$, we conclude that (B.2.1) is equivalent to the existence of a $\lambda \geq 0$ such that $\widehat{B} \succeq \lambda \widehat{A}$, which is exactly what is stated by the Inhomogeneous $\mathcal{S}$-Lemma.

Justifying Claim: We clearly have $[\bar{x};1]^T \widehat{A}[\bar{x};1] = f_A(\bar{x}) > 0$. Further, if $(*)$ is valid, then so is (B.2.1), since $f_A(x) = \widehat{f}_A([x;1])$, $f_B(x) = \widehat{f}_B([x;1])$. We see that all we need is to show that the validity of implication (B.2.1) implies the validity of implication $(*)$. Thus, assume that (B.2.1) is valid, and let us prove that $(*)$ takes place. Let $[x;t]$ be such that $[x;t]^T \widehat{A}[x;t] \geq 0$; we should prove that then $[x;t]^T \widehat{B}[x;t] \geq 0$. The case of $t \neq 0$ is trivial due to

$$[x;t]^T \widehat{A}[x;t] \geq 0 \Rightarrow \underbrace{[t^{-1}x;1]^T \widehat{A}[t^{-1}x;1]}_{f_A(x)} \geq 0 \Rightarrow \underbrace{[t^{-1}x;1]^T \widehat{B}[t^{-1}x;1]}_{f_B(x)}$$
$$\Rightarrow [x;t]^T \widehat{B}[x;t] \geq 0.$$

In order to prove that $[x;0]^T \widehat{A}[x;0] \geq 0$ implies that $[x;0]^T \widehat{B}[x;0] \geq 0$, it suffices to verify that the point $y = [x;0]$ can be represented as the limit of a sequence $y^i = [x^i;t^i]$ with $t^i \neq 0$ and $[y^i]^T \widehat{A} y^i \geq 0$. Indeed, in this situation, due to the already proved part of $(*)$, we would have $[y^i]^T \widehat{B} y^i \geq 0$ for all i, and passing to limit as $i \to \infty$, we would get the required relation $y^T \widehat{B} y \geq 0$.

To prove the aforementioned approximation result, let us pass to the coordinates z_j of a point z in the eigenbasis of $\widehat{A}$, so that

$$y^T \widehat{A} y = \sum_j \lambda_j y_j^2 \geq 0,$$

where $\lambda_1 \geq \lambda_2 \geq \dots$ are the eigenvalues of $\widehat{A}$. Observe that $\lambda_1 > 0$, since, as we remember, there exists $\bar{y}$ such that $\bar{y}^T \widehat{A} \bar{y} > 0$. It follows that replacing the first coordinate in y with $(1 + 1/i)y_1$ and keeping the remaining coordinates intact, we get points $\widehat{y}^i$ such that $\widehat{y}^i \to y$, $i \to \infty$, and $[\widehat{y}^i]^T \widehat{A} \widehat{y}^i > 0$. Since the latter inequalities are strict, we can perturb slightly the points $\widehat{y}^i$ to get a sequence $\{y^i\}$ that still converges to y, still satisfies $[y^i]^T \widehat{A} y^i > 0$, and, in addition, is comprised of points with nonzero t-coordinates. $\qquad \square$

B.3 APPROXIMATE $\mathcal{S}$-LEMMA

Theorem B.3.1. [11] Let $\rho > 0$, $A, B, B_1, ..., B_J$ be symmetric $m \times m$ matrices such that $B = bb^T$, $B_j \succeq 0$, $j = 1, ..., J \geq 1$, and $B + \sum\limits_{j=1}^{J} B_j \succ 0$.

Consider the optimization problem

$$\text{Opt}(\rho) = \max_x \left\{ x^T A x : x^T B x \leq 1, x^T B_j x \leq \rho^2, j = 1, ..., J \right\} \qquad (\text{B.3.1})$$

along with its semidefinite relaxation

$$
\begin{aligned}
\text{SDP}(\rho) &= \max_X \left\{ \text{Tr}(AX) : \text{Tr}(BX) \leq 1, \text{Tr}(B_j X) \leq \rho^2, \right. \\
&\qquad \left. j = 1, ..., J, X \succeq 0 \right\} \\
&= \min_{\lambda, \{\lambda_j\}} \left\{ \lambda + \rho^2 \sum_{j=1}^{J} \lambda_j : \lambda \geq 0, \lambda_j \geq 0, j = 1, ..., J, \right. \qquad (\text{B.3.2}) \\
&\qquad \left. \lambda B + \sum_{j=1}^{J} \lambda_j B_j \succeq A \right\}.
\end{aligned}
$$

Then there exists $\bar{x}$ such that

$$
\begin{aligned}
(a) &\quad \bar{x}^T B \bar{x} &\leq&\quad 1 \\
(b) &\quad \bar{x}^T B_j \bar{x} &\leq&\quad \Omega^2(J)\rho^2, j = 1, ..., J \qquad (\text{B.3.3}) \\
(c) &\quad \bar{x}^T A \bar{x} &=&\quad \text{SDP}(\rho),
\end{aligned}
$$

where $\Omega(J)$ is a universal function of J such that $\Omega(1) = 1$ and

$$\Omega(J) \leq 9.19\sqrt{\ln(J)}, \ J \geq 2. \qquad (\text{B.3.4})$$

In particular,

$$\text{Opt}(\rho) \leq \text{SDP}(\rho) \leq \text{Opt}(\Omega(J)\rho). \qquad (\text{B.3.5})$$

Proof. 1^0. First of all, let us derive the "in particular" part of the statement from its general part. Indeed, given that $\bar{x}$ satisfying (B.3.3) does exist, observe that

$\bar{x}$ is a feasible solution to the problem defining $\mathrm{Opt}(\Omega(J)\rho)$, whence $\mathrm{Opt}(\Omega(J)\rho) \geq \bar{x}^T A\bar{x} = \mathrm{SDP}(\rho)$. The first inequality in (B.3.5) is evident.

2^0. The problem

$$\mathrm{SDP}(\rho) = \max_X \left\{ \mathrm{Tr}(AX) : \mathrm{Tr}(BX) \leq 1, \mathrm{Tr}(B_j X) \leq \rho^2, 1 \leq j \leq J, X \succeq 0 \right\} \tag{B.3.6}$$

clearly is strictly feasible and solvable; by this reason, the semidefinite dual of this problem is solvable with the optimal value $\mathrm{SDP}(\rho)$, which is nothing but the second equality in (B.3.2).

3^0. Consider the case $J = 1$, where we should prove that here $\Omega(J) = 1$. This can be immediately derived from the following nice fact:

> **Theorem** [95] *Let A, B, B_1 be three $m \times m$ symmetric matrices with $m \geq 3$ such that certain linear combination of the matrices is $\succ 0$. Then the joint range $\mathcal{I} = \{(x^T Ax, x^T Bx, x^T B_1 x)^T : x \in \mathbb{R}^m\} \subset \mathbb{R}^3$ of the associated quadratic forms is a closed convex set.*

We, however, prefer to present an alternative straightforward proof.

    $3^0.0$. Since $B \succeq 0$, $B_1 \succeq 0$ and $B + B_1 \succ 0$, problem (B.3.6) clearly is solvable. All we need is to prove that this problem admits an optimal solution X_* of rank ≤ 1. Indeed, such a solution is representable in the form $\bar{x}\bar{x}^T$ for certain vector $\bar{x}$; from the constraints of (B.3.6) it follows that $\bar{x}$ satisfies (B.3.3.a–b) with $\Omega(1) = 1$, and from the optimality of $X_* = \bar{x}\bar{x}^T$ — that $\bar{x}$ satisfies (B.3.3.c) as well. Now, when proving that (B.3.6) admits an optimal solution with rank ≤ 1, we may assume that $B_1 \succ 0$. Indeed, assuming that in the latter case the statement we are interested in is true, we would conclude that whenever $\epsilon > 0$, the optimization problem

$$\mathrm{Opt}_\epsilon = \max_X \left\{ \mathrm{Tr}(AX) : \mathrm{Tr}(BX) \leq 1, \mathrm{Tr}([B_1 + \epsilon I]X) \leq \rho^2, X \succeq 0 \right\} \tag{P_ϵ}$$

has an optimal solution X_*^ϵ of rank ≤ 1. Since $B + B_1 \succ 0$, the matrices X_*^ϵ are bounded, so that, by compactness argument, there exists a matrix X_* of rank ≤ 1 such that

$$\mathrm{Tr}(AX_*) \leq \lim_{\epsilon \to +0} \sup \mathrm{Opt}_\epsilon, \mathrm{Tr}(BX_*) \leq 1, \mathrm{Tr}(B_1 X_*) \leq \rho^2.$$

We see that X_* is a feasible solution to (B.3.6), and all we need is to prove that this solution is optimal, that is, to prove that $\mathrm{SDP}(\rho) \leq \liminf_{\epsilon \to +0} \mathrm{Opt}_\epsilon$. To this end, let Y_* be an optimal solution to (B.3.6). For every γ, $0 < \gamma < 1$, the matrix γY_* clearly is feasible for problems (P_ϵ) with all small enough ϵ, whence $\gamma \mathrm{SDP}(\rho) \leq \limsup_{\epsilon \to +0} \mathrm{Opt}_\epsilon$. Since $\gamma < 1$ is arbitrary, we get $\mathrm{SDP}(\rho) \leq \liminf_{\epsilon \to +0} \mathrm{Opt}_\epsilon$, as required.

    Thus, we may focus on the case when $B_1 \succ 0$, and all we need is to prove that in this case (B.3.6) has an optimal solution of rank ≤ 1.

    $3^0.1$. Passing in the optimization problem in (B.3.1) from variables x to variables $B_1^{1/2}x$, we may assume w.l.o.g. that $B_1 = I$; in the latter case, passing to the orthonormal eigenbasis of A, we may further assume that A is diagonal: $A = \text{Diag}\{\lambda_1, ..., \lambda_m\}$ with $\lambda_1 \geq \lambda_2 \geq ... \geq \lambda_m$.

$3^0.2.$ Problem (B.3.6) clearly is solvable; all we need to prove is that this problem has an optimal solution X_* that is a matrix of rank ≤ 1.

$3^0.3.$ Assuming that $\text{SDP}(\rho) \leq 0$, the optimization problem in (B.3.6) clearly has an optimal rank 0 solution $X_* = 0$, and we are done. Thus, assume that $\text{SDP}(\rho) > 0$, which implies that $\lambda_1 > 0$. Note that since A is diagonal and $B_1 = I$, we have

$$\text{SDP}(\rho) \leq \max_X \left\{ \text{Tr}(AX) : \text{Tr}(X) \leq \rho^2, X \succeq 0 \right\} = \lambda_1 \rho^2. \tag{B.3.7}$$

$3^0.4.$ It is possible that $\lambda_1 = \lambda_2$. We clearly have $\text{SDP}(\rho) \leq \lambda_1 \rho^2$; on the other hand, there exists a vector $\bar{x}$, $\|\bar{x}\|_2 = \rho$, that is in the linear span of the first two basic orths and is orthogonal to b. The rank 1 matrix $X_* = \bar{x}\bar{x}^T$ clearly is feasible for (B.3.6), and for this matrix $\text{Tr}(AX_*) = \lambda_1 \rho^2$, so that X_* is an optimal solution to (B.3.6) by (B.3.7). Thus, (B.3.6) has a rank 1 optimal solution, and we are done.

$3^0.5.$ From now on we assume that $\lambda_1 > \lambda_2$. There are two possible cases: one where (B.3.6) has an optimal solution X_* with $\text{Tr}(BX_*) < 1$ ("Case I") and another one where $\text{Tr}(BX_*) = 1$ for every optimal solution X_* to (B.3.6). Assume, first, that we are in Case I, and let X_* be an optimal solution to (B.3.6) with $\text{Tr}(BX_*) < 1$. Let $X_* = V^T V$, let v_i be the columns of V and p_i be the Euclidean norms of the vectors v_i, so that $v_i = p_i f_i$, $\|f_i\|_2 = 1$. We clearly have

$$
\begin{align}
(a) \quad & \text{SDP}(\rho) = \sum_i \lambda_i (X_*)_{ii} = \sum_i \lambda_i p_i^2, \\
(b) \quad & \rho^2 \geq \text{Tr}(B_1 X_*) = \text{Tr}(X_*) = \sum_i p_i^2, \\
(c) \quad & 1 \geq \text{Tr}(BX_*) = \text{Tr}(bb^T V^T V) = \|\sum_i b_i v_i\|_2^2.
\end{align} \tag{B.3.8}
$$

We claim that in fact $p_i = 0$ for $i > 1$, so that X_* is a rank 1 optimal solution, as required. Indeed, assuming that there exists $i_* > 1$ with $p_{i_*} > 0$ and given ϵ, $0 \leq \epsilon < p_{i_*}^2$, let us pass from V to a new matrix V_+ as follows: we replace in V the column $v_{i_*} = p_{i_*} f_{i_*}$, with its multiple $v_{i_*}^+ = \gamma f_{i_*}$, where $\gamma > 0$ is such that $\|v_{i_*}^+\|_2^2 = p_{i_*}^2 - \epsilon$, and replace column $v_1 = p_1 f_1$ in V with the column $v_1^+ = \theta f_1$, where $\theta > 0$ is such that $\|v_1^+\|_2^2 = p_1^2 + \epsilon$; all remaining columns in V_+ are exactly the same as in V. Setting $X_+ = [V_+]^T V_+$, we clearly have $X_+ \succeq 0$, $\text{Tr}(B_1 X_+) = \text{Tr}(X_+) = \text{Tr}(X_*) \leq \rho^2$ and $\text{Tr}(AX_+) = \text{Tr}(AX_*) + (\lambda_1 - \lambda_{i_*})\epsilon > \text{Tr}(AX_*)$ (recall that $\lambda_1 \geq \lambda_2 \geq ... \geq \lambda_m$ and $\lambda_1 > \lambda_2$). On the other hand, for small $\epsilon > 0$, X_+ is close to X_*, so that for small enough $\epsilon > 0$ we have $\text{Tr}(BX_+) < 1$ due to $\text{Tr}(BX_*) < 1$. Thus, for small $\epsilon > 0$ X_+ is a feasible solution to (B.3.6) that is better than X_* in terms of the objective, which is a contradiction. Thus, $p_i = 0$ for $i > 1$, as claimed.

$3^0.6.$ It remains to consider Case II. Let X_* be an optimal solution to (B.3.6), and let V, v_i, p_i be defined exactly as in $3^0.5$, so that (B.3.8) takes place. Since we are in Case II, the vector $e = \sum_i b_i v_i$ is of Euclidean norm 1. Let $I = \{i : b_i \neq 0\}$. We claim that all vectors v_i, $i \in I$, are proportional to e. Indeed, assume that $i \in I$ and v_i is not proportional to e, so that the vectors v_i and $w_i = e - b_i v_i$ are nonzero and are not proportional to each other. Let v_i^+ be the vector of exactly the same Euclidean norm as v_i and of the direction opposite to the one of the vector $b_i w_i$, let V_+ be the matrix obtained from V by replacing the column v_i with the column v_i^+, and let $X_+ = [V_+]^T V_+$. By construction, the Euclidean norms of the columns in V_+ are the same as those of columns in V, whence

$$\text{Tr}(AX_+) = \text{Tr}(AX_*) = \text{SDP}(\rho), \ \text{Tr}(B_1 X_+) = \text{Tr}(B_1 X_*) \leq \rho^2. \tag{B.3.9}$$

At the same time, by construction

$$\begin{aligned}
\text{Tr}(BX_+) &= \|V_+ b\|_2^2 = \|b_i v_i^+ + w_i\|_2^2 = b_i^2 \|v_i^+\|_2^2 + 2(v_i^+)^T(b_i w_i) + \|w_i\|_2^2 \\
&= b_i^2 \|v_i\|_2^2 - 2\|b_i v_i\|_2 \|w_i\|_2 + \|w_i\|_2^2 \\
&< b_i^2 \|v_i\|_2^2 + 2b_i v_i^T w_i + \|w_i\|_2^2 = \|b_i v_i + w_i\|_2^2 = 1,
\end{aligned}$$

where the strict inequality is given by the fact that $b_i \neq 0$ and the nonzero vectors v_i and w_i are not proportional to each other. Invoking (B.3.9), we conclude that X_+ is an optimal solution to (B.3.6) with $\text{Tr}(BX_+) < 1$, which is impossible, since we are in Case II.

Thus, all v_i, $i \in I$, are proportional to e. Replacing in V columns v_i, $i \notin I$, with columns of the same Euclidean norms proportional to e, we get a matrix V_+ such that (a) all columns in V_+ are proportional to e, (b) the columns in V_+ are of the same Euclidean norms as the corresponding columns in V, and (c) $V_+ b = Vb$. From (b), (c) it follows that $X_+ = [V_+]^T V_+$ is a feasible solution to (B.3.6) with the same value of the objective as the one at X_*, i.e., is optimal for the problem, while (a) implies that $V_+ = ef^T$ for certain f, so that X_+ is a rank 1 solution, and we are done.

4^0. Now consider the case of $J > 1$. Let X_* be an optimal solution to the semidefinite program defining $\text{SDP}(\rho)$, and let

$$\widehat{A} = X_*^{1/2} A X_*^{1/2}.$$

Let also

$$\widehat{A} = U\Lambda U^T$$

be the eigenvalue decomposition of $\widehat{A}$, so that U is orthogonal and Λ is diagonal. Consider the random vector

$$\xi = X_*^{1/2} U\zeta,$$

where $\zeta \in \mathbb{R}^m$ is random vector with independent coordinates taking values ± 1 with probabilities 0.5. We have

$(a) \qquad \xi^T A \xi \;=\; \zeta^T U^T X_*^{1/2} A X_*^{1/2} U \zeta = \zeta^T U^T \widehat{A} U \zeta = \zeta^T \Lambda \zeta$
$\qquad\qquad\qquad =\; \mathrm{Tr}(\Lambda) = \mathrm{Tr}(U \Lambda U^T) = \mathrm{Tr}(\widehat{A}) = \mathrm{Tr}(A X_*)$
$\qquad\qquad\qquad =\; \mathrm{SDP}(\rho),$

$(b) \quad \mathbf{E}\{\xi^T B \xi\} \;=\; \mathrm{Tr}(B \mathbf{E}\{\xi \xi^T\}) = \mathrm{Tr}(B X_*^{1/2} U \mathbf{E}\{\zeta \zeta^T\} U^T X_*^{1/2}) \qquad \text{(B.3.10)}$
$\qquad\qquad\qquad\quad =\; \mathrm{Tr}(B X_*) \le 1,$

$(c) \quad \mathbf{E}\{\xi^T B_j \xi\} \;=\; \mathrm{Tr}(B_j \mathbf{E}\{\xi \xi^T\}) = \mathrm{Tr}(B_j X_*^{1/2} U \mathbf{E}\{\zeta \zeta^T\} U^T X_*^{1/2})$
$\qquad\qquad\qquad\quad =\; \mathrm{Tr}(B_j X_*) \le \rho^2$

(we have used the fact that X_* is an optimal solution to the problem defining $\mathrm{SDP}(\rho)$).

$5^0.$ We need the following

Lemma B.3.2. One has

$$\mathrm{Prob}\{\xi^T B \xi > 1\} \le 2/3. \qquad\qquad \text{(B.3.11)}$$

Proof. Recalling that $B = bb^T$, we have

$$\xi^T B \xi = \zeta^T U^T X_*^{1/2} bb^T X_*^{1/2} U \zeta = (\beta^T \zeta)^2,$$

where $\beta = U^T X_*^{1/2} b$. From (B.3.10.b) it follows that $\mathbf{E}\{(\beta^T \zeta)^2\} = \|\beta\|_2^2 \le 1$; the fact that in this situation one has $\mathrm{Prob}\{|\beta^T \zeta| > 1\} \le 2/3$ is proved in Lemma A.1 in [11]. $\qquad\qquad\qquad\qquad\qquad\qquad\qquad\qquad\qquad\qquad\square$

$6^0.$ We next need the following fact.

Lemma B.3.3. Let $e_1, ..., e_m$ be deterministic vectors such that

$$\sum_{i=1}^m \|e_i\|_2^2 \le 1.$$

Then

$$\forall (t > 1) : \mathrm{Prob}\left\{ \|\sum_{j=1}^m \zeta_i e_i\|_2 \ge t \right\} \le \phi(t) = \inf_{r:1 < r < t} \frac{r^2 \exp\{-(t-r)^2/16\}}{r^2 - 1}.$$

$$\text{(B.3.12)}$$

Proof uses the following fundamental fact:

Talagrand Inequality [see, e.g., [67]] *Let $\eta_1, ..., \eta_m$ be independent random vectors taking values in unit balls of the respective finite-dimensional vector spaces $(E_1, \|\cdot\|_{(1)}), ..., (E_m, \|\cdot\|_{(m)})$, and let $\eta = (\eta_1, ..., \eta_m) \in E = E_1 \times ... \times E_m$. Let us equip E with the norm*
$$\|(z^1, ..., z^m)\| = \sqrt{\sum_{i=1}^m \|z^i\|_{(i)}^2},$$ *and let Q be a closed convex subset of E. Then*
$$\mathbf{E}\left\{\exp\{\frac{\mathrm{dist}_{\|\cdot\|}^2(\eta, Q)}{16}\}\right\} \le \frac{1}{\mathrm{Prob}\{\eta \in Q\}}.$$

Let us specify the spaces $(E_i, \|\cdot\|_{(i)})$, $i = 1, ..., m$, as $(\mathbb{R}, |\cdot|)$, and let $\eta_i = \zeta_i$, $i = 1, ..., m$. Let, further,

$$Q_1 = \{u \in \mathbb{R}^m : \|\sum_{i=1}^{m} u_i e_i\|_2 \leq 1\}.$$

Observe that Q_1 is a closed convex set in $\mathbb{R}^m$ and that this set contains the unit $\|\cdot\|_2$-ball; indeed,

$$\|\sum_{i=1}^{m} u_i e_i\|_2 \leq \sum_{i=1}^{m} |u_i| \|e_i\|_2 \leq \|u\|_2 \sqrt{\sum_{i=1}^{m} \|e_i\|_2^2} \leq \|u\|_2.$$

Observe, further, that

$$\mathbf{E}\{\|\sum_i \zeta_i e_i\|_2^2\} = \sum_i \|e_i\|_2^2 \leq 1,$$

whence, by Tschebyshev Inequality,

$$\mathrm{Prob}\left\{\|\sum_i \zeta_i e_i\|_2 > r\right\} \equiv \mathrm{Prob}\{\zeta \notin rQ_1\} \leq \frac{1}{r^2} \; \forall r > 1. \tag{B.3.13}$$

For $t > r > 1$ we have

$$\|\sum_i u_i e_i\|_2 > t \Rightarrow u \notin \tfrac{t}{r}(rQ_1) \Rightarrow u \notin (rQ_1) + (\tfrac{t}{r} - 1)(rQ_1)$$
$$\Rightarrow \; \mathrm{dist}_{\|\cdot\|_2}(z, rQ_1) \geq (\tfrac{t}{r} - 1)r = t - r,$$

where the concluding inequality follows from the fact that Q_1 contains the unit Euclidean ball centered at the origin. We now have for $t > r > 1$:

$$
\begin{aligned}
\mathrm{Prob}\{\|\sum_i \zeta_i e_i\|_2 > t\} \;\; &\leq \;\; \mathrm{Prob}\left\{\frac{\mathrm{dist}_{\|\cdot\|_2}^2(\zeta, rQ_1)}{16} \geq \frac{(t-r)^2}{16}\right\} \\
&\leq \;\; \exp\{-\frac{(t-r)^2}{16}\} \mathbf{E}\{\frac{\mathrm{dist}_{\|\cdot\|_2}^2(\zeta, rQ_1)}{16}\} \\
&\qquad\qquad \text{[Tschebyshev Inequality]} \\
&\leq \;\; \frac{\exp\{-(t-r)^2/16\}}{\mathrm{Prob}\{\zeta \in rQ_1\}} \\
&\qquad\qquad \text{[Talagrand Inequality]} \\
&\leq \;\; \frac{\exp\{-(t-r)^2/16\}}{1 - 1/r^2},
\end{aligned}
$$

where the concluding $\leq$ is due to

$$\mathrm{Prob}\{\zeta \notin rQ_1\} = \mathrm{Prob}\{\|\sum_i \zeta_i e_i\|_2 > r\} \leq \frac{1}{r^2}. \qquad\qquad \square$$

$\mathbf{7^0}$. Given integer $J > 1$, let $\Omega(J) = \inf\{t \geq 1 : \phi(t) > 1/(3J)\}$. Note that from (B.3.12) it follows immediately that

$$J > 1 \Rightarrow \Omega(J) \leq C\sqrt{\ln(J)} \tag{B.3.14}$$

where C is an absolute constant (computer says that it can be set to the value 9.19). Denoting by e_i^j the columns of the matrix $\rho^{-1}B_j^{1/2}X_*^{1/2}U$, we have

$$\sum_{i=1}^{m} \|e_i^j\|_2^2 = \mathbf{E}\{\|\rho^{-1}B_j^{1/2}X_*^{1/2}U\zeta\|_2^2\} \leq 1, \tag{B.3.15}$$

where the concluding inequality is nothing but (B.3.10.c). Taking into account that

$$\mathrm{Prob}\{\xi^T B_j \xi > a\} = \mathrm{Prob}\{\zeta^T[B_j^{1/2}X_*^{1/2}U]^T[B_j^{1/2}X_*^{1/2}U]\zeta > a\}$$
$$= \mathrm{Prob}\{\|B_j^{1/2}X_*^{1/2}U\zeta\|_2^2 > a\} = \mathrm{Prob}\{\|\sum_i \zeta_i e_i^j\|_2^2 > \rho^{-2}a\}$$

and invoking Lemma B.3.3, we get

$$\mathrm{Prob}\{\xi^T B_j \xi > \rho^2 t^2\} = \mathrm{Prob}\{\|\sum_i \zeta_i e_i^j\|_2 > t\} \leq \phi(t),$$

whence

$$t > \Omega(J) \Rightarrow \mathrm{Prob}\{\xi^T B_j \xi > \rho^2 t^2\} < \frac{1}{3J}. \tag{B.3.16}$$

Invoking Lemma B.3.2, it follows that when $t > \Omega(J)$, one has

$$\mathrm{Prob}\{\xi : \xi^T B \xi > 1 \text{ or } \exists j : \xi^T B_j \xi > \rho^2 t^2\} < \frac{2}{3} + J\frac{1}{3J} = 1,$$

that is, there exists a realization $\tilde{\xi}$ of ξ such that

$$\tilde{\xi}^T B \tilde{\xi} \leq 1, \tilde{\xi}^T B_j \tilde{\xi} \leq t^2 \rho^2 \,\forall j, 1 \leq j \leq J.$$

Since $t > \Omega(J)$ is arbitrary, there exists a realization $\bar{x}$ of ξ such that

$$\bar{x}^T B \bar{x} \leq 1, \bar{x}^T B_j \bar{x} \leq \Omega^2(J)\rho^2 \,\forall j, 1 \leq j \leq J.$$

Since $\bar{x}$ is a realization of ξ, we have also

$$\bar{x}^T A \bar{x} = \mathrm{SDP}(\rho)$$

by (B.3.10.a). Thus, $\bar{x}$ satisfies (B.3.3). $\qquad\square$

B.4 MATRIX CUBE THEOREM

B.4.1 Matrix Cube Theorem, Complex Case

The "Complex Matrix Cube" problem is as follows:

CMC: *Let m, $p_1, q_1, ..., p_L, q_L$ be positive integers, and $A \in \mathbf{H}_+^m$, $L_j \in \mathbb{C}^{p_j \times m}$, $R_j \in \mathbb{C}^{q_j \times m}$ be given matrices, $L_j \neq 0$. Let also a partition $\{1, 2, ..., L\} = I_\mathrm{S}^\mathrm{r} \cup I_\mathrm{S}^\mathrm{c} \cup I_\mathrm{f}^\mathrm{c}$ of the index set $\{1, ..., L\}$ into three non-overlapping sets be given, and let $p_j = q_j$ for $j \in I_\mathrm{S}^\mathrm{r} \cup I_\mathrm{S}^\mathrm{c}$. With these data, we associate a parametric family of "matrix boxes"*

$$\mathcal{U}[\rho] = \left\{ A + \rho \sum_{j=1}^{L} [L_j^H \Theta^j R_j + R_j^H [\Theta^j]^H L_j] : \begin{array}{c} \Theta^j \in \mathcal{Z}^j, \\ 1 \leq j \leq L \end{array} \right\} \subset \mathbf{H}^m,$$

$$\tag{B.4.1}$$

where $\rho \geq 0$ is the parameter and

$$
\mathcal{Z}^j = \begin{cases}
\{\Theta^j = \theta I_{p_j} : \theta \in \mathbb{R}, |\theta| \leq 1\}, & j \in I_{\mathsf{S}}^{\mathsf{r}} \\
\quad [\text{``real scalar perturbation blocks''}] \\
\{\Theta^j = \theta I_{p_j} : \theta \in \mathbb{C}, |\theta| \leq 1\}, & j \in I_{\mathsf{S}}^{\mathsf{c}} \\
\quad [\text{``complex scalar perturbation blocks''}] \\
\{\Theta^j \in \mathbb{C}^{p_j \times q_j} : \|\Theta^j\|_{2,2} \leq 1\}, & j \in I_{\mathsf{f}}^{\mathsf{c}} \\
\quad [\text{``full complex perturbation blocks''}] .
\end{cases}
\tag{B.4.2}
$$

Given $\rho \geq 0$, check whether

$$
\mathcal{U}[\rho] \subset \mathbf{H}_+^m. \tag{$\mathcal{A}(\rho)$}
$$

Remark B.4.1. In the sequel, we always assume that $p_j = q_j > 1$ for $j \in I_{\mathsf{S}}^{\mathsf{c}}$. Indeed, one-dimensional complex scalar perturbations can always be regarded as full complex perturbations.

Our main result is as follows:

Theorem B.4.2. [The Complex Matrix Cube Theorem [12]] Consider, along with predicate $\mathcal{A}(\rho)$, the predicate

$$
\begin{aligned}
&\exists Y_j \in \mathbf{H}^m, \, j = 1, ..., L \text{ such that :} \\
&(a) \quad Y_j \succeq L_j^H \Theta^j R_j + R_j^H [\Theta^j]^H L_j \quad \forall(\Theta^j \in \mathcal{Z}^j, 1 \leq j \leq L) \\
&(b) \quad A - \rho \sum_{j=1}^{L} Y_j \succeq 0.
\end{aligned}
\tag{$\mathcal{B}(\rho)$}
$$

Then:

(i) Predicate $\mathcal{B}(\rho)$ is stronger than $\mathcal{A}(\rho)$ — the validity of the former predicate implies the validity of the latter one.

(ii) $\mathcal{B}(\rho)$ is computationally tractable — the validity of the predicate is equivalent to the solvability of the system of LMIs

$$
\begin{aligned}
(s.\mathbb{R}) \qquad & Y_j \pm \left[L_j^H R_j + R_j^H L_j \right] \succeq 0, \, j \in I_{\mathsf{S}}^{\mathsf{r}}, \\[2mm]
(s.\mathbb{C}) \qquad & \begin{bmatrix} Y_j - V_j & L_j^H R_j \\ R_j^H L_j & V_j \end{bmatrix} \succeq 0, \, j \in I_{\mathsf{S}}^{\mathsf{c}}, \\[2mm]
(f.\mathbb{C}) \quad & \begin{bmatrix} Y_j - \lambda_j L_j^H L_j & R_j^H \\ R_j & \lambda_j I_{p_j} \end{bmatrix} \succeq 0, \, j \in I_{\mathsf{f}}^{\mathsf{c}} \\[2mm]
(*) \qquad & A - \rho \sum_{j=1}^{L} Y_j \succeq 0.
\end{aligned}
\tag{B.4.3}
$$

in the matrix variables $Y_j \in \mathbf{H}^m$, $j = 1, ..., k$, $V_j \in \mathbf{H}^m$, $j \in I_{\mathsf{S}}^{\mathsf{c}}$, and the real variables λ_j, $j \in I_{\mathsf{f}}^{\mathsf{c}}$.

(iii) "The gap" between $\mathcal{A}(\rho)$ and $\mathcal{B}(\rho)$ can be bounded solely in terms of the maximal size

$$
p^{\mathsf{s}} = \max \left\{ p_j : j \in I_{\mathsf{S}}^{\mathsf{r}} \cup I_{\mathsf{S}}^{\mathsf{c}} \right\} \tag{B.4.4}
$$

of the scalar perturbations (here the maximum over an empty set by definition is 0). Specifically, there exists a universal function $\vartheta_{\mathbb{C}}(\cdot)$ such that

$$\vartheta_{\mathbb{C}}(\nu) \leq 4\pi\sqrt{\nu}, \ \nu \geq 1, \tag{B.4.5}$$

and

$$\text{if } \mathcal{B}(\rho) \text{ is not valid, then } \mathcal{A}(\vartheta_{\mathbb{C}}(p^{\mathrm{s}})\rho) \text{ is not valid.} \tag{B.4.6}$$

(iv) Finally, in the case $L = 1$ of single perturbation block $\mathcal{A}(\rho)$ is equivalent to $\mathcal{B}(\rho)$.

Remark B.4.3. From the proof of Theorem B.4.2 it follows that $\vartheta_{\mathbb{C}}(0) = \frac{4}{\pi}$, $\vartheta_{\mathbb{C}}(1) = 2$. Thus,

- when there are no scalar perturbations: $I_{\mathrm{S}}^{\mathrm{r}} = I_{\mathrm{S}}^{\mathrm{C}} = \emptyset$, the factor ϑ in the implication

$$\neg\mathcal{B}(\rho) \Rightarrow \neg\mathcal{A}(\vartheta\rho) \tag{B.4.7}$$

  can be set to $\frac{4}{\pi} = 1.27...$

- when there are no complex scalar perturbations (cf. Remark B.4.1) and all real scalar perturbations are non-repeated ($I_{\mathrm{S}}^{\mathrm{C}} = \emptyset$, $p_j = 1$ for all $j \in I_{\mathrm{S}}^{\mathrm{r}}$), the factor ϑ in (B.4.7) can be set to 2.

The following simple observation is crucial when applying Theorem B.4.2.

Remark B.4.4. Assume that the data $A, R_1, ..., R_L$ of the Matrix Cube problem are affine in a vector of parameters y, while the data $L_1, ..., L_L$ are independent of y. Then (B.4.3) is a system of LMIs in the variables Y_j, V_j, λ_j and y.

B.4.2 Proof of Theorem B.4.2.(i)

Item (i) is evident.

B.4.3 Proof of Theorem B.4.2.(ii)

The equivalence between the validity of $\mathcal{B}(\rho)$ and the solvability of (B.4.3) is readily given by the following facts:

Lemma B.4.5. Let $B \in \mathbb{C}^{m\times m}$ and $Y \in \mathbf{H}^m$. Then the relation

$$Y \succeq \theta B + \bar{\theta} B^H \quad \forall(\theta \in \mathbb{C}, |\theta| \leq 1) \tag{B.4.8}$$

is satisfied if and only if

$$\exists V \in \mathbf{H}^m : \ \begin{bmatrix} Y - V & B^H \\ B & V \end{bmatrix} \succeq 0. \tag{B.4.9}$$

Lemma B.4.6. Let $L \in \mathbb{C}^{\ell\times m}$ and $R \in \mathbb{C}^{r\times m}$.

(i) Assume that L, R are nonzero. A matrix $Y \in \mathbf{H}^m$ satisfies the relation

$$Y \succeq L^H U R + R^H U^H L \quad \forall(U \in \mathbb{C}^{\ell\times r} : \|U\|_{2,2} \leq 1) \tag{B.4.10}$$

if and only if there exists a positive real λ such that

$$Y \succeq \lambda L^H L + \lambda^{-1} R^H R. \tag{B.4.11}$$

(ii) Assume that L is nonzero. A matrix $Y \in \mathbf{H}^m$ satisfies (B.4.10) if and only if there exists $\lambda \in \mathbb{R}$ such that

$$\begin{bmatrix} Y - \lambda L^H L & R^H \\ R & \lambda I_r \end{bmatrix} \succeq 0. \tag{B.4.12}$$

Lemmas B.4.5, B.4.6 $\Rightarrow$ Theorem B.4.2.(ii). All we need to prove is that a collection of matrices Y_j satisfies the constraints in $\mathcal{B}(\rho)$ if and only if it can be extended by properly chosen V_j, $j \in I_f^C$, and λ_j, $j \in I_S^C$, to a feasible solution of (B.4.3). This is immediate, since matrices Y_j, $j \in I_f^C$, satisfy the corresponding constraints $\mathcal{B}(\rho).a$ if and only if these matrices along with some matrices V_j satisfy (B.4.3.s.$\mathbb{C}$)) (Lemma B.4.5), while matrices Y_j, $j \in I_S^C$, satisfy the corresponding constraints $\mathcal{B}(\rho).a$ if and only if these matrices along with some reals λ_j satisfy (B.4.3.f.$\mathbb{C}$) (Lemma B.4.6.(ii)). $\square$

Proof of Lemma B.4.5. "if" part: Assume that V is such that

$$\begin{bmatrix} Y - V & B^H \\ B & V \end{bmatrix} \succeq 0.$$

Then, for every $\xi \in \mathbb{C}^n$ and every $\theta \in \mathbb{C}$, $|\theta| = 1$, we have

$$0 \leq \begin{bmatrix} \xi \\ -\bar{\theta}\xi \end{bmatrix}^H \begin{bmatrix} Y - V & B^H \\ B & V \end{bmatrix} \begin{bmatrix} \xi \\ -\bar{\theta}\xi \end{bmatrix} = \xi^H(Y - V)\xi + \xi^H V\xi$$
$$- \xi^H[\theta B + \bar{\theta} B^H]\xi,$$

so that $Y \succeq \theta B + \bar{\theta} B^H$ for all $\theta \in \mathbb{C}$, $|\theta| = 1$, which, by evident convexity reasons, implies (B.4.8).

"only if" part: Let $Y \in \mathbf{H}^m$ satisfy (B.4.8). Assume, on the contrary to what should be proved, that there does not exist $V \in \mathbf{H}^m$ such that $0 \preceq \begin{bmatrix} Y - V & B^H \\ B & V \end{bmatrix}$, and let us lead this assumption to a contradiction. Observe that our assumption means that the optimization program

$$\min_{t,V} \left\{ t : \begin{bmatrix} tI_m + Y - V & B^H \\ B & V \end{bmatrix} \succeq 0 \right\} \tag{B.4.13}$$

has no feasible solutions with $t \leq 0$; since problem (B.4.13) is clearly solvable, its optimal value is therefore positive. Now, our problem is a conic problem on the (self-dual) cone of positive semidefinite Hermitian matrices; since the problem clearly is strictly feasible, the Conic Duality Theorem says that dual problem

$$\max_{\substack{Z \in \mathbf{H}^m, \\ W \in \mathbb{C}^{m \times m}}} \left\{ -2\Re\left\{\mathrm{Tr}(W^H B)\right\} - \mathrm{Tr}(ZY) : \begin{array}{ll} \begin{bmatrix} Z & W^H \\ W & Z \end{bmatrix} \succeq 0, & (a) \\ \mathrm{Tr}(Z) = 1 & (b) \end{array} \right\} \tag{B.4.14}$$

is solvable with the same — positive — optimal value as the one of (B.4.13). In (B.4.14), we can easily eliminate the W-variable; indeed, constraint (B.4.14.a), as it is well-known, is equivalent to the fact that $Z \succeq 0$ and $W = Z^{1/2}XZ^{1/2}$ with $X \in \mathbb{C}^{m \times m}$, $\|X\|_{2,2} \leq 1$. With this parameterization of W, the W-term in the objective of (B.4.14) becomes $-2\Re\{\mathrm{Tr}(X^H Z^{1/2} B Z^{1/2})\}$; as it is well-known, the maximum of the latter expression in X, $\|X\|_{2,2} \leq 1$, is $2\|\sigma(Z^{1/2}BZ^{1/2})\|_1$. Since the optimal value in (B.4.14) is positive, we arrive at the following intermediate conclusion:

(*) *There exists* $Z \in \mathbf{H}^m$, $Z \succeq 0$, *such that*

$$2\|\sigma(Z^{1/2}BZ^{1/2})\|_1 > \mathrm{Tr}(ZY) = \mathrm{Tr}(Z^{1/2}YZ^{1/2}). \qquad \text{(B.4.15)}$$

The desired contradiction is now readily given by the following simple observation:

Lemma B.4.7. Let $S \in \mathbf{H}^m$, $C \in \mathbb{C}^{m \times m}$ be such that

$$S \succeq \theta C + \bar{\theta} C^H \quad \forall(\theta \in \mathbb{C}, |\theta| = 1). \qquad \text{(B.4.16)}$$

Then $2\|\sigma(C)\|_1 \leq \mathrm{Tr}(S)$.

To see that Lemma B.4.7 yields the desired contradiction, note that the matrices $S = Z^{1/2}YZ^{1/2}$, $C = Z^{1/2}BZ^{1/2}$ satisfy the premise of this lemma by (B.4.8), and for these matrices the conclusion of the lemma contradicts (B.4.15).

Proof of Lemma B.4.7: As it was already mentioned,

$$\|\sigma(C)\|_1 = \max_X \left\{ \Re\{\mathrm{Tr}(XC^H)\} : \|X\|_{2,2} \leq 1 \right\}.$$

Since the extreme points of the set $\{X \in \mathbb{C}^{m \times m} : \|X\|_{2,2} \leq 1\}$ are unitary matrices, the maximizer X_* in the right hand side can be chosen to be unitary: $X_*^H = X_*^{-1}$; thus, X_* is a unitary similarity transformation of a diagonal unitary matrix. Applying appropriate unitary rotation $A \mapsto U^H A U$, $U^H = U^{-1}$, to all matrices involved, we may assume that X_* itself is diagonal. Now we are in the situation as follows: we are given matrices C, S satisfying (B.4.16) and a *diagonal* unitary matrix X_* such that $\|\sigma(C)\|_1 = \Re\{\mathrm{Tr}(X_* C^H)\}$. In other words,

$$\|\sigma(C)\|_1 = \Re\left\{ \sum_{\ell=1}^m (X_*)_{\ell\ell} \overline{C_{\ell\ell}} \right\} \leq \sum_{\ell=1}^m |C_{\ell\ell}| \qquad \text{(B.4.17)}$$

(the concluding inequality comes from the fact that X_* is unitary). On the other hand, let e_ℓ be the standard basic orths in $\mathbb{C}^m$. By (B.4.16), we have

$$\theta C_{\ell\ell} + \overline{\theta C_{\ell\ell}} = e_\ell^H [\theta C + \bar{\theta} C^H] e_\ell \leq e_\ell^H S e_\ell = S_{\ell\ell} \quad \forall(\theta \in \mathbb{C}, |\theta| = 1),$$

whence, maximizing in θ, $2|C_{\ell\ell}| \leq S_{\ell\ell}$, $\ell = 1, ..., m$, which combines with (B.4.17) to imply that $2\|\sigma(C)\|_1 \leq \mathrm{Tr}(S)$. $\qquad \square$

Proof of Lemma B.4.6 (cf. section 5.3).

(i), "if" part: Let (B.4.11) be valid for certain $\lambda > 0$. Then for every $\xi \in \mathbb{C}^m$ one has

$$
\begin{aligned}
\xi^H Y \xi \;\; &\geq \;\; \lambda \xi^H L^H L \xi + \lambda^{-1} \xi^H R^H R \xi \geq 2 \sqrt{\xi^H L^H L \xi} \sqrt{\xi^H R^H R \xi} \\
&= \;\; 2 \| L \xi \|_2 \| R \xi \|_2 \\
\Rightarrow \;\; \forall (U, \| U \|_{2,2} &\leq 1): \\
\xi^H Y \xi \;\; &\geq \;\; 2 | [L\xi]^H U [R\xi] | \geq 2 \Re \{ [L\xi]^H U [R\xi] \} \\
&= \;\; \xi^H [L^H U R + R^H U^H L] \xi,
\end{aligned}
$$

as claimed.

(i), "only if" part: Assume that Y satisfies (B.4.10) and $L \neq 0, R \neq 0$; we prove that then there exists $\lambda > 0$ such that (B.4.11) holds true. First, observe that w.l.o.g. we may assume that L and R are of the same sizes $r \times n$ (to reduce the general case to this particular one, it suffices to add several zero rows either to L (when $\ell < r$), or to R (when $\ell > r$)). We have the following chain of equivalences:

$$
\begin{aligned}
& (B.4.10) \\
\Leftrightarrow \;\; & \forall \xi \in \mathbb{C}^m: \quad \xi^H Y \xi \geq 2 \| L\xi \|_2 \| R\xi \|_2 \\
\Leftrightarrow \;\; & \forall (\xi \in \mathbb{C}^n, \eta \in \mathbb{C}^r): \quad \| \eta \|_2 \leq \| L\xi \|_2 \Rightarrow \xi^H Y \xi - \eta^H R \xi - \xi^H R^H \eta \geq 0 \\
\Leftrightarrow \;\; & \forall (\xi \in \mathbb{C}^m, \eta \in \mathbb{C}^r): \\
& \xi^H L^H L \xi - \eta^H \eta \geq 0 \Rightarrow \xi^H Y \xi - \eta^H R \xi - \xi^H R^H \eta \geq 0 \\
\Leftrightarrow \;\; & \exists (\lambda \geq 0): \quad \begin{bmatrix} Y & R^H \\ R & \end{bmatrix} - \lambda \begin{bmatrix} L^H L & \\ & -I_r \end{bmatrix} \succeq 0 \quad [\mathcal{S}\text{-Lemma}] \\
\Leftrightarrow (a) \;\; & \begin{bmatrix} Y - \lambda L^H L & R^H \\ R & \lambda I_r \end{bmatrix} \succeq 0.
\end{aligned}
$$

$$ \text{(B.4.18)} $$

(Note that $\mathcal{S}$-Lemma clearly holds true in the Hermitian case, since Hermitian quadratic forms on $\mathbb{C}^m$ can be treated as real quadratic forms on $\mathbb{R}^{2m}$.)

Condition (B.4.18.a), in view of $R \neq 0$, clearly implies that $\lambda > 0$. Therefore, by the Schur Complement Lemma (SCL), (B.4.18.a) is equivalent to $Y - \lambda L^H L - \lambda^{-1} R^H R \succeq 0$, as claimed.

(ii): When $R \neq 0$, (ii) is clearly equivalent to (i) and thus is already proved. When $R = 0$, it is evident that (B.4.12) can be satisfied by properly chosen $\lambda \in \mathbb{R}$ if and only if $Y \succeq 0$, which is exactly what is stated by (B.4.10) when $R = 0$. $\square$

B.4.4 Proof of Theorem B.4.2.(iii)

In order to prove (iii), it suffices to prove the following statement:

Lemma B.4.8. Assume that $\rho \geq 0$ is such that the predicate $\mathcal{B}(\rho)$ is not valid. Then the predicate $\mathcal{A}(\vartheta_{\mathbb{C}}(p^s)\rho)$, with appropriately defined function $\vartheta_{\mathbb{C}}(\cdot)$ satisfying (B.4.5), is also not valid.

We are about to prove Lemma B.4.8. The case of $\rho = 0$ is trivial, so that from now on we assume that $\rho > 0$ and that all matrices L_j, R_j are nonzero (the latter assumption, of course, does not restrict generality). *From now till the end of* *section B.4.4.3, we assume that we are under the premise of Lemma B.4.8, i.e., the predicate $\mathcal{B}(\rho)$ is not valid.*

B.4.4.1 First step: duality

Consider the optimization program

$$
\min_{\substack{t,\{Y_j\in\mathbf{H}^m\}_{j\in I_S^{\mathbf{r}}},\\ \{U_j,V_j\in\mathbf{H}^m\}_{j\in I_S^{\mathbf{c}}},\\ \{\lambda_j,\nu_j\in\mathbb{R}\}_{j\in I_f^{\mathbf{c}}}}} t : \left\{
\begin{array}{ll}
Y_j \pm \underbrace{[L_j^H R_j + R_j^H L_j]}_{2A_j, A_j = A_j^H} \succeq 0,\, j\in I_S^{\mathbf{r}}, & (a)\\[2ex]
\left[\begin{array}{cc} U_j & R_j^H L_j \\ L_j^H R_j & V_j \end{array}\right] \succeq 0,\, j\in I_S^{\mathbf{c}}, & (b)\\[2ex]
\left[\begin{array}{cc} \lambda_j & 1 \\ 1 & \nu_j \end{array}\right] \succeq 0,\, j\in I_f^{\mathbf{c}}, & (c)\\[2ex]
tI + A - \rho\left[\sum_{j\in I_S^{\mathbf{r}}} Y_j + \sum_{j\in I_S^{\mathbf{c}}}[U_j + V_j]\right.\\[2ex]
\left. + \sum_{j\in I_f^{\mathbf{c}}}[\lambda_j L_j^H L_j + \nu_j R_j^H R_j]\right] \succeq 0 & (d)
\end{array}\right\}. \qquad (B.4.19)
$$

Introducing "bounds" $Y_j = U_j + V_j$ for $j\in I_S^{\mathbf{c}}$ and $Y_j \succeq \lambda_j L_j^H L_j + \nu_j R_j^H R_j$ for $j\in I_f^{\mathbf{c}}$ and then eliminating the variables $U_j, j\in I_S^{\mathbf{c}}, \nu_j, j\in I_f^{\mathbf{c}}$, we convert (B.4.19) into the equivalent problem

$$
\min_{\substack{t,\{Y_j\in\mathbf{H}^m\}_{j=1}^L,\\ \{V_j\in\mathbf{H}^m\}_{j\in I_S^{\mathbf{c}}},\\ \{\lambda_j\in\mathbb{R}\}_{j\in I_f^{\mathbf{c}}}}} t : \left\{
\begin{array}{l}
Y_j \pm [L_j^H R_j + R_j^H L_j] \succeq 0,\, j\in I_S^{\mathbf{r}},\\[2ex]
\left[\begin{array}{cc} Y_j - V_j & R_j^H L_j \\ L_j^H R_j & V_j \end{array}\right] \succeq 0,\, j\in I_S^{\mathbf{c}},\\[2ex]
\left[\begin{array}{cc} Y_j - \lambda_j L_j^H L_j & R_j^H \\ R_j & \lambda_j I_{p_j} \end{array}\right] \succeq 0,\, j\in I_f^{\mathbf{c}},\\[2ex]
tI + A - \rho\sum_{j=1}^L Y_j \succeq 0
\end{array}\right\}.
$$

By (already proved) item (ii) of Theorem B.4.2, predicate $\mathcal{B}(\rho)$ is valid if and only if the latter problem, and thus problem (B.4.19), admits a feasible solution with $t \leq 0$. We are in the situation when $\mathcal{B}(\rho)$ is *not* valid; consequently, (B.4.19) does not admit feasible solutions with $t \leq 0$. Since the problem clearly is solvable, it means that the optimal value in the problem is positive. Problem (B.4.19) is a conic problem on the product of cones of Hermitian and real symmetric positive semidefinite matrices. Since (B.4.19) is strictly feasible and bounded below, the Conic Duality Theorem implies that the conic dual problem of (B.4.19) is solvable with the same positive optimal value. Taking into account that the cones associated with (B.4.19) are self-dual, the dual problem, after straightforward simplifications, becomes the conic problem

$$
\max \quad -2\rho\left[\sum_{j\in I_S^{\mathbf{r}}} \mathrm{Tr}([P_j - Q_j]A_j) + \sum_{j\in I_S^{\mathbf{c}}} \Re\{\mathrm{Tr}\left(S_j R_j^H L_j\right)\} + \sum_{j\in I_{\mathbf{f}}^{\mathbf{c}}} w_j\right]
$$
$$
-\mathrm{Tr}(ZA)
$$

subject to

$(a.1)$ $\qquad\qquad\qquad\qquad P_j, Q_j \ \succeq\ 0,\, j \in I_S^{\mathbf{r}},$

$(a.2)$ $\qquad\qquad\qquad\qquad P_j + Q_j \ =\ Z,\, j \in I_S^{\mathbf{r}};$

(b) $\qquad\qquad\quad \begin{bmatrix} Z & S_j^H \\ S_j & Z \end{bmatrix} \ \succeq\ 0,\, j \in I_S^{\mathbf{c}};$

(c) $\quad \begin{bmatrix} \mathrm{Tr}(L_j Z L_j^H) & w_j \\ w_j & \mathrm{Tr}(R_j Z R_j^H) \end{bmatrix} \ \succeq\ 0,\, j \in I_{\mathbf{f}}^{\mathbf{c}};$

(d) $\qquad\qquad\qquad\quad Z \succeq 0, \mathrm{Tr}(Z) \ =\ 1.$

$$(B.4.20)$$

in matrix variables $Z \in \mathbf{H}_+^m$, $P_j, Q_j \in \mathbf{H}^m$, $j \in I_S^{\mathbf{r}}$, $S_j \in \mathbb{C}^{m\times m}$, $j \in I_S^{\mathbf{c}}$, and real variables w_j, $j \in I_{\mathbf{f}}^{\mathbf{c}}$. Using (B.4.20.c), we can eliminate the variables w_j, thus arriving at the following equivalent reformulation of the dual problem:

$$
\text{maximize} \quad 2\rho\left[- \sum_{j\in I_S^{\mathbf{r}}} \mathrm{Tr}([P_j - Q_j]A_j) - \sum_{j\in I_S^{\mathbf{c}}} \Re\{\mathrm{Tr}\left(S_j R_j^H L_j\right)\} \right.
$$
$$
\left. + \sum_{j\in I_{\mathbf{f}}^{\mathbf{c}}} \underbrace{\sqrt{\mathrm{Tr}(L_j Z L_j^H)}}_{\|L_j Z^{1/2}\|_2} \underbrace{\sqrt{\mathrm{Tr}(R_j Z R_j^H)}}_{\|R_j Z^{1/2}\|_2} \right] - \mathrm{Tr}(ZA)
$$

subject to $\qquad\qquad\qquad\qquad\qquad\qquad\qquad\qquad\qquad\qquad\qquad$ $(B.4.21)$

$(a.1)$ $\qquad\qquad\quad P_j, Q_j \ \succeq\ 0,\, j \in I_S^{\mathbf{r}},$

$(a.2)$ $\qquad\qquad\quad P_j + Q_j \ =\ Z,\, j \in I_S^{\mathbf{r}};$

(b) $\qquad\quad \begin{bmatrix} Z & S_j^H \\ S_j & Z \end{bmatrix} \ \succeq\ 0,\, j \in I_S^{\mathbf{c}};$

(c) $\qquad Z \succeq 0, \mathrm{Tr}(Z) \ =\ 1.$

Next we eliminate the variables S_j, Q_j, R_j. It is clear that

1. (B.4.21.a) is equivalent to the fact that $P_j = Z^{1/2}\widehat{P}_j Z^{1/2}$, $Q_j = Z^{1/2}\widehat{Q}_j Z^{1/2}$ with $\widehat{P}_j, \widehat{Q}_j \succeq 0$, $\widehat{P}_j + \widehat{Q}_j = I_m$. With this parameterization of P_j, Q_j, the corresponding terms in the objective become $-2\rho\mathrm{Tr}([\widehat{P}_j - \widehat{Q}_j](Z^{1/2}A_j Z^{1/2}))$. Note that the matrices A_j are Hermitian (see (B.4.19)), and observe that if $A \in \mathbf{H}^m$, then

$$
\max_{P,Q\in\mathbf{H}^m} \{\mathrm{Tr}([P - Q]A) : 0 \preceq P,Q, P+Q = I_m\} = \|\lambda(A)\|_1 \equiv \sum_\ell |\lambda_\ell(A)|
$$

(w.l.o.g., we may assume that A is Hermitian and diagonal, in which case the relation becomes evident). In view of this observation, partial optimization in P_j, Q_j in (B.4.21) allows to replace in the objective of the problem the terms $-2\rho\mathrm{Tr}([P_j - Q_j]A_j)$ with $2\rho\|\lambda(Z^{1/2}A_j Z^{1/2})\|_1$ and to eliminate the constraints (B.4.21.a).

2. Same as in the proof of Lemma B.4.5, constraints (B.4.21.b) are equivalent to the fact that $S_j = -Z^{1/2}U_j Z^{1/2}$ with $\|U_j\|_{2,2} \leq 1$. With this parameterization, the corresponding terms in the objective become $2\rho\Re\{\mathrm{Tr}(U_j(Z^{1/2}R_j^H L_j Z^{1/2}))\}$, and the maximum of this expression in U_j, $\|U_j\|_{2,2} \leq 1$, is $2\rho\|\sigma(Z^{1/2}R_j^H L_j Z^{1/2})\|_1$. With this observation, partial optimization in S_j in (B.4.21) allows to replace in the objective the terms $-2\rho\Re\{\mathrm{Tr}(S_j R_j^H L_j)\}$ with $2\rho\|\sigma(Z^{1/2}R_j^H L_j Z^{1/2})\|_1$ and to eliminate the constraints (B.4.21.b).

After the above reductions, problem (B.4.21) becomes

$$
\text{maximize} \quad 2\rho\left[\sum_{j\in I_{\mathsf{S}}^{\mathrm{r}}} \|\lambda(Z^{1/2}A_j Z^{1/2})\|_1 + \sum_{j\in I_{\mathsf{S}}^{\mathsf{C}}} \|\sigma(Z^{1/2}R_j^H L_j Z^{1/2})\|_1 \right.
$$
$$
\left. + \sum_{j\in I_{\mathsf{f}}^{\mathsf{C}}} \|L_j Z^{1/2}\|_2 \|R_j Z^{1/2}\|_2 \right] - \mathrm{Tr}(ZA) \tag{B.4.22}
$$

subject to $\quad Z \succeq 0, \mathrm{Tr}(Z) = 1.$

Recall that we are in the situation when the optimal value in problem (B.4.20), and thus in problem (B.4.22), is positive. Thus, we arrive at an intermediate conclusion as follows.

Lemma B.4.9. Under the premise of Lemma B.4.8, there exists $Z \in \mathbf{H}^m$, $Z \succeq 0$, such that

$$
2\rho\left[\sum_{j\in I_{\mathsf{S}}^{\mathrm{r}}} \|\lambda(Z^{1/2}A_j Z^{1/2})\|_1 + \sum_{j\in I_{\mathsf{S}}^{\mathsf{C}}} \|\sigma(Z^{1/2}R_j^H L_j Z^{1/2})\|_1 \right.
$$
$$
\left. + \sum_{j\in I_{\mathsf{f}}^{\mathsf{C}}} \|L_j Z^{1/2}\|_2 \|R_j Z^{1/2}\|_2 \right] > \mathrm{Tr}(Z^{1/2}AZ^{1/2}). \tag{B.4.23}
$$

Here the Hermitian matrices A_j are given by

$$
2A_j = L_j^H R_j + R_j^H L_j, \ j \in I_{\mathsf{S}}^{\mathrm{r}}. \tag{B.4.24}
$$

B.4.4.2 Second step: probabilistic interpretation of (B.4.23)

The major step in completing the proof of Theorem B.4.2.(iii) is based on a probabilistic interpretation of (B.4.23). This step is described next.

Preliminaries. Let us define a standard Gaussian vector ξ in $\mathbb{R}^n$ (notation: $\xi \in \mathcal{N}_{\mathbb{R}}^n$) as a real Gaussian random n-dimensional vector with zero mean and unit covariance matrix; in other words, ξ_ℓ are independent Gaussian random variables with zero mean and unit variance, $\ell = 1, ..., n$. Similarly, we define a standard Gaussian vector ξ in $\mathbb{C}^n$ (notation: $\xi \in \mathcal{N}_{\mathbb{C}}^n$) as a complex Gaussian random n-dimensional vector with zero mean and unit (complex) covariance matrix. In other words, $\xi_\ell = \alpha_\ell + i\alpha_{n+\ell}$, where $\alpha_1, ..., \alpha_{2n}$ are independent real Gaussian random variables with zero means and variances $\frac{1}{2}$, and i is the imaginary unit.

We shall use the facts established in the next three propositions.

Proposition B.4.10. Let ν be a positive integer, and let $\vartheta_{\mathbf{S}}(\nu)$, $\vartheta_{\mathbf{H}}(\nu)$ be given by the relations

$$\vartheta_{\mathbf{S}}^{-1}(\nu) = \min_{\alpha}\left\{\mathbf{E}_{\xi}\left\{|\sum_{\ell=1}^{\nu}\alpha_{\ell}\xi_{\ell}^2|\right\} : \alpha \in \mathbb{R}^{\nu}, \|\alpha\|_1 = 1\right\} \quad [\xi \in \mathcal{N}_{\mathbb{R}}^{\nu}],$$

$$\vartheta_{\mathbf{H}}^{-1}(\nu) = \min_{\alpha}\left\{\mathbf{E}_{\chi}\left\{|\sum_{\ell=1}^{\nu}\alpha_{\ell}|\chi_{\ell}|^2|\right\} : \alpha \in \mathbb{R}^{\nu}, \|\alpha\|_1 = 1\right\} \quad [\chi \in \mathcal{N}_{\mathbb{C}}^{\nu}].$$

(B.4.25)

Then

(i) Both $\vartheta_{\mathbf{S}}(\cdot)$, $\vartheta_{\mathbf{H}}(\cdot)$ are nondecreasing functions such that

$$\begin{array}{llll}
(a.1) & \vartheta_{\mathbf{S}}(1) & = & 1,\ \vartheta_{\mathbf{S}}(2) = \frac{\pi}{2}, \\
(a.2) & \vartheta_{\mathbf{S}}(\nu) & \leq & \frac{\pi}{2}\sqrt{\nu},\ \nu \geq 1; \\
(b.1) & \vartheta_{\mathbf{H}}(1) & = & 1,\ \vartheta_{\mathbf{H}}(2) = 2, \\
(b.2) & \vartheta_{\mathbf{H}}(\nu) & \leq & \vartheta_{\mathbf{S}}(2\nu) \leq \pi\sqrt{\nu/2},\ \nu \geq 1.
\end{array}$$

(B.4.26)

(ii) For every $A \in \mathbf{S}^n$, one has

$$\mathbf{E}_{\xi}\left\{|\xi^T A\xi|\right\} \geq \|\lambda(A)\|_1 \vartheta_{\mathbf{S}}^{-1}(\mathrm{Rank}(A)) \quad [\xi \in \mathcal{N}_{\mathbb{R}}^n], \tag{B.4.27}$$

and for every $A \in \mathbf{H}^n$ one has

$$\mathbf{E}_{\chi}\left\{|\chi^H A\chi|\right\} \geq \|\lambda(A)\|_1 \vartheta_{\mathbf{H}}^{-1}(\mathrm{Rank}(A)) \quad [\chi \in \mathcal{N}_{\mathbb{C}}^n]. \tag{B.4.28}$$

Proof. 1^0. Observe that $\vartheta_{\mathbf{S}}(\cdot)$ satisfies (B.4.27). Indeed, since $\xi \in \mathcal{N}_{\mathbb{R}}^n$ implies that $U\xi \in \mathcal{N}_{\mathbb{R}}^n$ for an orthogonal matrix U, it suffices to verify (B.4.27) for a diagonal matrix $A = \mathrm{Diag}\{\lambda_1, ..., \lambda_{\nu}, 0, ..., 0\}$, where $\nu = \mathrm{Rank}(A)$, in which case (B.4.27) is readily given by the definition of $\vartheta_{\mathbf{S}}(\cdot)$. By construction, $\vartheta_{\mathbf{S}}(\cdot)$ is nondecreasing. To check that $\vartheta_{\mathbf{S}}(\cdot)$ satisfies (B.4.26.a), let $\alpha \in \mathbb{R}^{\nu}$, $\|\alpha\|_1 = 1$, let $\beta = [\alpha; -\alpha] \in \mathbb{R}^{2\nu}$, and let $\xi \in \mathcal{N}_{\mathbb{R}}^{2\nu}$. Let also

$$p_{\nu}(u) = (2\pi)^{-\nu/2}\exp\{-u^T u/2\} : \mathbb{R}^{\nu} \to \mathbb{R}$$

be the density of $\eta \in \mathcal{N}_{\mathbb{R}}^{\nu}$. Setting

$$J = \int\left|\sum_{i=1}^{\nu} u_i^2 \alpha_i\right| p_{\nu}(u)du,$$

we have

$$\mathbf{E}\left\{\left|\sum_{i=1}^{2\nu}\xi_i^2\beta_i\right|\right\} \leq \mathbf{E}\left\{\left|\sum_{i=1}^{\nu}\xi_i^2\alpha_i\right| + \left|\sum_{i=\nu+1}^{2\nu}\xi_i^2\alpha_{i-\nu}\right|\right\} = 2J. \tag{B.4.29}$$

On the other hand, setting $\eta_i = (\xi_i - \xi_{i+\nu})/\sqrt{2}$, $\zeta_i = (\xi_i + \xi_{i+\nu})/\sqrt{2}$, we get

$$\left|\sum_{i=1}^{2\nu}\xi_i^2\beta_i\right| = \left|\sum_{i=1}^{\nu}2\alpha_i\eta_i\zeta_i\right| = 2\left|\widehat{\eta}^T\zeta\right|, \quad \widehat{\eta} = [\alpha_1\eta_1; ...; \alpha_{\nu}\eta_{\nu}],\ \zeta = [\zeta_1; ...; \zeta_{\nu}].$$

(B.4.30)

Note that $\zeta \in \mathcal{N}_{\mathbb{R}}^{\nu}$ and $\widehat{\eta}$, ζ are independent. Setting $\widetilde{\eta} = [|\alpha_1 \eta_1|; ...; |\alpha_\nu \eta_\nu|]$, we have

$$
\begin{aligned}
\mathbf{E}\left\{|\widehat{\eta}^T \zeta|\right\} &= \mathbf{E}\left\{\|\widehat{\eta}\|_2\right\} \int |t| p_1(t) dt \\
&\qquad [\text{since } \widehat{\eta}, \zeta \text{ are independent and } \zeta \in \mathcal{N}_{\mathbb{R}}^{\nu}] \\
&= \mathbf{E}\left\{\|\widehat{\eta}\|_2\right\} \frac{2}{\sqrt{2\pi}} = \frac{2}{\sqrt{2\pi}} \mathbf{E}\left\{\|\widetilde{\eta}\|_2\right\} \\
&\geq \frac{2}{\sqrt{2\pi}} \|\mathbf{E}\left\{\widetilde{\eta}\right\}\|_2 = \frac{2}{\sqrt{2\pi}} \sqrt{\sum_{i=1}^{\nu} \alpha_i^2 \left(\frac{2}{\sqrt{2\pi}}\right)^2} \geq \frac{2}{\pi\sqrt{\nu}}.
\end{aligned}
\tag{B.4.31}
$$

Combining (B.4.29), (B.4.30) and (B.4.31), we get $2J \geq \frac{4}{\pi\sqrt{\nu}}$, i.e., $\frac{1}{J} \leq \frac{\pi\sqrt{\nu}}{2}$, which yields (B.4.26.$a$.2). Relation (B.4.26.$a$.1) is given by the following computation:

$$
\begin{aligned}
\frac{1}{\vartheta_{\mathbf{S}}(2)} &= \min_{\substack{\alpha \in \mathbb{R}^2, \\ \|\alpha\|_1 = 1}} \left\{\int |\alpha_1 u_1^2 + \alpha_2 u_2^2| p_2(u) du\right\} = \min_{\theta \in [0,1]} \underbrace{\int |\theta u_1^2 - (1-\theta) u_2^2| p_2(u) du}_{f(\theta)} \\
&= \frac{1}{2} \int |u_1^2 - u_2^2| p_2(u) du \\
&\qquad [\text{since } f(\theta) \text{ is convex and symmetric w.r.t. } \theta = 1/2] \\
&= \left[\int |t| p_1(t) dt\right]^2 = \frac{2}{\pi}.
\end{aligned}
$$

$\mathbf{2^0}$. From the definition of $\vartheta_{\mathbf{H}}(\cdot)$ it is clear that this function is nondecreasing. To establish (B.4.28), by the same reasons as in the case of (B.4.27), it suffices to verify (B.4.28) when $A = \text{Diag}\{\lambda_1, ..., \lambda_\nu, 0, ..., 0\}$, where $\nu = \text{Rank}(A)$, in which case (B.4.28) is readily given by the definition of $\vartheta_{\mathbf{H}}(\cdot)$.

It remains to verify (B.4.26.b). The relation $\vartheta_{\mathbf{H}}(1) = 1$ is evident. Further, we clearly have

$$
\vartheta_{\mathbf{H}}^{-1}(2) = \min_{\beta \in [0,1]} \psi(\beta), \quad \psi(\beta) = \mathbf{E}_\chi\left\{|\beta|\chi_1|^2 - (1-\beta)|\chi_2|^2|\right\}, \quad \chi \in \mathcal{N}_{\mathbb{C}}^2.
$$

The function $\psi(\beta)$ is convex in $\beta \in [0,1]$ and is symmetric: $\psi(1-\beta) = \psi(\beta)$. It follows that its minimum is achieved at $\beta = \frac{1}{2}$; direct computation demonstrates that $\psi(1/2) = 1/2$, which completes the proof of (B.4.26.b.1).

It remains to prove the first inequality in (B.4.26.b.2). Given $\alpha \in \mathbb{R}^\nu$, $\|\alpha\|_1 = 1$, let $\widetilde{\alpha} = [\alpha; \alpha] \in \mathbb{R}^{2\nu}$. Now, if $\chi = \eta + \imath\omega$ is a standard Gaussian vector in $\mathbb{C}^\nu$, then the vector $\xi = 2^{1/2}[\eta; \omega]$ is a standard Gaussian vector in $\mathbb{R}^{2\nu}$. We now have

$$
\begin{aligned}
\mathbf{E}_\chi\left\{|\sum_{\ell=1}^{\nu} \alpha_\ell |\chi_\ell|^2|\right\} &= \mathbf{E}_\chi\left\{|\sum_{\ell=1}^{\nu} \alpha_\ell [\eta_\ell^2 + \omega_\ell^2]|\right\} = \frac{1}{2} \mathbf{E}_\xi\left\{|\sum_{\ell=1}^{2\nu} \widetilde{\alpha}_\ell \xi_\ell^2|\right\} \\
&\geq \frac{1}{2} \|\widetilde{\alpha}\|_1 \vartheta_{\mathbf{S}}^{-1}(2\nu) = \vartheta_{\mathbf{S}}^{-1}(2\nu),
\end{aligned}
$$

whence $\vartheta_{\mathbf{H}}^{-1}(\nu) \geq \vartheta_{\mathbf{S}}^{-1}(2\nu)$, and the desired inequality follows. $\qquad\square$

Proposition B.4.11. For every $A \in \mathbb{C}^{n \times n}$ one has

$$
\mathbf{E}_\eta\left\{|\eta^H A \eta|\right\} \geq \|\sigma(A)\|_1 \frac{1}{4} \vartheta_{\mathbf{H}}^{-1}(2\text{Rank}(A)) \quad [\eta \in \mathcal{N}_{\mathbb{C}}^n].
\tag{B.4.32}
$$

Proof. Let $\widehat{A} = \begin{bmatrix} & A \\ A^H & \end{bmatrix}$, so that $\widehat{A} \in \mathbf{H}^{2n}$, $\mathrm{Rank}(\widehat{A}) = 2\mathrm{Rank}(A)$ and the eigenvalues of $\widehat{A}$ are $\pm\sigma_\ell(A)$, $\ell = 1, ..., n$. Let also $\chi = [\eta; \omega]$ be a standard Gaussian vector in $\mathbb{C}^{2n}$ partitioned into two n-dimensional blocks, so that η, ω are independent standard Gaussian vectors in $\mathbb{C}^n$. We have

$$
\begin{aligned}
\chi^H \widehat{A} \chi &= 2\Re\{\eta^H A \omega\} \\
&= \Re\Big\{ \big[(\eta+\omega)^H A(\eta+\omega) - \eta^H A\eta - \omega^H A\omega \big] \\
&\qquad + \imath \big[(\eta - \imath\omega)^H A(\eta - \imath\omega) - \eta^H A\eta - \omega^H A\omega \big] \Big\} \\
&\qquad\qquad\qquad\qquad\qquad\qquad \text{[polarization identity]}
\end{aligned}
\tag{B.4.33}
$$

$$
\begin{aligned}
\Rightarrow \mathbf{E}_\chi\left\{ |\chi^H \widehat{A}\chi| \right\} \leq\ & \mathbf{E}_{\eta,\omega}\left\{ |(\eta+\omega)^H A(\eta+\omega)| \right\} \\
&+ \mathbf{E}_{\eta,\omega}\left\{ |(\eta - \imath\omega)^H A(\eta - \imath\omega)| \right\} \\
&+ 2\mathbf{E}_\eta\left\{ |\eta^H A\eta| \right\} + 2\mathbf{E}_\omega\{ |\omega^H A\omega| \}.
\end{aligned}
$$

Since η, ω are independent standard Gaussian vectors in $\mathbb{C}^n$, the vectors $2^{-1/2}(\eta+\omega)$ and $2^{-1/2}(\eta - \imath\omega)$ also are standard Gaussian. Therefore (B.4.33) implies that

$$
\mathbf{E}_\chi\left\{ |\chi^H \widehat{A}\chi| \right\} \leq 8\mathbf{E}_\eta\left\{ |\eta^H A\eta| \right\}.
\tag{B.4.34}
$$

Since $\widehat{A}$ is a Hermitian matrix of rank $2\mathrm{Rank}(A)$ and $\|\lambda(\widehat{A})\|_1 = 2\|\sigma(A)\|_1$, the left hand side in (B.4.34), by (B.4.28), is $\geq 2\|\sigma(A)\|_1 \vartheta_{\mathbf{H}}^{-1}(2\mathrm{Rank}(A))$, and (B.4.34) implies (B.4.32). $\qquad\square$

Proposition B.4.12. (i) Let $L \in \mathbb{C}^{p \times n}, R \in \mathbb{C}^{q \times n}$, and let χ be a standard Gaussian vector in $\mathbb{C}^n$. Then

$$
\mathbf{E}_\chi\left\{ \|L\chi\|_2 \|R\chi\|_2 \right\} \geq \frac{\pi}{4} \|L\|_2 \|R\|_2.
\tag{B.4.35}
$$

(ii) Let $L \in \mathbb{R}^{p \times n}, R \in \mathbb{R}^{q \times n}$, and let ξ be a standard Gaussian vector in $\mathbb{R}^n$. Then

$$
\mathbf{E}_\xi\left\{ \|L\xi\|_2 \|R\xi\|_2 \right\} \geq \frac{2}{\pi} \|L\|_2 \|R\|_2.
\tag{B.4.36}
$$

Proof. (i): There is nothing to prove when L or R are zero matrices; thus, assume that both L and R are nonzero.

Let us demonstrate first that it suffices to verify (B.4.35) in the case when both L and R are rank 1 matrices. Let $L^H L = U^H \mathrm{Diag}\{\lambda\} U$ be the eigenvalue decomposition of $L^H L$, so that U is a unitary matrix and $\lambda \geq 0$. We have

$$
\begin{aligned}
\mathbf{E}\left\{\|L\xi\|_2\|R\xi\|_2\right\} &= \mathbf{E}\left\{\sqrt{\xi^H L^H L\xi}\|R\xi\|_2\right\} \\
&= \mathbf{E}\left\{\left((U\xi)^H \mathrm{Diag}\{\lambda\}\underbrace{(U\xi)}_{\chi}\right)^{1/2}\underbrace{\|RU^H\chi\|_2}_{\phi(\chi)\geq 0}\right\} \\
&= \mathbf{E}\left\{\phi(\chi)\sqrt{\sum_{\ell=1}^n \lambda_\ell|\chi_\ell|^2}\right\} = \Phi(\lambda),
\end{aligned}
\tag{B.4.37}
$$

$$
\Phi(x) = \mathbf{E}\left\{\phi(\chi)\sqrt{\sum_{\ell=1}^n x_\ell|\chi_\ell|^2}\right\}.
$$

The function $\Phi(x)$ of $x \in \mathbb{R}^n_+$ is concave; therefore its minimum on the simplex

$$
S = \{x \in \mathbb{R}^n_+ : \sum_\ell x_\ell = \sum_\ell \lambda_\ell\}
$$

is achieved at a vertex, let it be e. Now let $\widehat{L} \in \mathbb{C}^{d\times n}$ be such that $\widehat{L}^H\widehat{L} = U^H \mathrm{Diag}\{e\}U$. Note that $\widehat{L}$ is a rank 1 matrix (since e is a vertex of S) and that

$$
[\|\widehat{L}\|_2^2 =] \quad \mathrm{Tr}(\widehat{L}^H\widehat{L}) = \sum_\ell e_\ell = \sum_\ell \lambda_\ell = \mathrm{Tr}(L^H L) \quad [= \|L\|_2^2].
$$

Since the unitary factor in the eigenvalue decomposition of $\widehat{L}^H\widehat{L}$ is U, (B.4.37) holds true when L is replaced with $\widehat{L}$ and λ with e, so that

$$
\mathbf{E}\left\{\|\widehat{L}\chi\|_2\|R\chi\|_2\right\} = \Phi(e) \leq \Phi(\lambda) = \mathbf{E}\left\{\|L\chi\|_2\|R\chi\|_2\right\}.
$$

Applying the same reasoning to the quantity

$$
\mathbf{E}\left\{\|\widehat{L}\chi\|_2\|R\chi\|_2\right\}
$$

with R playing the role of L, we conclude that there exists a rank 1 matrix $\widehat{R}$ such that

$$
\|\widehat{R}\|_2 = \|R\|_2
$$

and

$$
\mathbf{E}\left\{\|\widehat{L}\chi\|_2\|\widehat{R}\chi\|_2\right\} \leq \mathbf{E}\left\{\|\widehat{L}\chi\|_2\|R\chi\|_2\right\}.
$$

Thus, replacing L and R with the rank 1 matrices $\widehat{L}$, $\widehat{R}$, we do not increase the left hand side in (B.4.35) and do not vary the right hand side, so that it indeed suffices to establish (B.4.35) in the case when L, R are rank 1 matrices. Note that so far our reasoning did not use the fact that χ is standard Gaussian.

Now let us look what inequality (B.4.35) says in the case of rank 1 matrices L, R. By homogeneity, we can further assume that $\|L\|_2 = \|R\|_2 = 1$. With this normalization, for rank 1 matrices L, R we clearly have $L\chi = z\ell$ and $R\chi = wr$ for unit deterministic vectors ℓ, r and a Gaussian random vector $[z; w] \in \mathbb{C}^2 = \mathbb{R}^4$ such that $\mathbf{E}\{|z|^2\} = \mathbf{E}\{|w|^2\} = 1$ (both z and w are just linear combinations, with appropriate deterministic coefficients, of the entries in χ). Since $\mathbf{E}\{|z|^2\} = \mathbf{E}\{|w|^2\} = 1$, we can express (z, w) in terms of a *standard* Gaussian vector $[\eta; \xi] \in$

$\mathbb{C}^2$ as $z = \eta$, $w = \cos(\theta)\eta + \sin(\theta)\xi$, where $\theta \in [0, \frac{\pi}{2}]$ is such that $\cos(\theta)$ is the absolute value of the correlation $\mathbf{E}\{z\overline{w}\}$ between z and w. With this representation, inequality (B.4.35) becomes

$$\phi(\theta) \equiv \int_{\mathbb{C}\times\mathbb{C}} |\eta||\cos(\theta)\eta + \sin(\theta)\xi|dG(\eta, \xi) \geq \frac{\pi}{4} \equiv \phi(\frac{\pi}{2}), \qquad (B.4.38)$$

where $G(\eta, \xi)$ is the distribution of $[\eta; \xi]$. We should prove (B.4.38) in the range $[0, \frac{\pi}{2}]$ of values of θ; in fact we shall prove this inequality in the larger range $\theta \in [0, \pi]$. Given $\theta \in [0, \pi]$, we set

$$u = \cos(\theta/2)\eta + \sin(\theta/2)\xi, \quad v = -\sin(\theta/2)\eta + \cos(\theta/2)\xi;$$

it is immediately seen that the distribution of (u, v) is exactly G. At the same time,

$$\eta = \cos(\theta/2)u - \sin(\theta/2)v, \quad \cos(\theta)\eta + \sin(\theta)\xi = \cos(\theta/2)u + \sin(\theta/2)v,$$

whence

$$\begin{aligned}\phi(\theta) &= \int_{\mathbb{C}\times\mathbb{C}} |\cos(\theta/2)u - \sin(\theta/2)v||\cos(\theta/2)u + \sin(\theta/2)v|dG(u, v) \\ &= \int_{\mathbb{C}\times\mathbb{C}} |\cos^2(\theta/2)u^2 - \sin^2(\theta/2)v^2|dG(u, v).\end{aligned}$$

We see that

$$\min_{\theta \in [0,\pi]} \phi(\theta) = \min_{0 \leq \alpha \leq 1} \psi(\alpha), \quad \psi(\alpha) = \int_{\mathbb{C}\times\mathbb{C}} |\alpha u^2 - (1 - \alpha)v^2|dG(u, v).$$

The function $\psi(\alpha)$ clearly is convex and $\psi(1 - \alpha) = \psi(\alpha)$ (since the distribution of $[u; v]$ is symmetric in u, v). Consequently, ψ attains its minimum when $\alpha = 1/2$, and ϕ attains its minimum when $\cos^2(\theta/2) = 1/2$, i.e., when $\theta = \pi/2$, which is exactly what is stated in (B.4.38).

(ii): Applying exactly the same reasoning as in the proof of (i), we conclude that it suffices to verify (B.4.36) in the case when L, R are real rank 1 matrices. In this case, the same argument as above demonstrates that (B.4.36) is equivalent to the fact that if ξ, η are independent real standard Gaussian variables and $G(\xi, \eta)$ is the distribution of $[\xi; \eta]$, then the function

$$\phi(\theta) = \int_{\mathbb{R}\times\mathbb{R}} |\xi||\cos(\theta)\xi + \sin(\theta)\eta|dG(\xi, \eta) \qquad (B.4.39)$$

of $\theta \in [0, \pi]$ achieves its minimum when $\theta = \frac{\pi}{2}$. To prove this statement, one can repeat word by word, with evident modifications, the reasoning we have used in the complex case. $\qquad \square$

B.4.4.3 Completing the proof of Theorem B.4.2.(iii)

We are now in a position to complete the proof of Theorem B.4.2.(iii). Let us set

$$
\begin{array}{rcl}
p_{\mathbb{R}}^{\mathrm{s}} & = & 2\max\left\{p_j : j \in I_{\mathrm{S}}^{\mathrm{r}}\right\}, \\
p_{\mathbb{C}}^{\mathrm{s}} & = & 2\max\left\{p_j : j \in I_{\mathrm{S}}^{\mathrm{C}}\right\}, \\
\vartheta_{\mathbf{S}} & = & \max\left[\vartheta_{\mathbf{H}}(p_{\mathbb{R}}^{\mathrm{s}}), 4\vartheta_{\mathbf{H}}(p_{\mathbb{C}}^{\mathrm{s}}), \tfrac{4}{\pi}\right];
\end{array}
\tag{B.4.40}
$$

here by definition the maximum over an empty set is 0, and $\vartheta_{\mathbf{H}}(0) = 0$. Note that by (B.4.26) one has

$$
\vartheta_{\mathbf{S}} \leq 4\pi\sqrt{p^{\mathrm{s}}}
$$

(cf. (B.4.4), (B.4.5)).

Let χ be a standard Gaussian vector in $\mathbb{C}^n$. Invoking Propositions B.4.10 — B.4.12, we have (for notation, see Lemma B.4.9):

$$
\begin{aligned}
&\|\lambda(Z^{1/2}A_j Z^{1/2})\|_1 \\
\leq\ & \vartheta_{\mathbf{H}}(\mathrm{Rank}(Z^{1/2}A_j Z^{1/2}))\mathbf{E}_\chi\left\{|\chi^H Z^{1/2}A_j Z^{1/2}\chi|\right\} \\
\leq\ & \vartheta_{\mathbf{S}}\mathbf{E}_\chi\left\{|\chi^H Z^{1/2}A_j Z^{1/2}\chi|\right\},\ j \in I_{\mathrm{S}}^{\mathrm{r}} \\
& \left[\begin{array}{l} \text{by Proposition B.4.10 since } A_j = A_j^H \text{ and} \\ \mathrm{Rank}(A_j) = \mathrm{Rank}([L_j^H R_j + R_j^H L_j]) \leq 2p_j \end{array}\right] \\
&\|\sigma(Z^{1/2}R_j^H L_j Z^{1/2})\|_1 \\
\leq\ & 4\vartheta_{\mathbf{H}}(2\mathrm{Rank}(Z^{1/2}R_j^H L_j Z^{1/2}))\mathbf{E}_\chi\left\{|\chi^H Z^{1/2}R_j^H L_j Z^{1/2}\chi|\right\} \\
\leq\ & \vartheta_{\mathbf{S}}\mathbf{E}_\chi\left\{|\chi^H Z^{1/2}R_j^H L_j Z^{1/2}\chi|\right\},\ j \in I_{\mathrm{S}}^{\mathrm{C}} \\
& \quad\quad [\text{by Proposition B.4.11 since } \mathrm{Rank}(R_j^H L_j) \leq p_j] \\
&\|L_j Z^{1/2}\|_2\|R_j Z^{1/2}\|_2 \\
\leq\ & \tfrac{4}{\pi}\mathbf{E}_\chi\left\{\|L_j Z^{1/2}\chi\|_2\|R_j Z^{1/2}\chi\|_2\right\} \\
\leq\ & \vartheta_{\mathbf{S}}\mathbf{E}_\chi\left\{\|L_j Z^{1/2}\chi\|_2\|R_j Z^{1/2}\chi\|_2\right\} \\
& \quad\quad\quad\quad\quad\quad [\text{by Proposition B.4.12.(i)}]
\end{aligned}
$$

and, of course,

$$
\mathbf{E}_\chi\left\{\chi^H Z^{1/2}A Z^{1/2}\chi\right\} = \mathrm{Tr}(Z^{1/2}A Z^{1/2}).
$$

In view of these observations, (B.4.23) implies that

$$
\begin{aligned}
\rho\vartheta_{\mathbf{S}}\Bigg[& \sum_{j\in I_{\mathrm{S}}^{\mathrm{r}}}\mathbf{E}_\chi\left\{|\chi^H Z^{1/2}[L_j^H R_j + R_j^H L_j]Z^{1/2}\chi|\right\} \\
& + \sum_{j\in I_{\mathrm{S}}^{\mathrm{C}}}\mathbf{E}_\chi\left\{2|\chi^H Z^{1/2}R_j^H L_j Z^{1/2}\chi|\right\} \\
& \quad + \sum_{j\in I_{\mathrm{f}}^{\mathrm{C}}}\mathbf{E}_\chi\left\{2\|L_j Z^{1/2}\chi\|_2\|R_j Z^{1/2}\chi\|_2\right\}\Bigg] > \mathbf{E}_\chi\left\{\chi^H Z^{1/2}A Z^{1/2}\chi\right\}
\end{aligned}
$$

(we have substituted the expressions for A_j, see (B.4.24)). It follows that there exists a realization $\widehat{\chi}$ of χ such that with $\xi = Z^{1/2}\widehat{\chi}$ one has

$$
\begin{aligned}
& \rho\vartheta_{\mathbf{S}}\Bigg[\sum_{j\in I_{\mathrm{S}}^{\mathrm{r}}}|\xi^H[L_j^H R_j + R_j^H L_j]\xi| + \sum_{j\in I_{\mathrm{S}}^{\mathrm{C}}}2|\xi^H R_j^H L_j \xi| + \sum_{j\in I_{\mathrm{f}}^{\mathrm{C}}}2\|L_j\xi\|_2\|R_j\xi\|_2\Bigg] \\
& > \xi^H A\xi.
\end{aligned}
\tag{B.4.41}
$$

Observe that

- The quantities $\xi^H[L_j^H R_j + R_j^H L_j]\xi$ are real; we therefore can choose $\theta_j = \pm 1$, $j \in I_S^r$, in such a way that with $\chi^j = \theta_j I_{p_j}$ one has

$$\xi^H[L_j^H \chi^j R_j + R_j^H [\chi^j]^H L_j]\xi = |\xi^H[L_j^H R_j + R_j^H L_j]\xi|, \ j \in I_S^r;$$

- For $j \in I_S^C$, we can choose $\theta_j \in \mathbb{C}$, $|\theta_j| = 1$, in such a way that with $\Theta^j = \theta_j I_{p_j}$ one has

$$\xi^H[L_j^H \Theta^j R_j + R_j^H [\Theta^j]^H L_j]\xi = 2|\xi^H R_j^H L_j \xi|, \ j \in I_S^C;$$

- For $j \in I_f^C$, we can choose $\Theta^j \in \mathbb{C}^{p_j \times q_j}$, $\|\Theta^j\|_{2,2} \leq 1$, in such a way that

$$\xi^H[L_j^H \Theta^j R_j + R_j^H [\Theta^j]^H L_j]\xi = 2\|L_j \xi\|_2 \|R_j \xi\|_2, \ j \in I_f^C.$$

With Θ^j's we have defined, (B.4.41) reads

$$\xi^H \underbrace{\left[A - \rho \vartheta_S \sum_{j=1}^{L} [L_j^H \Theta^j R_j + R_j^H [\Theta^j]^H L_j] \right]}_{C} \xi < 0,$$

so that C is not positive semidefinite; on the other hand, by construction $C \in \mathcal{U}[\vartheta_S \rho]$. Thus, the predicate $\mathcal{A}(\vartheta_S \rho)$ is not valid; recalling the definition of ϑ_S, this completes the proof of Lemma B.4.8, and thus the proof of Theorem B.4.2.(iii). □

B.4.5 Proof of Theorem B.4.2.(iv)

The fact that $\mathcal{A}(\rho)$ is equivalent to $\mathcal{B}(\rho)$ in the case of $L = 1$ is evident when the only perturbation block in question is a real scalar one, is readily given by Lemma B.4.5 when the block is a complex scalar one, and is readily given by Lemma B.4.6 when the block is full. □

B.4.6 Matrix Cube Theorem, Real Case

The Real Matrix Cube problem is as follows:

RMC: Let m, $p_1, q_1, ..., p_L, q_L$ be positive integers, and $A \in \mathbf{S}^m$, $L_j \in \mathbb{R}^{p_j \times m}$, $R_j \in \mathbb{R}^{q_j \times m}$ be given matrices, $L_j \neq 0$. Let also a partition $\{1, 2, ..., L\} = I_S^r \cup I_f^r$ of the index set $\{1, ..., L\}$ into two non-overlapping sets be given. With these data, we associate a parametric family of "matrix boxes"

$$\mathcal{U}[\rho] = \left\{ A + \rho \sum_{j=1}^{L} [L_j^T \Theta^j R_j + R_j^T [\Theta^j]^T L_j] : \Theta^j \in \mathcal{Z}^j, 1 \leq j \leq L \right\}$$
$$\subset \mathbf{S}^m,$$

(B.4.42)

where $\rho \geq 0$ is the parameter and

$$
\mathcal{Z}^j = \begin{cases}
\{\theta I_{p_j} : \theta \in \mathbb{R}, |\theta| \leq 1\}, j \in I_{\mathrm{s}}^{\mathrm{r}} \\
\quad [\text{"scalar perturbation blocks"}] \\
\{\Theta^j \in \mathbb{R}^{p_j \times q_j} : \|\Theta^j\|_{2,2} \leq 1\}, j \in I_{\mathrm{f}}^{\mathrm{r}} \\
\quad [\text{"full perturbation blocks"}]
\end{cases}
\qquad (B.4.43)
$$

Given $\rho \geq 0$, check whether

$$
\mathcal{U}[\rho] \subset \mathbf{S}_+^m \qquad\qquad \mathcal{A}(\rho)
$$

Remark B.4.13. In the sequel, we always assume that $p_j > 1$ for $j \in I_{\mathrm{s}}^{\mathrm{r}}$. Indeed, non-repeated ($p_j = 1$) scalar perturbations always can be regarded as full perturbations.

Consider, along with predicate $\mathcal{A}(\rho)$, the predicate

$$
\begin{aligned}
&\exists Y_j \in \mathbf{S}^m, j = 1, ..., L : \\
&(a) \quad Y_j \succeq L_j^T \Theta^j R_j + R_j^T [\Theta^j]^T L_j \ \forall \, (\Theta^j \in \mathcal{Z}^j, 1 \leq j \leq L) \\
&(b) \quad A - \rho \sum_{j=1}^L Y_j \succeq 0.
\end{aligned}
\qquad \mathcal{B}(\rho)
$$

The Real case version of Theorem B.4.2 is as follows:

Theorem B.4.14. [The Real Matrix Cube Theorem [10, 12]] One has:

(i) Predicate $\mathcal{B}(\rho)$ is stronger than $\mathcal{A}(\rho)$ — the validity of the former predicate implies the validity of the latter one.

(ii) $\mathcal{B}(\rho)$ is computationally tractable — the validity of the predicate is equivalent to the solvability of the system of LMIs

$$
\begin{aligned}
(s) &\qquad Y_j \pm \left[L_j^T R_j + R_j^T L_j\right] \ \succeq \ 0, j \in I_{\mathrm{s}}^{\mathrm{r}}, \\[2mm]
(f) &\qquad \begin{bmatrix} Y_j - \lambda_j L_j^T L_j & R_j^T \\ R_j & \lambda_j I_{p_j} \end{bmatrix} \ \succeq \ 0, j \in I_{\mathrm{f}}^{\mathrm{r}} \\[2mm]
(*) &\qquad\qquad\qquad A - \rho \sum_{j=1}^L Y_j \ \succeq \ 0.
\end{aligned}
\qquad (B.4.44)
$$

in matrix variables $Y_j \in \mathbf{S}^m$, $j = 1, ..., L$, and real variables λ_j, $j \in I_{\mathrm{f}}^{\mathrm{r}}$.

(iii) "The gap" between $\mathcal{A}(\rho)$ and $\mathcal{B}(\rho)$ can be bounded solely in terms of the maximal rank

$$
p^{\mathrm{s}} = \max_{j \in I_{\mathrm{s}}^{\mathrm{r}}} \mathrm{Rank}(L_j^T R_j + R_j^T L_j)
$$

of the scalar perturbations. Specifically, there exists a universal function $\vartheta_{\mathbb{R}}(\cdot)$ satisfying the relations

$$
\vartheta_{\mathbb{R}}(2) = \frac{\pi}{2}; \vartheta_{\mathbb{R}}(4) = 2; \vartheta_{\mathbb{R}}(\mu) \leq \pi\sqrt{\mu}/2 \ \forall \mu \geq 1
$$

such that with $\mu = \max[2, p^{\mathrm{s}}]$ one has

$$
\text{if } \mathcal{B}(\rho) \text{ is not valid, then } \mathcal{A}(\vartheta_{\mathbb{R}}(\mu)\rho) \text{ is not valid.} \qquad (B.4.45)
$$

(iv) Finally, in the case $L = 1$ of single perturbation block $\mathcal{A}(\rho)$ is equivalent to $\mathcal{B}(\rho)$.

The proof of the Real Matrix Cube Theorem repeats word by word, with evident simplifications, the proof of its complex case counterpart and is therefore omitted. Note that Remark B.4.4 remains valid in the real case.

B.5 PROOFS FOR CHAPTER 10

B.5.1 Proof of Theorem 10.1.2

Let

$$\mathrm{Erf}(t) = \tfrac{1}{\sqrt{2\pi}} \int\limits_{t}^{\infty} \exp\{-s^2/2\}ds,$$

$$\mathrm{ErfInv}(r) : \tfrac{1}{\sqrt{2\pi}} \int\limits_{\mathrm{ErfInv}(r)}^{\infty} \exp\{-s^2/2\}ds = r.$$

Theorem B.5.1. Let $\zeta \sim \mathcal{N}(0, I_m)$, and let Q be a closed convex set in $\mathbb{R}^m$ such that

$$\mathrm{Prob}\{\zeta \in Q\} \geq \chi > \frac{1}{2}. \tag{B.5.1}$$

Then

(i) Q contains the centered at the origin $\|\cdot\|_2$-ball of the radius

$$r(\chi) = \mathrm{ErfInv}(1 - \chi) > 0. \tag{B.5.2}$$

(ii) If Q contains the centered at the origin $\|\cdot\|_2$-ball of a radius $r \geq r(\chi)$, then

$$\forall \alpha \in [1, \infty) : \mathrm{Prob}\{\zeta \notin \alpha Q\} \leq \mathrm{Erf}\left(\mathrm{ErfInv}(1 - \chi) + (\alpha - 1)r\right)$$
$$\leq \mathrm{Erf}\left(\alpha \mathrm{ErfInv}(1 - \chi)\right) \leq \tfrac{1}{2} \exp\left\{-\tfrac{\alpha^2 \mathrm{ErfInv}^2(1-\chi)}{2}\right\}. \tag{B.5.3}$$

In particular, for a closed and convex set Q, $\zeta \sim \mathcal{N}(0, \Sigma)$ and $\alpha \geq 1$ one has

$$\mathrm{Prob}\{\zeta \notin Q\} \leq \delta < \tfrac{1}{2} \Rightarrow$$
$$\mathrm{Prob}\{\zeta \notin \alpha Q\} \leq \mathrm{Erf}(\alpha \mathrm{ErfInv}(\delta)) \leq \tfrac{1}{2} \exp\{-\tfrac{\alpha^2 \mathrm{ErfInv}^2(\delta)}{2}\}. \tag{B.5.4}$$

Proof. (i) is immediate. Indeed, assuming the opposite and invoking the Separation Theorem, Q is contained in a closed half-space $\Pi = \{x : e^T x \leq r\}$ with a unit vector e and certain $r < \mathrm{ErfInv}(\chi)$, and therefore $\mathrm{Prob}\{\eta \notin Q\} \geq \mathrm{Prob}\{\eta \notin \Pi\} = \mathrm{Erf}(r) > \chi$, which is a contradiction.

(ii): This is an immediate corollary of the following fact due to C. Borell [31]:

(!) For every $\alpha > 0$, $\epsilon \geq 0$ and every closed set $X \subset \mathbb{R}^k$ such that $\mathrm{Prob}\{\zeta \in X\} \geq \alpha$ one has

$$\mathrm{Prob}\left\{\mathrm{dist}(\zeta, X) > \epsilon\right\} \leq \mathrm{Erf}(\mathrm{ErfInv}(1 - \alpha) + \epsilon)$$

where $\mathrm{dist}(a, X) = \min\limits_{x \in X} \|a - x\|_2$.

The derivation of (ii) from (!) is as follows. Since Q contains the centered at the origin $\|\cdot\|_2$-ball B_r of the radius r, the set αQ, $\alpha \geq 1$, contains $Q + (\alpha - 1)Q \supset Q + (\alpha - 1)B_r$ and thus contains the set $\{x : \operatorname{dist}(x, Q) \leq \epsilon = (\alpha - 1)r\}$. Invoking (!), we arrive at the first inequality in (B.5.3); the second inequality is due to $r \geq r(\chi) = \operatorname{ErfInv}(1 - \chi)$, and the third is well known.

> Here is the demonstration of the inequality $\operatorname{Erf}(s) \leq \frac{1}{2}\exp\{-s^2/2\}$, $s \geq 0$. This is equivalent to $\frac{1}{2} \geq \int_s^\infty \exp\{(s^2 - r^2)/2\}(2\pi)^{-1/2}dr$, that is, $\frac{1}{2} \geq \int_0^\infty \exp\{(s^2 - (s+t)^2)/2\}(2\pi)^{-1/2}dt$, which indeed is true, since the latter integral is $\leq \int_0^\infty \exp\{-t^2/2\}(2\pi)^{-1/2}dt = 1/2$.

$\hfill\square$

B.5.2 Proof of Proposition 10.3.2

Items 1, 2, 3 are evident.

Item 4: Let $f(x, y) \in \mathcal{CF}_{r+s}$ ($x \in \mathbb{R}^r$, $y \in \mathbb{R}^s$), The functions $p(x) = \int f(x, y)dP_2(y)$, $q(x) = \int f(x, y)dQ_2(y)$ clearly belong to $\mathcal{CF}_r$ and $p \leq q$ due to $P_2 \preceq_c Q_2$. We have $\int f(x, y)d(P_1 \times P_2)(x, y) = \int p(x)dP_1(x) \leq \int p(x)dQ_1(x) \leq \int q(x)dQ_1(x) = \int f(x, y)d(Q_1 \times Q_2)(x, y)$, where the first $\leq$ follows from $P_1 \preceq_c Q_1$, and the second $\leq$ is given by $p \leq q$. The resulting inequality shows that $P_1 \times P_2 \preceq_c Q_1 \times Q_2$. $\hfill\square$

Item 5: Given $f \in \mathcal{CF}_m$, let us set $g(u_1, ..., u_k) = f(\sum_{i=1}^k S_i u_i)$, so that $g \in \mathcal{CF}_{kn}$. By item 4, we have $[\xi_1; ...; \xi_k] \preceq_c [\eta_1; ...; \eta_k]$, whence $\mathbf{E}\{g(\xi_1, ..., \xi_k)\} \leq \mathbf{E}\{g(\eta_1, ..., \eta_k)\}$, or, which is the same, $\mathbf{E}\{f(\sum_i S_i \xi_i)\} \leq \mathbf{E}\{f(\sum_i S_i \eta_i)\}$. since the latter inequality holds true for all $f \in \mathcal{CF}_m$, we see that $\sum_i S_i \xi_i \preceq_c \sum_i S_i \eta_i$. $\hfill\square$

Item 6: Let $f \in \mathcal{CF}_1$; we should prove that

$$\int f(s)dP_\xi(s) \leq \int f(s)dP_\eta(s). \qquad (*)$$

When we add to f an affine function, both quantities we are comparing change by the same amount (recall that $\mathcal{R}_1$ is comprised of probability distributions with zero mean). It follows that w.l.o.g. we can assume that $f(-1) = 0$ and $f'(-1+0) = 0$, so that f is nonnegative to the left of the point -1. Replacing f in this domain by 0, we preserve convexity, keep the quantity $\int f(s)dP_\xi(s)$ intact and do not increase $\int f(s)dP_\eta(s)$; it follows that it suffices to prove our inequality when f, in addition to $f'(-1+0) = f(-1) = 0$, is identically zero to the left of -1. Now, either f is identically zero on $[-1, 1]$, or $f(1)$ is positive. In the first case, the left hand side in $(*)$ is 0, while the right hand side is nonnegative (since f is nonnegative due to $f(-1) = f'(-1+0) = 0$), and $(*)$ holds true. When $f(1) > 0$, we can, by scaling f, reduce the situation to the one where $f(1) = 1$. In this case, $f(s) \leq (s+1)/2$ on $[-1, 1]$ by convexity, whence, recalling that ξ is supported on $[-1, 1]$ and has zero mean, we have

$$\int f(s)dP_\xi(s) \leq \int \frac{1}{2}(1+s)dP_\xi(s) = \frac{1}{2}$$

On the other hand, let $\alpha = f'(1+0)$, so that $\alpha > 0$. Besides this, f is nonnegative and $f(s) \geq 1 + \alpha(s-1)$ for all s, whence

$$f(s) \geq \max[0, 1 - \alpha + \alpha s] \ \forall s.$$

Consequently, setting $\sigma = \sqrt{\pi/2}$, we have

$$\int f(s)dP_\eta(s) \geq \int \max[0, 1 - \alpha + \alpha s] \frac{1}{\sqrt{2\pi}\sigma} \exp\{-s^2/(2\sigma^2)\}ds$$
$$\geq \min_{\alpha \geq 0} \int [\max[0, 1 - \alpha + \alpha s] \underbrace{\frac{1}{\sqrt{2\pi}\sigma} \exp\{-s^2/(2\sigma^2)\}}_{p(s)} ds.$$

The function $g(\alpha) = \int \max[0, 1 - \alpha + \alpha s]p(s)ds$ clearly is convex in $\alpha \geq 0$, and $g'(\alpha) = \int_{\frac{\alpha-1}{\alpha}}^{\infty} (s-1)p(s)ds$. Taking into account that $\int_0^{\infty} p(s)ds = \int_0^{\infty} sp(s)ds = 1/2$, we conclude that $g'(1) = 0$, that is, $\alpha = 1$ is a minimizer of g so that $g(\alpha) \geq g(1) = 1/2$ whenever $\alpha > 0$. Thus, the right hand side in $(*)$ is $\geq 1/2$ and thus $(*)$ is true. $\quad\square$

Item 7: Denoting by μ, ν the probability distributions of ξ, η respectively, we should verify that

$$\int f(s)d\mu(s) \leq \int f(s)d\nu(s)$$

for every $f \in \mathcal{CF}_1$. Since both ξ and η are symmetrically distributed w.r.t. 0, it suffices to verify this inequality for the case of an even $f \in \mathcal{CF}_1$ (pass from the original $f(x)$ to $\frac{1}{2}(f(x) + f(-x))$). An even convex function is monotone on the nonnegative ray, and it remains to use Proposition 4.4.2. $\quad\square$

Item 8: Due to absolute symmetry, the distribution of ξ is the limit, in the sense of weak convergence, of a sequence of convex combinations of uniform distributions on the vertices of cubes $\{u : \|u\|_\infty \leq r\}$, $r \leq 1$. By item 6, all these distributions are dominated by $\mathcal{N}(0, (\pi/2)I_n)$. It remains to apply item 2. $\quad\square$

Item 9: Since $0 \preceq \Sigma \preceq \Theta$, there exists a nonsingular transformation $x \mapsto Ax : \mathbb{R}^r \to \mathbb{R}^r$ such that the random vectors $\widetilde{\xi} = A\xi$ and $\widetilde{\eta} = A\eta$ are, respectively, $\mathcal{N}(0, \text{Diag}\{\lambda\})$ and $\mathcal{N}(0, \text{Diag}\{\mu\})$; since $\Sigma \preceq \Theta$, we have $\lambda \leq \mu$, whence, by item 3, $\widetilde{\xi} \preceq_c \widetilde{\eta}$, which, of course, is equivalent to $\xi \preceq_c \eta$. $\quad\square$

B.5.3 Proof of Theorem 10.3.3

Let $\zeta \in \mathcal{R}_L$ and $\eta \sim \mathcal{N}(0, I_L)$, and let $\zeta \preceq_c \eta$. Let, further, $Q \subset \mathbb{R}^n$ be a closed convex set such that $\chi \equiv \text{Prob}\{\eta \in Q\} \in (1/2, 1)$. All we need is to prove that whenever $\gamma > 1$, one has

$$\text{Prob}\{\zeta \notin \gamma Q\} \leq \inf_{1 \leq \beta < \gamma} \frac{1}{\gamma - \beta} \int_\beta^{\infty} \text{Erf}(r\text{ErfInv}(1 - \chi))dr. \tag{B.5.5}$$

Indeed, since Q is convex and $\text{Prob}\{\eta \in Q\} > 1/2$, the origin is in the interior of Q. Let $\beta \in [1, \gamma)$, let

$$\theta(x) = \inf\{t : t^{-1}x \in Q\}$$

be the Minkowski function of Q, and let $\delta(x) = \max[\theta(x) - \beta, 0]$. We clearly have $\delta(\cdot) \in \mathcal{CF}_n$, so that

$$\int \delta(x)dP_\zeta(x) \le \int \delta(x)dP_\eta(x). \qquad (a)$$

For $r \ge \beta$ let $p(r) = \text{Prob}\{\eta \notin rQ\} = \text{Prob}\{\delta(\eta) > r - \beta\}$. By Theorem B.5.1, for $r \ge \beta$ we have

$$p(r) \le \text{Erf}(r\text{ErfInv}(1 - \chi)). \qquad (b)$$

We have

$$\int \delta(x)dP_\eta(x) = -\int_\beta^\infty (r - \beta)dp(r) = \int_\beta^\infty p(r)dr \le \int_\beta^\infty \text{Erf}(r\text{ErfInv}(1 - \chi))dr,$$

whence

$$\int \delta(x)dP_\zeta(x) \le \int_\beta^\infty \text{Erf}(r\text{ErfInv}(1 - \chi))dr$$

by (a). Now, when $\zeta \notin \gamma Q$, we have $\delta(\zeta) \ge \gamma - \beta$. Invoking Tschebyshev Inequality, we arrive at

$$\text{Prob}\{\zeta \notin \gamma Q\} \le \frac{\mathbf{E}\{\delta(\zeta)\}}{\gamma - \beta} \le \frac{1}{\gamma - \beta}\int_\beta^\infty \text{Erf}(r\text{ErfInv}(1 - \chi))dr.$$

The resulting inequality holds true for all $\beta \in [1, \gamma)$, and (B.5.5) follows. $\qquad \square$

B.5.4 Conjecture 10.1

The validity of Conjecture 10.1 with $\kappa = \frac{3}{4}$ and $\Upsilon = 4\sqrt{\ln \max[m, 3]}$ is given by the following statement (to be applied to the matrices $B_\ell = A^{-\frac{1}{2}}A_\ell A^{-\frac{1}{2}}$; we assume w.l.o.g. that $A \succ 0$):

Proposition B.5.2. Let $B_1, ..., B_L$ be deterministic symmetric $m \times m$ matrices such that $\sum_{\ell=1}^L B_\ell^2 \preceq I$ and let ζ_ℓ be random perturbations satisfying Assumption A.I or A.II (see p. 235). Then with $\Upsilon = 4\sqrt{\ln(\max(3, m))}$ one has

$$\text{Prob}\left\{-\Upsilon I \preceq \sum_{\ell=1}^L \xi_\ell B_\ell \preceq \Upsilon I\right\} \ge \frac{3}{4}. \qquad (B.5.6)$$

Proof is readily given by the following deep result from Functional Analysis due to Lust-Piquard [78], Pisier [93] and Buchholz [35], see [111, Proposition 10]:

Let ϵ_ℓ, $\ell = 1, ..., L$, be independent random variables taking values ± 1 with probabilities $1/2$, and let $Q_1, ..., Q_L$ be deterministic matrices.

Then for every $p \in [2, \infty)$ one has

$$\mathbf{E}\left\{|\sum_{\ell=1}^{L} \epsilon_\ell Q_\ell|_p^p\right\}$$
$$\leq \left[2^{-1/4}\sqrt{p\pi/e}\right]^p \max\left[|\sum_{\ell=1}^{L} Q_\ell Q_\ell^T|_{\frac{p}{2}}, |\sum_{\ell=1}^{L} Q_\ell^T Q_\ell|_{\frac{p}{2}}\right]^{\frac{p}{2}}, \qquad (B.5.7)$$

where $|A|_p = \|\sigma(A)\|_p$, $\sigma(A)$ being the vector of singular values of a matrix A.

Observe, first, that (B.5.7) remains valid when the random variables ϵ_ℓ in the left hand side are replaced with ζ_ℓ. Indeed, assume first that we are in the case of **A.I** — ζ_ℓ are independent with zero means and take values in $[-1, 1]$. It is immediately seen that if μ is a random variable with zero mean taking values in $[-1, 1]$ and ν is a random variable taking values ± 1 with probabilities $1/2$, then $\nu \succeq_c \mu$ (see Definition 10.3.1). Applying Proposition 10.3.2.5, we conclude that if $\zeta = [\zeta_1; ...; \zeta_L]$ and $\epsilon = [\epsilon_1; ...; \epsilon_L]$ with ϵ_i as in (B.5.7), then $\epsilon \succeq_c \zeta$, which, by definition of $\succeq_c$, implies that $\mathbf{E}\{f(\epsilon)\} \geq \mathbf{E}\{f(\xi)\}$ for every convex function f, in particular, for the function $f(z) = |\sum_{\ell=1}^{L} z_\ell Q_\ell|_p^p$. In the case of **A.II**, let $\epsilon_{\ell i}$, $1 \leq \ell \leq L$, $1 \leq i \leq N$, be independent random variables taking values ± 1 with probability $1/2$, and let $Q_{\ell i} = \frac{1}{N} Q_\ell$, $1 \leq i \leq N$. By (B.5.7) we have

$$\mathbf{E}\{|\sum_{\ell=1}^{L} \overbrace{\sum_{i=1}^{N} \epsilon_{\ell i} Q_{\ell i}}^{\zeta_\ell^N Q_\ell} |_p^p\}$$
$$\leq \left[2^{-1/4}\sqrt{p\pi/e}\right]^p \max\left[|\sum_{\ell=1}^{L}\sum_{i=1}^{N} Q_{\ell i} Q_{\ell i}^T|_{\frac{p}{2}}, |\sum_{\ell=1}^{L}\sum_{i=1}^{N} Q_{\ell i}^T Q_{\ell i}|_{\frac{p}{2}}\right]^{\frac{p}{2}}$$
$$= \left[2^{-1/4}\sqrt{p\pi/e}\right]^p \max\left[|\sum_{\ell=1}^{L} Q_\ell Q_\ell^T|_{\frac{p}{2}}, |\sum_{\ell=1}^{L} Q_\ell^T Q_\ell|_{\frac{p}{2}}\right]^{\frac{p}{2}}.$$
$$(B.5.8)$$

The random variables $\zeta_\ell^N = \frac{1}{\sqrt{N}}\sum_{i=1}^{N}\epsilon_{\ell i}$, $\ell = 1, ..., L$, are independent, and their distributions, by the Central Limit Theorem, converge to the standard Gaussian distribution, so that (B.5.8) implies the validity of (B.5.7) when ϵ_ℓ are replaced with independent $\mathcal{N}(0, 1)$ random variables.

Applying (B.5.7) to matrices B_ℓ in the role of Q_ℓ and random variables ζ_ℓ in the role of ϵ_ℓ, we get

$$\mathbf{E}\left\{|\sum_{\ell=1}^{L} \zeta_\ell B_\ell|_p^p\right\} \leq \left[2^{-1/4}\sqrt{p\pi/e}\right]^p m,$$

Taking into account that $|A|_p \geq \|A\|$ ($\|A\|$ is the maximal singular value of A) and applying Tschebyshev Inequality, we get

$$\forall(\alpha > 0, p \geq 2): \mathrm{Prob}\left\{\|\sum_{\ell=1}^{L} \zeta_\ell B_\ell\| > \alpha\right\} \leq \left[\frac{2^{-1/4} m^{1/p}\sqrt{p\pi/e}}{\alpha}\right]^p.$$

Setting $\bar{m} = \max[m, 3]$, $p = 2\ln\bar{m}$ and $\alpha = 4\sqrt{\ln\bar{m}}$, we get

$$\mathrm{Prob}\left\{\|\sum_{\ell=1}^{L} \zeta_\ell B_\ell\| > 4\sqrt{\ln\bar{m}}\right\} \leq \left[\frac{2^{-1/4}\sqrt{2\pi\ln\bar{m}}}{4\sqrt{\ln\bar{m}}}\right]^p \leq \left[\frac{\sqrt{2\pi}}{2^{9/4}}\right]^{2\ln 3} \leq 1/4. \quad \square$$

Appendix C
Solutions to Selected Exercises

C.1 CHAPTER 1

Exercise 1.1: Setting $c_j^+ = c_j^n + \sigma_j/2$, $c_j^- = c_j^n - \sigma_j/2$, and similarly for $A_{ij}^{\pm}$, $b_i^{\pm}$, the RC is equivalent to

$$\min_{u \geq 0, v \geq 0} \left\{ \sum_j [c_j^+ u_j - c_j^- v_j] : \sum_j [A_{ij}^+ u_j - A_{ij}^- v_j] \leq b_i^- , 1 \leq i \leq m \right\};$$

the robust optimal solution is $u_* - v_*$, where u_*, v_* are the components of an optimal solution to the latter problem.

Exercise 1.2: The respective RCs are (equivalent to)

$$\begin{array}{ll} [a^n; b^n]^T [x; -1] + \rho \|P^T [x; -1]\|_q \leq 0, \ q = \frac{p}{p-1} & (a) \\ [a^n; b^n]^T [x; -1] + \rho \|(P^T [x; -1])_+\|_q \leq 0, \ q = \frac{p}{p-1} & (b) \\ [a^n; b^n]^T [x; -1] + \rho \|P^T [x; -1]\|_\infty \leq 0 & (c) \end{array}$$

where for a vector $u = [u_1; ...; u_k]$ the vector $(u)_+$ has the coordinates $\max[u_i, 0]$, $i = 1, ..., k$.

Comment to (c): The uncertainty set in question is nonconvex; since the RC remains intact when a given uncertainty set is replaced with its convex hull, we can replace the restriction $\|\zeta\|_p \leq \rho$ in (c) with the restriction $\zeta \in \text{Conv}\{\zeta : \|\zeta\|_p \leq \rho\} = \{\|\zeta\|_1 \leq \rho\}$, where the concluding equality is due to the following reasons: on one hand, with $p \in (0, 1)$ we have

$$\begin{array}{l} \|\zeta\|_p \leq \rho \Leftrightarrow \sum_i (|\zeta_i|/\rho)^p \leq 1 \Rightarrow |\zeta_i|/\rho \leq 1 \forall i \Rightarrow |\zeta_i|/\rho \leq (|\zeta_i|/\rho)^p \\ \Rightarrow \sum_i |\zeta_i|/\rho \leq \sum_i (|\zeta_i|/\rho)^p \leq 1, \end{array}$$

whence $\text{Conv}\{\|\zeta\|_p \leq \rho\} \subset \{\|\zeta\|_1 \leq \rho\}$. To prove the inverse inclusion, note that all extreme points of the latter set (that is, vectors with all but one coordinates equal to 0 and the remaining coordinate equal $\pm \rho$) satisfy $\|\zeta\|_p \leq 1$.

Exercise 1.3: The RC can be represented by the system of conic quadratic constraints

$$\begin{array}{l} [a^n; b^n]^T [x; -1] + \rho \sum_j \|u_j\|_2 \leq 0 \\ \sum_j Q_j^{1/2} u_j = P^T [x; -1] \end{array}$$

in variables $x, \{u_j\}_{j=1}^J$.

C.2 CHAPTER 2

Exercise 2.1: W.l.o.g., we may assume $t > 0$. Setting $\phi(s) = \cosh(ts) - [\cosh(t) - 1]s^2$, we get an even function such that $\phi(-1) = \phi(0) = \phi(1) = 1$. We claim that $\phi(s) \leq 1$ when $-1 \leq s \leq 1$.

Indeed, otherwise ϕ attains its maximum on $[-1, 1]$ at a point $\bar{s} \in (0, 1)$, and $\phi''(\bar{s}) \leq 0$. The function $g(s) = \phi'(s)$ is convex on $[0, 1]$ and $g(0) = g(\bar{s}) = 0$. The latter, due to $g'(\bar{s}) \leq 0$, implies that $g(s) = 0$, $0 \leq s \leq \bar{s}$. Thus, ϕ is constant on a nontrivial segment, which is not the case.

For a symmetric P supported on $[-1, 1]$ with $\int s^2 dP(s) \equiv \bar{\nu}^2 \leq \nu^2$ we have, due to $\phi(s) \leq 1$, $-1 \leq s \leq 1$:

$$\int \exp\{ts\} dP(s) = \int_{-1}^{1} \cosh(ts) dP(s)$$
$$= \int_{-1}^{1} [\cosh(ts) - (\cosh(t) - 1)s^2] dP(s) + (\cosh(t) - 1) \int_{-1}^{1} s^2 dP(s)$$
$$\leq \int_{-1}^{1} dP(s) + (\cosh(t) - 1)\bar{\nu}^2 \leq 1 + (\cosh(t) - 1)\nu^2,$$

as claimed in (2.4.33). Setting $h(t) = \ln(\nu^2 \cosh(t) + 1 - \nu^2)$, we have $h(0) = h'(0) = 0$, $h''(t) = \frac{\nu^2(\nu^2 + (1-\nu^2)\cosh(t))}{(\nu^2 \cosh(t) + 1 - \nu^2)^2}$, $\max_t h''(t) = \begin{cases} \nu^2, & \nu^2 \geq \frac{1}{3} \\ \frac{1}{4}\left[1 + \frac{\nu^4}{1-2\nu^2}\right] \leq \frac{1}{3}, & \nu^2 \leq \frac{1}{3} \end{cases}$, whence $\Sigma_{(3)}(\nu) \leq 1$.

Exercise 2.2: Here are the results:

n	ϵ	t_{tru}	t_{Nrm}	t_{Bll}	t_{BllBx}	t_{Bdg}
16	5.e-2	3.802	3.799	9.791	9.791	9.791
16	5.e-4	7.406	7.599	15.596	15.596	15.596
16	5.e-6	9.642	10.201	19.764	16.000	16.000
256	5.e-2	15.195	15.195	39.164	39.164	39.164
256	5.e-4	30.350	30.396	62.383	62.383	62.383
256	5.e-6	40.672	40.804	79.054	79.054	79.054

n	ϵ	t_{tru}	$t_{\text{E.2.4.11}}$	$t_{\text{E.2.4.12}}$	$t_{\text{E.2.4.13}}$	t_{Unim}
16	5.e-2	3.802	6.228	5.653	5.653	10.826
16	5.e-4	7.406	9.920	9.004	9.004	12.502
16	5.e-6	9.642	12.570	11.410	11.410	13.705
256	5.e-2	15.195	24.910	22.611	22.611	139.306
256	5.e-4	30.350	39.678	36.017	36.017	146.009
256	5.e-6	40.672	50.282	45.682	45.682	150.821

Exercise 2.3: Here are the results:

n	ϵ	t_{tru}	t_{Nrm}	t_{Bll}	t_{BllBx}	t_{Bdg}	$t_{\text{E.2.4.11}}$	$t_{\text{E.2.4.12}}$
16	5.e-2	4.000	6.579	9.791	9.791	9.791	9.791	9.791
16	5.e-4	10.000	13.162	15.596	15.596	15.596	15.596	15.596
16	5.e-6	14.000	17.669	19.764	16.000	16.000	19.764	19.764
256	5.e-2	24.000	26.318	39.164	39.164	39.164	39.164	39.164
256	5.e-4	50.000	52.649	63.383	62.383	62.383	62.383	62.383
256	5.e-6	68.000	70.674	79.054	79.054	79.054	79.053	79.053

Exercise 2.4: In the case of (a), the optimal value is $t_a = \sqrt{n}\text{ErfInv}(\epsilon)$, since for a feasible x we have $\xi^n[x] \sim \mathcal{N}(0, n)$. In the case of (b), the optimal value is $t_b = n\text{ErfInv}(n\epsilon)$. Indeed, the rows in B_n are of the same Euclidean length and are orthogonal to each other, whence the columns are orthogonal to each other as well. Since the first column of B_n is the all-one vector, the conditional on η distribution of $\xi = \sum_j \hat{\zeta}_j$ has the mass $1/n$ at the point $n\eta$ and the mass $(n-1)/n$ at the origin. It follows that the distribution of ξ is the convex combination of the Gaussian distribution $\mathcal{N}(0, n^2)$ and the unit mass, sitting at the origin, with the weights $1/n$ and $(n-1)/n$, respectively, and the claim follows.

The numerical results are as follows:

n	ϵ	t_a	t_b	t_b/t_a
10	1.e-2	7.357	12.816	1.74
100	1.e-3	30.902	128.155	4.15
1000	1.e-4	117.606	1281.548	10.90

C.3 CHAPTER 3

Exercise 3.1: *A possible model is as follows:* let us define the normal range $\mathcal{Z}$ of the uncertain data — the vector of prices c — as the box $\{c : 0 \leq c \leq \bar{c}\}$, where $\bar{c}$ is the vector of current prices, so that all "physically possible" price vectors form the set $\mathcal{Z} + \mathbb{R}_n^+$. To account for volatilities, is natural to measure deviations of the price vector from its normal range in the norm $\|c\| = \max_j |c_j|/d_j$. We now can model the decision making problem in question by the GRC of the uncertain LO problem

$$\min_x \left\{ c^T x : Px \geq b, x \geq 0 \right\}$$

the uncertain data being c. By Proposition 3.2.1, the GRC of this uncertain problem is the semi-infinite LO program

$$
\begin{aligned}
\min_{x,t} \quad & t \\
\text{subject to} \quad & \\
& Px \geq b & (a) \\
& x \geq 0 & (b) \\
& c^T x \leq t \ \forall c \in \mathcal{Z} & (c) \\
& \Delta^T x \leq \alpha \ \forall(\Delta \geq 0 : \|\Delta\| \leq 1), & (d)
\end{aligned}
$$

which is equivalent to the LO program

$$\min_x \left\{ \bar{c}^T x : \begin{array}{c} Px \geq b, x \geq 0 \\ d^T x \leq \alpha \end{array} \right\}.$$

With the given data, the meaningful range of sensitivities (the one where the GRC is feasible and the constraint $d^T x \leq \alpha$ is binding) is $[0.16, 0.32]$, and in this range the cost of the monthly supply at the current prices varies from 8000 to 6400.

C.4 CHAPTER 4

Exercise 4.1: In the notation of section 4.2, we have

$$
\begin{aligned}
\Phi(w) &\equiv \ln\left(\mathbf{E}\{\exp\{\textstyle\sum_\ell w_\ell \zeta_\ell\}\}\right) = \sum_\ell \lambda_\ell (\exp\{w_\ell\} - 1) \\
&= \max_u [w^T u - \phi(u)], \\
\phi(u) &= \max_w [u^T w - \Phi(w)] = \begin{cases} \sum_\ell [u_\ell \ln(u_\ell/\lambda_\ell) - u_\ell + \lambda_\ell], & u \geq 0 \\ +\infty, & \text{otherwise.} \end{cases}
\end{aligned}
$$

Consequently, the Bernstein approximation is

$$\inf_{\beta > 0} \left[z_0 + \beta \sum_\ell \lambda_\ell (\exp\{w_\ell/\beta\} - 1) + \beta \ln(1/\epsilon) \right] \leq 0,$$

or, in the RC form,

$$z_0 + \max_u \left\{ w^T u : u \in \mathcal{Z}_\epsilon = \{u \geq 0, \sum_\ell [u_\ell \ln(u_\ell/\lambda_\ell) - u_\ell + \lambda_\ell] \leq \ln(1/\epsilon)\} \right\} \leq 0.$$

Exercise 4.2: $w(\epsilon)$ is the optimal value in the chance constrained optimization problem

$$\min_{w_0}\left\{w_0 : \mathrm{Prob}\{-w_0 + \sum_{\ell=1}^{L} c_\ell\zeta_\ell \le 0\} \ge 1 - \epsilon\right\},$$

where ζ_ℓ are independent Poisson random variables with parameters λ_ℓ.

When all c_ℓ are integral in certain scale, the random variable $\zeta^L = \sum_{\ell=1}^{L} c_\ell\zeta_\ell$ is also integral in the same scale, and we can compute its distribution recursively in L:

$$p_0(i) = \left\{\begin{array}{ll} 1, & i = 0 \\ 0, & i \ne 0 \end{array}\right. , p_k(i) = \sum_{j=0}^{\infty} p_{k-1}(i - c_\ell j)\frac{\lambda_k^j}{j!}\exp\{-\lambda_k\};$$

(in computations, $\sum_{j=0}^{\infty}$ should be replaced with $\sum_{j=0}^{N}$ with appropriately large N). With the numerical data in question, the expected value of per day requested cash is $c^T\lambda = 7,000$, and the remaining requested quantities are listed below:

	ϵ					
	1.e-1	1.e-2	1.e-3	1.e-4	1.e-5	1.e-6
$w(\epsilon)$	8,900	10,800	12,320	13,680	14,900	16,060
CVaR	9,732 +9.3%	11,451 +6.0%	12,897 +4.7%	14,193 +3.7%	15,390 +3.3%	16,516 +2.8%
BCV	9,836 +10.5%	11,578 +7.2%	13,047 +5.9%	14,361 +5.0%	15,572 +4.5%	16,709 +4.0%
B	10,555 +18.6%	12,313 +14.0%	13,770 +11.8%	15,071 +10.2%	16,270 +9.2%	17,397 +8.3%
E	8,900 +0.0%	10,800 +0.0%	12,520 +1.6%	17,100 +25.0%	—	—

"BCV" stands for the bridged Bernstein-CVaR, "B" — for the Bernstein, and "E" — for the $(1 - \epsilon)$-reliable empirical bound on $w(\epsilon)$. The BCV bound corresponds to the generating function $\gamma_{16,10}(\cdot)$, see p. 97. The percents represent the relative differences between the bounds and $w(\epsilon)$. All bounds are right-rounded to the closest integers.

Exercise 4.3: The results of computations are as follows (as a benchmark, we display also the results of Exercise 4.2 related to the case of independent $\zeta_1, ..., \zeta_L$):

	ϵ					
	1.e-1	1.e-2	1.e-3	1.e-4	1.e-5	1.e-6
Exer. 4.2	8,900	10,800	12,320	13,680	14,900	16,060
Exer. 4.3, lower bound	11,000 +23.6%	15,680 +45.2%	19,120 +55.2%	21,960 +60.5%	26,140 +75.4%	28,520 +77.6%
Exer. 4.3, upper bound	13,124 +47.5%	17,063 +58.8%	20,507 +66.5%	23,582 +72.4%	26,588 +78.5%	29,173 +81.7%

Percents display relative differences between the bounds and $w(\epsilon)$

Exercise 4.4. Part 1: By Exercise 4.1, the Bernstein upper bound on $w(\epsilon)$ is

$$\begin{aligned} B_\lambda(\epsilon) &= \inf\left\{w_0 : \inf_{\beta>0}\left[-w_0 + \beta\sum_\ell \lambda_\ell(\exp\{c_\ell/\beta\} - 1) + \beta\ln(1/\epsilon)\right] \le 0\right\} \\ &= \inf_{\beta>0}\left[\beta\sum_\ell \lambda_\ell(\exp\{c_\ell/\beta\} - 1) + \beta\ln(1/\epsilon)\right] \end{aligned}$$

The "ambiguous" Bernstein upper bound on $w(\epsilon)$ is therefore

$$\begin{aligned} B_\Lambda(\epsilon) &= \max_{\lambda\in\Lambda}\inf_{\beta>0}\left[\beta\sum_\ell \lambda_\ell(\exp\{c_\ell/\beta\} - 1) + \beta\ln(1/\epsilon)\right] \\ &= \inf_{\beta>0}\beta\left[\max_{\lambda\in\Lambda}\sum_\ell \lambda_\ell(\exp\{c_\ell/\beta\} - 1) + \ln(1/\epsilon)\right] \end{aligned} \qquad (*)$$

where the swap of $\inf_{\beta>0}$ and $\max_{\lambda \in \Lambda}$ is justified by the fact that the function $\beta \sum_\ell \lambda_\ell(\exp\{c_\ell/\beta\} - 1) + \beta \ln(1/\epsilon)$ is concave in λ, convex in β and by the compactness and convexity of Λ.

Part 2: We should prove that if Λ is a convex compact set in the domain $\lambda \geq 0$ such that for every affine form $f(\lambda) = f_0 + e^T\lambda$ one has

$$\max_{\lambda \in \Lambda} f(\lambda) \leq 0 \Rightarrow \text{Prob}_{\lambda \sim P}\{f(\lambda) \leq 0\} \geq 1 - \delta, \tag{!}$$

then, setting $w_0 = B_\Lambda(\epsilon)$, one has

$$\text{Prob}_{\lambda \sim P}\left\{\lambda : \text{Prob}_{\zeta \sim P_{\lambda_1} \times \ldots \times P_{\lambda_L}}\left\{\sum_\ell \zeta_\ell c_\ell > w_0\right\} > \epsilon\right\} \leq \delta. \tag{?}$$

It suffices to prove that under our assumptions on Λ inequality (?) is valid for all $w_0 > B_\Lambda(\epsilon)$. Given $w_0 > B_\Lambda(\epsilon)$ and invoking the second relation in (∗), we can find $\bar\beta > 0$ such that

$$\bar\beta\left[\max_{\lambda \in \Lambda} \sum_\ell \lambda_\ell(\exp\{c_\ell/\bar\beta\} - 1) + \ln(1/\epsilon)\right] \leq w_0,$$

or, which is the same,

$$[-w_0 + \bar\beta \ln(1/\epsilon)] + \max_{\lambda \in \Lambda} \sum_\ell \lambda_\ell[\bar\beta(\exp\{c_\ell/\bar\beta\} - 1)] \leq 0,$$

which, by (!) as applied to the affine form

$$f(\lambda) = [-w_0 + \bar\beta \ln(1/\epsilon)] + \sum_\ell \lambda_\ell[\bar\beta(\exp\{c_\ell/\bar\beta\} - 1)],$$

implies that

$$\text{Prob}_{\lambda \sim P}\{f(\lambda) > 0\} \leq \delta. \tag{∗∗}$$

It remains to note that when $\lambda \geq 0$ is such that $f(\lambda) \leq 0$, the result of Exercise 4.1 states that

$$\text{Prob}_{\zeta \sim P_{\lambda_1} \times \ldots \times P_{\lambda_m}}\left\{-w_0 + \sum_\ell \zeta_\ell c_\ell > 0\right\} \leq \epsilon.$$

Thus, when $\omega_0 > B_\Lambda(\epsilon)$, the set of λ's in the left hand side of (?) is contained in the set $\{\lambda \geq 0 : f(\lambda) > 0\}$, and therefore (?) is readily given by (∗∗).

Exercise 4.5: The necessary and sufficient condition for x to satisfy (4.6.2) clearly is

$$\forall(P \in \mathcal{P}) :$$
$$\text{Prob}_{\eta \sim P}\Big\{\underbrace{\sup_{\xi \in \mathcal{Z}_\xi}\left\{[a^0]^Tx - b^0 + \sum_{\ell=1}^L \xi_\ell[[a^\ell]^Tx - b^\ell]\right\}}_{g(x)} + \sum_{\ell=1}^L \eta_\ell[[a^\ell]^Tx - b^\ell] \leq 0\Big\}$$
$$\geq 1 - \epsilon.$$

Now, by conic duality a pair x, t can be extended by properly chosen y to a solution of $(4.6.4.a - d)$ if and only if $t \geq g(x)$ (cf. proof of Theorem 1.3.4). Recalling the origin of f, we conclude that if x can be extended to a solution of (4.6.4), then x satisfies (4.6.2).

Exercise 4.7: Let us fix the true vector of expected returns μ. The RC mentioned in the exercise is the (random) optimization problem

$$\max_{x,t}\left\{t - \text{ErfInv}(\epsilon)\sigma(x) : \begin{array}{l} \nu^Tx \geq t \,\forall\nu \in \mathcal{M}(\widetilde\zeta) \\ x \geq 0, \sum_\ell x_\ell = 1 \end{array}\right\}.$$

Let $\widetilde{\zeta}$ be such that $\mathcal{M}(\widetilde{\zeta})$ contains a lower bound $\nu = \nu(\widetilde{\zeta})$ for μ; note that this happens with probability $\geq 1 - \delta$. Then the t-component of a feasible solution (x, t) to the RC satisfies the relation $t \leq \nu^T x \leq \mu^T x$, whence the optimal value $\mathrm{VaR}(\widetilde{\zeta})$ and the x-component $x_* = X(\widetilde{\zeta})$ of an optimal solution (x_*, t_*) to the problem satisfy the relation

$$\mathrm{VaR}(\widetilde{\zeta}) = t_* - \mathrm{ErfInv}(\epsilon)\sigma(x_*) \leq \mu^T x_* - \mathrm{ErfInv}(\epsilon)\sigma(x_*),$$

and (4.6.6) follows. $\qquad\qquad\qquad\qquad\qquad\qquad\qquad\qquad\qquad\qquad\qquad\qquad\qquad$ $\square$

Exercise 4.8: We have $\widehat{\mu} = \mu + \Sigma\eta$ with $\eta \sim \mathcal{N}(0, I_n)$. Therefore the set $\mathcal{M}$ given by (4.6.9) can be represented as
$$\mathcal{M} = \mu + [\Sigma\eta + \mathcal{O}].$$
Such a set contains vector that is $\leq \mu$ if and only if the set $\Sigma\eta + \mathcal{O}$ contains a nonpositive vector, i.e., if and only if $\mathcal{O} + \mathbb{R}^n_+$ contains the vector $-\Sigma\eta$. In the case of (4.6.10), the latter condition is indeed satisfied with probability $\geq 1 - \delta$. $\qquad\qquad\qquad\qquad$ $\square$

Exercise 4.9: The RC associated with the combined Ball-Box approximation is

$$\max_{x,u,v} \left\{ \sum_{\ell=1}^{n} \widehat{\mu}_\ell x_\ell - \rho_2 N^{-1/2}\sigma(u) - \rho_\infty N^{-1/2} \sum_{\ell=1}^{n} \sigma_i v_i : \begin{array}{l} x \in \Delta_n \\ u + v = x, u, v \geq 0 \end{array} \right\}, \quad \text{(C.4.1)}$$

cf. Proposition 2.3.3.

Exercise 4.10: The solution x_* to the *random* problem (C.4.1) depends on $\widetilde{\zeta}$: $x_* = x_*(\widetilde{\zeta})$. Therefore from the fact that the value of the problem's objective at every *fixed* point $x \in \Delta_n$ with probability $\geq 1 - \delta$ is a lower bound on the value of the objective of the "true" problem (4.6.5) at x it does *not* follow that the same is true at $x_*(\widetilde{\zeta})$. It can happen (and in fact indeed happens) that the "bad event" — the objective of the soft RC approximation *as evaluated* at $x_* = x_*(\widetilde{\zeta})$ is greater than the "true" objective at the same point — has probability significantly greater than δ. Whenever this bad event happens, the optimal value in the approximation, which is exactly the value of its objective at x_*, is greater than $\mathrm{VaR}_\epsilon[V^{x_*}]$, (which is nothing but the value of the true objective at x_*), that is, our target relation $\mathrm{VaR} \leq \mathrm{VaR}_\epsilon[V^{x_*}]$ takes place with probability less than the desired probability $1 - \delta$. We could save the day by ensuring that the objective of the approximation underestimates, with probability $\geq 1 - \delta$, the true objective *everywhere on* Δ_n, but this is much more than what is ensured by the soft approximation.

Exercises 4.6 — 4.11, numerical results. The results of our experiments are presented in table C.1. In the table:
• "Inv" is the empirical mean of the capital invested in "true assets" (those with $\ell \geq 2$);
• $M(\cdot)/S(\cdot)/P(\cdot)$ are the empirical mean/standard deviation/probability computed over a sample of 100 collections of historical data and associated portfolio selections;
• "Id" is the ideal portfolio given by the optimal solution to (4.6.5), while "Bl" "Bx," "BB," "S" and "CS" stand for portfolios yielded, respectively, by the Ball, Box, combined Ball-Box, soft, and corrected soft approximations.

C.5 CHAPTER 5

Exercise 5.1: The proof is correct up to the fact that the mere existence of similar ellipsoids, centered at the origin, with the similarity ratio $\sqrt{L}$ which "bracket" $\mathcal{Z}$ is not enough; we need a *tractable* approximation, so that we need an explicit description of these ellipsoids, e.g., a description by explicitly given quadratic inequalities. Whether such a description can be found efficiently, it depends on how $\mathcal{Z}$ itself is given. For

Portf	Inv	M(VaR)	S(VaR)	M(VaR$_\epsilon[V^x]$)	S(VaR$_\epsilon[V^x]$)	P(VaR > VaR$_\epsilon[V^x]$)
Id	1.000	1.053	0.000	1.053	0.000	0.000
Bl	1.000	1.032	0.001	1.044	0.001	0.000
Bx	0.000	1.000	0.000	1.000	0.000	0.000
BB	1.000	1.032	0.001	1.044	0.001	0.000
S	1.000	1.062	0.002	1.042	0.001	1.000
CS	1.000	1.046	0.002	1.052	0.000	0.000
				Data (4.6.7.a)		
Portf	Inv	M(VaR)	S(VaR)	M(VaR$_\epsilon[V^x]$)	S(VaR$_\epsilon[V^x]$)	P(VaR > VaR$_\epsilon[V^x]$)
Id	1.000	1.053	0.000	1.053	0.000	0.000
Bl	0.990	1.002	0.001	1.013	0.006	0.000
Bx	1.000	1.040	0.003	1.053	0.000	0.000
BB	1.000	1.040	0.003	1.053	0.000	0.000
S	1.000	1.046	0.003	1.053	0.000	0.020
CS	1.000	1.040	0.003	1.053	0.000	0.000
				Data (4.6.7.b)		
Portf	Inv	M(VaR)	S(VaR)	M(VaR$_\epsilon[V^x]$)	S(VaR$_\epsilon[V^x]$)	P(VaR > VaR$_\epsilon[V^x]$)
Id	1.000	1.018	0.000	1.018	0.000	0.000
Bl	0.680	1.001	0.001	1.008	0.006	0.000
Bx	0.000	1.000	0.000	1.000	0.000	0.000
BB	0.620	1.001	0.001	1.008	0.006	0.000
S	1.000	1.020	0.002	1.012	0.001	1.000
CS	1.000	1.011	0.002	1.018	0.000	0.000
				Data (4.6.7.c)		

Table C.1 Numerical results for Exercises 4.6 — 4.11 on data sets (4.6.7.a-c).

example, when $\mathcal{Z}$ is "black-box-represented," that is, it is given by a membership oracle (a "black box" capable to check whether a given point ζ belongs to $\mathcal{Z}$) or a separation oracle (a membership oracle that in the case of $\zeta \notin \mathcal{Z}$ returns vector e such that $e^T\zeta > \max_{\zeta' \in \mathcal{Z}} e^T\zeta'$), we do *not* know how to "round" $\mathcal{Z}$ within the factor $\vartheta = (1 + \epsilon)\sqrt{d}$ efficiently — i.e., how to find a pair of similar, with the similarity ratio ϑ, and centered at the origin ellipsoids which bracket $\mathcal{Z}$ in a polynomial in L and $\ln(1/\epsilon)$ number of calls to the oracle, with polynomial in L and $\ln(1/\epsilon)$ number of additional arithmetic operations per call. The best known so far ϑ for which ϑ-rounding of a black-box-represented solid $\mathcal{Z} = -\mathcal{Z}$ can be found efficiently, is $\vartheta = O(1)L$, and with this ϑ in the role of $\sqrt{d}$ the tightness factor of the safe tractable approximation of (5.3.3), (RC$_\rho$) developed at the end of section 5.3 jumps from L to $O(1)L^{3/2}$. To get better results, we need "more informative" representation of $\mathcal{Z}$. We could require, e.g., that along with the membership (or separation) oracle we have at our disposal an "inclusion" oracle — one that, given on input a centered at the origin ellipsoid E (represented by an explicit quadratic inequality) reports whether this ellipsoid is contained in $\mathcal{Z}$, and if it is not the case, returns a point from $E\backslash\mathcal{Z}$. In this case, we can approximate efficiently to whatever accuracy the largest volume ellipsoid contained in $\mathcal{Z}$ and thus round efficiently $\mathcal{Z}$ within factor $\vartheta = (1 + \epsilon)\sqrt{L}$ for whatever $\epsilon > 0$ (for justification of this and the subsequent claims, see [8, section 4.9]). Note that the inclusion oracle is easy to implement when, e.g., $\mathcal{Z}$ is given as intersection of finitely many ellipsoids represented by explicit quadratic inequalities. Indeed, in this case building an inclusion oracle reduces immediately to building a similar oracle for a pair of ellipsoids given by explicit quadratic inequalities, and the latter problem is easy. Note that in our context the inclusion oracle can be replaced with a "covering" one — a routine that, given on input a centered at the origin ellipsoid E, reports whether $E \supset \mathcal{Z}$, and if it is not the case, returns a point from $\mathcal{Z}\backslash E$. In this case, we can approximate efficiently to whatever accuracy the smallest volume ellipsoid containing $\mathcal{Z}$, thus once again arriving at an efficient ϑ-rounding of $\mathcal{Z}$ with $\vartheta = (1 + \epsilon)\sqrt{L}$, $\epsilon > 0$. In order for a covering oracle to be readily available, it suffices to assume that $\mathcal{Z}$ is given as a convex hull of the union of finitely many ellipsoids.

Exercise 5.2: Let $\mathcal{S}[\cdot]$ be a safe tractable approximation of $(C_{\mathcal{Z}_*}[\cdot])$ tight within the factor ϑ. Let us verify that $\mathcal{S}[\lambda\gamma\rho]$ is a safe tractable approximation of $(C_{\mathcal{Z}}[\rho])$ tight within the factor $\lambda\vartheta$. All we should prove is that (a) if x can be extended to a feasible solution to $\mathcal{S}[\lambda\gamma\rho]$, then x is feasible for $(C_{\mathcal{Z}}[\rho])$, and that (b) if x cannot be extended to a feasible solution to $\mathcal{S}[\lambda\gamma\rho]$, then x is not feasible for $(C_{\mathcal{Z}}[\lambda\vartheta\rho])$. When x can be extended to a feasible solution of $\mathcal{S}[\lambda\gamma\rho]$, x is feasible for $(C_{\mathcal{Z}_*}[\lambda\gamma\rho])$, and since $\rho\mathcal{Z} \subset \lambda\gamma\rho\mathcal{Z}_*$, x is feasible for $(C_{\mathcal{Z}}[\rho])$ as well, as required in (a). Now assume that x cannot be extended to a feasible solution of $\mathcal{S}[\lambda\gamma\rho]$. Then x is not feasible for $(C_{\mathcal{Z}_*}[\vartheta\lambda\gamma\rho])$, and since the set $\vartheta\lambda\gamma\rho\mathcal{Z}_*$ is contained in $\vartheta\lambda\rho\mathcal{Z}$, x is not feasible for $(C_{\mathcal{Z}}[(\vartheta\lambda)\rho])$, as required in (b). □

Exercise 5.3: 1) Consider the ellipsoid

$$\mathcal{Z}_* = \{\zeta : \zeta^T[\sum_i Q_i]\zeta \le M\}.$$

We clearly have $M^{-1/2}\mathcal{Z}_* \subset \mathcal{Z} \subset \mathcal{Z}_*$; by assumption, $(C_{\mathcal{Z}_*}[\cdot])$ admits a safe tractable approximation tight within the factor ϑ, and it remains to apply the result of Exercise 5.2.

2) This is a particular case of 1) corresponding to $\zeta^T Q_i \zeta = \zeta_i^2$, $1 \le i \le M = \dim\zeta$.

3) Let $\mathcal{Z} = \bigcap_{i=1}^{M} E_i$, where E_i are ellipsoids. Since $\mathcal{Z}$ is symmetric w.r.t. the origin, we also have $\mathcal{Z} = \bigcap_{i=1}^{M} [E_i \cap (-E_i)]$. We claim that for every i, the set $E_i \cap (-E_i)$ contains an ellipsoid F_i centered at the origin and such that $E_i \cap (-E_i) \subset \sqrt{2}F_i$, and that this ellipsoid F_i can be easily found. Believing in the claim, we have

$$\mathcal{Z}_* \equiv \bigcap_{i=1}^{M} F_i \subset \mathcal{Z} \subset \sqrt{2}\bigcap_{i=1}^{M} F_i.$$

By 1), $(C_{\mathcal{Z}_*}[\cdot])$ admits a safe tractable approximation with the tightness factor $\vartheta\sqrt{M}$; by Exercise 5.2, $(C_{\mathcal{Z}}[\cdot])$ admits a safe tractable approximation with the tightness factor $\vartheta\sqrt{2M}$.

It remains to support our claim. For a given i, applying nonsingular linear transformation of variables, we can reduce the situation to the one where $E_i = B + e$, where B is the unit Euclidean ball, centered at the origin, and $\|e\|_2 < 1$ (the latter inequality follows from $0 \in \operatorname{int}\mathcal{Z} \subset \operatorname{int}(E_i \cap (-E_i))$). The intersection $G = E_i \cap (-E_i)$ is a set that is invariant w.r.t. rotations around the axis $\mathbb{R}e$; a 2-D cross-section H of G by a 2D plane Π containing the axis is a 2-D solid symmetric w.r.t. the origin. By the results on inscribed/cisrumscribed ellipsoids mentioned at the end of chapter 5, there exists an ellipsis I, centered at the origin, that is contained in H and is such that $\sqrt{2}I$ contains H. This ellipsis can be easily found, see Solution to Exercise 5.1. Now, the ellipsis I is the intersection of Π and an ellipsoid F_i that is invariant w.r.t. rotations around the axis $\mathbb{R}e$, and F_i clearly satisfies the required relations $F_i \subset E_i \cap (-E_i) \subset \sqrt{2}F_i$. [1]

---

[1] In fact, the factor $\sqrt{2}$ in the latter relation can be reduced to $2/\sqrt{3} < \sqrt{2}$, see Solution to Exercise 7.1.

C.6 CHAPTER 6

Exercise 6.1: With y given, all we know about x is that there exists $\Delta \in \mathbb{R}^{p \times q}$ with $\|\Delta\|_{2,2} \leq \rho$ such that $y = B_n[x; 1] + L^T \Delta R[x; 1]$, or, denoting $w = \Delta R[x; 1]$, that there exists $w \in \mathbb{R}^p$ with $w^T w \leq \rho^2 [x; 1]^T R^T R[x; 1]$ such that $y = B_n[x; 1] + L^T w$. Denoting $z = [x; w]$, all we know about the vector z is that it belongs to a given affine plane $\mathcal{A}z = a$ and satisfies the quadratic inequality $z^T \mathcal{C}z + 2c^T z + d \leq 0$, where $\mathcal{A} = [A_n, L^T]$, $a = y - b_n$, and

$$[\xi; \omega]^T \mathcal{C}[\xi; \omega] + 2c^T [\xi; \omega] + d \equiv \omega^T \omega - \rho^2 [\xi; 1]^T R^T R[\xi; 1], \ [\xi; \omega] \in \mathbb{R}^{n+p}.$$

Using the equations $\mathcal{A}z = a$, we can express the $n+p$ z-variables via $k \leq n+p$ u-variables:

$$\mathcal{A}z = a \Leftrightarrow \exists u \in \mathbb{R}^k : z = Eu + e.$$

Plugging $z = Eu + e$ into the quadratic constraint $z^T \mathcal{C}z + 2c^T z + d \leq 0$, we get a quadratic constraint $u^T F u + 2f^T u + g \leq 0$ on u. Finally, the vector Qx we want to estimate can be represented as Pu with easily computable matrix P. The summary of our developments is as follows:

> (!) *Given y and the data describing $\mathcal{B}$, we can build k, a matrix P and a quadratic form $u^T F u + 2f^T u + g \leq 0$ on $\mathbb{R}^k$ such that the problem of interest becomes the problem of the best, in the worst case, $\| \cdot \|_2$-approximation of Pu, where unknown vector $u \in \mathbb{R}^k$ is known to satisfy the inequality $u^T F u + 2f^T u + g \leq 0$.*

By (!), our goal is to solve the semi-infinite optimization program

$$\min_{t,v} \left\{ t : \|Pu - v\|_2 \leq t \, \forall (u : u^T F u + 2f^T u + g \leq 0) \right\}. \tag{*}$$

Assuming that $\inf_u \left[u^T F u + 2f^T u + g \right] < 0$ and applying the inhomogeneous version of $\mathcal{S}$-Lemma, the problem becomes

$$\min_{t,v,\lambda} \left\{ t \geq 0 : \left[\begin{array}{c|c} \lambda F - P^T P & \lambda f - P^T v \\ \hline \lambda f^T - v^T P & \lambda g + t^2 - v^T v \end{array} \right] \succeq 0, \lambda \geq 0 \right\}.$$

Passing from minimization of t to minimization of $\tau = t^2$, the latter problem becomes the semidefinite program

$$\min_{\tau,v,\lambda,s} \left\{ \tau : \begin{array}{l} v^T v \leq s, \lambda \geq 0 \\ \left[\begin{array}{c|c} \lambda F - P^T P & \lambda f - P^T v \\ \hline \lambda f^T - v^T P & \lambda g + \tau - s \end{array} \right] \succeq 0 \end{array} \right\}.$$

In fact, the problem of interest can be solved by pure Linear Algebra tools, without Semidefinite optimization. Indeed, assume for a moment that P has trivial kernel. Then (*) is feasible if and only if the solution set S of the quadratic inequality $\phi(u) \equiv u^T F u + 2f^T u + g \leq 0$ in variables u is nonempty and bounded, which is the case if and only if this set is an ellipsoid $(u - c)^T Q(u - c) \leq r^2$ with $Q \succ 0$ and $r \geq 0$; whether this indeed is the case and what are c, Q, r, if any, can be easily found out by Linear Algebra tools. The image PS of S under the mapping P also is an ellipsoid (perhaps "flat") centered at $v_* = Pc$, and the optimal solution to (*) is (t_*, v_*), where t_* is the largest half-axis of the ellipsoid PS. In the case when P has a kernel, let E be the orthogonal complement to $\text{Ker}P$, and $\widehat{P}$ be the restriction of P onto E; this mapping has a trivial kernel. Problem (*) clearly is equivalent to

$$\min_{t,v} \left\{ t : \|\widehat{P}\widehat{u} - v\|_2 \leq t \, \forall (\widehat{u} \in E : \exists w \in \text{Ker}P : \phi(\widehat{u} + w) \leq 0) \right\}.$$

The set
$$\widehat{U} = \{\widehat{u} \in E : \exists w \in \mathrm{Ker}P : \phi(\widehat{u} + w) \leq 0\}$$
clearly is given by a single quadratic inequality in variables $\widehat{u} \in E$, and $(*)$ reduces to a similar problem with E in the role of the space where u lives and $\widehat{P}$ in the role of P, and we already know how to solve the resulting problem.

C.7 CHAPTER 7

Exercise 7.1: In view of Theorem 7.2.1, all we need to verify is that $\mathcal{Z}$ can be "safely approximated" within an $O(1)$ factor by an intersection $\widehat{\mathcal{Z}}$ of $O(1)J$ ellipsoids *centered at the origin*: there exists $\widehat{\mathcal{Z}} = \{\eta : \eta^T \widehat{Q}_j \eta \leq 1, 1 \leq j \leq \widehat{J}\}$ with $\widehat{Q}_j \succeq 0$, $\sum_j \widehat{Q}_j \succ 0$ such that
$$\theta^{-1}\widehat{\mathcal{Z}} \subset \mathcal{Z} \subset \widehat{\mathcal{Z}},$$
with an absolute constant θ and $\widehat{J} \leq O(1)J$. Let us prove that the just formulated statement holds true with $\widehat{J} = J$ and $\theta = \sqrt{3}/2$. Indeed, since $\mathcal{Z}$ is symmetric w.r.t. the origin, setting $E_j = \{\eta : (\eta - a_j)^T Q_j(\eta - a_j) \leq 1\}$, we have
$$\mathcal{Z} = \bigcap_{j=1}^{J} E_j = \bigcap_{j=1}^{J} (-E_j) = \bigcap_{j=1}^{J} (E_j \cap [-E_j]);$$

all we need is to demonstrate that every one of the sets $E_j \cap [-E_j]$ is in between two proportional ellipsoids centered at the origin with the larger one being at most $2/\sqrt{3}$ multiple of the smaller one. After an appropriate linear one-to-one transformation of the space, all we need to prove is that if $E = \{\eta \in \mathbb{R}^d : (\eta_1 - r)^2 + \sum_{j=2}^{k} \eta_j^2 \leq 1\}$ with $0 \leq r < 1$, then we can point out the set $F = \{\eta : \eta_1^2/a^2 + \sum_{j=2}^{k} \eta_j^2/b^2 \leq 1\}$ such that
$$\frac{\sqrt{3}}{2} F \subset E \cap [-E] \subset F.$$

When proving the latter statement, we lose nothing when assuming $k = 2$. Renaming η_1 as y, η_2 as x and setting $h = 1 - r \in (0, 1]$ we should prove that the "loop" $\mathcal{L} = \{[x; y] : [|y| + (1 - h)]^2 + x^2 \leq 1\}$ is in between two proportional ellipses centered at the origin with the ratio of linear sizes $\theta \leq 2/\sqrt{3}$. Let us verify that we can take as the smaller of these ellipses the ellipsis
$$\mathcal{E} = \{[x; y] : y^2/h^2 + x^2/(2h - h^2) \leq \mu^2\}, \mu = \sqrt{\frac{3 - h}{4 - 2h}},$$
and to choose $\theta = \mu^{-1}$ (so that $\theta \leq 2/\sqrt{3}$ due to $0 < h \leq 1$). First, let us prove that $\mathcal{E} \subset \mathcal{L}$. This inclusion is evident when $h = 1$, so that we can assume that $0 < h < 1$. Let $[x; y] \in \mathcal{E}$, and let $\lambda = \frac{2(1-h)}{h}$. We have
$$y^2/h^2 + x^2/(2h - h^2) \leq \mu^2 \Rightarrow \begin{cases} y^2 \leq h^2[\mu^2 - x^2/(2h - h^2)] & (a) \\ x^2 \leq \mu^2 h(2 - h) & (b) \end{cases};$$
$$(|y| + (1 - h))^2 + x^2 = y^2 + 2|y|(1 - h) + (1 - h)^2 \leq y^2 + \left[\lambda y^2 + \frac{1}{\lambda}(1 - h)^2\right]$$
$$+ (1 - h)^2 = y^2 \frac{2 - h}{h} + \frac{(2 - h)(1 - h)}{2} + x^2$$
$$\leq \left[\mu^2 - \frac{x^2}{h(2 - h)}\right](2h - h^2) + \frac{(2 - h)(1 - h)}{2} + x^2 \equiv q(x^2),$$

where the concluding $\leq$ is due to (a). Since $0 \leq x^2 \leq \mu^2(2h - h^2)$ by (b), $q(x^2)$ is in-between its values for $x^2 = 0$ and $x^2 = \mu^2(2h - h^2)$, and both these values with our μ are equal to 1. Thus, $[x; y] \in \mathcal{L}$.

It remains to prove that $\mu^{-1}\mathcal{E} \supset \mathcal{L}$, or, which is the same, that when $[x; y] \in \mathcal{L}$, we have $[\mu x; \mu y] \in \mathcal{E}$. Indeed, we have

$$[|y| + (1-h)]^2 + x^2 \leq 1 \Rightarrow |y| \leq h \ \& \ x^2 \leq 1 - y^2 - 2|y|(1-h) - (1-h)^2$$
$$\Rightarrow x^2 \leq 2h - h^2 - y^2 - 2|y|(1-h)$$

$$\Rightarrow \mu^2 \left[\frac{y^2}{h^2} + \frac{x^2}{2h - h^2} \right] = \mu^2 \frac{y^2(2-h) + hx^2}{h^2(2-h)} \leq \mu^2 \frac{h^2(2-h) + 2(1-h)\overbrace{[y^2 - |y|h]}^{\leq 0}}{h^2(2-h)} \leq \mu^2$$
$$\Rightarrow [x; y] \in \mathcal{E},$$

as claimed.

C.8 CHAPTER 8

Exercise 8.1: 1) We have

$$\begin{aligned}
\text{EstErr} &= \sup_{v \in V, A \in \mathcal{A}} \sqrt{v^T (GA - I)^T (GA - I) v + \text{Tr}(G^T \Sigma G)} \\
&= \sup_{A \in \mathcal{A}} \sup_{u : u^T u \leq 1} \sqrt{u^T Q^{-1/2} (GA - I)^T (GA - I) Q^{-1/2} u + \text{Tr}(G^T \Sigma G)} \\
&\qquad\qquad\qquad\qquad\qquad\qquad\qquad [\text{substitution } v = Q^{-1/2} u] \\
&= \sqrt{\sup_{A \in \mathcal{A}} \|(GA - I) Q^{-1/2}\|_{2,2}^2 + \text{Tr}(G^T \Sigma G)}.
\end{aligned}$$

By the Schur Complement Lemma, the relation $\|(GA - I)Q^{-1/2}\|_{2,2} \leq \tau$ is equivalent to the LMI $\left[\begin{array}{c|c} \tau I & [(GA-I)Q^{-1/2}]^T \\ \hline (GA-I)Q^{-1/2} & \tau I \end{array} \right]$, and therefore the problem of interest can be posed as the semi-infinite semidefinite program

$$\min_{t, \tau, \delta, G} \left\{ t : \begin{array}{c} \sqrt{\tau^2 + \delta^2} \leq t, \ \sqrt{\text{Tr}(G^T \Sigma G)} \leq \delta \\ \left[\begin{array}{c|c} \tau I & [(GA-I)Q^{-1/2}]^T \\ \hline (GA-I)Q^{-1/2} & \tau I \end{array} \right] \succeq 0 \ \forall A \in \mathcal{A} \end{array} \right\},$$

which is nothing but the RC of the uncertain semidefinite program

$$\left\{ \min_{t, \tau, \delta, G} \left\{ t : \begin{array}{c} \sqrt{\tau^2 + \delta^2} \leq t, \ \sqrt{\text{Tr}(G^T \Sigma G)} \leq \delta \\ \left[\begin{array}{c|c} \tau I & [(GA-I)Q^{-1/2}]^T \\ \hline (GA-I)Q^{-1/2} & \tau I \end{array} \right] \succeq 0 \end{array} \right\} : A \in \mathcal{A} \right\}.$$

In order to reformulate the only semi-infinite constraint in the problem in a tractable form, note that with $A = A_{\text{n}} + L^T \Delta R$ we have

$$\begin{aligned}
\mathcal{N}(A) &:= \left[\begin{array}{c|c} \tau I & [(GA-I)Q^{-1/2}]^T \\ \hline (GA-I)Q^{-1/2} & \tau I \end{array} \right] \\
&= \underbrace{\left[\begin{array}{c|c} \tau I & [(GA_{\text{n}}-I)Q^{-1/2}]^T \\ \hline (GA_{\text{n}}-I)Q^{-1/2} & \tau I \end{array} \right]}_{\mathcal{B}_{\text{n}}(G)} + \mathcal{L}^T(G) \Delta \mathcal{R} + \mathcal{R}^T \Delta^T \mathcal{L}(G),
\end{aligned}$$

$$\mathcal{L}(G) = \left[0_{p \times n}, LG^T \right], \mathcal{R} = \left[RQ^{-1/2}, 0_{q \times n} \right].$$

Invoking Theorem 8.2.3, the semi-infinite LMI $\mathcal{N}(A) \succeq 0 \ \forall A \in \mathcal{A}$ is equivalent to

$$\exists \lambda : \left[\begin{array}{c|c} \lambda I_p & \rho \mathcal{L}(G) \\ \hline \rho \mathcal{L}^T(G) & \mathcal{B}_{\text{n}}(G) - \lambda \mathcal{R}^T \mathcal{R} \end{array} \right] \succeq 0,$$

and thus the RC is equivalent to the semidefinite program

$$\min_{\substack{t,\tau,\\ \delta,\lambda,G}} \left\{ t : \begin{array}{c} \sqrt{\tau^2+\delta^2} \le t, \; \sqrt{\mathrm{Tr}(G^T\Sigma G)} \le \delta \\ \left[\begin{array}{c|c|c} \lambda I_p & & \rho L G^T \\ \hline & \tau I_n - \lambda Q^{-1/2} R^T R Q^{-1/2} & Q^{-1/2}(A_{\mathrm n}^T G^T - I_n) \\ \hline \rho G L^T & (GA_{\mathrm n} - I_n)Q^{-1/2} & \tau I_n \end{array} \right] \succeq 0 \end{array} \right\}.$$

2): Setting $v = U^T\widehat{v}$, $\widehat{y} = W^T y$, $\widehat{\xi} = W^T\xi$, our estimation problem reduces to the exactly the same problem, but with $\mathrm{Diag}\{a\}$ in the role of $A_{\mathrm n}$ and the diagonal matrix $\mathrm{Diag}\{q\}$ in the role of Q; a linear estimate $\widehat{G}\widehat{y}$ of $\widehat{v}$ in the new problem corresponds to the linear estimate $U^T\widehat{G}W^T y$, of exactly the same quality, in the original problem. In other words, the situation reduces to the one where $A_{\mathrm n}$ and Q are diagonal positive semidefinite, respectively, positive definite matrices; all we need is to prove that in this special case we lose nothing when restricting G to be diagonal. Indeed, in the case in question the RC reads

$$\min_{\substack{t,\tau,\\ \delta,\lambda,G}} \left\{ t : \begin{array}{c} \sqrt{\tau^2+\delta^2} \le t, \; \sigma\sqrt{\mathrm{Tr}(G^T G)} \le \delta \\ \left[\begin{array}{c|c|c} \lambda I_n & & \rho G^T \\ \hline & \tau I_n - \lambda\mathrm{Diag}\{\mu\} & \mathrm{Diag}\{\nu\}G^T - \mathrm{Diag}\{\eta\} \\ \hline \rho G & G\mathrm{Diag}\{\nu\} - \mathrm{Diag}\{\eta\} & \tau I_n \end{array} \right] \succeq 0 \end{array} \right\} \quad (*)$$

where $\mu_i = q_i^{-1}$, $\nu_i = a_i/\sqrt{q_i}$ and $\eta_i = 1/\sqrt{q_i}$. Replacing the G-component in a feasible solution with EGE, where E is a diagonal matrix with diagonal entries ± 1, we preserve feasibility (look what happens when you multiply the matrix in the LMI from the left and from the right by $\mathrm{Diag}\{I, I, E\}$). Since the problem is convex, it follows that whenever a collection $(t, \tau, \delta, \lambda, G)$ is feasible for the RC, so is the collection obtained by replacing the original G with the average of the matrices $E^T G E$ taken over all 2^n diagonal $n \times n$ matrices with diagonal entries ± 1, and this average is the diagonal matrix with the same diagonal as the one of G. Thus, when $A_{\mathrm n}$ and Q are diagonal and $L = R = I_n$ (or, which is the same in our situation, L and R are orthogonal), we lose nothing when restricting G to be diagonal.

Restricted to diagonal matrices $G = \mathrm{Diag}\{g\}$, the LMI constraint in $(*)$ becomes a bunch of 3×3 LMIs

$$\left[\begin{array}{c|c|c} \lambda & 0 & \rho g_i \\ \hline 0 & \tau - \lambda\mu_i & \nu_i g_i - \eta_i \\ \hline \rho g_i & \nu_i g_i - \eta_i & \tau \end{array} \right] \succeq 0, \; i = 1, ..., n,$$

in variables λ, τ, g_i. Assuming w.l.o.g. that $\lambda > 0$ and applying the Schur Complement Lemma, these 3×3 LMIs reduce to 2×2 matrix inequalities

$$\left[\begin{array}{c|c} \tau - \lambda\mu_i & \nu_i g_i - \eta_i \\ \hline \nu_i g_i - \eta_i & \tau - \rho^2 g_i^2/\lambda \end{array} \right] \succeq 0, \; i = 1, ..., n.$$

For given τ, λ, every one of these inequalities specifies a segment $\Delta_i(\tau, \lambda)$ of possible value of g_i, and the best choice of g_i in this segment is the point $g_i(\tau, \lambda)$ of the segment closest to 0 (when the segment is empty, we set $g_i(\tau, \lambda) = \infty$). Note that $g_i(\tau, \lambda) \ge 0$ (why?). It follows that $(*)$ reduces to the convex (due to its origin) problem

$$\min_{\tau,\lambda \ge 0} \left\{ \sqrt{\tau^2 + \sigma^2 \sum_i g_i^2(\tau, \lambda)} \right\}$$

with easily computable convex nonnegative functions $g_i(\tau, \lambda)$.

C.9 CHAPTER 9

Exercise 9.1: Note that 1), 2) are nothing but the real case counterparts of Lemma B.4.6. For the sake of completeness, we present the corresponding proofs.

1) Let $\lambda > 0$. For every $\xi \in \mathbb{R}^n$ we have $\xi^T[pq^T + qp^T]\xi = 2(\xi^T p)(\xi^T q) \leq \lambda(\xi^T p)^2 + \frac{1}{\lambda}(\xi^T q)^2 = \xi^T[\lambda pp^T + \frac{1}{\lambda} qq^T]\xi$, whence $pq^T + qp^T \preceq \lambda pp^T + \frac{1}{\lambda} qq^T$. By similar argument, $-[pq^T + qp^T] \preceq \lambda pp^T + \frac{1}{\lambda} qq^T$. 1) is proved.

2) Observe, first, that if $\lambda(A)$ is the vector of eigenvalues of a symmetric matrix A, then $\|\lambda(pq^T + qp^T)\|_1 = 2\|p\|_2\|q\|_2$. Indeed, there is nothing to verify when $p = 0$ or $q = 0$; when $p, q \neq 0$, we can normalize the situation to make p a unit vector and then to choose the orthogonal coordinates in $\mathbb{R}^n$ in such a way that p is the first basic orth, and q is in the linear span of the first two basic orths. With this normalization, the nonzero eigenvalues of A are exactly the same as the eigenvalues of the 2×2 matrix $\begin{bmatrix} 2\alpha & \beta \\ \beta & 0 \end{bmatrix}$, where α and β are the first two coordinates of q in our new orthonormal basis. The eigenvalues of the 2×2 matrix in question are $\alpha \pm \sqrt{\alpha^2 + \beta^2}$, and the sum of their absolute values is $2\sqrt{\alpha^2 + \beta^2} = 2\|q\|_2 = 2\|p\|_2\|q\|_2$, as claimed.

To prove 2), let us lead to a contradiction the assumption that $Y, p, q \neq 0$ are such that $Y \succeq \pm[pq^T + qp^T]$ and there is no $\lambda > 0$ such that $Y - \lambda pp^T - \frac{1}{\lambda} qq^T \succeq 0$, or, which is the same by the Schur Complement Lemma, the LMI

$$\begin{bmatrix} Y - \lambda pp^T & q \\ q^T & \lambda \end{bmatrix} \succeq 0$$

in variable λ has no solution, or, equivalently, the optimal value in the (clearly strictly feasible) SDO program

$$\min_{t,\lambda} \left\{ t : \begin{bmatrix} tI + Y - \lambda pp^T & q \\ q^T & \lambda \end{bmatrix} \succeq 0 \right\}$$

is positive. By semidefinite duality, the latter is equivalent to the dual problem possessing a feasible solution with a positive value of the dual objective. Looking at the dual, this is equivalent to the existence of a matrix $Z \in \mathbf{S}^n$ and a vector $z \in \mathbb{R}^n$ such that

$$\begin{bmatrix} Z & z \\ z^T & p^T Zp \end{bmatrix} \succeq 0, \quad \mathrm{Tr}(ZY) < 2q^T z.$$

Adding, if necessary, to Z a small positive multiple of the unit matrix, we can assume w.l.o.g. that $Z \succ 0$. Setting $\bar{Y} = Z^{1/2} Y Z^{1/2}$, $\bar{p} = Z^{1/2} p$, $\bar{q} = Z^{1/2} q$, $\bar{z} = Z^{-1/2} z$, the above relations become

$$\begin{bmatrix} I & \bar{z} \\ \bar{z}^T & \bar{p}^T \bar{p} \end{bmatrix} \succeq 0, \mathrm{Tr}(\bar{Y}) < 2\bar{q}^T \bar{z}. \tag{$*$}$$

Observe that from $Y \succeq \pm[pq^T + qp^T]$ it follows that $\bar{Y} \succeq \pm[\bar{p}\bar{q}^T + \bar{q}\bar{p}^T]$. Looking at what happens in the eigenbasis of the matrix $[\bar{p}\bar{q}^T + \bar{q}\bar{p}^T]$, we conclude from this relation that $\mathrm{Tr}(\bar{Y}) \geq \|\lambda(\bar{p}\bar{q}^T + \bar{q}\bar{p}^T)\|_1 = 2\|\bar{p}\|_2\|\bar{q}\|_2$. On the other hand, the matrix inequality in $(*)$ implies that $\|\bar{z}\|_2 \leq \|\bar{p}\|_2$, and thus $\mathrm{Tr}(\bar{Y}) < 2\|\bar{p}\|_2\|\bar{q}\|_2$ by the second inequality in $(*)$. We have arrived at a desired contradiction.

3) Assume that x is such that all $L_\ell(x)$ are nonzero. Assume that x can be extended to a feasible solution $Y_1, ..., Y_L, x$ of (9.2.2). Invoking 2), we can find $\lambda_\ell > 0$ such that $Y_\ell \succeq \lambda_\ell R_\ell^T R_\ell + \frac{1}{\lambda_\ell} L_\ell^T(x)L_\ell(x)$. Since $\mathcal{A}_\mathrm{n}(x) - \rho\sum_\ell Y_\ell \succeq 0$, we have $[\mathcal{A}_\mathrm{n}(x) - \rho\sum_\ell \lambda_\ell R_\ell^T R_\ell] - \sum_\ell \frac{\rho}{\lambda_\ell} L_\ell^T(x)L_\ell(x) \succeq 0$, whence, by the Schur Complement Lemma, $\lambda_1, ..., \lambda_L, x$ are feasible

for (9.2.3). Vice versa, if $\lambda_1, ..., \lambda_L, x$ are feasible for (9.2.3), then $\lambda_\ell > 0$ for all ℓ due to $L_\ell(x) \neq 0$, and, by the same Schur Complement Lemma, setting $Y_\ell = \lambda_\ell R_\ell^T R_\ell + \frac{1}{\lambda_\ell} L_\ell^T(x) L_\ell(x)$, we have

$$\mathcal{A}_n(x) - \rho \sum_\ell Y_\ell \succeq 0,$$

while $Y_\ell \succeq \pm \left[L_\ell^T(x) R_\ell + R_\ell^T L_\ell(x) \right]$, that is, $Y_1, ..., Y_L, x$ are feasible for (9.2.2).

We have proved the equivalence of (9.2.2) and (9.2.3) in the case when $L_\ell(x) \neq 0$ for all ℓ. The case when some of $L_\ell(x)$ vanish is left to the reader.

Exercise 9.2: *A solution might be as follows.* The problem of interest is

$$\min_{G,t} \left\{ t : t \geq \| (GA - I)v + G\xi \|_2 \, \forall (v \in V, \xi \in \Xi, A \in \mathcal{A}) \right\}$$

$$\Updownarrow$$

$$\min_{G,t} \left\{ t : u^T(GA - I)v + u^T G\xi \leq t \forall \left(u, v, \xi : \begin{array}{c} u^T u \leq 1 \\ v^T P_i v \leq 1, \\ 1 \leq i \leq I \\ \xi^T Q_j \xi \leq \rho_\xi^2, \\ 1 \leq j \leq J \end{array} \right) \forall A \in \mathcal{A} \right\}. \quad (*)$$

Observing that

$$u^T[GA - I]v + u^T G\xi = [u; v; \xi]^T \left[\begin{array}{c|c} & \frac{1}{2}[GA - I] \;\big|\; \frac{1}{2}G \\ \hline \frac{1}{2}[GA - I]^T & \\ \frac{1}{2}G^T & \end{array} \right] [u; v; \xi],$$

for A fixed, a sufficient condition for the validity of the semi-infinite constraint in $(*)$ is the existence of nonnegative μ, ν_i, ω_j such that

$$\left[\begin{array}{c|c} \mu I & \\ \hline & \sum_i \nu_i P_i \\ & \sum_j \omega_j Q_j \end{array} \right] \succeq \left[\begin{array}{c|c} & \frac{1}{2}[GA - I] \;\big|\; \frac{1}{2}G \\ \hline \frac{1}{2}[GA - I]^T & \\ \frac{1}{2}G^T & \end{array} \right]$$

and $\mu + \sum_i \nu_i + \rho_\xi^2 \sum_j \omega_j \leq t$. It follows that the validity of the semi-infinite system of constraints

$$\mu + \sum_i \nu_i + \rho_\xi^2 \sum_j \omega_j \leq t, \; \mu \geq 0, \nu_i \geq 0, \omega_j \geq 0$$

$$\left[\begin{array}{c|c} \mu I & \\ \hline & \sum_i \nu_i P_i \\ & \sum_j \omega_j Q_j \end{array} \right] \succeq \left[\begin{array}{c|c} & \frac{1}{2}[GA - I] \;\big|\; \frac{1}{2}G \\ \hline \frac{1}{2}[GA - I]^T & \\ \frac{1}{2}G^T & \end{array} \right] \qquad (!)$$

$$\forall A \in \mathcal{A}$$

in variables $t, G, \mu, \nu_i, \omega_j$ is a sufficient condition for (G, t) to be feasible for $(*)$. The only semi-infinite constraint in (!) is in fact an LMI with structured norm-bounded uncertainty:

$$
\left[
\begin{array}{c|c|c}
\mu I & & \\ \hline
 & \sum_i \nu_i P_i & \\ \hline
 & & \sum_j \omega_j Q_j
\end{array}
\right]
- \left[
\begin{array}{c|c|c}
 & \frac{1}{2}[GA - I] & \frac{1}{2}G \\ \hline
\frac{1}{2}[GA-I]^T & & \\ \hline
\frac{1}{2}G^T & &
\end{array}
\right] \succeq 0 \ \forall A \in \mathcal{A}
$$

$$\Updownarrow$$

$$
\underbrace{\left[
\begin{array}{c|c|c}
\mu I & -\frac{1}{2}[GA_{\mathrm n} - I] & -\frac{1}{2}G \\ \hline
-\frac{1}{2}[GA_{\mathrm n} - I]^T & \sum_i \nu_i P_i & \\ \hline
-\frac{1}{2}G^T & & \sum_j \omega_j Q_i
\end{array}
\right]}_{\mathcal{B}(\mu, \nu, \omega, G)}
$$

$$
+ \sum_{\ell=1}^L [\mathcal{L}_\ell(G)^T \Delta_\ell \mathcal{R}_\ell + \mathcal{R}_\ell^T \Delta_\ell^T \mathcal{L}_\ell(G)] \succeq 0
$$
$$
\forall (\|\Delta_\ell\|_{2,2} \leq \rho_A, 1 \leq \ell \leq L),
$$
$$
\mathcal{L}_\ell(G) = \tfrac{1}{2}\left[L_\ell G^T, 0_{p_\ell \times n}, 0_{p_\ell \times m} \right], \quad \mathcal{R}_\ell = [0_{q_\ell \times n}, R_\ell, 0_{q_\ell \times m}].
$$

Invoking Theorem 9.1.2, we end up with the following safe tractable approximation of $(*)$:

$$
\min_{t, G, \mu, \nu_i, \omega_j, \lambda_\ell, Y_\ell} \quad t
$$
$$
\text{s.t.}
$$
$$
\mu + \sum_i \nu_i + \rho_\xi^2 \sum_j \omega_j \leq t, \ \mu \geq 0, \nu_i \geq 0, \omega_j \geq 0
$$
$$
\left[
\begin{array}{c|c}
\lambda_\ell I & \mathcal{L}_\ell(G) \\ \hline
\mathcal{L}_\ell^T(G) & Y_\ell - \lambda_\ell \mathcal{R}_\ell^T \mathcal{R}_\ell
\end{array}
\right] \succeq 0, \ 1 \leq \ell \leq L
$$
$$
\mathcal{B}(\mu, \nu, \omega, G) - \rho_A \sum_{\ell=1}^L Y_\ell \succeq 0.
$$

C.10 CHAPTER 12

Exercise 12.1: For every $i \in \{1, \ldots, m\}$, the condition

$$
\forall \delta w, \ \|\delta w\|_\infty \leq \rho \|w\|_2 : y_i((w + \delta w)^T x_i + b) \geq 0
$$

is equivalent to

$$
y_i(w^T x_i + b) \geq \rho \|w\|_2,
$$

which is similar to the condition (12.1.2). Finding the maximally robust classifier is processed the same way as in the case of data uncertainty.

Exercise 12.2: For given vectors u, v, we have

$$
\max_{\Delta \in \mathcal{D}} u^T \Delta v = \max_{\alpha \geq 0 \, : \, \|\alpha\|_q \leq 1} \ \max_{\delta_i \, : \, \|\delta_i\|_p \leq \alpha_i} \sum_{i=1}^m v_i(u^T \delta_i)
$$

$$
= \max_{\alpha \geq 0 \, : \, \|\alpha\|_q \leq 1} \sum_{i=1}^m \max_{\delta \, : \, \|\delta\|_p \leq 1} \alpha_i |v_i| \cdot |u^T \delta|
$$

$$
= \|u\|_{p^*} \cdot \max_{\alpha \, : \, \|\alpha\|_q \leq 1} \sum_{i=1}^m |v_i| |\alpha_i|
$$

$$
= \|u\|_{p^*} \|v\|_{q^*},
$$

as claimed.

Exercise 12.3: 1): If $i \in \mathcal{J}$, then we need to compute

$$\max_{y_i \in \{-1, 1\}} [1 - y_i(w^T x_i + b)]_+ = 1 + |w^T x_i + b|.$$

Hence the robust counterpart expresses as

$$\min_{(w,b)} \left\{ \sum_{i \in \mathcal{I}} [1 - y_i(z_i^T w + b)]_+ + \sum_{i \in \mathcal{J}} |z_i^T w + b| \right\},$$

which can be expressed easily as a Linear Optimization problem. The above corresponds to a regularized version of the SVM, with weighted ℓ_1-norm, where the weights involve the data points.

2): With the notation $\Theta_k := \{\theta \in \{0,1\}^m : \mathbf{1}^T \theta = k\}$, this problem can be formulated as

$$\min_x \left\{ \max_{\theta \in \Theta_k} \sum_{i=1}^m \left[(1 - \theta_i) p_i^+ + \theta_i p_i^- \right] : p_i^\pm = [1 \mp y_i(z_i^T w + b)]_+, \, 1 \le i \le m \right\}.$$

Using the fact that we can replace equality constraints by convex inequalities without loss of generality in the above, we obtain the formulation

$$\min_x \left\{ \mathbf{1}^T p^+ + \max_{\theta \in \Theta_k} \sum_{i=1}^m \theta_i(p_i^- - p_i^+) : p_i^\pm \ge [1 \mp y_i(z_i^T w + b)]_+, \, 1 \le i \le m \right\},$$

or, equivalently:

$$\min_x \left\{ \mathbf{1}^T p^+ + \sum_{i=1}^k (p^- - p^+)_{[i]} : p_i^\pm \ge [1 \mp y_i(z_i^T w + b)]_+, \, 1 \le i \le m \right\},$$

which can be cast into a linear optimization format via

$$\min_{x, \mu} \left\{ \mathbf{1}^T p^+ + k\mu + \sum_{i=1}^m [p_i^- - p_i^+ - \mu]_+ : p_i^\pm \ge [1 \mp y_i(z_i^T w + b)]_+, \, 1 \le i \le m \right\}.$$

Exercise 12.4: 1): The solution involves addressing the following problem, where i, y_i and x_i are given:

$$\max_{\delta \in \Theta_k} [1 - y_i(w^T(x_i + \delta_i) + b)]_+,$$

where $\Theta_k := \{\delta \in \{-1, 0, 1\}^n : \|\delta\|_1 \le k\}$. In turn, we are led to a problem of the form

$$\max_{\delta \in \Theta_k} \delta^T r,$$

where $r \in \mathbf{R}^n$ is given. Without loss of generality, we can replace Θ_k by its convex hull. Using duality, we can then easily convert the above problem to

$$\min_u \{ \|r - u\|_1 + k\|u\|_\infty \}.$$

Thus, the robust counterpart to our problem takes the form

$$\min_{w,b} \sum_{i=1}^m [1 - y_i(x_i^T w + b) + \phi(w)]_+,$$

where

$$\phi(w) := \min_u \{k\|u\|_\infty + \|w - u\|_1\}.$$

2): The same approach leads us to the problem

$$\phi(r, x) := \max_\delta \left\{\delta^T r : 0 \le \delta + x \le \mathbf{1}, \ \|\delta\|_1 \le k\right\},$$

where $r \in \mathbf{R}^n$ is given. Using duality again, we obtain that the value of the above problem is

$$\phi(r, x) = \min_u \left\{\mathbf{1}^T (r - u) - (r - u)^T x + k\|u\|_\infty\right\}.$$

The robust counterpart thus expresses as

$$\min_{w,b} \sum_{i=1}^m [1 - y_i(z_i^T w + b) + \phi(y_i w, x_i)]_+,$$

where the function ϕ is given above.

Exercise 12.5: Let $z_i = [y_i x_i; y_i]$, $i = 1, \ldots, m$, and $Z = [z_1, \ldots, z_m]$. For the new data point/label pair (x_{m+1}, y_{m+1}), we can write $[y_{m+1} x_{m+1}; y_{m+1}] = Zu + [v; 0]$, where $u \in \{0, 1\}^m$, $\sum_{i=1}^m u_i = 1$, and $v \in \mathbf{R}^n$, $\|v\|_2 \le 1$. Denoting by $\mathcal{U}$ the set of such allowable data point/label pair, we have

$$\max_{(x_{m+1}, y_{m+1}) \in \mathcal{U}} [1 - y_{m+1}(x_{m+1}^T w + b)]_+ = [1 - \min_i y_i(x_i^T w + b) + \rho\|w\|_2]_+,$$

which means that the robust counterpart can be written as

$$\min_{w,b} \left\{\sum_{i=1}^m [1 - y_i(w^T x_i + b)]_+ + [1 - \min_i y_i(x_i^T w + b) + \rho\|w\|_2]_+\right\}.$$

The above can be expressed as the second-order cone optimization problem

$$\min_{w,b,\tau} \left\{\sum_{i=1}^m [1 - y_i(w^T x_i + b)]_+ + [1 - \tau + \rho\|w\|_2]_+ : \tau \le y_i(x_i^T w + b), \ 1 \le i \le m\right\}.$$

C.11 CHAPTER 14

Exercise 14.1: From state equations (14.5.1) coupled with control law (14.5.3) it follows that

$$w^N = W_N[\Xi]\zeta + w_N[\Xi],$$

where $\Xi = \{U_t^z, U_t^d, u_t^0\}_{t=0}^N$ is the "parameter" of the control law (14.5.3), and $W_N[\Xi]$, $w_N[\Xi]$ are matrix and vector affinely depending on Ξ. Rewriting (14.5.2) as the system of linear constraints

$$e_j^T w^N - f_j \le 0, \ j = 1, \ldots, J,$$

and invoking Proposition 3.2.1, the GRC in question is the semi-infinite optimization problem

$$\begin{aligned}
&\min_{\Xi, \alpha} \quad \alpha \\
&\text{subject to} \\
&\quad e_j^T[W_N[\Xi]\zeta + w_N[\Xi]] - f_j \le 0 \ \forall(\zeta : \|\zeta - \bar\zeta\|_s \le R) \quad (a_j) \\
&\quad e_j^T W_N[\Xi]\zeta \le \alpha \ \forall(\zeta : \|\zeta\|_r \le 1) \quad\quad\quad\quad\quad (b_j) \\
&\quad\quad\quad\quad\quad\quad\quad\quad\quad\quad\quad\quad\quad\quad 1 \le j \le J.
\end{aligned}$$

This problem clearly can be rewritten as

$$\min_{\Xi,\alpha} \quad \alpha$$
subject to
$$R\|W_N^T[\Xi]e_j\|_{s_*} + e_j^T[W_N[\Xi]\bar\zeta + w_N[\Xi]] - f_j \le 0, \; 1 \le j \le J$$
$$\|W_N^T[\Xi]e_j\|_{r_*} \le \alpha, \; 1 \le j \le J$$

where

$$s_* = \frac{s}{s-1}, \; r_* = \frac{r}{r-1}.$$

Exercise 14.2: The AAGRC is equivalent to the convex program

$$\min_{\Xi,\alpha} \quad \alpha$$
subject to
$$R\|W_N^T[\Xi]e_j\|_{s_*} + e_j^T[W_N[\Xi]\zeta + w_N[\Xi]] - f_j \le 0, \; 1 \le j \le J$$
$$\|[W_N^T[\Xi]e_j]_{d,+}\|_{r_*} \le \alpha, \; 1 \le j \le J$$

where

$$s_* = \frac{s}{s-1}, \; r_* = \frac{r}{r-1}$$

and for a vector $\zeta = [z; d_0; ...; d_N] \in \mathbb{R}^K$, $[\zeta]_{d,+}$ is the vector obtained from ζ by replacing the z-component with 0, and replacing every one of the d-components with the vector of positive parts of its coordinates, the positive part of a real a being defined as $\max[a, 0]$.

Exercise 14.3: 1) For $\zeta = [z; d_0; ...; d_{15}] \in \mathcal{Z} + \mathcal{L}$, a control law of the form (14.5.3) can be written down as

$$u_t = u_t^0 + \sum_{\tau=0}^{t} u_{t\tau}d_\tau,$$

and we have

$$x_{t+1} = \sum_{\tau=0}^{t}\left[u_\tau^0 - d_\tau + \sum_{s=0}^{\tau} u_{\tau s}d_s\right] = \sum_{\tau=0}^{t} u_\tau^0 + \sum_{s=0}^{t}\left[\sum_{\tau=s}^{t} u_{\tau s} - 1\right]d_s.$$

Invoking Proposition 3.2.1, the AAGRC in question is the semi-infinite problem

$$\min_{\{u_t^0,u_{t\tau}\},\alpha} \quad \alpha$$
subject to
(a_x) $\qquad |\theta\left[\sum_{\tau=0}^{t} u_\tau^0\right]| \le 0, \; 0 \le t \le 15$
(a_u) $\qquad |u_t^0| \le 0, \; 0 \le t \le 15$
(b_x) $\qquad |\theta \sum_{s=0}^{t}\left[\sum_{\tau=s}^{t} u_{\tau s} - 1\right]d_s| \le \alpha$
$\qquad\qquad \forall(0 \le t \le 15, [d_0; ...; d_{15}] : \|[d_0; ...; d_{15}]\|_2 \le 1)$
(b_u) $\qquad |\sum_{\tau=0}^{t} u_{t\tau}d_\tau| \le \alpha$
$\qquad\qquad \forall(0 \le t \le 15, [d_0; ...; d_{15}] : \|[d_0; ...; d_{15}]\|_2 \le 1)$

We see that the desired control law is linear ($u_t^0 = 0$ for all t), and the AAGRC is equivalent to the conic quadratic problem

$$\min_{\{u_{t\tau}\},\alpha}\left\{\alpha : \begin{array}{l} \sqrt{\sum_{s=0}^{t}\left[\sum_{\tau=s}^{t} u_{\tau s} - 1\right]^2} \le \theta^{-1}\alpha, \; 0 \le t \le 15 \\ \sqrt{\sum_{\tau=0}^{t} u_{\tau t}^2} \le \alpha, \; 0 \le t \le 15 \end{array}\right\}.$$

2) In control terms, we want to "close" our toy linear dynamical system, where the initial state is once and for ever set to 0, by a linear state-based non-anticipative control law in such a way that the states $x_1, ..., x_{16}$ and the controls $u_1, ..., u_{15}$ in the closed loop

system are "as insensitive to the perturbations $d_0, ..., d_{15}$ as possible," while measuring the changes in the state-control trajectory

$$w^{15} = [0; x_1; ...; x_{16}; u_1, ..., u_{15}]$$

in the weighted uniform norm $\|w^{15}\|_{\infty,\theta} = \max[\theta\|x\|_\infty, \|u\|_\infty]$, and measuring the changes in the sequence of disturbances $[d_0; ...; d_{15}]$ in the "energy" norm $\|[d_0; ...; d_{15}]\|_2$. Specifically, we are interested to find a linear non-anticipating state-based control law that results in the smallest possible constant α satisfying the relation

$$\forall \Delta d^{15} : \|\Delta w^{15}\|_{\infty,\theta} \le \alpha \|\Delta d^{15}\|_2,$$

where Δd^{15} is a shift of the sequence of disturbances, and Δw^{15} is the induced shift in the state-control trajectory.

3) The numerical results are as follows:

θ	α
1.e6	4.0000
10	3.6515
2	2.8284
1	2.3094

Exercise 14.4: 1) Denoting by x_γ^{ij} the amount of information in the traffic from i to j travelling through γ, by q_γ the increase in the capacity of arc γ, and by $O(k), I(k)$ — the sets of outgoing, resp., incoming, arcs for node k, the problem in question becomes

$$\min_{\substack{\{x_\gamma^{ij}\}, \\ \{q_\gamma\}}} \left\{ \sum_{\gamma \in \Gamma} c_\gamma q_\gamma : \begin{array}{l} \sum_{(i,j)\in \mathcal{J}} x_\gamma^{ij} \le p_\gamma + q_\gamma \; \forall \gamma \\[2mm] \sum_{\gamma \in O(k)} x_\gamma^{ij} - \sum_{\gamma \in I(k)} x_\gamma^{ij} = \left\{ \begin{array}{ll} d_{ij}, & k = i \\ -d_{ij}, & k = j \\ 0, & k \notin \{i,j\} \end{array} \right. \\[4mm] \hspace{3cm} \forall((i,j) \in \mathcal{J}, k \in V) \\[2mm] q_\gamma \ge 0, x_\gamma^{ij} \ge 0 \; \forall((i,j) \in \mathcal{J}, k \in V) \end{array} \right\}. \quad (*)$$

2) To build the AARC of $(*)$ in the case of uncertain traffics d_{ij}, it suffices to plug into $(*)$, instead of decision variables x_γ^{ij}, affine functions $X_\gamma^{ij}(d) = \xi_\gamma^{ij,0} + \sum_{(\mu,\nu)\in\mathcal{J}} \xi_\gamma^{ij\mu\nu} d_{\mu\nu}$ of $d = \{d_{ij} : (i,j) \in \mathcal{J}\}$ (in the case of (a), the functions should be restricted to be of the form $X_\gamma^{ij}(d) = \xi_\gamma^{ij,0} + \xi_\gamma^{ij} d_{ij}$) and to require the resulting constraints in variables $q_\gamma, \xi_\gamma^{ij\mu\nu}$ to be valid for all realizations of $d \in \mathcal{Z}$. The resulting semi-infinite LO program is computationally tractable (as the AARC of an uncertain LO problem with fixed recourse, see section 14.3.1).

3) Plugging into $(*)$, instead of variables x_γ^{ij}, affine decision rules $X_\gamma^{ij}(d)$ of the just indicated type, the constraints of the resulting problem can be split into 3 groups:

(a) $\sum_{(i,j)\in\mathcal{J}} X_\gamma^{ij}(d) \le p_\gamma + q_\gamma \; \forall \gamma \in \Gamma$

(b) $\sum_{\substack{(i,j)\in\mathcal{J} \\ \gamma\in\Gamma}} \mathcal{R}_\gamma^{ij} X_\gamma^{ij}(d) = r(d)$

(c) $q_\gamma \ge 0, X_\gamma^{ij}(d) \ge 0 \; \forall((i,j) \in \mathcal{J}, \gamma \in \Gamma)$.

In order to ensure the feasibility of a given candidate solution for this system with probability at least $1 - \epsilon, \epsilon < 1$, when d is uniformly distributed in a box, the linear equalities (b) *must* be satisfied for all d's, that is, (b) induces a system $A\xi = b$ of linear equality constraints on the vector ξ of coefficients of the affine decision rules $X_\gamma^{ij}(\cdot)$. We can use this system of linear equations, if it is feasible, in order to express ξ as an affine function of a shorter vector η of "free" decision variables, that is, we can easily find H and h in such a way that $A\xi = b$ is equivalent to the existence of η such that $\xi = H\eta + h$. We can now plug $\xi = H\eta + h$ into $(a), (c)$ and forget about (b), thus ending up with a system of constraints of the form

$$(a') \quad a_\ell(\eta, q) + \alpha_\ell^T(\eta, q)d \leq 0, \ 1 \leq \ell \leq L = \mathrm{Card}(\Gamma)(\mathrm{Card}(\mathcal{J}) + 1),$$
$$(b') \quad q \geq 0$$

with a_ℓ, α_ℓ affine in $[\eta; q]$ (the constraints in (a') come from the $\mathrm{Card}(\Gamma)$ constraints in (a) and the $\mathrm{Card}(\Gamma)\mathrm{Card}(\mathcal{J})$ constraints $X_\gamma^{ij}(d) \geq 0$ in (c)).

In order to ensure the validity of the uncertainty-affected constraints (a'), as evaluated at a candidate solution $[\eta; q]$, with probability at least $1 - \epsilon$, we can use either the techniques from chapters 2, 4, or the techniques from section 10.4.1.

Bibliography

[1] Barmish, B.R., Lagoa, C.M. The uniform distribution: a rigorous justification for the use in robustness analysis, *Math. Control, Signals, Systems* **10** (1997), 203–222.

[2] Bendsøe, M. *Optimization of Structural Topology, Shape and Material.* Springer-Verlag, Heidelberg, 1995.

[3] Ben-Tal, A., Nemirovski, A., Stable Truss Topology Design via Semidefinite Programming. *SIAM J. on Optimization* **7**:4 (1997), 991–1016.

[4] Ben-Tal, A., Nemirovski, A. Robust Convex Optimization. *Math. of Oper. Res.* **23**:4 (1998), 769–805.

[5] Ben-Tal, A., Nemirovski, A. Robust solutions of uncertain linear programs. *OR Letters* **25** (1999), 1–13.

[6] Ben-Tal, A., Kočvara, M., Nemirovski, A., and Zowe, J. Free Material Design via Semidefinite Programming. The Multiload Case with Contact Conditions. *SIAM J. on Optimization* **9** (1999), 813–832, and *SIAM Review* **42** (2000), 695–715.

[7] Ben-Tal, A., Nemirovski, A. Robust solutions of Linear Programming problems contaminated with uncertain data. *Math. Progr.* **88** (2000), 411–424.

[8] Ben-Tal, A., Nemirovski, A. *Lectures on Modern Convex Optimization: Analysis, Algorithms and Engineering Applications.* SIAM, Philadelphia, 2001.

[9] Ben-Tal, A., Nemirovski, A. Robust Optimization — methodology and applications. *Math. Progr. Series B* **92** (2002), 453–480.

[10] Ben-Tal, A., Nemirovski, A. On tractable approximations of uncertain linear matrix inequalities affected by interval uncertainty. *SIAM J. on Optimization* **12** (2002), 811–833.

[11] Ben-Tal, A., Nemirovski, A., Roos, C. Robust solutions of uncertain quadratic and conic-quadratic problems, *SIAM J. on Optimization* **13** (2002), 535–560.

[12] Ben-Tal, A., Nemirovski, A., Roos, C. Extended matrix cube theorems with applications to μ-theory in control. *Math. of Oper. Res.* **28** (2003), 497–523.

[13] Ben-Tal, A., Goryashko, A., Guslitzer, E. Nemirovski, A. Adjustable robust solutions of uncertain linear programs. *Math. Progr.* **99** (2004), 351–376.

[14] Ben-Tal, A., Golany, B., Nemirovski, A., Vial, J.-P. Supplier-retailer flexible commitments contracts: A robust optimization approach. *Manufacturing & Service Operations Management* **7:3** (2005), 248–271.

[15] Ben-Tal, A., Boyd, S., Nemirovski, A. Extending scope of robust optimization: Comprehensive robust counterparts of uncertain problems. *Math. Progr. Series B* **107:1-2** (2006), 63–89.

[16] Ben-Tal, A., Nemirovski, A. Selected topics in robust convex optimization. *Math. Progr. Series B* **112:1** (2008), 125–158.

[17] Ben-Tal, A., Margalit, T., Nemirovski, A. Robust modeling of multi-stage portfolio problems. In: H. Frenk, K. Roos, T. Terlaky, S. Zhang, eds. *High Performance Optimization*, Kluwer Academic Publishers, 2000, 303–328.

[18] Ben-Tal, A., El Ghaoui, L., Nemirovski, A. Robust semidefinite programming. In: R. Saigal, H. Wolkowitcz, L. Vandenberghe, eds. *Handbook on Semidefinite Programming*. Kluwer Academic Publishers, 2000, 139–162.

[19] Ben-Tal, A., Nemirovski, A., Roos, C. Robust versions of convex quadratic and conic-quadratic problems. In: D. Li, ed. *Proceedings of the 5th International Conference on Optimization Techniques and Applications (ICOTA 2001)*, **4** (2001), 1818–1825.

[20] Ben-Tal, A., Golany, B., Shtern, S. (2008). Robust multi-echelon, multi-period inventory control. Submitted to *Operations Research*.

[21] Bernussou, J., Peres, P.L.D., Geromel, J.C. A linear-programming oriented procedure for quadratic stabilization of uncertain systems. *Syst. Control Letters* **13** (1989), 65–72.

[22] Bertsimas, D., Sim, M. Tractable approximations to robust conic optimization problems. *Math. Progr. Series B* **107:1-2** (2006), 5–36.

[23] Bertsimas, D., Pachmanova, D. Sim, M. Robust linear optimization under general norms. *OR Letters* **32:6** (2004), 510–516.

[24] Bertsimas, D., Sim, M. The price of robustness. *Oper. Res.* **32:1** (2004), 35–53.

[25] Bertsimas, D., Sim, M. Robust discrete optimization and network flows. *Math. Progr. Series B* **98** (2003), 49–71.

[26] Bertsimas, D., Popescu, I. Optimal inequalities in probability theory: A convex optimization approach. *SIAM J. on Optimization* **15:3** (2005), 780–804.

[27] Bertsimas, D., Popescu, I., Sethuraman, J. Moment problems and semidefinite programming. In: H. Wolkovitz, R. Saigal, eds. *Handbook on Semidefinite Programming*, Kluwer Academic Publishers, 2000, 469–509.

[28] Bertsimas, D., Fertis, A. 2008. On the equivalence between robust optimization and regularization in statistics. To appear in *Oper. Res.*

[29] Bhattacharrya, C., Grate, L., Mian, S., El Ghaoui, L., Jordan, M. Robust sparse hyperplane classifiers: Application to uncertain molecular profiling data. *Journal of Computational Biology* **11:6** (2003), 1073–1089.

[30] Blanchard, O.J. The production and inventory behavior of the American automobile industry. *Journal of Political Economy* **91:3** (1983), 365–400.

[31] Borell, C. The Brunn-Minkowski inequality in Gauss space. *Inventiones Mathematicae* **30:2** (1975), 207–216.

[32] Boyd, S., El Ghaoui, L., Feron, E., Balakrishnan, V. *Linear Matrix Inequalities in System and Control Theory*. SIAM, Philadelphia, 1994.

[33] Boyd, S., Vandenberghe, L. *Convex Optimization*. Cambridge University Press, 2004.

[34] Brown, R.G., Hwang, P.Y.C. *Introduction to Random Signals and Applied Kalman Filtering*. 3rd ed. John Wiley & Sons, New York, 1996.

[35] Buchholz, A., Operator Khintchine inequality in the non-commutative probability. *Mathematische Annalen* **391** (2001), 1–16.

[36] Cachon, G.P., Taylor, R., Schmidt, G.M. (2005). In search of the bullwhip effect. Manufacturing & Service Operations Management **9:4** (2007), 457–479.

[37] Calafiore, G., Campi, M.C. Uncertain convex programs: Randomized solutions and confidence levels. *Math. Progr.* **102:1** (2005), 25–46.

[38] Calafiore, G., Campi, M.C. Decision making in an uncertain environment: The scenario-based optimization approach. In: J. Andrysek, M. Karny, J. Kracik, eds. *Multiple Participant Decision Making*. Advanced Knowledge International, 2004, 99–111.

[39] Calafiore, G., Topcu, U., El Ghaoui, L. 2009. Parameter estimation with expected and residual-at-risk criteria. To appear in *Systems and Control Letters*.

[40] Charnes, A., Cooper, W.W., Symonds, G.H. Cost horizons and certainty equivalents: An approach to stochastic programming of heating oil. *Management Science* **4** (1958), 235–263.

[41] Chen, F., Drezner, Z., Ryan, J., Simchi-Levi, D. Quantifying the bullwhip effect in a simple supply chain: The impact of forecasting, lead times and information. *Management Science* **46** (2000), 436–443.

[42] Dan Barb, F., Ben-Tal, A., Nemirovski, A. Robust dissipativity of interval uncertain system. *SIAM J. Control and Optimization* **41** (2003), 1661–1695.

[43] Dantzig, G.B. Linear Programming. In: J.K. Lenstra, A.H.G. Ronnooy Kan, A. Schrijver, eds. *History of Mathematical Programming. A Collection of Personal Reminiscences.* CWI, Amsterdam and North-Holland, New York 1991.

[44] De Farias, V.P., Van Roy, B. On constraint sampling in the linear programming approach to approximate dynamic programming. *Math. of Oper. Res.* **29:3** (2004), 462–478.

[45] Dentcheva, D., Prékopa, A., Ruszczynski, A. Concavity and efficient points of discrete distributions in probabilistic programming. *Mathematical Programming* **89** (2000), 55–77.

[46] Dhaene, J., Denuit, M., Goovaerts, M.J., Kaas, R., Vyncke, D. The concept of comonotonicity in actuarial science and finance: theory. *Insurance: Mathematics and Economics* **31** (2002), 3–33.

[47] Diamond, P., Stiglitz, J.E. Increases in risk and in risk aversion. *Journal of Economic Theory* **8** (1974), 337–360.

[48] Eldar, Y., Ben-Tal, A., Nemirovski, A. Robust mean-squared error estimation in the presence of model uncertainties. *IEEE Trans. on Signal Processing* **53** (2005), 168–181.

[49] El Ghaoui, L., Lebret, H. Robust solution to least-squares problems with uncertain data. *SIAM J. of Matrix Anal. Appl.* **18** (1997), 1035–1064.

[50] El Ghaoui, L., Oustry, F., Lebret, H. Robust solutions to uncertain semidefinite programs. *SIAM J. on Optimization* **9** (1998), 33–52.

[51] El Ghaoui, L., Lanckriet, G.R.G., Natsoulis, G. Robust classification with interval data. Technical Report # UCB/CSD-03-1279, EECS Department, University of California, Berkeley, Oct. 2003.
http://www.eecs.berkeley.edu/Pubs/TechRpts/2003/5772.html

[52] Falk, J.E., Exact solutions to inexact linear programs. *Oper. Res.* **24:4** (1976), 783–787.

[53] Genin, Y., Hachez, Y., Nesterov, Yu., Van Dooren, P. Optimization problems over positive pseudopolynomial matrices. *SIAM J. Matrix Anal. Appl.* **25** (2003), 57–79.

[54] Lanckriet, G.R.G., El Ghaoui, L., Bhattacharyya, C., Jordan, M.I. A robust minimax approach to classification. *J. Mach. Learn. Res.* **3** (2003), 555–582.

[55] Goulart, P.J., Kerrigan, E.C., Maciejowski, J.M. Optimization over state feedback policies for robust control with constraints. *Automatica* **42:4** (2006), 523–533.

[56] Grotschel, M., Lovasz, L., Schrijver, A. *Geometric Algorithms and Combinatorial Optimization.* Springer-Verlag, Berlin, 1987.

[57] Forrester, J.W. *Industrial Dynamics.* MIT Press, 1973.

[58] Hadar, J., Russell, W. Rules for ordering uncertain prospects. *American Economic Review* **59** (1969), 25–34.

[59] Hammond, J. Barilla SpA (A). Harvard Business School. Case No. 9-694-046 (1994).

[60] Hanoch, G., Levy, H. The efficiency analysis of choices involving risk. *Review of Economic Studies* **36** (1969), 335–346.

[61] Håstad, J. Some optimal inapproximability results. *J. of ACM* **48** (2001), 798–859.

[62] Hildebrand, R. An LMI description for the cone of Lorentz-positive maps. *Linear and Multilinear Algebra* **55:6** (2007), 551–573.

[63] Hildebrand, R. An LMI description for the cone of Lorentz-positive maps II. To appear in *Linear and Multilinear Algebra.* 2008. E-print: http://www.optimization-online.org/DB_HTML/2007/08/1747.html

[64] Holt, C.C., Modigliani, F., Shelton, J.P. The transmission of demand fluctuations through distribution and production systems: The tv-set industry. *Canadian Journal of Economics* **14** (1968), 718–739.

[65] Huber, P.J. *Robust Statistics.* John Wiley & Sons, New York 1981.

[66] Iyengar, G., Erdogan, E. Ambiguous chance constrained problems and robust optimization. *Math. Progr. Series B* **107:1-2** (2006), 37–61.

[67] Johnson, W.B., Schechtman, G. Remarks on Talagrand's deviation inequality for Rademacher functions. Banach Archive 2/16/90. *Springer Lecture Notes* 1470 (1991), 72–77.

[68] Khachiyan, L.G. The problem of calculating the volume of a polyhedron is enumerably hard. *Russian Math. Surveys* **44** (1989), 199–200.

[69] Kirkwood, C.W. *System Dynamics Methods: A Quick Introduction.* Arizona State University, 1998.

[70] Kouvelis, P., Yu, G. *Robust Discrete Optimization and its Applications.* Kluwer Academic Publishers, London, 1997.

[71] Lagoa, C.M., Li, X., Sznaier, M. Probabilistically constrained linear programs and risk-adjusted controller design. *SIAM J. on Optmization* **15** (2005), 938–951.

[72] Lebret, H., Boyd, S. Antenna array pattern synthesis via convex optimization. *IEEE Trans. on Signal Processing* **45:3** (1997), 526–532.

[73] Lee, H., Padmanabhan, V., Whang, S. Information distortion in a supply chain: The bullwhip effect. *Management Science* **43** (1997), 546–558.

[74] , Lee, H., Padmanabhan, V., Whang, S. The bullwhip effect in supply chains. *MIT Sloan Management Review* **38** (1997), 93–102.

[75] Lee, H., Padmanabhan, V., Whang, S. Comments on information distortion in a supply chain: The bullwhip effect. *Management Science* **50** (2004), 1887–1893.

[76] Lewis, A.S. Robust regularization. Technical Report, Department of Mathematics, University of Waterloo, 2002.

[77] Love, S. *Inventory Control*. McGraw-Hill, 1979.

[78] Lust-Piquard, F., Inégalités de Khintchine dans C_p $(1 < p < \infty)$. *Comptes Rendus de l'Académie des Sciences de Paris, Série I* **393:7** (1986), 289–292.

[79] Miller, L.B., Wagner, H. Chance-constrained programming with joint constraints. *Oper. Res.* **13** (1965), 930–945.

[80] Morari, M., Zafiriou, E. *Robust Process Control*. Prentice-Hall, 1989.

[81] Nemirovski, A., Roos, C., Terlaky, T. On maximization of quadratic form over intersection of ellipsoids with common center. *Math. Progr.* **86** (1999), 463–473.

[82] Nemirovski, A. On tractable approximations of randomly perturbed convex constraints. *Proceedings of the 42nd IEEE Conference on Decision and Control Maui, Hawaii. December 2003*, 2419–2422.

[83] Nemirovski, A., Shapiro, A. Convex approximations of chance constrained programs. SIAM J. on Optimization **17:4** (2006), 969–996.

[84] Nemirovski, A., Shapiro, A. Scenario approximations of chance constraints. In: G. Calafiore, F. Dabbene, eds. *Probabilistic and Randomized Methods for Design under Uncertainty*. Springer, 2006.

[85] Nemirovski, A. Sums of random symmetric matrices and quadratic optimization under orthogonality constraints. *Math. Progr. Series B* **109** (2007), 283–317.

[86] Nemirovski, A., Onn, S., Rothblum, U. Accuracy certificates for computaTIONAL problems with convex structure. E-print:
http://www.optimization-online.org/DB_HTML/2007/04/1634.html

[87] Nesterov, Yu. Squared functional systems and optimization problems. In: H. Frenk, T. Terlaky and S. Zhang, eds. *High Performance Optimization*, Kluwer, 1999, 405–439.

[88] Nesterov, Yu. Semidefinite relaxation and nonconvex quadratic optimization. *Optim. Methods and Software* **9** (1998), 141–160.

[89] Nikulin, Y. Robustness in combinatorial optimization and scheduling theory: An extended annotated bibliography. Working paper (2006). Christian-Albrechts University in Kiel, Institute of Production and Logistics. E-print: http://www.optimization-online.org/DB_HTML/2004/11/995.html

[90] Nilim, A., El Ghaoui, L. Robust control of Markov decision processes with uncertain transition matrices. *Oper. Res.* **53:5** (2005), 780–798.

[91] Packard, A., Doyle, J.C. The complex structured singular value. *Automatica* **29** (1993), 71–109.

[92] Pinter, J. Deterministic approximations of probability inequalities, *ZOR Methods and Models of Operations Research, Series Theory* **33** (1989), 219–239.

[93] Pisier, G. Non-commutative vector valued L_p spaces and completely p-summing maps. - *Astérisque* **247** (1998).

[94] Pólik, I., Terlaky, T. A survey of the S-Lemma. - *SIAM Review* **49:3** (2007), 371–418.

[95] Polyak, B.T. Convexity of quadratic transformations and its use in control and optimization, *J. on Optimization Theory and Applications* **99** (1998), 553–583.

[96] Prékopa, A. On probabilistic constrained programming. In: *Proceedings of the Princeton Symposium on Mathematical Programming*. Princeton University Press, 1970, 113–138.

[97] Prékopa, A. *Stochastic Programming*, Kluwer, Dordrecht, 1995.

[98] Prékopa, A., Vizvari, B., Badics, T. Programming under probabilistic constraint with discrete random variables. In: L. Grandinetti et al., eds. *New Trends in Mathematical Programming*. Kluwer, 1997, 235–257.

[99] Ringertz, J. On finding the optimal gistribution of material properties. *Structural Optimization* **5** (1993), 265–267.

[100] Rockafellar, R.T., *Convex Analyis*. Princeton University Press, 1970.

[101] Rothschild, M., Stiglitz, J.E. Increasing risk I: a definition. *Journal of Economic Theory* **2:3** (1970), 225–243.

[102] Rothschild, M., Stiglitz, J.E. Increasing Risk II: its economic consequences. *Journal of Economic Theory* **3:1** (1971), 66–84.

[103] Saab, J., Corrêa, E. Bullwhip effect reduction in supply chain management: One size fits all? *Int. J. Logistics Systems and Management* **1:2-3** (2005), 211–226.

[104] Shapiro, A. Stochastic programming approach to optimization under uncertainty. *Math. Program. Series B* **112** (2008), 183220.

[105] Shivaswamy, P.K., Bhattacharyya, C., Smola, A.J. Second order cone programming approaches for handling missing and uncertain data. *J. Mach. Learn. Res.* **7** (2006), 1283–1314.

[106] Singh, C. Convex programming with set-inclusive constraints and its applications to generalized linear and fractional programming. *J. of Optimization Theory and Applications* **38:1** (1982), 33–42.

[107] Sterman, J.B. Modeling managerial behavior: Misperceptions of feedback in a dynamic decision making experiment. *Management Science* **35** (1989), 321–339.

[108] Stinstra, E., den Hertog, D. Robust optimization using computer experiments. CentER Discussion Paper 2005–90, July 2005. CentER, Tilburg University, P.O. Box 90153, 5000 LE Tilburg, The Netherlands. To appear in *European Journal of Operational Research*. http://arno.uvt.nl/show.cgi?fid=53788

[109] Soyster, A.L. Convex programming with set-inclusive constraints and applications to inexact linear programming. *Oper. Res.* (1973), 1154–1157.

[110] Special Issue on Robust Optimization. *Math. Progr. Series B* **107:1-2** (2006).

[111] Tropp, J.A. The random paving property for uniformly bounded matrices. *Studia Mathematica* **185** (2008), 67–82.

[112] Terwiesch, C., Ren, J., Ho, T.H., Cohen, M. Forecast sharing in the semiconductor equipment supply chain. *Management Science* **51** (2005), 208–220.

[113] Tibshirani, R. Regression shrinkage and selection via the LASSO. *Journal of the Royal Statistical Society, Series B* **58:1** (1996), 267–288.

[114] Trafalis, T.B., Gilbert, R.C. Robust classification and regression using support vector machines. *European Journal of Operational Research* **173:3** (2006), 893–909.

[115] H. Wolkowicz, R. Saigal, L. Vandenberghe, eds. *Handbook of Semidefinite Programming*. Kluwer Academic Publishers, 2000.

[116] Xu, H., Mannor, S., Caramanis, C. 2008. Robustness, risk, and regularization in support vector machines. In preparation.

[117] Youla, D.C., Jabr, H.A., Bongiorno Jr., J.J. Modern WienerHopf design of optimal controllers, Part II: The multivariable case. *IEEE Trans. Automat. Control* **21:3** (1976), 319338.

[118] Zhang, X. Delayed demand information and dampened bullwhip effect. *Operations Research Letters* **33** (2005), 289–294.

Index

www.ingramcontent.com/pod-product-compliance
Ingram Content Group UK Ltd.
Pitfield, Milton Keynes, MK11 3LW, UK
UKHW010827161224
452264UK00001B/22